Behavior Analysis and Learning

Using a behavioral perspective, *Behavior Analysis and Learning* provides an advanced introduction to the principles of behavior analysis and learned behaviors, covering a full range of principles from basic respondent and operant conditioning through applied behavior analysis into cultural design. The text uses Darwinian, neurophysiological, and biological theories and research to inform B. F. Skinner's philosophy of radical behaviorism.

The seventh edition expands the focus on neurophysiological mechanisms and their relation to the experimental analysis of behavior, providing updated studies and references to reflect current expansions and changes in the field of behavior analysis. By bringing together ideas from behavior analysis, neuroscience, epigenetics, and culture under a selectionist framework, the text facilitates understanding of behavior at environmental, genetic, neurophysiological, and sociocultural levels. This "grand synthesis" of behavior, neuroscience, and neurobiology roots behavior firmly in biology. The text includes special sections, "New Directions," "Focus On," "Note On," "On the Applied Side," and "Advanced Section," which enhance student learning and provide greater insight on specific topics. This edition was also updated for more inclusive language and representation of people and research across race, ethnicity, sexuality, gender identity, and neurodiversity.

Behavior Analysis and Learning is a valuable resource for advanced undergraduate and graduate students in psychology or other behavior-based disciplines, especially behavioral neuroscience. The text is supported by Support Material that features a robust set of instructor and student resources: www.routledge.com/9781032065144.

Erin B. Rasmussen is Professor of Psychology at Idaho State University, USA. She has published studies in the experimental analysis of behavior, behavioral pharmacology and toxicology, and behavioral economics. Her current research examines the roles of diet and sociocultural factors that potentiate food reinforcement value and subsequent obesity. She teaches upper-division undergraduate and graduate courses on behavior analysis, behavioral pharmacology, research writing, and food and behavior. She served as president of the Association for Behavior Analysis International.

Casey J. Clay is Director of Behavior Programs at the Thompson Autism Center at Children's Hospital Orange County in California, USA. He has published in the area of social interaction for children with autism, clinical behavior assessment and intervention, and methods of training behavior assessment and intervention. His current research involves the use of virtual reality for training skills to mitigate and intervene in challenging behavior in clinical settings.

W. David Pierce (1945–2020) was a Professor Emeritus of Sociology at the University of Alberta, Canada. He has investigated a biobehavioral model of activity anorexia, exercise-induced taste aversion, and behavioral-neurometabolic determinants of obesity. His research also focused on wheel-running reinforcement, the response deprivation hypothesis, and extensions of behavioral momentum theory.

Carl D. Cheney is Professor Emeritus of Psychology at Utah State University, USA. He taught behavior analysis and physiological psychology, and published widely in the experimental analysis of behavior—conducting basic analyses of predator–prey interactions, diet selection in domestic livestock, as well as reinforcement-schedule alterations and transitions in humans. His current research is focused on encouraging the wider use of behavior analysis in all appropriate situations.

Behavior Analysis and Learning

A Biobehavioral Approach

Seventh Edition

Erin B. Rasmussen, Casey J. Clay, W. David Pierce, and Carl D. Cheney

NEW YORK AND LONDON

Designed cover image: © Getty Images

Seventh edition published 2023
by Routledge
605 Third Avenue, New York, NY 10158

and by Routledge
4 Park Square, Milton Park, Abingdon, Oxon, OX14 4RN

Routledge is an imprint of the Taylor & Francis Group, an informa business

© 2023 Erin B. Rasmussen, Casey J. Clay, W. David Pierce, and Carl D. Cheney

The right of Erin B. Rasmussen, Casey J. Clay, W. David Pierce, and Carl D. Cheney to be identified as authors of this work has been asserted in accordance with sections 77 and 78 of the Copyright, Designs and Patents Act 1988.

All rights reserved. No part of this book may be reprinted or reproduced or utilised in any form or by any electronic, mechanical, or other means, now known or hereafter invented, including photocopying and recording, or in any information storage or retrieval system, without permission in writing from the publishers.

Trademark notice: Product or corporate names may be trademarks or registered trademarks and are used only for identification and explanation without intent to infringe.

First edition published by Prentice Hall 1995
Sixth edition published by Routledge 2017

ISBN: 978-1-032-06514-4 (hbk)
ISBN: 978-1-003-20262-2 (ebk)

DOI: 10.4324/9781003202622

Typeset in Univers
by Newgen Publishing UK

Access the Support Material: www.routledge.com/9781032065144

CONTENTS

Preface	vi
1 A Science of Behavior: Perspective, History, and Assumptions	**1**
2 The Experimental Analysis of Behavior	**37**
3 Reflexive Behavior and Respondent Conditioning	**71**
4 Reinforcement and Extinction of Operant Behavior	**108**
5 Schedules of Reinforcement	**149**
6 Aversive Control of Behavior	**187**
7 Operant–Respondent Interrelations: The Biological Context of Conditioning	**233**
8 Stimulus Control	**271**
9 Choice and Preference	**311**
10 Conditioned Reinforcement	**358**
11 Correspondence Relations: Imitation and Rule-Governed Behavior	**393**
12 Verbal Behavior	**427**
13 Applied Behavior Analysis	**469**
14 Three Levels of Selection: Evolution, Behavior, and Culture	**509**
Glossary	543
References	583
Author Index	651
Subject Index	670

PREFACE

Instructors who use *Behavior Analysis and Learning: A Biobehavioral Approach* will notice a proverbial passing of the torch to the next generation of behavior analysts with the seventh edition. In 2020, the behavior science community was shocked at the unexpected death of David Pierce. He was a vibrant, creative, and productive colleague. Dave's contributions to the experimental analysis of behavior (EAB), including a cogent development of an animal model of activity-based anorexia, were impactful. Many of these research endeavors were conducted with his dear friend and collaborator W. Frank Epling (1943–1998), an original author of this text. Further editions of this book will continue to owe a great deal to both of them, as well as to Carl Cheney, who was an author since the third edition. Carl also has made a lifetime of research contributions to EAB using approximately 20 different species of animals and has taught thousands of students over the years.

The new set of authors, Erin B. Rasmussen and Casey J. Clay, were invited on the basis of their accomplishments and commitment to the science of behavior analysis. They represent both the basic experimental analysis research side and the behavioral engineering applied side. Rasmussen began her training in animal research in EAB as an undergraduate in Carl's lab at Utah State University. She completed her doctoral training in EAB at Auburn University with an emphasis in behavioral pharmacology and toxicology and has contributed numerous peer-reviewed research articles with both non-humans and humans to the field to date. Clay is a board-certified behavior analyst-doctoral level who also had humble beginnings in rat and pigeon labs at USU; however, his interests turned to human behavior, specifically individuals with developmental disabilities who engaged in severe challenging behavior. He continued his training at Northeastern University and through the New England Center for Children and earned his master's degree. He later returned to USU for doctoral training and continues to publish and engage in clinical work.

Indeed, *Behavior Analysis and Learning* has come a long way since its inception in 1995. Rasmussen fondly recalls the first edition of the text, which was published in 1995, the year she started graduate school. While instructors used the text for courses there, she used it as a supplementary resource to assist her understanding of the many complicated empirical articles that were required for conceptualizing her thesis and dissertation. The chapter on Choice in particular, effectively broke down the complexities of the generalized matching relation in a manner that made it easy to learn. No other textbook published at the time (or perhaps even now) described matching to students with such clarity. Therefore, the value of this text both inside and outside of a classroom has been historically significant to her and having the opportunity to contribute to it as an author has been a gratifying and exciting experience.

Clay recalls using this text in teaching his courses as a professor at the University of Missouri. Students were amazed at how the principles of behavior science could be applied universally and across species. He took great joy in explaining to students how an operant chamber with some modifications could be used to teach pigeons how to talk to each other. Students in his courses were also charmed when an invited guest would nonchalantly attend classes and sit in the back of the room and make comments on the lecture only to be eventually introduced as Dr. Carl Cheney—the author of the textbook they were using for the semester. Clay is also excited by the opportunity to contribute as an author and hopes to fill the large shoes offered as he attempts to stand on the shoulders of giants.

Indeed, the seventh edition of this text is transgenerational. On one hand, this version holds strongly to the foundational principles of behavior analysis. As with any solid science grounded in empiricism, the foundations of EAB have not changed much over the years—behavior is still a function of its consequences; reinforcement still strengthens behavior; resurgence of behavior still occurs during extinction; differential reinforcement of successive approximations still shapes new responses; and the operant chamber is still a major apparatus for exploring environment–behavior interactions.

On the other hand, our expansion and understanding of behavior has grown and become more complex. Our views of verbal behavior and transfer of function by way of relational frame theory, which was built on the science of stimulus equivalence, has advanced our understanding of the rapid expansion of verbal relational classes. We have strongly extended our characterization of choice with behavioral economics, as evidenced by the explosion of research on delay discounting and the conceptualization of reinforcer pathologies. We also have expanded on the third level of selectionism—cultural behavior. At the cultural level, we have observed rapid changes worldwide in a relatively small window of time. The internet, the streaming of media, smart phones, social media apps, and gaming systems serve as strong reinforcers to everyone from young children to older adults. We offer some application and analyses, including insights from fields outside of behavior analysis, on how these environmental influences might impact behavior in our sociocultural environment. We also have observed how millennial parenting practices, such as "helicopter parenting" might affect the behavior of offspring in the long run. We also continue to discuss how eco-friendly behavior and cultural practices might affect climate change predictions. Importantly, we have also been through a global pandemic since the sixth edition of the text, and with that, swift sociocultural movement; we would be remiss to exclude these topics and their influence on behavior from the seventh edition.

In this edition of the text, we also have continued to expand the presentation of neural mechanisms as another level to consider for the experimental analysis of behavior. The contributions of neuroscience substantially improve our ultimate explanation of behavior and how it can be influenced. We also expand coverage of behavior-gene interactions, including epigenetics. We recognize the importance of *heterogeneous reductionism* (Marr, 1977; Marr & Zilio, 2013)—the explaining of phenomena on multiple levels (e.g., neural and the behavioral)—as creating a more complete science.

We also continue to examine behavior through a selectionist lens—the evolutionary, behavioral, and cultural levels—and are impressed by the growing number of evidenced-based studies that support this position. We continue to promote the broad practice of applied behavior analysis and the growing literature illustrating diverse applications of behavior science. Several professional organizations exist whose members express, either in research or application, the philosophy of radical behaviorism in the analysis of behavior. The discovery of the way behavior works upon the world is illuminated by EAB in learning and genetics laboratories, free-ranging animal environments, psychopharmacology, educational instruction methods for classrooms, training centers for explosive-sniffing dogs, language and verbal development, treatment of substance abuse disorders, care and treatment of zoo animals, early intensive behavioral intervention for children with autism and developmental disabilities, computer labs and virtual human learning environments, applications to business and organizations, and university behavior laboratories investigating contingencies of reinforcement with a variety of organisms.

To ensure we stay current with the scientific analysis of behavior–environment relationships, we have added over 100 new references to the seventh edition, some from traditional sources like the *Journal of the Experimental Analysis of Behavior* and other citations from more general biological journals, such as *Science* and *Nature*. We also bring in impactful references from outside the field, such as clinical psychology, substance abuse, parenting, and health psychology. Refinements of technology, research design, and data analysis have vastly expanded the field of behavior analysis and therefore the topics and sources to present to students and readers. We have been driven not only by the breadth of findings from the scientific literature, but also media coverage of topics relevant to behavior analysis. We suggest instructors recommend their students access original sources to appreciate them more fully.

Chapter titles and their order have remained virtually the same as the sixth edition. The addition of more recent citations, however, has necessitated the removal of some references—in many cases older papers. Some of these removed papers gave a historical impression of how behavior analysis was conducted along with the ethical practices of the time in which they were published. Because some of these ethical practices give the reader an inaccurate impression of the field today (especially with work using aversive stimuli), we focus instead on more contemporary research and applied practices. However, we consider many early papers in the experimental analysis of behavior to remain as relevant today as ever, even when ethical practices were somewhat different. When these papers were retained in the seventh edition, we include balanced discussions about these ethical issues to assist our readers with understanding how our field has evolved over the years.

One may also notice the use of more sensitive and inclusive language and representation. We believe this practice is aligned with the values of a more diverse scientific and global community. In addition, we believe that all students need to see themselves represented in textbooks. To this end, we use gender-inclusive pronouns and language that involves all couples and sexualities. We also include greater ethnicity, race, and gender representation in photos and research throughout the text. Finally, we also describe the neurodiversity movement in a balanced manner in this edition and strive to use respectful terms for individuals with developmental disabilities. It is possible we missed some opportunities for inclusive language, and if so, we apologize and invite you to contact us so that subsequent editions can be amended.

In conclusion, the authors want to provide the information necessary for readers who may never take another course on behavior to better understand and apply evidence-based principles to manage the behavior of organisms. We are not promoting a theory or school of psychology, but rather showing through empirical evidence that environmental events influence behavior in predictable and replicable ways. We have always wanted to expand our realm of influence as behavior scientists to other fields, such as biology, wildlife, pharmacology, speech pathology, education, and health. Behavior is everywhere and making the principles by which behavior operates explicit is the purpose of this text. We hope you enjoy the seventh edition of *Behavior Analysis and Learning: A Biobehavioral Approach*.

Erin B. Rasmussen
Casey J. Clay
W. David Pierce
Carl D. Cheney

CHAPTER 1

A Science of Behavior: Perspective, History, and Assumptions

1. Inquire about *learning*, a science of behavior and behavior analysis.
2. Discover how *selection by consequences* extends to evolution and behavior.
3. Explore new directions in behavior analysis and behavioral neuroscience.
4. See how early learning is retained by *epigenetic* mechanisms.
5. Investigate the early beginnings of behavior analysis and learning.
6. Analyze feeling and thinking as complex behavior.

Learning refers to the acquisition, maintenance, and change of an organism's *behavior* as a result of lifetime experience and events. The **behavior** of an organism is everything it does, including private and covert actions like thinking and feeling (see "Science and Behavior: Some Assumptions" section of this chapter). Learning also involves **neuroplasticity**—alterations in the brain that accompany behavior change and participate in the regulation of behavior. While our focus in this book is centered on the study of behavior for its own sake, the links to the brain and neural processes are increasingly important to the field of learning and behavior analysis as we discuss throughout the book.

An important aspect of human learning concerns the experiences arranged by other people. From earliest history, people have acted to influence the behavior of other individuals. At the level of the individual, rational argument, rewards, bribes, threats, and force are used in attempts to promote learning or change the behavior of people. At the level of culture, people are required to learn socially relevant behavior of which the culture agrees and approves. As long as a person conforms, no one pays much attention. People may also praise or reward the socially appropriate behavior of the individual (e.g., "such an upstanding citizen!"). But when the conduct substantially departs from cultural norms, people may get upset and socially reject the behavior of the non-conformist or even the non-conformist themself—ensuring that most of us will comply with society's wishes (Williams & Nida, 2011). At the societal level (e.g., community, city, state, or nation), there are codes of conduct and laws that people must learn; those who break moral codes or civil laws may face penalties ranging from minor fines to capital punishment. Human learning, then, and the regulation of human conduct, is relevant

DOI: 10.4324/9781003202622-1

to families, communities, cultures, and societies. Without regulation of behavior, anarchy and confusion eventually destroy the civil order of society.

Human behavior has been attributed to a great variety of causes. The causes of behavior have been located both inside a person and outside, that is, from the environment. Internal causes have ranged from metaphysical entities like the soul to hypothetical constructs (e.g., impulsivity). Internal causes also include structures of the nervous system, such as parts of the brain. Suggested external causes of behavior have included the effect of the moon and tides, the arrangement of stars, and the whims of gods. Unfortunately, some of these bizarre, prescientific attempts to explain human behavior remain popular today, despite the lack of evidence for their role on behavior.

One such example is astrology. In an interview with Ashley Lutz for *Business Insider* (2012), Susan Miller, a successful astrologer with a business degree from NYU, said, "What I do is scientific. Astrology involves careful methods learned over the years and years of training and experience." Her website has six million visitors every month and she has built an empire based on her "scarily accurate" predictions, said the *Insider*. Miller states "one unlikely group of customers … are professional men from 25 to 45 years old. In these uncertain economic times, astrology is more important than ever!" Many people faced with the unpredictability of daily existence turn to the theory of celestial alignment (astrology) to inform and guide their actions in business, life, and personal relationships.

The trouble with astrology and other spurious accounts of human behavior is that they are not scientifically supported. These theories do not hold up to objective testing, replication, and close scrutinizing by researchers who follow the scientific method. Indeed, many astrological predictions that are made are general and **unfalsifiable**, meaning they are not capable of being proven false. Consider this astrological prediction for the new year: "Career wise, the year has many new opportunities to offer." This statement could mean many things, such as a person may find a new job, they may find something new in terms of opportunities (e.g., a task or leadership position) in their current career, or they find another way to enjoy what is not changeable in their current job (e.g., a new friendship with a colleague), or something else entirely. There is nothing specific about this prediction that can be proven false. Related to this issue is **confirmation bias**; that is, people may seek information that confirms a statement or belief and ignore other information that suggests otherwise. For example, if after this prediction, a person lost their job, but found a new one, they might say "See? I found a new career opportunity, just like my horoscope said", and ignore the issue of losing the job in the first place.

Over the last century, a science-based model of learning and behavior has developed and undergone careful and rigorous scientific testing. Behavioral theory states that all behavior is due to complex interactions between genetic influences and environmental experiences. The theory is based on observation and controlled experimentation, and it provides a natural-science account of the learning and behavior of most, if not all, organisms, including humans. This book is concerned with such an account.

SCIENCE OF BEHAVIOR

The **experimental analysis of behavior** is a natural-science approach to understanding behavior regulation. Experimental analysis is concerned with controlling and changing the factors that affect the behavior of humans and other animals in order to observe systematic and predictable changes. For example, a behavioral researcher in a classroom or a home environment may use a computer to arrange feedback for a student's mathematical performance—for example, the visual presentation of a star when a correct answer is given. The relevant condition manipulated or changed by the experimenter may involve presenting feedback on some days and withholding it on others—this latter condition is called a **control**. In this case, the researcher would probably observe more accurate mathematical performance on days with the visual presentation of the star. This simple experiment illustrates one of the most basic principles of behavior—the principle of **reinforcement**.

The principle of reinforcement (and other behavior principles) provides a scientific account of how people and animals learn complex actions. When a researcher identifies a condition or set of conditions that govern behavior, this is called an analysis of behavior. Thus, the experimental analysis of behavior involves discovery of the basic processes and principles through experimentation that regulate the behavior of organisms. Experiments are used to test the adequacy of the analysis. The manner in which this is done will be discussed in this book

---

NOTE ON: EXPERIMENTAL ANALYSIS OF BEHAVIOR IN THE NATURAL WORLD

Experimental analysis can occur outside of the laboratory and in a more naturalistic setting. Consider when, for example, a researcher notices that more seagulls fly and congregate along a shoreline when people are on the beach than when the beach is deserted. After checking that changes in climate, temperature, time of day, and other conditions do not affect the behavior of the seagulls, the researcher offers the following analysis: People feed the birds and this reinforces flocking to the beach. When the beach is abandoned, the seagulls are no longer fed for congregating on the shoreline. Therefore, the reason for flocking could be food. It could also be the mere presence of people (and not food), but the former is a reasonable first guess. This guess can only be tested by an experiment. Pretend that the behavior analyst owns the beach and has complete control over it. The experiment involves changing the usual relationship between the presence of people and food across different conditions. In a beginning condition, people are allowed on the beach, but are not allowed to feed the birds. In a second condition, the opposite holds—food is placed on the beach when people are not around. Over time and repeated days of each condition, the behavior analyst notes that there are fewer and fewer seagulls on the beach when people are present, and more and more gulls when the shoreline has food, but is deserted. The behaviorist concludes that birds came to the beach because the birds were fed, or reinforced, for doing so only when people were present. This is one example of an experimental analysis of behavior.

Behavior Analysis: A Science of Behavior

Experimental analysis is the fundamental method used to establish the principles for a **science of behavior**. A science of behavior is informed by a philosophy of naturalism and is called *behavior analysis*. This term implies a more general scientific approach that includes assumptions about how to study behavior, techniques for carrying out the analysis, a systematic body of knowledge, and practical implications for society and culture.

Behavior analysis is a comprehensive, natural-science approach to the study of the behavior of organisms. Primary objectives are the discovery of principles and laws that govern behavior, the extension of these principles across species, settings, and different levels of analysis (e.g., the level of one individual, a group, or the level of an entire culture), and the development of applied technology for the management of behavior. The discipline of behavior analysis used to be known as behavioral psychology, but was changed to behavior analysis to emphasize the focus on behavior and the analysis of conditions that change it. A good example of a principle of behavior that we will cover in greater depth later in this textbook is called **discrimination**. The principle of discrimination states that an organism will respond differently to two situations if its behavior has been reinforced in one setting but not in the other (*differential reinforcement*). For example, a deer may choose a pond from which to drink, in which the water is fresh and plentiful. Another pond nearby may have equally fresh and plentiful water, but on occasion, there is a cougar that frequents the area. The deer learns to discriminate between the two ponds and chooses the one without the predator.

The principle of discrimination may be extended to human behavior and social reinforcement. You may discuss dating with Carmen, but not Tracey, because Carmen has shown interest in such conversation while Tracey rolls her eyes at the topic and wants to instead discuss politics (*differential reinforcement*). In a classroom setting, the principle of discrimination can be used to improve teaching and learning. A child is given a series of multiplication problems from the 2-times tables such as 2 × 4 =_?. Correct answers result in the next question, while incorrect responses lead to corrective feedback from the teacher, and repetition of the question. In this way, most children learn their 2-times table. The use of behavior principles to solve practical problems is called **applied behavior analysis** and is discussed at some length in Chapter 13.

As you can see, behavior analysis has a strong focus on behavior–environment relations. The focus is on how organisms alter their behavior to meet the ever-changing demands of the environment. When an organism learns new ways of behaving in reaction to the changes in its environment, this is called *conditioning*. The two basic kinds of conditioning are called respondent and operant.

Two Types of Conditioning

A **reflex** involves behavior *elicited* by a biologically relevant stimulus. When a stimulus (S) automatically elicits (→) a stereotypical response (R) or **respondent**, the S → R *relationship* is called a reflex. The reflex is present at birth, in the sense that those animals that quickly and reliably responded to particular stimuli were more likely than other organisms to survive and reproduce. For instance, animals that startle to a sudden noise may also be able to escape a predator, hence the startle reflex may have provided an adaptive advantage over organisms that did not startle to the noise. Thus, reflexes are selected across the evolutionary history of the species. Of course, different species of organisms exhibit different sets of reflexes.

Respondent Conditioning

Respondent conditioning occurs when a feature of the environment (generally called a **stimulus**) without a known effect on behavior is **correlated** with an **unconditioned stimulus** (**US**). The US is a stimulus that naturally *elicits* a response based on an organism's biological history. For example, a puff of air (US) in the eyes elicits blinking (**unconditioned response** or

UR). You can see that this is an inherited response—it is present at birth and occurs without the organism's learning; indeed, the term "unconditioned" means *unlearned*. If I say the word "paper," it does not elicit eye blinking, and has no stimulus function with respect to the eye-blinking response before conditioning. Indeed "paper" is a neutral, non-functional stimulus. However, the word "paper" comes to predict the air puff (US) because we pair the two stimuli together multiple times; pretty soon the word "paper" by itself (without the US) will produce the blink response. When "paper" elicits the eye blink, it has acquired a **conditioned-stimulus (CS)** function. One method to ensure that a feature of the environment predicts the US is called pairing or *temporal contiguity*; the neutral stimulus precedes the presentation of the US close in time. For example, respondent conditioning occurs when the buzz of bees (initially neutral) is paired with painful stings, which are also startling to the individual receiving them (US). After this conditioning, a buzzing bee (CS) usually causes people to startle at the sound, which also occasions them to escape it; this startle response is the **conditioned response (CR)**.

You may have read about the Russian physiologist *Ivan Petrovich Pavlov* who made explicit this form of conditioning at the turn of the 20th century. He observed that dogs salivated when food was placed in their mouths. This relation between the food stimulus and salivation is an *unconditioned reflex*, and it occurs because of the animals' evolutionary history. However, when Pavlov rang a bell just before feeding the dogs, the animals began to salivate at the sound of the bell. In this way, a new feature (the sound of the bell) that predicted the presentation of food came to control the CR of salivation. As shown in Figure 1.1, the CR is now elicited by the new *conditioned stimulus* (CS).

Respondent (also called classical or Pavlovian) conditioning is one way in which organisms meet the challenge of change in their environments. A grazing animal that startles in conditions to the sound of rustling grass before a predator's attack, but not to grass blowing in the wind, gains a survival advantage. The animal is able to efficiently consume food, running away only when certain stimuli may signal a life-threatening situation. All species that have been tested, including humans, show this kind of conditioning. In terms of human behavior, many of what we call our likes and dislikes are based on "evaluative" conditioning. Evaluative conditioning of humans replicates many of the respondent-conditioning effects found in animals, although some differences have been noted (De Houwer, Thomas, & Baeyens, 2001). Generally, when good or bad things happen to us, we usually have an emotional reaction. These emotional responses can be conditioned to other people who are present when the positive or negative events occur. For example, if you were to get in a car wreck with a person you have never met,

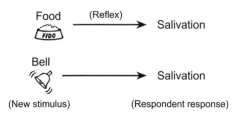

Fig. 1.1
Simple respondent conditioning: In a reflex for a dog, food in the mouth (US) produces salivation (UR) as reflexive behavior. Next, a bell rings (neutral stimulus) just before feeding the dog; after several pairings of the neutral bell and food the dog begins to salivate at the sound of the bell, even when the food is not presented. The bell has now become a conditioned stimulus (CS) that causes a CR.

you are likely to associate the person with the wreck and therefore feel ill feelings toward that person, even if that person is a decent human being. Thus, respondent conditioning plays an important role in our social relationships—determining, to a great extent, how we evaluate and come to "feel" about our friends as well as our enemies. Respondent conditioning is covered in more detail in Chapter 3.

Operant Conditioning

Operant conditioning involves the *regulation of behavior by outcomes or consequences*. B. F. Skinner called this kind of behavior regulation **operant conditioning** because, in a given situation or setting (S^D), behavior (R) *operates* on the environment to produce effects or outcomes (S^r). An **operant** is any behavior that operates on the environment to produce an effect. The effect or outcome in turn changes the likelihood that the operant will occur again in a similar situation. During operant conditioning, an organism emits operant behavior that produces an effect that increases (or decreases) the frequency of the response in a given situation. In the laboratory, a hungry rat in a chamber may receive food if it presses a lever when a light is on. If lever pressing increases in the presence of the light, then operant conditioning has occurred and food functions as *reinforcement* (S^r) for this operant response (Figure 1.2).

In this example, the light (S^D) eventually *sets the occasion for* lever pressing in the sense that *the operant is likely to occur* when the light is on and is unlikely to occur when it is off. Basically, the frequency of lever pressing increases in the presence of the light (S^D). Turning the light on, however, does not force or elicit lever pressing as with a respondently conditioned stimulus; it simply increases the probability of the lever-pressing response when the light is on. In essence, operants are voluntary, probabilistic behaviors that are **emitted** rather than elicited (as with respondent conditioning). The control of the bar press by the light stimulus is based on the past history of reinforcement for lever pressing in the presence of the light. It is also controlled by the absence of food delivery (no reinforcement) when the light is off.

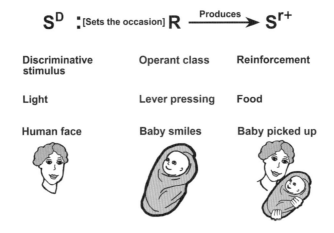

Fig. 1.2
Two examples of simple operant conditioning: In an operant chamber, lever pressing produces food for a hungry rat. The consequences of lever pressing (presentation of food) increase its frequency in that setting. In another example, a baby smiles at a human face and is picked up. The consequence of smiling (social attention) increases the frequency of this behavior in the presence of human faces.

Most of what we commonly call voluntary, willful, or purposive action is analyzed as operant behavior. *Operant conditioning* occurs, for example, when an older baby smiles at a human face and is then picked up. If smiling at faces increases in frequency because of such social attention, then smiling is an operant and the effect is a result of conditioning. The presentation of a human face (S^D) *sets the occasion for* infant smiling only after a history of operant conditioning. When a face appears, the frequency of smiling increases; also, smiling has a low frequency of occurrence when no one is around. In a more complex example using video games, the presence of targets on the screen (S^D) *sets the occasion for* pressing a sequence of buttons (operant) that results in hitting a target (S^r) and increasing the probability of the response sequence. Other examples of operant behavior include driving a car to work to get paid, talking on the phone, taking notes during class, walking to the store to buy groceries, reading a book for pleasure, writing a term paper for grades, or conducting an experiment to resolve a scientific question. In each case, we say the operant is *selected by its consequences or outcomes*.

Selection as a Causal Process

The behavioral view of psychology and behavior analysis is heavily influenced by B. F. Skinner (1938), who viewed psychology as the study of the behavior of organisms. From this point of view, psychology is a subfield of biology. The main organizing principle of contemporary biology is evolution through natural selection. Skinner generalized this concept to a broader principle of **selection by consequences**. Selection by consequences applies at three levels: (1) the selection over generations for genes related to survival and reproduction (natural or Darwinian selection); (2) the selection for behavior within the lifetime of an individual organism (selection by operant conditioning); and (3) the selection for behavior patterns (practices, traditions, or rituals) of groups of human beings that endure beyond the lifetime of a single individual (cultural selection). In all three cases, it is the outcomes or consequences arranged by the environment that select for (or against) the frequency of genetic, behavioral, and cultural forms (see Chapter 14). It is important to note that *consequences* refers to all events that affect behavior—both good and bad—not just ones we traditionally view as negative.

Selection by consequences is a form of causal explanation. In science we talk about at least two kinds of causation: immediate and remote. **Immediate causation** is the kind of mechanism studied by physics and chemistry—the "billiard ball" type of processes where we try to isolate a chain of events that directly, and close together in time, result in some effect. For example, chemical reactions are explained by describing molecular interactions. In the study of behavior, an immediate causal explanation might refer to the physiology or biochemistry of the organism. For example, the bar pressing of a rat for food or a gambler playing roulette could each involve the release of endogenous opiates and dopamine in the hypothalamus and the part of the brain responsible for reward called the nucleus accumbens (Shizgal & Arvanitogiannis, 2003). When this neurotransmitter is released, rewarding processes happen at the neural level.

In contrast, **remote causation** is typical of sciences like evolutionary biology, geology, and astronomy. In this case, we explain some phenomenon by pointing to remote events that made it likely or events that caused a chain of other events that led to the outcome. Thus, the causal explanation of a species characteristic (e.g., size, coloration, or exceptional vision) involves the working of natural selection on the gene pool of the parent population.

8 A Science of Behavior: Perspective, History, and Assumptions

An evolutionary account of species coloration, for example, would involve showing how this characteristic improved the reproductive success of organisms in a given ecological environment. Thus, natural selection for coloration explains the current frequency of the characteristic in the population.

On the behavioral level, the principle of selection by consequences is a form of explanation that is more remote causation. When a rat learns to press a lever for food, we explain the rat's behavior by pointing to its past consequences (the function of behavior). Thus, the current frequency of bar pressing is explained by the **contingency** between bar pressing and food in the past. The rat's behavior has been *selected by its history of reinforcement*. Thus, the history of reinforcement is what explains why the rat presses the lever. Though, at the same time, when the rat earns a food reinforcer, there are immediate causes (dopamine release) that also strengthen the response in the moment. Therefore, there are also immediate causes happening.

Both immediate and remote causal explanations are acceptable in science. **Behavior analysts** have emphasized functional analysis and selection by consequences (remote causation), but are also interested in direct analysis of physiological and neurochemical processes (immediate causation). Ultimately, both types of causal explanation will provide a more complete account of learning and behavior.

FOCUS ON: BEHAVIOR ANALYSIS AND NEUROSCIENCE

Behavior analysis is becoming more involved with the scientific analysis of the brain and nervous system or neuroscience. Researchers who primarily study the behavior of organisms and learning are often interested in the brain processes that participate in the regulation of behavior (Schaal, 2013; see also special issue on "Relation of Behavior and Neuroscience" (2005) in *Journal of the Experimental Analysis of Behavior, 84*, pp. 305–667). The word *participate* is used because the brain shows **neuroplasticity**, or changes in the interconnections of neurons or nerve cells (Kandel, 2006) and glia or non-neuronal cells (Fields, 2009) as an organism interacts with the world in which it lives—altering gene transmission, gene expression, and neural pathways related to learning and memory (McClung & Nestler, 2008; also see "New Directions: Epigenetics and Retention of Early Learning" in this chapter). The brain is not a static structure that determines behavior, but a malleable organ constantly adjusting to the behavioral requirements of everyday life or the laboratory (Draganski et al., 2004). For example, brain mechanisms (neurons or groups of neurons) obviously participate in the regulation of behavior (bar pressing) by its consequences (food). Describing how neurons assemble, code for, and respond to stimulation and reinforcement is an important and exciting addition to behavior analysis.

Currently, neuroscientists are mapping neurons to behavior in simple organisms like the fruit fly, *Drosophila* (Vogelstein et al., 2014). Flies are genetically engineered to selectively express a light-sensitive protein in defined sets of neurons (1054 neuron lines), which researchers activate with the presentation of light (optogenetic stimulation). Larvae are placed in plastic dishes and light stimulation is applied to the genetically engineered neurons, allowing observation and control of defined behavioral sequences (e.g., "wiggle escape" or "turn-turn-turn"). One finding is that the relation between a specific line of neurons and evoked behavior is probabilistic—repeatedly activating the same neurons did not always produce the identical behavioral sequence; thus, the *topography of response* varies even though the identical brain pathway is activated. The researchers note that optogenetic mapping of neurons to behavior would allow for an atlas of connectivity–activity maps to further investigate how neurons participate in the regulation of complex behavior.

At the practical level, knowing the reinforcement contingencies for lever pressing is sufficient by itself to allow us to predict and control the rat's behavior. We can get the rat to increase or decrease its lever pressing by providing or denying food reinforcement for this behavior—there is no need to look at neural systems. However, we gain a more complete account of how a rat's behavior increases when the action of neurons (and neural systems) is combined with the analysis of behavior. For example, in some cases it may be possible to "sensitize" or "desensitize" a rat to the behavioral contingencies by drugs that activate or block the action of specialized neurons (e.g., Bratcher, Farmer-Dougan, Dougan, Heidenreich, & Garris, 2005). Research at the neural level could, in this way, add to the practical control or regulation of behavior by its consequences.

Neural processes also may participate as immediate consequences (local contingencies) for behavior that had long-range benefits for organisms—remote contingencies, as in evolution and natural selection (Tobler, Fiorillo, & Schultz, 2005). The so-called *neural basis of reward* involves the interrelationship of the endogenous opiate and dopamine systems (as well as other neural processes) in the regulation of behavior and learning (Fiorillo, Tobler, & Schultz, 2003; Puig, Rose, Schmidt, & Freund, 2014). For example, rats that are food restricted and allowed to run in activity wheels increase running over days—up to 20,000 wheel turns. Wheel running leads to the release of neural opiates and endocannabinoids that reinforce this behavior (Pierce, 2001; Rasmussen & Hillman, 2011; Smith & Rasmussen, 2010). If wheel running is viewed as food-related travel, one function of neural reinforcement is to promote locomotion under conditions of food scarcity. The long-range or remote contingency (travel produces food: travel → food) is supported proximally by the release of endogenous opiates and endocannabinoids (physical activity → release of endogenous opiates and cannabinoids) that "keep the rat going" under conditions of food scarcity (e.g., famine or drought).

The integration of the science of behavior with neuroscience (**behavioral neuroscience**) is a growing field of inquiry. Areas of interest include the effects of drugs on behavior (behavioral pharmacology), neural imaging and complex stimulus relations, choice and neural activity, and the brain circuitry of learning, addiction, and obesity. We shall examine some of this research in subsequent chapters in sections that focus on behavior analysis and neuroscience ("Focus On" sections) or in sections that emphasize applications ("On the Applied Side" sections).

The Evolution of Learning

When organisms were faced with unpredictable and changing environments in their evolutionary past, natural selection favored those species whose behavior could be conditioned. Species that condition are more flexible, in the sense that they can learn new requirements and relationships in a dynamic environment (see section on "Behavioral Flexibility" in Chapter 14 for evidence by Mery and Kawecki (2002) on the link between learning ability and improved fitness in the fruit fly, *Drosophila melanogaster*). Such behavioral flexibility must reflect underlying structural changes of the organism. During embryonic development, genes are sequenced to form the anatomical and physiological characteristics of the individual, allowing for different degrees of functional flexibility (Mukherjee, 2016, pp. 185–199). Thus, differences in the structure of organisms based on genetic control give rise to differences in the regulation of behavior. Processes of learning, like operant and respondent conditioning, lead to greater (or lesser) reproductive success. Those organisms that changed their behavior as a result of experiences during their lifetimes survived and had offspring (passing on the genome), while those that were less flexible did not. Simply stated, this means that *the capacity for learning is inherited*.

The evolution of learning processes had an important payoff: Behavior that was closely tied to survival and reproduction could be influenced by experience. Specific physiological processes, orchestrated by genes and proteins at the cellular level, typically regulate behavior related to survival and reproduction. However, for behaviorally flexible organisms, this control by physiology may be modified by experiences during the lifetime of the individual. The extent of such modification depends on the amount and scope of behavioral flexibility. For example, sexual behavior is closely tied to reproductive success and is regulated by distinct physiological processes. For many species, sexual behavior is rigidly controlled by genetically driven mechanisms. In humans, however, sexual behavior is also influenced by socially mediated experiences. It is these experiences, not genes, which come to dictate when sexual behavior occurs, how it is performed, and who can be a sexual partner. Powerful religious or social controls also can make people abstain from sex. This example demonstrates that even the biologically relevant behavior of humans is partly determined by life experience.

The Biological Context of Behavior

As we have emphasized, behavior analysts recognize and promote the importance of biology, genes, and evolution, but focus more on the interplay of behavior and environment. To maintain this focus, the evolutionary history and biological status of an organism are examined as part of the *context of behavior* (see Morris, 1988). This contextualist view is seen in an analysis of imprinting in a duckling. Imprinting is viewed as behavior in which a very young animal fixates its attention on the first object with which it has visual, auditory, or tactile experience, and subsequently follows the object. Generally, this object is the mother, but other objects (other animals or even inanimate objects) can be the source of imprinting.

The duckling's evolutionary history, in terms of providing the *capacity for reinforcement* by proximity to a duck-sized object, is the context for the regulation of its behavior. In other words, this capacity for reinforcement is present at birth, but to what it will "imprint on" is contextual. Of course, the anatomy and neurophysiology of the duckling allow for this capacity. The way the environment is arranged during critical periods of development (i.e, early in life), however, determines the behavior of the individual organism on a specific occasion. Laboratory experiments in behavior analysis identify the general principles that govern the behavior of organisms, the specific events that regulate the behavior of different species, and the arrangement of these events during the lifetime of an individual.

NEW DIRECTIONS: EPIGENETICS AND RETENTION OF EARLY LEARNING

Did you know that diet plays a large role in the genetic expression of queen bees versus worker bees? Both queen and worker honeybees have the same genome and as larvae are fed a rich diet of royal jelly, which is produced by nurse bees. However, bees that are weaned early from royal jelly, and have it replaced with nectar and pollen, become worker bees. Worker bees (which are

Fig. 1.3
Epigenetics at work. The image shows structural changes in worker and queen bees that occur as a result of differential exposure to royal jelly. Queen bees are much larger than worker bees, but also lay eggs.
Source: Used with permission—May, 2022. ©Arizona Board of Regents / ASU Ask A Biologist https://askabiologist.asu.edu/bee-colony-life

sterile) forage for food, collect pollen, maintain the hive, and fight off predators. In contrast, a larval bee that continues an extended diet of royal jelly becomes an adult queen bee (see Figure 1.3). Queens can lay up to 2,000 eggs per day, are up to five times larger than worker bees, and live up to 20 times longer (Chittka & Chittka, 2010).

This ability for environmental experiences to make structural and functional changes in genetic expression is called *epigenetics* (Roth & Sweatt, 2011) and it has rocked the scientific world in recent years. Epigenetics is a branch of biology concerned with the expression of heritable, functional changes to the genome that do not involve alterations of the gene itself (sequence of deoxyribonucleic acid, or DNA code). All cells in the body have a nucleus which includes chromatin, a combination of DNA and histone protein in a spool-like structure (see illustration at https://en.wikipedia.org/wiki/Epigenetics#/media/File:Epigenetic_mechanisms.jpg). Biochemical markings of the chromatin control accessibility to genes and gene expression, allowing cells to adapt to an ever-changing environment, beginning with cell differentiation and fetal development in utero and continuing throughout the organism's lifetime. The molecular biology of epigenetic (outside of genetic) processes is beyond the scope of this textbook (see Tammen, Friso, & Choi, 2013 for overview), but the basics can be outlined briefly. There are two primary epigenetic mechanisms called DNA methylation (adding methyl groups to DNA) and histone modification (e.g., acetylation, adding acetyl groups to histone tails), both of which determine whether "packaged" cellular DNA is available for gene transcription by messenger RNA (mRNA) and subsequent translation by mRNA into proteins. DNA methylation increases the affinity between DNA and histone (alkaline proteins of eukaryotic cell nuclei) "spools," limiting accessibility to the genetic code and silencing the gene transcription machinery; therefore, DNA methylation provides an epigenetic mark (signal) for *gene silencing*. Histone acetylation, in contrast, usually decreases the affinity between histone and DNA, allowing for mRNA transcription and subsequent translation into proteins; thus, histone acetylation is an epigenetic mark for *gene activation*. Thus, an active chromatin structure ("packaged" genes available for activation) allows mRNA access to the genetic material for transcription and subsequent translation into proteins, which in turn controls the cell structure and function of the organism—including the cells or neurons in its brain (Day & Sweatt, 2011).

Evidence in rats indicates that epigenetic changes underlie the effects of maternal caretaking of pups on the adult behavior of these offspring (see Roth, 2012). Rodent mothers (dams) differ in the amount of grooming and licking they provide to pups within the first 10 days after birth. Compared to low nurturing mothers, dams that provided high levels of grooming and licking produced adult offspring with lower indicators of physiological stress and less fear responses to a novel environment.

Subsequent research showed that maternal care influenced DNA transcription of the glucocorticoid receptor gene (GR) in the hippocampus (see Roth, 2012). Notably, increased GR gene transcription helps to moderate the animal's neural and behavioral responses to stressful (aversive) situations with higher GR expression linked to less severe stress responses. Thus, adult male rats from high-caregiving dams were shown to have less DNA-methlylation markers (lower silencing of GR gene) and greater histone-acetylation markers (higher transcription of GR gene) in the hippocampus than dams providing lower amounts of grooming and licking of pups after birth. Further research subsequently established causal connections among postnatal maternal caretaking, epigenetic alterations of gene expression, and differences in adult offspring responses to stressful situations.

Non-genetic factors including learning experiences (e.g., conditioning) can result in epigenetic changes by histone acetylation and DNA methylation, which in turn affect brain and behavior via mRNA transcription and mRNA translation to proteins; although still controversial (Francis, 2014), it appears that cell division passes on epigenetic markings over one's lifetime and even from one generation to the next via noncoding mRNAs of sex cells or gametes (Dias & Ressler, 2014; Gapp et al., 2014; see Jablonka & Raz, 2009 for a complete discussion of transgenerational epigenetic inheritance). One implication is that learning sometimes can be transmitted epigenetically from one generation to the next with no change in the genes themselves. Also, in the future, it may be possible to produce lasting reversal of epigenetic changes by targeted early behavioral interventions (as in autism; see Chapter 13) or reverse epigenetic effects in later life by arranging new (re)programmed learning experiences (Tammen et al., 2013). Generally, evolution has provided animals with epigenetic mechanisms that allow for retention of learning experiences (changes in behavior due to the prevailing environmental contingencies) over an organism's lifetime and perhaps beyond.

The Selection of Operant Behavior

Early behaviorists like John B. Watson (1913) used the terminology of stimulus–response (S–R) psychology. From this perspective, stimuli force responses much like meat in a dog's mouth elicits (or forces) salivation. In fact, Watson based his stimulus–response theory of behavior on Pavlov's conditioning experiments. Stimulus–response theories are mechanistic in the sense that an organism is compelled to respond when a stimulus is presented. This is similar to a physical account of the motion of billiard balls. The impact of the cue ball (stimulus) determines the motion and trajectory (response) of the target ball. Although stimulus–response conceptions are useful for analyzing reflexive behavior and other rigid response patterns, the push–pull model is not as useful when applied to voluntary actions or *operants*. To be fair, Watson talked about "habits" in a way that sounds like operant behavior, but he lacked the experimental evidence and vocabulary to distinguish between respondent and operant conditioning.

It was B. F. Skinner (1935) who made the distinction between two types of conditioned responses, corresponding to the difference between operant and respondent behavior. In 1938, he introduced the term "operant" in his classic book, *The Behavior of Organisms*. Eventually, Skinner rejected the mechanistic (S–R) model of Watson and based operant conditioning on Darwin's principle of selection. The basic idea is that an individual *emits behavior that produces effects or consequences*. Based on these consequences, those performances that are appropriate to the environmental requirements increase, becoming more frequent in the population or class of responses for the situation; at the same time, less appropriate forms of response decline or become extinct.

Skinner recognized that operants are selected by their consequences (behavioral selection). He also noted that operant behavior naturally varies in form and frequency. Even the simple movement of opening the door to your house is not done exactly the same way each time, an observation consistent with recent optogenetic studies of variation of neuron firing in fruit flies (Vogelstein et al., 2014). Pressure on the doorknob, strength of pull, and the hand used change from one occasion to the next. If the door sticks and becomes difficult to open, a more forceful response may eventually occur (an effect of extinction). This energetic response may succeed in opening the door and become the most likely performance for the situation. Other forms of response may occur at different frequencies depending on how often they succeed in opening the door (reinforcement). Thus, *operants are selected by their consequences*.

Similarly, it is well known that babies produce a variety of sounds called "babbling." These natural variations in sound production are important for language learning. When sounds occur, parents usually react to them. If the infant produces a familiar sound (such as "ma"), parents often repeat it more precisely. Unfamiliar sounds are usually ignored. Eventually, the baby begins to produce sounds (we say talk) like other people in their culture or verbal community. Selection of verbal behavior by its social consequences is an important process underlying human communication and language (Skinner, 1957).

Culture and Behavior Analysis

Although much of the basic research in the experimental analysis of behavior is based on laboratory animals, contemporary behavior analysts are also concerned with human behavior. The behavior of people occurs in a social environment. Society and culture refer to aspects of the social environment, the social context, which regulates human conduct. One of the primary tasks of behavior analysis is to show how individual behavior is acquired, maintained, and changed through interaction with others. An additional task is to account for the practices of the group, community, or society that affect an individual's behavior (Lamal, 1997).

Culture is usually defined in terms of the ideas and values of a society. However, behavior analysts define **culture** as all the conditions, events, and stimuli arranged by other people that regulate human action (Glenn, 2004; Skinner, 1953). The principles and laws of behavior analysis provide an account of how culture regulates an individual's behavior. A person in an English-speaking culture learns to speak in accord with the verbal practices of that community. People in the community provide reinforcement for a certain way of speaking. In this manner, a person comes to talk like and share the language of other members of the public and, in doing so, contributes to the perpetuation of the culture. The customs or practices of a culture are therefore maintained through the social conditioning of individual behavior.

Another objective is to account for the evolution of cultural practices. Behavior analysts suggest that the principles of variation and selection by consequences occur at the biological, behavioral and cultural levels (Wilson, Hayes, Biglan, & Embry, 2014). Thus, cultural practices increase (or decrease) based on consequences produced in the past. A cultural practice of making containers to hold water is an advantage to the group because it allows for the transportation and storage of water. This practice may include using shells or hollow leaves, or making fired-clay containers. The cultural form selected (e.g., clay jars) is the one that proves most efficient and least costly. In other words, the community values and uses those containers that last the longest, hold the most, and are easily stored. People manufacture and use clay pots, while production and use of less efficient containers declines.

Behavior analysts are interested in cultural evolution because cultural changes alter the social conditioning of individual behavior. Analysis of cultural evolution suggests how the social environment is arranged and rearranged to support specific forms of human behavior. On a more practical level, behavior analysts suggest that the solution to many social problems requires a technology of cultural design. Behavioral technology also has been used to manage environmental pollution, encourage energy and resource conservation, increase recycling, enhance green technology and use, and regulate overpopulation (e.g., Austin, Hatfield, Grindle, & Bailey, 1993; Bostow, 2011; Lehman & Geller, 2004; Parece, Grossman, & Geller, 2013; Wilson, Hayes, Biglan, & Embry, 2014). Behavioral scientists have also analyzed the important cultural contingencies that have led to and maintained systemic racism, poverty, and violence and abuse (e.g., Biglan, 1995; Biglan, 2015; Isaacs, 1982; see review Matsuda, Garcia, Catagnus, & Brandt, 2020).

FOCUS ON: BURRHUS FREDERIC SKINNER

B. F. Skinner (1904–1990) was the intellectual force behind behavior analysis. He was born and named Burrhus Frederic Skinner on March 20, 1904 in Susquehanna, Pennsylvania. When he was a boy, Skinner spent much of his time exploring the countryside with his younger brother. He had a passion for English literature and mechanical inventions. His hobbies included writing stories and designing perpetual-motion machines. He wanted to be a novelist, and went to Hamilton College in Clinton, New York, where he graduated with a degree in English. After graduation in 1926, Skinner reported that he was not a great writer because he had nothing to say. He began reading about behaviorism, a new intellectual movement at that time, and as a result went to Harvard in 1928 to learn more about a science of behavior. He earned his master's degree in 1930 and his PhD the following year.

Fig. 1.4
A photograph of B. F. Skinner.
Source: Reprinted with permission from the B. F. Skinner Foundation.

Skinner (Figure 1.4) began writing about the behavior of organisms in the 1930s, when the discipline was in its infancy, and he continued to publish papers until his death in 1990. During his long career, he wrote about and researched topics ranging from utopian societies to the philosophy of science, teaching machines, pigeons that controlled the direction of missiles, air cribs for infants, and techniques for improving education. Some people considered him a genius, while others were critical of his theory of behavior.

Skinner was always a controversial figure. He proposed a natural-science approach to human behavior. According to Skinner, the behavior of organisms, including humans, was determined by observable and measurable processes. Although common sense suggests that we do things because of our feelings, thoughts, and intentions, Skinner stated that behavior resulted from genes and environment. This position bothered many people who believed that humans have some degree of self-determination and free will. Even though he was constantly confronted with arguments against his position, Skinner maintained that the scientific facts required that feelings, thoughts, and intentions—though very real private events—were not *causes* of behavior. He said that these internal (private) events were not explanations of behavior; rather these events were additional behaviors that, though not observable, could still be explained by genes and environment:

> The practice of looking inside the organism for an explanation of behavior has tended to obscure the variables which are immediately available for a scientific analysis. These variables lie outside the organism in its immediate environment and in its environmental history. They have a physical status to which the usual techniques of science are adapted, and they make it possible to explain behavior as other subjects are explained in science. These independent variables [causes] are of many sorts and their relations to behavior are often subtle and complex, but we cannot hope to give an adequate account of behavior without analyzing them.
>
> (Skinner, 1953, p. 31)

One of Skinner's most important achievements was his theory of operant behavior. The implications of behavior theory were outlined in his book, *Science and Human Behavior* (Skinner, 1953). In this book, Skinner discussed basic operant principles and their application to human behavior. Topics include self-control, thinking, the self, social behavior, government, religion, and culture. Skinner advocated the principle of positive reinforcement and argued against the use of punishment. He noted how governments and other social agencies often resort to punishment for behavior control. Although punishment works in the short run, he noted that it has many negative side effects. Positive reinforcement, Skinner believed, is a more effective means of behavior change—people act well and are happy when behavior is maintained by positive reinforcement.

People have misunderstood many of the things that Skinner said and did (Catania & Harnard, 1988). One popular misconception is that he raised his children in an experimental chamber—the so-called "baby in a box." Some critics claimed that Skinner used his daughter as an experimental subject to test his theories. A popular myth was that this experience drove his child crazy. His daughter, Julie, was confronted with this myth and recalls the following:

> I took a class called "Theories of Learning" taught by a nice elderly gentleman. He started with Hull and Spence, and then reached Skinner. At that time I had read little of Skinner, and I could not judge the accuracy of what was being said about Skinner's theories. But when a student asked whether Skinner had any children, the professor thought Skinner had children. "Did he condition his children?" asked another student. "I heard that one of the children was crazy." "What happened to his children?" The questions came thick and fast.
>
> What was I to do? I had a friend in the class, and she looked over at me, clearly expecting action. I did not want to demolish the professor's confidence by telling who I was, but I couldn't just sit there. Finally, I raised my hand and stood up. "Dr. Skinner has two daughters and I believe they turned out relatively normal," I said, and sat down.
>
> (Vargas, 1990, pp. 8–9)

In truth, the "box" that Skinner designed for his children had nothing to do with an experiment. The *air crib* is an enclosed bed that allows air temperature to be controlled. Because of this feature no blankets are needed, so the baby is free to move and there is no danger of suffocating. The air crib was designed to keep the child warm, dry, and safe. Most importantly, the infant spent no more time in the air crib than other children do in ordinary beds (Skinner, 1945). It was like a climate-controlled playpen. Indeed, the air crib was something that was advertised and sold in *Ladies Home Journal* at the time.

Although Skinner did not experiment with his children, he was always interested in the application of conditioning principles to human issues. His writings on applied behavioral technology led to the field of applied behavior analysis or ABA (see Rutherford, 2009 who provides an historical account of the transition from Skinner's work in the laboratory to applications of behavior principles in everyday life). Applied behavior analysis is concerned with the extension of behavior principles to socially important problems. One area of application that Skinner wrote about extensively was teaching and learning. Although Skinner recognized the importance of behavior principles for teaching people with learning disabilities, he claimed that the same technology could be used to improve our general educational system. In his book *The Technology of Teaching*, Skinner (1968) offered a personalized system of positive reinforcement for the academic performance of students. In this system, teaching involves arranging materials, designing the classroom, and programming lessons to shape and maintain the performance of students. Learning is defined objectively in terms of answering questions, solving problems, using grammatically correct forms of the language, and writing about the subject matter.

In the later part of his life, Skinner worked with Margaret Vaughan (Skinner & Vaughan, 1983) on positive approaches to the problems of old age. Their book *Enjoy Old Age: A Program of Self-Management* is written for the elderly reader and provides practical advice on how to deal with daily life. For example, the names of people are easy to forget, and even more so in old age. Skinner and Vaughan suggest you can improve your chances of recalling a name by reading a list of people you are likely to meet before going to an important occasion. If all else fails "you can always appeal to your age. You can please the friend whose name you have momentarily forgotten by saying that the names you forget are always the names you most want to remember" (Skinner & Vaughan, 1983, p. 52).

Skinner, who held the Edgar Pierce Chair in Psychology, officially retired from Harvard University in 1974. Following his retirement, he continued an active program of research and writing. Each day he walked two miles to William James Hall, where he lectured, supervised graduate students, and conducted experiments. Eight days before his death on August 18, 1990, B. F. Skinner received the first (and only) Citation for Outstanding Lifetime Contribution to Psychology from the American Psychological Association (Schlinger, 2011). The citation for the award, published in the *American Psychologist*, read "Few individuals have had such a dynamic and far-reaching impact on the discipline" (1990, p. 1205). In a study of renowned psychologists by Haggbloom (2002), Skinner ranked as the most eminent psychologist of the 20th century. Skinner's contributions to psychology and a science of behavior are documented in the film *B. F. Skinner: A Fresh Appraisal* (1999). Murray Sidman, a distinguished researcher in the experimental analysis of behavior, narrated the film (available from the bookstore of the Cambridge Center for Behavioral Studies, www.behavior.org).

A BRIEF HISTORY OF BEHAVIOR ANALYSIS

Contemporary behavior analysis is based on ideas and research that became prominent at the turn of the 20th century. The Russian scientist Ivan Petrovich Pavlov discovered the conditioned reflex (a reflex that only occurs under a particular set of conditions, such as the pairing of stimuli), and this was a significant step toward a scientific understanding of behavior.

Ivan Petrovich Pavlov (1849–1936)

Pavlov (Figure 1.5) was born the son of a village priest in 1849. He attended seminary school to follow his father into the priesthood. However, after studying physiology he decided on a career in the biological sciences. Although his family protested, Pavlov entered the University of St. Petersburg where he graduated in 1875 with a degree in physiology. After completing his studies in physiology, Pavlov was accepted as an advanced student of medicine. He distinguished himself and obtained a scholarship to continue his studies of physiology in Germany. In 1890, Pavlov was appointed to two prominent research positions in Russia. He was Professor of Pharmacology at the St. Petersburg Medical Academy and Director of the Physiology Department. For the next 20 years, Pavlov studied the physiology of digestion, and in 1904 he won the Nobel Prize for this work, the year that B. F. Skinner was born.

18 A Science of Behavior: Perspective, History, and Assumptions

Fig. 1.5
A photograph of Ivan Petrovich Pavlov.

Source: Reprinted with permission from the Archives of the History of American Psychology, Center for the History of Psychology, The University of Akron.

Ivan Pavlov initially worked on the physiology of salivation and digestion; later he began investigations of "psychic secretions" involving the salivary reflex and its role in digestion. The animals were brought into the laboratory and put in restraining harnesses. As shown in Figure 1.6, food was then placed in the dogs' mouths and the action of the salivary glands was observed and measured.

The analysis of the salivary reflex was based on prevailing notions of animal behavior. At this time, many people thought that animals, with the exception of humans, were complex biological machines. The idea was that a specific stimulus elicited a particular response in much the same way that turning a key starts an engine. In other words, animals reacted to the environment in a simple cause–effect manner. Humans, on the other hand, were seen as different from other animals in that their actions were purposive. Humans were said to anticipate future events. Pavlov noticed that his dogs began to salivate at the sight of an experimenter's lab coat *before* food was placed in the animal's mouth. This suggested that the dogs "anticipated" the delivery of food. Pavlov recognized that such a result challenged conventional wisdom.

Pavlov made an important observation in terms of the study of behavior. He reasoned that anticipatory reflexes were learned or conditioned. Further, Pavlov concluded that these conditioned reflexes were an essential part of the behavior of organisms. Although some behaviors were described as innate (present at birth) reflexes, other actions were based on conditioning that occurred during the animal's life. These **conditioned reflexes** (termed conditional reflexes in Pavlov, 1960) were present to some degree in all animals, but were most prominent in humans.

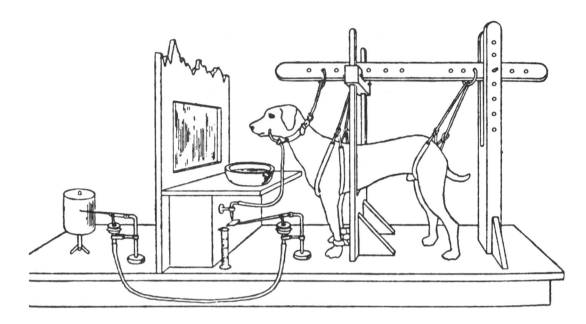

Fig. 1.6
A dog in the experimental apparatus used by Pavlov.

The question was how to study conditioned reflexes systematically. Pavlov's answer to this question represents a major advance in the experimental analysis of behavior. If dogs reliably salivate at the sight of a lab coat, Pavlov reasoned, then any arbitrary stimulus that preceded food might also be conditioned and evoke salivation. Pavlov replaced the experimenter's lab coat with a stimulus that he could systematically manipulate and reliably control. In some experiments, a metronome (a device used to keep the beat while playing the piano) was presented to a dog just before it was fed. This procedure resulted in the dog eventually salivating to the sound of the metronome. If a particular beat preceded feeding while other rhythms did not, the dog salivated most to the sound associated with food.

Although Pavlov was a physiologist and believed in mental associations and subjective experience (Specter, 2014), his research was directed at observable responses and stimuli and he foreshadowed the modern study of behavioral neuroscience, the objective and direct study of brain and behavior emphasized in this book. He discovered many principles of the conditioned reflex including spontaneous recovery, discrimination, generalization, and extinction. The later part of his career involved an experimental analysis of neurosis in animals. He continued these investigations until his death in 1936.

John Broadus Watson (1878–1958)

Pavlov's research became prominent in North America, and the conditioned reflex was incorporated into a more general theory of behavior by the famous behaviorist John B. Watson (Figure 1.7). Watson argued that there was no need to invent unobservable mental associations to account for human and animal behavior. He proposed that psychology should be a science based on observable behavior. Thoughts, feelings, and intentions had no place in a

Fig. 1.7

An image of John Watson.

Source: Reprinted with permission from the Archives of the History of American Psychology, Center for the History of Psychology, The University of Akron.

scientific account, and researchers should direct their attention to muscle movements and neural activity. Although this was an extreme position, Watson succeeded in directing the attention of psychologists to behavior–environment relations, although his status as the originator or "founder" of behaviorism is doubtful (Malone, 2014).

Watson graduated from Furman University in 1899, when he was 21 years old. After spending a year as a public-school teacher, he was admitted to graduate studies at the University of Chicago. There he studied philosophy with John Dewey, the famous educator. While a graduate student at the University of Chicago, he also studied psychology with James Angell and biology and physiology with Henry Donaldson and Jacques Loeb (Pauley, 1987). In 1903, he obtained his doctorate for research with laboratory rats. The experiments concerned learning and correlated changes in the brains of these animals.

Watson published his most influential work in 1913 in the *Psychological Review*, "Psychology as the Behaviorist Views It." This paper outlined Watson's views on **behaviorism** as the only way to build a science of psychology, avoiding the philosophical speculation of mind–body dualism and focusing research on objective behavior (Watson, 1913; also see Malone & Garcia-Penagos, 2014 for Watson's contributions to behaviorism).

In his 1913 paper, Watson rejected as scientific data what people said about their thoughts and feelings by pointing to the unreliability of psychological inferences about another person's mind. He also noted that the psychology of mind (because it is not observable) had little practical value for behavior control and public affairs. Modern behavior analysts, informed by the writings of Watson and Skinner, study what people *say* as verbal behavior regulated by the social environment (behavior of other people), not reports on the mind or mental states. Behavior analysts study thinking and remembering as private behavior often related to challenges or problems faced by the person. A person may "think about the house key" and check their coat pocket just before leaving for work, especially if they have been locked out in the past. Notice that thinking about the key is not treated as a *cause* of behavior, but as more behavior to be explained by its interaction with the environment (history of being locked out of the house). In other words, private behavior falls under the same environmental control as overt (observable) behavior.

In addition to his influence on modern behaviorism, Watson is best known for his early studies of conditioning. Perhaps Watson's most famous experiment was one he conducted with Rosalie Rayner, a student at Johns Hopkins University, on the study of fear conditioning with Little Albert. Rayner later went on to publish studies on developmental psychology with Watson, but also independently.

Little Albert (see Figure 1.8) was a normal, healthy 9-month-old child who attended a day-care center. Watson and his assistant used classical-conditioning procedures to condition Little Albert to fear a white rat. When first exposed to the animal, the child looked at the rat and tried to touch it. The unconditioned stimulus was the sound of a hammer hitting an iron rail. This sound made Little Albert jump, cry, and fall over. After only six presentations of the rat (CS) and the noise (US) the furry animal alone also produced the fear responses. The next phase of the experiment involved a series of tests to see if the child's fear reaction transferred or *generalized* to similar stimuli. Albert was also afraid (showed fear behaviors) when presented with a white rabbit, a dog, and a fur coat. While the conditioning of fear in infants today would likely be seen as unethical, at the time when it was conducted, views on research ethics with humans were not well developed.

At this point, Watson and his student Rayner discussed a number of techniques that could be used to eliminate the child's fear. (If fear behavior was learned, the opposite behavior could also be learned.) Unfortunately, Little Albert was removed from the day-care center before counterconditioning could be carried out.

NOTE ON: LITTLE ALBERT AND WHAT HAPPENED TO HIM

Not surprisingly, Watson received bad press for his fear conditioning experiment with Little Albert (see Figure 1.8), and introductory psychology textbook authors and instructors have commented on the methods and ethics of the study (Kalat, 2014). A long-standing mystery has been the identity of Little Albert and what happened to him after the experiment ended. Did the child grow up with a strange phobic reaction to white, furry objects? In a 2009 article in the *American Psychologist* on "Finding Little Albert ..." Hall P. Beck and his associates claimed to have found the identity of the boy's mother and discovered that Albert was actually Douglas Merritte (Beck, Levinson, & Irons, 2009). With further investigation by Alan J. Fridlund, Beck and colleagues reported that Douglas had hydrocephalus, a neurological impairment, and the boy was not a healthy, normal child as Watson had stated in his 1920 experiment (Fridlund, Beck, Goldie, & Irons, 2012; Watson & Rayner, 1920). These authors erroneously suggested that Watson had known about the child's neurological condition and had intentionally misrepresented his health, a conclusion that further damaged the already soiled reputation of Watson, behaviorism, and the Little Albert experiment.

Fig. 1.8
Rosalie Rayner holds Little Albert as he plays with a white lab rat before fear conditioning. Watson, left, observes the boy.
Source: Archives of the History of American Psychology, The University of Akron. Published with permission.

In a 2014 evidence-based rebuttal published in the *American Psychologist*, however, these conclusions and charges by Beck's team were severely challenged by Russell Powell and his colleagues in their article, "Correcting the Record on Watson, Rayner and Little Albert ..." (Powell, Digdon, Harris, & Smithson, 2014). Powell's team was able to identify another child by the name of Albert Barger with characteristics that matched closely those originally described for Little Albert, involving normal health and development. Furthermore, these investigators by an extensive historical analysis established that the weight of the evidence indicated that Albert Barger was in fact the Little Albert in Watson's experiment and not Douglas Merritte as claimed by the Beck team. With regard to the claim of misrepresentation by John Watson, Powell and his colleagues concluded "there is no evidence that he [Watson] committed fraud in his scientific endeavors" (p. 23). Given the life of most textbooks, it will take time to correct the accusations directed at Watson. Many introductory psychology textbooks probably will continue to report the bogus claims of neurological impairment of Little Albert and fraudulent science by John B. Watson, thereby maintaining this myth about behaviorism and its renowned advocate.

As for whether Little Albert (Albert Barger) grew up with a conditioned phobic reaction to furry objects, Powell and colleagues were able to establish that he showed some aversion to and dislike of dogs, but there was no clear evidence that Albert showed a generalized avoidance of furry animals or other objects related to his participation in Watson's experiment. Thus, the speculation by Watson and Rayner of lasting effects of their fear conditioning is not well supported by the extensive follow-up inquiry on Albert Barger's life history.

Watson had many professional interests, and he investigated and wrote about ethology, comparative animal behavior, neural function, physiology, and philosophy of science. Based on his provocative views and charisma, he was elected president of the American Psychological Association in 1915 when he was only 37 years old. After leaving Johns Hopkins University in

1920, he became successful in industry by applying conditioning principles to advertising and public relations (Buckley, 1989). Watson implemented the use of "subliminal" suggestion and the pairing of hidden symbols in advertising—techniques that are still used today.

Edward Lee Thorndike (1874–1949)

Watson's behaviorism emphasized the conditioned reflex. His analysis focuses on the events that precede action and is usually called a stimulus–response approach. Another American psychologist, Edward Lee Thorndike (Figure 1.9), was more concerned with how success and failure affect the behavior of organisms. His research emphasized the events that followed behavior. In other words, Thorndike was the first scientist to systematically study operant behavior, although he called the changes that occurred **trial-and-error learning** (Thorndike, 1898).

Thorndike was always intrigued by animal behavior. While he was at Harvard, his landlady became upset because he was raising chickens in his bedroom. By this time, William James, his mentor at Harvard, and Thorndike were good friends, and Thorndike moved his experiments to the basement of James's house when he could not get laboratory space at Harvard. (This is not something you could do today, as all laboratory animal research with protected species must be approved and conducted in regulated facilities.) He continued his research and supported himself by tutoring students for two years at Harvard. Then Thorndike moved to Columbia University, where he studied with James McKeen Cattell, the famous expert on intelligence testing. Thorndike took two of his "smartest" chickens with him to Columbia, but soon switched to investigating the behavior of cats.

Fig. 1.9
An image of Edward Thorndike.

Source: Reprinted with permission from the Archives of the History of American Psychology, The University of Akron. Published with permission.

At Columbia University, Thorndike began his famous experiments on trial-and-error learning in cats. Animals were placed in what Thorndike called a "puzzle box" and food was placed outside the box (Chance, 1999). A cat that struggled to get out of the box would accidentally step on a treadle, pull a string, and lift a latch. These responses resulted in opening the puzzle-box door. Thorndike found that most cats took less and less time to solve the problem after they were repeatedly returned to the box (i.e., repeated trials). From these and additional observations, Thorndike developed the first formulation of the **law of effect**, which in essence states that behaviors that result in a satisfaction or pleasure get "stamped in" and those that result in discontent and are not successful, get "stamped out." We can see some problems with the law of effect, of course. Not everything that results in satisfaction or success gets repeated. For example, a college student may find candy satisfying, but is it a strong enough stimulus to alone increase the behavior of studying for a big exam? It might be the exam that does this, which is not all that pleasant. Indeed, there are also situations in which we have to repeatedly do things that we do not enjoy. Many of us, for example, have experienced going to places of employment that are labor-intensive or have toxic working environments, yet we show up and do the work. Therefore, the law of effect has many exceptions, rendering it not a law at all.

Today, Thorndike's law of effect is restated more carefully and scientifically as the principle of reinforcement. This principle states that all operants may be followed by outcomes that increase or decrease the probability of response in the same situation. The focus is not on the aspects of the stimulus, but rather how the *stimulus impacts the probability of response*. Notice that references to "stamping in" and "pleasure" are not necessary for this definition and that nothing is lost by this more accurate restatement of the law of effect.

B. F. Skinner and the Rise of Behavior Analysis

The works of Pavlov, Watson, Thorndike, and many others have influenced contemporary behavior analysis. Although the ideas of many scientists and philosophers have had an impact, B. F. Skinner (1904–1990) is largely responsible for the development of modern behavior analysis. In the "Focus On: Burrhus Frederic Skinner" section, we described some details of his life and some of his accomplishments. An excellent biography is available (Bjork, 1993), and Skinner himself wrote a three-volume autobiography (Skinner, 1976; Skinner, 1979; Skinner, 1983).

Behavior analysis has continued to grow and thrive since the early days. In 1964, the number of behavior analysts had grown so much that the American Psychological Association established Division 25 called The Experimental Analysis of Behavior, which has several thousand members. Subsequently, the Association for Behavior Analysis (ABA) (now called Association for Behavior Analysis International—ABAI) was founded in the late 1970s, and is still growing today with thousands of members worldwide. This association holds an annual international convention in the USA, a convention every other year in a non-US country, an annual conference on autism, and special conferences on particular topics, such as culture or substance abuse. It also has nearly 100 affiliated regional chapters (over 30,000 members) all over the world, many with their own regional conferences. ABAI publishes six journals with their most impactful journal being *Perspectives on Behavior Science* (formerly *The Behavior Analyst*).

A continuing issue in the field of behavior analysis is the continuum of applied behavior analysis and basic research. During the 1950s and 1960s, no clear distinction existed between applied and basic investigations (see Rutherford, 2009 for more on the transition from the operant laboratory to applications of behavior principles in everyday life). This was because applied behavior analysts were trained as basic researchers. The first applications of behavior principles therefore came from the same people who were conducting laboratory experiments. The applications of behavior principles were highly successful, and this led to a greater demand for people trained in applied behavior analysis. Soon applied researchers were no longer working in the laboratory or reading the basic journals. Today, Masters programs that focus on the training of applied behavior analysts far outnumber those that train basic scientists in behavior analysis (e.g., PhDs in the experimental analysis of behavior). The demand for effective treatments of applied behavioral problems has been met with extensive training of those that can help. The difference in these applied programs as opposed to the basic, though, is that many of these programs do not contain the basic laboratory experience including animals (e.g., pigeons, rats). However, applied behavior analysts are naturally more concerned with procedures to improve or help those with severe disorders, and are less connected to exploring basic phenomena directly with traditional animal laboratory models.

The large influx of applied behavior analysts has been observed greatly in areas such as the treatment of autism and developmental disabilities, behavioral safety, and behavioral medicine. Many of the college degree programs for applied behavior analysts offer the credential of Board Certified Behavior Analyst, which means that specific standards of training and ethics have been met. These training requirements involve extensive fieldwork experience and supervision while working in applied settings (e.g., autism services). This certification is also frequently required for legal licensure to practice in several states in the US.

More recent trends in behavior analysis also show the integration of the basic science in behavior analysis with other areas. For instance, there has been a trend towards infusing behavioral economics, behavioral neuroscience, and behavior analysis in examining problems such as substance abuse. The reinforcer pathology model (Bickel et al., 2011), for example, focuses on variations in behavior sensitivity to delays and effort to stimuli that contribute to the overconsumption of drugs and food (e.g., DeHart et al., 2020). Neural structures, such as reward areas and the prefrontal cortex, that play a role in these reinforcer pathologies, have been identified as important neural mechanisms. Another area is in obesity. The role of the D2 dopamine receptor subtype is altered with chronic exposure to high-fat, high-sugar diets. These types of diets also alter reward processes, contributing to food functioning more highly as a reinforcer (see Johnson & Kenny, 2010; Robertson & Rasmussen, 2017).

The principles of behavior analysis are also foundational to many empirically supported treatments in clinical psychology. The earlier days of behavior therapy, in which environmentally based treatments were used to reduce a variety of problems and disorders, such as enuresis (bed wetting), bruxism (teeth grinding), anxiety-related disorders (including post-traumatic stress disorder), and depression were effective and successful, and many are still used today. More recently, the third-wave behavioral therapy Acceptance and Commitment Therapy (ACT) (see Hayes et al., 2009), which is based on changing the relations of emotions and contexts with the words we tell ourselves through conditioning, is one of the most successful clinical therapies of modern times. It is important to mention that the foundations for all of these therapies were derived from studies from the laboratory. One area in particular—relational frame theory—contributes to ACT (see Hayes, 2004).

Other changes in the field relate to inclusion and diversity. Before the late 1800s and early 1900s, our understanding of medicine was superstitious and often flat-out inaccurate. The main belief in disease was that illnesses and mental health problems were caused from an imbalance of blood. This is why leeches and blood-letting was a common practice for treating a variety of conditions at the time. Then, the scientific method became the foundation of medicine and fundamentally changed our understanding of medicine (see *The Great Flu Pandemic* for this historical account). Other areas, including psychology, also began systematically using the foundation of the scientific method for determining causes. During this era when we read the history books of various sciences, we see that it was mostly white men who were participating as scientists during this time. This was true even in psychology with a few notable exceptions, such as Rosalie Rayner and Mary Cover Jones. Today, women are participating much more as both scientists and practitioners and there is much more demographic variation in terms of who our scientists and practitioners are from the perspectives of ethnicity, race, sexual orientation, gender identity, ablebodiedness, and the like. Indeed, the Black Applied Behavior Analyst, Inc. association, which was organized in 2019 as one of the ABAI affiliate chapters, has the primary aim to reduce barriers in the field related to race and to promote diversity and inclusion; the group has over 1,000 members to date.

The early years of behavior analysis are sometimes difficult to understand through our modern lens. Back in the early days of behavior analysis (1940s–1960s), family members sought behavior analysts to help loved ones with challenging behaviors, including aggression and self-injurious behavior. The goal was to increase safety and reduce harm, but sometimes the goal was to help behavior conform to what was viewed at the time as "normal." One especially controversial study was published in 1974 by Rekers and Lovaas. In this study, a male child was showing what was regarded at the time as gender-non-conforming behavior, such as preferences for cross-gender clothing and cosmetics and behavioral mannerisms that were described as feminine. At the time, this type of behavior for a young boy was viewed as abnormal by the American Psychiatric Association (see Kawa & Giordano, 2012 for a history of the *Diagnostic and Statistical Manual*). Because parents wanted their children to fit in, they sought help for changing this behavior. The applied behavior analysts used social reinforcement to increase more masculine behavior and punishment to reduce feminine behavior.

Today, because of an extensive body of research, we know a great deal more about behavior related to gender identity and sexual orientation and what we accept and affirm for every individual includes a much broader range of behavior. The Association for Behavior Analysis International, the American Psychological Association, and the American Psychiatric Association have taken strong positions on treatments that aim to change behavior related to gender orientation or sexual orientation, such as conversion therapy. Not only are these therapies often ineffective in the long run, they can also be harmful to the individual. In short, these types of therapies are viewed as unethical and their practice is condemned. Presently, there is also discussion on the neurodiversity movement, in which more variation in behavior in individuals who are diverse from a neurological standpoint (e.g., autistic individuals) is accepted. These types of cultural changes affect treatment goals for behavioral change.

Science is self-correcting. When we discover new information, it is up to us to look at our current views and determine if they reflect what the new data offer. We do not hold tightly to old, outdated views, no matter how much we may feel they are right. Being a scientist means looking at perspectives with as much objectivity and humanity as possible, to consider alternate perspectives, and to do that with humility—we recognize that we do not know all of the answers. Openness is key, especially if there are views that can be potentially harmful or unfair to others that are in less dominant positions. When we practice openness and humility, we help society progress.

Our goal in this textbook also is to further the integration of applied and basic areas of behavior analysis and to encourage closer ties to other biological sciences. We have written this book assuming that an acquaintance with basic research is important, even for those who are primarily concerned with behavioral applications and translational studies. We also assume that those who have an interest in laboratory science should also appreciate the manner and complexities of the issues that surround the application of behavioral principles to socially and culturally relevant problems. Students can study this text for a basic grounding in behavior science, or for a solid foundation in human behavior and application.

SCIENCE AND BEHAVIOR: SOME ASSUMPTIONS

All scientists make assumptions about their subject matter. These assumptions are based on prevailing views in the discipline and guide scientific research. In terms of behavior analysis, researchers assume that the behavior of organisms is lawful and deterministic. This means that it is possible to study the interactions between an organism and its environment in an objective manner. To carry out the analysis, it is necessary to *isolate* behavior–environment relations. The scientist must identify events that reliably precede the onset of some action and the specific effects or consequences that follow behavior. If behavior systematically changes with variation in the environmental conditions, then behavior analysts assume that they have explained the action of the organism. There are other assumptions that behavior analysts make about their science.

The Private World

Contemporary behavior analysts include internal, *private events* as part of an organism's environment. Part of each person's experiences is private with stimulation only available to that person. Consider physiological sensations that relate to internal functioning, like an upset stomach, full bladder, or low blood sugar. Consider also the emotions of feeling sad or worried (e.g., muscle tension or "knots" in your stomach). These private events have the same status as external, public stimuli such as light, noise, odor, and heat; they are very real to the observer. Both public and private events regulate behavior. Although this is so, behavior analysts usually emphasize the external, public environment. This is because public events are the only stimuli available for behavior change, a major goal of behavior analysis. It is difficult (if not impossible) for two more people to observe objectively what someone is privately experiencing. We have to rely on a response (often a verbal report) of these types of phenomena, which are then also subjective to consequences from the environment, such as social disapproval.

It is important to note that, just like public events, private events also are regulated by the environment. The objective procedures of psychological experiments or clinical treatments often are giving instructions and observing how the person acts. From a behavioral view, the instructions are external, public stimuli that regulate both verbal and nonverbal behavior. Even when a drug is administered to a human participant, the chemical alters the person's biochemistry, which can sometimes be felt (private event), but the *direct injection* of the drug or swallowing of the pill is an external, public event that subsequently regulates behavior. To make this clear, without the drug administration neither the biochemistry nor the behavior of the person would change.

Many psychological studies involve giving information or a condition to a person to change or activate cognitive processes, which in turn activates behavior. In the cognitive view, thoughts are used to *explain* behavior. The problem is that the existence of thoughts (or feelings) is often *inferred* from the behavior to be explained, leading to circular reasoning. For example, a child who peers out of the window at around the time her mother usually comes home from work is said to do this because of an "expectation." The expectation of the child is said to explain why the child peers out of the window. This is an explanatory fiction because the cognition (expectation) is inferred from the behavior it is said to explain, when actually, the explanation is observable and testable and can be found in the environment: The boy looks out the window because around this same time every day, his mother comes home. You can test this by having his mother come home at a different time; his window peering would likely change. Cognitions could explain behavior if the existence of thought processes were based on some evidence other than behavior. In most cases, however, there is no independent evidence that cognitions caused behavior, and the explanation is not scientifically valid. One way out of this problem of logic is not to use thinking and feeling as causes of behavior. Thinking and feeling are treated *as* behavior to be explained and the explanations come from the environment—just like observable behavior.

Feelings and Behavior

Many people assume that their feelings and thoughts explain why they act as they do. Contemporary behavior analysts agree that people feel and think, but they do not consider these events as *causes* of behavior. They note that these terms are more correctly used as verbs rather than nouns. Instead of talking about thoughts, behavior analysts point to the action word "thinking." And instead of analyzing feelings as things we possess, the behavioral scientist focuses on the *action* of feeling or sensing. In other words, thinking and feeling are activities of the organism that require explanation.

Feelings: Real, But Not Causes

Because feelings often occur at the same time as we act, they are often mistaken as causes of behavior. Although feelings and behavior necessarily go together, it is the environment that determines how we act, and at the same time, how we feel. Feelings are real private activities, but they are the result of the environmental events that regulate behavior. Thus, a behavioral approach requires that the researcher trace feelings back to the interaction between behavior and environment.

Pretend that you are in an elevator between the 15th and 16th floors when the elevator suddenly stops, and the lights go out. You hear a sound that appears to be the snapping of elevator cables. Suddenly, the elevator lurches and then drops 2 feet. You call out, but nobody comes to your rescue. After about an hour, the elevator starts up again, and you get off on the 16th floor. Six months later, a good friend invites you to dinner. You meet downtown, and you discover that your friend has made reservations at a restaurant called The Room at the Top, which is located on the 20th floor of a skyscraper. Standing in front of the elevator, a sudden feeling of panic overwhelms you. You make a socially appropriate excuse like, "I don't feel well," and you leave. What is the reason for your behavior and the accompanying feeling?

There is no question that you feel anxious, but this feeling is not why you decide to go home. Both the feelings of anxiety and your decision to leave are easily traced to the negative experience in the elevator that occurred six months ago. It is this prior conditioning that behavior analysts emphasize. Notice that the behavioral position does not deny your feelings. These are real events. However, it is your previous interaction with the broken elevator that changed both how you feel and how you act.

Reports of Feelings

You may still wonder why behavior analysts study overt behavior instead of feelings—given that both are changed by experience. The answer concerns the accessibility of feelings and overt behavior. Much of the behavior of organisms is directly accessible to the observer or scientist. This public behavior provides a relatively straightforward subject matter for scientific analysis. In contrast, feelings are largely inaccessible to the scientific community. Of course, the person who feels anxiety has access to this private stimulation, but the problem is that reports of feelings are highly unreliable.

This unreliability occurs because we learn to talk about our feelings (and other internal, private events) as others have trained us to do. During socialization, people teach us how to describe ourselves, but when they do this, they have no way of accurately knowing what is going on inside us. Parents and teachers rely on public cues to train self-descriptions. They do this by commenting on and correcting verbal reports when behavior or events suggest a feeling. A preschooler is taught to say "I feel happy" when the parents guess that the child is happy. The parents may base their judgment on smiling, excitement, and affectionate responses from the child. Another way in which this training occurs is that the child may be asked "Are you happy?" in a circumstance where the parents expect the child to feel this way (e.g., on their birthday). When the child appears to be sad, or circumstances suggest this should be so, saying "I am happy" is not reinforced by the parents. Eventually, the child says "I am happy" in some situations and not in others.

Perhaps you have already noticed why reports of feelings are not good scientific evidence. Reports are only as good as the training of *correspondence* between public conditions and private events. In addition to inadequate training, there are other problems with accurate descriptions of feelings. Many of our internal functions are poorly correlated (or uncorrelated) with public conditions, and this means that we cannot be taught to describe such events accurately. Although a doctor may ask for the general location of a pain (e.g., the abdomen), they are unlikely to ask whether the hurt is in the liver or the spleen. This report is simply inaccessible to the patient because there is no way to teach the correspondence between exact location of damage and public conditions. Generally, we are able to report in a limited way on private events, but the unreliability of such reports makes them questionable as scientific observations. Based on this realization, behavior analysts focus on the study of behavior rather than feelings.

Thinking as Behavior

Behavior analysts have also considered "thinking" and its role in a science of behavior. In contrast to views that claim a special inner world of thought, behavior analysts suggest that human thought may be analyzed as human behavior.

Fig. 1.10
An image of Darrian Robinson moving a piece on a chessboard. Ms. Robinson, who learned to play chess when she was 9 years old, became one of the highest-ranking chess players in the United States Chess Federation by the time she was 19. Ms. Robinson illustrates that thinking is operant behavior.
Source: *Chicago Tribune*, July 20, 2014.

A number of behavioral processes, such as generalization, discrimination, matching to sample, and stimulus equivalence (see later chapters), give rise to behavior that, in a particular situation, may be attributed to higher mental functions. From this perspective, thinking is treated as **private behavior** (see Moore, 2003, and Tourinho, 2006 on private events); that is, behavior only accessible to the person doing it.

One of the more interesting examples of thinking, which involves private behavior, might be observed in a game of chess (Figure 1.10). We may ask another person, "What is the player thinking about?" A response like "She is probably thinking of moving the castle" refers to thinking that precedes the move itself. Sometimes this prior behavior is observable—the player may place a hand on the castle in anticipation of the move, showing a high probability of moving the castle given the layout of the chessboard. At other times, such behavior is private and cannot be observed by others. An experienced chess player may think about the game, evaluating the consequences of moving the castle to different locations. Of course, evaluating the consequences of chess moves itself depends on the actual consequences of such moves in the past (*history of reinforcement*). How often did the move result in a capture or checkmate?

Presumably, the private behavior of playing chess is overt when a person learns to play the game. For example, first the basic rules of the game are explained and a novice player is shown how the pieces move and capture. In moving the pieces from place to place, the player is asked to describe the relations between the opposing chess pieces. This establishes the behavior of viewing and describing the layout of the board and possible moves. As the player receives additional corrective feedback and reinforcement, viewing the layout becomes more skillful. The novice begins to see relations and moves that were not previously apparent. During the first few games, new players are often given instructions like "Don't move your knight there, or you'll lose it." Additionally, the player may be told, "A better move would have been …," and a demonstration of the superior move is usually given. After playing a number of games, the student is asked to explain why a particular move was made, and the explanation is discussed and evaluated. Eventually, the teacher stops prompting the player and encourages the person to play chess in silence. At this point, viewing the layout of the board (e.g., white controls the center of the board) and describing the possible consequences of moves (e.g., moving the knight to this square will split the two rooks) becomes covert and private.

As skill at chess further improves, the player no longer relies on viewing the board and its layout, but increasingly relies on thinking about the game. Blindfolded chess masters report that their thinking does not use mental images as surrogates for viewing the board, but rather *abstractions with minimal or no physical features*. Mechner (2010) suggests that "visualizing the board" involves private verbalizing rather than mental imaging—something like verbal chess notation. In chess notation, a letter and the coordinate of the destination square indicate each move of a piece. For example, a sequence of moves might be as follows: move a bishop to position e5 (Be5), move a knight to f3 (Nf3), and move a pawn to c5 (c5 with no initial).

FOCUS ON: THINKING ALOUD ABOUT CHESS MOVES

The current layout of the chessboard provides players with the opportunity to think about moving pieces to gain either a momentary or long-run advantage in the game. One way to study "thinking about chess moves" is to have tournament and expert players think aloud (saying what they are thinking about) when presented with chessboard problems with known best solutions (Moxley, Ericsson, Charness, & Krampe, 2012). Although the "think aloud" technique is related to the study of information processing and cognition, thinking aloud also is useful in behavior analysis. From a behavioral viewpoint, "thinking aloud" is considered a verbal measure of private behavior (thinking about X) that precedes public actions (making a choice), allowing for an analysis of the correspondence between saying (what I am thinking) and doing (making the choice), and its behavioral consequences (see Austin & Delaney, 1998). Moxley et al. (2012) used a chess computer program to assign "move strength scores" to the first move mentioned and to the final move actually made after further deliberation (more thinking aloud time), regarding chessboard problems (positions of pieces) that varied in degree of difficulty (easy and hard).

Figure 1.11 shows the estimated move strength scores for first move mentioned and final move chosen after further deliberation for experts and tournament players on easy and hard chessboard problems. For easy problems (top panel), experts do much better (higher move strength) on first moves mentioned than tournament players (two black bars), an effect called intuitive thinking; both types of players (tournament and expert) show improved move strength after further deliberation (move chosen), but improvement is greater for tournament players. Thus, thinking more (deliberating) about easy chess problems generally benefits all chess players (experts and tournament) by gaining a subsequent advantage (move strength) in the game, but tournament players gain the most from continuing to think about the problem.

As for the hard chessboard problems (bottom panel), experts again do better in move strength than tournament players (two black bars), indicating a difference in intuitive thinking favoring experts; tournament players show a small, but statistically unreliable, increase in move strength on the move chosen after deliberation, which suggests that further thinking about difficult chess problems did not gain an advantage for tournament players. In contrast, experts confronted with difficult problems showed a reliable improvement in move strength after

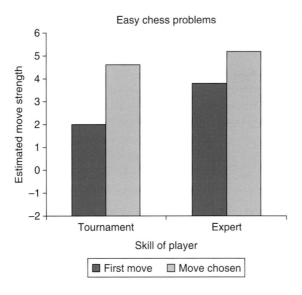
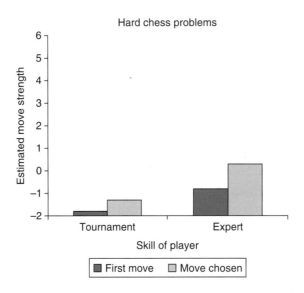

Fig. 1.11

The top panel of this figure shows a graph that depicts the estimated move strength in chess as a function of the skill of the chess player—tournament level or expert level for each chess problem. This is for easy chess problems. The strength of the first mentioned move (black) and the move chosen after deliberation (gray) is shown on the y-axis. For easy chess problems, the strength of the first move is higher for the expert player than the tournament player, but the deliberated move for both types of players is quite similar. The bottom panel shows data for hard chess problems. The move strength is very low for both types of chess players for both types of moves compared with easy chess problems. But the expert has higher move strength in the deliberated moves compared to the tournament chess player. See text for discussion of these findings.

Source: Based on a graph from J. H. Moxley, K. A. Ericsson, N. Charness, & R. T. Krampe (2012). The role of intuition and deliberative thinking in experts' superior tactical decision-making. *Cognition, 124*, pp. 72–78, 75, but omitting the results for medium difficulty chess problems. Published with permission of Elsevier B. V. all rights reserved.

deliberation (move chosen). Thus, thinking longer, and presumably in different ways, about difficult chessboard problems results in better moves (move strength) for expert players, but not for tournament players. The next step, from a behavioral view, is to use the "thinking aloud" method to analyze the verbal behavior of experts to extract rules and strategies that could be taught to tournament players, showing that once these "ways of thinking" are learned, move scores improve, as does general expertise in chess.

Verbal concepts drawn from a player's pre-existing behavior repertoire are used for conceptualizing moves (a form of personal notation), not mental representations of the external world as suggested by cognitive psychology. To illustrate this, verbal concepts linked to vision might include the vocabulary of color (black or white), movement in space (coordinates), and distance to and between objects (knight to bishop on e5). Chess masters appear to use their full range of verbal concepts to "conceptualize certain of [the chess position] relational and dynamic attributes" (Mechner, 2010, p. 376). The term "conceptualize" is used by Mechner to refer to verbal "discrimination between [stimulus] classes and generalization within classes," not some inferred mental entity.

The function of thinking as *private behavior* is to increase the effectiveness of practical action. People conceptualize the game privately without committing themselves publicly. An advantage is that thinking about a sequence of moves (high probability response) is not revealed to your opponent; also thinking may be revoked if the judged consequences are not reinforcing based on your recent *history of reinforcement*. Once a conceptualization is implemented, the player faces the objective consequences. If conceptualizing guides actions resulting in checkmate, then this kind of thinking is strengthened. On the other hand, conceptualizing moves that result in loss of the game weakens this sort of thinking in the future. Overall, thinking is complex operant behavior (mostly verbal) controlled by its consequences. Thinking about a move that guides effective action is likely to occur again, while thinking that prompts ineffective performance declines.

In this section, we have discussed thinking as private behavior. There are many other ways in which the term "thinking" is used. When a person remembers, we sometimes talk about thinking in the sense of searching and recalling. Solving problems often involves private behavior that furthers a solution. In making a decision, people are said to think about the alternatives before a choice is made. The creative artist is said to think of novel ideas. In each of these instances, it is possible to analyze thinking as private behavior regulated by specific features of the environment. The remainder of this book discusses the behavioral processes that underlie all behavior, including thinking.

CHAPTER SUMMARY

This chapter has introduced the idea that many behaviors are acquired during one's lifetime as a result of experience. At birth we emit behavior with very little organized activity. However, as our behaviors cause consequences, some responses are strengthened while others are weakened. The consequences or outcomes of behavior function to select and establish a behavior repertoire. Several prominent persons were introduced to illustrate the history of the science of behavior analysis. In particular, B. F. Skinner was described as a major force behind the experimental and applied analyses of behavior, which is the topic of this book. This approach is related to biology in that behavior is considered to be

> a product of genes interacting with the organism's environment over a lifetime. In this regard, we saw that responses to environmental contingencies alter gene expression (epigenetic effects), neural pathways (neuroplasticity), and retention of learning. Behavior analysis currently is extended to the understanding of feelings and to complex behavior involving problem solving and thinking.

Key Words

- Applied behavior analysis
- Behavior
- Behavior analysis
- Behavior analysts
- Behavioral neuroscience
- Behaviorism
- Conditioned reflexes
- Confirmation bias
- Contingency
- Correlation
- Culture
- Discrimination
- Experimental analysis of behavior
- Immediate causation
- Law of effect
- Learning
- Neuroplasticity
- Operant
- Operant conditioning
- Private behavior
- Reflex
- Reinforcement
- Remote causation
- Respondent
- Respondent conditioning
- Science of behavior
- Selection by consequences
- Trial-and-error learning
- Unconditioned response
- Unconditioned stimulus

On the Web

www.bfskinner.org As you learned in this chapter, B. F. Skinner established the natural science of behavior used in this textbook. The B. F. Skinner Foundation was established in 1987 to educate the public about Skinner's work and accomplishments, and to promote an understanding of contingencies of reinforcement in regulation of human behavior.

www.behavior.org The website for the Cambridge Center for Behavioral Studies is useful to learn more about behavior analysis, behaviorism, and applied behavior analysis. The Center publishes several journals, is host to the Virtual Behavioral Community, and offers recent publications through an online bookstore.

http://web.utk.edu/~wverplan William S. Verplanck, a major contributor to the experimental analysis of behavior and behaviorism, died in September 2002. This award-winning website provides information on his past activities, publications, and addresses, plus interesting issues in psychology and the study of behavior.

www.abainternational.org Go to the webpage for the Association of Behavior Analysis International to find out about the annual convention, the official journals (*Perspectives on Behavior Science* and *Verbal Behavior*), and membership in the Association.

http://psych.athabascau.ca/html/aupr/ba.shtml Find the resource website for the behavior analysis and learning program at Athabasca University in Canada. Many useful links are given for students who want to explore a range of issues in the field. Many of these sites can be accessed to supplement your learning from each chapter in the textbook.

www.simplypsychology.org/classical-conditioning.html This webpage tells you about John Watson and includes a video presentation of his fear conditioning experiment with the infant known as Little Albert.

www.youtube.com/watch?v=JTBg6hqeuTg This brief TEDx talk by geneticist Dr. Courtney Griffins introduces the science of epigenetics and how the environment during an organism's lifetime controls which genes are active or inactive (gene transcription). Once you watch the presentation, return to the section on "Epigenetics and Retention of Early Learning" in this chapter and read the material again.

www.nature.com/scitable/topicpage/translation-dna-to-mrna-to-protein-393 Students that need a quick reference to DNA transcription by mRNA and translation of mRNA into proteins would benefit from this website by Nature Education.

Brief Quiz

1. _____ is the alteration (or maintenance) of an organism's behavior due to _____.
    a. behavior; causes
    b. learning; lifetime events
    c. culture; social norms
    d. evolution; genes

2. The experimental analysis of behavior is:
    a. a natural-science approach to understanding behavior regulation
    b. concerned with controlling and changing factors that affect behavior
    c. concerned with the principle of reinforcement
    d. all of the above

3. A _____ is behavior that is elicited by a biologically relevant stimulus, while a _____ is behavior that is controlled by its consequences.
    a. reflex; respondent
    b. respondent; voluntary
    c. reflex; operant
    d. operant; respondent

4. Selection by consequences occurs at three levels. What are these?
    a. natural selection, behavior selection, and cultural selection
    b. artificial selection, culture, and linguistic selection
    c. natural selection, artificial selection, and cultural selection
    d. artificial selection, natural selection, and linguistic selection

5. Skinner emphasized that behavior (including human behavior) resulted from _____.
    a. genes
    b. environment
    c. self-determination
    d. both (a) and (b)

6. Which of the following statements is true of Pavlov and his contributions?
    a. he won the Nobel Prize
    b. he investigated the salivary reflex
    c. he discovered the conditioned (or conditional) reflex
    d. all of the above

7 How are thinking and feeling treated from a behavioral perspective?
   a   as more private behavior to be explained
   b   as the causes of overt behavior
   c   the relation between the mental and the physical
   d   the mind and its regulation of behavior

Answers to Brief Quiz: 1, b (p. 1); 2, d (p. 2); 3, c (p. 4); 4, a (p. 7); 5, b (p. 15); 6, d (p. 18); 7, a (p. 28).

The Experimental Analysis of Behavior

1 Learn about a *functional* analysis of behavior.
2 Inquire about the method of experimental analysis of behavior.
3 Focus on drugs and behavioral baselines.
4 Learn how to design and assess behavioral experiments.
5 Conduct a behavior analysis of perception.

The experimental analysis of behavior (EAB) refers to a field of study that analyzes behavior–environment relations. This method is called functional analysis. **Functional analysis** involves *classifying behavior* according to its response functions and analyzing the *environment* in terms of stimulus functions. The term *function* refers to the *characteristic effect produced* by either a behavioral or an environmental event. Once a reliable classification has been established, the researcher uses experimental methods to show a causal relation between the environmental event and a specified response. Because of this objective method, behavior analysts need not restrict their findings to one or a few species. The principles of behavior–environment relations hold for all animals. Based on this assumption, and for convenience, researchers often use nonhuman subjects (e.g., rats and pigeons) as their "tools" for discovering principles of behavior, in the same way that geneticists use the fruit fly (*Drosophila melanogaster*) to establish laws of genetics.

FUNCTIONAL ANALYSIS OF BEHAVIOR

There are two ways to classify the behavior of organisms: structurally and functionally. In the **structural approach**, behavior is categorized by age to infer stages of development. For example, many developmental psychologists are interested in the intellectual growth of children. These researchers often use what a person does at a given point in the life cycle to test theories of intellectual development. Children may be said to show "object permanence" when they look for a familiar object that has just been hidden. The structure of the response or what the child does (e.g., looking for and finding the hidden object) is used to make an *inference* about cognitive development. The form or structure of behavior is emphasized because it is said to reveal the underlying "stage" of intellectual growth—knowing that objects exist

DOI: 10.4324/9781003202622-2

38 The Experimental Analysis of Behavior

even when they are out of sight. The *structural approach* draws inferences from behavior about *hypothetical* cognitive abilities such as object permanence. These cognitive abilities in turn are used to explain the child's behavior in finding hidden objects. The child is said to find hidden objects *because she has developed the concept of object permanence*. You may remember from the previous chapter that one problem with using structure to explain behavior is circular reasoning—the behavior of finding the hidden object is used to infer cognitive abilities (object permanence), and these presumed abilities are then used to explain the behavior. Notice that nothing about the child's behavior is actually explained from a causal standpoint with the structural, developmental approach.

In the previous chapter, we noted that behavior analysts study behavior *for its own sake* and at its own level. To keep attention focused on behavior, structure and function are interrelated. A particular form of response by a child, such as opening and looking inside a handbag for a hidden object, is traced to its characteristic effects or consequences (Figure 2.1). The form, structure, or **topography**, of response occurs because that way of doing it has been highly efficient, relative to other ways of opening the bag. Thus, the current topography (structure) of a response is determined by the function (effects or consequences) of this behavior.

In the example of a child who finds a hidden object, a *functional analysis* suggests that this behavior has resulted in specific consequences—the child usually has discovered the hidden toy or object. Rather than infer the existence of some intellectual stage of development or cognitive ability (object permanence), the behavior analyst suggests that a particular **history of**

Fig. 2.1
Picture shows a toddler finding an object hidden in a purse. The conditioning of finding hidden objects (object permanence) may begin early in life when a child pulls at their mother's clothing to uncover the breast for feeding, and is further refined by finding things under blankets and is directly trained as gift-opening on birthdays or holidays and finding chocolate eggs on Easter.
Source: Shutterstock.

reinforcement is responsible for the child's capability. Presumably, from a behavioral perspective, a child who demonstrates object permanence (searching for objects when they are not in sight) has had numerous opportunities to search for and find missing or hidden objects. One advantage of this functional account is that it is testable.

A mother who breastfeeds her newborn often removes some of their clothing just before feeding the baby. After some experience, the baby may tug at the mother's blouse when they are hungry. This is one potential instance of the early conditioning of searching for hidden objects. A few months later, the infant may inadvertently cover up a favorite rattle. In this situation, getting the toy reinforces pulling back the cover when things are hidden. As children get older, they are directly taught to find hidden objects. This occurs when children are given presents to open at birthdays and when they hunt for eggs at springtime holidays, such as Easter. A functional analysis of object permanence accounts for the behavior by pointing to its usual effects or consequences. Object permanence occurs because searching for out-of-sight objects has usually resulted in finding them. Also, children who do not have these or similar experiences (playing peek-a-boo) will perform poorly on a test of object permanence, but should be able to learn this behavior by systematic instruction of the component skills (Bruce & Muhammad, 2009).

Response Functions

Behavior is often considered as *performance* that follows a specific stimulus and at some point, results in a particular consequence. (One three-term notation system used to denote this arrangement is A: B → C, which stands for antecedent stimulus (A), behavior (B), and consequence (C).) Although we shall use the term *response* throughout this book, this term does not always refer to a discrete movement like a muscle twitch or a lever press. *A response is an integrated set of movements, or a behavioral performance*, which is functionally related to environmental events or happenings in the world. For example, the performance of taking notes during a lecture requires a rather complex set of behaviors in which an individual uses a writing utensil (or a computer) to write or type a series of words that characterize what the professor just stated. It involves listening and perhaps reading what is listed on a PowerPoint presentation, thinking quickly about what to write, and then writing it. This is all occasioned by something the professor says (the stimulus).

Functionally, we speak of two basic types or classes of behavior: respondent and operant (Figure 2.2). These behavioral classes were briefly discussed in Chapter 1 and are discussed further throughout the book, but here we emphasize the functional classification of behavior. The term **reflex** is used to refer to behavior that is invariantly caused by the presentation of a stimulus (or event) that *precedes* the response. We say that the *presence of the stimulus induces or **elicits** the response*; that is the stimulus reliably and often involuntarily activates the response. The notation used with elicited behavior is S → R. The stimulus S causes (arrow) the response R. An example of a reflex is the pupillary response to light. The constriction (and dilation) of the eye pupil is reflexive behavior. It occurs when a bright light is directed into (or away from) the eye. Salivation to food is another reflex; food elicits salivation in the mouth. The stimulus S (food) elicits the response R (salivation). For the moment you may consider reflexes to be the activity of smooth muscles or glands (i.e., eye blinks and salivation), and many of them indeed are. But, as we show in Chapters 3 and 7, the modern view of behavior theory substantially expands the organism's reflexive repertoire.

Functional response classes

Response class	Function	Controlling event
Respondent	Elicited	Stimulus preceding the response
Operant	Emitted	Stimulus following the response

Fig. 2.2
The figure summarizes the concepts used for a functional classification of behavior. The response classes are either reflexive or operant. Reflexive behavior is elicited by a stimulus preceding the response; operant behavior is emitted and increased (or decreased) by a stimulus that has followed the response in the past.

There is another large class of behavior that does not depend on an eliciting stimulus. This behavior is **emitted** at some frequency based on an organism's genetic endowment. For example, human infants randomly emit vocal sounds usually referred to as "babbling." These sounds contain the basic elements of all human languages. English-speaking parents attend to, repeat back, and reinforce babbling that "sounds like" English and the baby soon begins to emit more English sounds, especially when the parents are present. The same happens for those who speak other languages: Arabic-speaking parents, for example, attend to and reinforce sounds that sound Arabic. When emitted, behavior is strengthened or weakened by the events that follow the response, it is called **operant** behavior. Thus, operants are emitted responses that occur more or less often depending on the prior consequences produced in a given situation. To make clear the subtle distinction between emitted behavior and operants, consider the action word *walking* versus the phrase *walking to the store*. Walking is emitted behavior, but it has no specified function. In contrast, walking to the store is an operant defined by getting items from the store. Pecking a disk or response key is emitted behavior by a pigeon, but it is an operant when pecking the key has resulted in food. Generally, operants are emitted responses occurring without an eliciting stimulus; effects or consequences of behavior control these responses. We should note that the distinction between elicited and emitted behavior currently is disputed because often there is blending of them in behavior, but most behavior analysts support understanding the differences and we maintain the distinction throughout this version of our textbook (see Domjan, 2016 for details about this dispute).

Operant and reflexive behaviors often occur at the same time when dealing with a single organism. When you step out of a movie theater in the middle of a bright afternoon you may show both types of responses. The change from dark to bright light will elicit pupil contraction, a type of reflex. At the same time, you may shade your eyes with a hand or put on a pair of sunglasses. This latter behavior is *operant* because it is strengthened by the removal of the brightness—the aversive stimulus. In another example, you find that you have failed an important exam. The bad news may elicit a number of emotional responses (conditioned reflexes called **respondents**, which will be discussed later), such as heart palpitations, changes in blood pressure, and perspiration. You probably interpret these physiological responses as dread or anxiety. The person standing next to you as you read the results of the exam asks, "How did you do on the test?" You say, "Oh, not too bad" and walk down the hall with your head down. Your reply and posture is *operant* behavior that avoids the embarrassment of discussing your poor performance. Although operant and reflexive (or respondent) behaviors often occur at the same moment, we will usually analyze them separately to simplify and clarify the environmental conditions that regulate such behavior.

Response Classes

When a person emits a relatively simple operant such as putting on a coat, the performance changes from one occasion to the next. The coat may be put on using either the left or right hand; it may be grasped at the collar or held up by a sleeve. Sometimes one arm is inserted first, while in other circumstances both arms may be used. Careful observation of this everyday action will reveal an almost infinite variety of responses. The important point is that each variation of the putting-on-a-coat response has the common effect of staying warm. To simplify the analysis, it is useful to introduce the concept of a class of responses. A **response class** refers to all the topographic forms (or variations) of the performance that have a similar function (e.g., to keep warm). In some cases, the responses in a class have close physical resemblance, but this is not always the case. A response class for "convincing an opponent" may include dramatic gestures, giving sound reasons, and paying attention to points of agreement. To get service from a restaurant server, you may call out as they pass, wave your hand in the air, or ask the bus-person to send the server to your table; these are all responses within a response class.

Responses within the response class tend be emitted in a particular order (Baer, 1982a), becoming arranged as a **response hierarchy** based on likelihood of occurrence. Often response–response relations (i.e., relative degree of effort) influence the ordering of responses in the class. In a study of effort and ordering of responses, Shabani, Carr, and Petursdottir (2009; Study 1) trained children to separately press each of three buttons (low, medium, and high effort) for identical rates of reinforcement (each press resulted in reinforcement). Next, when all three buttons were available at the same time, children mainly pressed the low-effort button for reinforcement. When pressing the low-effort button no longer resulted in reinforcement (extinction), children primarily pressed the medium-effort button; reinstatement of continuous reinforcement on the low-effort alternative resulted in the children selecting and again pressing this button while reducing responses to the higher effort alternatives. In Study 2, when both low- and medium-effort responses were not functional, children pressed the high-effort button. Generally, the results suggested that the three button-press responses were part of a response class and were arranged in hierarchical order in terms of degree of effort (see also Mendres and Borrero, 2010 for modification of a response class by positive and negative reinforcement; see Beavers, Iwata, & Gregory, 2014 for influence of response–reinforcement relations—reinforcement rate, quality, delay and magnitude—on emergence of response-class hierarchies).

FUNCTIONAL ANALYSIS OF THE ENVIRONMENT

In Chapter 1, we noted that behavior analysts use the term **environment** to refer to events and stimuli that change behavior. These events may be external to the organism or may arise from internal physiology. The sound of a jet aircraft passing close overhead or an upset stomach may both be classified as aversive by their common effects on behavior—attempts to escape them. Both events strengthen any behavior that removes them. In the case of a passing jet, people may cover their ears; a stomach pain may be removed by taking antacid medication.

The location of the source of a stimulus, internal versus external, is not a critical distinction for a functional analysis. There are, however, methodological problems with stomach pains that are not raised by external, public events like loud sounds. Internal, private sources of stimulation must be indirectly observed with the aid of instruments or inferred from observable behavior–environment interactions (see Chapter 1 on private events). Evidence for stomach pain, beyond the verbal report, may include the kinds of foods recently eaten, the health of the person when the food was ingested, and current external signs of discomfort.

Stimulus Functions

All events and stimuli, whether internal or external, may *acquire* the capacity to affect behavior. When the occurrence of an event changes the behavior of an organism, the event has a **stimulus function**. Both the conditioning of reflexes (respondent) and operant conditioning are ways to create stimulus functions (see Figure 2.3). During respondent conditioning, an arbitrary event is paired with a reflex. More specifically, though, the arbitrary stimulus, such as a tone is paired with a reflexive stimulus, such as food. Here, food is also called an *unconditioned stimulus* because based on the animal's evolutionary history, food elicits a particular response like salivation. No learning is necessary; this unconditioned stimulus is something that is present at birth. But, after repeated pairings of the tone and food, the tone on its own now elicits salivation. Once the tone effectively elicits salivation without the food present, the tone is said to have a **conditioned-stimulus function** for salivation. In the absence of a conditioning history, the tone is neutral in terms of eliciting salivation, but the sound has no specified function and does not affect behavior. After conditioning, though, it has a condition stimulus function.

Functional stimulus classes

Type of conditioning	Stimulus function	Temporal location	Effect on behavior
Respondent	Unconditioned Conditioned	Before Before	Elicits response Elicits response
Operant	Discriminative Reinforcement	Before After	Occasions response Increases response

Fig. 2.3
The table summarizes the concepts used for a functional classification of the environment. The type of conditioning or arrangement of behavior–environment relations is either respondent or operant. For respondent conditioning, the stimulus that comes before behavior (temporal location) can be either an unconditioned or conditioned stimulus. An unconditioned stimulus elicits a response based on the genetic endowment of the animal (food elicits salivation). A conditioned stimulus elicits a response based on a history of association with the unconditioned stimulus (tone elicits salivation). For operant conditioning, a stimulus that comes after a response (key pecking produces food) subsequently increases the rate of response (key pecking for food increases). A discriminative stimulus (key light) precedes reinforcement of the response (operant) and eventually *sets the occasion for* the response, making the response more likely when the stimulus is presented (key light increases pecking for food).

Similarly, operant conditioning generally results in establishing or changing the functions of stimuli. Any stimulus (or event) that follows a response and increases its frequency is said to have a **reinforcement function**. An event or stimulus that has this function is called a reinforcing stimulus or reinforcer (S^r). When an organism's behavior is reinforced, those events that reliably precede responses come to have a **discriminative function**. Events with a discriminative function *set the occasion for* behavior in the sense that an operant is more likely when the event occurs (see Chapter 1). Events, settings, and situations that precede operant behavior and increase its probability are called **discriminative stimuli**. Discriminative stimuli (S^D) acquire this function because they predict (have been followed by) reinforcement of operant behavior. In the laboratory, a pigeon's key pecks may be followed by food when the key is illuminated red, but not reinforced when the key is blue. After some time, the red key color is said to *set the occasion for* the response. In everyday language, the red key "tells" the bird when pecking will be reinforced. More technically, the red key is a discriminative stimulus, as the probability of reinforcement for pecking is higher when the key is red than when it is blue. The bird is said to discriminate or make a *differential response* to red and blue.

The concept of stimulus function is an important development in the analysis of behavior. The nervous system of humans and other animals has evolved in such a way that it can sense those aspects of the environment that have been important for survival. Of all the stimuli that can be physically measured and sensed by an organism at any one moment, only some affect behavior (have a stimulus function). Imagine you are sitting on a park bench with a friend on a nice sunny day. The physical environment includes heat, wind current, sounds and smells from traffic, birds, insects, rustling leaves, tactile pressure from sitting, and the sight of kids playing ball, people walking in the park, and the color of flowers, grass, and trees. Although all of these (and many more) physical events are present, only some affect your behavior—in the sense that you turn your face to the sun, comment on the beauty of the flowers, wrinkle your nose to the odor of exhaust, and look in the direction of a passing fire truck. The remaining parts of physical environment, at this moment in time, either have no function or serve as the context for those events that do.

Stimulus Classes

In a preceding section, we noted that responses that produce similar effects are many and varied. To encompass responses that vary in form, but have a common consequence or function, behavior analysts use the term *response class*. Stimuli that regulate operant and respondent behavior also vary from one time to the next in terms of how they might affect behavior. When stimuli vary across a physical dimension, but have common effects on behavior, they are said to be part of the same **stimulus class**. Bijou and Baer (1978) have used the concept of stimulus class in an analysis of child development and have made the point that:

> A mother's face has a fair consistency to it, we may think, in that we know our mother's face from anyone else's face. But careful observations show that it is sometimes shiny, sometimes dusty, sometimes wet; occasionally creased into its facial lines, but sometimes smooth; the eyes range between fully open and fully closed, and assume a wide range of angles of regard; sometimes hairs fall in front of the face, sometimes not. Then let us remember that whenever we speak of a stimulus, we will almost surely mean a class of stimuli.
>
> (Bijou & Baer, 1978, p. 25)

It is important to note that a stimulus class is defined by its *common effect on behavior*. A stimulus class cannot be defined by the apparent similarity of the stimuli. Consider the words "boring" and "uninteresting." In common English, we say that they have the same meaning. In behavior analysis, because these words have a similar effect on the behavior of the person who reads or hears them (e.g., if someone says a particular professor is boring or uninteresting, you might avoid taking their class), they belong to the same stimulus class even though they have completely different physical dimensions (i.e., letters on the page). Other stimuli may appear physically similar but belong to different stimulus classes. For example, mushrooms and toadstools look somewhat similar, but for an experienced woodsperson these stimuli have different functions—you pick and eat mushrooms but avoid toadstools.

Classes of Reinforcing Stimuli

The concept of stimulus class may also be used to categorize the consequences of behavior. When behavior operates on the environment to produce effects, it is an operant; the effects that increase the frequency of response are a class of reinforcing stimuli. Some consequences strengthen behavior when they are presented, such as money for a job well done, and others strengthen it when they are removed, such as scratching an itch. In this case, we can divide the general class of reinforcing stimuli into two subsets. Those events that increase behavior when presented are called **positive reinforcers**, and those that increase behavior when removed are **negative reinforcers**. For example, a smile and a pat on the back may increase the probability that a child will complete their homework; thus, the smile and pat are positive reinforcers. The same child may stop dawdling and start working on a school project when a parent scolds the child for wasting time and the nagging stops when they get going. In this case, reinforcement for working is based on the removal of scolding, and the reprimand is a negative reinforcer.

Reinforcement Dynamics: Motivating Operations

The relations between stimulus and response classes depend on the broader **context of behavior**. Behavior–environment relations are always probabilistic and conditional—that is, they depend on other circumstances. One of the most common ways to change behavior–environment relations is to have the person (or other organism) experience a period of deprivation or satiation. For example, a pigeon will peck a key for food only if it is deprived of food for some period of time. More specifically, the peck-for-food contingency depends on level of food deprivation.

An important distinction is made between the discriminative and motivational functions of stimuli (Michael, 1982a). The term **establishing operation (EO)** refers to any environmental change that has two major effects: first, the change should increase the momentary effectiveness of reinforcement supporting operant behavior, and second, the change should increase momentarily the responses that had in the past produced such reinforcement (see also Michael, 1993, 2000). For example, the most commonly described establishing operation is deprivation for primary reinforcement. The procedure involves withholding reinforcement for some period of time (see Chapter 5). This establishing operation of deprivation has two effects. First, food becomes an effective reinforcer for any operant that produces it. The deprivation procedure *establishes* the reinforcement function of food. Second, behavior that has previously resulted in getting food becomes more likely—in the wild, a bird may start to forage in places where it has previously found food.

Establishing operations regularly occur in everyday life and depend on a person's conditioning history. For example, television commercials are said to influence a person's attitude toward a product. One way to understand the effects of TV commercials is to analyze them as establishing operations (technically, *conditioned establishing operations* or CEOs). In this case, an effective commercial alters the reinforcement value of the product and increases the likelihood of purchasing the item or using it if available. For example, dairy farmers advertise the goodness of ice-cold milk. Those who are influenced by the commercial are likely to go to the fridge and have a glass of milk. Of course, this immediate effect of the commercial depletes the amount of milk you have on hand, and eventually you buy more milk. In this analysis, television commercials are examples of conditioned visuals, instructions or rules, which function as CEOs increasing the reinforcement value or effectiveness of milk and evoking behavior (buying and drinking milk) that has produced this reinforcement in the past. A study by Harris, Bargh, and Brownell (2009) showed that for children, commercials that advertise food increase the amount of eating snack foods (e.g., Goldfish crackers) compared to commercials that advertise for goods that are not food. These establishing operations for food can increase consumption without awareness.

Establishing operations are ubiquitous and can be observed even with games. In the game *Monopoly*, the rules of the game concerning going to and getting out of jail are CEOs, establishing the "get out of jail free" cards as reinforcement for exchanging them when landing in jail. When you draw one from the "Chance" pile, it emits saving these cards for later; they have value. Notice that outside of the game's rules (CEOs), "get out of jail free" has no reinforcement and evocative functions.

In contrast to the establishing operation, an **abolishing operation** (**AO**) decreases the effectiveness of behavioral consequences, and momentarily reduces behavior that has resulted in those consequences in the past (Michael, 1982a). For example, a person who has spent an 8-hr period at work constantly talking to others may not feel like talking with family members when they come home from work. Therefore, frequent talking at work serves as an abolishing operation for social reinforcement at home. In another example, drinking alcohol can reduce the effectiveness of money loss as a punisher (Rasmussen & Newland, 2009). Indeed, this is a behavioral process capitalized on by casinos; offering free drinks to gamblers suppresses sensitivity to losing money—hence alcohol is an abolishing operation for punishment.

Both the establishing operation and abolishing operations fall under the umbrella term of **motivational operation** (**MO**). MOs refer to any event that alters the effectiveness of behavioral consequences and changes the frequency of behavior maintained by those consequences (see Laraway, Snycerski, Michael, & Poling, 2003). Figure 2.4 shows that the motivational operation (MO) must be in place before the three-term contingency of reinforcement [$S^D: R \rightarrow S^R$] is functional. In the operant laboratory, a period of food deprivation (MO) increases the effectiveness of food as a reinforcer for lever-pressing. At work, an 8-hour period of socially interacting with colleagues and customers (MO) reduces the effectiveness of social reinforcement at home, in terms of talking and interacting with loved ones.

Motivational operations can have diverse effects on behavior. Rispoli and colleagues (2011) examined the effects of presession delivery of items on problem behavior and

$$MO[S^D : R \rightarrow S^r]$$

MO	$[S^D : R \rightarrow S^r]$
Deprive of food	Light ON: Press lever → food pellet
8 hrs social interaction	Home: Interacting → social reinforcement

Fig. 2.4

The figure depicts the motivational operation (MO) that functions to alter the effectiveness of the contingency of reinforcement ($S^D: R \rightarrow S^r$). In the laboratory, food deprivation (MO) increases the effectiveness of food reinforcement and also increases the probability of responses previously reinforced by food in this situation (bar pressing for a rat). A great deal of social interaction at work (MO) can reduce the social reinforcement value of loved ones at home and decrease the probability that an individual will talk out loud and interact when at home. The first example is an EO and the second is an AO.

academic engagement in the classroom. The authors first established that problem behavior was a function of gaining access to preferred items for a 5- and 7-year-old boy with autism. Next, they tested two conditions. In one condition they offered the items for up to 45 min before classroom sessions began; in the other condition there was not presession access to the preferred items. The researchers found lower levels of problem behavior and higher levels of academic engagement in the condition in which the boys had presession access to the items. Researchers concluded presession access to items decreased the value of the items during subsequent classroom sessions, which resulted in a corresponding decrease in the frequency of problem behavior previously correlated with obtaining these items. Therefore, preexposure to preferred items has an AO function on problem behavior in the classroom, but an EO function on academic behavior. In another study, events such as delaying planned activities or sleep deprivation had multiple motivating effects on problem behavior of boys with developmental disabilities (Horner, Day, & Day, 1997). For one boy, sleep deprivation reduced the effectiveness of staff praise as reinforcement for problem behavior (AO effect) and increased the effectiveness of food items as reinforcers (EO effect).

A recent study of alcohol and social bonding indicates that consumption of alcohol has motivational effects (MO) on human social interaction in addition to the reinforcing effects of alcohol itself (Sayette et al., 2012). In this study, researchers created groups of three strangers who drank cranberry juice (control drink), tonic water mixed with cranberry juice (placebo drink), or vodka mixed with cranberry juice (alcohol drink) in a social setting. Social interactions in these groups were video recorded, and facial expressions as well as speech behaviors were systematically coded. Alcohol consumption had behavior-enhancing social effects (MO) compared with control and placebo groups. Alcohol use increased interpersonal smiling, speaking to others, participants' ratings of social bonding, and ratings of the reinforcing effects of the group. The motivational effects of alcohol as an EO on social contingencies of reinforcement (social reinforcement from interaction with others) may explain its continued use across diverse cultures.

Motivational operations also involve aversive events that alter the effectiveness of negatively reinforcing events (recall that negative reinforcers are those that strengthen behavior through removal of an aversive stimulus). A toothache is an aversive event that functions as an EO, often making social conversation aversive; the greater the pain from the toothache the more reinforcing is escape from social settings and conversations. In addition, the toothache as an EO affects behavior. We are more likely to talk less to others and retreat to a less social environment, such as going to our bedroom and closing the door. In another example, a stressful workplace may function as an EO, establishing reduction or escape from work as reinforcement, and making behavior like absenteeism more likely. Extending this analysis, the clinical problem of depression is treated as a mood disorder, but motivational events play a crucial role. When much of the world becomes aversive (filled with stressful and painful happenings), removal of most life events becomes reinforcing and generalized escape (getting away from life) is generated—mood changes often accompany the generalized escape responses of depression but, from a behavioral perspective, do not cause them.

When the intensity of aversive events is decreased, we may say an AO is in place. Removal or escape from aversive events is less reinforcing and evokes less escape behavior. Sunglasses function as an AO, reducing or abolishing the glare of the sun as negative reinforcement and evoking less squinting of the eyes and looking away from the sun. Another example involves vaccination and prevention of disease. Vaccination for the flu is a health prevention that functions as an AO (especially for those who already have experienced a severe case of the flu), decreasing the aversiveness of being around large crowds of possibly sick people and making escape from social contact with others less reinforcing during the period of contagion. As you can see, the regulation of most human behavior involves numerous and varied motivational operations (both EO and AO). Furthermore, behavior change to solve personal and social problems is more effective when a functional analysis identifies and alters these motivational events (see Langthorne and McGill, 2009 for a more complete analysis of motivational operations in applied behavior analysis).

ON THE APPLIED SIDE: FUNCTIONAL ANALYSIS OF CHALLENGING BEHAVIOR

Functional analysis (FA) has been used in therapeutic settings with atypically developing people. FA involves analyzing and testing potential functions of challenging behavior. The basic idea is that behavior problems (e.g., head hitting, aggression, and stereotypy) of people diagnosed with developmental delays and autism are maintained by the operating contingencies of reinforcement. A person diagnosed with autism may be hitting teachers and caretakers possibly because this behavior has resulted in positive reinforcement from social attention or perhaps in negative reinforcement by escape from demands and requests of those in charge. Applied behavior analysts use FA to obtain evidence of the operating contingencies (positive social reinforcement vs. negative reinforcement by escape) and design a tailored, individualized program or intervention to ameliorate the behavioral problem (Dixon, Vogel, & Tarbox, 2012).

Iwata and his associates used FA in their classic article, "Toward a Functional Analysis of Self-Injury" (Iwata, Dorsey, Slifer, Bauman, & Richman, 1982/1994). The study concerned nine children and adolescents with developmental disabilities and self-injurious behavior (SIB) and the study involved determining the possible functions of SIB for these children. The experimental design included alternating 15-min periods of four conditions: academic, alone, social disapproval, and play.

The first three conditions assessed three general functions of SIB, involving positive, negative, and automatic reinforcement. The play condition served as the control for the experiment.

Briefly, in the academic condition the subject was given various tasks and prompted to complete the sequence of required actions. If the child displayed SIB, the experimenter turned away and removed the task demands for a short period—this condition was designed to assess the effects of negative reinforcement by escape from demands on the number of occurrences of the target behavior (SIB). In the alone condition, the subject was left alone in the therapy room with no toys or any other items, a so-called "deprived" environment to determine the extent to which the SIB might be a function of self-stimulation or automatically reinforcing. Again, the researchers monitored the occurrence of the target behavior, SIB, under these conditions. For the social disapproval condition, the experimenter and the child entered the therapy room arranged with a variety of toys and the youngster was asked to play with the toys while the experimenter did some work. If the child displayed SIB, the experimenter presented statements of disapproval like "Don't do that, you will hurt yourself" while also delivering physical attention such as pats on the shoulder. Thus, the third condition assessed the positive reinforcement function of SIB, gaining access to attention from others. In the play (control) condition the child was placed in the therapy room with toys; there were no demands and no attention provided by the experimenter.

The results of FA showed SIB increased mostly in the attention condition (social disapproval), indicating a positive reinforcement function for this behavior. Other children, however, were most likely to emit SIB in the escape from demands condition (academic) involving negative reinforcement, while the SIB of still others occurred mostly in the alone "deprived" environment condition, suggesting an automatic reinforcement function. The FA study showed that SIB has different operant functions depending on the unique learning history of each child and that an FA is necessary to design an individualized program of behavior change—reducing the occurrence of SIB in each subject.

Since these early studies, there now have been literally hundreds of investigations of FA using a variety of problem behaviors. A review by Beavers, Iwata, and Lerman (2013), combining studies from an earlier summary, identified 981 FA graphs of individual problem behaviors, with 94% showing clear regulation by contingencies across diverse response topographies (self-injury, aggression, vocalization, and others). About 30% of these behaviors were maintained by social-negative reinforcement (escape from demands and requests) and another third were regulated by social-positive reinforcement (social attention and caretaker access to tangible items or activities). Responding was maintained by automatic reinforcement (reinforcement gained just from doing the activity) in about 16% of the cases and by multiple reinforcement contingencies especially for aberrant behavior—involving multiple responses of dissimilar topography (e.g., aggression and self-injury).

Figure 2.5 depicts the percentage of cases (differentiated results) for three response topographies (self-injury, aggression, and stereotypy) as a function of the source of reinforcement maintaining the problem behavior. Self-injury (black bars) is almost equally maintained by social-negative reinforcement (escape from demands), social-positive reinforcement (attention and access to tangible items/activities) and by automatic reinforcement from just doing the activity. On the other hand, aggression is predominantly regulated by escape contingencies and social reinforcement (attention and access to tangible items/activities), but not by automatic reinforcement from the activity itself. Unlike aggressive behavior, stereotypy (the excessive repetition of a response or routine) is most often maintained by automatic reinforcement from engaging in the activity itself, occasionally by escape from demands, and rarely, if ever, by social-positive reinforcement (attention and tangible). Functional analysis shows that different response topographies (problem behavior)

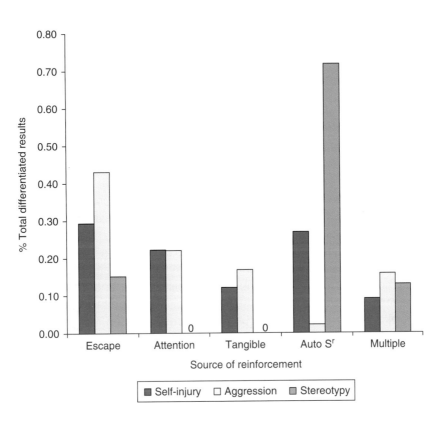

Fig. 2.5
The percentage of self-injury, aggression, and stereotypy as a function of the source of reinforcement involving escape, attention, tangible, automatic reinforcement, and multiple sources of reinforcement.

Source: The data are based on a review of 435 studies and were recalculated from a table of "summary of functional analysis outcomes" from G. A. Beavers, B. A. Iwata, & D. Lerman (2013). Thirty years of research on the functional analysis of problem behavior. *Journal of Applied Behavior Analysis, 46*, pp. 1–21, using differentiated results (results showing a difference).

of people with developmental disabilities and autism are regulated by distinct contingencies of reinforcement. Most importantly, behavioral programs and interventions are more effective once the operating contingencies have been identified by functional analysis.

TACTICS OF BEHAVIORAL RESEARCH

To discover elementary relationships between functional stimuli, responses, and consequences, behavior analysts have relied on experimental methods developed in biology, medicine, and behavior analysis (Bernard, 1927; Bushell & Burgess, 1969; Johnston & Pennypacker, 1993; Sidman, 1960). In 1865, the French physician Claude Bernard outlined the central objectives for experimental analysis. He stated that:

> We can reach knowledge of definite elementary conditions of phenomena only by one road, viz., by *experimental analysis*. Analysis dissociates all the complex phenomena successively into more simple phenomena, until they are reduced, if possible, to just two

elementary conditions. Experimental science, in fact, considers in a phenomenon only the definite conditions necessary to produce it.

(Bernard, 1927, p. 72)

In his book *An Introduction to the Study of Experimental Medicine*, Bernard provided a classic example of experimental analysis:

> One day, rabbits from the market were brought into my laboratory. They were put on the table where they urinated, and I happened to observe that their urine was clear and acid. This fact struck me, because rabbits, which are herbivora, generally have turbid and alkaline urine; while on the other hand carnivora, as we know, have clear and acid urine. This observation of acidity in the rabbits' urine gave me an idea that these animals must be in the nutritional condition of carnivora. I assumed that they had probably not eaten for a long time, and that they had been transformed by fasting, into veritable carnivorous animals, living on their own blood. Nothing was easier than to verify this preconceived idea or hypothesis by experiment. I gave the rabbits grass to eat; and a few hours later, their urine became turbid and alkaline. I then subjected them to fasting and after twenty-four hours, or thirty-six hours at most, their urine again became clear and strongly acid; then after eating grass their urine became alkaline again, etc. I repeated this very simple experiment a great many times, and always with the same result. I then repeated it on a horse, an herbivorous animal that also has turbid and alkaline urine. I found that fasting, as in rabbits, produced prompt acidity of the urine, with such an increase in urea that it spontaneously crystallizes at times in the cooled urine. As a result of my experiments, I thus reached the general proposition which then was still unknown, to wit, that all fasting animals feed on meat, so that herbivora then have urine like that of carnivora.
>
> But to prove that my fasting rabbits were really carnivorous, a counter proof was required. A carnivorous rabbit had to be experimentally produced by feeding it with meat, so as to see if its urine would then be clear, as it was during fasting. So I had rabbits fed on cold boiled beef (which they eat very nicely when they are given nothing else). My expectation was again verified, and as long as the animal diet was continued, the rabbits kept their clear and acid urine.

(Bernard, 1927, pp. 152–153)

Bushell and Burgess (1969) outlined the basic tactics of experimental analysis used by Bernard in the rabbit experiment. The following account is loosely based on their outline. Notice that Bernard made an observation that, as a physiologist, seemed unusual and puzzling—namely, that the rabbits from the market had urine that was characteristic of that of carnivores. Only a trained physiologist familiar with carnivores and herbivores would notice the anomaly of the urine. Most of us would run and get a cloth to wipe it up. The point is that a researcher must have a thorough familiarity with the subject matter to find a significant problem.

Once Bernard had identified the *problem*, he stated it in terms of a conjecture. The problem statement related type of diet to the chemistry of the urine. A period of fasting results in the animal living off its own body stores, and this produces acidity of the urine. On the other hand, when herbivores eat their usual diet of grass, their urine is alkaline. Thus, there is a clear relationship between type of diet and the nature of the animal's urine.

Experimentally, Bernard's statement suggests that we change, manipulate, or control the type of diet and measure the chemistry of the urine. The condition changed or controlled by the experimenter (i.e., type of diet) is called the **independent variable** (variable X), because it is free (or independent) to vary at the discretion of the researcher. Bernard manipulated the animal's diet and measured the effect on the urine. The measured effect is called the **dependent variable** (variable Y), because its change *depends* on the value of the independent variable set by the experimenter. Whether the urine is acid or alkaline (dependent variable) depends on the nature of the diet (independent variable). Figure 2.6 explains the terms used in this section.

The purpose of any experiment is to establish a cause-and-effect relationship between the independent (X) and dependent (Y) variables. To establish such a relationship, the researcher must show that changes in the independent variable are functionally related to changes in the dependent variable. In addition, the experimenter must show that the changes in the independent variable *preceded* changes in the dependent variable (X precedes Y). Both of these conditions are seen in Bernard's experiment.

In Figure 2.7, you can see that changes between fasting and grass diet reliably alter the chemistry of the rabbits' urine. Thus, changes in the type of diet (the X variable) may be said

Independent variable	Dependent variable
• What is changed in an experiment	• What is measured in an experiment
• X variable	• Y variable
• Commonly called a *cause*	• Commonly called an *effect*
• In Bernard's experiment, type of diet	• In Bernard's experiment, chemistry of urine
• In behavioral experiments, environmental change	• In behavioral experiments, behavior of the organism

Fig. 2.6
The figure shows scientific terms used to discuss cause-and-effect relationships.

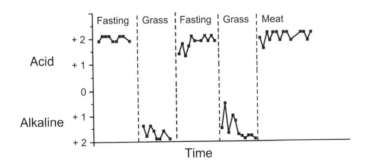

Fig. 2.7
The graph shows the results of Bernard's experiment. Notice that the change in diet (independent variable) reliably changes the chemistry of the urine (dependent variable). Each time the diet is changed, the urine changes from acid to alkaline or vice versa.

Source: Based on results reported by D. Bushell, Jr., & R. L. Burgess (1969). Characteristics of the experimental analysis. In R. L. Burgess & D. Bushell, Jr. (Eds.), *Behavioral sociology: The experimental analysis of social processes* (pp. 145–174). New York: Columbia University Press, p. 133. Published with permission of Robert Burgess.

to covary with degree of acidity of the urine (the Y variable of Figure 2.7). Recall that Bernard manipulated or controlled the type of diet and then measured its effects on the urine. This procedure of manipulating the independent variable ensures that a change in X (type of diet) precedes the change in Y (chemistry of urine). At this point, Bernard has shown two of the three important conditions for causation—first, covariation of X and Y, and second, the independent variable precedes a change in the dependent variable.

The central question in all experiments is whether the values of the dependent variable (effect) are uniquely *caused* by manipulations of the independent variable. The problem is that many other factors may produce changes in the dependent variable, and the researcher must rule out the possibility of the *other variables being the cause*. In the Bernard experiment, the initial change from fasting to grass diet may have been accompanied by an illness caused by contaminated grass. Suppose that the illness changed the chemistry of the animals' urine. In this case, changes from fasting to grass, or from grass to fasting, will change the chemistry of the urine, but the changes are caused by the unknown illness rather than the type of diet. The unknown illness (other variable) is said to *confound* the effects of type of diet on the acidity of the urine. At this point, stop reading and look again at Bernard's description of his experiment and at Figure 2.7. Can you determine how Bernard eliminated this rival hypothesis?

One procedure for eliminating rival explanations is the systematic introduction and elimination of the grass diet. Notice that Bernard withholds and gives the grass diet and then *repeats* this sequence. Each time he introduces and removes the grass, a rapid change occurs from alkaline to acid (and vice versa). This rapid and systematic change makes it unlikely that illness accounts for the results. How can an animal recover from and contract an illness so quickly? Another procedure would be to use different batches of grass, because it is unlikely that they would all be contaminated. However, the most convincing feature of Bernard's experiment, in terms of *eliminating rival explanations*, is his final procedure of introducing a meat diet. The meat diet is totally consistent with Bernard's claim that the animals were living off their body stores, and counteracts the rival explanation that the animals were ill. More generally, the reversal of conditions (direct replication) and the addition of the meat diet (systematic replication) help to eliminate most other explanations, yielding high *internal validity* of the experiment or a strong attribution that the type of diet caused the observed changes in urine. (Before his death, Dr. Oliver Sacks, the brilliant neurologist, used Claude Bernard's *single-subject method* to analyze the causes of brain and spinal cord problems of his patients, although often medicine had not advanced enough for a viable treatment or cure; Groopman, 2015.)

The Reversal Design and Behavior Analysis

Bernard's experimental design for physiology is commonly used to study behavior–environment relations. The design is called an **A-B-A-B reversal**, and is a powerful tool used to show causal relationships among stimuli, responses, and consequences. The reversal design is ideally suited to show that specific features of the environment control the behavior of a single organism. This kind of research is often called a *single-subject experiment* and involves several distinct phases. The A phase or **baseline** measures behavior before the researcher introduces an environmental change. During baseline, the experimenter takes repeated measures of the behavior under study and these measures establish a criterion which any subsequent changes, caused by the independent variable, may be compared to and assessed. Following the baseline phase, an environmental condition is changed (B phase) and behavior is repeatedly measured. If the independent variable or environmental condition has an effect, then the behavioral measure (dependent variable) will change (increase or decrease).

At the same time, as we have indicated, the researcher must rule out rival explanations for the change in behavior, such as simple coincidence or chance. To do this, the baseline phase is reintroduced (A) and behavior is once more measured repeatedly. Notice that under removal of the cause or independent variable, behavior should return to pretreatment or baseline levels. Finally, the independent variable is re-inserted and the behavior is carefully measured again (B). According to the logic of the design, behavior should return to a level observed in the initial B phase of the experiment. This second application of the independent variable helps to ensure that the behavioral effect is caused by the manipulated condition, and not by some extraneous, confounding factor (other variable problem).

The Reversal Design Applied to Behavior Analysis
An example of the reversal design, as used in behavior analysis, is seen in an experiment conducted in an elementary school classroom (Hasazi & Hasazi, 1972). The teacher reported a challenge with an 8-year-old boy named Bob who had difficulty in adding numbers that yielded two-digit sums. Given the problem 5 + 7, Bob would write 21, reversing the digits. We call this kind of behavior dyslexia, which is categorized as a type of learning disability. The researchers designed an experiment to manipulate the contingency between Bob's digit-reversal responses and teacher attention (reinforcement). Basically, digit-reversal responses were expected to occur at high frequency when followed by "extra help" (attention) from the teacher, but should decrease when no longer supported by teacher-arranged consequences.

For each day of the study, at the same time in the morning, the teacher gave Bob 20 arithmetic problems with two-digit sums. Thus, the maximum number of digit reversals for a day was 20 responses. For the initial baseline (Baseline 1), the teacher used her usual method for checking problems. After Bob had completed the 20 additions, he raised his hand and the teacher came to his desk to check the worksheet. The teacher marked correct answers with "C" and digit reversals with an "X" for incorrect, explaining, "This one is incorrect. You see [pointing], you reversed the numbers in the answer." Next, the teacher gave Bob "extra help" on incorrect answers by taking him through the adding process and providing verbal and physical prompts to obtain the correct answer.

For the first experimental phase (Experimental 1), the researchers changed the contingency between digit-reversal responses and teacher attention. All digit reversals were now marked with a "C," assuming that incorrect responses (and the mark "X") had acquired a conditioned reinforcement function. Also, the teacher no longer made comments about digit reversals, no longer supplied "extra help" for reversals, and Bob's statements about reversal errors were ignored. As in the past, correct sums were marked with a "C" and followed by the usual teacher consequences of a smile, a pat on the back, and the comment "This one is *very* good." As you can see from Figure 2.8, the experimental procedures no longer reinforced Bob's digit reversals with teacher attention. Technically, digit-reversal behavior was placed on *extinction* (no longer reinforced) while correct sums continued to be reinforced (*differential reinforcement* for correct sums).

The Baseline 2 phase reinstated the teacher's usual method of checking problems as in the initial baseline—assuming that digit reversals would increase once more. The Experimental 2 phase replicated the procedures used in the first experimental phase; the teacher again used *differential reinforcement* of correct sums—no longer giving attention to Bob's digit reversals,

54 The Experimental Analysis of Behavior

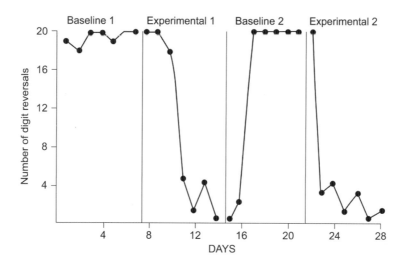

Fig. 2.8
An A-B-A-B design is depicted. The A phase or baseline provides a measure of digit-reversal responses when the teacher used her usual method of "extra help" for digit reversals. Next, the reinforcement contingency for digit reversals (B) is changed, placing these responses on extinction while maintaining reinforcement of correct sums. For the third phase (A reinstated), the teacher again followed digit reversals with "extra help." Finally, the experimental contingencies were once more in effect (B reinstated)—extinction for digit reversals and reinforcement of correct sums.

Source: Adapted from J. E. Hasazi & S. E. Hasazi (1972). Effects of teacher attention on digit-reversal behavior in an elementary school child. Journal of Applied Behavior Analysis, 5, pp. 157–162. Copyright 1972 held John Wiley & Sons Ltd. Published with permission.

and supplying attention only for correct sums. Digit-reversal responses were expected to decrease when placed on *extinction* once again. The results of this experiment are shown in Figure 2.8.

The experimental design is an A-B-A-B reversal. The A phase provides a measure of digit-reversal responses when the teacher used her usual method of "extra help" for digit reversals. Next, the researchers manipulated the reinforcement contingency for digit reversals (B), placing these responses on extinction while maintaining reinforcement of correct sums. During the third phase (A reinstated), the usual contingency arranged by the teacher was again in effect—supporting digit reversals with "extra help." Finally, the experimental contingencies were once more in effect (B reinstated)—extinction for digit reversals, and reinforcement of correct sums. The independent variable for this experiment is the contingency of reinforcement—arranging reinforcement versus extinction for digit reversals. The dependent variable is the number of digit-reversal responses that Bob produced during each phase of the experiment (maximum of 20). As you can see in Figure 2.8 the dependent variable reliably changes in the expected direction with changes in the contingency of reinforcement (i.e., teacher attention for digit reversals). Clearly, in this experiment with this child, digit reversals were operant responses that were inadvertently maintained by teacher-arranged reinforcement.

Limitations to Use of the Reversal Design
The A-B-A-B reversal design is the most fundamental research design used in behavior analysis. There are, however, difficulties that may make this design inappropriate for a given research

question. One major problem is that some behavior, once changed, may not return to baseline levels. Consider what might happen if you used a reinforcement technique to teach a person to read. You could measure reading level, introduce your teaching technique, and after some time withdraw reinforcement for reading. It is very unlikely that the student will again become unable to read. In behavioral terms, the student's reading is maintained by other sources of reinforcement, such as getting information that enables the student to behave effectively (e.g., reading a menu, traffic signs, and books).

Another difficulty is that it is sometimes unethical to reverse the effects of a behavioral procedure. Suppose that a behavioral program to eliminate the use of the drug fentanyl works, but the doctors who run the program are not absolutely certain that the decline in drug use is caused by reinforcement procedures. It would be highly unethical to remove treatment (i.e., implement the second A condition) and reinsert the therapy to be certain about causation. This is because removing the treatment could lead to an increase in drug use. Therefore, in these situations, an A-B design is preferable (also see multiple baseline design in Chapter 13). Nonetheless, when these and other difficulties are not encountered, the A-B-A-B reversal design is a standard mode of analysis.

Throughout this book, we address research that uses the reversal design, modified reversal designs (e.g., adding other control conditions), and other forms of designs for experimental and functional analysis (see Perone & Hursh, 2013 for a detailed overview of single-subject experimental designs). We have concentrated on the reversal design in this chapter because it demonstrates the basic logic of behavioral experimentation. *The task of all behavioral experiments is to establish with high certainty the cause-and-effect relations that govern the behavior of organisms.* Based on these causal relations, behavior analysts search for general principles that organize experimental findings (e.g., principle of reinforcement).

NEW DIRECTIONS: OPERANT BASELINES FOR BEHAVIORAL NEUROSCIENCE

In a given setting, behavior reinforced in a particular way (e.g., every 10 responses produce food) becomes very stable (low variability) over repeated experimental sessions. An animal might show a run of responses followed by a break (or time without responding) and then another run. This pattern might be repeated over and over again after long exposure to the reinforcement procedure (called **steady-state performance**). Stable performance under a contingency of reinforcement can be used as a baseline for the effects of other independent variables. When behavior is very stable under a given arrangement of the environment, it is possible to investigate other conditions that disrupt, increase, or decrease the steady-state performance of animals. Recognizing this advantage, behavioral neuroscientists often use steady-state operant behavior as baselines (control conditions) to investigate the effects of drugs on the brain and behavior (see Winger & Woods, 2013 for research in behavioral pharmacology, which we treat as in intersection of behavior analysis and behavioral neuroscience in this textbook).

Regarding drugs and baselines, the more stable the baseline, the easier it is to detect the effects of small doses of the drug. If an animal's average number of responses for 20 experimental sessions is ten per minute, with a range of ± 1 response per minute (more stable baseline), a smaller dose of a drug would show an effect than if the baseline had the same average with a range of ± 5 responses per minute (less stable baseline). Notice that the same drug dose that produces a detectable effect for the stable baseline is claimed to be ineffective when inserted on the less stable baseline. The point is that we can detect small effects of drugs (and other variables) if the operant baseline is very stable during steady-state performance.

Operant baselines are said to show sensitivity to drugs. **Baseline sensitivity** means that a low dose of a drug such as amphetamine (a dopamine agonist) can cause substantial changes in baseline behavior. In contrast, the same operant baseline may not show sensitivity to doses of morphine (an opioid agonist). One implication of this kind of finding is that the effectiveness of the reinforcement contingency on behavior may involve the dopamine system more than the endogenous opiates. Based on this inference, the behavioral neuroscientist can further explore how the dopamine system participates in the control of the behavior and what neural structures are involved. Subsequent research could involve anatomical and physiological studies as well as further experiments using behavioral baselines.

Behavioral pharmacologists have used operant baselines to investigate the role of drugs that selectively "release" punished behavior—an area called anti-punishment effects (e.g., Morgan, Carter, DuPree, Yezierski, & Vierck, 2008). In a series of classic experiments, rats in an operant chamber were trained to respond to presentations of sweetened condensed milk (Geller & Seifter, 1960; Geller, Kulak, & Seifter, 1962). Next, when a clicker sounded, each lever press resulted in the milk and also a slight, uncomfortable electric shock to the floor grid. Data on typical performance by the rats showed that responding was greatly reduced during periods of punishment. A series of sedative or tranquilizing drugs were then administered, and the most interesting findings were that the tranquilizers did not affect overall responding for milk, but they increased responding during the clicker/shock periods. The clicker period is when there is a conflict between responding for milk (positive reinforcement) and receiving the electric shock (punishment). Apparently, the class of drugs called tranquilizers prevented the usual effects of punishment, while the other classes of drugs did not have this effect.

Drug effects on extinction (withholding reinforcement for previously reinforced behavior) have been investigated using operant baselines that require a set number of responses for reinforcement. In these studies, as outlined by Leslie (2011), rats receive reinforcement (food pellets) after lever pressing a given number of times; the lever is then withdrawn for a short period and reinserted, allowing the rat to complete the ratio requirement again. After behavior stabilizes on these ratio schedules (baseline), extinction is programmed and lever pressing no longer is reinforced with food. One question about extinction that behavioral pharmacologists try to answer is: which neural systems are involved? Researchers administered the anti-anxiety drug chlordiazepoxide (CDP) before beginning extinction and in another group of rats, the glutamatergic agonist D-cycloserine (DCS) after the extinction procedure. Both drugs facilitated operant extinction but for different reasons. The drug CDP potentiates the GABAergic (gamma-aminobutyric acid) inhibitory system resulting in its relaxing effects; but further research indicates that the anti-anxiety action is not central to CDP effects on operant extinction, and neither is its nonspecific effects on activity level. The drug appears to directly affect the extinction process but how this occurs is not yet known. As for the drug DCS, which targets the NMDA (N-methyl-D-aspartate) glutamate receptor, evidence suggests that activation of the NMDA receptor furthers the retention of new "inhibitory learning" that occurs during extinction—remembering not to press the lever for food when it is available.

One practical implication of this research is that the reinforcement and extinction procedures can be extended to "memory assessment" and used as an animal model for Alzheimer's Disease (AD). Once the animal model is well established by linking the conditioning procedures with known genetic markers for AD in transgenic mice (mice bred with these markers), it may be possible to investigate both drug interventions to improve or ameliorate learning and memory deficits of the disease (Leslie, 2011).

SINGLE-SUBJECT RESEARCH

Single-subject research is a well-founded scientific strategy. A single individual (rat, pigeon, or human) is exposed to the values of the independent variable, and the experiment may be conducted with several subjects. Each subject is a replication of the experiment; if there are four subjects, the investigation is repeated four separate times. Thus, every additional individual in a single-subject experiment constitutes a **direct replication** of the research and adds to the **generality** of the research findings—the more we replicate the methods and findings, the more our confidence increases that we can say something more general or universal in terms of a statement. Direct replication involves manipulating the independent variable in the same way for each subject in the experiment.

Another way to increase the generality of a finding is by **systematic replication** of the experiment. Systematic replication uses procedures that are slightly different, but are logically related to the original research question (see Sidman, 1960 for a detailed discussion of direct and systematic replication). For example, in Bernard's research with the rabbits, changing the diet from fasting to grass altered the chemistry of the urine and may be considered an experiment in its own right. Feeding the animals meat may be viewed as a second experiment—systematically replicating the initial research using a grass diet. Given Bernard's hypothesis that all fasting animals become carnivores, it logically follows that meat should change the chemistry of the urine from alkaline to acid.

In a behavioral experiment, such as the teacher attention and digit-reversal study (Hasazi & Hasazi, 1972), the researchers could have established generality by using a different teacher and a different kind of reinforcement (e.g., tactile contact such as hugging). Here the central idea is that the **contingency of reinforcement** is the critical factor that maintained the digit-reversal behavior of the child. The observed change in digit reversals did not depend on the specific teacher or the nature of the reinforcer (positive attention). In fact, many behavioral experiments have shown that contingencies of reinforcement generalize across species, type of reinforcement, diverse settings, and different operants.

Generality and Single-Subject Research

A common concern about single-subject experiments is that generalizations are not possible because the scores of a few individuals are not representative of the larger population. Many social scientists believe that experiments must include a large group of individuals to make general statements (called the *statistical groups design*). This position is valid if the social scientist is interested in descriptions of what the average individual does. For example, single-subject research is not appropriate for questions like "What sort of advertising campaign is most effective for getting people in Los Angeles to recycle garbage?" In this case, the independent variable might be a type of advertising and the dependent variable the number of citizens in Los Angeles who recycle their waste. The central question is concerned with how many people recycle, and a group statistical experiment is the appropriate way to approach the problem.

Behavior analysts are less interested in aggregate or group effects. Instead, the analysis usually focuses on the behavior of the *single individual*. These researchers are concerned with predicting, controlling, and interpreting the behavior of each organism. The generality of the effect in a behavioral experiment is established by **replication** of the effect, more than by *statistical significance*. A similar strategy is sometimes used in analytical sciences like chemistry. The process of electrolysis can be observed in an unrepresentative sample of water from a location such as Pocatello, ID. A researcher who follows the procedures for electrolysis will observe the same result in each batch (or unit) of water, whether the batch is taken from Portneuf River or from the Ganges. Importantly, the researcher may claim—on the basis of a single experiment—that electrolysis occurs in all water, at all times, and in all places. Of course, only repeated *replication* (both direct and systematic) of the electrolysis experiment will increase confidence in this empirical generalization. Today, behavioral researchers use both single-subject, statistical groups experimental designs (especially repeated measures designs), and even correlational studies (which simply examine the extent to which two or more variables are co-related) to establish the scientific validity of their research claims to a wider scientific audience. All of these methods are important and contribute in meaningful ways to understanding a behavioral problem or phenomenon.

Assessment of Experimental Control and Behavior Change

As mentioned earlier, single-subject experiments require a pre-intervention baseline period of measurement. This baseline serves as a comparison or reference for any subsequent change in behavior produced by the independent variable. The baseline is essential to know if your independent variable (which is not yet implemented) has any effect. To construct an appropriate baseline, it is necessary to define the *response class* objectively and clearly.

Definition and Measurement of the Response

In the animal laboratory, the response of pressing a lever is most often defined by the closure of an electrical switch. There is no dispute about the state of the switch: it is either on or off. An animal may press the lever in many different ways. The left or right paw may be used as well as the hind foot, nose, or mouth. The point is that no matter how the response is made, all actions that result in a switch closure define the response class. Once the response class is defined, the number of times the response occurs can be counted and a baseline constructed.

Outside of the laboratory, response classes are usually more difficult to define. Imagine that you are asked to help manage the disruptive behavior of a child in a classroom setting. The teacher complains that the child's behavior interferes with her teaching. On the surface, measuring the disruptive behavior of the child seems easy. Further reflection, however, suggests that it is not easy to define the operant class. What exactly does the teacher mean when she says "the child's behavior is disruptive"? After talking to the teacher and observing the child in the classroom, several "disruptive" responses may be identified. The child is often *out of her seat* without permission and at times when a lesson is being taught. Another behavior that occurs is *talking loudly* to other children during study periods. Both of these responses are more clearly defined than the label "disruptive," but objective measurement may still be difficult. Notice that each response is partially defined by prior events (permission) and the current situation (study periods). In addition, terms like *loud* and *out of seat* are somewhat subjective. How loud is loud, and is sitting on the edge of the desk out of seat? The answer is to keep refining the response definition until it is highly objective. When two observers can agree most of the time on whether a response has occurred, a baseline can be established.

In addition to defining the response class, assessment of behavior change requires *measurement of the response*. During the baseline, repeated measures of the target behavior are

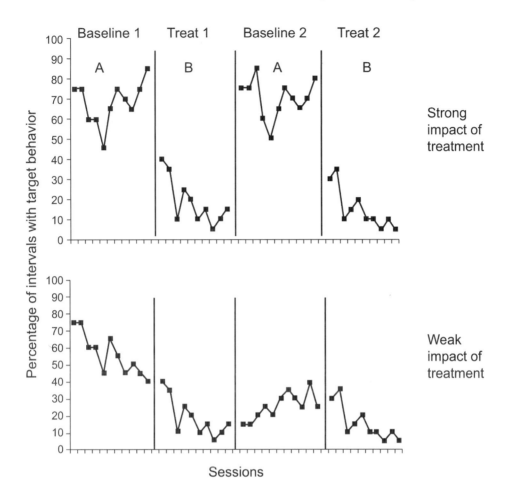

Fig. 2.9
Compare your assessment of the treatment effect in the Strong (top) and Weak (bottom) impact graphs. What visual properties of the two graphs lead you to assess the treatment (environmental manipulation) as effective in the top graph but not in the bottom one?

taken and plotted to assess *response variability*. Figure 2.9 portrays an idealized experiment to modify the out-of-seat behavior of the child in the foregoing classroom example. Pretend that the teacher is requested to pay attention and give tokens to the child only when she is sitting quietly in her seat during instruction sessions. At the end of the school day, the tokens may be exchanged for small prizes. For each 5-min interval of an instruction session, the teacher records whether an out-of-seat or target response has occurred (see Chapter 13 on interval recording), and the percentage of intervals with target responses is calculated. Does this procedure alter the child's behavior?

Assessment of Experimental Control of the Response
The upper and lower panels of Figure 2.9 show two possible results of an A-B-A-B reversal study of this classroom intervention or treatment (attention and tokens for sitting quietly). Compare your assessment of the treatment effect in the two panels. You probably judge that

the reinforcement procedure was effective in the top panel (strong impact) but possibly not in the lower one (weak impact). What do you suppose led to your conclusion?

A standardized visual assessment of treatment effectiveness uses several features of the graphed data (Kratochwill et al., 2010). These features are displayed in Figure 2.10, which shows four graphic displays of the classroom intervention for the child's out-of-seat behavior. The display labeled "Level" shows the **change in level** or average for the baseline and treatment (reinforcement) phases, with ten instruction sessions per phase. Compared with the baseline phases, the teacher's reinforcement for sitting quietly (treatment) is expected to decrease the level of out-of-seat responses. Also, when the teacher removes the reinforcement contingency for sitting quietly (Baseline 2), it is expected that target responses will return to the initial baseline level (Baseline 1). These shifts in level in the appropriate direction are large and convincing for the high-impact graph but not for the low-impact plot.

Changes in level produced by the treatment must also be assessed in terms of the **range of variability** of the dependent variable (percentage of intervals with out-of-seat responses). The range of variability is the difference between the highest and lowest values (percentages), and is shown in Figure 2.10 for each phase of the study in the display labeled "Variability." In terms of the high-impact results (top), you can see that the scores for both baselines lie outside the range of scores in the treatment phases. Inspection of the *range of variability* for the low-impact data shows that the baseline phases overlap the score variation in the treatments. This overlap of the percentage scores from phase to phase makes the results less convincing, in the sense of attributing any change in level to the effects of the treatment (reinforcement for sitting quietly in the seat). One reasonable strategy would be to increase the power of the intervention. In this case, the attempt is to produce a larger shift in behavior, relative to the baseline. For example, the small prizes earned at the end of the school day may be changed to more valuable items. Notice that this tactic leads to refinement of the procedures used in the experiment. This increases the experimental control over the participant's behavior—a primary objective of the experimental analysis of behavior.

You also may have taken into account the **immediacy of change** from baseline to treatment and from treatment to return to baseline (Baseline 2). You are using an assumption that the cause of a change in behavior must immediately precede the change. In behavior analysis, immediacy is assessed using the last three data points of the baselines and the first three data points for the treatment phases (ovals and hexagons of the "Immediacy" chart in Figure 2.10). You also probably assessed the immediacy of change from Treatment 1 to the return to baseline (Baseline 2), encompassed by the squares. Notice that, for the high-impact results (top display of chart), the change in the dependent variable is almost immediate with the changes in the independent variable (from baseline to treatment or treatment to baseline). Now inspect the low-impact display and you are less convinced that the changes in the behavioral measures are caused by the teacher's reinforcement of sitting quietly in the seat.

A fourth visual feature used for assessment involves **trend**—a systematic rise or decline in the values of the scores. The trend is assessed by visually depicting a "best-fit" straight line moving through the data points for each phase of the experiment. These lines are shown in Figure 2.10 in the chart labeled "Trend." For the high-impact findings (top display) there is an upward trend for the baseline phases, indicating that the out-of-seat responses are increasing. There is, however, a downward trend for each treatment phase, suggesting that the teacher's reinforcement of sitting quietly decreases the out-of-seat target behavior.

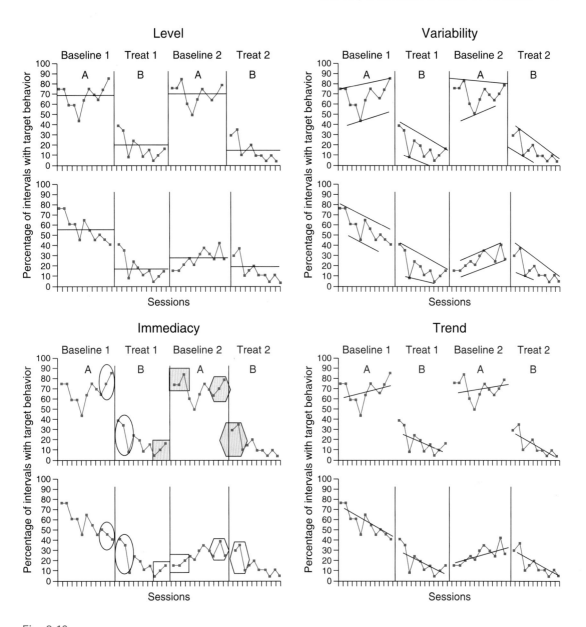

Fig. 2.10
Portrayed in the four panels are the standardized features for visual assessment of treatment (environmental manipulation) effectiveness. The first panel (Level) emphasizes the change in level or average for the Strong (top) and Weak (bottom) impact graphs from Figure 2.9. A second property is the range of variability of the scores for each phase of the experiment, as portrayed in panel two (Variability) by the range lines around the data. Immediacy of the change in the dependent variable is also used to assess the effectiveness of treatment and is portrayed in the third panel (Immediacy). Note that both change in level and immediacy of change are assessed relative to the range of variability. Finally, the fourth panel (Trend) depicts the changes in trend or drift in the scores for the Strong and Weak impact graphs by including trend lines. Drift from baseline to treatment, in the direction of treatment (downward), is a distinct problem for the A-B-A-B reversal design. Stable baselines without trend are required to attribute causation to the treatment or independent variable.

Now inspect the low-impact display for trend. Notice that the downward trend in the initial baseline (Baseline 1) is carried into the first treatment phase (Treatment 1). A drift in baseline measures can be problematic when the treatment is expected to produce a change in the same direction as the trend. In this case, the child is decreasing out-of-seat behavior before the intervention and continuing to decline in the treatment phase. Perhaps the child's parents are receiving more complaints from the school, and as the complaints mount, they put more pressure on the child to "sit quietly in class." You can hypothesize other reasons for the downward drift. The trends for the return to baseline (Baseline 2) and Treatment 2 are seemingly in the appropriate directions for the teacher's intervention, but could also reflect some kind of cyclical variation in the response measure. In summary, the trend data for the low-impact results do not suggest effective control by the teacher's reinforcement contingency. Generally, single-subject research requires a large shift in the level or direction of behavior relative to baseline. This shift must be clearly observed when the independent variable is introduced and withdrawn.

ADVANCED SECTION: PERCEIVING AS BEHAVIOR

Most of us believe that we accurately perceive the world around us and are able to report on happenings with some reliability. In everyday language and in psychology, perception is an inferred, underlying cognitive process that determines behavior. In contrast, behavior analysis suggests that *perceiving is behavior* that must be accounted for by environment–behavior relationships.

In the traditional account of perception, the person is said to transform the sensations from the sense organs (eyes, ears, and nose) by mentally organizing the input into a meaningful representation of the situation. From a behavioral perspective, the difficulty with this view of perception is that the mental organization and representation of sensory input are not directly observable. There is no objective way of obtaining information about such hypothetical events except by observing the behavior of the organism. Such **hypothetical constructs** are not always undesirable in science, but when used to account for behavior, these terms usually lack explanatory power because they are grounded in the very behavior that they are used to explain. This problem of explanatory power is seen in the traditional perception account of the well-known Stroop effect (Stroop, 1935).

Perception: The Stroop Effect

Figure 2.11 gives an example of a variation of the Stroop effect that you can try for yourself. First, look at the dots about the line at the top of the figure. Now, as fast as you can, say out loud whether each dot is positioned above or below the line—go! How many errors did you make and about how long did it take? OK, now look at the words ABOVE or BELOW the line at the bottom of the figure. As fast as you can, say whether each word is positioned above or below the line—go! How many errors did you make this time and how long did it take? Most people do pretty well when the problem involves dots (top), but poorly when they have to say the position of the words (bottom). Why do you think it is harder to do the problem with the words? Read on to find out.

One account involves perception and cognition, as in the following:

[T]he highly practiced and almost automatic perception of word meaning [ABOVE or BELOW] facilitates reading. However, this same perception automatically makes it difficult to ignore meaning and pay attention only to [the position of the word] stimulus. Thus, the Stroop effect is a failure of selective perception.

(Darley, Glucksberg, & Kinchla, 1991, p. 112)

Stroop Effect: A Behavior Analysis

From a behavior analytic perspective, the foregoing account restates the fact that your performance is better with the dots than with the words. The meanings and attention referred to in the passage are inferences from behavior with no independent evidence for their occurrence. Without evidence, the selective-perception explanation is not satisfying to the behavior analyst. The question is: How do environment–behavior relations regulate performance on this task?

The first thing to notice is that all of us have extensive experience of identifying the position of objects as above or below some reference point (the line in Figure 2.11). Thus, the position of the object comes to set the occasion for the perceptual response of reporting "above" or "below." We also have an extensive history of reinforcement for reading the words ABOVE and BELOW; in books these words correspond to the position of objects in pictures, as in "the airplane is above the ground" or "the sun is below the horizon." Because of this learning,

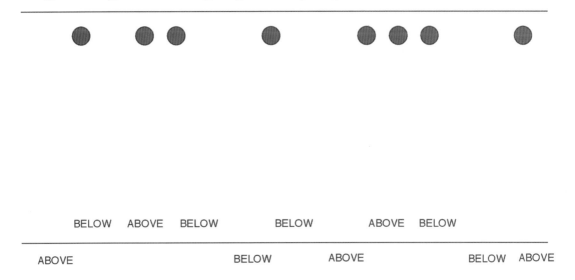

Fig. 2.11
A variation of the Stroop effect using dots above and below a line (top) and words for position (bottom). The bottom problem is more difficult in the sense that the position and the word (ABOVE vs. BELOW) compete for the response "saying the position of the word."

the physical position (location of object X) and written words for position ("above/below") come to control the response class ("object X is above [or below] the reference point"). When written words for location are presented in positions that do not correspond to the word (word = ABOVE; position = below), the two properties of the complex stimulus (word/position) compete for the respective responses. Based on the simultaneous control of behavior by two aspects of the blended stimulus, the time to complete the task increases and errors occur. Consider what you might do if you were driving and came to an intersection with a red hexagon sign that had the word PROCEED painted on it. You would probably wonder what to do and show "brake and go" responses. Instead of using an account based on selective perception, the behavior analysts would point to *response competition and reinforcement history* as reasons for your hesitation.

The Effect of the World on What We See

There are other interesting implications of a functional analysis of perceiving. For example, you walk into a room and look around, believing that you are taking in reality. But what do you see? Seeing itself is something an organism is prepared to do based on its biology and genetic endowment, but seeing a particular object on a given occasion may be analyzed as respondent or operant behavior. In this behavioral view, seeing an object or event is behavior elicited by the event, has a high probability due to past consequences, or becomes likely due to motivating conditions (e.g., hunger, thirst, or aversive stimulation).

Imagine that you have gone camping with several friends. After dinner you decide to entertain your friends by telling a horror story about an axe murder that took place in the same area a few years ago. One of your companions is finishing dinner, and the fried egg on her plate begins to look like a giant dead eye about to explode with yellow "glop." As the night gets darker, another camper hears ominous sounds and begins to see figures moving in the brush. In everyday words, your friends are imagining these events. Behaviorally, the frightening story may be analyzed as a motivating condition that momentarily increases the probability of seeing things that appear to be threatening.

B. F. Skinner (1953) has described other conditions that affect seeing as a conditioned response. One of these conditions is previous experience. He stated that:

> Conditioned seeing explains why one tends to see the world according to one's previous history. Certain properties of the world are responded to so commonly that "laws of perception" have been drawn up to describe the behavior thus conditioned. For example, we generally see completed circles, squares, and other figures. An incomplete figure presented under deficient or ambiguous circumstances may evoke seeing a completed figure as a conditioned response. For example, a ring with a small segment missing when very briefly exposed may be seen as a completed ring. Seeing a completed ring would presumably not be inevitable in an individual whose daily life was concerned with handling incomplete rings.
>
> (Skinner, 1953, pp. 267–268)

Skinner later points out that reinforcement can also affect what is seen:

> Suppose we strongly reinforce a person when he finds a four-leaf clover. The increased strength of "seeing a four-leaf clover" will be evident in many ways. The person will be more inclined to look at four-leaf clovers than before. He will look in places where he has

found four-leaf clovers. Stimuli that resemble four-leaf clovers will evoke an immediate response. Under slightly ambiguous circumstances he will mistakenly reach for a three-leaf clover. If our reinforcement is effective enough, he may even see four-leaf clovers in ambiguous patterns in textiles, wallpaper, and so on. He may also "see four-leaf clovers" when there is no similar visual stimulation—for example, when his eyes are closed or when he is in a dark room. If he has acquired an adequate vocabulary for self-description, he may report this by saying four-leaf clovers "flash into his mind" or he "is thinking about" four-leaf clovers.

(Skinner, 1953, p. 271)

You should realize that no one knows what a person "sees" at any moment. What we know is what the perceiver *says* they see or reports by providing a response (e.g., pressing a left or right key to a visual stimulus). But once the person tells us they see (or do not see) something, it is now public, and therefore becomes amenable to environmental influence. In other words, this statement or report is now an operant and is therefore subject to consequences.

Psychologists have talked about conditioned seeing as "perceptual set," "search image," or "mind-set". In fact, many psychologists do not consider perceiving as operant behavior. These researchers prefer to study perception as a cognitive process that underlies behavior (e.g., Langer, Djikic, Pirson, Madenci, & Donohue, 2010 on perception and visual acuity). Although the issue is not resolved here, Skinner makes it clear that analyzing seeing as behavior is one way to understand such processes. Thus, perceiving may be treated like *detecting signals* rather than mental states and processes (Green & Swets, 1966).

FOCUS ON: PERCEPTION, SIGNAL DETECTION, AND THE PAYOFF MATRIX

Perception often is considered to involve transformation of an environmental stimulus into sensation, neural activity, and interpretation. The environmental stimulus is said to stimulate the sensory organs (e.g., the retina in the eye, or the cochlea in the inner ear), which transform the input into neural activity (coded stimulus) that in turn activates information processing by the brain, yielding a mental representation or percept of the object. One can, however, separate perception per se from a perceptual response. Notably, the *perceptual response* is necessary to infer the presence of a perception (interpretation of sensory input such as "I see a red dot" when a red dot is presented). Basically, there is no way for us to know the perception of an environmental stimulus except by observing and measuring the perceptual response of the organism—*perception is always inferred from behavior*. By declaring that perceptual responses are behavioral responses (operant), it follows that perception per se is a function of its consequences. The theory of signal detection explicitly demonstrates this (see also Lynn & Barrett, 2014 for a behavioral economic analysis of signal detection).

Signal detection theory (SDT) is based on the idea of a viewer detecting a signal presented in the midst of a great deal of background noise (Goldiamond, 1962; Green & Swets, 1966; Nevin, 1969). Noise can be static on a radio, television, oscilloscope, or cell phone and the signal is the song, picture, airplane, or caller's voice, respectively. In actuality, the signal/noise metaphor can refer to any situation where a weak message is muddled by a confusing context. In this case, and almost every other situation where perception is involved, the perceptual response of reporting detection of a signal is an *operant*. Notice that as an operant the perceptual response must be a function of its consequences. In each instance of signal detection, the observer chooses between reporting the signal as "present" or "absent" under conditions of uncertainty and their choice is a function of signal probability, signal intensity, and consequences for the response. So, reporting the presence of a weak signal imbedded in strong noise (like guessing) depends mainly on the payoffs for being correct and incorrect.

When radar was invented, the screen often contained a great deal of static or visual noise. It was difficult for the observer to detect the actual signal amidst all the clutter. Of course, with radar it was critical to separate "blips" of the airplane (signal) from clouds, flights of birds, or just static in the system (noise). The observer faced a challenge of reporting the targets accurately among all the foils on the screen (for a related problem of vigilance and safety see Lattal, 2012). It became even more critical during and after the World War II when enemy aircraft or missiles could strike our homeland. Under wartime conditions, the radar observer viewed a monitor and there "might" be a target (enemy aircraft). They had to ask themselves: "If I report this as enemy aircraft, the military will activate defensive or retaliatory action, but if I don't report it, because it 'might' be a flock of seagulls, we could be bombed. What should I do?"

The observer actually has only the two possible perceptual responses ("Yes, it's a target" or "No, it's noise") and the situation has two possibilities (Yes = signal is present or No = signal is absent). These possibilities compose a 2 x 2 *outcome matrix* with responses (Yes or No) on the columns and signal presentations (Yes or No) on the rows, as shown in Figure 2.12 (top

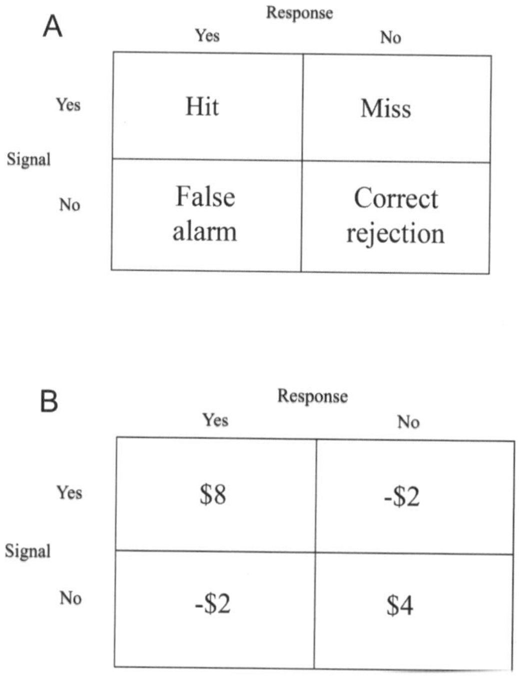

A

	Response Yes	Response No
Signal Yes	Hit	Miss
Signal No	False alarm	Correct rejection

B

	Response Yes	Response No
Signal Yes	$8	-$2
Signal No	-$2	$4

Fig. 2.12
The top panel shows a signal detection 2 x 2 outcome matrix (A) of the possible outcomes (hit, miss, false alarm, or correct rejection) when the signal is present (yes) or absent (no) and the perceptual response is "the signal occurred" (yes) or "the signal did not occur" (no). The bottom panel shows a payoff matrix (B) of possible gains and losses for the different signal-detection outcomes (hit, miss, false alarm, or correct rejection). See text for an analysis of control of the perceptual response by the payoffs and signal strength.

panel A). But just because there are only two possible responses does not mean that they are equally likely, even when 50% of the presentations are signals and 50% are not (only noise). If a weak signal (just detectable target in noisy background) is in fact present and the observer responds "Yes," the outcome is a *Hit*; if no target is present, but the observer reports "Yes," the outcome is a *False Alarm*; if a signal occurs but the response is "No," the outcome is a *Miss*; and if no signal is reported as "No," a *Correct Rejection* is the outcome. So far so good; but, when a *payoff matrix* (in dollars for human research) is set for each of the four outcomes as in Figure 2.12 (bottom panel B), perceptual responses come under the control of the payoffs (see Stuttgen, Yildiz, & Gunturkun, 2011 for experimental analysis of signal detection in pigeons).

In a situation where signal and noise are equally likely events (50/50), let's consider what will happen to "Yes" and "No" perceptual responses (observer reports) based on the stipulated payoffs in dollars shown in Figure 2.12 (lower panel B). For correct detections, notice that the positive payoff ($8) for a "Yes" response in the *Hit* cell is higher than for a "No" response in the *Correct Rejection* ($4) category; also, for error responses, notice that the losses in dollars ($2) are equal in both categories or cells (*False Alarm* and *Miss*). Under these contingencies and weak signals, the observer would respond "Yes" when a signal is discriminated as present (certain it's a signal) and "No" when discriminated as absent (certain it's noise). If a discriminative response is not possible (uncertainty), the observer would respond "Yes" (guess signal present) about twice as often as "No" (guess no signal), based on the relative payoffs ($8 vs. $4). Of course, the payoff matrix would have minimal effects if strong signals were presented (detectable target with little noise)—responses will mainly depend on the signal not the payoffs.

Considerations of signal intensity and payoffs have other implications. One factor that immediately stands out is that to maximize *Hits* (say Yes, there is a signal), one will also give many *False Alarms*, depending on the signal-to-noise ratio. One would detect 100% of the targets by reporting "Yes" always, but *False Alarms* are often costly and discouraged (as in wartime). Because of these types of consequences, the chances of a Yes would go down, and with it, the number of *Hits* would be reduced, as observers would not choose "Yes" as often when in doubt. In addition, if *Hits* were more highly valued (reinforced differentially) than *Misses*, many more "Yes" responses would occur; if, however, there were strong penalties for *False Alarms*, fewer real targets would be detected. As we have seen, the situation is always a problem when the signal is weak and uncertain (low signal/noise ratio). The observer asks, "Was something there or not? I'm not sure. What should I say?" This uncertainty makes the "perceptual response" an operant that depends on its consequences. If *False Alarms* are punished (high cost), I must tolerate occasionally missing a real target and not reporting "Yes" as often.

The signal-detection dilemma is not only a good example of how perceptual behavior is operant, it also provides a plausible account of hallucinations or the reporting of things not present. Parents sometimes teach children to "pretend" this is a piece of chewing gum—taking a piece from an empty hand and chewing on a nonexistent stick of gum. We, and kids, pretend because we receive some attention or other payoffs for this perceptual behavior. Professional actors, who use pantomime for entertainment, receive money and applause for perceptual responses to things that are not present. Claiming to see "pink elephants" when none exist is not surprising when you realize others have taken seriously (attended to) reports of such "visions" in the past. The famous Rorschach test requires an observer "to report" a signal when in fact none exists.

SDT describes an important and useful way to organize information when required to make decisions in the presence of complex data with high uncertainty. Daniel Levitin devotes an appendix in his book *The Organized Mind* (2014) to constructing 2 x 2 tables for the reader to better assess Bayesian probabilities for signal-detection type problems. He quotes President Obama saying "Nothing comes to my desk that is perfectly solvable" as an example of having to decide based on imperfect information some issue with only negative outcomes, which is far worse than a payoff matrix providing both gains and losses.

> **CHAPTER SUMMARY**
>
> In summary, this chapter introduces the science of behavior analysis. In particular, it is the assessment of the antecedents and consequences of behavior as aspects of the environment that can be manipulated and affect the behavior in question. Innumerable formal and informal studies have determined that the events that follow a specific response will influence whether that response is likely to occur again. If a response has a destructive, obnoxious, painful, or otherwise unpleasant outcome, an organism will likely not repeat it. We are built by natural evolutionary processes to behave, to walk, to pick up things, to vocalize, etc. We look around and we see different sights; our head turning has a function—it moves our eyes so that we see in different directions. Behavior analysts work to discover the functions of behavior and also to provide functions that end up creating novel behaviors.
>
> A functional analysis is conducted in several uniform and proven effective sets of procedures. A major tactic is the A-B-A-B reversal process whereby the researcher determines if a certain functional effect (an applied consequence) does indeed control the appearance of a behavior. If a rat gets pellets for lever pressing, the rat presses the lever, and when pellets stop coming it stops pressing. The behavior of organisms can be studied objectively and scientifically, and that is why several issues are described concerning replication, validity, generalization, and assessment. The experimental analysis of behavior is a systematic set of tactics for the exploration of the controlling variables of behavior.

Key Words

A-B-A-B reversal design
Abolishing operation (AO)
Baseline
Baseline sensitivity
Change in level (baseline to treatment)
Conditioned-stimulus function
Context of behavior
Contingency of reinforcement
Dependent variable
Direct replication
Discriminative function
Discriminative stimuli

Elicited (behavior)
Emitted (behavior)
Environment
Establishing operation (EO)
Functional analysis
Generality
History of reinforcement
Hypothetical construct
Immediacy of change (baseline to treatment)
Independent variable
Motivational operation (MO)
Negative reinforcer

Operant
Positive reinforcer
Range of variability (in assessment)
Reinforcement function
Replication (of results)
Respondent
Response class
Response hierarchy

Single-subject research
Steady-state performance
Stimulus class
Stimulus function
Structural approach
Systematic replication
Topography
Trend (in baseline)

On the Web

http://onlinelibrary.wiley.com/journal/10.1002/(ISSN)1938–3711 The homepage for the *Journal of the Experimental Analysis of Behavior* (*JEAB*)—a journal that illustrates the experimental method discussed in Chapter 2. Early issues of the journal are helpful in terms of basic design (A-B-A-B reversal). You may be able to get access to the journal's issues via your university online library.

http://onlinelibrary.wiley.com/journal/10.1002/(ISSN)1938–3703 The webpage is for the *Journal of Applied Behavior Analysis* (*JABA*)—a journal devoted to the application of behavior principles. The articles often illustrate the basic designs used in behavior analysis, especially in earlier issues of the journal. You may be able to get access to the journal's issues via your university online library.

www.michaelbach.de/ot Visit the website of Michael Bach and enjoy the 92 visual illusions he has assembled. Most of these illusions are used to reveal how humans perceive the world. A behavior analysis turns perception on its head, asking how the world controls what we see. Try to explain your favorite illusion by referring to a previous history of reinforcement for "seeing X" and the current contingencies set up by the illusion. You might also use the concepts of *signal detection theory* addressed at the end of the chapter. Discuss the illusions with your classmates and instructor.

www.dharma-haven.org/science/myth-of-scientific-method.htm Here is a website maintained by Terry Halwes, who argues that scientists deviate in important ways from the logical hypothesis-testing view taught in most scientific methods books. He states that "the procedure that gets taught as 'The Scientific Method' is entirely misleading. Studying what scientists actually do is far more interesting."

Brief Quiz

1. In terms of finding an object that is missing or hidden:
    a. a structural account points to stages of development and object permanence
    b. a behavioral account points to a particular history of reinforcement
    c. the form or structure of behavior is used by behavior analysts to infer mental stages
    d. both (a) and (b) are true
2. The term _____ refers to behavior that is elicited, and the term _____ refers to behavior that is emitted.
    a. operant; respondent
    b. respondent; operant
    c. reflexive; flexible
    d. flexible; reflexive

3 Any stimulus (or event) that follows a response and increases its frequency is said to have:
   a   a reinforcement function
   b   a discriminative function
   c   a conditioned-stimulus function
   d   a consequence function
4 In functional analysis, positive and negative reinforcers are examples of:
   a   response classes
   b   stimulus classes
   c   conditioned stimuli
   d   unconditioned stimuli
5 In terms of behavior–environment relations, establishing operations:
   a   are used to construct the foundations of behavior
   b   increase the momentary effectiveness of reinforcement
   c   increase momentarily responses that produce reinforcement
   d   both (b) and (c)
6 The variable manipulated by the experimenter is the _____, and the measured effect is the _____.
   a   dependent; independent
   b   extraneous; dependent
   c   independent; dependent
   d   independent; extraneous
7 In terms of the reversal design and behavioral experiments:
   a   the A phase is called the baseline
   b   the B phase is the experimental manipulation
   c   the design is used in single-subject experiments
   d   all of the above are true
8 Baseline sensitivity means that:
   a   behavior is sensitive to a low dose of drug
   b   behavior is sensitive to a high dose of drug
   c   behavior is sensitive to both high and low doses of drug
   d   behavior is sensitive to stimuli that accompany the drug dose
9 The presence of trend in baseline measures:
   a   refers to a systematic rise or decline in the baseline values
   b   is a problem when the treatment-expected change is in the direction of the trend
   c   can be helpful when inferring that an independent variable has produced an effect
   d   is characterized by both (a) and (b)
10 **ADVANCED SECTION:** In terms of the Stroop effect, behavior analysts point to _____ and _____ as reasons for hesitation.
   a   response competition; learning
   b   learning; reinforcement
   c   response competition; history of reinforcement
   d   history of reinforcement; memory

Answers to Brief Quiz: 1, d (p. 37); 2, b (p. 39); 3, a (p. 43); 4, b (p. 44); 5, d (p. 44); 6, c (p. 51); 7, d (p. 52); 8, a (p. 56); 9, d (p. 60); 10, c (p. 64).

Reflexive Behavior and Respondent Conditioning

1. Learn about fixed-action patterns (FAPs) and modal action patterns (MAPs).
2. Investigate the primary laws of the reflex and the process of habituation.
3. Study Pavlov's experiments on respondent conditioning of salivation.
4. Learn about the complexities of higher-order conditioning.
5. Discover the conditioning basis of drug tolerance and overdose.

A biological imperative faced by all creatures is to survive long enough to reproduce. Because of this necessity, behavior related to survival and reproduction often appears to be built into the organism. Thus, organisms are born with a range of behavior that aids survival and reproduction. Creatures that fly to avoid predators are likely to be born with the capacity or ability to fly. Thus, flying does not need to be learned; it results from the organism's evolutionary history as a species. The complex array of motor movement and coordination involved in flying could be learned, but it is much more dependable when this behavior is primarily based on genetic endowment.

For most animals, survival at birth depends on being able to breathe, digest food, and move about. When a worm is dangled over a young robin's head, this stimulus elicits opening of the mouth and chirping. The behavior of the chick is the result of biological mechanisms and is elicited by the sight of the dangling worm. The relation between the dangling worm (stimulus) and the open mouth (response) is called a *reflex*. Presumably, in the evolutionary history of robins, chicks that presented a gaping mouth and chirped were fed more often than those that did not, contributing to survival and reproduction. There are, however, learned modifications of such initial behaviors. For example, Tinbergen and Kuenen (1957) observed that if feeding did not follow, chicks stopped gaping to a realistic, artificial parental stimulus. In humans, reflexive crying to discomfort or hunger by an infant ensures more effective care from the child's parents. Parents engage in a variety of caretaking behaviors, which may result in cessation of crying. Usually, parental responses such as changing a soiled diaper, feeding, or burping the infant stop the ongoing fussing (see escape in Chapter 6).

DOI: 10.4324/9781003202622-3

PHYLOGENETIC BEHAVIOR

Stimulus–response relations that predominantly are based on genetic endowment are described as **phylogenetic**, and are present on the basis of the evolutionary history of a species (species history). Behavior that is involuntary and invariant that aids survival and procreation is often (but not always) unlearned. This is because past generations of organisms that engaged in such behavior survived and reproduced—passing on their genes over generations. Thus, species history provides an organism with a basic repertoire of responses that are elicited by specific environmental conditions. Darwin said these physical and behavioral characteristics were naturally selected, as they occurred through no immediate human action or intervention.

Sequences of Behavior

Fixed-action patterns or **FAPs** are sequences of behavior (a series of connected movements) that are phylogenetic in origin. All members of a particular species (though it may be sex-dependent) engage in the FAP when the appropriate *releasing stimuli* are presented. Fixed-action patterns have been observed and documented in a wide range of animals and over a large number of behaviors related to survival and reproduction. To illustrate, Tinbergen (1951) noted that the male stickleback fish responds with a stereotyped sequence of aggressive displays and movements when other male sticklebacks intrude on his territory during the mating season. The female spider *Cupiennius salei* constructs a cocoon and deposits her eggs in it by engaging in a fixed sequence of responses (Eibl-Eibesfeldt, 1975). A graylag goose presented with an egg outside its nest will spontaneously roll the egg into the nest by reaching over the egg (with her bill) and pulling it carefully toward the nest. If the egg is removed, the bird continues with the fixed sequence of egg-retrieval actions. Basically, the bird continues behaving as if the egg is present even though it has been removed. The following passage describes the fixed-action pattern that the squirrel (*Sciurus vulgaris* L.) engages in while storing nuts for the winter:

> The squirrel ... buries nuts in the ground each fall, employing a quite stereotyped sequence of movement. It picks a nut, climbs down to the ground, and searches for a place at the bottom of a tree trunk or a large boulder. At the base of such a conspicuous landmark it will scratch a hole by means of alternating movements of the forelimbs and place the nut in it. Then the nut is rammed into place with rapid thrusts of the snout, covered with dirt with sweeping motions and tamped down with the forepaws.
>
> (Eibl-Eibesfeldt, 1975, p. 23)

Ethologists refer to such predictable and stereotypic behaviors as *fixed-action patterns* to suggest that these behaviors are built in and immutable. These researchers are looking for heritable genetic factors, which appear to account for behavior of all members of the species. On the other hand, the behavior science model used in this textbook considers all behaviors as flexible and adaptable, at least to some degree. So, given the adaptive ability of most animals, we refer to this behavior as a **modal action pattern** or **MAP**. Although the topographic features of these reflexive sequences may appear similar across most individuals and situations (modal), the concept of MAP denotes the numerous idiosyncratic differences or variations in behavior—implying some degree of flexibility rather than rigid genetic control. For example, robins (*Turdus migratorius*) build nests that appear very similar in construction. It is clear, however, they do not all build in the same location, or use the same materials. There

Fig. 3.1
A photo of a young woman yawning. Yawning is considered a common modal action pattern (MAP) that appears to be phylogenetic in origin. One possibility is that yawning functions to reinstate surveillance and monitoring of the environment following periods of low vigilance and reduced alertness.
Source: Shutterstock.

is substantial individual variation in all phases of nest construction, suggesting modification by the environment (ontogeny).

One common and often overlooked stereotypic MAP in humans is yawning (Figure 3.1), which involves gaping of the mouth, an extended intake of air, and a short period of expiration (Provine, 2005). In humans, yawning (without intake of air) begins in utero by the 15th week of pregnancy and continues to occur in infants, children, and adults of all ages. The behavior is mostly fixed, but shows flexibility in duration, frequency, and form. Once started, a yawn progresses through a sequence of responses, which is difficult to stop—almost like a sneeze. Yawning typically occurs in bouts, with a highly variable inter-yawn interval, averaging about a minute. Yawns last about 6 s, but show considerable variation about this value. Also, the pattern of yawning for each person (within individuals) is quite stable over several weeks of observation and there is no compensation between yawning frequency and duration. Those who yawn for shorter durations do not do it more frequently than others with longer durations; also, people with longer duration yawns do not yawn less often than those with shorter durations (Provine, 2005). Although the precise stimulus that sets off yawning is hard to define, research by Provine and Hamernik (1986) indicates that time of exposure to unchanging, monotonous (highly boring) visual stimulation (e.g., unchanging color-bar test patterns without sound) substantially increase yawning compared to more dynamic visual/auditory stimulation (music videos with accompanying audio). People yawn most when sleepy, especially in the first hour after waking, but also in the hour before bedtime as well. Yawning after waking is often accompanied by stretching of the arms, while yawning before bedtime usually only involves yawning without stretching. As you can see, yawning involves stereotypic behavior of phylogenetic origin that often occurs during periods of low vigilance and reduced alertness—perhaps acting to momentarily reinstate surveillance and monitoring of the environment. [Note: arousal or physiological activation of the brain is a disputed cause of yawning (Gallup, 2011; Guggisberg, Mathis, Schnider, & Hess, 2010); *contagious yawning* is reviewed by Demuru & Palagi, 2012; Provine, 2005; see also Gallup, Swartwood, Militello, & Sacket, 2015 for birds.]

Reaction chains are similar to FAPs, but with one major difference—each set of responses in a reaction chain requires an appropriate stimulus to set it off. Recall that once a fixed-action pattern (FAP) begins, the animal usually continues the sequence even when the stimuli that set off the behavior are removed. In the previous squirrel and nuts example, the animal continues to dig a hole and bury the non-existent nut, even if the nut is removed. In contrast, a reaction chain requires the presence of a specific stimulus to activate each link in the sequence of behavior. An organism's performance produces stimuli that set off the next series of responses in the chain; these behaviors in turn produce the stimuli followed by another set of responses. Presenting a stimulus that ordinarily occurs in the middle part of the sequence activates the chain at that point rather than at the beginning. Also, unlike FAPs, if the stimuli that activate behavior are removed, the sequence is disrupted.

The courtship ritual of the male and female stickleback fish (*Gasterosteus aculeatus*) is a reaction chain (Figure 3.2). Reaction chains often show behavioral flexibility similar to MAPs, but here we describe an idealized behavioral sequence. During the mating season, the reaction

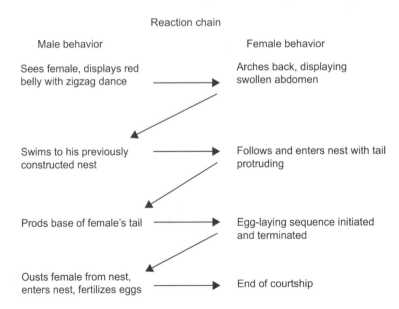

Fig. 3.2
The reaction chain for courtship of male and female sticklebacks is shown. This sequence of reflexive responses is initiated in the breeding season when the male stickleback spies a female and begins his zigzag dance, displaying his red belly. See text for a description of this reaction chain.

chain is initiated when a male stickleback sees a female and begins a zigzag dance, displaying his red underbelly. For a receptive female with eggs, the male's display functions as a stimulus that causes the female to arch her back and reveal her swollen abdomen. This initiates swimming by the male to a previously constructed nest. The female follows and enters the nest with her tail protruding, causing the male to nibble at the base of her tail and initiate the egg-laying sequence. Once the female has deposited her eggs, the male ousts her from the nest, enters the nest, and fertilizes the eggs, thereby completing the reaction chain.

This courtship chain may be terminated at any point if the behavioral displays by either fish (male or female) are inadequate or inappropriate to function as a stimulus for the next link in the chain. Thus, red-bellied males that skillfully perform the zigzag dance are more likely to attract females to their nests and fertilize eggs than males that execute the dance less skillfully. **Sexual selection**, by providing more chances to mate, ensures that genes related to skillful execution of the dance are more represented in the next generation and that the offspring have a high probability of successful courtship and reproduction.

Reflexive Behavior

The principles that describe the reflex (and its conditioning) are similar across different kinds of reflexes. For example, the laws that govern pupil contraction when a light is shone in the eye, or the principles that describe the relation between a sudden loud noise and a startle response, also hold for the salivation produced when you eat a meal. Early work by Sherrington (1906) focused on the reflex and the relations discovered over a century ago generalize to a remarkable variety of stimulus–response relations. When food is placed in a dog's mouth, the salivary glands produce saliva. This relationship between food in the mouth and salivation is a reflex, which is based on the heredity and is not learned. Many reflexes serve defensive, protective, or survival functions. Frequently, such reflexes are not learned because they must function before adequate experience is provided.

All organisms are born with a built-in set of reflexes; some are conserved across species, but others are particular to a species. Thus, humans are born with an array of responses elicited by specific stimuli. For example, tactile stimulation of the human infant's cheek evokes the rooting response—turning toward the stimulation with the mouth open, which makes it more likely that the infant will contact milk through the mother's then receives the nipple. Also, as we have noted, in young robins the so-called "begging" reflex (open mouth and chirping) serves a similar function—getting fed. Because these relationships are relatively invariant and biologically based, we refer to the eliciting or activating event as the **unconditioned stimulus** (**US**). The related behavior following the stimulus is called the **unconditioned response** (**UR**). The term *unconditioned* is used because the reflex does not depend on an organism's experience or conditioning during its lifetime (i.e., learning).

When an unconditioned stimulus elicits an unconditioned response (US → UR), the relation is called a **reflex**. Reflexive behavior is automatic in the sense that a physically healthy organism always produces the unconditioned response when presented with an unconditioned stimulus. You do not choose to salivate or not when you have food in your mouth; the US (which is "food in the mouth") draws out the UR of salivation; that is, salivation is said to be **elicited** by the US. This is the way the animal is built by evolution. However, there are times and conditions described below where the US does not elicit the UR. When repeated presentations of the US leads to a reduction of the UR, we call the process *habituation*.

Laws of the Reflex

Around 350 BC, Aristotle developed principles of association that were rediscovered by psychologists and by Pavlov (a physiologist) in the 1900s (Hothersall, 1990, p. 22). Sherrington (1906) studied many different types of reflexes, and formulated the laws of reflex action. These laws are general in that they hold for all eliciting or unconditioned stimuli (e.g., food in the mouth, a touch of a hot surface, a sharp blow just below the knee, or a light shining in the eye) and the corresponding unconditioned responses (salivation, quick finger withdrawal, an outward kick of the leg, pupil contraction, respectively).

The unconditioned stimuli (US) that elicit unconditioned responses (UR) may vary in intensity. For example, light shining in the eye may be bright enough to hurt or so faint that it is difficult to detect. A tap below the knee, causing a kick, may vary from a modest to a heavy blow. The intensity of the eliciting US has direct effects on the elicited reflex. Three **primary laws of the reflex** describe these effects.

1. The **law of the threshold** is based on the observation that at very weak intensities a stimulus will not elicit a response, but as the intensity of the eliciting stimulus increases, there is a point at which the response is elicited. Thus, *there is a point below which no response is elicited and above which a response always occurs*. This point is called the threshold.
2. The **law of intensity–magnitude** describes the relationship between the intensity of the eliciting stimulus and the size or magnitude of the elicited response. *As the intensity of the US increases, so does the magnitude of the elicited UR*. Once a stimulus reaches threshold, a light tap on the patellar tendon (just below the kneecap) will evoke a slight jerk of the lower leg; a stronger tap will produce a more vigorous kick of the leg (the patellar reflex). Of course, there are upper limits to the magnitude of the tap.
3. The **law of latency** concerns the time between the onset of the eliciting stimulus and the appearance of the reflexive response. Latency is a measure of the amount of time that passes between these two events. *As the intensity of the US increases, the latency to the appearance of the elicited UR decreases*. Thus, a strong puff of air will elicit a quick blink of the eye. A weaker puff will also elicit an eye blink, but the onset of the response will be delayed.

These three laws of the reflex are basic properties of all reflexes. They are called primary laws because, taken together, they define the relations between values of the eliciting stimulus (US) and measured properties of the unconditioned response (UR). Reflexes, however, have other characteristics and one of these, *habituation*, has been shown in animals as simple as protozoa and as complex as humans.

Habituation

One of the more documented *secondary properties* of the reflex is called **habituation**. Habituation is observed to occur when an unconditioned stimulus repeatedly elicits an unconditioned response and the response gradually declines in magnitude. When the UR is repeatedly elicited by the US, the reflex may eventually fail to occur at all. For example, Wawrzyncyck (1937) repeatedly dropped a 4-g weight onto a slide on which the protozoa *Spirostomum ambiguum* were mounted. The dropped weight initially elicited a contraction or startle response that steadily declined to near zero with repeated stimulation.

Fig. 3.3
A volcano repeatedly threatening to erupt, without a major eruption, results in habituation of people's startle/panic responses, including running away.
Source: Shutterstock.

An interesting report of human habituation, in a dangerous setting, appeared in the July 1997 issue of *National Geographic* (Figure 3.3). The small island of Montserrat has been home to settlers since 1632. Unfortunately, the relatively silent volcano on the island reawakened in July 1995. Suddenly, the quiet existence that had characterized living on Montserrat was rudely interrupted. Before the major eruption of the volcano, a large group of inhabitants refused to evacuate the island, even though these people suffered through several previous small volcanic explosions:

> Gerard Dyer and his wife, Judith, [have] been staying with friends in St. John's, about as far north of the volcano as you can get.... People could get passes to visit the unsafe zone, which is how Gerard came to be working on the flanks of Soufriere Hills that bright morning.
> "If you have animals and crops, you can't just leave them," said Gerard as we walked back to his truck. "You have to come look after them and hope nothing happen." As he spoke, the volcano made a crackling sound like distant thunder—blocks of solid lava rolling down the side of the dome. Gerard didn't even look up.
> Montserratians have become so used to the volcano's huffing and puffing that the initial terror has gone. As one woman said, "At first when there was an ashfall, everybody run. Now when the ash falls, everybody look."
>
> (Williams, 1997, p. 66)

In this example, Gerard has been repeatedly exposed to the sound (US) of minor volcanic explosions. At first, this sound elicited a startle/panic response, accompanied by running, but these URs habituated to near zero with repeated eruptions of the volcano. A similar process is observed when people live under an airport flight path; initially the sound of a jet taking off or landing is bothersome, but after some time the sound is barely noticed. There are a number of general properties that characterize habituation (Rankin et al., 2009; Thompson & Spencer, 1966); some are outlined here. First, the decrease in the habituated response is large initially, but gets progressively smaller as habituation continues. Second, if the unconditioned stimulus is withheld for some time, the habituated response recovers, a process called *spontaneous recovery*. Third, when habituation is repeatedly produced, each series of stimulus presentations generates progressively more rapid habituation. In other words, habituation occurs more quickly on a second series of US presentations than on the first, and then even faster on a third set. This quicker onset of habituation with repeated series of US presentations may define the simplest form of learning and remembering (Tighe & Leaton, 1976).

On a daily basis, animals are exposed to aversive events that activate a complex stress response by the hypothalamic-pituitary-adrenal (HPA) axis, a response that seemingly shows habituation to repeated presentation of the stressful event. This is an important function, because an organism learns that while some stimuli are loud (inducing startle), many are not dangerous. Habituation allows the HPA stress response to adapt to non-threatening stimuli; otherwise, organisms would be in a chronic state of hyperarousal, which would require a great deal of energy. In a review of the HPA axis and stress neurobiology, Grissom and Bhatnagar (2009) concluded that HPA activity to stressors shows many of the features of response habituation to repeated presentations of a stimulus. However, they also stated that the HPA-axis response to repeated stress is more complicated than response habituation. The decline in HPA activity to stressful events does involve habituation, but also negative feedback mechanisms in the brain regulated by the release of glucocorticoids (a class of steroid hormones) as well as more complex respondent and operant learning, involving previous exposures to stressful stimuli.

Habituation is a conserved behavioral process that has come about because of a phylogenetic history. Those animals that habituated were more likely to survive and produce offspring—passing on their genes to the next generation. A herbivore that runs away each time the grass rustles gets less to eat than one that is not scared away. A rustling sound of grass may indicate the presence of a predator, or simply the wind blowing. Repeated, unnecessary activation of respondent mechanisms also causes stress to the animal, which is not good in terms of health and physiology.

ONTOGENETIC BEHAVIOR

In addition to phylogenetic history, the behavior of an organism is affected by environmental experience. Each organism has a unique **ontogenetic** history or lifetime of conditioning. Changes in behavior, as a result of such experiences, are called *learning*, consisting of moment-to-moment interactions of the organism's behavior with the environment. Events of the physical world include social and cultural impact on the behavior of organisms. Learning builds on *phylogenetic* history to determine when, where, and what kind of behavior will occur at a given moment.

For example, salivation is involved in the digestion of food. People do not learn to salivate to the taste of food; this is a phylogenetic characteristic of the species. After some experience learning that McDonald's goes with food, you may salivate to the sight of the golden arches of McDonald's, especially if you are hungry and like to eat hamburgers. Salivating at the specific sight of McDonald's arches occurs because of *respondent conditioning*—you were not born that way. It is, however, important to note that respondent conditioning and other learning processes evolved because they provided some sort of reproductive advantage. Those organisms whose behavior came under the control of important environmental events presumably gained an advantage over those whose behavior did not. Through Darwinian evolution and selection, respondent conditioning became a means of behavioral adaptation. In other words, *organisms with a capacity for respondent or associative learning were more likely to survive and reproduce*—increasing their genes in the population (for possible steps in the evolution of non-associative and associative learning, see Pereira & van der Kooy, 2013).

Respondent Conditioning

Respondent conditioning involves the transfer of the control of a reflex to an additional stimulus (or set of stimuli) by stimulus–stimulus (or S–S) association. In Chapter 1, we saw that the sound of a bell could come to elicit salivation after the bell had been associated with food. This kind of conditioning occurs in all species, including humans, and is common in everyday life. Imagine that you are out for an early morning walk and pass a bakery where you smell fresh donuts. When this happens, your mouth begins to water and your stomach starts to growl. These conditioned responses occur because, in the past, the smell has been associated with food in the mouth.

Figure 3.4 shows the classical conditioning of salivation described by Pavlov (1960). The upper panel first shows the basic reflex of food (US) eliciting salivation (UR). Then, when an arbitrary stimulus such as a light (which is initially neutral in terms of eliciting salivation; as such, it is called a neutral stimulus or NS) is presented just before the US (food) is placed in a dog's mouth, which elicits the UR. After several presentations of the light with the food,

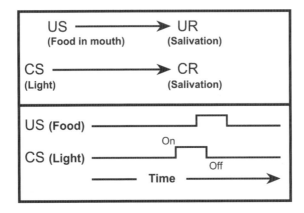

Fig. 3.4

Simple respondent conditioning. An arbitrary stimulus such as a light is presented as a neutral stimulus just before food (US) is placed in a dog's mouth. After several pairings of light and food, the light is presented alone. If the light now elicits salivation, it is called a conditioned stimulus (CS), and salivation to the light is a conditioned response (CR).

the light is presented alone. This is shown in the lower panel. If the light now elicits salivation during a test phase, the light is now called a **conditioned stimulus** (**CS**), and salivation to the light is called the **conditioned response** (**CR**).

Notice that a new and previously neutral feature of the environment (a light) has now come to regulate the behavior (salivation) of the organism. Thus, Pavlovian (classical or respondent) conditioning involves the transfer of behavior control to new and often arbitrary aspects of the environment by mere association. To experience this sort of conditioning, try the following: read the word *lemon* and consider the last time you ate a slice of lemon. Many people salivate at this CS because the word has been correlated with the sour taste of the fruit. This shift in controlling stimulus from food to word is possible because inputs to the visual system end up activating the neurons innervating the salivary gland.

Because the CR is a response elicited by the CS, it is often called a **respondent**. The terms *conditioned response* and *respondent* are interchangeable throughout this text. The process of pairing close together in time the NS with the US so that the CS comes to elicit the conditioned response (CR) is called **respondent conditioning**.

Note that the association that is formed is between the CS and US (i.e., the word *lemon* and the real fruit in the mouth) because of the CS–US contingency (time-based or temporal pairing)—not because of some cognitive (internal mental) association of events. This is an important point. The word "association" is sometimes taken to mean an internal mental process that a person or other animal performs. We hear people say, "The dog salivates when the bell is sounded because it has associated the sound with the food." In contrast, a behavior analyst points to the association or correlation of stimuli (CS and US) that occurred in the past. In other words, the association is between events—it does not refer to mental associations to explain the conditioning. The word *lemon* (CS) elicits salivation (CR) because the word has been followed closely in time by the lemon (US), which produced salivation (UR).

The basic *measures* of behavior for respondent conditioning are magnitude (amount of salivation) and latency (time to salivation) of response following presentation of the US or CS. Magnitude and latency make sense as behavioral measures because respondent conditioning often involves the actions of smooth muscles and glands or responses such as eye blinks and skin resistance (the UR or CR) that vary on these two dimensions. Other respondents, however, such as attending or orienting to a stimulus or reflexive movement toward a place or location, are elicited and often confused with operant behavior. When the CS controls behavior based on a respondent-conditioning procedure, the behavior is classified as a respondent, no matter what its form or appearance. Behavior controlled by its consequences is operant, even when this behavior involves actions of the smooth muscles, glands, or eye blinking (see more in Chapter 7).

Relative Nature of Stimuli

The conditioned stimulus (CS) is defined as an event that initially does not elicit the target response (i.e., the NS), while the US does (US → UR). Consider a sweet-flavored liquid as an NS for inducing nausea; prior to conditioning, it does not induce nausea at all. However, when it is repeatedly paired with a drug that induces sickness (e.g., lithium chloride, which is a US that produces nausea), the sweet-flavored liquid will now serve as a CS for nausea. Indeed, animals show avoidance of the sweet-flavored solution in this situation, an effect known as **conditioned taste aversion** (**CTA**). In this situation, note that the sweet liquid functions as two different types of stimuli, depending on the kind of response it elicits. It can be a CS, depending on its pairing with a US such as lithium chloride. It can also be an NS prior to condition. How we define it depends solely on *what kind of response* it elicits: a CR or nothing.

You may have noted to yourself that sweet-flavored liquid is not completely neutral, however. Because it does not naturally induce nausea, we call it an NS. However, it indeed naturally induces salivation and tastes good (US). As such, we could use the sweet liquid in a different procedure where the CS is a particular place or location and the sweet-flavored solution is the US. In this case, we repeatedly give the animal the liquid in one distinct chamber (with stripes) but not in another (white). During test trials, the animal shows a preference by a choice test for the location paired with the solution, an effect known as **conditioned place preference** (**CPP**).

It is important to note that the tasty solution functioned as an NS (before conditioning) and as CS for nausea for the CTA conditioning. And in the CPP procedure, that same stimulus served as a US. Also, note that the liquid was something to be avoided in the CTA procedure and played a role in approach behavior in the CPP procedure. Therefore, the same stimulus or event may function as either a CS, a US, or neutral, depending on its relation to other stimuli in the situation and the type of response it produces (or does not produce).

Contiguity and Contingency of Stimuli

When stimuli or events occur near in time or are paired together, they often become similar in function. While close temporal proximity between CS and US is usually required, conditioning of biologically significant relations between CS and US may be extended in time. For example, in CTA there can be a substantial delay between the CS-taste onset and US-sickness, but strong conditioned taste aversion occurs. In typical laboratory preparations, an NS is followed by an effective US and acquires some of the behavior functions related to the US, an equivalence relation referred to as *stimulus to stimulus (S–S) conditioning* (Rescorla & Wagner, 1972).

For most practical purposes, the conditioning procedure should arrange close temporal proximity, pairing, or contiguity of the CS and US, but research shows that it is the **contingency**, predictiveness, or correlation between the stimuli that is critical. Thus, *the US should occur more frequently when the CS is present than when it is absent—the CS should predict or signal the US* (Rescorla, 1988). The importance of contingency cannot be overstated, as these stimulus relations occur everywhere from moment to moment. Although many nonhuman studies of respondent conditioning are cited in this textbook, these principles also account for an enormous amount of human behavior.

In humans, conditioning by contingency happens all the time. When the features of a person are reliably followed by sexual gratification, these features take on the positive aspects of sexual stimulation. The correlation of human features with pleasant events (and other positive stimuli) in our life enhances many social relationships. Mother–infant bonding is a product of good USs from the mother (such as mother's milk) being associated with mom's facial features, her odor, her stroking, the sound of her voice, and perhaps her heartbeat, many aspects that can become very potent signals for the mother (see Blass, Ganchrow, & Steiner, 1984 for conditioning in infants 2–48 hours of age; see Sullivan and Hall, 1988 for odor (CS) conditioning by presentations of milk and stroking of fur (US) in 6-day-old rat pups). Affectionate and sexual conditioning occurs between lovers and may result in exclusive commitment to that person. Advertising often involves enhancing product appeal by linking the item with something already attractive. Thus, Bierley, McSweeney, and Vannieuwkerk

(1985) showed that human preference ratings were greater for stimuli (colored geometric figures) that predicted pleasant music (music from *Star Wars*) than for stimuli that predicted the absence of music. These preferences also transferred to other stimuli resembling those used during conditioning.

The attractiveness of the human face has been an important component in human dating and sexual attraction, and a prominent feature in the reproductive success of humans (Rhodes, 2006). Research indicates that attractive faces function as reinforcers or unconditioned stimuli (US), suggesting that attractive human faces are capable of conditioning other objects or events (CS) predictive of them. Additional research, using brain-imaging technology, has revealed neural activity to presentations of attractive faces in the reward areas of the brain. In the advertising industry, many television commercials and magazine advertisements involve arranging a contingency between products (e.g., cars, clothing, and beer) and attractive people, suggesting that the US value of attractive human faces can transfer to commercial products by conditioning procedures.

Given that attractive human faces appear to have US value, Bray and O'Doherty (2007) asked whether human participants would show conditioning when neutral stimuli (fractal or self-similar shapes) were followed by presentation of attractive and unattractive faces. The study also used functional magnetic resonance imaging (fMRI) of the participants' brains to reveal neural-reward areas active during conditioning. The results showed that participants' positive ratings or evaluations of the fractal shapes (CS) increased from preconditioning to postconditioning when paired with attractive female faces (US). Also, the fMRI imaging showed increased brain activity in the ventral striatum on attractive-face presentations (compared with unattractive-face presentations). These findings indicate that a neutral stimulus acquires CS value when it signals pleasant social stimuli (attractive female faces), and this conditioning works on the reward centers in the brain related to the dopamine neurons in the ventral striatum. The results also suggest how commercial advertisements linking products with attractive social stimuli act on the brain's reward system to influence consumer preference and the purchasing of goods and services.

Respondent Acquisition

When a neutral stimulus (NS) is repeatedly paired with an unconditioned stimulus (US), the NS becomes a conditioned stimulus (CS) and elicits the conditioned response (CR). The process of this elicitation of the CR to the presentation of the CS is called **respondent acquisition**. In one experiment, Anrep (1920) demonstrated the conditioning of the salivary reflex to a tone stimulus. The acquisition procedure involved turning on the tone for a brief period, and then placing food in a dog's mouth. Anrep measured the CR as the number of drops of saliva during 30-s intervals wherein the tone occurred without food. Figure 3.5A (acquisition) shows that the amount of salivation to the tone increases rapidly during the first 25 trials and then levels off, or reaches its maximum, called the **asymptote**. In other words, with repeated presentations of the CS and US, the magnitude of the conditioned response increases. Once the conditioned reflex reaches asymptote, however, further CS–US presentations have little to no additional effects.

It is important to note that the asymptote for the conditioned response depends on the intensity of the unconditioned stimulus. As the intensity of the US increases, the magnitude of the UR also increases up to a point. The magnitude of the UR limits the maximum associative

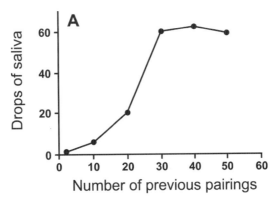

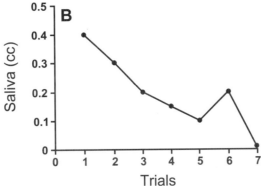

Fig. 3.5
The acquisition and extinction of salivation. The acquisition curve (A) is taken from an experiment by Anrep (1920), who paired a tone (CS) with food placed in a dog's mouth (US). The extinction curve (B) is from Pavlov (1960, p. 53), who presented the CS (sight of food) in the absence of the US (food in the mouth). Results are portrayed as a single experiment.

strength of the CR. Thus, the more food a dog is given the greater the amount of salivation. If a dog is given 60 g of meat, there will be more salivation than if it is presented with 30 g. A tone associated with 60 g of food will elicit salivation as a CR at a higher level (at asymptote) than a tone associated with 30 g of food. These relations, then, are limited by an organism's physiology. If a dog is given a much larger US—say, 450 g of steak, it will probably salivate at maximum strength, and a change to an even larger US—900 g will have no further effect. Similar limits are observed for reflexes such as variation in pupil size in response to light, magnitude of the knee jerk in response to a tap, and the degree of startle in response to noise.

Conditioned and Unconditioned Responses

Notice that the conditioned response of salivation appears identical to the unconditioned response. When conditioning to the tone has occurred, the sound of the tone will elicit salivation. This response to the tone seems to be the same as the salivation produced by food in the dog's mouth. In fact, early theories of learning held that the tone substituted for the food stimulus. This implies that the CS–CR relationship is the same as the US–UR relation. If the

CS–CR and the US–UR relationships are the same, then both should follow similar laws and principles. And the *laws of the reflex* govern the US–UR relationship, as you have seen.

If the CS–CR and US–UR relationships are the same, then the law of intensity–magnitude should hold for conditioned stimuli and responses. Thus, a rise in the intensity of the CS should increase the magnitude of the CR. In addition, the CS–CR relation should follow the law of latency. An increase in the intensity of the CS should decrease the latency between the CS onset and the conditioned response. Research has shown that these, and other laws of the reflex, typically do *not* hold for the CS–CR relationship (Millenson & Hendry, 1967). Generally, a change in the intensity of the conditioned stimulus decreases the strength of the conditioned response. In the experiment by Anrep (1920), the tone occurred at a particular intensity, and after conditioning, it elicited a given magnitude and latency of salivation. If Anrep had increased the intensity of the tone, there would have been less salivation and it would have taken longer to occur. Thus, the CS–CR relation is specific to the original conditioning and does not follow the laws of the reflex. One reason for this is that the CS–CR relationship involves processes such as respondent discrimination.

Respondent Extinction

Pavlov (1960) reported a very simple experimental procedure that is called **respondent extinction**. The procedure involves repeatedly presenting the CS without the US. Figure 3.5B (extinction) shows the decline in salivation when Pavlov's assistant repeatedly presented the CS but no longer fed the dog. As you can see, the amount of salivation declines and reaches a minimal value by the seventh trial. This minimum level of the CR is often similar to the value obtained during the first trial of acquisition, and probably reflects the **baseline level** of this behavior. Baseline refers to the strength of the target response (e.g., salivation) before any known conditioning has occurred. Extinction can be valuable as a therapeutic procedure for reducing or removing unwanted emotional responses such as claustrophobia or arachnophobia. Simply present the CS (the spider or the small space, respectively) but do not present the fearful or frightening US component. The phobic CR will then gradually decrease with repeated trials.

A distinction should be made between extinction as a procedure and extinction as a behavioral process. The procedure involves presenting the CS but not the US after conditioning has occurred. As a behavioral process, extinction refers to the decline in the strength of the conditioned response when an extinction procedure is in effect. In both instances, the term *extinction* is used correctly. Extinction is the procedure of breaking the CS–US association, resulting in the decline of the CR.

In some cases, the decline in the strength of the CR is often rapid, such as with salivation. But other types of conditioned responses may vary in *resistance to extinction*. Even with salivation, Pavlov noted that as the time between trials increased, the CR declined more slowly. A test trial (also called a probe trial) is any instance in which the CS is given in the absence of the unconditioned stimulus. Of course, repeated test trials are the same as extinction. The slower extinction of salivation with longer intervals between test trials may reflect what is called spontaneous recovery.

Spontaneous Recovery

Spontaneous recovery is the observation of an increase in the CR after respondent extinction has occurred. Recall that after repeated presentations of the CS without the US, the

CR declines to near-baseline level. Following extinction of the response, after some time has passed, the CS might again elicit the CR, and the more time that elapses between the first and second extinction sessions, the more likely the spontaneous recovery (Brooks & Bouton, 1993).

The typical effect is seen in Figure 3.6, which shows the course of extinction and spontaneous recovery from another experiment by Pavlov (1960). In this experiment, the CS was the sight of meat powder, and the US was food in the dog's mouth. As you would expect, the sight of meat powder eventually elicited a conditioned response of salivation. When extinction began, the dog responded with 1 mL of salivation at the first sight (trial 1) of the CS. By the fifth extinction trial, the animal showed almost no salivation to the sight of food powder, but after 20 min of rest without stimulus presentations, the CS again elicited a conditioned response. Note, however, that the amount of salivation on the spontaneous-recovery trial is much less than the amount elicited on the first extinction trial.

A behavioral analysis of spontaneous recovery suggests that the CS–CR relation is weakened by extinction, but the context or features of the situation in general maintain some level of control over the conditioned response (see Bouton, 2014 for more on context and behavior change). During respondent conditioning, many stimuli (additional CSs) not specified by the researcher as the CS, but present in the experimental situation, come to regulate behavior. For example, background odors, general illumination of the room, the presence of particular researchers, the passage of time, and all the events that signal the start of a conditioning series come to exert some control over the conditioned response. Each time a recovery test is made, some part of the situation that has not yet been extinguished evokes the CR. This gradual decline in **contextual stimulus** control through repeated extinction also accounts for progressively less recovery of the conditioned response. The role of extinction of contextual CSs in spontaneous recovery is investigated in behavioral neuroscience (Zelinski, Hong, Tyndall, Halsall, & McDonald, 2010).

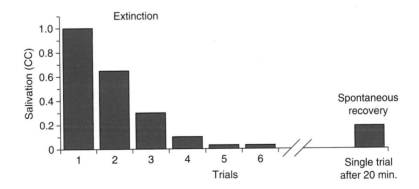

Fig. 3.6

Extinction and spontaneous recovery of salivation elicited by the sight of meat powder.

Source: I. P. Pavlov (1960). *Conditioned reflexes: An investigation of the physiological activity of the cerebral cortex* (G. V. Anrep, trans.). New York: Dover (original work published in 1927); with data replotted from Bower, G. H., & Hilgard, E. R. (1981). *Theories of learning*. Englewood Cliffs, NJ: Prentice-Hall, p. 51.

NEW DIRECTIONS: NEUROSCIENCE AND LEARNING IN HONEYBEES

A 2007 scientific report describes the integrated use of behavior and neurophysiological techniques in the honeybee to provide a more complete account of associative conditioning in human learning—(Giurfa, 2007). Bees are excellent models for this type of research for several reasons. These insects have rich behavioral repertoires, a social lifestyle, and well-developed motor and sensory systems; they travel relatively large distances for food, water, and information; they also see color, are able to learn, remember, and communicate, as well as learn relational concepts such as same/different and above/below (Avargues-Weber & Giurfa, 2013). Honeybees also are easy to train individually and to observe communally. They exhibit both operant and respondent conditioning and, because these forms of learning rely on the association of specific stimuli, it is possible to study where and how such stimuli activate brain areas and neural pathways. Thus, the location and interactions of the neural pathways involved and how they are modified by experience are available for examination in the everyday honeybee. Such research is also worthwhile in terms of learning about honeybees themselves, as they are currently experiencing a global decrease in population due to viruses such as Deformed Wing (Brettel & Martin, 2017) and anthropogenic causes, such as pesticides, habitat destruction, air pollution, and climate change.

In the appetitive-learning situation (positive US or reinforcer), one specific odor CS is correlated with a sucrose US, which elicits a proboscis extension UR, and ultimately a CR to the CS (Bitterman, Menzel, Fietz, & Schafer, 1983). One goal of this work is to identify the dopaminergic-neuronal circuit involved and whether that circuit differs in the appetitive and aversive learning procedures (Vergoz, Roussel, Sandoz, & Giurfa, 2007). Another objective is to identify the role of dopamine and octopamine, the latter being a major neurotransmitter in the honeybee which is especially involved in appetitive learning of directional flying related to energy regulation (Hammer & Menzel, 1998). Virtually all addictive drugs increase dopamine levels in specific parts of the brain, and these brain areas appear to be part of a neural-reward system. It is the presence of this common neurotransmitter (dopamine) in humans and other animals, including insects, which allows research findings with lower organisms such as the honeybee to be extended to humans in the search for the neural underpinnings of associative learning, and eventually the treatment of compulsions and addictions based on respondent and operant learning (Volkow, Fowler, & Wang, 2003).

To research the neuroscience of honeybee behavior, individual bees are tethered (with a really tiny harness) and receive implanted electrodes in the nervous system. There are many different protocols that use bees, but most of them involve sucrose as the reinforcer (or US), and flying to a visually cue-signaled food source. (A variant of this procedure is being used to train bees to locate hidden mines in a field, or hidden drugs in airport luggage as reported by Shaw et al., 2005.) Honeybees in particular, and insects in general, make convenient subjects for neuroscience research, just as the common fruit fly (*Drosophila melanogaster*) is an excellent model for genetic research. Eric Kandel received a Nobel Prize primarily for his work on the neuronal basis of memory using the marine snail *Aplysia*, illustrating how fundamental, behavioral neuroscience often extends to a more complete understanding of human behavior.

Respondent Generalization and Discrimination

Generalization

Pavlov conducted a large number of conditioning experiments and discovered many principles that remain useful today. One of his important findings concerned the principle of **respondent generalization**. Respondent generalization occurs when an organism shows a conditioned response to values of the CS that were not trained during acquisition. For example, respondent acquisition occurs when a specific stimulus, such as a 60-dB tone at a known frequency (e.g., 375 Hz), is associated with a US (e.g., food). After several CS–US presentations, the CS elicits a conditioned response, in this case salivation. If a 60-dB tone of 375 Hz is now presented without the US (a test trial), the animal will salivate at maximum level. To show generalization, the researcher varies some property of the conditioned stimulus. For example, a 60-dB tone of 75, 150, 225, 300, 375, 450, 525, 600, and 675 Hz is presented on test trials, and the magnitude of the conditioned response is measured (this would be like holding the volume of the stimulus constant, but varying the frequency or pitch). Figure 3.7 shows possible results of such an experiment. As you can see, the amount of salivation declines as the test stimulus departs in both directions from the value used in training. In other words, the more the test stimulus approximates the trained stimulus, the more response there is; the less it approximates the trained stimulus, the less response there is. This graph, which plots stimulus value against magnitude of response, is called a **generalization gradient**.

Generalization is an adaptive process, allowing the organism to respond similarly even when conditions do not remain exactly the same from trial to trial. Consider a situation in which a predator's approach (US) is associated with the sound of snapping twigs, rustling grass, and waving shrubs (CS). An organism that runs away (CR) only in the presence of these exact stimulus conditions would probably not last long. This is because the events that occurred during conditioning are never precisely repeated—each approach of a predator produces variations in sounds, sights, and smells. Even in the laboratory, where many features of the environment are controlled, there is some variation in stimuli from one trial to the next. When a bell is presented and followed by food, the dog may change its orientation to the bell and thereby alter the sound; room humidity and other factors may also produce slight variations in tonal quality. Because of generalization, a CS–CR relationship can be strengthened even though the stimulus conditions are never exactly the same from trial to trial. Thus, stimulus

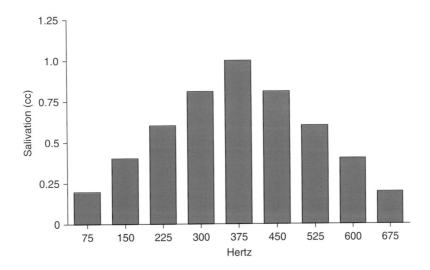

Fig. 3.7
A hypothetical generalization gradient for the salivary response. In this idealized experiment, training would occur at 375 Hz and then CSs ranging from 75 to 675 Hz would be presented.

generalization is likely an adaptive process, allowing organisms to respond to the variations of life.

Discrimination

Another conditioning principle that Pavlov discovered is called differentiation or discrimination. **Respondent discrimination** occurs when an organism shows a conditioned response to one value of the stimulus, but not to other values. A discrimination-training procedure involves presenting both positive and negative conditioning trials. For example, a positive trial occurs when a CS$^+$ such as a 60-dB tone is associated with an unconditioned stimulus (US) like food. On negative trials, a 40-dB tone is presented (CS$^-$) but never paired with the US (e.g., food). Because of stimulus generalization, the dog may salivate to both the 60-dB (CS$^+$) and 40-dB (CS$^-$) tones on the early trials. If the procedure is continued, however, the animal no longer salivates to the CS$^-$ (40-dB tone), but shows a reliable response to the CS$^+$ (60-dB tone). Once a *differential response* occurs, we may say that the dog discriminates between the tones.

Respondent discrimination is another adaptive learning process. It would be a chaotic world if an animal spent its day running away from most sounds, sights, and smells—generalizing to everything. Such an animal would not survive and reproduce as there would be no time for other essential activities such as eating, drinking, and procreating. Discrimination allows an organism to budget its time and responses in accord with the requirements of the environment. In the predator example, noises that are reliably associated with an animal that considers you a main course should become CS$^+$ for flight or fight. Similar noises made by the wind or harmless animals are CS$^-$ for such behavior. Notice, however, that there is a fine line between discrimination and generalization in terms of survival.

Pre-Exposure to Stimuli

Recall that people who are first exposed to the rumblings of a volcano startle and run away, but these reflexes show habituation after repeated exposure to volcano sounds without a major eruption. This observation suggests that familiar events or stimuli do not elicit as intense a reaction as novel ones. The same is true for respondent conditioning, where novelty of the CS or US increases its effectiveness. Repeated exposure (familiarity) to the CS or US, however, reduces its effectiveness and interferes with respondent acquisition.

Considerable research indicates that pre-exposure to the CS weakens subsequent conditioning with the US—the **CS-pre-exposure effect** (Lubow & Moore, 1959). **Latent inhibition** denotes the inhibition of learning of the CS–US relation by pre-exposure of the CS, as revealed by an acquisition test following the conditioning phase.

Recall that conditioned taste aversion (CTA) occurs when we present a distinctive flavor or taste, which is followed by drug-induced illness. Compared to animals without pre-exposure to the taste (novel taste), those with pre-exposure to the flavor (familiar taste) show weaker CTA on a test for acquisition. You may have had similar experiences. For example, you eat a juicy steak with your meal, adding sauce béarnaise to the steak for the first time. Later, you become extremely sick (perhaps with a flu virus) and subsequently avoid the béarnaise sauce, eating your steak instead with good old familiar steak sauce. Now consider what would happen if you had eaten sauce béarnaise repeatedly with your meals without any ill effects, but now had the steak dinner with the sauce and became extremely sick. Given the repeated pre-exposure to the sauce béarnaise without illness, it is unlikely that you will condition to the taste of the sauce. In other words, you will not show avoidance of the sauce for your next steak dinner. *Latent inhibition* and CTA have been studied extensively, and the neurophysiological underpinnings are now a central focus of research in associative or respondent conditioning (Lubow, 2009).

Other research has focused on the novelty of the US by giving pre-exposure to the US before using it in subsequent conditioning. Animals are first given repeated exposures to the US by itself and then a series of pairing of a neutral stimulus with a US until a CS develops (i.e., conditioning occurs). Compared with animals given pairing using a novel US, those pre-exposed to the US show weaker and slower conditioning on the acquisition test—a result called the **US-pre-exposure effect**.

When a sweet saccharin stimulus is followed by injections of lithium chloride (US), animals become sick and avoid the sweet taste (CS), reducing their consumption of the flavored solution on an acquisition test. However, when the drug (US) is injected repeatedly before being used in conditioning (pre-exposure), animals show less avoidance of the sweet taste CS than other animals without a history of pre-exposure to the drug (Hall, 2009). One would guess that the weaker conditioning following pre-exposure to the drug is due to simple habituation of the unconditioned stimulus. However, this is not the case. Research has shown that the *context* in which the drug US is injected can function as a CS. Even when the drug is injected alone, aspects of the context or background acquire CS functions. When the sweet saccharin CS is subsequently conditioned in the same context, the contextual cues that signal the upcoming injection can blunt or block conditioning to the sweet saccharin CS, resulting in weak or no avoidance of the sweet solution on the acquisition test (De Brugada, Hall, & Symonds, 2004; see Revillo, Arias, & Spear, 2013 for evidence against contextual conditioning during US-pre-exposure in rat pups).

TEMPORAL RELATIONS AND CONDITIONING

Delayed Conditioning

There are several ways to arrange the temporal relationship between the presentation of a CS and the unconditioned stimulus (US). So far, we have described a procedure in which the CS is presented a few seconds before the US occurs. This procedure is called **delayed conditioning** (the presentation of the US is slightly delayed relative to the CS) and is shown in Figure 3.8A.

Delayed conditioning is considered the most effective way to condition simple autonomic reflexes such as salivation. In the diagram, the CS is turned on, and 3 s later the US is presented and there is some overlap in the presentation of both stimuli. The interval between the onset of the CS and the onset of the US (called the CS–US interval) determines the effectiveness of conditioning. For autonomic responses, such as salivation, blood pressure, skin temperature, hormone levels, and sweat secretion, a CS–US interval of between 5 and 30 s appears to be most effective. A brief CS–US interval of about 0.5 s seems to be optimal for the conditioning of quick skeletal responses, such as a knee jerk, eye blinks, and retraction of a limb from a hot surface. In human eye-blink conditioning, a delay of 0.4 s between the CS and the US produces the fastest conditioning in young adults, but a longer delay of about 1 s is more effective with older people (Solomon, Blanchard, Levine, Velazquez, & Groccia-Ellison, 1991). Currently, delayed eye-blink conditioning serves as a model system for analysis of the neural mechanisms participating in respondent or associative learning (Freeman & Steinmetz, 2011).

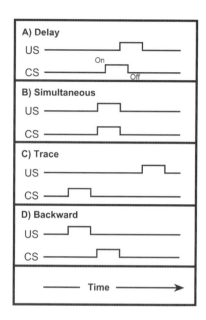

Fig. 3.8
Several temporal arrangements between CS and US commonly used for simple respondent conditioning. Time is shown in the bottom panel of the figure and moves from left to right. The other panels depict the temporal arrangement between US and CS for four basic respondent conditioning arrangements: delay, simultaneous, trace, and backward. For example, delayed (or delay) conditioning is shown in panel (A), where the CS is turned on and, a few seconds later, the US is presented. See text for other temporal arrangements.

Simultaneous Conditioning

Another temporal arrangement is called **simultaneous conditioning**, where the CS and US onsets begin and end at the same time (i.e., they are presented at the same time). This procedure is shown in Figure 3.8B, where the CS and US are presented at the same moment. For example, at the same time that the bell rings (CS), food is placed in the dog's mouth (US). Compared with delayed conditioning, where the CS precedes the US briefly, simultaneous conditioning produces a weaker conditioned response (White & Schlosberg, 1952). A way to understand this weaker effect is to note that the CS does not predict or signal the impending occurrence of the US in simultaneous conditioning. Based on this observation, many researchers have emphasized the *predictiveness of the CS* as the central feature of classical conditioning (see Rescorla, 1966, 1988). From this viewpoint, the CS works because it provides information that "tells" the organism a US will follow. In simultaneous conditioning, however, there is no predictive information given by the CS, and yet some conditioning occurs. One possibility is that predictiveness of the CS is *usually required* for conditioning, but *contiguity* or close temporal proximity of CS and US also plays a role (Papini & Bitterman, 1990; also see "Backward Conditioning" in this chapter).

Trace Conditioning

The procedure for trace conditioning is shown in Figure 3.8C. The CS is presented for a brief period, both onset and offset, and after some time the US occurs after the offset of the CS.

For example, a light is flashed for 2 s, and 20 s later food is placed in a dog's mouth. The term **trace conditioning** comes from the idea of a "memory trace," the small, or "trace" amount of conditioning that has occurred. Generally, as the time between the CS and US increases, the conditioned response becomes weaker (Lucas, Deich, & Wasserman, 1981). For eye-blink conditioning, the CR does not occur when the CS and US are separated by only a few seconds. When compared with delayed conditioning with the same interval between the onset of the CS followed by the US, trace conditioning is not as effective—producing a weaker conditioned response. Contemporary research has extended trace conditioning to taste aversion learning (see Chapter 7) and to neural changes that help to bridge stimulus associations over the trace interval (Cheng, Disterhoft, Power, Ellis, & Desmond, 2008; Waddell, Anderson, & Shors, 2011; see Raybuck & Lattal, 2014 for the neural circuitry of trace-fear conditioning).

Backward Conditioning

As shown in Figure 3.8D, **backward conditioning** stipulates that the US comes on and goes off before the CS comes on (US followed by CS). The general consensus has been that backward conditioning is unreliable, and many researchers question whether it occurs at all (see Spetch, Wilkie, & Pinel, 1981 for supportive review of evidence). It is true that backward conditioning usually does not produce a conditioned response because the to-be CS does not predict the US. If you place food in a dog's mouth and then ring a bell, the bell does not predict food and elicit salivation when presented later. Most conditioning experiments have used arbitrary stimuli such as lights, tones, and shapes as the conditioned stimuli. However, Keith-Lucas and Guttman (1975) found backward conditioning when they used a biologically significant CS.

These researchers reasoned that following an unsuccessful attack by a predator, the sights, sounds, and smells of the attacker (to-be CSs) would be associated with pain from the attack (US). Consider a situation in which a grazing animal is unaware of the approach of a lion. The attack (US) comes swiftly and without warning (no CS), as depicted in Figure 3.9. The prey animal survives the onslaught and manages to run away. In this case, the pain inflicted by the attack is a US for flight that precedes the sight of the predator (CS). For such a situation, backward conditioning would have adaptive value as the prey animal would learn to avoid lions.

Keith-Lucas and Guttman (1975) designed an experiment to test this adaptive-value hypothesis. Rats were placed in an experimental chamber and fed a sugar pellet in a particular location. While eating the pellet, the rats were given a one-trial presentation of a small electric shock (US). After the shock, the chamber was made completely dark for 1, 5, 10, or 40 s. When the light in the chamber came back on, a toy hedgehog (CS) was presented to the rat. To make this experiment clear, eating sugar pellets was viewed as the laboratory equivalent of grazing, the shock represented an attack, and the appearance of the toy hedgehog substituted for the predator. Two control groups were run under identical conditions, except that one group saw the hedgehog but did not get shocked (no US), and the other group received the shock but did not see a hedgehog (no to-be CS).

After these conditioning trials, the number of fear responses were measured. Compared with the control groups, backward conditioning to the hedgehog was found after a delay of 1, 5, and 10 s but not after 40 s. Relative to control animals, experimental subjects showed

Fig. 3.9

An image of a lion attacking a wildebeest from the rear. If the prey escapes the attack, backward conditioning (US → CS) would result in avoidance of lions.

Source: Shutterstock.

greater avoidance (fear) of the hedgehog, spent less time in the presence of the hedgehog, and ate less food. Presumably, the shock (US) elicited a fear–flight reaction (UR), and backward conditioning transferred this reaction to the toy hedgehog (CS) at briefer intervals between the US and CS. This experiment shows that backward conditioning can occur in some conditions—perhaps those in which a noxious or harmful US is used. Despite this study, most contemporary researchers suggest that backward conditioning (US followed by CS) does not result in reliable conditioning (see Arcediano & Miller, 2002 for timing and backward conditioning).

SECOND-ORDER RESPONDENT CONDITIONING

So far in this chapter, we have described the classic respondent conditioning paradigm, often referred to as first-order conditioning. To briefly review, in **first-order conditioning** a non-functional or neutral event is presented before an unconditioned stimulus. After several conditioning trials, the ability of the US stimulus to elicit a UR is now extended to the no-longer neutral stimulus; it is now a CS that elicits a CR. Second-order conditioning extends the controllability to elicit a conditioned response to other events that have not been directly associated with the unconditioned stimulus. These events gain this eliciting power through the association of an established conditioned stimulus. Thus, **second-order conditioning** involves presentation of a second CS_2 along with an already conditioned CS_1, rather than following a CS by a US (Rizley & Rescorla, 1972; see also Witnauer & Miller, 2011 for determinants of second-order conditioning). Such higher-order conditioning is important because it extends the range of behavioral effects produced by respondent conditioning, especially with regard to evaluative (e.g., approach or avoid) conditioning in humans (see Hofmann, De Houwer, Perugini, Baeyens, & Crombez, 2010 for a meta-analytic review).

Higher-order conditioning may cause phobic reactions (i.e., an intense and seemingly irrational fear) in people. Consider a person who refuses to sit with friends in the backyard on a nice summer day. The sight of flowers greatly upsets her and she says, "With so many flowers there are probably bees." A possible interpretation is that a bee (CS_1) previously has stung (US) the person, and she has noticed that bees hover around flowers (CS_2). The "phobic" fear of flowers occurs because of the association of bees (CS_1) with flowers (CS_2). Thus, phobic reactions and other emotional responses may sometimes involve higher-order respondent conditioning (see Martin & Pear, 2006 on systematic desensitization and the fear hierarchy).

ON THE APPLIED SIDE: DRUG USE, ABUSE, AND RESPONDENT CONDITIONING

Basic research on simple and complex (i.e., including contextual effects) respondent conditioning has major applied importance. One example of this involves factors that affect drug use and abuse. Several experiments have shown that conditioned stimuli (CS) that have been paired with drugs (USs) can produce drug-like effects in both humans and other animals. In addition, with chronic drug use, the CRs that elicit *an opposing process* from the drug. For example, when animals are injected with insulin (US), the unconditioned response is a reduction in blood sugar levels (UR). If a CS is paired with insulin chronically, the response to a stimulus (CS) that has been followed by insulin is exactly the opposite—blood sugar levels increase (Siegel, 1975)!

Similar counteractive effects have been found with drugs other than insulin. For example, amphetamine reduces appetite, but a CS correlated with chronic amphetamine increases food intake (Poulos, Wilkinson, & Cappell, 1981). Pentobarbital is a sedative, but the response to a conditioned stimulus associated with chronic pentobarbital exposure counteracts the drowsiness ordinarily associated with the drug (Hinson, Poulos, & Cappell, 1982).

Effects such as these suggest that respondent conditioning plays a major role in drug tolerance. Here is how it works. Each time a drug (US) is administered, the effects of the drug are the unconditioned responses. For example, the drug morphine causes USs such as analgesia (pain relief), constricted pupils, and euphoria. However, there are also *opposing processes*, that is physiological processes that counteract the main processes, such that the net effect is homeostasis. With chronic drug use, these opposing processes become larger, often overwhelming the initial effects of the drug; these include pain, dilated pupils, and feelings of extreme unhappiness. Indeed, these are also considered withdrawal effects.

These opposing withdrawal effects can also be elicited by CSs that predict the drug. For example, stimuli associated with an injection (such as the needle piercing the skin) are paired repeatedly with the drug (US), the conditioned opposing process (CR) gains in strength and increasingly opposes the unconditioned (UR) effects of the drug. This means that not only are withdrawal effects triggered by the CS, but also larger and larger amounts of the US will be needed for the user to experience the same degree of drug effect. In everyday life, conditioned stimuli arise from the time of day that a drug is taken, the way it is administered (e.g., using a needle), the location (in a tavern or at home), and social events (a party or dance). Notice that tolerance, which is a reduction in the effect of the drug (UR), is not due to habituation, but rather it is the result of the counteractive effects or opposing process (CR) to the injection and setting (CS). When more of a drug (US) is needed to obtain the same drug effects (UR), we talk about drug **tolerance** (Baker & Tiffany, 1985). Thus, the counteractive effects of CSs are major components of conditioned drug tolerance and conditioned drug withdrawal.

Heroin Overdose and Context

To consider drug tolerance as a conditioned response helps to explain instances of drug overdose. Individuals with heroin use disorder are known to survive a drug dose that would kill a person who did not regularly use the drug. Despite this high level of tolerance, over 14,000 Americans per year have died of heroin overdose since 2015; about the same number have overdosed per year from prescription opioid drugs since 2009 (National Institute on Drug Abuse, 2021). These victims typically die from drug-induced respiratory depression. Surprisingly, many of these individuals with substance use disorders (SUD) die from a dose similar to the amount of opioid they usually took each day. Siegel, Hinson, Krank, and McCully (1982) proposed that these deaths resulted from "a failure of tolerance. The individual (with an opioid use disorder), who can usually tolerate extraordinarily high doses, is not tolerant on the occasion of the overdose" (p. 436). They suggested that when a drug is administered in the usual context (CS), the CRs that counteract the drug allow for a large dose, i.e., induce tolerance. When the situation in which the drug is taken is changed, the CSs are not present, the opposing conditioned response (CR) does not occur, and the drug is sufficient to kill the person. Siegel and associates designed a number of animal experiments to test these ideas. One is described below.

In one study, rats were injected with heroin every other day for 30 days. The amount of heroin was gradually increased to a dose level that would produce tolerance to the drug. On non-heroin days, these rats were injected with dextrose solution (i.e., sugar and water). Both heroin and dextrose injections were given in one of two distinctive contexts—the ordinary colony room that the rats lived in, or a different room with constant white noise. A control group of rats was injected only with the dextrose solution in the two situations. The researchers predicted that experimental animals would develop a tolerance to the drug; this tolerance would occur if aspects of the room in which heroin injections were given became CSs that elicited opposing responses (CRs) to the drug.

To test this assumption, Siegel and colleagues (1982) on the test day doubled the amount of heroin given to experimental animals. The same high dose of heroin was given to the control group, who had no history of tolerance. Half of the experimental animals received this larger dose in the room where the drug was usually administered. The other rats that were given chronic heroin exposure were injected with the higher dose in the room where they usually received a dextrose injection.

Figure 3.10 shows the results of this experiment. As you can see, the large dose of heroin killed almost all of the animals in the control group. For the two groups of animals with a history of heroin exposure, one group (same room) received the higher dose in the room where they were usually injected with heroin. Only 32% of the rats died in this condition, presumably because the CSs set off the opposing conditioned responses. This inference is supported by the mortality rate of rats in the different room group. These rats were injected with the double

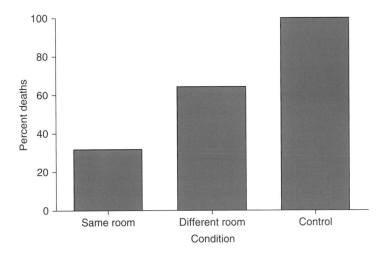

Fig. 3.10
Results of the experiment by Siegel, Hinson, Krank, and McCully (1982). The same room group of rats received the higher dose in the room where they usually were injected with heroin, and only 32% died. Twice as many animals in the different room condition died from the larger dose presumably because they were injected in a room where heroin had not been given. Heroin killed almost all of the animals in the control group.

Source: Adapted from S. Siegel, R. E. Hinson, M. D. Krank, & J. McCully (1982). Heroin "overdose" death: The contribution of drug-associated environmental cues. *Science, 216*, pp. 436–437. Copyright 1982 held by the American Association for the Advancement of Science. Published with permission.

dose of heroin in a room that had never been associated with heroin administration. Twice as many animals in this condition died from the larger dose (64%) when compared with the same room group; this group represents a typical metabolic tolerance to the drug that develops with chronic exposure. It seems that the effects of context during this kind of respondent conditioning can be a matter of life or death—tolerance to heroin (and perhaps other drugs) is relative to the situation in which the conditioning has occurred, and in humans involves conditioning of both external and internal events (Siegel, 2005).

What happens when the drug-related CS is presented without the drug US, as in the classical extinction procedure? In this case the elicited respondents often are called "cravings," and the process is known as **conditioned withdrawal**. When the CS elicits the opposing processes of the drug and there is no US delivered, the net effects do not result in homeostasis. Indeed, these craving and withdrawal effects are conditioned responses. A person with a heroin use disorder can have withdrawal symptoms immediately terminated by a heroin injection. If you are accustomed to having a cigarette after a meal, the craving you experience can be alleviated with a smoke.

NOTE ON: PHYSIOLOGY AND THE CONTROL OF PREPARATORY RESPONSES

The physiological concept of homeostasis helps to clarify the control by the CS over conditioned responses opposite to those induced by the US. **Homeostasis** is the tendency for a system to remain stable and resist change. In terms of a biological system, homeostasis refers to the regulation of the system by negative feedback loops. For example, the body maintains a temperature within a very small window around the average of 98.7 degrees Fahrenheit (37.06 degrees Celsius). If the environment warms up or cools down, physiological mechanisms (sweating or shivering) involving the sympathetic and parasympathetic nervous systems are activated to reduce the drift from normal body temperature. In terms of drug exposure, when a drug (US) is administered it upsets the stability of the system—increasing heart rate or reducing respiration (UR). If some aspect of the environment is consistently present when the drug is delivered (e.g., drug paraphernalia, a person, or the room), that stimulus becomes a conditioned stimulus (CS) capable of eliciting a conditioned response, which is often preparatory and compensatory (CR). If the US drug causes the heart rate to increase, the **conditioned compensatory response** (CR) is homeostatic in nature. For example, it would be a decrease in heart rate—thus, the learned component ($CS_{needle} \rightarrow CR_{heart\ rate\ decrease}$) counteracts the opposing response to the drug ($US_{drug} \rightarrow UR_{heart\ rate\ increase}$). This counteracting, homeostatic effect may be so great that it nullifies the responses to the drug, and the user no longer experiences the typical high—a process called *tolerance*. The onset of tolerance can be dangerous for the drug user. As we have seen, if a larger dose of the drug is taken to overcome tolerance and the compensatory conditioned response is not produced, an overdose can occur. Furthermore, if the preparatory stimuli (CSs) elicit conditioned responses and the drug (US) is not delivered, a condition called *craving* or *withdrawal* occurs.

Conditioned Immunosuppression

Conditioned immunosuppression is another example of environmental influences altering what are generally considered to be internal and automatically controlled processes. In this procedure, a CS is followed by a US drug that suppresses immune-system function, such as the production of antibodies. Drugs like cyclophosphamide are commonly administered to suppress rejection of a transplanted organ, so this drug function is important with those who have received donated organs. After several presentations of the CS–US contingency, the CS is presented alone and the immune-system reaction is measured. Ader and Cohen (1981) were the first to systematically investigate and describe this phenomenon. Clearly, the next question is whether the immune system can also be conditioned to increase an immune reaction. It appears that it can. In a human study, a flavor (CS) followed by adrenaline injection (US) subsequently raised natural killer (NK) cell production (Buske-Kirschbaum, Kirschbaum, Stierle, Jabaij, & Hellhammer, 1994; also Hadamitzky, Engler, & Schedlowski, 2013 reported learned immunosuppression using a conditioned taste aversion (CTA) procedure).

The issue of conditioned enhancement of the immune system also speaks to the findings of **placebo effects** (Vits & Schedlowski, 2014). How can an inert substance, a placebo, have any effect on a person's physiological well-being? Many studies have shown that people who receive a sugar pill do as well as those in the legitimate treatment group (Brody, 2000). How can this be possible when the placebo, by definition, cannot directly cause any change? One conclusion is that respondent conditioning is occurring. A verbal CS, say the patient's "belief" (resulting from experience with doctors and medication) that she is receiving treatment, is activated and elicits the CRs of improvement. Active placebos, which cause some subjective interoreceptive (internal) feelings, have been shown to be just as effective in improving symptoms of depression than SSRI medication (see Kirsch, 2014). Even sham (fake) arthroscopic surgery for arthritis can be functional as actual surgery with fewer side effects and less cost (Moseley et al., 2002).

One thing that these types of studies indicate is that there is much greater two-way interaction between the environment and physiological mechanisms than had been suspected. Organisms are adaptive and they learn. It appears that organs (e.g., salivary glands) and organ systems (e.g., the immune system) can also alter function as a result of experience. A review in the *Philosophical Transactions of the Royal Society* (Vits et al., 2011) presents a plethora of research on the topic of behavioral conditioning of the immune system including ongoing clinical applications.

NEW DIRECTIONS: EPIGENETIC "MEMORY" AND TRAINED IMMUNITY

A grand synthesis of brain and behavior is gradually developing in the fields of biology and psychology. Components that have previously been studied separately are becoming integrated as research methods, instrumentation, findings, and theory are expanding and overlapping. For example, many neuroanatomical-brain circuits underlying various activities have been mapped, genes responsible for features of appearance and some behaviors are known, and adaptive processes are unfolding between neurophysiology and performance. It is noteworthy that many previously nonbehavioral disciplines within biology are now using behavioral terms and concepts to better conceptualize their findings. The discovery of epigenetic influence and transmission (Bonasio, Tu, & Reinberg, 2010), as well as experience-dependent epigenetic changes to the brain (Sweatt, 2009), has opened another fertile area for understanding and analyzing complex biological processes on both macro and micro levels (see Chapter 1 "New Directions: Epigenetics and Retention of Early Learning").

The journal *Science* published two studies of "trained immunity" involving epigenetic "memory" (histone modification) in innate immune cells exposed to certain pathogens or their antigens—substances provoking an adaptive immune response (Cheng et al., 2014; Saeed et al., 2014). Macrophages are a type of white blood cell of the immune system, which engulf and dispose of dead cells and pathogens. Under certain conditions, these cells can become tolerant of pathogens or "trained" by epigenetic regulators to react against pathogens. Immunological researchers assume both innate and adaptive arms to the human immune system. The innate arm is unconditioned, broad, and general in its response to invading organisms; the adaptive, conditioned arm establishes immune cell "memories" via vaccination or natural-exposure conditioning for specific pathogens. But now it seems even the adaptive component of specifically "trained" macrophages actually spreads to fighting other additional microorganisms—a generalized response not previously observed.

In one of the two epigenetic studies from *Science* (Cheng et al., 2014), the researchers identified the HIF1α gene, which transcribes for a protein involved in glucose metabolism, and is critical for *generalized trained immunity*. Mice were exposed initially to a fungal polysaccharide antigen that induced a trained immune response. Subsequently, the mice were challenged with a bacterial pathogen, a secondary infection causing sepsis—an inflammation throughout the body often leading to death. Following exposure to the antigen, normal mice with active HIF1α gene transcription were resistant to the pathogen; however, other mice trained on the antigen, but with the HIF1α gene *deleted* from their immune cells (no protein regulating glucose metabolism), were not protected from the secondary bacterial infection. Thus, a generalized immune response to an invading bacterial pathogen seems to require a history of trained immunity, epigenetic cell "memory" for this training, and active HIF1α gene transcription related to glucose metabolism.

As for respondent conditioning, the injected inert pathogen or antigen in a vaccine might be considered a generalized stimulus (perhaps conditioned), established on the basis of its similar biological structure to the original pathogen or US, eliciting an adaptive conditioned immune response (macrophages engulf and destroy) similar to the innate UR reaction. In this view, mechanisms of immunity are selected by consequences and those that work are retained—suggesting the paradigm may have operant properties as well as respondent ones. The use of behavior analysis terms, concepts, and paradigms in treating complex epigenetic molecular processes may require further refinements, adjustments, or replacements in the future as behavioral and other biological disciplines continue to merge based on overlapping research interests.

NUANCES AND COMPLEXITIES OF RESPONDENT CONDITIONING

We so far have examined CS–US relations in isolation, ignoring for the most part the complexities in which these events occur. In the laboratory, we use simple stimuli such as tones and lights. In the real world, stimuli in the environment are more complicated. When one enters a bakery or restaurant, for example, the odor of food probably becomes a CS for salivation, having been followed by donuts or burgers and fries (US). Other related stimuli, however, such as the name, the order clerk, the location of the store, and the outdoor signs are also correlated with eating. These additional features of the food experience may become conditioned stimuli that function as the context (compound CS), which evokes salivation. Differences in conditioning procedures related to compound stimuli result in the behavioral processes called blocking and overshadowing.

To investigate these complexities of respondent behavior, researchers have arranged situations involving **compound stimuli**. In these cases, and to keep things somewhat simple, two conditioned stimuli (tone and light) are presented together before (delayed) or during (simultaneous) a US. This arrangement of two controllable stimuli (compound CSs) presented together in a respondent contingency can be shown to acquire the capacity to elicit a single conditioned response. While only two dimensions of the environment are examined (visual and auditory), a number of interesting findings have emerged that deepen our understanding of how respondent conditioning works with the complex environment. These phenomena follow.

Overshadowing

Pavlov (1960) first described **overshadowing**. A compound stimulus is arranged consisting of two or more simple stimuli presented at the same time. For example, a faint light and loud tone (compound CS) may be presented at the same time and followed by an unconditioned stimulus (US) such as food. Pavlov found that the *most salient element* of the compound stimulus came to elicit the conditioned response. In this case the loud tone and not the faint light would become a CS for salivation. The tone is said to *overshadow* conditioning to the light. This happens even though the weak light could function as a CS if it was originally presented by itself and followed by a US.

The conditioning of events that happened on 9/11/2001 in the United States might be an example of overshadowing. There were three locations (CSs) that were tragically attacked by terrorists (USs) that day, leading to the loss of over 3,000 lives: the World Trade Center (WTC), the Pentagon, and Flight 93 that crashed over Somerset County, PA. All three of these events were horrifying, but the one that elicits the most intense emotional images for most people in America is the one that occurred at the WTC. One reason is because it was the most

Fig. 3.11
A firefighter searches through debris after the World Trade Center was attacked by terrorists on 9/11/2001.
Source: AP Photo/Graham Morrison.

salient—many of us saw it unfold on television in real time and it was intense and horrific. (There are other reasons, too, such as the repeated airing of the images.) Some people may have less intense emotional responses to the attacks that happened in Pennsylvania or even the Pentagon, even though the details of those were also horrifying. In contrast to the Twin Towers (as the WTC was called), though, it was simply less salient.

Blocking

Kamin (1969) reported another complexity of conditioning with compound stimuli. This effect is called **blocking**, and describes a situation in which CS_1 when followed by the US blocks a subsequent CS_2–US association. In blocking, a CS_1 is followed first by the US ($CS_1 \rightarrow US$) until the conditioned response (CR) reaches maximum strength. Following this conditioning of CS_1, a second stimulus or CS_2 is presented at the same time as the original CS_1, and this compound stimulus ($CS_1 + CS_2$) is then followed by the unconditioned stimulus (Compound $\rightarrow$ US). On test trials, CS_1 still evokes the CR but the second stimulus or CS_2 does not! For example, a tone (CS_1) may be associated with food (US) until the tone reliably evokes salivation (Tone $\rightarrow$ Food). Next, the tone and a light are presented together as a compound CS and both are associated with the food-US (Tone/Light $\rightarrow$ Food). On test trials, the tone will elicit salivation but the light will not. This seems counterintuitive because CS_2 has been paired with two stimuli

(CS$_1$ and US) that elicit a response—surely CS$_2$ should elicit a conditioned response, too. But indeed, the previously conditioned tone blocks conditioning of the light stimulus.

In his classic studies of blocking, Kamin (1969) used a procedure called **conditioned suppression** (see Estes & Skinner, 1941). In conditioned suppression, a CS such as a tone is followed by an aversive US such as an electric shock, which causes freezing in place (a reflexive defensive response) as a UR. After several conditioning trials, the CS becomes a conditioned aversive stimulus (CS^{ave}) for freezing. Once the CS^{ave} has been conditioned, its effects may be observed by changes in an organism's operant behavior. For example, a rat may be trained to press a lever for food. After a stable rate of response is established, the CS^{ave} is introduced without the US. When this occurs, the animal's lever pressing is disrupted, presumably by the defensive freezing elicited by the CS^{ave}. Basically, we could say that the CS^{ave} induces the conditioned freezing response in the animal and it stops pressing the bar. Conditioned suppression historically has been a widely used procedure in respondent conditioning, and as you will see later it is important in the study of human emotions.

Using a conditioned-suppression procedure, Kamin (1969) discovered the phenomenon of blocking. Two groups of rats were used: a blocking group and a control group. In the blocking group, rats were presented with a tone (CS^{ave}) that was associated with small electric shocks for 16 trials (CS$_1$ → US). Following this, the rats received eight trials during which the tone was presented with light as a compound stimulus; this compound stimulus, was followed by shock (Compound CS → US). The control group did not receive the 16 light-shock conditioning trials but did have the eight trials of the compound tone and light followed by shock (Compound CS → US). Both groups were tested for conditioned suppression of lever pressing in the presence of the tone presented alone, which functioned as a CS+ in both groups for conditioned suppression. But the more interesting test was when the light was presented alone. In the control group, the light was presented alone and suppression of bar pressing for food was observed, indicating the light functioned as a CS. However, when the light alone was presented for the blocking group, conditioned suppression was not observed. Indeed, the light did not take on aversive properties and it did not function as a CS—it did not elicit the CR. In other words, prior conditioning with the tone alone blocked or prevented conditioning to the light. Functionally, the light acted as a CS^{ave} in the control group but not in the blocking group.

Recall that the best case for conditioning to occur is when a stimulus predicts the US. Blocking has been interpreted as a case of redundant stimuli. When a CS already predicts a US effectively, other elements of the compound stimulus are redundant. The most salient and predictive features of a compound stimulus will gain the eliciting properties, while other features do not. All stimulus manipulations are conducted in some place, be it the laboratory or an everyday setting like a classroom, and noticeable elements of that environment often signal the stimuli of interest. It is the repeated, consistent, and predictable nature of the specific CS–US contingency that tends to restrict the connection to only the stimuli that best predict the US.

ADVANCED SECTION: RESCORLA–WAGNER MODEL OF CONDITIONING

The Rescorla–Wagner model provides a set of elegant mathematical predictions on how much conditioning will occur to a simple (or compound) CS overall and on each trial. There are two central ideas of the Rescorla–Wagner model. First, a conditioned stimulus acquires **associative strength** as it is paired with a US. We use the term "associative strength" to describe the relation between the CS and the magnitude of the CR, in other words, how strong the CR is when the CS is presented. A neutral stimulus before conditioning, for example, has no associative strength. After 20 trials in which it is paired with a US, it is now a CS that can elicit a strong CR; therefore, it has a large amount of associative strength. In general, the *total* amount of associative strength for a given CS increases over conditioning trials and reaches some maximum level. This is the **maximum associative strength** for the CS. Thus, a tone (CS) associated with 1 g of food has maximum associative strength when salivation (CR) to the tone is about the same as salivation (UR) to the 1 g of food (US). This magnitude sets the upper limit for the CR. The CS cannot elicit a greater response than the one produced by the US.

Second, there are predictions about the amount of associative strength the CS acquires on *each* trial. The amount of gain or increment depends on several factors. One obvious factor is the salience of the CS. Weaker CSs (e.g., a 5-dB tone) will not condition as strongly as a stronger CS (e.g., a 50-dB tone). A second factor involves the intensity of the US; stronger USs will result in the conditioning of stronger CSs. A third factor is *the disparity between the CS and US* in terms of their relative ability to elicit a response. The difference between a neutral stimulus which does not elicit a response compared to a US, which does elicit a response, has a high amount of disparity. The higher the disparity, the more conditioning will take place on that trial. Therefore, the first trial of a CS–US pairing will result in the greatest amount of learning (i.e., the CS will acquire the greatest amount of associative strength). In comparison, on the tenth trial, when the CS has acquired some associative strength through nine previous CS–US pairings, the disparity between the CS and US is likely very small. Therefore, less associative strength will be acquired on the 10th trial.

For example, assume a 10-trial experiment in which 1 g of meat evokes 2 mL of saliva and a tone is paired with the meat. In terms of change in associative strength, the most gain will occur on the first trial, there will be less gain by the fifth trial, and there will likely be almost no gain in associative strength by the tenth trial.

The **change in associative strength** of a conditioned stimulus (CS_1) is also affected by the strength and predictability of other conditioned stimuli (CS_2, CS_3, etc.) that elicit the conditioned response in that situation. Because there is a maximum associative strength set by the US, it follows that the associative strength of each CS would proportionately combine in some way as they are paired with a US. If a weak light and strong tone are presented together as a compound stimulus before a US, each will acquire associative strength as a CS presented alone that is in proportion to their relative intensity; in other words, a weak light, which contributes 30% of the compound stimulus, will by itself produce a CR that is approximately 30% of the possible maximal value of the CR (overshadowing).

Similarly, if a tone by itself has been frequently paired with meat, it would evoke almost maximum salivation. If a light is now introduced and presented along with the tone, it would show little control over salivation, as most of the possible associative strength has accrued to the tone (blocking).

The Rescorla–Wagner model of respondent conditioning describes a large number of findings and has stimulated a great deal of research. The model makes intuitive predictions that have been confirmed in a variety of experimental settings. Since the early 1970s, scores of experiments have been conducted to test some of the implications of the model.

The Rescorla–Wagner Equation

An equation suggested by Rescorla and Wagner (1972) but simplified here for presentation is shown in Equation 3.1:

$$\Delta V = S(US - V_{SUM})$$ (Equation 3.1)

The symbol ΔV stands for the amount of *change in associative strength* (or change in value of the stimulus, *V*) *of any CS that occurs on any one trial*. The symbol *S* is a constant that varies between 0 and 1, and may be interpreted as the **salience** (e.g., dim light versus bright light) of the CS based on the sensory capacities of the organism and how intense the stimulus is. A larger salience coefficient makes the associative strength of the CS rise more quickly to its maximum. *US* is the magnitude of the US, which represents a maximal value of a stimulus, in terms of eliciting a response. V_{SUM} is the total amount of associative strength gained by the CS from other trials ($V_{SUM} = CS_2 + CS_3 + \ldots CS_N$). Note that the ($US - V_{SUM}$) represents the disparity between the US and the CS. Therefore, for a given trial, the CS–US disparity is multiplied by the constant of the salience of the to-be CS. To understand this more thoroughly, let's work through the equation with some numbers as we explore acquisition of a CS.

Acquisition

Figure 3.12 is a table of values for an idealized experiment on the acquisition of a CR based on Equation 3.1. Figure 3.13 is the graph of the associative strength *V* based on the data in the table. In this hypothetical experiment, a tone CS is repeatedly followed by an US such as food. In the figure, the salience is not that strong—*S* is set at 0.25. The asymptote (or maximum possible strength) of the US is 10 units of the conditioned response (e.g., salivation). The value of V_{SUM} is assumed to be zero, because before conditioning takes place, the CS is neutral. The value of ΔV for trial 1 is given by the equation when we substitute $S = 0.25$, $US = 10$, and the value of V_{SUM} is zero before conditioning begins. Based on Equation 3.1, the increase in associative strength from no conditioning to the first trial is:

Trial	ΔV	V	
1	2.5	2.5	V = Associate strength or value
2	1.9	4.4	$S = 0.25$
3	1.4	5.8	$V_{max} = 10$
4	1.1	6.9	$V_{sum} = 0$
5	0.8	7.7	
6	0.6	8.3	
7	0.4	8.7	
8	0.3	9.0	
9	0.25	9.3	
10	0.18	9.4	

Fig. 3.12
A table of values for a 10-trial acquisition experiment based on solving Rescorla–Wagner Equation 3.1. The symbols V_{sum} and ΔV_x refer to the total amount of associative strength and the change in associative strength for a given trial, respectively. The values of US and *S* are also given in the table. See text for details.

$$\Delta V_1 = 0.25 (10 - 0) = 2.50$$

Therefore, on trial 1, the CS acquires 2.5 units of associative strength. Notice that the value of V_{SUM} has changed from 0 to 2.50 (check this with the rounded tabled values of Figure 3.12).

On each subsequent trial, the associative strength of the CS is 0.25 (salience) of the remaining distance to the asymptote or maximum. Thus, for trial 2 we substitute the value 2.50 for V_{SUM} and obtain an increase of 1.88 units of associative strength for the tone. Therefore, for trial 2, the ΔV is 1.88.

$$\Delta V_2 = 0.25 (10 - 2.50) = 1.88.$$

Now the associative strength of the CS (V_{SUM}) after the second trial is 2.50 + 1.88, or 4.38. This means that roughly one-half of the potential maximal associative strength ($US = 10$) of the CS has been acquired by trial 2.

The change in associative strength for trial 3 uses $V = 4.38$ from the second trial and obtains the value:

$$\Delta V_3 = 0.25 (10 - 4.38) = 1.40.$$

And the new estimate of V is 4.38 + 1.40, or 5.78 (used to obtain ΔV_{SUM} on the fourth trial). Estimates of ΔV and V_{SUM} for all 10 trials of the experiment are obtained in the same way, using Equation 3.1.

As you can see in Figure 3.13, the equation yields a negatively accelerating curve for the associative strength, V, which approaches *but never quite reaches* maximum associative strength (the same magnitude of the US). You can see from the horizontal and perpendicular lines that the largest increase in associative strength is on the first trial, and this change corresponds to the difference in associative strength between trial 0 and trial 1 (2.5-unit increase). The change in associative strength (ΔV) becomes smaller and smaller over trials (check this out in the table of Figure 3.12). Notice how the values of ΔV and V_{SUM} depend on the salience, S, of the CS (tone). If the salience of the tone were greater, say $S = 0.90$ rather than $S = 0.25$, a new set of estimates would be given by Equation 3.1 for ΔV and V_{SUM}. Conditioning would happen more rapidly (in fewer trials). Calculate these values for the first five trials of ΔV yourself and check them at the end of the chapter.

Extinction

Equation 3.1 can also be used to account for respondent extinction. As before, assume that a tone is paired with food until the tone (CS) elicits a conditioned response close to maximum; there is no US presentation in extinction so the value of the US = 0. We start with a value of V_{SUM} that is greater than 0—whatever the intensity of the CS is (which is determined by the strength of the CR; here it is 10) since the procedure is respondent extinction, the CS will lose associative strength. If $S = 0.25$ and the US = 0, the decline in associative strength on the first extinction trial is:

$$\Delta V = 0.25 (0-10) = -2.50.$$

Thus, the value of the tone (CS) after the first extinction trial is 10 – 2.50, or 7.50 ($V = 7.50$). Other values of the CS during extinction are determined in a similar fashion (compare with respondent acquisition). Figure 3.14 shows that the predicted extinction curve is the *exact opposite of the acquisition curve* shown in Figure 3.13. It is important to note that the actual

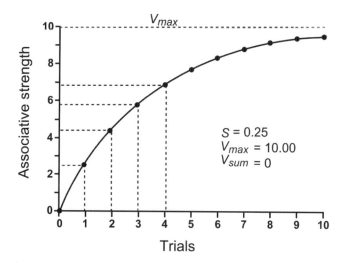

Fig. 3.13
The acquisition curve predicted by the Rescorla–Wagner equation (our Equation 3.1). Gain in associative strength, from trial to trial, declines as the CR comes closer to the asymptote. The asymptote or upper-flat portion of the curve is set in the equation by the value of the US. The curve is based on the data in Figure 3.12.

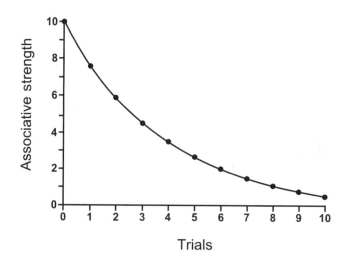

Fig. 3.14
The extinction curve predicted by the Rescorla–Wagner model. Notice that V_{MAX}, or the asymptote, is zero because extinction is in effect.

associative strength of the tone before extinction is never exactly equal to the V_{MAX}, but for simplicity we have assumed that it is in Figure 3.14.

As you can see, the Rescorla–Wagner equation describes many of the basic aspects of respondent conditioning such as acquisition and extinction. It can also describe other second-order conditioning effects (e.g., latent inhibition) not discussed here as well as conditioning with compound stimuli, such as overshadowing and blocking.

CHAPTER SUMMARY

This chapter has introduced reflexive behavior, which is based on species history or phylogeny. It has explored reflexive sequences or patterns set off by a releasing stimulus and reaction chains where each response requires an appropriate stimulus to keep the sequence going. Reflexive behavior obeys the three laws of the reflex (threshold, magnitude, and latency) and secondary principles such as reflex habituation. Next, the discussion turned to the ontogeny of behavior and respondent conditioning, which involved presenting a CS (tone) followed by a US (food in mouth). The importance of respondent contingency (correlation of CS and US) and contiguity (timing) was described throughout the chapter. Respondent behavior is elicited by the US, and conditioning establishes this function. Opposing functions, such as the conditioned compensatory response, can also be respondently conditioned, such as the case of chronic drug use and conditioned tolerance) for the CS. A sweet flavor stimulus may be a CS for conditioned taste aversion and a US for conditioned taste preference. Both respondent acquisition and extinction were described, and research examples were provided. Spontaneous recovery that occurs during respondent behavior was also discussed.

When a single CS is trained, organisms show generalization of respondent behavior to stimuli, depending on how physically similar they are to the trained CS. Organisms also show discrimination (a differential response) when the US is followed by one stimulus but withheld when other values of the stimulus array are presented. In addition to simple conditioning effects, temporal relationships between the CS and US are important as in delayed, simultaneous, trace, and backward conditioning—and the phenomenon known as second-order conditioning. The implications of respondent conditioning were extended to an analysis of drug use and abuse, with some attention to context and drug tolerance. Finally, more advanced issues of complex conditioning and compound-stimulus effects such as overshadowing and blocking were introduced. The Rescorla–Wagner model of conditioning was described, and expressed as a mathematical equation. This equation predicts the increase in the respondent over trials (acquisition) and the decrease during extinction.

Key Words

Associative strength
Backward conditioning
Baseline level
Blocking
Change in associative strength
Compound stimuli
Conditioned compensatory response
Conditioned place preference (CPP)
Conditioned response (CR)
Conditioned stimulus (CS)
Conditioned suppression
Conditioned taste aversion (CTA)
Conditioned withdrawal
Contextual stimulus
Contingency
CS-pre-exposure effect
Delayed conditioning
Elicited (behavior)
First-order conditioning
Fixed-action pattern (FAP)
Generalization gradient (respondent)
Habituation

Homeostasis
Latent inhibition
Law of intensity–magnitude
Law of latency
Law of the threshold
Maximum associative strength
Modal action pattern (MAP)
Ontogenetic
Overshadowing
Phylogenetic
Placebo effect
Primary laws of the reflex
Reaction chain
Reflex
Rescorla–Wagner model
Respondent

Respondent acquisition
Respondent conditioning
Respondent discrimination
Respondent extinction
Respondent generalization
Salience
Second-order conditioning
Sexual selection
Simultaneous conditioning
Spontaneous recovery (respondent)
Tolerance (to a drug)
Trace conditioning
Unconditioned response (UR)
Unconditioned stimulus (US)
US-pre-exposure effect

On the Web

www.flyfishingdevon.co.uk/salmon/year3/psy337DrugTolerance/drugtolerance.htm — Introduction Learn more about drug use, abuse, and tolerance from the website of Paul Kenyon. Some neat data on conditioning and tolerance are provided and discussed.

www.youtube.com/watch?v=LcojyGx8q9U One application of respondent conditioning is called *systematic desensitization*—an effective treatment for anxiety and phobia. This video clip outlines the basic procedure of graded exposure to the fear stimulus, a snake.

Brief Quiz

1. Behavior relations based on the genetic endowment of the organism are described as:
    a   operants
    b   reflexes
    c   ontogenetic
    d   phylogenetic

2. Complex sequences of released behaviors are called:
    a   traits
    b   reaction chains
    c   fixed-action patterns
    d   second-order conditioned reflexes

3. Reflexive behavior is said to be _____ and _____.
    a   built in; flexible
    b   involuntary; elicited
    c   respondent; emitted
    d   voluntary; inflexible

4 Primary laws of the reflex do not include:
   a   the law of latency
   b   the law of threshold
   c   the law of habituation
   d   the law of magnitude
5 A diminution in the UR due to repeated presentation of the US is called:
   a   habituation
   b   extinction
   c   forgetting
   d   sensitization
6 Respondent conditioning might also be called:
   a   S–R conditioning
   b   S–S pairing
   c   CS–CR association
   d   R–S learning
7 To reduce an unwanted CR, one should:
   a   present the CS without the CR
   b   present the CR without the US
   c   present the US without the CS
   d   present the CS without the US
8 Drug tolerance induced by a particular stimulus or setting has been shown to be a result of:
   a   generalization
   b   metabolization
   c   discrimination
   d   conditioned compensatory response
9 Which of the following is not a traditional way of relating the CS and a US?
   a   trace
   b   simultaneous
   c   delayed
   d   overshadowing
10 The Rescorla–Wagner theory suggests that a CS becomes effective:
   a   gradually
   b   through backward conditioning
   c   by conditioned inhibition
   d   following tolerance

Answers to Brief Quiz: 1, d (p. 72); 2, c (p. 72); 3, b (p. 75); 4, c (p. 76); 5, a (p. 76); 6, b (p. 79); 7, d (p. 84); 8, d (p. 96); 9, d (p. 98); 10, a (p. 100).

Answers to Rescorla–Wagner:
$\Delta V_1 = 9$; $\Delta V_2 = 0.9$, $\Delta V_3 = 0.09$, $\Delta V_4 = 0.009$; $\Delta V_5 = 0.0009$

CHAPTER 4

Reinforcement and Extinction of Operant Behavior

1 Learn about operant behavior and the basic contingencies of reinforcement.
2 Discover whether reinforcement undermines intrinsic motivation.
3 Learn how to carry out experiments on operant conditioning.
4 Delve into reinforcement of variability, problem solving, and creativity.
5 Investigate operant extinction and resistance to extinction.

A hungry lion returns to the waterhole where it has successfully ambushed prey. A person playing slot machines wins a jackpot and is more likely to play again. Students who ask questions and are told "That's an interesting point worth discussing" are prone to ask more questions. When a professor ignores questions or gives confusing answers, students eventually stop asking questions. In these examples (and many others), the consequences that follow behavior determine whether they will be repeated.

Recall that operant behavior is said to be emitted (Chapter 2). When operant behavior is selected by reinforcing consequences, it increases in frequency. Behavior not followed by reinforcing consequences decreases in frequency. This process of reinforcement is a part of **operant conditioning**, or the manner of learning in which behavioral change is based on *ontogeny* or life experience. Reinforcement is just one process that is operant in nature. It is important, however, to recognize that the ability to learn through operant conditioning is an evolved process. The ability to change our behavior to a rapidly changing and dynamic world full of reinforcers and punishers is something that species have evolved to do. Biologically, operant (and respondent) conditioning is a *general behavior-change process* based on *phylogeny* or species history. In other words, those organisms whose behavior changed dynamically on the basis of consequences encountered during their lifetimes, were more likely to survive and reproduce than animals that did not evolve such a capacity. Adaptation by operant learning is a mechanism of survival that furthers reproductive success. Therefore, operant learning is both an ontogenic and phylogenic process.

DOI: 10.4324/9781003202622-4

OPERANT BEHAVIOR

Operant behavior is often referred to as "choice" behavior or any behavior that an organism chooses to do voluntarily. Operant behavior is sometimes described as intentional, free, or willful. Examples of operant behavior include conversations with others, driving a car, taking notes, reading a book, and painting a picture. From a scientific perspective, though, operant behavior is deterministic and lawful and may be analyzed in terms of its relations to environmental events. Formally, responses that produce a change in the environment and change in frequency due to an environmental event are called operants. The term **operant** comes from the verb *to operate* and refers to behavior that operates on the environment to produce effects or consequences. The consequences of operant behavior are many and varied and occur across all sensory dimensions. When you turn on a light, make a phone call, drive a car, or open a door, these operants result in visual clarity, conversation, reaching a destination, or entering a room. A **reinforcer** is defined as any consequence or effect that increases the probability of the operant that produced it. For example, suppose that your car will not start, but when you jiggle the ignition key it fires right up. Based on past reinforcement, the operant—jiggling the key—is likely to be repeated the next time the car does not start.

Operants are defined by two parts: the behavior *and* the consequences or effects they produce. Opening the door to reach the other side is the operant, not just the physical movement of manipulating the door. Operants are a class of responses that may vary in **topography**. Topography refers to the physical form or characteristics of the response. Consider the number of different ways you could open a door—you may turn the handle, push it with your foot, or (if your arms are full of books) ask someone to open it for you. All of these responses vary in form or topography and result in the consequence of reaching the other side of the door. Because these responses result in the same consequence or effect, they are members of the same **operant class**. Thus, the term *operant* refers to a class of related responses that may vary in topography, but produce a common environmental consequence.

Discriminative Stimuli

Operant behavior is **emitted** in the sense that because operants are voluntary, by-choice behaviors, they are probabilistic, meaning that sometimes they occur (when they are previously reinforced) and sometimes they do not. Their probability depends highly on the outcomes or consequences they produce. This emitted behavior is in contrast to reflexive responses, which are said to be **elicited** or caused by a preceding stimulus (see Chapters 1 and 2). Reflexes are tied to the physiology of an organism and, under appropriate conditions, *always* occur when the eliciting stimulus is presented. For example, Pavlov showed that dogs automatically salivated when food was placed in their mouths. Dogs do not learn the relationship between food and salivation; this reflex is a characteristic of the species.

Stimuli may also precede operant behavior, though. These events, however, do not force the occurrence of the response that follows them. A stimulus that precedes an operant, and *sets the occasion for behavior*, is called a **discriminative stimulus**, or S^D (pronounced ess-dee). For example, when a traffic light changes from red to green, we choose to press the gas pedal and go. That is, when the S^D of a green light is present, pressing the gas pedal is reinforced with our ability to safely drive forward.

Sometimes S^Ds are not easily observed—when an S^D is not easily observed, we sometimes invent a stimulus inside the person and call it an impulse ("He ate the extra piece of cake on an impulse"). Recall that such inferred mental events are not acceptable explanations in behavior analysis, unless backed up by independent scientific evidence. It is a scientist's job to understand what the conditions are and test them.

Discriminative stimuli change the probability that an operant is emitted based on a history of differential reinforcement. **Differential reinforcement** involves reinforcing an operant in one situation (S^D) but not in another (S^Δ). The probability of emitting an operant in the presence of an S^D may be very high, but these stimuli do not have a one-to-one relationship with the response that follows them. For example, a cell phone ringtone associated with your good friend increases the chances that you emit the operant, answering the phone, but the ringtone does not force you to do so. Similarly, a nudge under the table may *set the occasion for* changing the conversation topic or ending your commentary abruptly. The events that occasion operant behavior may be private as well as public. Thus, a private event such as a headache may set the occasion for taking an aspirin, which will alleviate the pain.

Discriminative stimuli are defined by *setting the occasion for* specific behavior. The probability of raising your hand in class is much greater when the instructor is present than when she is absent. Thus, the presence of an instructor is an S^D for asking questions in class. The teacher functions as an S^D only when their presence changes the student's behavior. The student who is having difficulty with a math problem may ask questions when the teacher enters the room. A student who is easily mastering the material, however, is unlikely to do this. Based on the contingencies, the teacher functions as an S^D (for asking questions) for the first student but not the second. This discussion should make it clear that a stimulus is defined as an S^D only when it changes the probability of operant behavior. You may typically stop when you pull up to a traffic sign that reads STOP; the sign is a discriminative stimulus. If, however, you are driving a badly injured friend to the hospital, the same sign may not function as an S^D. Thus, discriminative stimuli are not defined by physical measures (e.g., color, size, tone); rather, they are defined as stimuli that precede and alter the probability of operant responses. But what makes an S^D effective at occasioning behavior?

The consequences that follow operant behavior establish the control exerted by discriminative stimuli. When an S^D is followed by an operant that produces positive reinforcement, the operant is more likely to occur the next time the stimulus is present. For example, a student may ask a particular teaching assistant questions because in the past that teaching assistant has provided clear and concise answers. In this example, the assistant is an S^D and asking questions is the operant that increases in their presence. When an operant does not produce reinforcement, the stimulus that precedes the response is called an **S-delta**, or **S^Δ** (pronounced ess-delta). In the presence of an S^Δ, there is a low probability of emitting the operant. For example, if a second teaching assistant answers questions in a confused and muddled fashion, the student is less likely to ask that person questions. That is, the behavior of asking questions is *not* reinforced. In this case the second teaching assistant becomes an S^Δ and the probability of asking questions is reduced in their presence. Students who are offered both of these teaching assistants will likely ask the first teacher questions and will likely avoid the other teacher.

Contingencies of Reinforcement

A **contingency of reinforcement** defines the relations among the events that set the occasion for behavior (S^D), the operant class (R for response), and the consequences (such as a reinforcer or S^r) that follow operant behavior. In a dark room (S^D), when you flip on a light switch (R), the light usually comes on (S^r). This behavior does not guarantee that the room lights up on a given occasion—the bulb may be burned out, or the switch broken. It is likely that the light comes on, but it is not certain. In behavioral terms, the probability of reinforcement is high, but it is not guaranteed. The probability may vary between 0 and 1. A high probability of reinforcement in the past for turning the switch to the "on" position (say, 0.99 or 99 times out of 100) establishes and maintains a high likelihood of this behavior on a given occasion.

Discriminative stimuli that precede behavior have an important role in the regulation of operant responses (Skinner, 1969). Signs that read OPEN, RESUME SPEED, or RESTAURANT, traffic lights, and a smile from across the room are examples of simple discriminative stimuli that may set the occasion for specific operants. These events regulate behavior because of a *history of reinforcement* in their presence. A smile from across a room may set the occasion for approaching and talking to the person who smiled. This is because, in the past, people who smiled have reinforced social interaction.

All three of these events—the S^D, the operant response, and the consequences of behavior—make up the *contingency of reinforcement*. Consider the example of this three-part contingency shown in Figure 4.1. The cell phone ring is a discriminative stimulus that sets the occasion for the operant class of answering the phone. This behavior occurs because, in the past, talking to the other party or having a conversation reinforced the operant of answering the phone. The probability of response is very high in the presence of the ring, but it is not certain. Perhaps you are in the process of leaving for an important meeting, or you are in the shower. Or perhaps the ringtone you hear is associated with a person whom you do not want to talk to right now. In that case, another contingency (punishment) will reduce the likelihood that you answer the phone.

Discriminative stimuli regulate behavior, but do not stand alone. The consequences that follow behavior determine the probability of response in the presence of the discriminative stimulus. For example, most people show a high probability of answering their phones when they ring. If you are in a dead zone in which the signal is faulty so that it rings but you cannot hear the other person when you answer it, the probability of answering the phone decreases as a function of no reinforcement. In other words, you stop answering a phone that does not work. (You may try texting instead.)

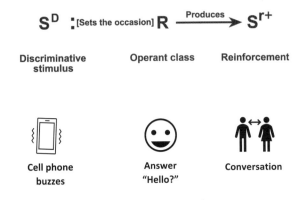

Fig. 4.1
The three-term contingency of reinforcement is illustrated. A discriminative stimulus (S^D) sets the occasion for operant behavior (R) that is followed by a reinforcing consequence (S^{r+}).

The three-term contingency (S^D: $R \rightarrow S^r$) is the basis for the neuroscience of habits, defined by neural patterns that accompany "cue–routine–reward" sequences (see Chapter 1 of *The Power of Habit*, Duhigg, 2012). For various reasons outlined in this chapter, behavior analysts prefer to use operant chambers and rate of response (rate of lever pressing for rats or rate of key pecking for birds) to investigate basic operant processes; however, some neuroscientists continue to use a more traditional method of maze-learning trials to study the neural circuitry of rats. When a rat is first placed in a maze, a click is presented and a barrier is removed so that the animals can run down the central route and discover a piece of chocolate if it turns left, but not right.

At first, the rat emits many exploratory responses like sniffing and rising as it traverses the central route of the maze and even makes wrong turns (right) at the choice point. The rat's brain activity, especially in the basal ganglia that is implicated in motor learning, is high and consistent throughout this exploratory learning period (also called **acquisition**). As the maze trials are repeated, however, the rat no longer emits exploratory responses. Upon hearing the click (Cue), the rat immediately runs down the central route to the choice point and turns left (Routine) to receive a piece of chocolate (Reward). This behavior pattern is correlated with an underlying shift in brain activity; prior to the sound of the click there is a spike in neural activity of the basal ganglia, followed by a low phase of neural responses during the routine, and ending with a second spike that accompanies chocolate consumption.

The brain pattern (high activity before cue, low brain activity during the routine, and high neural activity at the reward) is called "chunking" and is indicative of an automatic behavioral routine called a habit at the neural level. Once the rat shows neural chunking, it can carry out the "maze-running for chocolate" habit automatically, with very little additional brain activity. Habits, in this sense, are a pervasive part of human behavior. Each morning we get out of bed and go through a routine automatically, often without much awareness or mindfulness. Consider, for example, the behavioral sequence of brushing your teeth; there is very little mental effort in remembering the steps; it is a behavioral habit that also likely has neural chunking. More complex sequences of behavior, such as getting dressed or driving a car, involve learning to emit different responses to a series of cues (S^Ds), ending in reinforcement (S^r). These "chains" of behavior also become automatic with repeated practice and reinforcement, presumably accompanied by chunking at the neural level (see "Stimulus Control of Behavior Sequences" in Chapter 8). This is one reason elite athletes are often cautioned not to "over think" a particular situation or performance.

NEW DIRECTIONS: NEUROBIOLOGY OF OPERANT LEARNING IN *DROSOPHILA*

The three-term operant contingency of reinforcement stipulates the discriminative stimulus (S^D), the operant (R), and the reinforcing stimulus (S^r). To study the genetic and neural mechanisms underlying operant-contingency learning, neurobiologists have broken the reinforcement contingency into two learning components—learning about the consequences of behavior (R $\rightarrow$ S^r) and the learning about the relation between the discriminative stimulus and reinforcement (S^D: S^r). In this section, we refer to learning about the consequences of one's behavior as *behavior-consequence learning* (BCL), while learning about the stimuli that predict reinforcement is termed *stimulus-relation learning* (SRL). Notice that in SRL the operant contingency (S^D: S^r) is similar to CS–US or S–S learning in the respondent conditioning model, suggesting a common learning mechanism at the neurobiological level.

To isolate these two aspects of operant-learning contingencies, neurobiologists have found it useful to study the flying behavior of invertebrate *Drosophila* or fruit flies. In the operant laboratory, rats usually press levers, pigeons peck keys, and food hoppers make distinctive sounds. These (or any other) operanda would function as stimuli signaling reinforcement, possibly confounding neurobiological experiments focused only on behavior-consequence learning (BCL). A slightly more intricate preparation has been designed for *Drosophila* to eliminate potential confounding factors (Figure 4.2). To investigate BCL, the flies are tethered to a torque meter, which measures the yaw-torque produced by the angular momentum of flying (left or right, which get recorded—see graph at the bottom of Figure 4.2) in an enclosed cylindrical drum (Brembs, 2011). To study BCL at the neurobiological level without contamination by SRL, positive torque values (e.g., right turns) produce hot temperatures (which fireflies avoid) with no change in visual or auditory stimulation within the drum.

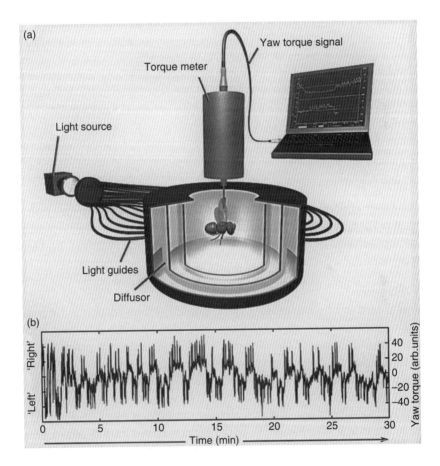

Fig. 4.2
A flight simulator for a fruit fly designed by Dr. Bjorn Brembs is shown. In the operant learning experiments, a fruit fly is tethered in a cylindrical drum uniformly illuminated from behind. The fly is able to emit left or right turns, which are measured as changes in angular momentum (yaw-torque signal) and fed into a computer and recorded as left or right activity on the graph below. Experimental contingencies, involving delivery of a heat beam, are used to separate behavior-consequence learning (R → S^r) from stimulus-relations learning (S^D: S^r) or to study combined learning (S^D: R → S^r) at the neurobiological level.

Source: ResearchGate Download Scientific Diagram.

Thus, there are no external cues such as levers or feeder sounds that signal the contingency, thereby isolating the behavior-consequence component for neurobiological analysis. In addition, to conduct a study of SRL, the angular speed of the rotating cylinder can be made proportional to the fly's yaw-torque, allowing the fruit fly to change flight direction based on visual patterns inside the drum or even use its yaw-torque to produce changes in visual signals while flying.

But what do fruit fly learning and human learning have in common? Recently, genetic mutation and gene expression studies in *Drosophila* have focused on the role of the Forkhead Box P (*FoxP*) gene and its family members on operant self-learning across species. One form of this gene (*FoxP2*) is necessary for normal human speech and language development and is implicated in vocal learning as well as other forms of motor learning (Mendoza et al., 2014). The similarity in structure and function among animals with one of the four members of the *FoxP* gene family (*FoxP1* to *FoxP4*) suggests that the ancestral form of the gene (*FoxP*) evolved as a central component of the neural circuitry activated in motor learning. In human speech, vocal learning by birds, and tethered flying by fruit flies, the animal emits highly variable, exploratory actions (such as babbling, subsong, and spontaneous directional turns) producing sensory feedback that eventually shapes the behavior of the organism, reducing its variability. One implication is that the *dFoxP* orthologue (a gene evolved from the ancestral *FoxP*), known to be important for vocal and motor learning, plays a central role in behavior-consequence learning (BCL) during tethered flying by *Drosophila*, but not for stimulus-relation learning (SRL) in the same situation. That is, this gene is specific to behavior-consequence learning only.

Four Basic Contingencies

There are four basic contingencies of reinforcement. Events that follow behavior may be either presented or removed (environmental outcome). These events can increase or decrease behavior (effect on behavior). The cells of the matrix in Figure 4.3 define the basic contingencies of reinforcement.

Positive Reinforcement

Positive reinforcement is one of the four basic contingencies of operant behavior. **Positive reinforcement** is portrayed in Figure 4.3 (cell 1), where a stimulus follows behavior and, as a result, the rate of that behavior increases. For example, a child is praised for sharing a toy (operant behavior), and the child begins to share toys more regularly (increase in response strength). Positive reinforcers often are consequences such as food, praise, and money. *These events, however, cannot be called or defined as positive reinforcers until they have been shown to increase behavior.* Just because something is pleasant does not mean it is a reinforcer; the proof is in the behavior. Also, reinforcers are individualized, meaning they vary from one individual to another, and dynamic, meaning they change with conditions. More on these processes will be described later in the chapter.

Negative Reinforcement

When a behavior *removes* an event, and the rate of response increases as a result, the contingency is called **negative reinforcement**, as shown in cell 3 of the matrix in Figure 4.3. Negative reinforcement is often behavior in which an organism escapes or avoids a situation.

	Effect on behavior	
Stimulus following behavior	Increase	Decrease
On/presented	1 Positive reinforcement	2 Positive punishment
Off/removed	3 Negative reinforcement	4 Negative punishment

Fig. 4.3
This figure shows the four basic contingencies of reinforcement. The stimulus following a response (consequence) can be either presented (turned on) or removed (turned off). The effect of these procedures is to increase or decrease rate of response. The cells of the matrix in this figure define the contingencies of reinforcement. A particular contingency of reinforcement depends on whether the stimulus following behavior is presented or removed and whether behavior increases or decreases in frequency.

Negative reinforcement plays a major role in the regulation of everyday human behavior. For example, you put on sunglasses because in the past this behavior *removed* the glare of the sun. You open your umbrella when it is raining because doing so has prevented you from getting wet. You leave the room when someone is rude or critical because this behavior has ended other similar conversations. Suppose that you live in a place with a very sensitive smoke detector. Each time you are cooking, the smoke detector goes off. You might remove the sound by tripping the breaker that controls the alarm. In fact, you will probably learn to do this each time before cooking. As a final example, a parent may pick up and rock their crying baby because, in the past, comforting the child has stopped the crying. In each of these instances, removing an aversive event strengthens an operant (see Chapter 6 on escape and negative reinforcement). Negative reinforcement is often confused with punishment, but the effects on behavior, as we will describe in the next section, are different.

Positive Punishment
Cell 2 of the matrix in Figure 4.3 depicts a situation in which an operant produces an event and the rate of operant behavior *decreases*. This contingency is called **positive punishment**. For example, bombing an enemy for attacking an ally is positive punishment if the enemy now stops hostile actions. Or, if speaking loudly in front of a sleeping baby wakes the baby up and the infant begins crying, the (addition of a) crying baby may reduce loud speaking during nap hours. If so, this is positive punishment.

An important note: In everyday life, people often talk about punishment (and reinforcement) without reference to behavior. For example, a mother scolds her child for playing with matches. The child continues to play with matches, and the parents may comment, "Punishment doesn't work with Nathan." In behavior analysis, positive punishment is defined functionally (i.e., by its effects); when behavior is not reduced by aversive events, punishment has not occurred. In other words, the parents are arranging an ineffective contingency.

The parents could identify an aversive event that reliably decreases behavior; however, this strategy may backfire. For example, as you will see in Chapter 6, punishment may produce serious emotional and aggressive behavior. Because of this, punishment should not be used alone and usually as a last resort for the modification of behavior problems.

Negative Punishment

Punishment can also be arranged by removing stimuli contingent on behavior (cell 4 in Figure 4.3). This contingency is called **negative punishment**. In this case, the removal of an event or stimulus *decreases* operant behavior. For example, two men are watching football on television and begin to argue with one another. The bartender says "That's enough fighting" and turns off the television. A person new to your social circle may tell a sexist joke and those in the circle stop talking to him (removal of social support). At school, a student who is passing notes is caught by the teacher and required to leave the room for a short period of time (also called a timeout, which involves removal from an otherwise reinforcing environment). In these examples, watching television, talking to others, and participating in classroom activities are assumed to be positively reinforcing events. When removal of these events is contingent on fighting, telling sexist jokes, or passing notes and behavior *decreases*, negative punishment has occurred.

FOCUS ON: DO REWARDS HARM INTRINSIC MOTIVATION?

Over the past 30 years, some social psychologists and educators have questioned the practice of using rewards in business, education, and behavior modification programs. The concern is that rewards (the terms "reward" and "reinforcement" are often used similarly in this literature) are experienced as controlling, thereby leading to a reduction in an individual's self-determination, intrinsic motivation, and creative performance (see, for example, Deci, Koestner, & Ryan, 1999). Thus, when a child who enjoys drawing is rewarded for drawing, with praise or with tangible rewards such as points or money, the child's motivation to draw is said to decrease. From this perspective, the child will come to draw less and enjoy it less once the reward is discontinued. In other words, the contention is that reinforcement undermines people's intrinsic motivation (see Figure 4.4).

This view unfortunately has been influential and has led to a decline in the use of rewards and incentive systems in some applied settings. In an article published in 1996 in *American Psychologist*, Robert Eisenberger and Judy Cameron provided an objective and comprehensive analysis of the literature concerned with the effects of reinforcement/reward on people's intrinsic motivation. Contrary to the belief of many psychologists, their findings indicated no inherent negative property of reward. Instead, their research demonstrates that reward has a much more favorable effect on interest in activities than is generally supposed (Eisenberger & Cameron, 1996).

Research Findings and Implications

To organize and interpret the diverse findings on rewards and intrinsic motivation, Cameron and her associates conducted several quantitative analyses of the literature (Cameron & Pierce, 2002; Cameron, Banko, & Pierce, 2001). Using a statistical procedure known as meta-analysis, Cameron et al. (2001) analyzed the results from 145 experiments on rewards and intrinsic

Fig. 4.4
The image depicts a boy receiving a monetary reward, perhaps for bringing home good grades. This kind of reward is said to undermine the child's intrinsic motivation for academic subjects. Research suggests, however, that tangible rewards tied to high performance can be used to increase intrinsic motivation as well as perceived competence and self-determination.
Source: Shutterstock.

motivation to answer the question "Do rewards have pervasive negative effects on intrinsic motivation?"

The findings indicated that rewards could be used effectively to enhance or maintain an individual's intrinsic interest in activities. Specifically, verbal rewards (praise, positive feedback) were found to increase people's performance and interest in tasks. In terms of tangible rewards such as money, the results showed that these consequences increased performance and interest for activities that were initially boring or uninteresting. Children who find little interest in reading or mathematics may gain intrinsic motivation from a well-designed reward program that *ties rewards to increasing mastery of the material*.

In a large-scale economic study of student achievement and financial incentives (Fryer, 2010), paying students for grades (output) had no reliable effect on academic achievement as would be expected if payments were given independent of daily mastery of the subject matter. In the same study, rewards tied to student input (doing homework and attending classes) modestly increased academic achievement, with no loss of intrinsic motivation. However, none of the financial interventions of this study was tied to mastery of the academic material (daily performance in the classroom), which would have produced much greater gains in student achievement, according to the studies by Cameron and associates cited previously.

For activities that people find inherently interesting and challenging, the results from the meta-analysis point to the *reward contingency* as a major determinant of intrinsic motivation. Cameron et al. (2001) found that tangible rewards loosely tied to performance produce a slight decrease in intrinsic motivation. One interpretation is that people rewarded simply for showing up or for doing a job, even an interesting one, repetitively carry out the task, put in low effort, and lose interest (see also the discussion of stereotypy and variability in the section on "Reinforcement and Problem Solving" in this chapter).

When tangible rewards are offered for achieving high performance or exceeding the performance of others, intrinsic interest is maintained or enhanced. In the work world, employees who are offered rewards for high performance assign high ratings to their perceived competence and self-determination (Eisenberger & Shanock, 2003)—findings contrary to the claim that rewards are perceived as controlling and reduce personal autonomy. Furthermore, rewards tied to progressively increasing levels of achievement or mastery instill higher intrinsic interest than rewards for meeting a set, unchanging standard of performance (Cameron, Pierce, Banko, & Gear, 2005). Overall, rewards do not have pervasive negative effects on intrinsic motivation. Rewards tied to high performance, achievement, and progressive mastery increase intrinsic motivation, perceived competence, and self-determination.

Identifying a Reinforcing Stimulus

How do we know if a given event or stimulus will function as reinforcement to a specific human or animal? To identify a *positive reinforcer*, you devise a test. The test is to find out whether a particular consequence *increases* behavior. If it does, the consequence is defined as a positive reinforcer. Such tests are common in science. For example, a litmus test in chemistry tells us whether the solution is acid or base. A potential $100 payout is defined as a positive reinforcer for many people (though, not all) because it increases the frequency of betting a dollar and pulling the handle on the slot machine. Notice that the test for a reinforcer is not the same as explaining the behavior. We explain behavior by pointing to the *contingencies of reinforcement* (S^D: R → S^r) and basic behavior principles, not by merely identifying a reinforcing stimulus. For example, we can explain a person's betting in a casino by analyzing the schedule of monetary reinforcement (involving large intermittent payoffs) that has strengthened and maintained gambling behavior. Our analysis subsequently would be tested by a series of experiments under controlled conditions, as well as by naturalistic observations of human gambling behavior (convergent evidence).

The Premack Principle

Another way to identify a positive reinforcer is based on the **Premack principle**. This principle states that *a higher-frequency behavior will function as reinforcement for a lower-frequency behavior*. For a person who spends little time practicing the piano but lots of time playing basketball, the Premack principle means that playing basketball (high-frequency behavior) reinforces practicing the piano. Generally, David Premack (1959) proposed that reinforcement involved a contingency between two sets of behaviors, operant behavior and reinforcing behavior (behavior^{operant} → behavior^{Sr}), rather than between an operant (behavior) and a following stimulus (R → S^r). Premack suggests it is possible to describe reinforcing events not just as stimuli, but also as *actions* of the organism. Thus, reinforcement can be viewed as eating rather than the presentation of food, drinking rather than provision of water, and reading rather than the effects of textual stimuli.

Reinforcement is dynamic, meaning certain conditions can change its efficacy. In his 1962 experiment, Premack deprived rats of water for 23 hours and then measured their behavior in a setting in which they could choose to run on an activity wheel or drink water. Of course, the animals spent more time drinking than running. Next, Premack arranged a contingency between running and drinking. The rats received a few seconds of access to drinking tubes when they ran on the wheels. Running on the wheel increased when it produced the opportunity to drink water—showing that drinking reinforced running. In other words, the rats ran on the wheel to get a drink of water. At this point in the experiment, Premack (1962) gave the rats free access to water. Now, when the rats were allowed to choose between drinking and running, they did little drinking and a lot more running. Under these conditions, Premack reasoned that wheel running could actually be used as a reinforcer for drinking because now running occurred at a higher frequency than drinking. To test this hypothesis, the running wheel was locked and the brake was removed if the rats licked the water tube for a few seconds. Based on this contingency, Premack showed that drinking water increased when it produced running. The animals drank water for opportunities to run on the wheels.

Overall, this experiment shows that wheel reinforcement and water reinforcement as reinforcers are dynamic. On one hand, drinking reinforces running when rats are motivated to drink. On the other hand, running reinforces drinking when running is the preferred activity (i.e., when rats are not water deprived). Thus, when behavior is measured in a situation that allows a choice among different activities, those responses that occur at a higher frequency may be used to reinforce those that occur at a lower frequency.

Premack's principle has obvious applied implications, and it provides another way to identify reinforcement in everyday settings. Behavior is measured in a situation where all relevant operant behaviors can occur without restriction; any behavior of relatively higher frequency will reinforce an operant of lower frequency. To illustrate, a child is observed in a situation at

Fig. 4.5
The image illustrates the Premack principle. In this example, homework is a low-frequency behavior and watching television is a high-frequency response. The parents have set the contingency that the child can watch television if they accurately complete their homework after coming home from school.

Source: Shutterstock

home where doing homework, watching television, playing with toys, and gaming may all freely occur. Once relative frequencies of behavior have been measured, the Premack principle holds that any higher-frequency (or longer-duration) behavior may serve as reinforcement for any behavior of lower frequency. If television watching is longer in duration than doing homework, watching television may be made contingent on completing homework assignments (see Figure 4.5). This contingency usually increases the number of homework assignments completed, as most parents know from experience.

The Premack principle states that a higher-frequency behavior can reinforce a lower-frequency operant. In a free-choice setting, several behaviors occur at different frequencies—yielding a **response hierarchy**. Any response in the hierarchy should reinforce any behavior below it; also, that response would be reinforced by any behavior above it. But this hierarchy can also change based on deprivation. Depriving a rat of water ensures that drinking occurs at a higher frequency than wheel running, and drinking then becomes behavior that would reinforce running (or other behavior such as lever pressing). On the other hand, restriction of running increases its frequency relative to drinking, and running should now reinforce drinking. Thus, **response deprivation**, or depriving an organism of the opportunity to engage in a behavior, leads to a reordering of the response hierarchy and determines which behaviors function as reinforcement at a given moment.

OPERANT CONDITIONING AND THE LAW OF EFFECT

Operant conditioning refers to an increase or decrease in operant behavior as a function of a contingency of reinforcement. In a simple demonstration of operant conditioning, an experimenter may alter the consequences that follow operant behavior. The effects of environmental consequences on behavior were first described in 1911 by the American psychologist E. L. Thorndike, who reported results from a series of animal experiments that eventually formed the basis of operant conditioning. Cats, dogs, and chicks were placed in situations in which they could perform complex sequences of behavior to obtain food. For example, hungry cats were confined to an apparatus that Thorndike called a puzzle box, shown in Figure 4.6. Food was placed outside the box, and if the cat managed to pull out a bolt, step on a lever, or emit some other behavior, the door would open and the animal could eat the food.

After some time in the box, the cat would accidentally pull the bolt or step on the lever and the door would open. Thorndike measured the time from closing the trapdoor until the cat managed to get it open. This measure, called **latency**, tended to decrease with repeated exposures to the box. In other words, the cats took less and less time to escape from the apparatus as they were given more trials. According to Thorndike, the puzzle-box experiment demonstrated learning by *trial and error*. The cats repeatedly tried to get out of the box and made fewer and fewer errors. Thorndike made similar observations with dogs and chicks and, on the basis of these observations, formulated the **law of effect**. A modern paraphrase of this law is the principle of reinforcement, namely that operants may be followed by the presentation of contingent consequences that increase the rate (frequency of response divided by time) of this behavior.

Skinner argued that simply measuring the time (or latency) taken to complete a task misses the moment-to-moment changes that occur across several operant classes. Responses that

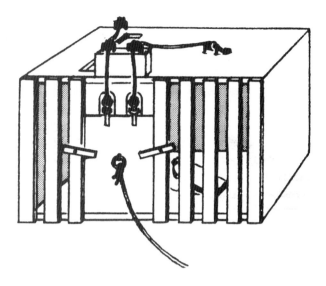

Fig. 4.6
Thorndike's puzzle box for cats is shown. Food was placed outside the box, and if the cat managed to pull out a bolt or step on a lever, the door would open and the animal could escape the box and eat the food. When the cats were given repeated trials in the box, they became faster and faster at getting out.

Source: Redrawn from E. L. Thorndike (1898), *Animal intelligence: Experimental studies*. New York: The MacMillan Co, p. 30.

resulted in escape and food were selected while other behavior that did not lead to those outcomes decreased in frequency. Eventually those operants that produced reinforcing consequences came to dominate the cat's behavior, allowing the animal to get out of the box in less and less time. Thus, latency was an indirect measure of a change in the animal's operant behavior. Today, **rate of response** or operant rate (the number of responses per a specified interval, such as a minute) is considered a better measure of operant behavior. Operant rate estimates the probability of response and provides a direct measure of the selection of behavior by its consequences.

FOCUS ON: BEHAVIORAL NEUROSCIENCE AND OPERANT CONDITIONING OF THE NEURON

How does the environment affect the organism itself during the process of operant conditioning? One possibility is that reinforcement and operant conditioning occur at the level of brain units or elements. Skinner (1953, pp. 93–95) addressed brain units when he stated that:

> [T]he element rather than the response [is] the unit of behavior. It is a sort of behavioral atom, which may never appear by itself upon any single occasion but is the essential ingredient or component of all observed instances [of behavior].

At the time that Skinner made this claim he had no way of knowing the basic element or "behavioral atom" of operant conditioning. Today, the evidence is strong that the basic units of reinforcement are not complex brain structures of whole responses but elements as small as the neuron itself.

It is possible to investigate the neuron and reinforcement by the method of ***in-vitro reinforcement*** or **IVR** (Stein & Belluzzi, 2014; Stein, Xue, & Belluzzi, 1994). The idea is that calcium bursts or firings (L-type Ca^{2+}) of a neuron are reinforced by dopamine (a neurotransmitter) binding to specialized receptors. Furthermore, the process of neuronal conditioning can be investigated "*in vitro*" using brain-slice preparations and drug injections that stimulate the dopamine receptor (dopamine agonists).

In these IVR experiments, a small injector tube (micropipette) is aimed at cells of the brain slice (hippocampal cells from pyramidal cell layer of CA1). During operant conditioning, micropressure injections of a dopaminergic drug (an agonist, which enhances dopamine binding) are applied to the cell for 50 ms following bursts of activity (amplified action potentials). When the computer identifies a predefined burst of activity for the target neuron, the pressure-injection pump delivers a tiny droplet of the drug to the cell. Drug-induced increases in bursting indicate operant conditioning if the *contingency* between neuron bursts and drug presentation is critical. To be sure that the drug is not just stimulating burst activity, the same drug is given independently of bursting on a *noncontingent* basis.

The results showed that the bursting responses of individual neurons increase in a dose-related manner by *response-contingent* injections of dopamine agonists. Also, noncontingent presentation of the same drug injections did not increase the bursting responses of the neurons. The findings indicate that reinforcement occurs at the level of individual neural units (Skinner's atoms of behavior), and suggest that subtypes of dopamine neurons (D1, D2, or D3 types) are involved in cellular and behavioral operant conditioning.

Additional IVR experiments indicate that bursting responses of CA1 pyramidal neurons also increase with injections of cannabinoid drugs, whereas the firings of CA3 neurons increase with drugs that stimulate the opiate receptors (Stein & Belluzzi, 1988; Xue, Belluzzi, & Stein, 1993). When these drug injections are administered independent of cellular activity, bursting responses do not increase and often are suppressed. Furthermore, contingent and noncontingent glutamate injections to the CA1 neurons over a range of doses fail to increase bursting or decrease this response. Thus, drug agonists that target specific receptors implicated in reward and addiction (e.g., dopamine, cannabinoid, and opioid) act as reinforcement for neural bursting, whereas glutamate, an excitatory transmitter not associated with behavioral reinforcement, fails to augment cellular activity or even suppresses it.

New research has revealed the conditioning of single neurons in the brains of live Japanese monkeys (*Macaca fuscata*) as the animals performed a visual fixation task (Kobayashi, Schultz, & Sakagami, 2010). Monkeys were mildly deprived of fluid and were seated in a primate chair with their head fixed, facing a computer monitor that presented images on each trial. The researchers isolated and monitored single neurons in the lateral prefrontal cortex (LPFC), an area of the brain associated with intentional, purposive action (operant behavior), and reinforced neuronal spikes that exceeded an established criterion. When the neuron emitted a spike in the established range, the monkeys received juice to drink as reinforcement. Control conditions involved all of the procedures of the experimental phase, including visual fixation and presentation of juice, but there was no contingency between neuron firing and reinforcement. Evidence indicated that individual neurons in the LPFC showed operant conditioning. Further experiments in the series indicated that LPFC neurons would respond in accord with the schedule of reinforcement. The researchers noted that alternative explanations such as simple reward prediction, attention, and arousal were unlikely to account for the findings. Also, the LPFC has few motor neurons, so it is unlikely that neuron activity in this region directly coincides with movement of the animal. One possibility is that LPFC neurons contribute to behavioral flexibility—a neural substrate for operant behavior. These neurons may "enhance the action signals [in other brain areas] to ensure the motor execution when operant control is required" (Kobayashi et al., 2010, p. 1854; see also Schafer & Moore, 2011; see also Ishikawa, Matsumoto, Sakaguchi, Matsuki, & Ikegaya, 2014 for "*in vivo*" rapid operant conditioning of selected hippocampal neurons in mice, using neurofeedback reinforcement that requires NMDA receptor activity—extending the "*in-vitro*" experiments described previously).

Operant conditioning is a major adaptive mechanism that change behavior on the basis of lifetime experiences (ontogeny). From a biological standpoint, operant conditioning allows for behavioral flexibility, survival, and reproduction. Evidence is accumulating that behavioral flexibility is based on **neuroplasticity**—alterations of neurons and neural interconnections during a lifetime by changes in the environmental contingencies (Caroni, Donato, & Muller, 2012). *In-vitro* reinforcement experiments show that endogenous brain chemicals binding to particular types of receptors increase the likelihood of neuronal activity. These molecular neural processes presumably underlie the large-scale changes in operant behavior that occur as humans and other animals interact with the world in which they live, from moment to moment, over a lifespan.

Procedures in Operant Conditioning

Operant Rate and Probability of Response

Rate of response refers to the number of operant responses that occur in some defined unit of time. For example, if you ask five questions during a 2-h class, your rate is 2.5 questions per hour. An animal that presses a lever 1,000 times in a 1-h session generates a rate of 1,000 bar presses per hour (or 16.7 responses per minute). Skinner (1938) proposed that rate of response is the basic datum (or measure) for operant analysis. **Operant rate** is a measure of the probability of behavior (the **probability of response**). In other words, an operant that occurs at a high rate in one situation has a high probability of being emitted in a similar situation in the future. This increased probability of response is observed as a change in operant rate. Of course, probability of response may decrease, and in this case is observed as a reduction in rate.

The Free-Operant Method

In the **free-operant method**, an animal may repeatedly respond over an extensive period of time. The organism is free to emit many responses or none at all. Using this method, responses can be made without interference from the experimenter. For example, a laboratory rat may press a lever for food pellets. Lever pressing is under the control of the animal, which may press the bar rapidly, slowly, or quit pressing. Importantly, this method allows the researcher to observe changes in the *rate of response*. This is important because rate of response is used as a measure of response probability. Rate of response must be free to vary if it is used to index the future probability of operant behavior.

The analysis of operant rate and probability of response is not easily accomplished when an organism is given a series of trials (as in the Thorndike experiments or with mazes). This is because the experimenter largely controls the animal's rate of behavior. For example, a rat that runs down a T-maze for food reward is picked up at the goal box and returned to the starting point. Because the experimenter sets the number of trials and response opportunities, changes in the rate of response and moment to moment changes in behavor (such as turning right or left or pausing) are aggregated into a single measure of response latency. Comparing the T-maze trial procedure with the free-operant method, it is clear that the free-operant method is more suitable for studying the *probability of response* in a given situation. The free-operant method is clearly demonstrated by the procedures used in operant conditioning.

The Operant Chamber

To study operant conditioning in a laboratory, a device called an **operant chamber** is used. Of course, operant conditioning is also investigated outside laboratories. Nonetheless, investigating the behavior of animals in operant chambers has resulted in the discovery of many principles of behavior. Figure 4.7 shows an operant chamber designed to accommodate a laboratory rat. The chamber is a small, enclosed box that contains a lever (or two) with a light above it, and a food magazine or cup connected to an external feeder. The feeder delivers a small food pellet (typically 45 mg) when electronically activated. It is also equipped with a fan to circulate air and a small speaker that plays white noise to mask extraneous sounds in the ambient environment. The operant chamber is housed inside a sound-attenuating cubicle,

Fig. 4.7

A standard operant chamber for a rat (note: image shows the right half of the chamber; chamber is actually twice as large as photo depicts). The chamber is a box that has a lever (often two) that the animal can press. There is a cue light above each lever that can be illuminated. A food magazine or cup is connected to a feeder. The feeder delivers a small, 45-mg food pellet to the cup. In this situation, the food pellet serves as reinforcement for lever pressing. There is also a houselight to illuminate the chamber, a speaker for white noise delivery, a fan to circulate air, and the entire chamber is enclosed in a sound-attenuating cubicle.

Source: Used with permission from E. Rasmussen, 5/2022.

which also blocks extra-experimental sounds. All of the inputs (e.g., responses) and outputs (e.g., pellet delivery, cue lights) are controlled by a computer.

In this situation, the food pellet serves as reinforcement for lever pressing. The operant chamber *structures the situation* so that the desired behavior will occur and incompatible behavior is reduced. Thus, lever pressing is highly likely, while behavior such as running away is minimized. A school classroom also attempts to structure the behavior of students with regard to learning. The classroom, unlike the operant chamber, often contains many distractions (e.g., looking out the window) that interfere with on-task behavior and concentrating on the material being presented. Just like other sciences that isolate variables in a laboratory, the operant chamber reduces all of the extraneous variables (sights, sounds, smells, etc.) of an environment so that *only* the behavior and contingency of reinforcement (or other variables) can be studied.

Changing Motivation with Deprivation
Because the delivery of food is used as reinforcement, an animal must be motivated to obtain food. An objective and quantifiable measure of motivation for food is percentage of free-feeding body weight (note that another way of quantifying deprivation is specifying the time since the rat last consumed the reinforcer). Prior to a typical experiment, an animal is brought from a commercial (or research) colony into a laboratory, placed in a cage, given free access to food, and weighed on a daily basis. The average weight is calculated, and this value is used as a baseline. Next, the daily food ration is gradually reduced until the animal reaches 85% of its free-feeding weight. The procedure of restricting access to food (the potentially reinforcing stimulus) is called a **deprivation operation** (see "Motivational Operations" in Chapter 2). At this point, the experimenter assumes, but does not know, that food is a reinforcing stimulus. This is because food delivery must increase the frequency of an operant before it can be defined as reinforcement.

The weight loss or deprivation criterion is not as bad as it sounds. Laboratory animals typically have food freely available 24 h a day, whereas animals in the wild must forage for their food. The result is that lab animals that free-feed (eat as much as they want whenever they want) tend to be heavier than their free-ranging counterparts. Alan Poling and his colleagues nicely demonstrated this point by showing that captured free-range pigeons gained an average 17% body weight when housed under free-feeding laboratory conditions (Poling, Nickel, & Alling, 1990). Notice that weight gain for these birds was roughly equal to the weight loss typically imposed on laboratory animals when using a deprivation operation. Importantly, research shows that maintaining animals at deprivation levels that are typical in operant studies enhances longevity and reduces health problems. It is actually beneficial for animals (including humans) to not feed freely.

At the physiological level, loss of body weight (food deprivation) activates hormones related to energy homeostasis, involving insulin, leptin, and ghrelin. Neuroendocrine research shows that variation in these hormones increase (or decrease) food-reinforced behavior and modulate neurotransmitters controlling the endogenous reward circuitry, especially the midbrain dopamine (DA) and opiodergic systems. Thus, changes in the feeding regimen of the rat impact operant control by reinforcement through a complex interplay of metabolism, neurotransmitters, and endogenous reward pathways (Figlewicz & Sipols, 2010).

REINFORCER HABITUATION ALSO CHANGES MOTIVATION

Deprivation shows us that a reinforcer's value is not constant; it is dynamic. Another event that may change the strength of an activity or reinforcer concerns **reinforcer habituation**. Have you ever noticed the first contact with a reinforcer is the strongest? In a single sitting, such as when you eat a piece of cheesecake, this is often the case. The first bite is delicious, but that last bite is not as subjectively satisfying. One reason, of course, is that deprivation decreases as you eat. But there is another process at work—reinforcer habituation. Recall that habituation (Chapter 3) refers to a decrease in the strength of a response with repeated exposure. When you eat a piece of cheesecake, each bite is an exposure and with each subsequent bite, its strength as a reinforcer decreases.

Frances McSweeney (see Figure 4.8) conducted a number of studies with rats on the phenomenon of reinforcer habituation (see review by McSweeney & Murphy, 2009) and showed that the presentation of a novel event can disrupt habituation (called *dishabituation*) of reinforcement, temporarily increasing the potency of food as a reinforcer. A paper by Aoyama and McSweeney (2001) is a good illustration. In this study, rats pressed levers for food and response rates were examined across the experimental session as a measure of reinforcer strength. The left panel of Figure 4.9 shows the decline of response rate across session. In the baseline condition (open circles), response rate was high initially and with each reinforcer consumed, rate decreased in a systematic manner; this was likely due to deprivation levels getting lower, but it could also be habituation to the food stimulus. To determine if habituation was also playing a role, another condition was implemented midway in the session. Habituation was disrupted briefly with the presentation of a novel event (dishabituation—see dotted vertical line)—in this case, simply withdrawing the lever from the operant chamber for 3 minutes. When the lever was withdrawn, response rate decreased during that time, since the rats could not press the lever, but then immediately afterward, when the lever was reintroduced, response rate increased even higher than baseline after the disruption. Note that this disruption did not affect deprivation levels; this temporary bump in reinforcer potency can be explained solely by the novel change in the chamber (the withdrawal of the lever). This circled area on the figure shows that habituation to reinforcement is also playing a role in the decline in response rate.

Fig. 4.8
Dr. Frances McSweeney, the behavior scientist who characterized habituation of reinforcement.
Source: Used with permission of Frances McSweeney, 2021.

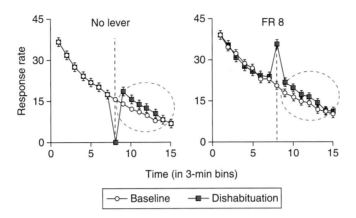

Fig. 4.9
On the left, a graph that shows two curves in which response rates decline within session. The baseline condition shows what a normal habituation curve looks like for food. In the dishabituation condition, the lever is withdrawn for a 3-min period (see dotted vertical line). After is it reintroduced, there is a temporary increase in response rate (see circled data), showing that the potency of food increases after the introduction of a novel event. This increase is also shown on the right figure in which a change in the schedule of reinforcement from 4 to 8 responses (FR 8) for a 3-min period (dishabituation) leads to a temporary increase in reinforcer potency (also circled).

Source: Based on data from Aoyama, K., & McSweeney, F. K. (2001). Habituation may contribute to within-session decreases in responding under high-rate schedules of reinforcement. *Animal Learning & Behavior*, *29*(1), 79–91. https://doi.org/10.3758/BF03192817.

With rats, dishabituation of food reinforcement has been accomplished in many ways besides removing the lever for a 3-min period. Dishabituation can be induced by simply changing the schedule of reinforcement delivery. In that same study by Aoyama and McSweeney, for example, another condition was implemented: the requirement for food was changed from 4 lever-presses to 8 lever-presses for a 3-min period. The right panel of Figure 4.9 shows that when this response requirement doubled, the response rate increased strongly during the dishabituation period (because the response requirement for food increased), but on return to the four-response contingency, reinforcer potency was temporarily heightened (see circled area). This dishabituation can occur also by introducing a light for a brief period of time, changing the location of a light, or presenting a brief tone (see review by Lloyd, Medina, Hawk, Fosco, & Richards, 2014). Interestingly, stimulants such as nicotine and methamphetamine can interfere with the habituation process of food; in other words, stimulants can disrupt habituation by creating intense novelty and stimulation (Lloyd, Hausknecht, & Richards, 2014).

Reinforcer habituation has also been studied with humans. In a study with children by Epstein et al. (2003), habituation to cheeseburgers was studied by examining saliva production to the smell of cheeseburgers. Children could also earn access to cheeseburgers by working on a computer. Predictably, saliva production and response rate decreased with exposure to the cheeseburger stimulus. When apple pie was introduced as a novel stimulus, though, salivation and response rate increased; that is, the pie disrupted habituation to the cheeseburger. This may be one reason that when we are full, we seem to be able to find room for dessert—dessert temporarily dishabituates food as a reinforcer. Reinforcer dynamics, such as habituation and dishabituation, play a large role in our eating behaviors and food choices.

The Operant Class

A lever press is an easily definable response— the press of a lever is detected by the computer when a switch closes that makes an electrical connection. Any behavior emitted by the rat that results in a switch closure defines the operant class. A lever press with the left or right paw, or even with its mouth, produces an identical electrical connection. Another advantage of lever pressing as an operant is that it may be emitted at a range of rates from very low to very high. This is an advantage because the primary focus of operant research is on the conditions that affect the rate (probability) of operant behavior.

Shaping: The Method of Successive Approximation

But how does the rat initially learn the lever press? Sometimes, a rat may happen to press the lever a few times and naturally discover the contingency between lever-press and food. More often, though, this takes a lot of time. Another process—**shaping** or the method of **successive approximations**—is implemented. We first establish an easily detectable cue— the "click of the feeder" as a stimulus that is associated with the delivery of the food. (This is called magazine training.) Once established, because it is paired with food (a *primary*, or unlearned, *reinforcer*), the click now takes on a reinforcing property—it can be used to strengthen behavior. We call this click a **conditioned** or learned reinforcer (more on this can be found in Chapter 10). Now we can activate the sound of a clicker to reinforce approximations to the lever press. We may start with activating the clicker when the rat is near the lever: when the rat nears the lever, we activate the click and the rat immediately approaches the food aperture and eats the food pellet. Then, afterward, the rat will likely return to "hovering" near the lever. Once the rat reliably does this, we can now stop reinforcing hovering near the lever and begin reinforcing a new approximation: moving its head closer to the lever. This process of only reinforcing one behavior and not others is called **differential reinforcement**. When the rat reliably moves its head closer to the lever, we begin differentially reinforcing the placement of one of its front paws near the lever. When the rat reliably places its paw near the lever, we stop reinforcing placement of the paw near the lever and then require contact of the paw with the lever to produce reinforcement. And then finally when the rat is reliably touching the lever, we require for reinforcement that the rat press the lever with enough force to close the circuit and activate the feeder. You can see that with shaping, the researcher or trainer is arranging the environment such that the organism is required to "level up" in their behavior. These steps are found in Figure 4.10.

Steps in shaping lever pressing

- Standing near the lever reinforced
- Facing lever reinforced
- Move head toward lever reinforced
- Touching lever with parts of body (nose) reinforced
- Touching lever with paws reinforced
- Touching lever with specified (right) paw reinforced
- Partially depressing lever with right paw reinforced
- Raising and depressing lever with right paw reinforced
- Raising and depressing lever completely with right paw reinforced

Fig. 4.10
Steps for shaping a lever press in an operant chamber.

In the preceding example, we took advantage of a rat's behavioral repertoire to train a lever press—something that the rat does not normally do. The animal's **repertoire** refers to the behavior it is capable of naturally emitting on the basis of species and environmental history. Rats naturally explore their surroundings, for example, so we begin with simply reinforcing the rat's exploring near the lever. Then, we systematically use **shaping** to establish the lever-press response. Shaping involves differentially reinforcing successive approximations to the final performance (i.e., pressing the lever).

An application of shaping to humans that is also a fun game to play with students is the hot/cold game. In this game an individual's behavior is shaped by a group of people playing the game. The individual is asked to leave the room or out of hearing distance while the group decides on the final performance of behavior they would like to see the individual demonstrate. After a definition of the final behavior is agreed upon by the group the individual returns to the group. The individual is instructed to "do anything," then the group either says "hotter" or "colder" depending on how close the individual's behavior is to the final performance. This feedback from the group serves as reinforcement of successive approximations until the individual performs the final behavior. Not surprisingly, the individual usually demonstrates the behavior in a relatively short amount of time without formal instructions or verbal descriptions given by the group. You may also realize this example can be applied by adults and teachers in teaching children new skills such as how to say specific words, how to play sports, or even tie their shoes.

Notice that shaping makes use of **behavioral variability**—the animal's (whether human or nonhuman) tendency to emit variations in response form in a given situation. The range of behavioral variation is related to an animal's capabilities based on genetic endowment, degree of neuroplasticity, and previous interactions with the environment. Behavioral variability allows for selection by reinforcing consequences, and is analogous to the role of genetic variability in natural selection (Neuringer, 2009). Shaping by successive approximation uses undifferentiated operant behavior (a lump of clay), which is moved toward "a functionally coherent unit" (pressing a lever for food) by a process of differential reinforcement. Shaping as an important behavioral procedure cannot be overemphasized. It is the process by which nearly all complex behaviors are acquired. The steps involve the explicit definition of the terminal behavior (final performance), description of the subject's or student's current repertoire (baseline level), and the contingent delivery of an effective reinforcer after each approximation toward the terminal performance (see also "Behavior Analysis in Education," Chapter 13).

NEW DIRECTIONS: SHAPING AND NEUROSCIENCE OF BIRDSONG

Successive approximation to a final performance occurs as part of the natural contingencies shaping the singing of songbirds. The learning of birdsong appropriate to one's species begins with auditory-vocal correspondence; the nestling hears the song of adult birds (tutors), which evokes species-specific, highly variable vocal responses or subsong, akin to babbling in human infants (Brainard & Doupe, 2002). Research indicates that this "sensory learning" phase appears to involve activation and encoding by the mirror-neuron system (Prather, Peters, Nowicki, & Mooney, 2008; see also Chapter 11 on correspondence and mirror neurons).

To achieve adult song, nestlings also must be able to hear themselves. Birds deafened after exposure to the songs of tutors, but before they practice singing on their own, show abnormal songs as adults. The songs of the nestlings must be perfected by self-initiated practice (operant) and auditory feedback (reinforcement)—called the "sensorimotor learning" phase. In this phase, immature birdsong is shaped toward the adult song (final performance) by hearing how closely vocal responses correspond with those of adult birds (auditory feedback). Using self-produced feedback from singing, the youngster's melody is fine-tuned toward an adult "crystallized" song with only small variations from one rendition to the next.

At the neural level, vocal learning in birds is viewed as a form of motor learning, involving the *anterior forebrain pathway* (AFP) composed of an area homologous to the mammalian basal ganglia, as well as an area of the thalamus. Basal ganglia pathways connect with the *posterior descending pathway* (PDP), especially the high vocal center (HVC), to regulate production, timing, and sequencing of song. To achieve mature birdsong, neural circuitry is required for production of vocalizations, circuitry for hearing and discriminating sounds of self and others, and connections between the two pathways (Brainard & Doupe, 2013). Some researchers propose an "error correction" model for acquisition (and maintenance) of birdsong, emphasizing the AFP and the lateral magnocellular nucleus of anterior nidopallium (LMAN). In this model, the LMAN, by its connection to the premotor nucleus (RA), allows for adjustments in song by hearing self-produced vocalizations (perhaps encoded by the nidopallium caudomediale or NCM, an analog of the mammalian audio-association cortex) compared with the "song template" of the adult birds. At the present time, however, the exact neural processes involved in the use of auditory feedback to shape mature birdsong are not well understood (Brainard & Doupe, 2013). An operant delayed matching to sample analysis (see Chapter 8), with adult song as the sample and hearing self-produced song as the matching response (via LMAN, NCM, and RA circuitry), would emphasize the "hits" (reinforcement) for selection of neural connections and pathways (plasticity) rather than appealing to an inferred cognitive process, involving comparison of errors between self-song and the memory song template (see "Focus On: Behavioral Neuroscience and Operant Conditioning of the Neuron" in this chapter).

A Model Experiment

In the previous discussion of operant behavior, some basic principles were illustrated using the laboratory rat. It is important to realize that these same principles can be extended to a variety of species, including humans, and even single-celled organisms. For example, recent reports indicate that the single-celled organism, slime mold, has behavior that is sensitive to consequences (Boisseau et al., 2016; Dussutour, 2021). The point is that the behavior of less complex organisms is influenced by the environment. This appears to be a universal process in nature thar occurs across most, if not all, species.

In the following demonstration of operant conditioning, pigeons are used as the experimental subjects. Pigeons are placed in an operant chamber and required to peck a small plastic disk or key illuminated by a white light. A peck at the key activates a microswitch and makes an electrical connection that controls a food hopper. Presentation of food functions as reinforcement for pecking. A food hopper filled with grain swings forward and remains available for a few seconds. The bird can eat the grain by sticking its head through an opening. Figure 4.11 shows an operant chamber designed for birds. Note that the chamber is very similar to the one used to study the operant behavior of rats.

Before an experiment, the bird is taken from its home colony and placed alone in a cage. Each pigeon is given free access to food and water. The bird is weighed each day for about a week, and its baseline weight is calculated. Next, the daily food ration is reduced until the bird

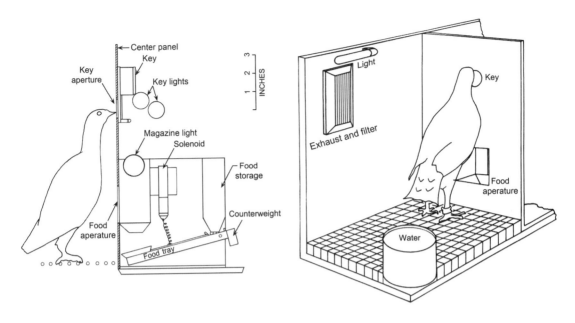

Fig 4.11

A schematic of an operant chamber for birds. The chamber contains a small plastic disk illuminated by a light. A peck at the disk activates a microswitch and makes an electrical connection. When reinforcement is scheduled to occur, the food hopper swings forward and remains available for a few seconds. The bird can eat grain from the hopper by sticking its head through the opening in the chamber wall. In principle, the chamber is similar to the one used to study the operant behavior of rats and has many of the same features (e.g., feeder, houselight, fan, speaker, sound-attenuating cubicle), though not all are seen here.

Source: Adapted from C. B. Ferster & B. F. Skinner (1957), *Schedules of reinforcement*. New York: Appleton-Century-Crofts.

reaches approximately 80% of free-feeding or **ad libitum weight**. After the deprivation procedure, the pigeon is placed in the operant chamber for magazine training.

When the bird is placed in the chamber for the first time, it may show a variety of emotional responses, including wing flapping and defecating. This is because the chamber, like any new environment, presents a number of new features that may initially function as aversive stimuli. For example, the operation of the feeder makes a loud sound that may startle the bird. Eventually, these emotional responses are extinguished by repeated exposure to the apparatus. As the emotional responses dissipate, the bird explores the environment and begins to eat from the food magazine. Since the sound of the hopper is associated with food, the sound becomes a conditioned positive reinforcer. At this point, the bird is said to be magazine trained.

A bird's operant level of key pecking is typically very low, and it is convenient to train these responses by the method of successive approximation. Shaping of key pecking in pigeons is similar to shaping lever pressing in rats; in both cases, shaping involves reinforcing closer and closer approximations to the final performance of pecking the key hard enough to operate the microswitch. As each approximation occurs, it is reinforced with the presentation of the food hopper. Earlier approximations are no longer reinforced and reduce in frequency (extinction). This process of reinforcing closer approximations, and withholding reinforcement for earlier approximations, eventually results in the pigeon pecking the key with sufficient force to operate

132 Reinforcement and Extinction of Operant Behavior

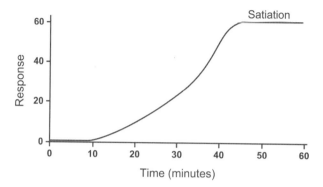

Fig. 4.12
This figure shows typical acquisition of key pecking on CRF or continuous reinforcement throughout a session. Rate of response is low when the animal is initially placed in the chamber. After this brief period, rate of response is high and stable. Finally, rate of response levels off toward the end of the session. This latter effect is caused by satiation, as is labeled on the curve.

the microswitch. The key peck that operates the microswitch to produce food is the *first definable response*. The switch closure and electrical connection define the operant class of pecking for food. At this point, a microcomputer is programmed so that each key peck results in the presentation of food for a few seconds. Because each response produces reinforcement, the schedule is called continuous reinforcement, or CRF.

Figure 4.12 shows the key pecking under continuous reinforcement across the time of the session (the bird has presumably been shaped to peck the key for food). Notice early in the session (the first 10 min), that there are hardly any responses; that is, the rate of response is low. This period is called the warm-up, and probably occurs because of the abrupt change from home cage to the operant chamber. Once lever-pressing begins and the pigeon makes contact with reinforcement, the rate begins to increase across the session and becomes high and stable.

Finally, toward the end of the session, the figure shows that rate of response plateaus, indicating that the bird stops pecking the key. This latter effect is called **satiation**, and it occurs because the bird has eaten enough food. More technically, *rate of response declines because repeated presentations of the reinforcer weaken its effectiveness*. A satiation operation (and likely habituation to reinforcement, too) decreases the effectiveness of reinforcement. This effect is opposite to deprivation in which withholding the reinforcer increases its effectiveness.

To be sure that an increase in the rate of response is caused by the contingency of reinforcement, it is necessary to create a second experimental condition in which the contingency between key pecking and food is broken. In other words, if food is no longer presented, the pigeon should give up pecking the key. If the peck–food contingency caused key pecking, then withdrawal of the contingency will result in a decline in key pecking toward the operant level.

Figure 4.13 presents records for periods in which pecking produces food (labeled as Reinforcement) or does not produce food (labeled as Extinction). The initial peck–food contingency produces a steady rate of response. Under Extinction, the rate of response declines and eventually key pecking stops and behavior levels off earlier in the session. Thus, key pecking clearly depends upon the contingency of reinforcement.

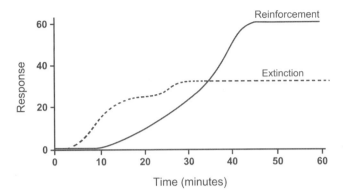

Fig. 4.13
This figure compares performances on CRF vs Extinction and shows responses across time. The curve labeled "Reinforcement" shows that responding is maintained throughout the session when reinforced. For the curve labeled "Extinction" shows that when responding is no longer reinforced, the rate of response declines earlier in the session and eventually responding stops.

FOCUS ON: REINFORCEMENT AND PROBLEM SOLVING

Reinforcement and Response Stereotypy

Barry Schwartz (1980, 1982a) carried out a series of experiments with pigeons to show that reinforcement produced *response stereotypy*, a pattern of responding that is repetitive, invariant, and highly practiced. Subsequently, Schwartz (1982b) used similar procedures with college students to demonstrate the presumed negative effects of reinforcement for response stereotypy in human problem solving. Indeed, response stereotypy can be thought of as the opposite to variation, and in some ways, creativity.

College students were given points on a counter when they completed a complex sequence of responses. The responses were left and right key presses that moved a light on a checkerboard-like matrix of 25 illuminated squares. Figure 4.14 shows the matrix, with the light in the top left square. The task required that the participant press the keys to move the light from the top left corner to the bottom right square. A press on the right key moved the light one square to the right. When the left-hand key was pressed, the light moved one square down. Schwartz required exactly four left (L) and four right (R) presses in any order (e.g., LRLRLRLR, LLLLRRRR, etc.). There were 72 different possible orders of left and right key presses that would move the light to the bottom right corner. When the light reached the bottom right corner, a point registered on the counter. The points were later exchanged for money. If the participant pressed any key a fifth time (e.g., RRRRR), all of the matrix lights were turned off and the trial ended without reinforcement.

In a series of experiments, Schwartz found that students developed a stereotyped pattern of responding. The point is that as soon as a student hit on a correct sequence, he or she repeated it and rarely tried another pattern, even though many other patterns were possible. In other experiments (Schwartz, 1982b), participants were explicitly reinforced for varying their response pattern, meaning that reinforcement would only be given if a response pattern differed from previous response patterns. When this was done, the students developed higher-order stereotypes meaning they produced a series of varied patterns, but repeated them.

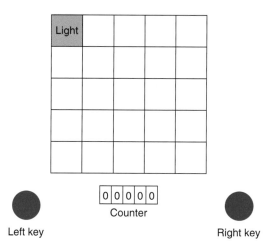

Fig. 4.14
The 5 × 5 matrix task used by Schwartz (1982b). A right press moved the light one square to the right; a left button press moved the light down one square. A counter is in the middle which tallies responses.

From these experiments, Schwartz concluded that reinforcement interfered with problem solving because it produced stereotyped response patterns.

Reinforcement and Response Variability

Allen Neuringer (Figure 4.15) is a behavior analyst who investigates variability, randomness, and behavior (see Neuringer & Jensen, 2013 for an overview of this research). He suggested that the contingencies of reinforcement of the Schwartz experiments produced response stereotypy, but this was not an inevitable outcome of reinforcement. In the Schwartz experiments, the response patterns were constrained by the requirement to emit *exactly four* pecks on each key in any order. This constraint means that of the 256 possible sequences of pecks, only 72 patterns resulted in reinforcement; also, a fifth peck to either left or right keys resulted in a *timeout from reinforcement* (negative punishment). This timeout contingency punished many instances of response variability, reducing its occurrence.

A classic study by Page and Neuringer (1985) eliminated the constraints imposed by Schwartz and tested the assumption that variability could actually increase with reinforcement. The experiment involved pigeons pecking left and right keys on the light matrix task. Each session consisted of numerous trials with the bird emitting 8 pecks (instead of 4 responses, like Schwartz's work) to the two keys on each trial—ending with food for 8 pecks that met the criterion for variability and a short timeout for other sequences. The contingency for the variability phase (VAR) involved *lags* in which the current sequence of 8 pecks had to differ from the pattern of pecks emitted on some previous trials. The number of previous trials defined the Lag value. For example, using a Lag 3 contingency, the current sequence of 8 pecks to left and right keys had to diverge from each of the patterns for the previous three trials for the current trial to end in food reinforcement; other "incorrect" sequences of pecks resulted in a timeout. The larger the lag, the more varied the responses. And here is something even more interesting: When a Lag 50 was in place (meaning that the current 8-response sequences had to be different from the last 50), the birds responded almost randomly (like coin flipping) on

Fig. 4.15
Dr. Allen Neuringer.
Source: Photograph by Vera Jagendorf, Portland, OR. Published with permission.

the left and right keys, not by "remembering" what they did over the past 50 trials. In a sense, they became random generators!

At this point in the experiment, the researcher introduced a critical control comparison phase where the birds' sequences were reinforced at the end of some of the 8-peck trials, but the presentation of food *did not* depend on the bird's variability. Reinforcements were now tied or *yoked* to an earlier pattern of reinforcement delivered in the VAR phase. If a bird's sequence had been reinforced on a given trial of the VAR session, then the equivalent trial in the yoked phase also ended with reinforcement—regardless of whether the lag contingency had been met. In other words, the contingency between behavioral sequence and reinforcement was broken. By yoking each bird's reinforcement in the two phases, the researchers ensured that the amount of reinforcement was identical for a given bird, and that the only difference was the requirement of the lag contingency. Would response variability be higher when variability produced reinforcement than when reinforcement occurred independently of variability? The answer was a clear yes. Birds were highly variable when reinforced for varying (VAR), but tended to repeat pecks to one of the keys when reinforcement did not require response variability (yoked). One conclusion is that variability is an operant dimension of behavior (much like force of response or speed of response) regulated by the contingency of reinforcement.

Reinforcement and Problem Solving: An Assessment

The current evidence indicates that *variability is an operant* that increases when reinforcement is contingent on behavioral variation (Lee, Sturmey, & Fields, 2007; Neuringer, 2002, 2009). To date, the reinforcement of variability has been shown in a number of species, including pigeons, dolphins, rats, and human adults and children (Goetz & Baer, 1973; van Hest, van Haaren, & van de Poll, 1989; Machado, 1997; Neuringer, 1986; Pryor, Haag, & O'Reilly, 1969; Stokes, Mechner, & Balsam, 1999). In addition, different experimental procedures have been used to produce variability with a number of different response forms (Blough, 1966; Eisenberger & Shanock, 2003; Goetz & Baer, 1973; Machado, 1989; Morgan & Neuringer, 1990; Odum, Ward, Barnes, & Burke, 2006; Pryor et al., 1969). Variability, constraints by task and contingencies, and artistic creativity have also been of interest (Stokes, 2001). More recently, an experiment with pigeons manipulated reinforcement magnitude (amount) for 4-peck sequences to left and right keys in a complex schedule arrangement. The results showed that large reinforcers disrupted the reinforcement of variability, inducing a high level of behavioral repetition as the time to reinforcement approached (Doughty, Giorno, & Miller, 2013).

In summary, on one hand, reinforcement produces behavioral inflexibility and rigidity. In contrast, when variation in response is part of the reinforcement contingency (such as with a Lag contingency), novel, even creative, sequences of behavior result (Neuringer, 2004; Neuringer & Jensen, 2013; Machado, 1989, 1992, 1997). Generally, a close analysis of the contingencies is required in problem-solving situations because "what you reinforce is what you get" (stereotypy or variability).

EXTINCTION

The procedure of *withholding reinforcement* for a previously reinforced response is called **extinction**. The first study of extinction goes back to the days of Skinner (1938). To produce extinction, you would disconnect the relation between reinforcement and behavior. For example, one could disable the food hopper after the bird had been reinforced for key pecking. It is important to note, though, that the procedure of *extinction is a contingency of reinforcement*. The contingency is defined as zero probability of reinforcement for the operant response. Extinction is also a behavioral process and, in this case, refers to a *decline in rate of response* caused by withdrawal of reinforcement. For example, you may raise your hand to ask a question and find that a certain professor ignores you. Asking questions may decline because the professor no longer reinforces this behavior.

Behavioral Side Effects of Extinction

As Chapter 5 will describe, side effects of reinforcement or behavioral effects that are not required for the contingency of reinforcement, but predictably develop. In addition to the main effect of a decline in the rate of response, extinction produces several behavioral side effects. In the section that follows, we consider the range of effects generated by the cessation of reinforcement. Many of the responses of organisms to the withdrawal of reinforcement are intuitive from an evolutionary perspective. Presumably, when things no longer worked in an ecological niche (extinction), natural selection favored organisms that repeated behavior that had "worked" in the past, made a greater range of responses in the situation (behavioral variability), emitted more forceful responses to the circumstances, and attacked other members of the species associated with the withdrawal of reinforcement.

Extinction Burst

When extinction is started, operant behavior will initially *increase* in frequency. Basically, organisms repeat behavior that has been reinforced in the past. A pigeon will initially increase the rate of key pecking, or you may raise your hand more often than you did in the past. You may explain your increased tendency to raise your hand by telling a friend, "The instructor doesn't see me; I have an important point to make." If the bird could talk it might also "explain" why it was pecking at an increased rate. The point is that an initial increase in the rate of response, or **extinction burst**, occurs when reinforcement is first withdrawn.

Extinction-Induced Variability

In addition to extinction bursts, operant behavior becomes increasingly variable as extinction proceeds (**extinction-induced variability**). Behavioral variation increases the chances that the organisms will reinstate reinforcement or contact other sources of reinforcement. You may wave your hand about in an attempt to catch the professor's eye (think of Hermione in the Harry Potter movies when Professor Snape wouldn't call on her). The pigeon may strike the key in different locations and with different amounts of force. A classic experiment by Antonitis (1951) demonstrated this effect. Rats were taught to poke their noses through a 50-cm-long slot for food reinforcement. When this occurred, a photocell was triggered and a photograph of the animal was taken. The position of the rat and the angle of its body were recorded at the moment of reinforcement. After the rat reliably poked its nose through the slot, it was placed on extinction.

Antonitis reported that reinforcement produced a stereotyped pattern of response. The rat repeatedly poked its nose through the slot at approximately the same location, and the position of its body was held at a particular angle. When extinction occurred, the nose poking and position of the body varied. During extinction, the animal poked its nose over the entire length of the slot. A similar study with humans was published by Jennifer Kinloch, Mary Foster, and James McKewan (2009). They had human participants press a space bar under a schedule of reinforcement called differential reinforcement of low rates (DRL), which requires a minimum pause between space-bar presses. For example, a DRL 3-s schedule means that at least 3 seconds must pass between two responses before reinforcement (points) were delivered. Under this schedule, mean interresponse times (also called IRTS; pauses between responses) hovered just above the DRL contingency (in this case, 3 s). When extinction was placed in effect, IRTs varied immensely from person to person.

Force of Response

Reinforcement may be made contingent on the **force of response** (or other properties) resulting in **response differentiation**. A classic study by Notterman (1959) measured the force that rats used to press a lever during periods of reinforcement and extinction. During reinforcement sessions, animals came to press the lever with a force that varied within a relatively narrow range (just enough force to activate the lever and produce reinforcement). When extinction occurred, the force of lever pressing became more variable. Interestingly, some responses were more forceful than any emitted during reinforcement or during operant level. This increase in response force is sometimes labeled as emotional behavior generated by extinction procedures and is implicated in extinction-induced aggressive behavior (see next section on emotional responses to extinction).

For example, imagine that you have pushed a button for an elevator but the elevator does not arrive, and you have an important appointment on the 28th floor. At first you increase the frequency of pressing the elevator button; you also change the way you hit the button. You probably feel angry and frustrated, and you may smash the button. These responses and accompanying feelings occur because of the change from reinforcement to extinction.

Emotional Responses
Consider what happens when someone puts money in a vending machine and is not reinforced with an item (e.g., a beverage). The person whose behavior is placed on extinction may hit the machine, curse, and engage in other emotional behavior. Soda machines once killed several US soldiers. Young soldiers at the peak of physical fitness are capable of emitting forceful operants. When some of the soldiers put money in soda machines that failed to operate, extinction-induced emotional behavior became so powerful that the men pulled over the 2-ton machines. Thus, their deaths were an indirect outcome of emotional behavior produced by extinction.

A variety of **emotional responses** occur under conditions of extinction. Birds flap their wings, rats bite the response lever, and humans may swear and kick at a vending machine. One important kind of emotional behavior that occurs during extinction is aggression. Azrin, Hutchinson, and Hake (1966) trained pigeons to peck a key for food. After training, a second immobilized pigeon was placed in the operant chamber. The "target" bird was restrained and placed on an apparatus that caused a switch to close whenever the bird was attacked. Attacks to the target reliably occurred when the contingencies of reinforcement were changed from CRF to extinction. Many of the attacks were vicious and unrelenting, lasting up to 10 min. (It is important to note that this experiment was conducted many decades ago, and current research ethics that are based on federal regulations would likely not allow this experiment to be conducted today.)

In children with severe behavior disorders, the modification of self-injurious behavior (SIB) maintained by social attention often uses extinction procedures. Two commonly observed side effects with the onset of extinction are the extinction burst (a sudden increase in the SIB) and extinction-induced aggression. Lerman, Iwata, and Wallace (1999) analyzed 41 sets of data on individuals who received extinction treatment for SIB, and found that bursts and aggression occurred in nearly 50% of the cases. When extinction was accompanied by other procedures, such as differential reinforcement of alternative behavior (DRA), bursting and aggression were substantially reduced. The recommendation when using extinction of inappropriate or harmful behavior is to also reinforce appropriate behavior. This will lessen the side effects of bursting and aggression (see Fritz, Iwata, Hammond, & Bloom, 2013 for analysis of severe problem behavior).

Discriminated Extinction
Suppose that a pigeon's key pecks are reinforced in the presence of a green light. When a red light comes on, however, pecking is not reinforced. During the course of training, the animal would emit emotional responses and extinction bursts when the red light (but not the green light) is turned on. After training, though, the bird would not emit this behavior; it would simply stop responding when the light changes from green to red. The red light becomes a *discriminative stimulus (S^Δ) that signals a period of extinction*. This effect is called **discriminated extinction**, and is commonly observed in human behavior. A sign on a vending machine that reads OUT OF ORDER is an S^Δ that signals extinction for putting money in the machine.

The procedures for rapid extinction of respondently conditioned responses (see Chapter 3) seem close to the operant procedure of discriminated extinction (see Lattal & Lattal, 2012 for a comparison of respondent and operant extinction procedures). Comparing the procedures, we assume that a respondently conditioned response to the light is similar to the emission of an operant on a given occasion. Discriminated extinction involves signaling extinction periods with an exteroceptive (externally presented) stimulus, such as a change in key color from green to red. This change from green to red in the operant procedure is like adding the tone during respondent extinction. When the key is green, a pigeon is trained to peck it for food. Every once in a while, the key color changes to red, and reinforcement for pecking no longer occurs. During these extinction periods, rate of response should decline. This decline would occur more rapidly when extinction is signaled by a change in color than when the key color remains the same. Finally, since the red key is consistently associated with extinction, it acquires a discriminative function (S^Δ), suppressing responding when it is presented. In this situation, if the red and green stimuli are alternated, the onset of red sets the occasion for any behavior except pecking; then with the onset of green, pecking is reinforced.

Resistance to Extinction

As extinction proceeds, emotional behavior subsides and the rate of response declines. When extinction has been in effect long enough, though, behavior may still return to baseline levels. This is because more than one extinction session is usually required. Extinction is typically measured as the number of responses emitted in some amount of time. For example, a bird's key pecks may be reinforced on CRF for ten consecutive daily sessions; following this, extinction is initiated. The pigeon's responses are recorded over three extinction sessions. The number of responses emitted by the bird or the rate of response during the last session may be used to index **resistance to extinction**. Operants are rapidly extinguished if a behavior has only been reinforced a few times, but when operants are reinforced many times, resistance to extinction increases. Experiments have shown that peak resistance to extinction occurs after a condition in which 50 to 80 responses have been reinforced (see Lattal & Lattal, 2012 for resistance to extinction in respondent and operant procedures).

Partial Reinforcement Effect (PRE)

Resistance to extinction is substantially increased when a partial or **intermittent schedule of reinforcement** has been used to maintain behavior. On an intermittent reinforcement schedule, only some responses are reinforced. For example, instead of reinforcing each response (CRF), the experimenter may program reinforcement after an average of 50 key pecks have been emitted. In this situation, the bird must emit an unpredictable number of pecks before food is presented. This intermittent schedule will generate many more responses during extinction than continuous reinforcement. When people are described as having a persistent or tenacious personality, their behavior may reflect the effects of intermittent reinforcement.

Nevin (1988) indicates that the **partial reinforcement effect (PRE)** is the result of two basic processes: reinforcement and discrimination. According to Nevin's analysis of behavioral momentum, reinforcement has the single effect of increasing resistance to change. Thus, *the higher the rate of reinforcement for an operant, the greater the resistance to change*. The implication is that behavior maintained by a CRF schedule is more resistant to change than behavior controlled by an intermittent schedule of reinforcement.

Extinction, however, occurs more rapidly on CRF compared with intermittent reinforcement. One reason for the discrepancy is that *discrimination* between reinforcement and extinction is more rapid on CRF than on intermittent reinforcement. In other words, an organism discriminates the difference between a high steady rate of reinforcement (CRF) and no reinforcement (extinction) more easily than the difference between a low intermittent rate of reinforcement and no reinforcement. In tests for resistance to extinction, the discrimination factor overrides the rate of reinforcement variable, and animals typically show greater resistance on intermittent than CRF schedules. If the effects of discrimination (between reinforcement and extinction) are controlled, behavior maintained by CRF is in fact more resistant to extinction than on intermittent schedules of reinforcement (see Nevin, 2012 for an analysis of resistance to extinction and behavioral momentum).

An additional reason for increased resistance to extinction following intermittent reinforcement involves *contact with the contingencies*. When a rat's bar pressing is reinforced for every 100 responses, it must emit 100 responses before contacting the change from reinforcement to extinction. In contrast, an animal reinforced for each response contacts the extinction contingency immediately. In the latter situation, since each response is not reinforced, the animal repeatedly encounters the change to extinction. If an animal on CRF emits 50 responses during extinction, it has contacted the extinction contingency 50 times. A rat on intermittent reinforcement may have to emit 5,000 responses to have equal experience with the change in contingencies.

Discriminative Stimuli and Extinction

Intermittent reinforcement is not the only factor that determines the return to baseline level during extinction. Resistance to extinction is also affected by discriminative stimuli that are conditioned during sessions of reinforcement. Pigeons were trained to peck a yellow triangle on an intermittent schedule of food reinforcement. After training, a red triangle was substituted for the yellow one and extinction was started. During 15 min of extinction in the presence of the red triangle, the rate of response substantially declined. At this point, the yellow triangle replaced the red one but extinction was continued. The effect of introducing the yellow triangle was that rapid responding began immediately, and the usual extinction curve followed. This effect is portrayed in Figure 4.16 in which responding in the presence of the yellow triangle is at a high rate during the first 30 min of intermittent reinforcement. When the red triangle and extinction were introduced, the rate of response declined. Finally, extinction was continued and the yellow triangle was reinstated. Notice that the rate of response immediately recovers and then declines toward extinction.

Spontaneous Recovery

An interesting phenomenon that occurs during extinction is **spontaneous recovery**. After a session of extinction, the rate of response may be close to near-zero level. At this point, the animal is taken out of the operant chamber and returned to a holding cage. The next day, the organism is again placed in the operant chamber and extinction is continued. Surprisingly, the animal begins to respond at a higher rate than the near-zero extinguished rate, and this defines spontaneous recovery. Over repeated sessions of extinction, the amount of spontaneous recovery decreases. If many sessions of extinction are provided, the rate of response no longer recovers.

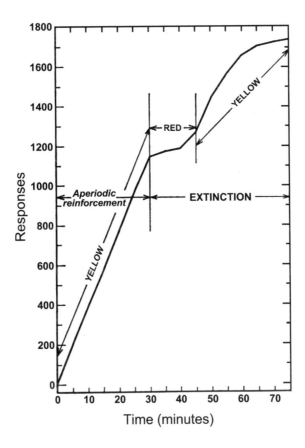

Fig. 4.16
A figure depicting responding during extinction as a function of discrimination is shown. Responses are shown across time in a session. Responding in the presence of the yellow triangle is high during the first 30 min of intermittent reinforcement. When the red triangle and extinction are introduced, rate of response declines. Extinction is continued and the yellow triangle is reinstated. When the yellow triangle is presented, rate of response recovers and then declines toward extinction.
Source: From B. F. Skinner (1950). Are theories of learning necessary? *Psychological Review*, 57, pp. 193–216.

Spontaneous recovery is really not spontaneous. Stimuli that have accompanied reinforced responding are usually presented at the beginning of extinction sessions. Skinner (1950) noted that handling procedures and the stimulation arising from being placed in an operant chamber set the occasion for responding at the beginning of each extinction session (habituation may also be involved; and extinction context may directly inhibit responses; see Todd, 2013). Skinner stated:

> No matter how carefully an animal is handled, the stimulation coincident with the beginning of an experiment must be extensive and unlike anything occurring in the latter part of an experimental period. Responses have been reinforced in the presence of, or shortly following, this stimulation. In extinction it is present for only a few moments. When the organism is again placed in the experimental situation the stimulation is restored ... The only way to achieve full extinction in the presence of the stimulation of starting an experiment is to start the experiment repeatedly.
>
> (Skinner, 1950, pp. 199–200)

Human behavior also shows apparent spontaneous recovery generated by stimuli based on previous conditioning. Imagine that you are stranded in a secluded mountain cabin during a weeklong snowstorm. Your limited cell service looks something like this: your cell phone rings, you answer, but all you get is dead air. You voice frustration at the phone and press the "end call" button repeatedly (extinction burst, variation in behavior). Next, you try to contact the cell phone company and discover that you are not able to dial out. Over the course of the first day your phone rings many times, you answer it, but it does not work. By the end of the day, you may not be inclined to answer the phone—you just let it keep on ringing. The next morning you are having breakfast and your phone rings again. What do you do? The best guess is that you will again answer the phone. You may say to yourself, "Perhaps cell service is better today." On this second day of extinction, you answer your phone but give up more quickly. On day 3, the phone rings at 10:00 a.m. and even though you doubt that it will work, you answer it "just to check it out." By day 4, you have had it with the "damn phone and the stupid cell service," and extinction is complete.

Reinstatement of Responding with Non-Contingent Reinforcment
Another kind of response recovery, called **reinstatement**, involves the recovery of behavior when the reinforcing stimulus is presented independently—without a **contingent response**—after a period of extinction (Bouton, 2004). In an operant model of reinstatement, Baker, Steinwald, and Bouton (1991) established lever-pressing with food reinforcement and then extinguished the lever-pressing response by withholding food. After extinction, reinstatement involved retracting the levers so that the rats could not press them, but dropping food into the food cup with no behavioral contingencies. Then, with extinction still in place, the response levers were made available again. On these tests, animals that were given non-contingent food deliveries showed more reinstatement of lever pressing than control animals. The results indicated that response-independent reinforcement activates contextual stimuli from the original learning situation that set the occasion for previously reinforced lever pressing.

Application to substance use disorder. At the practical level, reinstatement is often observed in the treatment of drug addiction. A person with a substance use disorder (SUD) spends a great amount of time drug seeking and administering the drug. In other words, drug seeking and the behavior of getting the drug into the body (called **self-administration**) are reinforced by the potent reinforcer of the drug itself. Drugs of abuse indeed can be powerful reinforcers. The person with the SUD may find, though, that this pattern of behavior is destructive to other aspects of their life, such as maintaining personal relationships, stable employment, or good health. They may seek help for the SUD and treatment may involve no longer using the drug (extinction) in a therapeutic setting, e.g., in a 30-day in-patient facility. When the client is returned to their former neighborhood and drug culture (original setting), drugs may be available on a response-independent basis—the client who is no longer using the drug may see it handed out on street corners or their friends may use the drug in front of them. These stimuli could activate the original setting events that have set the occasion for obtaining and using drugs in the past, and this could reinstate drug use.

One practical implication of this research is that extinction of drug use in a treatment setting (extinction context) may inadvertently exacerbate reinstatement of use when the person returns to the home environment (original context for reinforcement of drug use). Even when the person with the SUD makes dramatic changes in lifestyle such as by changing city, friends, and work, even small stimuli that a person may be unaware of (e.g., driving by an old street corner in which drug deals were made) could trigger drug use. Researchers have proposed a variety of cue-exposure treatments to prevent relapse of drug use (Havermans & Jansen, 2003), but the use of contingency management for drug abstinence likely would be more efficacious in the long run (see Chapter 13 on contingency management and substance abuse).

Extinction and Forgetting

During extinction, operant behavior decreases over time. People often talk about the weakening of behavior as loss of memory or forgetting. An important question concerns the procedural differences between forgetting and extinction. Extinction is a procedure in which a previously reinforced response no longer produces reinforcement. The opportunity to emit the operant *remains available* during extinction. Thus, the pigeon may still peck the illuminated key, or the rat may continue to press the response lever. In contrast, forgetting is said to occur after the mere passage of time. An organism that has learned a response is tested for retention after some amount of time has passed. In this case, there is no apparent opportunity to emit the behavior.

A classic experiment by Skinner (1938) illustates this. This experiment assessed the behavioral loss that occurs after the passage of time. In this experiment, four rats were trained to press a lever, and each animal received 100 reinforced responses. After 45 days of rest, each animal was placed in the operant chamber and responding was extinguished. The number of responses emitted during extinction was compared with the performance of four other rats selected from an earlier experiment. These animals were similar in age, training, and number of reinforced responses to the experimental subjects. The comparison animals had received extinction one day after reinforced lever pressing.

Figure 4.17 shows the results of Skinner's experiment. The results are presented as the cumulative average number of responses emitted by each group of animals. The group that received extinction one day after response strengthening emitted an average of 86 responses in 1 h. The group that was extinguished after 45 days made an average of 69 responses in 1 h. Notice that both groups of animals show a similar number of responses during the first few minutes of extinction. In other words, the animals in both groups immediately began to press the lever when placed in the operant chamber. This shows that the rats that received extinction after 45 days had not forgotten what to do to get food (Skinner, 1938).

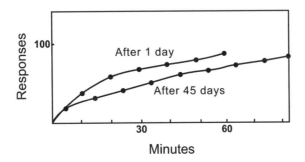

Fig. 4.17
The figure shows two extinction curves of four rats that are averaged together. Minutes are on the x-axis and responses on the y-axis. One curve represents responding after Day 1 of training and the other after 45 days after training. Both curves show an increase in response rate early in the session and response rate slows and levels off with time. For the Day 45 curve, though, response rate declines earlier in the session and there are fewer responses overall compared to Day 1.

Source: Curves are taken from B. F. Skinner (1938), *The behavior of organisms*. New York: Appleton-Century-Crofts.

Following the first few minutes of extinction, there is a difference in the cumulative-average number of responses for the two groups. Resistance to extinction is apparently reduced by the passage of time. Rats that were required to wait 45 days before extinction generated fewer responses per hour than those that were given extinction 1 day after reinforcement. Although the curves rise at different rates, animals in both groups appear to stop responding after approximately 90 unreinforced lever presses. Overall, the results suggest that the passage of time affects resistance to extinction, but a well-established performance is not forgotten (for an account of remembering in elephants, see Markowitz, Schmidt, Nadal, & Squier, 1975; see also Dale, 2008).

ON THE APPLIED SIDE: EXTINCTION OF TEMPER TANTRUMS

Williams (1959) has shown how extinction effects play an important role in the modification of human behavior. In this study, a 20-month-old child was having temper tantrums when put to bed. If the parents stayed up with the child, he did not scream and cry, and he eventually went to sleep. A well-known source of reinforcement for children is parental attention, and Williams reasoned that this was probably maintaining the bedtime behavior. In this analysis, when the parents left the bedroom, the child began screaming and crying. These tantrums were reinforced by the return of the parents to the bedroom. The parental behavior stopped the tantrum, and withdrawal of screaming by the child reinforced the parental behavior of returning to the bedroom. Based on these contingencies, the parents were spending a good part of each evening in the child's room waiting for him to go to sleep. At this point, the parents were advised to implement extinction by leaving the room and closing the door after the child was put to bed. Figure 4.18 demonstrates the rapid decline in duration of crying when this was done (first extinction).

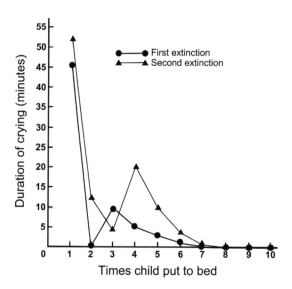

Fig. 4.18

The graph shows first and second extinction procedures for a child's temper tantrums. The x-axis is times the child was put to bed and the y-axis is duration of crying in minutes. The most crying took place on the first time the child was put to bed with rapid extinction afterward with both first and second extinction trials, though there was some variation with the second extinction trial.

Source: Adapted from C. D. Williams (1959). The elimination of tantrum behavior by extinction procedures. *Journal of Abnormal and Social Psychology, 59*, p. 269.

When extinction was first attempted, the child screamed and cried for 45 min (an extinction burst). However, on the next night he did not cry at all. On the third night, the child emitted tantrums for 10 min. By the end of 10 days, the boy was smiling at his parents when they left the room. Unfortunately, his aunt, who reinforced crying by staying in the room with him, put the boy to bed and his temper tantrums reoccurred. A second extinction procedure was then implemented. The duration of crying was longer for the second than for the first period of extinction. The higher probability of response during the second extinction phase is presumably caused by the intermittent reinforcement of tantrums. Recall that intermittent reinforcement increases resistance to extinction. Fortunately, the boy's tantrums were not reinforced again, and his tantrums eventually declined to a zero rate. At a two-year follow-up, the parents reported that his bedtime tantrums had been completely eliminated.

CHAPTER SUMMARY

In this important chapter, we have addressed the idea that behavior is a function of the outcomes (or consequences) they produce. Operants are responses that operate on the environment to produce changes or effects and, as a result, have an increased (or decreased) probability of occurrence. The measure of the probability of response is most often the rate of operant behavior. If the rate of the particular response (or class of behavior) increases as a result of some specific outcome, then that outcome is defined as a positive reinforcer. The exact definition of a positive reinforcer, as a stimulus or event that increases or maintains the rate of the response upon which it is contingent, is fundamental to a science of behavior. The delivery of the reinforcer is contingent upon the response (depends on the response), and no matter what the stimulus or event, it increases the frequency of operant behavior.

Other situations exist that require formal definitions. For example, when you encounter a disturbing event, you may turn away or cover your ears—behavior that is strengthened by the removal or reduction in occurrence of the event. This is a type of response strengthening called negative (subtracts the event) reinforcement (increases the escape response). Other outcomes reduce the rate of the response and are called punishers. The procedure of making a punisher contingent on a response and the response decreases is called punishment.

If some behavior has a low probability of occurrence, the response can be shaped by differential reinforcement of successive approximations. In this way, new behavior is generated from the variability existing in the response repertoire. When reinforcement is no longer delivered, the rate of the response eventually declines to near-zero levels. This is the process of extinction. For example, if a rat presses a lever and food pellets are delivered, it will continue to press the lever as long as some responses are followed by food. When no pellets are delivered, however, the rate of lever pressing eventually declines to zero. This simple demonstration of reinforcement followed by extinction illustrates the central point of the chapter—behavior is a function of the outcomes it produces.

Behavior analysis is a scientific discipline based on changing or manipulating outcomes and thereby shaping, directing, and altering the behavior of organisms. Precise procedures and apparatus have been invented to systematize this analysis.

Key Words

Ad libitum weight
Behavioral variability
Conditioned reinforcer
Contingency of reinforcement
Contingent response
Continuous reinforcement (CRF)
Deprivation operation
Differential reinforcement
Discriminated extinction
Discriminative stimulus (S^D)
Dishabituation
Elicited (behavior)
Emitted (behavior)
Emotional response
Extinction
Extinction burst
Extinction-induced variability
Force of response
Free-operant method
Intermittent schedule of reinforcement
In-vitro reinforcement (IVR)
Latency
Law of effect
Negative punishment
Negative reinforcement
Neuroplasticity

Operant
Operant chamber
Operant class
Operant conditioning
Operant rate
Partial reinforcement effect (PRE)
Positive punishment
Positive reinforcement
Positive reinforcer
Premack principle
Probability of response
Rate of response
Reinforcer habituation
Reinstatement (of behavior)
Repertoire (of behavior)
Resistance to extinction
Response deprivation
Response deprivation hypothesis
Response differentiation
Response hierarchy (free-choice setting)
Satiation
S-delta (S^Δ)
Shaping
Spontaneous recovery (operant)
Successive approximation
Topography

On the Web

http://psych.athabascau.ca Click the sidebar on Positive Reinforcement. The purpose of this site is to teach the concept of positive reinforcement and also to provide an idea of the kind of self-instructional exercises that are used at Athabasca University in Alberta, Canada.

www.wagntrain.com/OC This is a website for all those seeking to use positive reinforcement in animal training. If you have a dog or cat, you can use the "clicker training" method of positive reinforcement to teach your animal new behavioral sequences and skills.

www.karawynn.net/mishacat Here is a website about the toilet training of Misha the cat. The trainer does not provide the general principles that shaped the cat's behavior from the litter box to the toilet, but many of the principles are outlined in Chapter 4. See if you can figure out how to train any cat to do what Mischa did.

Brief Quiz

1. The term *operant* comes from the verb _____ and refers to behavior that _____.
   a. opponent; opposes its consequences in a given environment
   b. opendum; opens the door to its effects on a given occasion
   c. operates; operates on the environment to produce effects
   d. opara; presents the opportunity to respond on a given occasion

2. What defines a contingency of reinforcement?
   a. discriminative stimulus
   b. operant
   c. reinforcement
   d. all of the above

3. Which of the following is *not* one of the four basic contingencies?
   a. positive reinforcement
   b. positive extinction
   c. negative punishment
   d. negative reinforcement

4. In terms of rewards and intrinsic motivation, Cameron et al. (2001) conducted a statistical procedure called _____, and one of the findings indicated that verbal rewards _____ performance and interest on tasks.
   a. multivariate analysis; decreased
   b. meta-analysis; decreased
   c. meta-analysis; increased
   d. multivariate analysis; increased

5. The Premack principle states that a higher-frequency behavior will:
   a. function as reinforcement for a lower-frequency behavior
   b. function as punishment for a high-frequency behavior
   c. function as intermittent reinforcement for a low-frequency behavior
   d. none of the above

6. To experimentally study the probability of response, a researcher uses _____ as the basic measure and follows the _____ method.
   a. latency; T-maze
   b. latency; free operant
   c. operant rate; T-maze
   d. operant rate; free operant

7. Shaping of behavior involves:
   a. the molding of a response class by the physical arrangement of the operant chamber
   b. differentially reinforcing closer and closer approximations to the final performance
   c. withholding and giving food for correct performance of a specified level of response
   d. none of the above

8. A classic experiment on the effects of extinction by Antonitis (1951) involved:
   a. nose poking by rats for food reinforcement
   b. photographs of the rats' position and body angle
   c. increased variability of nose poking during extinction
   d. all of the above

9 In terms of response stereotypes, variability, and reinforcement, the work by Barry Schwartz shows that reinforcement can produce _____ patterns of behavior, while the work of Neuringer and his colleagues indicates that reinforcement can produce _____.
   a   stereotyped; response variability
   b   response variability; stereotyped
   c   stereotyped; response stability
   d   response stability; response variability

10 Which of the following is involved in the partial reinforcement effect?
   a   longer extinction on intermittent reinforcement compared with CRF
   b   the higher the rate of reinforcement, the greater the resistance to change
   c   discrimination between reinforcement and extinction is more rapid on CRF
   d   all of the above

Answers to Brief Quiz: 1, c (p. 109); 2, d (p. 110); 3, b (Figure 4.3); 4, c (p. 116); 5, a (p. 118); 6, d (p. 123); 7, b (p. 128); 8, d (p. 137); 9, a (p. 133); 10, d (p. 139).

Schedules of Reinforcement

1. Learn about the basic schedules of reinforcement.
2. Investigate rates of reinforcement and resistance to change.
3. Inquire about behavior during transition between schedules of reinforcement.
4. Discover how schedules of reinforcement are involved in cigarette smoking.
5. Distinguish molecular and molar accounts of performance on schedules.

The events that precede operant behavior and the consequences that follow may be arranged in many different ways. A **schedule of reinforcement** describes this arrangement. In other words, a schedule of reinforcement is a statement of the conditions under which a behavior is reinforced. In the laboratory, sounding a buzzer in an operant chamber may be a signal (S^D) that sets the occasion for lever pressing (operant) to produce food (reinforcer). A similar schedule operates when a dark room (S^D) sets the occasion for a person to turn on a lamp, which is followed by illumination of the room.

At first glance, a rat pressing a lever for food and a person turning on a light to see appear to have little in common. Humans are very complex organisms—they build cities, develop economies, write books, go to college, use computers, conduct scientific experiments, and do many other things that rats cannot do. In addition, pressing a lever for food appears to be very different from switching on a light. Nonetheless, foundational processes involved in schedules of reinforcement have been found to be remarkably similar across different organisms, behavior, and reinforcers. When the same schedule of reinforcement is in effect, a child who solves math problems for teacher approval may generate a pattern of behavior comparable to a bird pecking a key for water.

IMPORTANCE OF SCHEDULES OF REINFORCEMENT

Schedules of reinforcement were a major discovery first described by B. F. Skinner in the 1930s. Subsequently, Charles Ferster and B. F. Skinner reported the first and most comprehensive study of schedules ever conducted (Ferster & Skinner, 1957). Their work on this topic is unsurpassed and represents the most extensive study of these independent variables of behavior

DOI: 10.4324/9781003202622-5

science. Today, few studies focus directly on simple, basic schedules of reinforcement—they are embedded in the foundation of such phenomena as substance abuse, impulsivity, and toxicant exposure. The lawful relations that have emerged from the analysis of reinforcement schedules, however, remain central to the science of behavior—being used in virtually every study reported in the *Journal of the Experimental Analysis of Behavior*. The knowledge that has accumulated about the effects of schedules is central to understanding behavior regulation.

Modern technology has made it possible to analyze performance on schedules of reinforcement in increasing detail. Nonetheless, early experiments on schedules remain important. The experimental analysis of behavior is a progressive science in which observations and experiments build on one another. In this chapter, we present early and later research on schedules of reinforcement. The analysis of schedule performance ranges from a global consideration of cumulative records to a detailed analysis of the time between responses.

BEHAVIOR ANALYSIS: A PROGRESSIVE SCIENCE

In the experimental analysis of behavior research findings are accumulated and integrated to provide a general account of the behavior of organisms. Often, simple animals in highly controlled settings are studied. The strategy is to build a comprehensive theory of behavior that rests on direct observation and experimentation.

The field of behavior analysis emphasizes a descriptive approach and discourages speculations that go substantially beyond the data. Such speculations include reference to the organism's memory, thought processes, expectations, and undocumented accounts based on presumed physiological states. For example, a behavioral account of schedules of reinforcement provides a detailed description of how behavior is altered by contingencies of reinforcement. One such account is based on evidence that a particular schedule sets up differential reinforcement of the time between responses (interresponse times, or IRT; see later in this chapter). An alternative account is that behavior is integrated into larger units of performance according to the molar or macro contingencies of reinforcement (overall rate of reinforcement). Both of these analyses contribute to an understanding of an organism's behavior in terms of specific environment—behavior relationships—without reference to hypothetical cognitive events or presumed physiological processes.

Recall that behavior analysts study the behavior of organisms, including people, for its own sake. Behavior is not studied to make inferences about hypothetical mental states or real physiological processes. Although most behaviorists acknowledge and emphasize the importance of biology and neurophysiological processes, they focus more on the interplay of behavior with the environment during the lifetime of an organism. Of course, *direct analysis of neurophysiology* of animals provides essential details about how behavior is changed by the operating contingencies of reinforcement and behavioral neuroscientists currently are providing many of these details, as we discuss throughout this textbook.

Contemporary behavior analysis continues to build on previous research. The extension of behavior principles to more complex processes and especially to human behavior is of primary importance. The analysis, however, remains focused on the environmental conditions that control the behavior of organisms. Schedules of reinforcement concern the arrangement of environmental events that regulate behavior. The analysis of schedule effects is currently viewed within a biological context. In this analysis, biological factors play several roles. One way in which biology affects behavior is through specific neurophysiological events (e.g., release of neurotransmitters in reward areas of the brain) that function as reinforcement and discriminative stimuli. Biological variables may also constrain or enhance environment–behavior relationships (see Chapter 7). As behavior analysis and the other biological sciences progress, an understanding of biological factors becomes increasingly central to a comprehensive theory of behavior.

CUMULATIVE RECORDS

A commonly used laboratory instrument that records the frequency of operant behavior in time (rate) is called a **cumulative recorder**. Figure 5.1 illustrates a version of this device, which was used in the early research of schedules of reinforcement. Each time a lever press occurs, the pen steps up one increment. When reinforcement occurs, this same pen makes a downward deflection. Once the pen reaches the top of the paper, it resets to the bottom and starts to step up again. Since the paper is drawn across the roller at a constant speed, the cumulative recorder depicts *a real-time measure of the rate of operant behavior*. The faster the lever presses, the steeper the slope or rise of the **cumulative record**.

A cumulative record of key pecking by a pigeon is shown in Figure 5.2. In this illustration, a bird responded 50 times to produce one food delivery. Notice that periods of responding are followed by reinforcement (indicated by the deflection of the pen). After reinforcement, the rate of response is zero, as indicated by the plateaus or flat portions of the cumulative record.

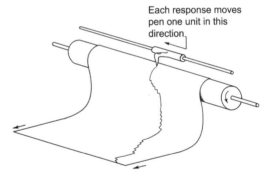

Each response moves pen one unit in this direction

Fi . 5.1
Photograph shows a laboratory instrument used to record operant responses, called a cumulative recorder. This device was used in the early research on operant conditioning. The recorder gives a real-time measure of the rate of operant behavior. The faster the lever presses, the steeper the slope or rise of the cumulative record. This occurs because paper is drawn across the roller at a constant speed and the pen steps up a defined distance for each response. Modern laboratories show cumulative records generated by computer software along with additional behavioral measures.
Source: Ralph Gerbrands Corporation, Arlington, MA.

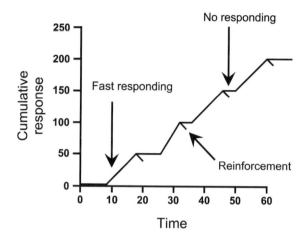

Fig. 5.2
A cumulative record of key pecking by a pigeon. In this illustration, a bird responded 50 times to produce one food delivery. Notice that 50 pecks are followed by reinforcement and that this is indicated by a downward deflection of the pen. Following reinforcement, the rate of response is zero, as indicated by the plateaus or flat portions of the record.
https://report.nih.gov/nihdatabook/category/10

In a modern operant laboratory, the cumulative record is generated by computer software that collects behavioral input from organisms. The record provides the experimenter with an immediate report of the animal's behavior. Researchers have discovered many basic principles of behavior by examining cumulative records (e.g., Ferster & Skinner, 1957). Today, computers allow researchers to collect, display, and record multiple measures of behavior (e.g., rate of response) that are later submitted to complex numerical and statistical analyses (Chen & Steinmetz, 1998; Gollub, 1991). In this book, we present examples of cumulative records and numerical analyses that have been important to the experimental analysis of behavior.

Schedules and Patterns of Response

Response patterns develop as an organism interacts with a schedule of reinforcement (Ferster & Skinner, 1957). These patterns come about after an animal has extensive experience with the *contingency of reinforcement* (S^D: R → S^r arrangement) defined by a particular schedule. This contingency is the schedule's **main effect**—the behavioral requirements that are necessary to produce the reinforcer. Subjects (often pigeons or rats) are exposed to a schedule of reinforcement and, following an acquisition period which examines behavior across multiple sessions, behavior typically settles into a consistent or **steady-state performance** (Sidman, 1960).

While an organism may learn a contingency quickly, it may take many experimental sessions before a consistent pattern across sessions emerges. What is interesting, though, is that the patterns that are generated after acquisition are orderly, predictable, and replicable. Even more interesting, these patterns are *not required* by the schedule contingencies themselves. They are **side-effect patterns** of behavior. The steady-state behavioral pattern generated when a fixed number of responses are reinforced illustrates one of these side-effect patterns. For example, a hungry rat might be required to press a lever 10 times to get a food

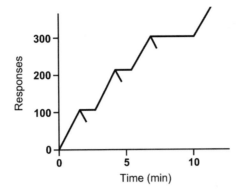

Fig. 5.3
A side-effect break-and-run pattern generated from the fixed ratio schedule of reinforcement.

pellet. This is the contingency or the main effect (also called direct effect) of the schedule—all that is required to produce reinforcement is 10 responses. Following reinforcement, the animal has to make another 10 responses to produce the next bit of food, then 10 more responses, and so forth. But here is what is remarkable: when organisms (rats, pigeons, or humans) are reinforced after a fixed number of responses, a *break-and-run side-effect pattern* of behavior often develops. Responses are made rapidly (and the higher the fixed ratio, the more rapid) and a characteristic pause in responding follows each reinforcer delivery, followed by another quick burst of responses (see "Fixed Ratio" section in this chapter for more details). This pattern repeats over and over again and occurs even when the ratio size of the schedule is changed. When examining this pattern on a cumulative record, it looks like the one shown in Figure 5.3.

Note that the schedule contingencies do not require this pattern for reinforcement—just 10 responses—as quickly or slowly as the organism chooses—is all that is required for food. Nonetheless, the side-effect break-and-run pattern emerges regardless.

NOTE ON: INNER CAUSES, SCHEDULES, AND RESPONSE PATTERNS

We sometimes speak of people being "highly motivated" when we observe them investing energy or time in some project. Motivation seems to explain why people behave as they do. People are sometimes said to be unmotivated when they put off or fail to do assignments or tasks; in contrast, we refer to individuals as highly motivated when they study hard and overachieve. From a behavioral perspective, there is no need to infer a hypothetical internal process of motivation or drive to understand this kind of behavior. Schedules of reinforcement generate unique and predictable patterns of behavior that are often interpreted as signs of high motivation; other schedules produce pausing and low rates of response used as indicators of low motivation or even clinical depression. In both cases, behavior is due to environmental contingencies rather than the *inferred inner cause* called motivation.

Similarly, habits or personality traits are said to be response dispositions or patterns that occur consistently across time and contexts (Neal, Wood, & Quinn, 2006). Reference is often made to internal dispositions that account for regular and frequent actions or habits. Instead of inferring dispositions as internal causes, one might say instead that habits or traits are *patterns of steady-state responding*; these regularities of behavior are maintained by the content and consistency of the schedules of reinforcement. Consistent or reliable schedules of reinforcement generate habitual, stable rates and patterns of responding. It is these characteristic patterns of behavior that people use to infer dispositional causes. A behavior analysis, though, indicates that the actual causes are often the behavioral contingencies and contexts rather than dispositional states within us (Phelps, 2015).

The stability of behavior and choice patterns generated by reinforcement contingencies, which allows people to infer others' dispositions and personality, also allows for reliable inferences of emotional states. Based on behavioral stability and consistency, computer algorithms are now able to recognize human faces and "read" emotions from a person's facial expressions. Our faces evolved as organs of emotional communication and there is money to be made with emotionally responsive algorithms. Algorithms with visual inputs are able to code facial expressions and, together with choice patterns and voice analysis, they can predict buying, voting, depression, attention, and additional affective behaviors (Khatchadourian, 2015). This is the case with social media as well.

Schedules and Natural Contingencies

In everyday life, behavior is often reinforced on an intermittent basis. On an intermittent schedule of reinforcement, an operant is reinforced occasionally rather than each time it is emitted. For example, not every cry from a child is reinforced with attention. Not every text or email sent to someone is answered. Each time a predator hunts, it is not always successful. When you call a number for customer service, sometimes you get through, but sometimes it takes time or you cannot reach another human. Buses do not immediately arrive when you go to a bus stop. Persistence is often essential for survival or achievement of success; thus, an account of perseverance on the basis of the maintaining schedule of reinforcement is a major discovery.

Consider a bird foraging for food. The bird turns over sticks or leaves and once in a while finds a seed or insect. These bits of food occur only every now and then, and the distribution of reinforcement is the schedule that maintains the animal's foraging behavior. If you were watching this bird hunt for food, you would probably see the animal's head bobbing up and down. You might also see the bird pause and look around, change direction, and move to a new spot. This sort of activity is often attributed to the animal's instinctive behavior patterns. Labeling the behavior as instinctive, however, does not explain it. Although evolution and biology certainly play a role in this foraging episode, perhaps as importantly, so does the schedule of food reinforcement.

Behavior analysts consider that there are always many schedules of reinforcement available at any one time and these schedules vary in terms of the amount and quality of reinforcement, as well as how immediate it is, and what kinds of responses are required to gain access to the sources of reinforcement, and any additional factors, such as schedules of punishment that are superimposed on top of the schedules of reinforcement (e.g., drinking beer is fairly reinforcing until one drinks too much and gets sick or has a hangover). The relative consequences compete for behavior allocation; that is, the organism chooses each moment how to allocate its behavior to these varying sources of reinforcement. For example, when a college student awakens in the morning, they have a choice to get out of bed and engage in the behaviors that will allow them to get to their 8:30 a.m. class on time; other alternative reinforcers might be to stay in bed and sleep or maybe get up and watch TV or hop on the PlayStation. The relative reinforcement associated with each option matters in terms of what is chosen. This is the study of choice and behavior analysts use the concurrent schedule of reinforcement, which arranges two or more schedules that are simultaneously available to begin understanding how choices are made. More on concurrent schedules and choice is discussed in Chapter 9.

Carl Cheney (an author of this textbook) and his colleagues examined foraging in the laboratory by using a concurrent schedule arrangement. In this arrangement, pigeons were able to choose between two food patches by pecking keys (Cheney, Bonem, & Bonem, 1985). Each food patches delivered food based on the concurrent progressive-ratio schedule using a two-key procedure (see "Progressive-Ratio Schedules" in this chapter; and see discussion of concurrent schedules in Chapter 9). As food reinforcers were obtained from one key, the density of food reinforcement on that key decreased, but more responses were required to produce bits of food—a progressively increasing ratio schedule (this simulates a depleting patch of food, or what happens when a patch is hunted or foraged). Concurrently (at the same time), the number of responses for each reinforcement decreased on the other key (repleting patch of food)—a progressively decreasing ratio schedule. As would be expected, this change in reinforcement density up and down generated switching back and forth between the two patches depending on response requirement and how much food was available; the pigeons preferred lower response requirements with higher amounts of food.

To change patches, however, the bird had to peck a center key under differing response requirements—simulating travel time and effort between patches (the side keys). Cheney and his colleagues found that the cost of hunting—represented by the increasing ratio schedule for pecking in a patch—resulted in pigeons changing patches and sampling more from the repleting patch. In addition, the effort (number of responses) required to change patches also played a role in changing patches; the higher the response requirement for changing, the less changing patches occurred; a lower response requirement increased the chances of sampling a different patch. This experiment depicts an animal model of foraging—using schedules of reinforcement to simulate natural contingencies operating in the wild.

In summary, schedules of reinforcement have main effects, which are the contingencies of behavior-reinforcement relations. They also have side effects, which produce unrequired, but reliable, consistent response patterns. These patterns are observed across different reinforcers, organisms, and operant responses. In our everyday experience, schedules of reinforcement are so common that we take such effects for granted. We wait for a taxi or Uber to arrive, line up at a store to have groceries scanned, or solve 10 math problems for homework. These common episodes of behavior and environment interaction illustrate schedules of reinforcement operating in our everyday lives. We can understand the main and side effects of schedules in everyday life by studying them in a laboratory.

SIMPLE SCHEDULES OF POSITIVE REINFORCEMENT

Continuous Reinforcement

Continuous reinforcement, or **CRF**, is probably the simplest schedule of reinforcement. On this schedule, every operant required by the contingency is reinforced. For example, every time a hungry pigeon pecks a key, food is presented. When every operant is followed by reinforcement, responses are emitted relatively quickly depending upon the time to consume the reinforcer. The organism continues to respond until it is satiated. Simply put, when the bird

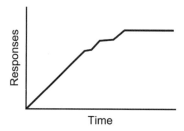

Fig. 5.4
Side-effect schedule patterning from a cumulative record is shown when behavior is placed under a continuous reinforcement schedule. Hatch marks indicating reinforcement are omitted since each response is reinforced. The flat portion of the record occurs when the animal stops making the response because of satiation.

is hungry (food deprived), it rapidly pecks the key and eats the food until it is full (satiated). If the animal is again deprived of reinforcement and exposed to a CRF schedule, the same pattern of responding followed by satiation is repeated. Figure 5.4 is a typical cumulative record of performance on continuous reinforcement. As mentioned in Chapter 4, the typical vending machine delivers products on a continuous (CRF) schedule. Having a conversation with a friend is another example in which each verbal response, such as "Hi, how are you today?" is likely reinforced with a verbal reinforcer, such as "I'm great! What's been happening with you?" and so forth.

Conjugate reinforcement is a type of CRF schedule in which properties of reinforcement, including the rate, amplitude, and intensity of reinforcement, are tied to particular dimensions of the response (see Weisberg & Rovee-Collier, 1998 for a discussion and examples). For example, loud, energetic, and high-rate operant crying by infants is often correlated with rapid, vigorous, and effortful caretaking (reinforcement) by parents. Basically, a repetitive "strong" response by the infant results in proportionally quick, "strong" caretaking (reinforcement) by the parents. Many repetitive behavior problems (stereotypy), such as head banging by atypically developing children, are automatically reinforced by perceptual and sensory effects (Lovaas, Newsom, & Hickman, 1987), in which high-rate, intense responding produces equally rapid, strong sensory reinforcement. This makes the behavior difficult to manage (see Rapp, 2008 for a brief review). Research with infants on conjugate schedules, involving leg thrusting for visual/auditory stimulation (e.g., stronger leg thrusts produce clearer image), has shown rapid acquisition with higher peak responding than on simple CRF schedules (Voltaire, Gewirtz, & Pelaez, 2005). Additional research has used college students responding to clarify fuzzy or pixilated pictures on a computer monitor; in this study, students' responding increased to enhance the clarity of a visual stimulus (a chosen image). In addition, when a schedule of negative punishment was implemented in which responses reduced the clarity of the images, under these conditions, rates of response decreased with conjugate negative punishment (MacAleese, Ghezzi, & Rapp, 2015).

CRF and Resistance to Extinction
Continuous reinforcement (CRF) generates weak resistance to extinction compared with intermittent reinforcement (Harper & McLean, 1992). Recall from Chapter 4 that *resistance to extinction* is a measure of persistence in behavior when reinforcement is discontinued. This

perseverance can be measured in several ways. The most obvious way to measure resistance to extinction is to count the number of responses and measure the length of time until behavior stabilizes under extinction. For example, consider a rat whose behavior was reinforced under a schedule of reinforcement that generates high rates of behavior. Then, extinction is placed in effect. At the beginning of extinction, the rat may press the lever with a rate of 20 times per minute with no explicit contingency of reinforcement in place. But once extinction is in effect for some time, measuring the time it takes and the number of responses made until a new pattern of stable behavior develops (much lower than early in the session, most likely) is the best gauge of resistance to extinction. It may take 30–50 minutes for stability to develop or longer (sometime, multiple sessions).

Although continuing extinction until low rates of behavior stabilize provides the best measure of behavioral persistence, this method requires considerable time and effort. Thus, arbitrary measures that take less time are usually used. Resistance to extinction may be *estimated* by counting the number of responses emitted over a fixed number of sessions. For example, after exposure to CRF, reinforcement could be discontinued and the number of responses made in three daily 1-h sessions are counted. Another index of resistance to extinction is based on how fast the rate of response declines during unreinforced sessions. The point at which no response occurs for 5 min may be used to index resistance. The number of responses and time taken to that point are used as indicators of behavioral persistence or resistance to extinction. The important criterion for any method is that it must be quantitatively related to extinction of responding.

Hearst (1961) compared resistance to extinction produced by CRF and intermittent schedules of reinforcement, which we will discuss next in this chapter. In this experiment, birds were trained on CRF and two schedules in which reinforcement for pecking a key was delivered intermittently. The number of extinction responses that the animals made during sessions of no reinforcement (extinction) was then counted. Basically, Hearst found that the birds made many more extinction responses after training on an intermittent schedule than after continuous reinforcement.

Response Stereotypy on CRF
On CRF schedules, the form or topography of response becomes stereotypical or invariant. In a classic study, Antonitis (1951) found that operant responses were repeated with very little change or variability in topography on a CRF schedule. In this study, rats were required to poke their noses anywhere along a 50-cm horizontal slot to get a food pellet (see Figure 5.5). Although not required by the contingency, the animals frequently responded at the same position on the slot. Only when the rats were placed on extinction did their responses become more variable. These findings are not limited to laboratory rats, and reflect a principle of behavior—reinforcement often narrows operant variability while extinction increases it; one might say that "failure creates innovation". More on this interesting idea is discussed in Chapter 4, "Focus On: Reinforcement and Problem Solving").

Under intermittent schedules, though, responses are less stereotyped. Research with pigeons suggests that response variability may be inversely related to the rate of reinforcement. In other words, as more responses are reinforced, less variation occurs in the operant. The reverse is also true: the less behavior is reinforced, the more variation in the operant occurs. In one study (Herrnstein, 1961a), pigeons' pecking was reinforced under an intermittent

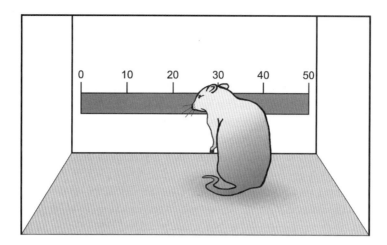

Fig. 5.5
The apparatus used by Antonitis (1951). Rats could poke their noses anywhere along the 50-cm horizontal slot to obtain reinforcement.

schedule. The birds pecked at a horizontal strip and pecks were occasionally reinforced with food. When some responses were reinforced, most of the birds pecked at the center of the strip—although they were not required to do so. During extinction, the animals made fewer responses to the center and more to other positions on the strip. Eckerman and Lanson (1969) replicated this finding in a subsequent study, also with pigeons. They varied the rate of reinforcement and compared where the pigeon pecked on a horizontal strip. Responses were stereotypical on CRF and became more variable when the birds were on extinction, but were also more varied under an intermittent schedule.

One interpretation of these findings is that organisms become more variable in their responding as reinforcement becomes less frequent or predictable. When a schedule of reinforcement is changed from CRF to an intermittent reinforcement, the rate of reinforcement declines and response variability increases. A further change in the rate of reinforcement occurs when extinction is started. In this case, the operant is no longer reinforced and response variation is maximal. The general principle appears to be "When things no longer work, try new ways of behaving." Or, as the saying goes, "If at first you don't succeed, try, try again." Though, this latter axiom is also related to resistance to extinction—the rate at which you keep trying has much to do with the kind of schedule—CRF vs. intermittent—that was previously maintaining the behavior.

When solving a problem, people usually use a solution that has worked in the past. When the usual solution does not work, most people—especially those with a history of reinforcement for response variability and novelty—try novel approaches to problem solving. Suppose that you are a camper who is trying to start a fire. Most of the time, you gather leaves and sticks, place them in a heap, strike a match, and start the fire. This time the fire does not start. What do you do? If you are like most of us, you try different ways to get the fire going, many of which may have worked in the past. You may change the kindling, add newspaper, use lighter fluid, or even build a shelter. Clearly, your behavior becomes more variable and inventive when reinforcement is withheld after a period of success. This increase in topographic variability during extinction after a period of reinforcement has been referred to as **resurgence** (Epstein, 1985), possibly contributing to the development of creative or original behavior on the one hand (Neuringer, 2009), and relapse of problem behavior during treatment on the other (Shahan & Sweeney, 2013).

In summary, CRF is the simplest schedule of positive reinforcement. On this schedule, every response produces reinforcement. Continuous reinforcement produces weak resistance to extinction and generates stereotypical response topographies. Resistance to extinction and variation in form of response both increase under intermittent schedules. Extinction can also induce variation in behavior.

INTERMITTENT SCHEDULES OF REINFORCEMENT: RATIO AND INTERVAL

Under intermittent schedules of reinforcement, some responses, as opposed to all of them, are reinforced. **Ratio schedules** are response based; that is, these schedules are set to deliver reinforcement following a set number of responses. The ratio specifies the number of responses required for reinforcement. **Interval schedules** produce reinforcement when *one* response is made after some amount of time has passed. Interval and ratio schedules may be fixed or variable. Fixed schedules mean a dimension is constant and stays the same across the session. Under variable schedules, a dimension varies across the session. Thus, there are four basic schedules—fixed ratio, variable ratio, fixed interval, and variable interval. In this section, we describe these four basic schedules of reinforcement (shown in Figure 5.6) and illustrate the typical side-effect response patterns that they produce. We also present an analysis of some of the reasons for the side effects produced by these basic schedules.

Ratio (Response-Based) Schedules

Fixed Ratio
A **fixed-ratio (FR)** schedule is programmed to deliver reinforcement after a fixed number of responses have been made. Continuous reinforcement (CRF) is defined as FR 1—the ratio is

		Reinforcement contingent on	
		Responses	Time*
Response/time requirement	Fixed	Fixed ratio (FR)	Fixed interval (FI)
	Variable	Variable ratio (VR)	Variable interval (VI)

*The first response after a given amount of time

Fig. 5.6
A table is shown of the four basic schedules of positive reinforcement.
Source: Adapted from C. B. Ferster, S. Culbertson, & M.C.P. Boren (1975). *Behavior principles*. Englewood Cliffs, NJ: Prentice Hall.

one reinforcer for one response. On fixed-ratio 25 (FR 25), 25 lever presses must be made before food is presented. This contingency is in effect across a session; therefore, an organism can earn many reinforcers in a session, depending on how many times they complete these 25-response operants.

An FR can assume any value. Of course, it is unlikely that very high values (say, FR 100,000,000) would ever be completed, but there are some reinforcers that are so potent that organisms will behave under FRs that require hundreds or even thousands of responses for small deliveries of reinforcement. Drugs in the opioid drug class, such as fentanyl, alfentanil, and remifentanil, are good examples (Ko et al., 2002). Not surprisingly, these drugs tend to also have high abuse liability with humans (see Jannetto et al., 2019 for discussion of this epidemic in the US and European Union). Indeed, the FR schedule is used to determine abuse liability for different types of drugs (see, for example, Hursh et al., 2005).

Ferster and Skinner (1957) first described the FR schedule and the characteristic effects, patterns, and rates, along with cumulative records of performance on about 15 other schedules of reinforcement. Their observations remain valid after literally thousands of replications: *FR schedules produce a rapid run of responses, followed by reinforcement, and then a pause in responding* (Ferster & Skinner, 1957). This is the break-and-run response described earlier and this is the FR side-effect schedule patterning (see Fig 5.7). The record looks somewhat like a set of stairs (except at very small FR values, as shown by Crossman, Trapp, Bonem, & Bonem, 1985). There is a steep period of responding (*run of responses*), followed by reinforcement (oblique line), and finally a flat portion (the pause)—a pattern known as **break and run**.

The flat parts of the cumulative record after the reinforcer is delivered are often called the **postreinforcement pause** (**PRP**), to indicate where it occurs. This pause is not required by the schedule, but is so predictable that it is regarded as a side effect of the schedule. One would think that the PRP occurs because the organism is consuming the food or because they become satiated and therefore not motivated to continue. That, however, is not correct. Skinner (1938) indicated that the length of the PRP depended on the preceding reinforcement

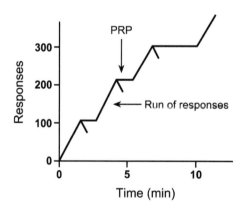

Fig. 5.7
A cumulative record is shown of a well-developed performance on an FR 100 schedule of reinforcement. The typical break-and-run pattern is presented. Reinforcement is indicated by the hatch marks. This is an idealized record that is typical of performance on many fixed-ratio schedules.

ratio, and called it a postreinforcement pause. He noted that on FR schedules one reinforcer never immediately follows another. Thus, the occurrence of reinforcement became a signal (or discriminative stimulus) for momentary extinction, and the animal paused. Subsequent research has shown that the moment of reinforcement contributes to the length of the PRP, but is not the only controlling variable (Schlinger, Derenne, & Baron, 2008).

Another general observation of the side-effect break-and-run pattern of the FR schedules reveals two interesting findings. One, as the ratio increases, the rate of response, which is indicated by the slope of the pattern, increases. This means that the rate is high (the higher the response rate, the steeper the slope). In addition, as the ratio requirement increases, longer PRPs also appear in the cumulative record. The larger the ratio, the larger the pause.

However, some ratios may be too high. Here, there may be almost no responding. If responding occurs at all, the animal responds at low rates. This is called *ratio strain* and is an indicator that the amount of effort required for reinforcement is too high to maintain responding. It is almost like the organism says, "I give up!"

These findings of FR size and the break-and-run side effects can provide clues about what actually controls the PRP. Detailed investigations of PRP on FR schedules indicate that the *upcoming ratio requirement* is perhaps a more critical determinant of the pause duration. Researchers found this when they alternated two different FR ratios within a session, each signaled by a different stimulus (called a multiple schedule). For example, in a classic study by Griffiths and Thompson (1973), an FR 30 was alternated with an FR 60 several times within a session. The pauses and rates that were generated during these FR schedules were predicted by the ratio of the upcoming schedule. In other words, when an FR 30 was completed (after the rat consumed the food pellet and the FR 60 was signaled by a stimulus) there was a long pause before the rat began pressing the lever under the FR 60. Indeed, it seems that whenever a larger ratio is signaled, the pause becomes longer. So, while calling this pause a "post"-reinforcement event accurately locates the pause, the upcoming requirements actually exert predominant control over the PRP. Thus, contemporary researchers often refer to the PRP as a **preratio pause** (e.g., Schlinger et al., 2008).

Conditioned reinforcers such as money, praise, and successful completion of a task also produce a pause when they are scheduled under fixed ratio. Consider what you might do if you had five sets of 10 math problems to complete for a homework assignment. A good bet is that you would solve 10 problems, and then take a break before starting on the next set. When constructing a sun deck, one of the authors bundled nails into lots of 50 each. This had an effect on the "nailing behavior" of friends who were helping to build the deck. The response pattern that developed was to put in 50 nails, then stop, take a drink, look over what has been accomplished, have a chat, and finally start nailing again. In other words, this simple scheduling of the nails generated a break-and-run pattern typical of FR reinforcement.

In industry, an FR schedule is often referred to as piece rate and the characteristic side effects can be observed on the job performances of the workers. For example, an artist may be commissioned to produce 10 or 50 pieces (FR 10 and FR 50, respectively) of art and will not get paid until they are complete. In some cases, the FRs, especially those with higher ratios can generate such large amounts of work that occur rapidly (i.e., a high rate of work), with only a pause before starting the next run, that it can exhaust a person or lead to lower-quality products, such as those produced in "fast fashion". The term "sweat shop" is also applied to piecemeal labor in which underprivileged people (usually women and children) are exploited with extremely low pay and poor working conditions. Though sweat shops are illegal in the US and other countries, they are still known to occur underground.

These examples of FR pausing suggest that the analysis of FR schedules has relevance for human behavior. We often talk about procrastination and people who put off or postpone doing things. It is likely that some of this delay in responding is similar to the pausing induced by the ratio schedule. A person who has a lot of upcoming work to complete (ratio size) may show a period of low or no productivity. Human procrastination may be modeled by animal performance on ratio schedules; translational research linking human productivity to animal performance on ratio schedules, however, has yet to be attempted (Schlinger et al., 2008).

In summary, fixed-ratio schedules have been used as a basis for understanding some of the dynamics of work and motivation. In addition to using FR schedules to manipulate the price (ratio size) of reinforcement to determine the value of reinforcers such as food or drugs of abuse, the FR schedule can be used to understand piecemeal work and procrastination.

Variable Ratio

Variable-ratio (VR) schedules are similar to fixed ratio except that the number of responses required for reinforcement is varied after each reinforcer is presented. A variable-ratio schedule is literally a series of fixed ratios with each FR of a different size. The *average number of responses* to reinforcement is used to define the VR schedule. A subject may press a lever for reinforcement 5 times, then 15, 7, 3, and 20 times. Adding these response requirements for a total of 50 and then dividing by the number of separate response runs (5) yields the schedule value, VR 10. The main effect or contingency of the schedule, then, is this average number of responses.

In general, ratio schedules produce a high rate of response. When VR and FR schedules are compared, responding is typically faster on VR. One reason for this is that pausing after reinforcement (PRP) is reduced or eliminated when the ratio contingency is changed from fixed to variable. This provides further evidence that the PRP does not occur because the animal is tired or is consuming the reinforcer (i.e., eating food). A rat or pigeon responding for food on VR does not pause as many times, or for as long, after reinforcement. When VR schedules are not excessive, PRPs do occur, although these pauses are typically much smaller than those generated by FR schedules (Mazur, 1983). Figure 5.8 portrays a typical side-effect

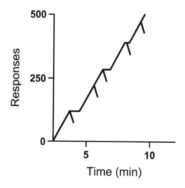

Fig. 5.8

A cumulative record is shown of the characteristic side effect responding on a VR schedule of reinforcement. Reinforcement is indicated by the hatch marks. Notice that PRPs are reduced or eliminated when compared with FR performance, resulting in a high, consistent pattern of response.

pattern of response on a VR schedule of positive reinforcement. The response is high and consistent with little, if any, pausing.

A VR schedule with a low mean ratio can contain some very small ratio requirements. For example, on a VR 10 schedule there cannot be many ratio requirements above 20 responses because, to offset those high ratios and average 10, there will have to be many very low ratios. It is the occasional occurrence of a reinforcer right after another reinforcer, *the short runs to reinforcement*, that reduces the likelihood of pausing on a VR schedule of reinforcement. Variable-ratio schedules with high mean ratios (e.g., VR 100) have some short ratios following one another, too, but typically generate longer PRPs because of the higher ratios.

Behavior maintained by VR schedules can be quite resistant to extinction. The change from VR reinforcement to extinction initially shows little or no change in rate of response. A pigeon on VR 110 shows a high steady rate of response (approximately 3 pecks per second). With the onset of extinction, the bird continues to respond at a similar high rate for about 3,000 responses, followed by a shift to a somewhat lower rate of response for 600 responses. The last part of the record shows long pausing and short bursts of responses at a rate similar to the original VR 110 performance. The pauses become longer and longer and eventually all responding stops, as it does on FR schedules (Ferster & Skinner, 1957, pp. 411–412).

Playing "hard to get" might be an example of a VR schedule. In this scenario, a potential love interest—Taylor—may only sometimes reinforce Joe's attempts at contacting them. For example, Taylor may answer Joe's every second, third, or fourth text message; Joe never knows which one it will be—only that it is inconsistent. Therefore, Joe's behavior of texting Taylor is reinforced under a VR 2 schedule. (This is frustrating by the way to Joe, but it indeed builds intrigue and perseverance.) Remember also that extinction under an intermittent schedule such as the VR is more resistant to extinction. This resistance can also be considered persistence. Taylor should remember, though, that if the ratio of texting back is too high (Taylor only responds after 7–10 of Joe's texts), this can induce ratio strain and may not maintain Joe's texting. (A word to Joe: Taylor may not be that into you. You are a valued person—consider removing yourself from Taylor's head games.) Joe might respond in kind by implementing extinction with this person's behavior—either signaled with one last text to say goodbye or with nothing at all—also called ghosting. (We think it is kinder to do the former, but understand there are conditions for the latter.)

Sometimes gambling is considered to be a VR schedule, as the moment-to-moment payoffs are varied; however, games of chance actually involve schedules that are referred to as random-ratio (RR) schedules of reinforcement. Research has shown that performance on RR schedules resembles that of a VR schedule, but these probabilistic schedules "lock you in" to high rates of response, as in gambling, by early runs of payoffs and by the pattern of unreinforced responses (Haw, 2008). The RR is different from the VR, however, in at least one important manner: with VR, each response gets the organism closer to the reinforcer delivery. With RR, each response has an equal probability of being reinforced. The distinction is made with the flipping of a coin with the hope of landing on heads. If you flip the coin and land on tails four times in a row, the probability is still 0.5 that on the fifth time you will land on heads. The previous four times does not change the probability or "get you any closer" to landing on heads; each response has an equal probability of reinforcement.

In everyday life, variability and probability are routine. Thus, ratio schedules involving probabilistic payoffs are more common than strict VR or FR contingencies from the laboratory. You may have to hit one nail three times to drive it in, and the next nail may take six swings of the hammer. It may, on average, take 70 casts with a fly rod to catch a trout, but any one strike is probabilistic. In baseball, the batting average reflects the player's schedule of reinforcement. A batter with a .300 average gets 3 hits for 10 times at bat on average, but nothing guarantees a hit for any particular time at bat. The schedule depends on a complex interplay among conditions set by the pitcher and the skill of the batter.

Interval (or Time-Based) Schedules

Fixed Interval

On **fixed-interval** (**FI**) schedules, a single response is reinforced after a fixed amount of time has passed. For example, on a fixed-interval 90-s schedule (FI 90 s), one bar press after 90 s results in reinforcement. Following reinforcement, another 90-s period goes into effect, and after this time has passed another response will produce reinforcement. It is important to note that responses made before the time period has elapsed do not produce reinforcement. Therefore, the main effect or contingency of the schedule is one response after 90-s elapses. [Note: there is a schedule called *fixed time* (*FT*) in which reinforcement is delivered without a response following a set, or fixed, length of time. This is also referred to as a *response-independent schedule*. Unless otherwise specified, one should always assume that a response is required on whatever schedule is in effect.]

When organisms are exposed to interval contingencies, and they have no way of "telling time" (i.e., using a stimulus that shows how much time has passed), they typically produce many more responses than the schedule requires. For the side-effect schedule patterning, fixed-interval schedules produce a pattern of responding that accelerates throughout the interval. First, there is a pause after reinforcement (PRP), then a few probe responses (which are extinguished), followed by a more rapid rate of response that is constant and high rate by the time the interval elapses. This accelerating side-effect pattern of response is called **scalloping**. Figure 5.9 is an idealized cumulative record of FI performance. Each interreinforcement interval (IRI) can be divided into three distinct classes—the PRP, followed by a period of a gradually increasing rate, and finally a high terminal rate of responding. Remember, though, that only one response is required for the reinforcer (as long as it occurs after the interval elapses)—the direct effect. Under the FI, you get these three characteristic side-effect patterns instead.

Suppose that you have volunteered to be in an operant experiment. You are brought into a small room, and on one wall there is a lever with a cup under it. Other than those objects, the room is empty. You are not allowed to keep your watch while in the room, and you are told, "Do anything you want." After some time, you press the lever to see what it does. Ten dollars fall into the cup. A good prediction is that you will press the lever again. You are not told this, but the schedule is FI 5 min. You have 1 h per day to work on the schedule. If you collect all 12 (60 min ÷ 5 min = 12) of the scheduled reinforcers, you can make $120 a day.

Assume you have been in this experiment for 3 months. Immediately after collecting a $10 reinforcer, there is no chance that a response will pay off (discriminated extinction). But, as you are standing around or doing anything else, the interval is timing out. You check out the contingency by making a probe response (you guess the time might be up). The next response occurs more quickly because even more time has passed. As the interval continues to time out, the probability of reinforcement increases and your responses are made faster and faster.

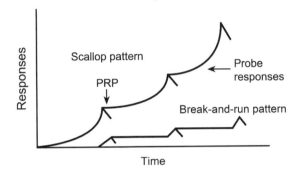

Fig. 5.9
A cumulative record of a fixed-interval schedule. Fixed-interval schedules usually produce a side-effect pattern that is called scalloping. There is a PRP following reinforcement, then a gradual increase in rate of response (acceleration) to the moment of reinforcement. A break-and-run pattern, though not as common, could also develop, as shown in the second curve.

This pattern of responding is described by the scallop shown in Figure 5.9 and is typical for FI schedules (Ferster & Skinner, 1957).

Following considerable experience with FI 5 min, you may get very good at judging the time period. In this case, you would wait out the interval and then emit a burst of responses. Perhaps you begin to pace back and forth during the session, and you find out that after 250 steps the interval has almost elapsed. This kind of *mediating behavior* may develop after experience with FI schedules (Muller, Crow, & Cheney, 1979). Other animals sometimes behave in a similar way and occasionally produce a *break-and-run* pattern of responding, similar to FR schedules (Ferster & Skinner, 1957).

In everyday life, FI schedules are arranged when people set timetables for trains and buses. Next time you are at a bus stop, take a look at what people do while they are waiting for the next bus. If a bus has just departed, people stand around and perhaps talk to each other for a while. Then, the operant of "looking for the bus" begins at a low rate of response. As the interval times out, the rate of looking for the bus increases and most passengers are now looking for the arrival of the next bus. The passengers' behavior approximates the scalloping pattern we have described in this section. These are not perfect examples of the FI scallop, however, as "looking up" does not contingently make the bus appear (though, it helps ensure that you will not miss boarding the bus when it comes). Schedules of reinforcement are a pervasive aspect of human behavior, but we seldom recognize the effects of these contingencies.

FOCUS ON: GENERALITY OF SCHEDULE EFFECTS

The **assumption of generality** implies that the effects of contingencies of reinforcement extend over species, reinforcement, and behavior. For example, a fixed-interval schedule is expected to produce the scalloping pattern for a pigeon pecking a key for food, and for a child solving math problems for teacher approval.

Fergus Lowe (1979) conducted numerous studies of FI performance with humans who pressed a button to obtain points later exchanged for money. Figure 5.10 shows typical performances on fixed-interval schedules by a rat and two human subjects. Building on

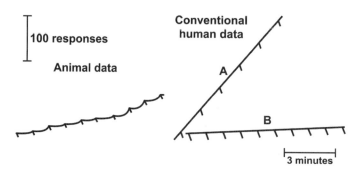

Fig. 5.10
Typical animal performances on FI schedules are shown along with the high- and low-rate performance usually seen with adult humans. The first cumulative record, labeled "Animal data" shows the FI scallop. Two additional records are labeled "Conventional human data". Pattern A shows a high consistent rate of responding and pattern B shows a low, slow rate of responding.

Source: Adapted from C. F. Lowe (1979). *Reinforcement and the organization of behavior.* Wiley: New York, p. 162. Copyright 1979 held by C. F. Lowe. Published with permission.

research by Harold Weiner (1969), Lowe argued that animals show the characteristic scalloping pattern, and humans generally do not. Humans often produce one of two patterns—an inefficient steady, high-rate of response or an efficient low-rate, break-and-run performance. (Note that they rarely, if ever, produce the main effect of the schedule, which is one response after a fixed period of time.) Experiments by Lowe and his colleagues focused on the conditions that produce the high- or low-rate patterns in humans.

Schedule performance in humans also reflects the influence of language. In conditioning experiments, people generate some verbal rule and proceed to behave according to the self-generated rule rather than the experimentally arranged FI contingencies. This is an example of rule-governed behavior, which we will explore later in Chapter 12. In most cases, people who follow self-generated rules (even if the rule is wrong) still satisfy the requirements of the schedule, obtain reinforcement, and continue to follow the rule. For example, one person may say, "I should press the button fast," and another says, "I should count to 50 and then press the button." Only when the contingencies are arranged so that self-generated rules conflict with programmed reinforcement do people reluctantly abandon the rule and behave in accord with the contingencies (Baron & Galizio, 1983). Humans also naturally find it easy to follow a self-instruction or rule and effortful to reject it (Harris, Sheth, & Cohen, 2007).

One implication of Lowe's analysis is that humans without language skills would show characteristic effects of schedules. Lowe et al. (1983) designed an experiment to show typical FI performance by children less than 1 year old. Figure 5.11 shows an infant (Jon) seated in a highchair and able to touch a round metal cylinder. Touching the cylinder produced a small bit of food (pieces of fruit, bread, or candy) on FI schedules of reinforcement. A second infant, Ann, was given 4 s of music played from a variety of music boxes on the same schedules. Both infants produced FI scalloping response patterns similar to the rat's performance in Figure 5.10. Thus, infants who are not verbally skilled behave in accord with the FI side effect. There is no doubt that humans become more verbal as they grow up; however, many other changes occur from infancy to adulthood. A possible confounding factor is the greater experience that adults compared to infants have with ratio-type contingencies of reinforcement. Infants rely on the

Fig. 5.11
Infant Jon in study of FI schedules had to touch the metal cylinder to receive small snack items like pieces of fruit, bread, and candy. The order of the FI values for Jon were 20, 30, 10, and 50 s.

Source: From C. Fergus Lowe, A. Beasty, & R. P. Benthall (1983). The role of verbal behavior in human learning: Infant performance on fixed-interval schedules. *Journal of the Experimental Analysis of Behavior, 39*, pp. 157–164. Reproduced with permission and copyright 1983 held by the Society for the Experimental Analysis of Behavior.

caregiving of other people. This means that most of an infant's reinforcement is delivered on the basis of time and behavior (interval schedules). A baby is fed when the parent has time to do so, although fussing may decrease the interval. As children get older, they begin to crawl and walk and reinforcement is delivered more and more on the basis of their own behavior (ratio schedules). When this happens, many of the contingencies of reinforcement change from interval to ratio schedules. The amount of experience with ratio-type schedules of reinforcement may contribute to the differences between adult human and animal/infant performance on fixed-interval schedules.

In fact, research by Wanchisen, Tatham, and Mooney (1989) has shown that rats perform like adult humans on FI schedules after a history of ratio reinforcement. The animals were exposed to variable-ratio (VR) reinforcement schedule and then were given 120 sessions on a fixed-interval 30-s schedule (FI 30 s). Two patterns of response developed on the FI schedule—a high-rate pattern with little pausing and a low-rate pattern with some break-and-run performance. These patterns of performance are remarkably similar to the schedule performance of adult humans (see Figure 5.10). One implication is that human performance on schedules may be explained by a special history of ratio-like reinforcement rather than self-generated rules. At this time, it is reasonable to conclude that both reinforcement history and verbal ability contribute to FI performance of adult humans (see Bradshaw & Reed, 2012 for appropriate human performance on random-ratio (RR) and random-interval (RI) schedules only for those who could verbally state the contingencies).

Variable Interval

On a **variable-interval** (**VI**) schedule, responses are reinforced after a variable amount of time has passed (see Figure 5.16). For example, under a variable interval 30-s (VI 30-s) schedule, consider that a program is set to deliver reinforcers 50 times within a session, but the time between reinforcer deliveries varies. For example, the first reinforcer may set up after 3 s passes; that is, the first bar-press a rat makes after 3 s produces the reinforcer. The second reinforcer may be delivered after a response occurs after 45 s, and so forth. If you average the 50 intervals across the session, the average variable ratio for reinforcement is 30 s. Notice that the time to each reinforcer changes in a manner that is unpredictable from moment to moment, but the *average* time is 30 s, and is therefore overall predictable. The *average amount of time* required for reinforcement is used to define the schedule.

Interval contingencies are common in the ordinary world of people and other animals. For example, people stand in line, sit in traffic jams, wait for elevators, time a boiling egg, and are put on hold. In everyday life, variable time periods occur more frequently than fixed ones. Waiting in line to get to a bank teller may take 5 min one day and 15 min the next time you go to the bank. A wolf pack may run down prey following a long or short hunt. A baby may cry for 5 s, 15 sec, or 2 min before a parent picks up the child. A cat waits varying amounts of time in ambush before a bird becomes a meal. Waiting for a bus is rarely reinforced on a fixed schedule, despite the efforts of transportation officials. The bus arrives around an average specified time and waits only a given time before leaving. A carpool is an example of a VI contingency with a limited hold. The car arrives more or less at a specified time, but waits for a rider only a limited (and usually brief) time. In the laboratory, this **limited-hold** contingency, where the reinforcer is only available for a set time after a variable interval, is in place. This limited hold increases the rate of responding by reinforcing short interresponse times (IRTs refer to the pauses between responses). In the case of the carpool, people on the VI schedule with limited hold are ready for pick-up and rush out of the door when the car arrives.

The main effect or contingency of the VI schedule is that a single response after a varied period of time produces the reinforcer. Only one response is required for reinforcement, as long as the time period has elapsed. However, consistent with the other schedules, the side-effect schedule patterning is what is more interesting. Figure 5.12 portrays the VI side-effect pattern of response generated on a VI schedule. On this schedule, rate of response is low to moderate and steady. The pause after reinforcement that occurs on FI does not usually appear in the VI record. Notably, this steady rate of response (as opposed to an extinction burst) is maintained during extinction. Ferster and Skinner (1957) described the cumulative record of a pigeon's performance for the transition from VI 7 min to extinction. The bird maintains a moderately stable rate (1.25 to 1.5 pecks per second) for approximately 8,000 responses. After this,

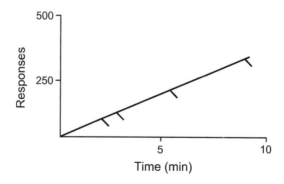

Fig. 5.12
Idealized cumulative pattern of response produced by a variable-interval schedule of reinforcement.

the rate of response continuously declines to the end of the record. Generally, VI response rates initially continue to be moderate and stable on extinction, showing an overall large output of behavior (resistance to extinction). Because the rate of response remains steady and moderate, VI performance is often used as a *baseline for evaluating other independent variables,* such as drug effects on behavior. Rate of response under VI schedules may increase or decrease as a result of experimental manipulations. For example, tranquilizing drugs such as chlorpromazine decrease the rate of response on VI schedules (Waller, 1961), while stimulants increase VI performance (Segal, 1962). Murray Sidman has commented on the usefulness of VI performance as a baseline:

> An ideal baseline would be one in which there is as little interference as possible from other variables. There should be a minimal number of factors tending to oppose any shift in behavior that might result from experimental manipulation. A variable-interval schedule, if skillfully programmed, comes close to meeting this requirement.
> (Sidman, 1960, p. 320)

In summary, VI contingencies are common in everyday life. The side-effect pattern is a moderate steady rate of response; the steadiness is resistant to extinction. Because of this characteristic pattern, VI performance is frequently used as a baseline to assess the effects of other variables, especially performance-altering drugs.

NOTE ON: VI SCHEDULES, REINFORCEMENT RATE, AND BEHAVIORAL MOMENTUM

Behavioral momentum refers to behavior that persists or continues following a change in environmental conditions (Nevin, 1992; Nevin & Grace, 2000; see PRE effect and behavioral momentum in Chapter 4). Furthermore, response rate declines more slowly relative to its baseline level in the presence of an S^D for high-density (compared to low-density) reinforcement (Shull, Gaynor, & Grimer, 2002). When you are working at the computer on a report, and keep working even though you are called to dinner, your behavioral persistence indicates behavioral momentum. Also, if you continue messaging on social media despite alternative sources of reinforcement (e.g., watching a favorite streamed series), that too shows behavioral momentum. In the classroom, students with a higher rate of reinforcement (correct answers) for solving math problems are less likely to be distracted by the sights and sounds outside the classroom window than other students with a lower rate of reinforcement for problem solving.

At the basic research level, Nevin (1974) used a multiple schedule of reinforcement to investigate behavioral momentum. The multiple schedule arranged two separate VI reinforcement components, each signaled by a distinct discriminative stimulus (S^D) and separated by a third darkened component, so it was easy to discriminate which VI schedule was in place. Rates of responding were higher in the richer VI component. But, when free food was provided in the third darkened component (disruption), responding decreased less in the VI condition with the higher rate of reinforcement. Thus, behavior in the component with the rich VI schedule (high rate of reinforcement) showed increased momentum. It continued to keep going despite the disruption by free food (see also Cohn, 1998; Lattal, Reilly, & Kohn, 1998).

Another study by John Nevin and associates compared the resistance to change (momentum) of behavior maintained on ratio and interval schedules of reinforcement (Nevin, Grace, Holland, & McLean, 2001). Pigeons pecked keys on a multiple schedule of random-ratio (RR) random-interval (RI) reinforcement to test relative resistance to change. On this multiple schedule, a distinctive discriminative stimulus (S^D) signaled each component schedule, either RR or RI, and the researchers ensured that the reinforcement rates for the RR component were equated with those of the RI segment. Disruptions by free feeding (allowing the rats to eat freely without restriction of schedules) between components, extinction, and pre-feeding (before the session) were investigated. The findings indicated that, with similar obtained rates of reinforcement, the *interval schedule is more resistant to change, and has higher momentum, than performance on ratio schedules*. Notice that resistance to change is exactly the opposite to the findings for rate of response on these schedules—ratio schedules maintain higher rates of response than interval schedules.

Researchers also use behavioral momentum theory to evaluate the long-term effects of reinforcement programs on targeted responses challenged by disruptions (Wacker et al., 2011). In one applied study, two individuals with severe developmental disabilities performed self-paced discriminations on a computer using a touch-sensitive screen and food reinforcement (Dube & McIlvane, 2001). Responses on two separate problems were differentially reinforced. On-task behavior with the higher reinforcement rate showed more resistance to change due to pre-feeding, free snacks, or alternative activities. Thus, the disruptive factors reduced task performance depending on the prior rates of on-task reinforcement. When performance on a task received a high rate of reinforcement, it was relatively impervious to distraction compared with performance maintained on a lower rate schedule. One applied implication is that children with attention deficit hyperactivity disorder (ADHD) who are easily distracted in the school classroom may suffer from low rates of reinforcement for on-task behavior and benefit more from a change in rates of reinforcement than from administration of stimulant medications with potential adverse effects.

Basic Schedules and Biofeedback

The major independent variable in operant conditioning is the program for delivering consequences, called the schedule of reinforcement. Regardless of the species, the shape of the response curve for a given schedule often approximates a predictable form. Fixed-interval scallops, fixed-ratio break and run, and other patterns were observed in a variety of organisms and were highly uniform and regular (see exceptions in "Focus On: Generality of Schedule Effects" in this chapter). The predictability of schedule effects has been extended to the phenomenon of biofeedback and the apparent willful control of physiological processes and bodily states.

Biofeedback usually is viewed as conscious, intentional control of bodily functions, such as brainwaves, heart rate, blood pressure, temperature, headaches, and migraines—using instruments that provide information or feedback about the ongoing activity of these systems. An alternative view is that biofeedback involves operant responses of bodily systems regulated by consequences, producing orderly changes related to the schedule of "feedback."

Early research showed schedule effects of feedback on heart rate (Hatch, 1980) and blood pressure (Gamble & Elder, 1990). Subsequently, behavioral researchers investigated five different schedules of feedback on forearm-muscle tension (Cohen, Richardson, Klebez, Febbo, & Tucker, 2001). The study involved 33 undergraduate students who were given extra class credit and a chance to win a $20 lottery at the end. Three electromyogram (EMG) electrodes were attached to the underside of the forearm to measure electrical activity produced by muscles while participants squeezed an exercise ball. They were instructed to contract their arm "in a certain way" to activate a tone and light; thus, their job was to produce the most tone/light presentations they could. Participants were randomly assigned to groups that differed in the schedule of feedback (tone/light presentations) for EMG electrical responses. Four basic schedules of feedback (FR, VR, FI, or VI) were programmed, plus CRF and extinction. Ordinarily, in basic animal research, sessions are run with the same schedule until some standard of stability is reached. In this applied experiment, however, 15-min sessions were conducted on three consecutive days with a 15-min extinction session added at the end.

Cumulative records were not collected to depict response patterns, presumably because the length and number of sessions did not allow for stable response patterns to develop. Instead, researchers focused on rate of EMG activation as the basic measure. As might be expected, ratio schedules (FR and VR) produced higher rates of EMG electrical responses than interval contingencies (FI or VI). Additionally, the VI and VR schedules showed the most resistance to extinction (see "Note On: VI Schedules, Reinforcement Rate, and Behavioral Momentum" in this chapter). CRF produced the most sustained EMG responding, while FR and VR schedules engendered more muscle pumping action of the exercise ball.

The EMG electrical responses used in this study were sensitive to the schedule of feedback, indicating the operant function of electrical activity in the forearm muscles. Together with studies of biofeedback and responses of the autonomic nervous system, the Cohen et al. (2001) experiment shows that responses of the somatic nervous system also are under tight operant control of the schedule of reinforcement (feedback). Further detailed analyses of biofeedback schedules on physiological responses clearly are warranted, but have been lacking in recent years. In this regard, we recommend the use of steady-state, single-subject designs that vary the interval or ratio schedule value over a wide range to help clarify how schedules of feedback regulate seemingly automatic bodily activity.

PROGRESSIVE-RATIO SCHEDULES

On a **progressive-ratio (PR) schedule** of reinforcement, the ratio requirements for reinforcement are increased systematically, typically after each reinforcer (Hodos, 1961). In an experiment, the first response requirement for reinforcement might be set at a small ratio value such as 5 responses. Once the animal emits 5 responses resulting in reinforcement, the next ratio requirement might increase by 10 responses (the step size). Now reinforcement occurs only after the animal has pressed the lever 15 times, followed by ratio requirements of 25 responses, 35, 45, 55, and so on (adding 10 responses on each step). The increasing ratios (5, 15, 25, 35, and so on) are the progression and give the schedule its name. At some point in the progression of ratios, the ratio becomes too high and the reinforcer is not potent enough to sustain the behavior; the animal stops responding. This is called ratio strain. The highest ratio value completed on the PR schedule is designated the **breakpoint**.

The type of progression on a PR schedule may be arithmetic, as when the difference between two ratio requirements is a constant value such as 10 responses. Another kind of progression is geometric, as when each ratio after the first is found by multiplying the previous one by a fixed number. A geometric progressive ratio might be 2, 6, 18, and so on, where 3 is the fixed value. The type of progression (arithmetic or geometric) is an important determinant of behavior on PR schedules. In one study of PR schedules, Peter Killeen and associates found that response rates on arithmetic and geometric PR schedules increased as the ratio requirement progressed and then at some point decreased (Killeen, Posadas-Sanchez, Johansen, & Thrailkill, 2009). Response rates maintained on arithmetic PR schedules decreased in a linear manner—as the ratio size increased, there was a linear decrease in response rates. Response rates on geometric PR schedules, however, showed a negative deceleration toward a low and stable response rate—as ratio size increased geometrically, response rates rapidly declined and then leveled off. Thus, the relationship between response rates and ratio requirements of the PR schedule depends on the type of progression—arithmetic or geometric. These relationships can be described by mathematical equations, and this is an ongoing area of research (Killeen et al., 2009).

Progressive-Ratio Schedules and Neuroscience

The PR schedule has also been used in applied research (Roane, 2008). Most of the applied research on PR schedules uses the giving-up or breakpoint as a way of measuring **reinforcement efficacy** or effectiveness, especially of drugs like cocaine. The breakpoint for a drug indicates how much operant behavior the drug will maintain at a given dose. For example, a rat might self-administer morphine on a PR schedule as the dose size is varied and breakpoints are determined for each dose size. It is also possible to determine the breakpoints for different kinds of drugs (e.g., stimulants or opioids), assessing the drugs' relative reinforcement effectiveness. In these tests, it is important to recognize that the time allotted to complete the ratio (e.g., 120 min) and the progression of the PR schedule (progression of ratio sizes) have an impact on the breakpoints—potentially limiting conclusions about which drugs are more "addictive" and how the breakpoint varies with increases in drug dose (dose–response curve).

Progressive-ratio schedules allow researchers to assess the reinforcing effects of drugs. Another way to talk about the reinforcing effects of a drug is abuse potential or drug abuse liability. While many illicit drugs, such as cocaine or heroin, have abuse liability, so do some prescribed drugs, such as opioids (e.g., oxycodone or oxycontin) that are prescribed for analgesia (pain control). A drug prescribed to control hyperactivity might have abuse liability in some conditions—an effect recommending against its use. The drug Ritalin® (methylphenidate) is commonly used to treat attention deficit hyperactivity disorder (ADHD) and is chemically related to Dexedrine® (d-amphetamine). Amphetamine is a drug of abuse, as are other stimulants, such as cocaine. Thus, people who take methylphenidate without a prescription might develop addictive behavior similar to behavior maintained by amphetamine. In one study, human volunteers with stimulant-use disorders were used to study the reinforcing efficacy of three doses of methylphenidate (16, 32, and 48 mg) and d-amphetamine (8, 16, and 24 mg), including a placebo (Stoops, Glaser, Fillmore, & Rush, 2004). For the sake of reference, a typical adult dose of methylphenidate for ADHD is 20–30 mg. The reinforcing efficacy of the drug was assessed by a modified PR schedule (over days) where participants had to press a key on a computer (50, 100, 200, 400, 800, 1,600, 3,200, and 6,400 times) to earn capsules of the drug; completion of the ratio requirement resulted in oral self-administration of the drug. Additional monetary contingencies were arranged to ensure continued participation in the study. As shown in Figure 5.13, the results indicated that the number of responses to the breakpoint increased at the intermediate dose of methylphenidate and d-amphetamine compared with the placebo control. Thus, at intermediate doses methylphenidate is similar in reinforcement effectiveness to d-amphetamine. One conclusion is that using Ritalin® to treat ADHD may be contraindicated for people who have a history of stimulant use disorder due to the potential for abuse (see Stoops, 2008 for a review of the relative reinforcing effects of

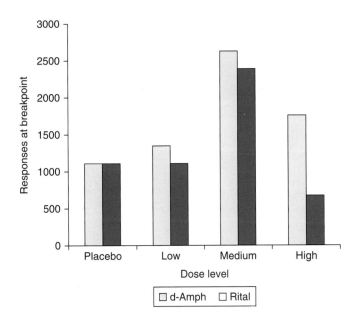

Fig. 5.13
Breakpoints produced by 10 volunteers with stimulant use disorders who self-administered low, medium, and high doses of two stimulants, Ritalin® (methylphenidate) and d-amphetamine, as well as a placebo control. The medium doses of each drug result in statistically higher breakpoints for the drugs compared to the placebo condition, but the two drugs do not reliably differ from each other.

Source: Based on data for average breakpoints in W. W. Stoops, P. E. A. Glaser, M. T. Fillmore, & C. R. Rush (2004). Reinforcing, subject-rated, performance and physiological effects of methylphenidate and d-amphetamine in stimulant abusing humans. *Journal of Psychopharmacology, 18*, pp. 534–543, 538.

stimulants in humans on PR schedules; also Bolin, Reynolds, Stoops, & Rush, 2013 provide an assessment of d-amphetamine self-administration on PR schedules, including verbal ratings of drug effects).

In another context, PR schedules have been used to study the reinforcement efficacy of palatable food on overeating and obesity. Leptin is a hormone mostly produced in the adipocytes (fat cells) of the white adipose (fat) tissue. The basic function of leptin is to signal when to stop eating—counteracting neuropeptides that stimulate feeding. Based on a genetic variation, the ob/ob (obese-prone) mouse is deficient in sensitivity to the leptin signal—they overeat and gain excessive body weight compared with the lean-prone littermates of the same strain (see Figure 5.14). Generally, overeating and obesity vary with genotype (obese-prone vs. lean-prone) of these rodents.

Researchers have investigated the reinforcing properties of food with other rodent strains such as these. For example, Rasmussen and Huskinson (2008) compared genetic variation in the reinforcing value of food in the Zucker rat strain. Obese-prone Zucker (two *fa* alleles) and lean-prone Zucker (at least one *FA* allele) rats were trained to lever-press for flavored sucrose pellets. Next, the rats were tested on a PR schedule, using a geometric progression of ratio values. The highest ratio completed was the breakpoint. After establishing the PR baselines, both obese-prone and lean-prone rats were administered low and high doses of an anorectic drug (rimonabant) and given more sessions on the PR schedule. This drug is important because it blocks the reinforcing properties of food by blocking the CB1

Fig. 5.14
Photograph is shown of the ob/ob obese-prone mouse and lean-prone littermate. The ob/ob genotype has a deficiency in leptin production that results in obesity when food is freely available.
Source: Public access photo.

endocannabinoid receptor. The results for breakpoints showed that under control and placebo conditions, obese-prone Zucker rats had higher breakpoints for sucrose compared to lean-prone Zuckers. The reinforcing efficacy of palatable food, then, was higher for obese-prone rats than lean-prone rats based solely on their genotypes. Also, rimonabant reduced breakpoints in a dose-dependent manner, but the obese-prone rats were more sensitive to the drug—that is, a lower dose of the drug (1 mg/kg) reduced breakpoints, but it took a dose that was three times larger to reduce breakpoints in the lean-prone group. This difference in drug sensitivity suggests that something is neurochemically different with obese-prone Zucker rats, in terms of endocannabinoid activity. One possibility is that obese Zuckers have higher amounts of endocannabinoids in their brains and bodies than lean Zuckers (DiMarzo, et al., 2001); this may explain not only why they eat more (since endocannbinoids are implicated in food reward), but why they have a greater sensitivity to drugs that work on this neurotransmitter system.

SCHEDULE PERFORMANCE IN TRANSITION

We have described typical performances generated by different schedules of reinforcement. The patterns of response on these schedules take a relatively long time to develop. Once behavior has stabilized, showing little change from day to day, the organism's behavior is said to have reached a *steady state*. The break-and-run pattern that develops on FR schedules is a steady-state performance and is only observed after an animal has considerable exposure to the contingencies. Similarly, the steady-state performance generated on other intermittent schedules takes time to develop. When an organism is initially placed on any schedule of reinforcement, typical behavior patterns are not consistent or regular. This early performance on a schedule is called a **transition state**. Transition states are the periods between initial steady-state performance and the next steady state (see Sidman, 1960 for steady-state and transition-state analysis).

Consider how you might get an animal to press a lever 100 times for each presentation of food (FR 100). First, you shape the animal to press the bar on CRF (see Chapter 4). After some arbitrary steady-state performance is established on CRF, you are faced with the problem of how to program the steps from CRF to FR 100. Notice that in this transition there is a large shift or step in the ratio of reinforcement to bar pressing. This problem has been studied using a *progressive-ratio schedule*, as we described earlier in this chapter. The ratio of responses following each run to reinforcement is programmed to increase in steps. Stafford and Branch (1998) employed the PR schedule to investigate the behavioral effects of step size and criteria for stability. If you simply move from CRF to the large FR value, the animal will probably show **ratio strain** in the sense that it pauses longer and longer after reinforcement. One reason is that the time between successive reinforcements contributes to the postreinforcement pause (PRP). The pause gets longer as the **interreinforcement interval** (IRI, or time between reinforcement) increases. Because the PRP makes up part of the IRI and is controlled by it, the animal eventually stops responding. Thus, there is a negative feedback loop between increasing PRP length and the time between reinforcers in the shift from CRF to the large FR schedule.

Transitions from one schedule to another play an important role in human development. Developmental psychologists have described periods of life in which major changes in behavior typically occur. One of the most important life stages in Western society is the transition from childhood to adolescence. Although this phase involves many biological and behavioral processes, one of the most basic changes involves schedules of reinforcement.

When a child reaches puberty, parents, teachers, peers, and others begin requiring more behavior and more skillful performance than they did during childhood. A young child's reinforcement schedules are usually simple, regular, and immediate. In childhood, food is given when the child tells a caregiver "I'm hungry" after playing a game or is scheduled at regular times throughout the day. On the other hand, a teenager is told to fix their own food and clean up the mess. Notice that the schedule requirement for getting food has significantly increased. The teenager may search through the refrigerator, open packages and cans, sometimes cook, get out plates, eat the food, and clean up. Of course, any part of this sequence may or may not occur depending on the disciplinary practices of the parents. Although most adolescents adapt to this transition state, others may show signs of *ratio strain* and extinction. Poor eating habits among teenagers, such as an overreliance on fast food or already-prepared meals, may reflect the change from regular to intermittent reinforcement.

Many other behavioral changes may occur during the transition from childhood to adolescence. Ferster, Culbertson, and Boren (1975) noted the transition to intermittent reinforcement that occurs in adolescence, such as getting a job so that independent transportation (e.g., a car, insurance, and fuel) is attainable. The workload for school also increases and includes longer term papers and presentations, as well as learning more complex subjects. This workload is even higher if one plans to prepare for college. If a student is an athlete, the time and energy used to develop athletic skills, and attend practices and games also increases.

There are other periods of life in which our culture demands large shifts in schedules of reinforcement. A current problem involves a rapidly aging population and the difficulties generated by forced or elected retirement. In terms of schedules, retirement is a large and rapid change in the contingencies of reinforcement. Retired people face significant alterations in social, monetary, and work-related consequences. For example, a professor who has enjoyed an academic career is no longer reinforced for research and teaching by the university community. Social consequences for these activities may have included approval by colleagues, academic advancement and income, the interest of students, and intellectual discussions. Upon retirement, the rate of social reinforcement is reduced or completely eliminated. It is, therefore, not surprising that retirement is an unhappy time of life for some people. Although retirement is commonly viewed as a problem of old age, a behavior analysis points to the abrupt change in rates and sources of reinforcement (Skinner & Vaughan, 1983).

ON THE APPLIED SIDE: SCHEDULES AND CIGARETTES

As we have seen, the use of drugs is operant behavior maintained in part by the reinforcing effects of the drug. One implication of this analysis is that reinforcement of an incompatible response (i.e., abstinence) can reduce the probability of taking drugs. The effectiveness of an abstinence contingency depends on the magnitude and schedule of reinforcement for nondrug use (e.g., Davis et al., 2016; Higgins, Bickel, & Hughes, 1994).

In applied behavior analysis, contingency management involves the systematic use of reinforcement to establish desired behavior and the withholding of reinforcement or punishment of undesired behavior (Higgins & Petry, 1999). An example of contingency management is seen in a study using reinforcement schedules to reduce cigarette smoking. Roll, Higgins, and Badger (1996) assessed the effectiveness of three different schedules of reinforcement for promoting and sustaining drug abstinence. These researchers conducted an experimental analysis of cigarette smoking because cigarettes can function as reinforcement, smoking can be reduced by reinforcement of alternative responses, and it is relatively more convenient to study cigarette smoking than illicit drugs. Furthermore, cigarette smokers usually relapse within several days following abstinence. This suggests that reinforcement factors regulating abstinence exert their effects shortly after the person stops smoking and it is possible to study these factors in a short-duration experiment.

Sixty adults, who smoked between 10 and 50 cigarettes a day, took part in the experiment. The smokers were not currently trying to give up cigarettes. Participants were randomly assigned to one of three groups: progressive reinforcement, fixed rate of reinforcement, and a control group. They were told to begin abstaining from cigarettes on Friday evening so that they could pass a carbon monoxide (CO) test for abstinence on Monday morning. Each person in the study went for at least 2 days without smoking before reinforcement for abstinence began. On Monday through Friday, participants agreed to take three daily CO tests. These tests could detect prior smoking.

Twenty participants were randomly assigned to the progressive reinforcement group. The progressive schedule involved increasing the magnitude of reinforcement for remaining drug free. Participants earned $3.00 for passing the first carbon monoxide test for abstinence. Each subsequent consecutive CO sample that indicated abstinence increased the amount of money participants received by $0.50. The third consecutive CO test passed earned a bonus of $10.00. To further clarify, passing the first CO test yielded $3.00, passing the second test yielded $3.50, passing the third test yielded $14.00 ($4.00 and bonus of $10.00), and passing the fourth test yielded $4.50. In addition, a substantial response cost was added for failing a CO test. If the person failed the test, the payment for that test was withheld and the value of payment for the next test was reset to $3.00. Three consecutive CO tests indicating abstinence following a reset returned the payment schedule to the value at which the reset occurred (Roll et al., 1996, p. 497), supporting efforts to achieve abstinence.

Participants in the fixed reinforcement group (N = 20) were paid $9.80 for passing each CO test. There were no bonus points for consecutive abstinences and no resets. The total amount of money available for the progressive and fixed groups was the same. Smokers in both the progressive and fixed groups were informed in advance of the schedule of payment and the criterion for reinforcement. The schedule of payment for the control group was the same as the average payment obtained by the first 10 participants assigned to the progressive condition. For these people, the payment was given no matter what their carbon monoxide levels were. The control group was, however, asked to try to cut their cigarette consumption, reduce CO levels, and maintain abstinence.

Smokers in the progressive and fixed reinforcement groups passed more than 80% of the abstinence tests, while the control group only passed about 40% of the tests. The effects of the schedule of reinforcement are shown in Figure 5.15(a). The figure indicates the percentage of participants who passed three consecutive tests for abstinence and then resumed smoking over the 5 days of the experiment. Only 22% of those on the progressive schedule resumed smoking, compared with 60% and 82% in the fixed and control groups, respectively. Thus, the progressive schedule of reinforcement was superior in terms of preventing the resumption of smoking (after a period of abstinence).

Figure 5.15(b) shows the percentage of smokers who gave up cigarettes throughout the experiment. Again, a strong effect of schedule of reinforcement is apparent. Around 50% of those on the progressive reinforcement schedule remained abstinent for the 5 days of the experiment, compared with 30% and 5% of the fixed and control participants, respectively.

In a subsequent experiment, Roll and Higgins (2000) found that a progressive reinforcement schedule with a response–cost contingency increased abstinence from cigarette use compared

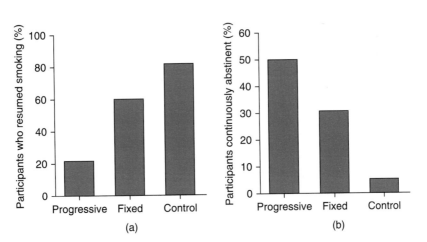

Figure shows the percentage of participants in each group who obtained three consecutive drug-free tests, but then resumed smoking (a). Also shown is the percentage of smokers in each group who were abstinent on all trials during the entire experiment (b).

Source: From J. M. Roll, S. T. Higgins, & G. J. Badger (1996). An experimental comparison of three different schedules of reinforcement of drug abstinence using cigarette smoking as an exemplar. *Journal of Applied Behavior Analysis*, 29, pp. 495–505. Published with permission from John Wiley & Sons, Ltd.

with a progressive schedule without the response cost or a fixed incentive-value schedule. Overall, these results indicate that a progressive reinforcement schedule, combined with an escalating response cost, is an effective short-term intervention for abstinence from smoking. Further research is necessary to see whether a progressive schedule maintains abstinence after the schedule is withdrawn. Long-term follow-up studies of progressive and other schedules are necessary to assess the lasting effects of reinforcement schedules on abstinence. What is clear, at this point, is that schedules of reinforcement may be an important component of smoking cessation programs (see more on contingency management in Chapter 13).

ADVANCED SECTION: SCHEDULE PERFORMANCE

Each of the basic schedules of reinforcement (FR, FI, VR, and VI) generates a unique pattern of responding. Ratio schedules produce a higher rate of response than interval schedules. A reliable pause after reinforcement (PRP) occurs on fixed-ratio and fixed-interval schedules, but not on variable-ratio or variable-interval schedules.

Rate of Response on Schedules

Some, though not all, of the determinants of rate differences on ratio vs. interval schedules have been resolved. One issue concerns molecular vs. molar determinants of schedule control. **Molecular accounts of schedule performance** focus on small moment-to-moment relations between behavior and its consequences. One important feature of a molecular account of behavior is the **interresponse time (IRT)** or the time or pause that occurs between successive responses. On the other hand, **molar accounts of schedule performance** are concerned with general, more large-scale factors that occur over the length of an entire session, such as the overall rate of reinforcement and the relation between response rate and reinforcement rate (called the feedback function).

Molecular Account of Rate Differences
The time between any two responses, or what is called the *interresponse time* (IRT), may be treated as an operant. Technically, IRTs are units of time and cannot be reinforced. The behavior, or pause, between any two responses is measured indirectly as IRT, and it is this behavior that contributes to reinforcement. Consider Figure 5.16, in which 30-s segments of performance on VR and VI schedules are presented. Responses are portrayed by the vertical marks, and the occurrence of reinforcement is denoted by the familiar symbol S^{r+}. As you can see, IRTs are much longer on VI than on VR. On the VR segment, 23 responses occur in 30 s, which gives an average time between responses, or IRT, of 1.3 s. The VI schedule generates longer IRTs with a mean of 2.3 s; longer IRTs, then, slow down response rate.

Generally, ratio schedules produce shorter IRTs, and consequently higher rates of response, than interval schedules. Skinner (1938) suggested that this came about because ratio and interval schedules reinforce short and long interresponse times, respectively. To understand this, consider the definition of an operant class. It is a class of behavior that may increase or decrease in frequency on the basis of contingencies of reinforcement. In other words, if it could

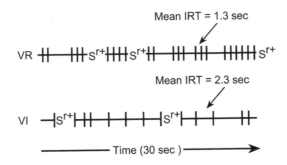

Fig. 5.16
Idealized distributions of response on VR and VI schedules of reinforcement. Responses are represented by the vertical marks, and S^r+ stands for reinforcement.

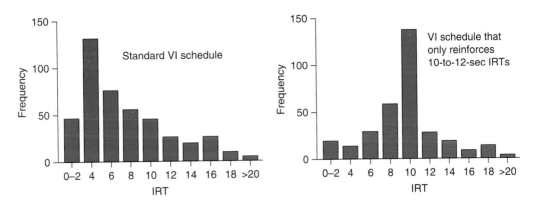

Fig. 5.17
Hypothetical distributions are shown of interresponse times (IRTs) for an animal responding on a standard VI schedule of reinforcement and on a VI that only reinforces IRTs that fall between 10 and 12 s.

be shown that the time between responses changes as a function of reinforcement, then the IRT is by definition an operant in its own right. To demonstrate that the IRT is an operant, it is necessary to identify an IRT of specific length (e.g., 2 s between any two responses) and then reinforce that interresponse time (really, you are reinforcing a response, and a pause of at least 2 s before the next response occurs), showing that it increases in frequency.

Computers and other electronic equipment have been used to measure the IRTs generated on various schedules of reinforcement. A response is made and the computer starts timing until the next response is emitted. Typically, these IRTs are slotted into time bins. For example, all IRTs between 0 and 2 s are counted, followed by those that fall in the 2- to 4-s range, and then the number of 4- to 6-s IRTs. This method results in a distribution of interresponse times. Several experiments have shown that the distribution of IRTs may in fact be changed by selectively reinforcing interresponse times of a particular duration (for a review, see Morse, 1966). Figure 5.17 shows the results of a hypothetical experiment in which IRTs of different duration are reinforced on a VI schedule. On the standard VI, most of the IRTs are 2–4 s long. When an additional contingency is added to the VI schedule that requires longer IRTs, for example, IRTs of 10–12 s in duration, the IRTs indeed increase. Also, a new distribution of IRTs is generated. Whereas on a VR schedule, the next response may be reinforced regardless of the IRT, on the VI schedule, the combination "response plus pause plus response" is required for reinforcement.

As an operant, the IRT is considered to be a property of the response that ends the time interval between any two responses. This response property may be increased by reinforcement. For example, a rat may press a lever (response) five consecutive times: R_1, R_2, R_3, R_4, and R_5. The time between lever presses R_1 and R_2 is the IRT associated with R_2. In a similar fashion, the IRT for R_5 is the elapsed time between R_4 and R_5. This series can be said to constitute a homogeneous chain.

In a classic experiment by Anger (1956), animals were placed on a VI 300-s schedule of reinforcement (Anger, 1956). On this schedule, the response that resulted in reinforcement had to occur 40 s or more after the previous response. If the animal made any faster-paced responses with IRTs of less than 40 s, the schedule requirements would not be met. In other words, only IRTs of more than 40 s were the operant that was reinforced. Anger found that this procedure shifted the distribution of IRTs toward 40 s. Thus, the IRT that is reinforced is more likely to be emitted than other IRTs.

Ratio schedules are different in terms of IRTs. Ratio schedules generate rapid sequences of responses with short IRTs (Gott & Weiss, 1972; Weiss & Gott, 1972). On a ratio schedule, consider what the probability of reinforcement is following a burst of very fast responses (short IRTs) or a series of responses with long IRTs. Recall that ratio schedules are based on the number of responses that are emitted—each response gets the organism closer to the delivery of reinforcement. Bursts of responses with short IRTs rapidly count down the ratio requirement and are more likely to be reinforced than sets of long IRT responses, in which slow responding (longer IRTs) is reinforced. Thus, ratio schedules, because of the way they are constructed, differentially reinforce short IRTs. According to the molecular IRT view of schedule control, this is why the rate of response is high on ratio schedules.

When compared with ratio schedules, interval contingencies generate longer IRTs and consequently a lower rate of response. Interval schedules pay off after some amount of time has passed and a response is made. As the IRTs become longer, the probability of reinforcement increases, as more and more of the time requirement on the schedule elapses. In other words, longer IRTs are differentially reinforced on interval schedules (Morse, 1966). In keeping with the molecular view, interval contingencies differentially reinforce long IRTs, and the rate of response is moderate on these schedules.

Molar Accounts of Rate Differences
Molar explanations of rate differences are concerned with the overall relation between responses and reinforcement. In molar terms, the *correlation between responses and reinforcement or feedback function* produces the difference in rate on interval and ratio schedules. Generally, if a high rate of response is correlated with a high rate of reinforcement in the long run, animals will respond rapidly. When an increased rate of response does not affect the rate of reinforcement, organisms do not respond faster (Baum, 1993).

Consider a VR 100 schedule of reinforcement. On this schedule, a subject could respond 50 times per minute and in a 1-h session obtain 30 reinforcements. On the other hand, if the rate of response now increases to 300 responses per minute (not outside the range of pigeons or humans), the rate of reinforcement would increase to 180 an hour. According to supporters of the molar view, this correlation between increasing rate of response and increased rate of reinforcement is responsible for rapid responding on ratio schedules.

A different correlation between rate of response and rate of reinforcement is set up on interval schedules. Recall that interval schedules program reinforcement after time has passed and one response is made. Suppose you are responding on a VI 3-min schedule for $5 as reinforcement. You have 1 h a day to work on the schedule. If you respond at a reasonable rate, say 30 lever presses per minute, you will get most or all of the 20 payouts (60 min of work on a VI 3 min will give you 20 opportunities for reinforcement). Now pretend that you increase your rate of response to 300 lever presses a minute. The only consequence is a sore wrist, and the rate of reinforcement remains at 20 per hour. In other words, you waste responses when you needlessly increase the response rate (and it may hurt!), so it does not pay to increase the rate of response on interval schedules—hence low to moderate response rates are maintained on interval schedules.

Postreinforcement (Preratio) Pause on Fixed Schedules

Fixed-ratio and fixed-interval schedules generate a pause that follows reinforcement. Accounts of pausing on fixed schedules also may be characterized as molecular and molar. Molecular accounts of pausing are concerned with the moment-to-moment relations that immediately precede reinforcement. Such accounts address the relation between the number of bar presses that produce reinforcement and the subsequent postreinforcement pause (or really, preratio pause) (PRP). In contrast, molar accounts of pausing focus on the overall rate of reinforcement and the average pause length for a session.

Research shows that the PRP is a function of the *interreinforcement interval* (*IRI*). As the IRI becomes longer, the PRP increases. On FI schedules, in which the experimenter controls the time between reinforcement, the PRP is approximately half of the IRI. For example, on an FI 300-s schedule (in which the time between reinforcements is 300 s), the average PRP will be 150 s. On FR schedules, the evidence suggests similar control by the IRI—the PRP becomes longer as the ratio requirement increases (Powell, 1968).

There is, however, a difficulty with analyzing the PRP on FR schedules. On ratio schedules, the IRI is partly determined by what the animal does. Thus, the animal's rate of pressing the lever affects the time between reinforcements. Another problem with ratio schedules for an analysis of pausing is that the rate of response goes up as the size of the ratio is increased (Boren, 1961). Unless the rate of response exactly coincides with changes in the size of the ratio, adjustments in ratio size alter the IRI. For example, on FR 10 a rate of 5 responses per minute produces an IRI of 2 min. This same rate of response produces an IRI of 4 min on an FR 20 schedule. Thus, the ratio size, the IRI, or both may cause changes in PRP.

Molar Interpretation of Pausing

We have noted that the average PRP is half of the IRI. Another finding is that the PRPs are normally distributed (bell-shaped curve) over the time between reinforcements. In other words, on an FI 320-s schedule, pauses will range from 0 to 320 s, with an average pause of around 160 s. As shown in Figure 5.18, these results can be accounted for by considering what would happen if the normal curve moved upward so that the mean pause was 225 s. In this case, many of the pauses would exceed the FI interval and the animal would get fewer reinforcers

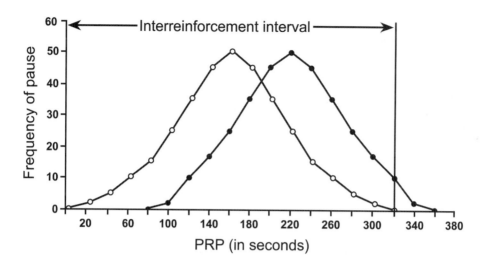

Fig. 5.18
The figure shows two possible distributions of PRPs on a fixed-interval 320-s schedule. The distribution given by the open circles has a mean of 160 s and does not exceed the interreinforcement interval (IRI) set on the FI schedule. The bell curve for the distribution with the dark circles has an average value at 225 s, and many pauses exceed the IRI.

for the session. An animal that was sensitive to *overall rate of reinforcement* or the long-range payoffs (also called maximization; see Chapter 9) should come to emit pauses that are on average half the FI interval, assuming a normal distribution. Thus, maximization of reinforcement provides a molar account of the PRP (Baum, 2002).

Molecular Interpretations of Pausing
There are two molecular accounts of pausing on fixed schedules that have some degree of research support. One account is based on the observation that animals often emit other behavior during the PRP (Staddon & Simmelhag, 1971). For example, rats may engage in grooming, sniffing, scratching, and stretching after the presentation of a food pellet. Because this other behavior reliably follows reinforcement, we may say it is induced by the schedule. *Schedule-induced* behaviors (see Chapter 6) may be viewed as operants that automatically produce reinforcement. For example, stretching may relieve muscle tension, and scratching may eliminate an itch. One interpretation is that pausing occurs because the animal is maximizing local rates of reinforcement. Basically, the rat gets food for bar pressing as well as the automatic reinforcement from the induced activities (see Shull, 1979). The average pause should therefore reflect the allocation of time to induced behavior and to the operant that produces scheduled reinforcement (food). At present, experiments have not ruled out or clearly demonstrated the induced-behavior interpretation of pausing (e.g., Derenne & Baron, 2002).

A second molecular account of pausing is based on the **run of responses** or amount of work that precedes reinforcement (Shull, 1979, pp. 217–218). This "work-time" interpretation holds that the previously experienced run of responses regulates the length of the PRP. Work time affects the PRP by altering the value of the next scheduled reinforcement. In other words, the more effort or time expended for the previous reinforcer, the lower the value of the next reinforcer and the longer it takes for the animal to initiate responding (pause length). This view suggests that the higher the response rate for reinforcement, the less valuable the next reinforcement is, and therefore the longer it takes to start working again.

Neither the induced-behavior nor the work-time accounts of pausing are sufficient to explain all that is known about patterning on schedules of reinforcement. A schedule of reinforcement is a procedure for combining a large number of different conditions that regulate behavior. Some of the controlling factors arise from the animal's behavior and the experimenter sets others via the programmed contingencies. This means that it is exceedingly difficult to unravel the exact processes that produce characteristic schedule performance. Nonetheless, the current interpretations of pausing point to some of the more relevant factors that regulate behavior on fixed schedules of reinforcement.

CHAPTER SUMMARY

A schedule of reinforcement describes the arrangement of discriminative stimuli, operants, and consequences. Such contingencies are central to the understanding of behavior regulation in humans and other animals. The research on schedules and performance patterns is a major component of the science of behavior, a science that progressively builds on previous experiments and theoretical analysis. Schedules of reinforcement have main, or direct effects (the actual behavioral contingencies of the reinforcement schedule) and the characteristic side-effect behavior patterns that are not necessary for reinforcement, but occur regardless. These side effects are best observed with cumulative records. Schedules of reinforcement generate consistent, steady-state performances involving runs of responses and pausing that are characteristic of the specific schedule (ratio or interval). In the laboratory, the arrangement of progressive-ratio schedules can help us assess the reinforcing efficacy of a particular stimulus like drugs of abuse or food. Intermittent reinforcement plays a role in most human behavior, especially social interaction.

In this chapter, we described continuous reinforcement (CRF) and resistance to extinction on this schedule. CRF also results in response stereotypy based on the high rate of reinforcement. Fixed-ratio (FR) and fixed-interval (FI) schedules were introduced, as well as the postreinforcement pausing (or preratio) (PRP) on these contingencies. Adult humans have not shown the side-effect pattern classic scalloping or break-and-run patterns on FI schedules, and the performance differences of humans relate to language or verbal behavior as well as histories of ratio reinforcement. Variable-ratio (VR) and variable-interval (VI) schedules produce less pausing and higher overall rates of response. Adding a limited hold to a VI schedule increases the response rate by reinforcing short interresponse times (IRTs). When rates of reinforcement are varied on VI schedules, the higher the rate of reinforcement the greater the behavioral momentum.

The study of behavior during the transition between schedules of reinforcement has not been well researched, due to the boundary problem of steady-state behavior. Transition states, however, play an important role in human behavior—as in the shift in the reinforcement contingencies from childhood to adolescence or the change in schedules from employment to retirement. Reinforcement schedules also have applied importance, and research shows that cigarette smoking can be regulated by a progressive schedule combined with an escalating response–cost contingency. Finally, in the Advanced Section of this chapter, we addressed molecular and molar accounts of response rate and rate differences on schedules of reinforcement. We emphasized the analysis of IRTs for molecular accounts, and the correlation of overall rates of response and reinforcement for molar explanations.

Key Words

Assumption of generality
Break and run
Breakpoint
Continuous reinforcement (CRF)
Cumulative record
Fixed interval (FI)
Fixed ratio (FR)
Interreinforcement interval (IRI)
Interresponse time (IRT)
Interval schedules
Limited hold
Main (or direct) effect of schedule
Molar account of schedule performance
Molecular account of schedule performance
Postreinforcement pause (PRP)

Preratio pause
Progressive-ratio (PR) schedule
Ratio schedules
Ratio strain
Reinforcement efficacy
Resurgence
Run of responses
Scalloping
Schedule of reinforcement
Side-effect schedule pattern
Steady-state performance
Transition state
Variable interval (VI)
Variable ratio (VR)

On the Web

www.thefuntheory.com/ Control of human behavior by programming for fun (called Fun Theory) is shown in these short videos; schedules of reinforcement (fun) are arranged for seatbelt use, physical activity, and cleaning up litter. See if you can think up new ways to use reinforcement schedules in programming fun to regulate important forms of human behavior in our culture.

www.youtube.com/watch?v=l_ctJqjlrHA This YouTube video discusses basic schedules of reinforcement, and B. F. Skinner comments on variable-ratio schedules, gambling, and the belief in free will.

www.pigeon.psy.tufts.edu/eam/eam2.html This module is available for purchase and demonstrates basic schedules of reinforcement as employed in a variety of operant and discrimination procedures involving animals and humans.

http://opensiuc.lib.siu.edu/cgi/viewcontent.cgi?article=1255&context=tpr&sei-redir=1-search="conjugateschedulereinforcement" A review of the impact of Ferster and Skinner's publication of *Schedules of Reinforcement* (Ferster & Skinner, 1957), from the study of basic schedules to the operant analysis of choice, behavioral pharmacology, and microeconomics of gambling. Contingency detection and causal reasoning by infants, children, and adults are addressed as areas influenced by schedules of reinforcement.

www.wadsworth.com/psychology_d/templates/student_resources/0534633609_sniffy2/sniffy/download.htm If you want to try out shaping and basic schedules with Sniffy the virtual rat, go to this site and use a free download for 2 weeks of fun. After this period, you will have to pay to continue your investigation of operant conditioning and schedules of reinforcement.

Brief Quiz

1. Schedules of reinforcement were first described by:
   a. Charles Ferster
   b. Francis Mechner
   c. B. F. Skinner
   d. Fergus Lowe
2. Infrequent reinforcement generates responding that is persistent. What is this called?
   a. postreinforcement pause
   b. partial reinforcement effect
   c. molar maximizing
   d. intermittent resistance
3. The main or direct effect of an FR 25 schedule is:
   a. break and run
   b. scallop
   c. 25 responses
   d. one first response after 25 s
4. Resurgence happens when:
   a. behavior is put on extinction
   b. reinforcement magnitude is doubled
   c. high-probability behavior persists
   d. response variability declines
5. Schedules that generate predictable break-and-run "stair-step" patterns are:
   a. fixed interval
   b. fixed ratio
   c. variable ratio
   d. random ratio
6. Variable-ratio schedules generate:
   a. postreinforcement pauses
   b. locked rates
   c. break-and-run performance
   d. high rates of response
7. Schedules that combine time and one response are called:
   a. partial reinforcement schedules
   b. complex schedules
   c. interval schedules
   d. fixed-time schedules
8. The side-effect shape of the response pattern generated by an FI is called a:
   a. scallop
   b. ogive
   c. break and pause
   d. accelerating dynamic

9 Human performance on FI differs from animal data due to:
   a   intelligence differences
   b   self-instruction
   c   contingency effects
   d   alternative strategies

10 Behavior is said to be in transition when it is between:
   a   the PRP and IRI
   b   stable states
   c   one schedule and another
   d   a response run

Answers to Brief Quiz: 1, c (p. 149); 2, b (p. 153); 3, c (p. 159); 4, a (p. 159); 5, b (p. 160); 6, d (p. 162); 7, c (p. 159); 8, a (p. 164); 9, b (p. 166); 10, b (p. 174).

Aversive Control of Behavior

1. Distinguish between positive and negative punishment.
2. Investigate negative reinforcement as the basis of escape and avoidance.
3. Discover how reduction in aversive stimuli regulates avoidance.
4. Inquire about learned helplessness induced by inescapable aversive stimuli.
5. Distinguish between displaced and counter-control aggression.
6. Learn about coercion and its negative side effects in our society.

Aversive stimuli are events or happenings that organisms escape from, evade, or avoid. Insect stings, physical attacks, foul odors, bright light, and very loud noises are common events that organisms are prepared to evade on the basis of phylogeny; that is, they are natural aversive stimuli. Escaping or avoiding these **primary aversive stimuli** was adaptive, presumably because those animals, which acted to remove or prevent contact with these events, more often survived and reproduced. In other words, organisms do not learn how to react to some aversive stimuli; they are biologically prepared to avoid or escape such events.

Other stimuli acquire aversive properties when associated with primary aversive events during an animal's lifetime. For people, **conditioned aversive stimuli** (S^{ave}) include verbal threats, public criticism, a failing grade, a frown, and verbal disapproval. To affect behavior, these events usually depend on a history of punishment. A 1-week-old infant is not affected by disapproval such as a headshake. By the time the child is 2 years old, however, the gesture may stop the toddler from tearing pages out of your favorite book. Animals also learn responses to conditioned stimuli as aversive events. People commonly shout "No!" when pets misbehave, and this auditory stimulus eventually reduces the probability of the response it follows (e.g., chewing on your new chair).

DOI: 10.4324/9781003202622-6

AVERSIVE CONTROL IN EVERYDAY LIFE

Aversive Control, Elephants, and Bees

Elephants are said to run away from mice, but research indicates that they are more likely to escape from the sounds of African bees (King, Douglas-Hamilton, & Vollrath, 2011). The "buzz" of bees is conditioned as an aversive stimulus when followed by bee stings inside the elephant's trunk—an apparently extremely painful event, with swelling that can last for weeks. In Kenya, farmers and elephants are often in conflict over crops that elephants raid and destroy. Rumors among game wardens suggested that elephants avoid trees with beehives, leading King and her colleagues to test the behavioral effects of a 4-min recording of bee sounds with 17 herds of elephants in Kenya's Buffalo Springs and Samburo National Reserves. The "buzzing" worked as 16 of the 17 herds took off running, and one herd even ran across a river to get away (Figure 6.1). On average, the elephants moved 64 m away from the speakers when

Fig. 6.1

A herd of African elephants is shown. These herds often invade the crops of farmers, eating the crops and destroying the property. Sounds of bees and the presence of beehives keep the elephants away, based on the elephants' conditioning history involving bee stings to the inside of the trunks.

Source: Shutterstock.

"buzzing" sounds were played, but only 20 m when the sound of random white noise was played. The equipment for playing bee sounds is too expensive (think of the extension cords that would be needed) for farmers. Beehives with real bees, however, are a feasible alternative that also provides farmers with extra food and income from the honey. The scientists placed beehives every 10 m along a 1700-m "beehive fence" on farms in northern Kenya, which were usually protected only by thorn-bush barriers. Over a period of 2 years, only one elephant broke through the beehive fence, compared with 31 invasions through only the thorny barriers. The evidence suggests that bees and their "buzz" are a deterrent for elephants approaching a space, presumably as a result of a conditioning history of bee stings and social learning.

Aversive Control of Human Behavior

Elephants' escape from, and avoidance of, bees illustrates that a large amount of animal behavior may be regulated by naturally occurring aversive stimuli. Humans also extensively use and arrange aversive stimuli to control the behavior of others at the individual, societal, and institutional levels.

In the physical world, punishment is a fact of life. With regard to the social world, Sidman (2001) has documented our excessive reliance on coercion to control human behavior. The excessive use and advocacy of punishment by some groups is illustrated by the beating of children as a form of discipline. In 2010, CBS News reported the beating to death of 7-year-old Lydia Schatz by her adopted parents. The beatings of Lydia and her 11-year-old sister, who recovered from her injuries, ironically took place in Paradise, California. The report stated:

> **CHICO, Calif. (CBS/KOVR)** Three years ago, Kevin Schatz and his wife Elizabeth did something so noble, a local television station featured them; the pair decided to adopt three children from Liberia. Now, they're accused of killing one of the children because she mispronounced a word.... Prosecutors say that the California couple used quarter-inch plastic tubing to beat their seven-year-old adopted daughter to death. Apparently, they got the idea from a fundamentalist Christian group, which promotes this as a way of training children to be obedient. Butte County District Attorney Mike Ramsey says for several hours the 7-year-old was held down by Elizabeth and beaten dozens of times by Kevin on the back of her body, which caused massive tissue damage.
>
> (Martinez, 2010; CBS News)

Subsequently, the couple pleaded guilty and Kevin Schatz was sentenced to 22 years in prison for murder, torture, and misdemeanor cruelty to a child. Elizabeth Schatz received a sentence of 13 years and 4 months for voluntary manslaughter, infliction of unlawful corporal punishment, and misdemeanor cruelty to a child (full story on CNN video; Tuckman, 2011). Sadly, this is only one story of many. In the USA in 2019, the Centers for Disease Control and Prevention report that 1 in 7 children have experienced child maltreatment (mostly neglect and physical abuse). That same year 1,840 children died (that's five children per day), and it is likely that this kind of abuse is underreported. During the early part of the COVID-19 pandemic, while emergency room visits related to child abuse decreased, the rates of hospitalization related to child abuse increased. The added stressors of school closures, remote learning, loss of income, and social isolation likely contributed to this (Swedo et al., 2020).

The use of punishment and aversive control in child rearing, unfortunately, has historically been a pervasive practice in American culture, and is well documented (see Gershoff, 2002 and Straus, 2001 on the use of corporal punishment in America; also Park, 2002 describes

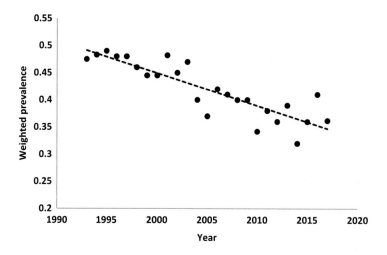

Fig. 6.2
The prevalence of parents in the USA who use spanking across the last 30 years. Use of spanking has decreased in recent years.

Source: Data based on a study by Mehus, C. J. & Patrick, M. E. (2021). Prevalence of spanking in US national samples of 35-year-old parents from 1993 to 2017. *Journal of the American Medical Association of Pediatrics*. 2021;175(1):92–93. doi:10.1001/jamapediatrics.2020.2197

the difficulty of isolating the effects of parental punishment from a "package" of disciplinary tactics). The use of physical punishment and specifically spanking, however, has declined quite a bit the last 30 years as a disciplinary tactic as more research on its harmful effects has emerged (Mehus & Patrick, 2021). Figure 6.2 shows this decline.

Aversive Control and Domestic Violence
In America, aversive control and punishment not only occur in parent–child relationships, but also are prevalent aspects of domestic violence between couples (see Figure 6.3). Domestic violence typically involves a person using physically coercive tactics that escalate in intensity to control the behavior of the victim. These manners of controlling others are illegal and harmful. Someone who engages in these types of behavior should seek therapy to learn prosocial ways to problem solve, communicate, and interact with their loved ones. Unfortunately, this does not happen often enough.

A person never deserves to be controlled or punished through violent or emotionally abusive means. Therefore, leaving the relationship is a reasonable behavioral option when domestic violence is present. However, many victims stay in abusive relationships (Miller, Lund, & Weatherly, 2012). It is important, then, to understand the contingencies of why a victim may choose to stay with, or leave, an abusive domestic partner. Research shows that the dynamics of domestic violence, also called the "cycle of violence," operate to increase the effectiveness of reinforcement for staying in the relationship while also increasing the punishment for leaving. This complexity may make it difficult for someone to leave an abusive situation.

Let us consider the conditions that might increase the odds of staying in an abusive relationship that are relevant to the abuser. Following an abusive episode, there may be a

Fig. 6.3
Domestic violence is an extreme form of aversive control.

"honeymoon" or period of reconciling and remorseful behavior on the part of the perpetrator. A behavioral analysis suggests that this erratic shift in behavior from abuse to intense affection and sorrow serves as intermittent reinforcement for staying in the relationship and increases the chances that staying becomes resistant to extinction—increasing the probability that the victim will choose to stay in the abusive relationship. Staying in the abusive relationship also may be maintained by negative reinforcement, which arises from a pattern of mounting conflict and tension followed by a "cooling off" period with reduction in the immediate threat (for more on negative reinforcement and conditioned reinforcement in the violence cycle, see Miller et al., 2012). Thus, intermittent positive reinforcement and negative reinforcement operate to establish and maintain the victim's staying in the abusive relationship.

Victims of domestic violence often report that the abuse did not happen suddenly overnight and the perpetrator was previously kind, courteous, and thoughtful. In these scenarios, abuse may begin with less severe forms of verbal abuse (e.g., name-calling, pushing) that may escalate in severity to slapping or striking, and finally to very severe forms, such as punching, burning, raping, or torturing the victim. One implication here is that the less intense forms of aversive control by the abuser may initially be effective in reducing behavior (e.g., getting the victim to comply with their demands), but as the victim is exposed, more severe forms are necessary for compliance. This represents one of the side effects of aversive stimuli that will be discussed later in the chapter. In addition, these severe forms may no longer be functional for just getting compliance—they may be used to harm the victim and cause suffering.

Although it may be hard to understand the reasons (positive and negative reinforcers) for staying in an abusive relationship, one must also consider the response costs for alternative behavior, such as leaving the relationship. These costs may make leaving the relationship difficult. Costs of leaving would include effort and ability to obtain food, shelter, and security;

this may be especially difficult when children live in the home. There are also "social costs" imposed by potential judgments and behavior of family, friends, or clergy. There is also fear about potential harm that could be caused by pursuit and intimidation by the abuser. Overall, domestic violence involves a complex interplay of behavioral contingencies both within and outside of the abusive relationship. Behavioral interventions often focus on changing the conditions that influence the victim's stay/leave behavior by altering contingencies and changing motivating operations (Miller et al., 2012). Sadly, the perpetrator's behavior is less often the focus of behavioral change in these situations; this would be an important area for behavior analysts to examine.

CONTINGENCIES OF PUNISHMENT

Given the acceptance and pervasiveness of punishment in our society, it is notable that today there is very little basic research on aversive control and how best to minimize its side effects and reduce its prevalence (Catania, 2008). Most studies of the basic principles of punishment (and negative reinforcement) were conducted in the 1950s to early 1980s, at which point studies of punishment almost stopped. A major reason for this was the 1985 amendment to the USDA's Animal Welfare Act—federal guidelines that had implications for the use of experiments with animals. This amendment outlined a number of protections for animals, one of which was that ethics committees at institutions that use animals in research (such as universities or medical schools) *must* evaluate research protocols before research can be conducted to ensure any animal pain or distress that occurs is justifiable, minimized, or mitigated (consider, for example, animal testing in treatments for cancer). There have also been changes in cultural values for the use of animals in research, including the animal rights activist movement. These two historical events made it challenging to conduct experiments on punishment, as the use of animals with punishment procedures was considered by many as inhumane. In this chapter, we present the accumulated findings on punishment and other forms of aversive control, assuming that this knowledge about the effects and side effects of aversive control is a better strategy for improving the human condition than not having any research or data. You will notice that many of these studies were conducted between the 1950s and 1980s.

While it is tempting to assume the term punishment means discipline of some sort, the term is used differently in behavior analysis. When a behavioral contingency results in a *decrease* in the rate of response, the contingency is defined as, and called, **punishment**. Any event or stimulus that decreases the rate of operant behavior is called a **punisher**. At this point, we now discuss contingencies of punishment; negative reinforcement is addressed later in this chapter (see also Hineline & Rosales-Ruiz, 2013 chapter on punishment and negative reinforcement in the *APA Handbook of Behavior Analysis*).

Many operations other than a punishment contingency reduce behavior frequency and the rate of response. These include satiation, extinction, behavioral contrast, exhaustion, restraint, precommitment, and richer alternative schedules of reinforcement. Each of these procedures is discussed throughout this textbook. Punishment is defined when an event is contingent on the occurrence of a specified response and the probability of that response is reduced. If a small electric shock or a loud tone is presented contingent on lever pressing, and lever pressing repeatedly has produced shocks or perhaps the loud tone, the rat is less likely to press the lever. The effect and the contingency are called positive punishment.

	Effect on behavior	
Stimulus Following behavior	Increase	Decrease
On/presented		2 Positive punishment
Off/removed	3 Negative reinforcement	4 Negative punishment

Fig. 6.4
Aversive contingencies of reinforcement and punishment (adapted from Figure 4.3 in Chapter 4). When a stimulus or event follows operant behavior, then the behavior increases or decreases in frequency. It is this relation between behavior and consequence that defines the contingency.

Positive Punishment

Positive punishment occurs when a stimulus is *presented* following an operant and the operant *decreases* in frequency. The contingency of positive punishment is shown in cell 2 of Figure 6.4. When a parent reprimands a child for running into the street and the child stops doing it, this is positive punishment. Of course, technically, reprimands function as punishment *only if* they decrease the probability of running into the street. This is an important point because in usual language people talk about punishment without considering its effects on behavior. For example, you may shout and argue with another person when she expresses a particular political position. Your shouting is positive punishment only if the other person stops (or decreases) talking about politics. In fact, the person may increase their rate of political conversation (as often happens in arguments). In this case, you actually have *reinforced* rather than punished arguing with you. Thus, positive punishment is defined as a decrease in operant behavior produced by the presentation of a stimulus that follows it. By this functional definition, punishment always works.

Overcorrection as Positive Punishment
In applied behavior analysis, **overcorrection** is a positive punishment procedure that uses "restitution" to reduce or eliminate destructive or aggressive behavior. The person emitting aggressive or destructive responses is required to "restore the disturbed situation to a greatly improved state" (Foxx & Azrin, 1973, p. 15). Thus, a child or teenager who throws a tantrum by throwing toys and objects all over their room and knocks over furniture can overcorrect the environmental effects of their actions by rearranging the toys and furniture of the entire room and apologizing to family members. Overcorrection may also involve *positive practice*, requiring the child or teen to intensively practice a corrected form of the action.

Overcorrection includes additional procedures such as differential reinforcement of alternative behavior or extinction, which contribute to the effectiveness of overcorrection as an intervention package. For example, Azrin, Besalel-Azrin, and Azrin (1999) recommend providing praise, attention, and other forms of reinforcement to the learner each time they might "spontaneously" perform appropriate behavior during similar activities. It is important to note that a distinction should be made on whether the challenging behavior is "deliberate" or the result of a skill deficit (Azrin, Besalel-Azrin, & Azrin, 1999). Positive practice may be effective because the extra effort is aversive and discourages future challenging behavior if the behavior is deliberate. If the behavior is the result of insufficient learning the intensive practice of the correct behavior should take the place of the challenging behavior. To date, a detailed component analysis of the "package" has not been a primary focus of applied behavioral research. However, critical analyses and reviews of overcorrection suggest that care should be taken when implementing overcorrection and it should only be implemented by experts given its punitive nature and potential for side effects (e.g., escape responses, emotional reactions, and elicited or operant aggression) (Miltenberger & Fuqua, 1981).

It is important to point out that behavior targeted for punishment is typically maintained by reinforcement before it is punished. When the punished behavior stops, the density of reinforcement in a person's life also declines. Thus, life often gets worse with the use of punishment in a treatment program. One strategy is to arrange alternative sources of high-density reinforcement with socially appropriate ways of getting them whenever punishment is used as part of a treatment program, including programs using overcorrection (Cautela, 1984).

Negative Punishment

Negative punishment is portrayed in cell 4 of Figure 6.4. When an ongoing stimulus is *removed* contingent on a response and this removal results in a *decrease* in the rate of behavior, the contingency is called **negative punishment** (or omission). In other words, if the organism responds, the stimulus is taken away and behavior decreases. A child is watching TV, but if they run around, the television is turned off. A driver has earned money and is fined for speeding (money is taken away). In these cases, stimuli (i.e., TV or money) are *removed* contingent on behavior, and the behavior decreases.

Negative punishment is often confused with extinction. *Extinction occurs when a previously reinforced response no longer produces reinforcement*. In this case, a response has produced reinforcement; extinction for that response is in effect when the response → reinforcer contingency is discontinued. A pigeon may peck a key for food, but when extinction is programmed, pecking no longer produces food reinforcement. Similarly, a child may be allowed to stream a favorite television show after completing homework assignments. But if the Wi-Fi is down, the contingency is no longer in effect and doing homework is under extinction.

In another example, ongoing reinforcement could be eating a meal with the family (food and good conversation are the reinforcers). Responses may involve talking to a sister, passing food around the table, or checking social media postings. Checking social media posts may result in a family member telling you to forgo your smartphone (negative punishment, assuming a decrease in social media checking). Giving up your phone reduces your tendency to check social media posts when you next have a meal with your family.

Timeout from Reinforcement as Negative Punishment

In behavioral terms, forgoing your phone is **timeout from positive reinforcement**, assuming that the procedure reduces checking social media posts during dinner. With timeout the teen loses access to positive reinforcement for a specified period (until the next family meal) for engaging in the undesirable behavior. In the classroom, timeout can involve either exclusion or non-exclusion (Cooper, Heron, & Heward, 2007). In timeout with non-exclusion, the student is not physically removed from the situation. Timeout by non-exclusion occurs when the teacher uses planned ignoring of the behavior, withdrawal of a specific positive reinforcer, or handing the person a timeout token exchanged later for a timeout period. Exclusion timeout, on the other hand, involves placing the person in a space separate from others (and the associated reinforcement), such as timeout room—a partitioned space for timeout. Another example might be placing the offending student in a barren school hallway.

For a timeout procedure to be effective in a classroom, the teacher must ensure that the classroom activities are reinforcing for the student in the first place, define the responses that lead to timeout, and decide on the method to use (non-exclusion or exclusion). In addition, the maximum duration of timeout (usually a brief period) must be specified, the exiting criteria should be established, and permission to use timeout must be obtained from the relevant parties, such as the school principal and parents. In addition, returning the child to the classroom for regular activities may be used as reinforcement for good behavior during timeout. As in all behavioral interventions, the teacher should keep precise records to evaluate the effectiveness of the procedure.

Response Cost as Negative Punishment

Response cost is another negative punishment procedure in which conditioned reinforcers (tokens) are removed contingent on behavior, and the behavior decreases. In humans, common response–cost contingencies involve the loss of money or privileges for disobedience, and this loss decreases rule breaking. For example, Bartlett and colleagues (2011) used response cost to decrease spitting for an 8-year-old boy with autism. The researchers first tested a non-contingent reinforcement condition in which the boy had access to preferred toys and reinforcers. Spitting was decreased in this condition relative to baseline; however, it was still occurring at socially unacceptable rates. The researchers then tested response cost condition in which the boy lost access to a preferred toy radio contingent on spitting. They found this condition was effective in decreasing spitting to near-zero levels. Importantly, researchers attempted a reinforcement-based procedure for socially appropriate behavior before implementing response cost.

Response cost can also be arranged using token reinforcers, which are subsequently subtracted or removed following a response. Conyers et al. (2004) conducted a study to reduce disruptive behaviors of 25 preschool children. They used a multielement design to compare differential reinforcement of other behavior (DRO) condition, in which children would earn tokens for not engaging in disruptive behavior, and response cost, in which children would lose tokens for engaging in disruptive behavior. They found both conditions were effective in decreasing disruptive behavior; however, response cost was more effective. The researchers recommended using a combination of both continencies to decrease emotional behavior side effects. Research in the laboratory with pigeons using light-emitting diodes (LEDs) as tokens exchanged for access to food has supported the human findings. Birds show suppression of behavior by response–cost contingencies, and effects similar to traditional punishers such as electric shock (Pietras & Hackenberg, 2005).

Conditions that Influence Punishment Effectiveness

Unlike reinforcement, contingencies of punishment do not teach or condition new behavior that replace the punished behavior. Contingencies of punishment alone eliminate or, more often, temporarily suppress the rate of operant behavior. In this section, we describe some of the conditions that increase the effectiveness of punishment as a behavior-reduction procedure.

One critical thing to remember about punishment is that the to-be punished response is always maintained by a schedule of reinforcement. Therefore, a schedule of punishment usually is superimposed on a schedule of reinforcement. This means that we are really investigating the effects of punishment applied to behavior maintained by some schedule of positive reinforcement, and the results may reflect both of these contingencies. Which will win in terms of influencing behavior the most—the schedule of reinforcement or punishment?

Intensity of Punishment
Consider the following scenario. Mike has bought a new home audio system and his friend Joe and Joe's 2-year-old daughter drop in for a visit. The child is eating a glob of peanut butter and makes a beeline for the new equipment. Nervously, Mike looks at his friend, who says, "Emily, don't touch—that's Mike's new audio system." The child continues to touch the control knobs on Mike's high-end system, and Joe says more emphatically, "Please leave that alone!" Emily is still smearing peanut butter on Mike's investment, so Joe glowers at his child and loudly says, "I said stop that!" Emily does not stop, and is now threatened with "If you don't stop, I will give you a spanking!" Emily still plays with the knobs of the audio system. In desperation, Joe gives Emily a light tap on the bottom, which she ignores. In this circumstance, presumed punishers are introduced at low intensity and gradually increased. Such actions teach the child to disregard early requests by the parents and that the "empty threats" are meaningless. Of course, the best solution for the audio system problem would be to wipe the child's hands or place the equipment out of reach—this is called response prevention.

Research from the laboratory shows us that if punishment is going to be used, it is most effective when introduced at a moderate to high intensity on the first occasion. Generally, higher-intensity punishment results in greater response suppression. Low-intensity positive punishment may leave behavior relatively unaffected, and the recipient only annoyed, while higher values of the punisher may permanently change behavior (Appel & Peterson, 1965). Several experiments have shown that intense punishment can completely eliminate responding (Appel, 1961; Storms, Boroczi, & Broen, 1962). One interesting implication is that once complete suppression of responding occurs, behavior is unlikely to recover for some time even when the punishment contingency is withdrawn. This is because the organism stops responding and never contacts the changed environment, which is usually desirable. On the other hand, an organism may respond again when reinforcement is available after punishment has been withdrawn. In these conditions, behavior of course may recover quickly to pre-punishment levels. These observations led Skinner and others to suggest that punishment by itself only produces a temporary suppression of behavior.

Skinner objected to the use of punishment for behavior regulation. He repeatedly argued against the use of punishment and instead for the use of positive reinforcement. Remember, however, that both reinforcement and punishment are defined functionally by a change in the behavior; if the rate of response does not change, neither reinforcement nor punishment can be said to have occurred.

Immediacy of Punishment
Punishment is most effective at reducing responses when it closely follows behavior (Azrin, 1956; Cohen, 1968). This effect easily can be missed because punishment often generates

emotional behavior that may disrupt operant responses. In other words, when it is first introduced, *positive punishment elicits reflexive behavior that prevents the occurrence of operant behavior*. Watch a child (or adult) who has just been chastised severely for making rude noises. You will probably see the child sit quietly, possibly cry, or look away from others. In common language, we may say that the child is pouting or humiliated, but in fact what is happening is that reflexive emotional behavior is disrupting all operant behavior. If punishment follows immediately for making rude noises (the target behavior), those noises (as well as many other operant responses) would decrease in frequency. Making noises, however, would be relatively unaffected if punishment did not closely follow the target response.

Some animal research also supports that punishment suppresses behavior through the elicitation of reflexes. Azrin (1956), for example, punished some rats' responses immediately while another group received *delayed* punishment. After the first hour of exposure to positive punishment, immediate versus delayed punishment made a large difference—responses that were punished after a delay recovered substantially, but when the punisher was delivered immediately, responses were often completely eliminated. One interpretation of this is that the introduction of immediate punishment generates responses that may disrupt operant behavior. With delayed punishment, though, the contingency of reinforcement for the target behavior (a bar press, for example) is contacted immediately and may in time override the punisher's effects. Therefore, punishment is most effective when it is delivered immediately after the response.

Schedule of Punishment
In general, positive punishment is most effective when it is delivered after each response (a continuous schedule of punishment) (Zimmerman & Ferster, 1963) rather than intermittently (Filby & Appel, 1966). Azrin, Holz, and Hake (1963) trained pigeons to peck a key on a VI 3-min schedule of food reinforcement. Once responding was stable, shocks were presented after 100, 200, 300, 500, or 1,000 key pecks. Rate of response substantially declined even when punishment was delivered after 1,000 responses. As the rate of punishment increased, the number of responses per hour declined. In other words, as more responses were punished, operant rate decreased. Continuous punishment (FR 1) produced the greatest response suppression. The maximal effect is similar to what happens when increasing the intensity of the punisher. Therefore, to maximize suppression of responses, frequent delivery at high intensity is most effective. Notably, with high intensity punishment, response suppression often occurs with few punishers, perhaps just one.

Rate of response patterns on various schedules of punishment (FR, FI, VI, and VR) are usually *opposite* to the patterns produced on similar schedules of positive reinforcement. For example, an FI schedule of punishment when superimposed on a VI schedule of reinforcement for key pecking by pigeons produces an *inverse scallop* (recall that FI reinforcement often yields a scalloping pattern). Each occurrence of the punisher is followed by an immediately high rate of pecking that gradually declines as the time to the next punishment approaches (Azrin, 1956).

In summary, punishment is most effective when it is intense, immediately following the response, and under a continuous schedule.

Reducing the Effectiveness of Positive Reinforcement
Punishment suppresses behavior more when the positive reinforcement maintaining the response is simultaneously reduced in effectiveness. Azrin, Holz, and Hake (1963) trained

pigeons to peck a key on a VI 3-min schedule of food reinforcement. After responding was stable, they introduced a shock for every 100th response. Birds were exposed to the schedule of reinforcement plus punishment at several levels of food deprivation. Recall that food deprivation is an *establishing operation* that should increase pecking of the key for food (and increase the reinforcement effectiveness of food). At lower deprivation, punishment virtually stopped the birds' responding. However, when they were more deprived of food, punishment was less effective and response rate was more likely to be maintained.

These findings may have practical implications. Punishment is often used in the hope of reducing the frequency of undesirable human behavior. But there are side effects of the punitive regulation of behavior, which will be discussed later in this chapter, suggesting that these techniques should be used with careful consideration and caution. The use of aversive stimuli, especially shock, has a controversial history and is a topic of importance in contemporary behavior analysis. We will describe these issues in a subsequent section of this textbook.

Arranging Response Alternatives
A straightforward way to make punishment more effective or unnecessary is to give a person another way to obtain reinforcement. When a reinforced response alternative is available, even moderate levels of punishment suppress behavior. Indeed, sometimes a punisher is not even required! To use a response alternative procedure, it is essential to identify the consequences that are maintaining the target behavior. Next, the person (or animal in a laboratory) is given another way to obtain the same or a preferred reinforcer. Herman and Azrin (1964) had people lever press under a VI schedule of reinforcement. Each lever press then produced an annoying buzzing sound, but the procedure only slightly reduced the rate of response. Finally, the people were given another response option that did not produce the buzzing sound; they quickly changed to that alternative, and punished responses were eliminated.

Imagine that there is a convenience store (a reinforcer) in the middle of the block directly behind your house. You often walk to the store, but if you turn left to go around the block you pass a chained dog that lunges and growls at you. On the other hand, if you turn right, you do not pass the dog. It is obvious that most people, after experience with these contingencies, would choose the unpunished route to the store. If, however, turning right leads to a path that does not get you to the store, you may continue to walk past the lunging dog. In reality, of course, you could walk on the other side of the street or drive to the store—these are also unpunished alternative responses.

NEW DIRECTIONS: EPIGENETICS IN RETENTION OF FEAR CONDITIONING

In Chapter 1, we briefly described the rapidly growing field of epigenetics and its relevance to the retention of early learning (see "New Directions: Epigenetic Mechanisms and Retention of Early Learning"). Recall that learning experiences are sometimes retained by epigenetic mechanisms at the cellular level. DNA methylation and histone acetylation tighten and loosen respectively the chromatin structure that envelopes the genes, allowing for differences in *gene expression* (transcription and translation) without any alteration of the DNA sequence. Differences in gene expression instigated by the environment (external and internal) operating on molecular and epigenetic mechanisms, change neural interconnections related to behavior, a process known as *neuroplasticity* (Johansen, Cain, Ostroff, & LeDoux, 2011). Here we present a non-technical description of some of the evidence for epigenetic effects on retention of fear learning. Our overview draws on a review from *Trends in Neuroscience* by Kwapis and Wood (2014).

It is often adaptive for organisms to retain or remember aversive learning experiences. To study the epigenetic regulation of learning and remembering, behavioral neuroscientists have adopted a simple, well-characterized protocol of respondent fear conditioning. In the training phase, animals (often mice) are presented with a tone (CS) followed by a small foot shock (US) in a specific location (white chamber or context) and freezing is used as the fear response (CR). The neural circuitry of basic fear conditioning has been researched (Johansen et al., 2011) and involves the responses of neurons in the amygdala to the tone (or context) and shock contingency. In addition, the CR of freezing in a specific context (lighting, shape, color, or texture of the chamber) involves the dorsal hippocampus and medial prefrontal cortex of the brain. Experimental disruption of these brain regions impairs contextual and higher-order conditioning. Because the neural pathways are well described, behavioral neurobiologists are able to conduct epigenetic studies of fear conditioning at the cellular and molecular levels.

After initial training by fear conditioning in the white chamber, the gene expression of experimental animals is manipulated while controls do not receive this treatment. Next, after a 24-h delay, animals are moved to a new context (gray box) to test for freezing to the tone stimulus (tone test), followed by a test for freezing in the original white chamber, but without the tone (context test). Based on these conditioning tests and gene expression manipulations, it is known that fear conditioning promotes epigenetic changes (markings) related to transcription of genes concerned with *retention of fear learning* and also inhibits expression of retention-limiting genes. Histone acetylation (carried out by histone acetyltransferase enzymes or HATs) increases retention and is the most widely researched epigenetic modification in fear conditioning. In tone and context tests, drugs that inhibit HATs and histone acetylation decrease retention of fear responses, whereas drugs that inhibit HDACs (enzymes that remove acetyl groups) increase retention (histone deacetylation and HDACs are discussed further in Chapter 14 in "New Directions: Epigenetic Reprogramming of Social Behavior in Carpenter Ants"). Additional studies show that manipulations of histone acetylation in the amygdala specifically affects freezing to the tone (CS) while targeted manipulations in the hippocampus enhance (or impair) the fear response to the context.

To investigate long-term retention of learning, behavioral neurobiologists use a fear conditioning situation that involves training, reactivation, and a test for remembering after a long delay (Maddox, Watts, Doyere, & Schafe, 2013). The animal initially receives training in a white novel chamber; a tone (CS) is associated with up to 3 presentations of foot shock (US). Twenty-four hours later, the animal is placed in a gray chamber and a single tone (CS) is presented without the shock to reactivate the fear conditioning. One hour after reactivation, HAT inhibitors are used to manipulate gene expression (drug vs. vehicle). Approximately 1 day later the animal is again placed in the gray chamber and tested 10 times for freezing to the CS-tone (remembering the fear after a 48-h delay).

HAT enzymes are found to enhance long-term remembering of the fear conditioning. Thus, after reactivation of a weak-fear experience (one training trial), drug inhibition of HAT enzymes in the lateral amygdala (LA) disrupts freezing to the tone one day later, indicating the animals no longer remembered the conditioned fear response. In other experiments, rats received 3 tone-shock presentations (strong conditioning) followed by 2 weeks in home cages without further training. Next, the rats were transferred to the gray chamber for the usual reactivation procedure and given intra-LA infusion of the HATs inhibitor. One day later and 21 days after aversive training, rats infused with the HATs inhibitor in the LA showed less freezing to the tone than control animals. Significantly, inhibition of HAT enzymes following reactivation of a fear experience impairs remembering, especially by neurons in the lateral amygdala (see Kwapis & Wood, 2014 for evidence on manipulation of HDACs).

Studies of reactivation and epigenetic manipulation to reduce fear responses after a long delay have applied importance. Post-traumatic stress disorder (PTSD) and other trauma and anxiety-related problems often use some form of CS-exposure therapy involving repeated presentation of the fear cue without the aversive stimulus. In fact, in the laboratory, arranging an extinction contingency—repeated presentation of CS-tone without the US—does reduce freezing to the tone. Research shows, however, that extinction does not wipe out the old learned fear involving the amygdala; rather, extinction establishes new learning, especially in the ventral segment of the medial prefrontal cortex (MPC), that the CS tone no longer predicts the US-shock (see Kwapis & Wood, 2014). Neural projections from the MPC to the amygdala help to inhibit the reactivation of the fear responses in new settings or fear responses when the aversive stimulus (shock) is encountered again. Evidence is mounting that HDAC inhibitors targeting the ventral segment of the MPC promote robust and persistent extinction learning that may outcompete the original fear learning (CS-tone predicts US-shock).

At the behavioral level, researchers may be able to produce similar effects to HDAC inhibitors by adding other procedures to extinction contingencies. One possibility is to use *positive counterconditioning* following reactivation of the original fear learning (Richardson, Riccio, Jamis, Cabosky, & Skoczen, 1982). In this procedure after fear training, the learning is reactivated (CS-fear presented) and, shortly after, rats are moved to feeding cages to drink a solution of sugar water (US-positive). Under this delayed US-positive procedure the animals showed substantially less fear responses than control rats, but only after reactivation of learning by a brief presentation of the fear stimulus (CS-aversive). One possibility is that positive counterconditioning works on epigenetic mechanisms related to retention of the original fear learning (amygdala) or perhaps on retention of new extinction learning (ventral segment of the MPC). In either case, treatment of PTSD and other anxiety disorders may benefit from a desensitization therapy combining extinction and positive counterconditioning.

Research in the field of epigenetics and retention of fear learning illustrates the ongoing synthesis of behavior analysis with neurobiological procedures and principles. This synthesis is providing a more complete understanding of the environmental and biological components of fear retention, which soon may provide new applications for prevention and treatment of stress-related behavior disorders.

USE OF PUNISHMENT IN TREATMENT

There are people with intellectual and developmental disabilities or psychological disorders with psychotic episodes who, for a variety of reasons, engage in self-destructive behavior. This behavior may escalate to the point at which the person is hitting, scratching, biting, or gouging themselves most of the day. In some cases, self-injurious acts are so frequent and intense that the person may break bones, destroy tissue, or is hospitalized. Occasionally physical injury is irreversible, as when a person repeatedly bangs their head on a wall, leading to a traumatic brain injury. In addition, individuals with these disorders may also act aggressively and direct these same types of behavior outwardly—towards caregivers. Although positive reinforcement programs have been used to alleviate these types of severe behavior problems, these contingencies are not always successful. In this situation, the choice is often one of four possibilities: 1) continue to let the individual engage in severe self-injury or aggression, which is harmful and unethical; 2) sedate the individual with CNS-depressant or anti-psychotic drugs, which lowers their quality of life; 3) use physical restraint to prevent the injurious behaviors, which in addition to being unpleasant, also reduces contact with potential reinforcement sources (and lowers quality of life); or 4) use a punishment procedure to target the self-injurious or aggressive behavior. Behavior therapists occasionally have resorted to this latter alternative of punishment as a way of reducing self-destructive behavior.

The Punishment Debate

The use of aversive stimuli such as misting, loud noises, and mild shock to reduce behavior has been highly controversial, resulting in a **use of punishment debate** (Feldman, 1990). Opponents of punishment argue that such procedures are morally wrong, advocating a total ban on their use (e.g., Sobsey, 1990). These researchers also suggest that punishment is not necessary because many positive methods are available to treat severe behavior problems. In fact, research shows that positive behavioral support is as effective as punishment for eliminating severe behavior problems in young children with autism (up to 90% reduction), especially when the treatment program is based on a behavioral assessment of the functions (e.g., social attention, automatic reinforcement, self-stimulation) of the self-injurious responses (Horner, Carr, Strain, Todd, & Reed, 2002). Thus, it is possible to use positive behavior management (without punishment) with young children who show severe problem behavior such as self-injurious behavior.

Proponents of response reduction by punishment include some therapists and parents who have not observed successful reductions in especially life-threatening or self-damaging behavior of the children with other treatments. These individuals advocate the individual's right to effective treatment (e.g., Van Houten et al., 1988). The proponents of effective treatment claim that a combination of positive behavioral support and punishment is the best (and perhaps the only) way to manage severely self-injurious behavior without sedation (see review by Minshawl, 2008). Notice, however, that the predominant strategy from all perspectives is positive behavior management, regardless of whether punishment is added to the overall program. No one in applied behavior analysis is advocating for the sole or predominant use of punishment without an overall positive behavioral program.

The use of Contingent Electric Skin Stimulation (CESS), or low-intensity skin shock is especially controversial. CESS used as positive punisher has been shown to rapidly reduce self-injury (Lovaas & Simmons, 1969) or aggression and physical harm to others (Israel et al., 2008). Besides the reduction of harm, the rationale for CESS, like other aversive stimuli, is that when less time is spent engaging in self-injury or aggression (or being chemically or physically restrained), more time can be spent making contact with positive reinforcers, which enhances the quality of life of the patient (although not specific to CESS, see for example, Hanley et al., 2005). Proponents of the use of CESS have argued that the temporary use of an effective punishment procedure, when no other procedure will work, can save a child from years of self-injury or sedation.

It is important to point out that very few applied behavior analysts use CESS and many do not support CESS as a way to reduce behavior. In addition, a number of entities and agencies, such as the International Association for the Scientific Study of Intellectual and Developmental Disabilities, have clear statements that oppose the use of CESS to treat even severe behaviors of those with intellectual and developmental disabilities (Zarcone, et al., 2020). One reason given for not using CESS (and other punishers) in applied settings is that aversive techniques may generate emotional distress, mental health problems such as anxiety and depression, substitution of other negative behaviors, and aggression (LaVigna & Donnellan, 1986; Meyer & Evans, 1989; Zarcone et al., 2020). In a treatment setting, these side effects imply that aversive therapy for self-injurious behavior may produce as many problems as it alleviates (Lerman & Vorndran, 2002; see also the section on "Side Effects of Aversive Procedures" below). This indeed is also the case for the use of CESS (Zarcone, et al., 2020). Salvy et al. (2004), however, found that an 85-v, 3.5-mA CESS (feels subjectively similar to a rubber band snap on the arm) eliminated severe head banging and injury in a preschool child; no negative side effects were reported. Moreover, just the placement of the device *without* the delivery of shock kept self-injury to near-zero levels, even at a 7-month follow-up. It is important to note, however, that many of the studies that report on the use of CESS *do not adequately assess or report on symptoms of mental health in a systematic way or at longer-term follow-up*; they may be noted anecdotally in more extreme situations. Therefore, it is difficult to state the extent to which these symptoms are present in more contemporary studies. Efforts are under way to weigh the costs and benefits of CESS and other alternatives in treating these extreme forms of self-injury.

Permanence of Punishment

One issue is whether punishment by itself, without additional procedures such as extinction or reinforcement of alternative behavior, can permanently eliminate undesirable behavior. In applied behavior analysis, the issue of permanence is cast as the maintenance of response suppression over extended periods. James O'Heare (2009; Weblog), a certified animal behavior consultant, has commented on the **permanence of punishment** and the maintenance of response suppression. He states:

> To suggest that punishment can eliminate a behavior is to conflate punishment with punishment plus other processes. If a behavior is exhibited then it has a reinforcement history; that contingency is in place. When we impose a punishment contingency on that behavior, we merely add that contingency to the prevailing reinforcement contingency. If the reinforcer is stronger than the punisher then the behavior will continue to occur, although perhaps at a reduced rate or frequency. If the punishment is stronger than the reinforcer then the behavior will be suppressed to some extent, depending on just how much stronger it is. Indeed, the behavior may be suppressed to a rate of zero. But as soon as the punishment contingency is discontinued, the existing reinforcement contingency prevails again and the behavior is expected to return to pre-punishment strength…. The main point here is to note that punishment alone does not eliminate the behavior; it merely suppresses it to some extent while that contingency is in effect. If the behavior is permanently eliminated, it is not because of the punishment contingency alone. What would have occurred is likely extinction simultaneously imposed, or more commonly the suppression allowed for the performance and reinforcement of other behaviors.
>
> (O'Heare, 2009; Weblog)

In applied settings, instances of punishment always involve other behavioral procedures as part of the treatment package. Overall, it is reasonable to conclude that punishment in combination with other positive procedures can have lasting suppressive effects on human behavior.

FOCUS ON: PHYSICAL PUNISHMENT AND PSYCHIATRIC DISORDERS

Worldwide the use of physical punishment to discipline children is controversial, even if it is not considered socially as maltreatment or abuse (World Health Organization, 2021). Physical punishment by parents includes the use of spanking, smacking, or slapping when children are judged to misbehave. These methods of correction involve the infliction of pain for the purpose of disciplining or reforming the behavior of the child. Thus, punishment is not functionally defined by its effects on behavior, but refers to painful procedures used to reform disobedient behavior, which is likely reinforcing to the person who is inflicting the punishment. One issue is whether the infliction of pain for disobedience is linked to psychiatric disorders, even when there is an absence of more extreme maltreatment involving physical abuse, sexual abuse, emotional abuse, physical and emotional neglect, and exposure to family violence.

One study used a nationally representative US sample to examine the long-term relation between physical punishment and mental health (Afifi, Mota, Dasiewicz, MacMillan, & Sareen, 2012). The data were drawn from a 2004–2005 national survey of nearly 35,000 adult respondents, with abused and maltreated respondents excluded from further analyses. The results showed that frequent or harsh use of physical punishment in childhood occurred in about 6% of the national sample, with males reporting harsher physical punishment than females. Also, respondents who indicated that they grew up in a dysfunctional family were more likely to report harsh physical punishment by adult caretakers or parents. Using statistical techniques and controlling for socioeconomic variables, it was found that those who experienced physical punishment in childhood were more likely than non-punished respondents to indicate a diagnosis of major depression, mania, mood disorder, specific phobia, anxiety disorders, or alcohol and drug dependence.

Compared with studies using experimental designs and direct measurement of behavior, this national survey has several limitations, as has been noted by the researchers. One problem is that the data are based on retrospective reports of what happened in childhood, and these verbal responses may be influenced by many factors that distort recall, including the interview situation itself (Loftus & Zanni, 1975). Another limitation is that the survey design (which is cross-sectional) does not allow for any inference of causation between harsh physical punishment and mental disorders—the best one can say is that there may be a link or correlation. However, the experimental analysis of punishment in nonhuman animals suggests that a causal relation is a strong possibility, especially when corporal punishment of children is arranged on the basis of whims and mistaken parenting beliefs of the caretakers; in other words, when punishment is not contingent on behavior.

CONTINGENCIES OF NEGATIVE REINFORCEMENT

When an organism emits a behavior in response to an aversive stimulus, the behavior may be viewed as either escape or avoidance. If the response is made while the punishing stimulus is occurring or present, it is an **escape** response. The vicious dog is growling at you, and you escape by crossing to the other side of the street. The bright sun is shining in your eyes, so you put on your sunglasses to block it. When the operant *prevents* the aversive stimulus, the behavior is **avoidance**. You turn right to go around the block and thereby do not walk past the dog. You may put on your sunglasses moments *before* you go out into the sun to prevent the sun's rays from hitting your eyes. In both cases, the removal or prevention of an event or stimulus *increases* operant behavior and the contingency is defined as **negative reinforcement** (cell 3 of Figure 6.4).

Any event or stimulus that increases operant rate by its removal (or prevention) is called a **negative reinforcer**. Notice that the same event—delivery of electric shock—is a punisher in a positive punishment procedure, and a negative reinforcer in a negative reinforcement procedure. Whether a stimulus is used in a punishment or negative reinforcement procedure, we refer to the event as an *aversive stimulus*—a stimulus that the organism escapes, or avoids.

In everyday life, the distinction between negative and positive reinforcement is occasionally confused, and continues to be a major topic of debate in behavior analysis (Baron & Galizio, 2005, 2006). For example, do you open a window on a hot day to get a cool breeze as positive reinforcement or to escape the heat as negative reinforcement? Putting on corrective lenses clarifies vision, but also removes a blurry view of the world. One issue is that physics tells us that there is no such thing as cold—there are only increases or decreases in heat. Thus, a person who places logs on a fire is adding heat and the behavior is controlled by positive reinforcement, not negative reinforcement (removing cold). On the other hand, there are many other instances where it may be difficult to tell the difference in everyday life. In the operant laboratory, however, the distinction between positive and negative reinforcement is reasonably easy to arrange, and experimental investigations of negative reinforcement are relatively clear-cut. When a response results in the removal of an ongoing event or postponement of a stimulus and the rate of response increases, negative reinforcement has occurred.

Escape Learning

In escape learning, an operant behavior changes the situation from one in which an aversive stimulus is present to one in which it is absent, for some period of time. A pigeon is exposed to continuous loud white noise and when the bird pecks a key the noise is turned off. If pecking the key increases because of the removal of the noise, then this defines the procedure as negative reinforcement. The noise would be considered a negative reinforcer. People reprimand others who are bothering them with statements such as "Stop that" or "Don't do that, it's not nice" and other forms of disapproval. Research shows that such reprimands are controlled by negative reinforcement in the form of escape from the aggravating behavior (Miller, Lerman, & Fritz, 2010).

In classic experiments on escape learning, a dog jumped across a barrier to escape electric shock. Figure 6.5 is a diagram of a shuttle-box apparatus used to train escape in dogs. The dog may engage in many responses, such as sniffing, grooming, or resting, but only jumping the barrier to the shock-free area (called the safety area) terminates the shock. Generally, organisms acquire escape responses more readily than avoidance responses. The reason is that in escape, but not avoidance, there is an *immediate change* from the presence or the absence of the aversive stimulus.

Another factor that affects how quickly an escape response occurs is its compatibility with reflexive behavior elicited by the negative reinforcer. Evolution has ensured that organisms respond to aversive stimuli. In the everyday world, an animal may only get one chance to save its life in the presence of an aversive event. Running like crazy makes good sense (in many cases) when a predator appears. Those animals that "ponder over" the situation are likely to

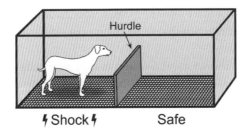

Fig. 6.5
A shuttle box is depicted that may be used to condition escape responses in dogs. The animal is placed in the left compartment at the start of a trial. A small electric shock is turned on, and the dog can escape the aversive stimulus by jumping the barrier to the safe area, on the right side of the box.

have contributed calories to the predator, but not genes to the next generation. Thus, natural selection has ensured that species-specific behavior often is elicited by aversive stimuli that also function as negative reinforcers. When rats are presented with electric foot shock, they typically show defensive responses. These species-typical responses include jumping to the onset of foot shocks and freezing in the post-shock interval. These species-typical behaviors elicited by electric shocks sometimes interfere with escape conditioning. For example, the operant of lever pressing is incompatible with freezing after the shock. If the animal is simply required to press the lever and hold it down, the escape response is more readily acquired; this is because freezing interferes less with holding down the lever. Generally, *negative reinforcement may elicit reflexive behavior that interferes with learning the behavior required for the removal of the negative reinforcer.*

Conditioning escape behavior is easier when the *operant is similar to reflexive behavior elicited by the aversive stimulus*. A rat can be readily trained to run on a wheel to escape electric shocks, but conditioning the animal to stand up is much more difficult (Bolles, 1970). Running is part of the species-typical responses to electric shock, but standing up is not. Although respondent and operant conditioning interact during escape training, behavior eventually comes under the control of the operant contingency. For example, rats that are trained to run on a wheel (or hold down a lever) to escape shock stop running (or lever holding) if this response does not terminate the negative reinforcer. The species-specific response does not predominate operant behavior required by the contingencies of reinforcement.

FOCUS ON: ESCAPE AND INFANT CAREGIVING

Parents frequently are confronted with a crying baby. Crying is a normal response of infants usually interpreted as a way in which the baby communicates their wants and needs and parents need to be responsive to an infant's cries. Excessive and persistent crying, however, is a major factor linked to infant abuse (e.g., shaking the baby), and a leading cause of severe injury and death. It appears that infant crying arranges an ongoing aversive stimulus for the parents or caregivers, which usually is removed by caretaking behavior (e.g., feeding, changing, playing, or rocking). The infant and caregiver are locked in a social interaction; this interaction involves escape contingencies where actions of the caregiver are negatively reinforced by the removal of crying, and the baby's vocalizations are positively reinforced by parental care. In this view, infant abuse is the result of inescapable crying where nothing the parent or caregiver does removes the aversive stimulus (Donovan, 1981; see also the section "Learned Helplessness" in this chapter).

To study experimentally the effects of infant crying on caretaking behavior, it is necessary to manipulate the contingencies of reinforcement. Thus, infant crying must be controlled by the researcher and its removal made contingent on specified actions of adult caregivers. One method is an experimental simulation of infant caretaking where the removal of recorded infant cries requires specific responses by the caregivers. In a study by Thompson, Bruzek, and Cotnoir-Bichelman (2011), infant crying was recorded as the parents rocked their baby before naps. Next, undergraduate students with some experience in infant caretaking were recruited as participants. They were placed in an experimental setting that included a crib, baby doll, blanket, toys, bottle, and a tape recording of infant cries controlled from an adjacent observation room. Participants were told that the study simulated a caretaking situation and they should "do what comes naturally." Negative reinforcement involved the presentation of recorded infant crying until the participant performed the target response (horizontal rocking, vertical rocking, feeding, or playing). An extension of the reversal design (A-B-A-B) compared negative reinforcement to an extinction phase that arranged response-independent crying; in addition, the target behavior differed by participant and phase of the experiment.

The results in Figure 6.6 show the cumulative number of seconds engaged in the target behavior and the alternative responses for three participants (P-7, P-8, and P-9). Notice the increase in time spent on the target response for each participant in the initial negative reinforcement (S^{r-}) crying phase. When the contingency is changed to another target response, cumulative time on this response increases while the previously reinforced response drops to near-zero level or reaches a plateau. During extinction the target response decreased, but responses that were previously effective in removing crying increased, indicating *resurgence* of caretaking responses based on a history of negative reinforcement (see also Bruzek, Thompson, & Peters, 2009).

Overall, the results indicate that infant caretaking is under the control of crying emitted by an infant. Furthermore, the findings suggest that negative reinforcement by escape is part of the early parent–infant relationship. Clearly, this need not be the case in practice. If the caretaker is attending to the child, signals from the infant would indicate child needs (hunger, wet diaper, etc.) prior to the crying. Caretakers who change a diaper or begin breastfeeding based on pre-crying signals (e.g., restlessness, vocalizations, lip-smacking) would reinforce the infant's signaling behavior as an early form of appropriate human communication.

Avoidance Learning

The top of Figure 6.7 shows a schematic of escape behavior. The black box represents the onset, duration, and offset of an aversive stimulus. Not that a response (R) cancels the aversive and a period of aversive-free time is in place (also called a safety period). When another aversive is presented, the response again terminates the aversive stimulus. Escape behavior, then, always makes contact with the aversive stimulus by cancelling it.

When an operant behavior *prevents the occurrence* of an aversive stimulus, the contingency is called avoidance. In avoidance learning, the animal's response cancels an impending aversive event so that nothing happens. The behavior, then, does not make contact with the aversive stimulus. This is depicted at the bottom of Figure 6.7. For example, you typically may

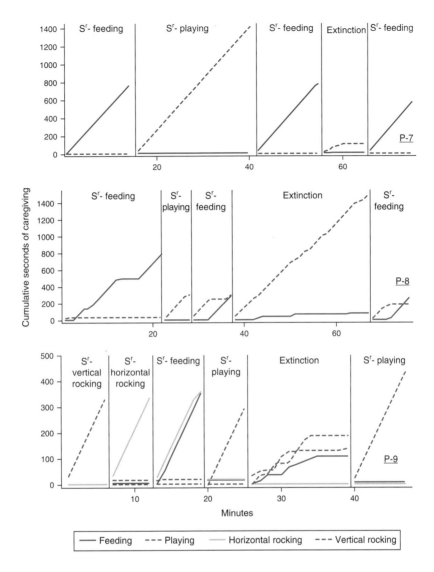

Fig. 6.6
Cumulative duration (seconds) of caregiving for participants (P-7, P-8, and P-9) is shown. Data are depicted in 1-min bins on the X-axis and breaks in the data plots indicate a new session.
Source: Reprinted from R. H. Thompson, J. Bruzek, & N. Cotnoir-Bichelman (2011). The role of negative reinforcement in infant caregiving: An experimental simulation. *Journal of Applied Behavior Analysis, 44*, pp. 295–304. Copyright 2011 held by John Wiley & Sons Ltd. Published with permission.

walk the shortest distance to the university, but recently an acquaintance, Kent, has joined you at the halfway mark, and continues to talk incessantly about topics you don't find very interesting. Given this history, now you walk a longer distance than necessary to the university using a route that does not take you past Kent's house, which prevents an interaction with Kent. In essence, you avoid Kent.

Similarly, during the annual migration, young wildebeests stop to drink at a river infested with large crocodiles. The crocodiles wait each year for this gourmet lunch and "pig out" on

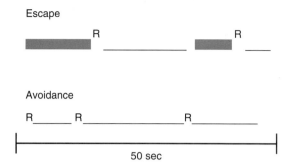

Fig. 6.7
A schematic of the difference between escape and avoidance.

rare wildebeest. Survivors of the crocodile picnic from last time may likely choose a different watering spot the next year. They too learn to avoid.

Discriminated Avoidance
Avoidance may involve responding when a warning signal precedes an aversive stimulus. Because the organism only responds when the warning signal occurs, the procedure is called **discriminated avoidance**. A parent may say to a child, "Nathan, keep the noise down or else you will have to go to bed." An antelope may smell a lion and change the direction in which it is traveling. In these cases, the child is told what to do and the antelope detects what direction to avoid an aversive stimulus.

In the operant laboratory, discriminated avoidance may acquire more slowly than one would guess. Rats, for example, will quickly learn to lever press for food, but may take a longer time to acquire lever pressing to avoid electric shock (Solomon & Brush, 1956). Pigeons are also slow at acquiring avoidance behavior when they are required to peck a key to avoid an aversive event. A major reason for the long acquisition is that, in the discriminated avoidance procedure, the warning stimulus (S^{ave}) is also a CS, which elicits respondently conditioned behaviors (like freezing) that interfere with the operant behavior that is defined as the avoidance response (Meyer, Cho, & Wesemann, 1960). As stated in the section on escape conditioning, other reflexive responses such as running and jumping are elicited by shock and are acquired much more readily than lever pressing. For example, Macphail (1968) reported that pigeons required 120 trials of signaled avoidance to run down a straight alley to avoid shock. Notice that running is not species-typical behavior for pigeons—they usually fly. Rats, on the other hand, required only two or three trials to learn to jump onto a platform when a warning stimulus occurred (Baum, 1965, 1969). For rats, jumping to safety may have some elements of respondent conditioning; this behavior is also compatible with the operant avoidance contingency.

In a series of experiments, Modaresi (1990) found that for rats, lever pressing to avoid shock is acquired more readily if the lever is high on the wall, and if lever pressing not only avoids the shocks, but also results in access to a platform to stand on. Additional experiments showed that these two aspects of the situation were in accord with the rats' species-specific behavior. Rats and other organisms naturally stretch upward and seek a safe area when painful aversive stimuli are delivered from the floor. Thus, to produce rapid acquisition of signaled avoidance responses, behavior that is also elicited by the negative reinforcer and the current situation is most effective.

A classic experiment by Solomon and Wynne (1953) exemplifies discriminated avoidance using an avoidance response that is more natural to the species. The contingencies were as follows. A dog was placed in a compartment of a shuttlebox similar to the one depicted in Figure 6.5. A light was illuminated in the compartment in which the dog was placed. In time, the light went off and the light in the other compartment was illuminated. If the dog did nothing, a shock was delivered 10 s after the light in its compartment went off (the light turning off signaled shock in 10 s). If the dog jumped the barrier to the other illuminated compartment before the 10 s elapsed, then it would avoid the impending shock. Under these conditions, dogs quickly (within 5 trials) learned to jump to avoid the shocks. Indeed, they would jump the barrier right before the 10-s warning signal elapsed.

Nondiscriminated (Sidman) Avoidance

In the Solomon and Wynne (1953) study, there was a warning stimulus for shock (when the light turned off in the compartment the dog was in and turned on in the other compartment). When there is no warning stimulus, the contingency is called **nondiscriminated avoidance**. Some people compulsively wash their hands to get rid of unseen germs. In this case, hand washing is the operant, but it is difficult to see what stimulus is being avoided. Perhaps it is a reduction of anxiety (a private event that is not observable to anyone but the person who is washing their hands) that negatively reinforces hand washing. As you will see in later chapters, negative reinforcement appears to underlie many anxiety-based behavior patterns.

This book was written on a computer, and an unexpected power failure could result in many hours of lost work. To avoid this event, the authors regularly emit the behavior of *saving the document*. This avoidance response saves the text to a hard drive or cloud and presumably is maintained because it has prevented computer crashes from costing the authors a day's or even a few minutes of work. Over time, however, clicking the save icon is so effective that loss of work rarely occurs and the rate of response begins to decline—we say we "forgot" to save, were careless, or it wasn't necessary. At this point, a computer crash or equivalent "shock" happens and suddenly reinstates the avoidance behavior. Thus, *avoidance is inherently cyclical*. It is a paradox that the more effective the avoidance response the fewer shocks are received, but the fewer shocks received the weaker the avoidance behavior. Like all operant behavior, avoidance responses must be negatively reinforced at least occasionally for the behavior to be maintained at high strength. In *Coercion and Its Fallout*, Murray Sidman pointed to the *avoidance paradox* when he compared contingencies of avoidance and positive reinforcement. He stated:

> The avoidance paradox reveals a critical difference between positive reinforcement and negative reinforcement by avoidance. With avoidance, success breeds failure; the behavior weakens and will stop unless another shock brings it back. With positive reinforcement, success breeds more of the same; the behavior continues. If the only reason a student studies is to keep from failing, an occasional failure or near-failure will be necessary to keep the studying going. A student who studies because of the options that new learning makes available will stop only if the products of learning become irrelevant. If citizens keep within the law only because that keeps them out of jail, they will eventually exceed the

speed limit, cheat on their income taxes, give or accept bribes, or worse. Citizens who keep within the law because of the benefits from participating in an orderly community will not face cyclic temptations to break the law.

(Sidman, 2001, p. 145)

The use of check stops by police and audits by the tax department ensure that drivers and taxpayers encounter or are threatened with occasional negative reinforcers. Without these occasional "shocks" there would be far fewer honest people in our society.

Murray Sidman (1953) was the first to investigate nondiscriminated avoidance, and the procedure is often called **Sidman avoidance** or free-operant avoidance. The procedure was similar to Solomon and Wynne's 1953 shuttle-box experiment, except there was no light that signaled an impending shock or safety from the shock. A **shock–shock interval** (**SSI**) was placed in effect. Here, if the dog did nothing, it would receive a shock at a predictable time, such as every 40 s. Second, at the same time, a **response–shock interval** (**RSI**) was also placed in effect. Here, when the dog jumped the barrier (the avoidance response), it could provide, for example, 60 s of safety from shock. Therefore, a dog could potentially avoid all shocks by repeatedly jumping the barrier about every 60 s, as long as its inter-jump interval was less than 60 s. Waiting too long (i.e., after the RSI elapses—60 s or more in this case) would result in a shock. The question was: Could the dogs respond to avoid shock without a signal?

The answer is: absolutely. What this might look like to an observer is that the animal is responding with no consequences at all. If the dogs spaced their responses such that the inter-jump intervals are just under the RSI, they could avoid most, if not all, of the shocks. This is indeed what they did. Avoidance responding, whether discriminated or non-discriminated, is learned more rapidly when the *RSI interval is longer than the SSI interval* (Sidman, 1962). In other words, when the operant delays the aversive stimulus (RSI) for a period greater than the time between shocks (SSI), conditioning is enhanced; this is because a larger time of safety is more reinforcing than a shorter time of safety.

Most of the research on avoidance learning has used shock as the aversive stimulus. Research has shown, however, that *timeout (TO) from positive reinforcement* (food) has functional properties similar to shocks. DeFulio and Hackenberg (2007) found that discriminated TO-avoidance parallel results obtained from signaled shock-avoidance contingencies. As for TO-avoidance with no warning stimulus, response rates to avoid timeout from response-independent food deliveries were a function of both magnitude (food quality) and rate of food pellet delivery, a finding similar to the manipulation of shock parameters in a Sidman avoidance procedure (Richardson and Baron, 2008).

Avoidance and Public-Health Vaccination Programs
In terms of avoidance behavior and community health, it seems there is always a minority of the population who refuse to vaccinate their children (see Figure 6.8) or who do not maintain a program of vaccination for childhood illnesses (Kristof, 2015 on "The Dangers of Vaccine Denial" in *The New York Times*). From a behavioral perspective, typical outbreaks of disease, such as the flu or measles, would function as negative reinforcement—maintaining nondiscriminated avoidance by vaccination. [Note: outbreaks are unpredictable events with no reliable warning for onset.] Parents who refuse to vaccinate their child may be from a generation that never has encountered an outbreak of diseases and has not learned the health benefits of avoidance of disease by vaccination, "seeing no reason for it." Also, health warnings to vaccinate are not

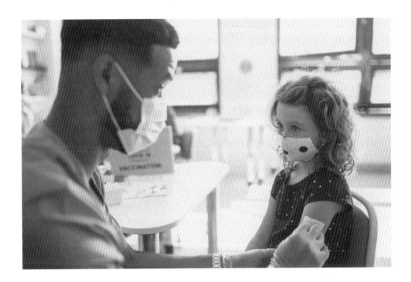

Fig. 6.8
A nurse gives a vaccine to a child.
Source: Shutterstock.

reliable or graphic enough signals for disease outbreaks and, without a history of discriminative avoidance, are not believed, in the sense that anti-vaccine people claim infringement on their freedom and fail to vaccinate their children.

An interesting phenomenon is vaccination rates for the COVID-19 virus. At the time of writing of this textbook version, the United States had just reached 1,000,000 COVID-related deaths (Centers for Disease Control, 2022a); the worldwide death rate had just reached 6.2 million (World Health Organization, 2022). Despite the vaccine's effectiveness at reducing deaths, intensive-care unit visits, and hospitalizations, and its high accessibility in most US pharmacies (available free of charge), only 66% of eligible individuals in the United States to date have gotten fully vaccinated. This could be partially due to the contingencies of COVID-19 not being a salient presence. For example, some people may not have been directly affected by a COVID-related death or hospitalization of a loved one or if they contracted COVID, they may have only experienced mild symptoms. In other words, there is no direct contact with the aversive stimulus that motivates getting the vaccine. Contributing to this is another possibility: the heightened spread of misinformation about the COVID-19 virus, which discourages vaccines. Such information has included that COVID-19 is a hoax or that the vaccine is unsafe or causes unfounded side effects, such as sterilization. Other misinformation includes government-related conspiracy theories, such as the vaccine contains a tracking device. These unfounded rumors may decrease the motivation for some to get the vaccine, which may also be reinforced by other like-minded members of their verbal community. Sadly, the reality is that lower vaccination rates mean that eradicating this virus will take a long time, even though we have the means to do it more rapidly.

Given that vaccination is avoidance behavior, the better we get at eliminating diseases by vaccination, the less probable another outbreak of disease will happen. As with other medical issues that involve avoidance, such as taking medication to avoid adverse health problems, one recommendation is that the situation be turned into escape behavior. If medical doctors or public-health officials can generate a more fear-based scenario (that is consistent with actual risk) to serve as a CS, from which the patient can escape by complying with vaccination suggestions, doing as prescribed may be enhanced (Cheney, 1996; Witte & Allen, 2000; but see Nyhan, Reifler, Richey, & Freed, 2014 for negative evidence about this strategy). This strategy can also backfire, however, if it is viewed as a fear tactic with no real contingency.

Aversive Stimulus Frequency and Avoidance Behavior

Avoidance behavior occurs when it reduces the frequency of an aversive event (Sidman, 1962). It does not need to necessarily eliminate the aversive stimulus completely, though—it just needs to *decrease the likelihood* of an aversive stimulus. In a classic experiment, Herrnstein and Hineline (1966) exposed 18 rats to a random sequence of small electric shocks. The animals could press a lever to reduce, but not eliminate, the probability of shock. Seventeen of the 18 rats in this experiment showed avoidance responding—they reliably pressed the lever.

This finding has generated a debate over the critical factors that regulate avoidance behavior. Essentially, the issue concerns molar versus molecular control of behavior in avoidance. From a **molecular account of schedule performance**, the moment-to-moment time between shocks (SSI) and the time from response to shock (RSI) represent the essential variables regulating avoidance responses (Dinsmoor, 1977, 2001a, 2001b). Nonetheless, the bulk of the evidence supports a **molar account of schedule performance** (Baum, 2001), suggesting that the molar variable, *overall reduction in shock frequency* (or sensitivity to rates of shock), establishes and maintains operant avoidance (Gardner & Lewis, 1976; Hineline, 1970).

Consider what happens when your friend persistently nags you to stop binge watching a series on Netflix and start working on your term paper. You may say, "Leave me alone, I'll get to it after this episode is over." This likely reduces the frequency of nagging, but does not eliminate it. In fact, your friend may now state, "I can't understand how you can just sit there glued to the TV when you have so much to do." Assuming that the nagging is a negative reinforcer, how can your vocal operant ("Leave me alone …") be maintained? The answer, of course, is that it has reduced the overall rate of nagging episodes from your roommate while you have been engrossed in *Stranger Things*.

The basic question concerning avoidance is how the lack of a contingent consequence—the lack of aversive stimulus—can maintain operant responding. How can "nothing" reinforce something? In discriminated avoidance the warning stimulus becomes aversive by being paired with the aversive stimulus (shock, for example) in the Solomon and Wynne (1953) shuttlebox study, and therefore a response that prevents the shock (avoidance) also terminates the conditioned aversive stimulus (light, which is a CS). This latter effect can be accounted for as an escape contingency, since the contingent consequence of the response is the safety signal of the light (CS). But when there is no change in the context as a result of a response, it is difficult to account for the operant behavior. This debate remains a topic of concern and research in the experimental analysis of behavior (Hineline and Rosales-Ruiz, 2013).

Avoidance as Impending Doom

Hackenberg and Hineline (1987) used a conditioned-suppression paradigm to show the interrelations between avoidance and behavior maintained by positive reinforcement.

Conditioned suppression is a procedure in which a conditioned aversive stimulus (S^{ave}, a tone that has signaled shock) is presented when an animal is responding for food reinforcement. The tone (S^{ave}) usually suppresses the operant behavior regulated by food. Hackenberg and Hineline (1987) introduced an interesting twist to show that a similar effect could be obtained when a period of avoidance either preceded or followed entire sessions of food reinforcement.

In their experiment, eight rats were trained to press a lever for food on a fixed-interval 3-min schedule (FI 3 min). After response rates were stable on the FI schedule, animals were exposed to 100 min of unsignaled shock avoidance. During this period, shocks occurred every 5 s (SSI = 5 s) unless the rat pressed a lever that postponed the shocks for 20 s (RSI = 20 s). These avoidance periods were presented to four rats just before the food reinforcement sessions. The other four animals were given the avoidance period immediately after they responded for food. The question was whether the avoidance periods would suppress responding during food reinforcement sessions.

The results indicated that operant responding for positive reinforcement was disrupted when avoidance periods either preceded or followed the food sessions. This suppression occurred even though the response rates of the rats remained high enough to obtain most of the available food. The avoidance periods had an effect that did not depend on interference with behavior maintained by positive reinforcement. When avoidance periods came after food reinforcement sessions, there was more disruption of food-related behavior than when avoidance periods preceded FI responding for food. In addition, when avoidance was discontinued, operant responses for food took longer to recover if the avoidance periods came after the sessions of positive reinforcement.

In everyday language, the rats "seemed worried" about their appointment with doom (remember that the animals had experienced these appointments in the past). This is not unlike a student who has difficulty studying because she is scheduled to have a wisdom tooth extracted a few hours later. People, and apparently rats, respond to *long-term aversive consequences in their environment*. This disruption of responding is severe when long-term aversive consequences are impending. Immediately-delivered aversive events can also suppress operant behavior but, all things being equal, do not appear to affect responses as strongly as long-delayed aversive consequences. By implication, a child who receives reprimands from a teacher for talking out of turn will show little disruption of play and school work. In contrast, a student who is regularly harassed by a bully (or an overly demanding parent) after school is over may show *general disruption* of school activities throughout the day.

Timeout from Avoidance

We all value a holiday from the stress of school requirements or the duties and obligations of the work world. One way to analyze work and holidays is to recognize that much of our behavior as students and employees is maintained by schedules of avoidance. We get our reports and essays in on time to avoid the reprimands, low grades, or low performance evaluations that we received from our boss or teacher in the past. Avoidance contingencies are so prevalent that we spend much of our day engaged in avoiding the aversive stimuli arranged by others and the natural world. Think about it.

Now consider how much we value holidays, leaves of absence, and other periods that temporarily suspend or remove the everyday aversive and behavioral requirements that pervade our lives. These periods of **timeout from avoidance** may be analyzed as negative reinforcement of behavior that terminates, prevents, or postpones the avoidance contingencies. For example, university professors are given sabbatical leave (suspension of teaching and administration duties) contingent on writing a sabbatical application outlining a program of research and academic inquiry. A strong record of publishing and obtaining research grants is part of the faculty assessment for granting sabbaticals. Professors with high publication records and large research grants are judged worthy of sabbaticals as well as monetary merit awards. Obviously, sabbatical leave, as timeout from avoidance, is part of the reinforcement contingencies that maintain much of the academic behavior of university professors.

Timeout from avoidance has been studied experimentally in a procedure developed by Perone and Galizio (1987). Rats could press either of two levers. Responses on the right lever postponed electric foot shocks arranged on a Sidman avoidance schedule. Pressing the left lever intermittently produced 2 min of timeout from avoidance. Insertion or withdrawal of the response levers, illumination of the chamber, and auditory white noise signaled periods of time in and timeout of avoidance. In a number of experiments using the two-lever procedure, timeout from avoidance maintained behavior on a variety of interval and ratio schedules of reinforcement (Foreman, 2009). Thus, taking a holiday from avoidance has proved to be effective as negative reinforcement.

But what is it about a holiday that makes it function as negative reinforcement? The timeout period involves three distinct changes in the environment. A stimulus change from avoidance to timeout occurs, the frequency of shocks (stressors) is reduced, and the response requirements for avoidance and timeout are suspended, resulting in reduced response effort. Research shows that the stimulus change from avoidance to timeout does not maintain responding on the timeout lever (Perone & Galizio, 1987). Additional research indicates that reduction in shock frequency plays a relatively minor role in maintaining timeout behavior, while reduction in response effort is the key factor (Courtney & Perone, 1992; Foreman, 2009). People may value holidays, leaves of absence, and sabbaticals because these are *periods of reduced response effort* during which many of the behavioral requirements of life are suspended.

SIDE EFFECTS OF AVERSIVE PROCEDURES

There are obvious ethical reasons for not using punishment contingencies to change behavior. These ethical issues arise even though punishment, by definition, always works and works quickly if used appropriately. There also are serious side effects that often arise when contingencies of punishment and negative reinforcement are employed. Skinner (1953, 1971) recognized these problems and has consistently argued against the use of punishment techniques:

> The commonest technique of control in modern life is punishment. The pattern is familiar: if a man does not behave as you wish, knock him down; if a child misbehaves, spank him; if the people of a country misbehave, bomb them. Legal and police systems are based on such punishments as fines, flogging, incarceration, and hard labor. Religious control is exerted through penances, threats of excommunication, and consignment to hell-fire. Education has not wholly abandoned the birch rod. In everyday personal contact we control through censure, snubbing, disapproval, or banishment. In short, the degree to which we use punishment as a technique of control seems to be limited only by the degree to which we can gain the necessary power. All of this is done with the intention of reducing

tendencies to behave in certain ways. Reinforcement builds up these tendencies; punishment is designed to tear them down.

(Skinner, 1953, pp. 182–183)

The use of punishment for the social regulation of behavior maintained by positive reinforcement has serious unintended effects, to which we shall now turn.

Behavioral Persistence by the Punisher

As we have seen, punishment may under some circumstances produce a rapid decline in behavior. Thus, individuals who use punishment effectively are more likely to use it on future occasions (Miller et al., 2010); that is, their behavior of using punishment is strengthened, or negatively reinforced, by the eradication of problem behavior of others. This is an important point—the "successful" use of punishment leads to further use of the technique, which produces the additional side effects of aversive control, such as counter-aggression, escape, and ennui.

Avoidance Prevents Extinction of the Conditioned Stimulus

Consider a person who has received a painful wasp sting. The sight and buzzing of the insect precede the sting and (for some people) become powerful conditioned stimuli (CS) that elicit anxiety. These CSs are likely to generalize to similar sights and sounds (i.e., the sight of other flying insects, or the buzzing of a harmless fly). The CS also has a dual function. In terms of Pavlovian associative conditioning, the stimulus elicits anxiety; in an operant sense, it functions as a conditioned aversive stimulus (S^{ave}), which strengthens behavior that removes it (negative reinforcement). To extinguish the effects of the CS, it must be presented in the absence of the unconditioned stimulus or US (respondent extinction). Under ordinary circumstances the CS would rapidly extinguish, as buzzing sounds and flying insects (CS) are rarely accompanied by pain (US). People who are afraid of wasps and bees, however, avoid places where these insects are found and immediately leave these locations if inadvertently encountered. Generally, *avoidance behavior maintained by operant conditioning prevents respondent extinction*. Indeed, it is well documented in research on the therapeutic treatment Acceptance and Commitment Therapy (ACT), that avoidance behavior can actually result in strengthening fears and anxiety (called experiential avoidance; Hayes & Wilson, 1994). Confronting the feared stimulus is critical to reduction of anxiety and fear in a variety of psychotherapies, including exposure therapy, systematic desensitization, and ACT.

One way to extinguish fear-based CSs is to expose the organism to the aversive stimulus while preventing an effective escape response. For example, if a person has a crippling fear of flying in airplanes, having them board a commercial jet until takeoff can ensure there will be no escape response. This is similar to what takes place in exposure therapy. That would likely be an initially terrifying situation for someone, but they can learn that flying on an airplane will likely result in nothing dangerous (air travel is, after all, the safest way to travel). A more gentle approach might be to use a fear hierarchy, such as what is used in systematic desensitization (e.g., Wolpe, 1961). Here, a series of stimuli rated to the feared event that generate lower anxiety (e.g., making a flight reservation) to highest anxiety (e.g., flying on the plane) are listed. Relaxation is taught for each step in the hierarchy starting with the lowest ranked stimulus until anxiety is low or manageable and eventually the client meets the target stimulus (e.g., flying in an airplane).

In therapeutic settings, fears that disrupt life can be alleviated by a professional. In many everyday settings and behaviors, however, escape and avoidance responses that are less disruptive to life can be resistant to extinction. This *persistence occurs when the stimulus conditions between danger (the presence of the punisher) and extinction (the absence of the punishment) setting is low.* In other words, extinction may not clearly be signaled by the CS. When the difference between the extinction setting and the situation associated with an aversive stimulus is slight, extinction is not discriminated, and avoidance responding continues. For example, in everyday life, a dentist's office might smell and look similar to the hospital where pain was once experienced, or flowers on a neighbor's patio might be similar to those in a garden where flying insects and buzzing sounds were once accompanied by a sting. It is not clear whether an actual aversive stimulus (such as pain) will be experienced in these conditions.

Learned Helplessness

A similar persistence effect occurs when animals are exposed first to inescapable, aversive stimulation and are later given an opportunity to escape. In the phenomenon called **learned helplessness**, an animal is first exposed repeatedly to inescapable and unpredictable aversive stimuli. Eventually, the animal gives up and stops attempting to avoid or escape the situation because nothing works. Next, the conditions are changed and an escape response, which under ordinary circumstances would be acquired easily, is made available. Surprisingly the animal does not make the escape response; it has learned to become helpless. In an early experiment, Seligman and Maier (1967) exposed dogs to inescapable and unpredictable electric foot shock. Following this, they attempted to teach the animals to avoid signaled shocks by jumping across a shuttle-box barrier. The dogs failed to avoid the shocks, and even after the shocks came on, they would not escape by crossing the barrier to safety. The researchers suggested that the dogs had learned to give up and become helpless when presented with inescapable aversive stimuli. Of course, dogs that are not first exposed to inescapable shock learn quickly to escape and avoid shocks in a shuttle box.

Learned helplessness has been found in a large number of experiments, and has been documented in other animals (e.g., Jackson, Alexander, & Maier, 1980; Maier & Seligman, 1976; Maier, Seligman, & Solomon, 1969; Overmier & Seligman, 1967; Seligman & Maier, 1967). In a report from *Current Biology*, escape-trained flies (*Drosophila*) that stopped walking for a brief period received blasts of heat, which could be terminated by resumption of walking (Yang, Bertolucci, Wolf, & Heisenberg, 2013). These flies show reliable resumption of walking after receiving heat pulses. Other inescapable-trained flies received the heat pulses in exactly the same sequence as flies from the escape-trained condition, but could do nothing to escape the heat blasts and gave up responding. Following training, both groups of flies (escape-trained and inescapable-trained) were given a test for locomotion. Flies that received inescapable heat pulses walked more slowly, rested more frequently, and appeared "depressed" compared to escape-trained insects.

Similar results have been reported for humans. Hiroto and Seligman (1975) exposed college students to a series of inescapable-loud noises. Following this procedure, the students had to solve a number of anagram problems. Students exposed to inescapable noise had more difficulty solving problems than students who were not exposed to the loud noise. Most control subjects solved all of the anagrams and reached solutions faster and faster. In contrast, students who were exposed to inescapable noise failed many problems and made slow improvements in performance.

The practical implication of these findings seems obvious. When people are exposed to inescapable failure (shocks), they may learn to give up and become helpless. A parent who spanks a child on the basis of his or her mood rather than for the child's misbehavior may create a socially withdrawn individual. The child has learned that "No matter what I do, I get a spanking." An individual who frequently "blows up" for no apparent reason might produce a similar set of responses in their partner.

Helplessness, Punishment, and Avoidance

Inescapable social "shocks" are not the only way to learn helplessness. Indiscriminant punishment and avoidance contingencies were brutally arranged by the Nazi guards of concentration camps to instill a kind of helplessness and docility in Jewish prisoners (Figure 6.9). Many people have questioned how so many people could have gone to their deaths without resisting the Nazi captors. The answer lies in the power of aversive control, which far exceeds what we can imagine.

The German jailors often used unpredictable and arbitrary slaughter of prisoners to maintain control after using death by execution for any minor act of resistance. In this situation, the Jewish captives learned to avoid death by doing what they were expected to do. Once this helpless avoidance had been set up, the SS guards could keep it going by occasionally

Fig. 6.9
A US postage stamp depicts Holocaust survivors after the Allied liberation in 1945. The survivors showed the profound effects of indiscriminant use of punishment and avoidance contingencies by the Nazi guards.
Source: Shutterstock.

selecting a few prisoners to shoot or exterminate on an arbitrary whim. These executions were unrelated to anything that the victims did or did not do—they were unavoidable. Murray Sidman explains that imposing indiscriminate death by execution on learned avoidance of death was the basis of the observed helplessness:

> If the shock had merely been painful [instead of death], the Jews might have resisted, welcoming death as the ultimate escape. With death itself as the shock, however, escape from death was the controlling contingency. That shock, delivered frequently with machinelike ruthlessness, was at first contingent on the prisoners' actions—when they resisted, for example, or failed to obey orders. Later, the shocks bore no relation to anything they actually did or failed to do. Because the original contingencies had generated required avoidance behavior—docility—the subsequent noncontingent shocks [arbitrary shooting of prisoners] kept that form of avoidance going. An outside observer, or a historian, could see that their quiet march to the ovens was futile. The change in the rules had come without notice, however, and those who were about to be murdered were simply doing what the original contingencies [of avoidance] had taught them was necessary for survival. Their deaths served to maintain the docility of those who remained.
>
> (Sidman, 2001, pp. 147–148)

It is important to recognize that helplessness had nothing to do with the Jewish people being unable to resist. Anyone exposed to similar kinds of coercive control would behave in a similar fashion, regardless of race, ethnicity, or religious orientation. Helplessness does not rest within the victims of violence, but with the powerful behavioral effects engendered by the aversive contingencies arranged by the Nazis.

Learned Helplessness and Depression
Seligman (1975) argued that the research on learned helplessness with animals provides a model for clinical depression. For example, there is evidence that helplessness is involved in the relationship between alcohol dependence and depression (Sitharthan, Hough, Sitharthan, & Kavanagh, 2001). More generally, thousands of people each year are diagnosed with depression. These individuals show insomnia, report feeling fatigued, may sleep often (avoidance of their difficult circumstances), have difficulty performing routine tasks, and may be suicidal. Clinical depression is severe, long lasting, and not easily traced to a recent environmental experience.

Although animal experiments may shed light on human depression, there are differences (Abramson, Seligman, & Teasdale, 1978; Peterson & Seligman, 1984). For the most part, Seligman points to differences that occur because of human verbal behavior. People talk about their problems and attribute them to either internal or external causes. When people attribute their difficulties to personal causes (e.g., "I am a failure"), these attributions could set the occasion for giving up (rule-governed behavior, as discussed in Chapter 11). In terms of treatment, Seligman suggested that depressed individuals be placed in situations in which they cannot fail. In this manner, the person may make contact with reinforcement, and eventually learn successful responses in the presence of negative life events.

Seligman (1991) also suggested how to prevent learned helplessness and depression. A person who has already learned to escape from punitive control may be "immunized" against the effects of inescapable aversive events. Such an effect is suggested by experiments in which animals initially learn some response (e.g., wheel running) to escape aversive stimuli.

First, the animals learn an effective escape response to negative reinforcement contingencies. Next, the animals are exposed to the typical learned-helplessness procedures of inescapable aversives. Finally, the subjects are tested in a situation where a new response produces escape from the aversive stimuli (e.g., switching sides in a shuttle box). The typical effect of *pre-exposure to escape* is that this experience blocks the learned helplessness usually brought on by inescapable aversive stimulation (Maier & Seligman, 1976; Williams & Lierle, 1986; but see Dos Santos, Gehm, & Hunziker, 2010 for negative evidence). This can also be considered building resilience.

The importance of "immunization" against learned helplessness can also be illustrated by a millennial parenting strategy referred to as *helicopter parenting*. This childrearing practice refers to a tendency for a parent to pay close attention to their children's activities in an attempt to prevent their child from experiencing disappointment or pain, such as failing, unfairness, or rejection from peers. In other words, well-meaning parents try to prevent their children from experiencing aversive stimuli throughout life. Research shows that while this may have positive benefits for children when they are young, in the long run, it actually may be harmful. College-aged students who experienced helicopter parenting as children are more likely to have poor psychological well-being, suffer from depression, and have greater difficulty adjusting to college (LeMoyne & Buchanan, 2011; Reed et al., 2016; Schiffren et al., 2014). These challenges are tied to a lack of autonomy and competence; in other words, individuals with helicopter parenting likely have an underdeveloped ability to control the aversive stimuli they encounter as college students. It may make sense, then, to allow children to experience difficulties as they develop, but provide positive support and coping skills while allowing them to label and experience their emotional responses. These types of strategies can build resilience and resistance to helplessness.

Learned Helplessness, Depression, and Neuroscience
In addition to behavioral approaches that immunize against learned helplessness, neuroscience research is currently analyzing the underlying brain mechanisms. The objective is to identify the brain structures, neuronal systems, and neurochemistry implicated in learned helplessness and depression, in the hope of discovering new medical treatments (see LoLordo & Overmier, 2011).

One promising brain structure, relating inescapable shocks (or stressors) to behavioral depression, is the medial prefrontal cortex, which is rich in 5-hydroxytryptamine (5-HT) receptors that are activated by the neurotransmitter serotonin (Amat et al., 2005). The 5-HT receptors modulate the release of many neurotransmitters and hormones related to stress and "reward-negative" reactions, including the neurotransmitter dopamine. In this regard, a line of research links dopamine and several brain sites to behavioral depression induced by learned helplessness. Drugs that target the dopaminergic pathways may eventually offer a treatment for clinical depression (Bertaina-Anglade, La Rochelle, & Scheller, 2006; Takamori, Yoshida, & Okuyama, 2001), especially when combined with behavioral interventions focused on overcoming and preventing learned helplessness.

At the cellular level, depressive behavior is poorly understood. Neurons in the lateral habenula (LHb) near the dorsal thalamus are implicated, as they allow communication between the forebrain and midbrain areas associated with learning about "reward-negative" events (see Mirrione et al. 2014 for brain imaging in learned helplessness). Importantly, LHb neurons project to, and modulate, dopamine nerve cells in the ventral tegmental area (VTA) of the midbrain. Dopamine neurons in the VTA in turn participate in the control of depressive behavior induced by inescapable shock. A study in *Nature* showed that learned-helplessness procedures increased excitatory synaptic responses of LHb neurons projecting to the VTA (Li et al., 2011). Furthermore, enhancement of LHb synaptic activation results from presynaptic release of a neurotransmitter, and correlates with an animal's depressive behavior. Repeated electrical stimulation of LHb afferent nerves depletes the release of the neurotransmitter, substantially decreases excitatory synaptic responses of LHb neurons in brain slices, and significantly reduces learned helplessness behavior in rats. Overall, the results indicate that transmitter release onto LHb neurons contributes to the rodent model of learned helplessness and depression. Also, the electrical stimulation method used to deplete transmitter release and reduce learned helplessness is a promising medical treatment, which could supplement behavioral therapy (Strosahl & Robinson, 2016), for patients diagnosed with clinical depression.

NEW DIRECTIONS: BEHAVIORAL NEUROSCIENCE OF SOCIAL DEFEAT

Psychiatric disorders that include depression, social phobia, and post-traumatic stress disorder (PTSD) have been linked to social withdrawal and to abnormalities of the dopaminergic system. To gain a better understanding of these links, Berton et al. (2006) used a social defeat procedure that profoundly alters the social interactions of rodents.

In their study, mice were given daily episodes of social defeat, followed by a period of protected exposure to the larger aggressor—both animals were placed in a cage separated by a barrier to allow for sensory contact. The test mice were subjected to defeat by different aggressors over a period of 10 days, and measures of social behavior were obtained. The researchers measured social approach to an unfamiliar mouse enclosed in a wire cage, using a video-tracking system. Control animals (undefeated) spent most of the time in close proximity to the unfamiliar mouse. Defeated mice displayed intense aversion responses and spent less time near the unfamiliar mouse in the cage, but not when the wire cage was empty. Thus, the response was to the social target (unfamiliar mouse), not the novel wire cage. When tested again after 4 weeks, mice with a history of social defeat still displayed avoidance of the social target. Not surprisingly, these avoidance responses were greater to the aggressor, but also generalized to unfamiliar mice that were physically distinct from the aggressor.

Next, the researchers showed that antidepressant drugs used with humans improved the social interaction of defeated mice, but anxiety-related drugs did not have this effect. One possibility is that antidepressant drugs operate on the dopaminergic (DA) pathways of the brain. To further characterize the neurobiological mechanisms of social aversion induced by defeat, Berton et al. (2006) targeted the dopamine neurons of the mesolimbic brain in the ventral tegmental area (VTA), as well as the projections of these neurons to the nucleus accumbens (NAc). Previous research has shown that these pathways are associated with emotionally salient stimuli and avoidance behavior. The neurotrophic factor BDNF (brain-derived neurotrophic factor) is a major regulator of the mesolimbic dopamine pathway—modulating the release of dopamine. BDNF is also involved with dopamine (DA) release in the NAc via the TrkB receptor (tropomyosin receptor kiase B) on the dopamine nerve terminals. The findings showed that BDNF levels in the NAc were increased by social defeat, and this effect occurred 24 h and even 4 weeks after the episodes of social defeat.

The source of the BDNF protein in the NAc is thought to be the VTA, where the messenger RNA (mRNA) for BDNF is expressed. Berton et al. (2006) deleted the gene encoding for BDNF in the VTA of adult mice and found an *antidepressant-like effect*; the deletion of the gene for BDNF and

DA release reduced the acquisition of social avoidance behavior in defeated mice. This finding and other control conditions indicated that *BDNF from the VTA neurons is required for a social target to become an aversive stimulus* that regulates the avoidance behavior of defeated mice. Subsequent research, using phasic optogenetic-light stimulation *in vivo*, has shown that optogenetic activation of the DA mesolimbic pathway increases BDNF in the NAc of socially stressed mice, but not non-stressed mice. This stress activation of BDNF signaling is mediated by corticotrophin-releasing factor (CRF, a stress hormone and neurotransmitter) acting on NAc neurons—providing a stress context-detecting mechanism for the brain's mesolimbic DA-reward circuit (Walsh et al., 2014).

One implication of the neuroscience of social defeat is that humans diagnosed with affective disorders may be showing avoidance responses acquired by a history of social punishment and defeat. These behavioral effects may involve BDNF, CRF, and the dopaminergic (DA) pathways. Behavior therapy (Strosahl & Robinson, 2016) when combined with specialized antidepressant drugs could be especially effective at reducing social aversion and increasing socially appropriate behavior. In addition, innovative neurochemical research is linking learned helplessness and social defeat procedures to common brain mechanisms, with the objective of yielding a more complete account of major depressive disorder (Amat, Aleksejev, Paul, Watkins, & Maier, 2010; Hammack, Cooper, & Lezak, 2012).

Aggression: A Prominent Side Effect

Displaced Aggression

When two rats are placed in the same setting and painful shocks are delivered, the animals may attack one another (Ulrich & Azrin, 1962; Ulrich, Wolff, & Azrin, 1964). The fighting generated by these contingencies is called **displaced aggression**, because the attack follows the presentation of aversive events. Attack occurs even though neither animal is responsible for the occurrence of the shocks. Displaced aggression has been documented in several species, including humans (Hutchinson, 1977), and has been found with stimuli other than electric shock (Azrin, Hake, & Hutchinson, 1965); indeed stimuli that are correlated with extinction (previous reinforcement that is no longer available) can induce displaced aggression. Most people recognize that they are more prone to aggression when exposed to aversive stimuli, whether painful, aversive, or correlated with extinction. When you are feeling good you may never shout at your partner, but you may do so if you have a severe toothache or headache.

In these early experiments (O'Kelly & Steckle, 1939), rats were placed in a small enclosure and electric shock occurred periodically, no matter what the animals did. When the rats were periodically shocked, they began to fight. Twenty-three years later, Ulrich and Azrin (1962) systematically investigated the fighting behavior of rats exposed to inescapable and intermittent shocks. These researchers began by testing whether two rats would fight when simply placed in a small operant chamber, noting that the animals showed a low probability of fighting when placed in a confined space without shocks. When random shocks were delivered, however, the rats would immediately face each other, assume a standing posture with mouth opened, and vigorously strike and bite one another (see Figure 6.10)—inflicting serious physical injury on each other unless precautions were taken. Notably, the amount of pain-elicited fighting between rats critically depended on the chamber size. In a small chamber about 90% of the shocks elicited a fighting response. Fighting decreased with larger floor space with only 2% of the shocks eliciting fighting in the largest chamber. Thus, confinement with another animal

Fig. 6.10
Two rats in the attack position induced by electric shock.

Source: Reprinted from R. E. Ulrich & N. H. Azrin (1962). Reflexive fighting in response to aversive stimulation. *Journal of the Experimental Analysis of Behavior, 5*, pp. 511–520. Copyright 1962 held by John Wiley & Sons Ltd. Published with permission.

combined with shocks produced high amounts of displaced aggression (see Ulrich, Hutchinson, & Azrin, 1965 for a review of the moderating factors related to pain-induced aggression).

In some studies, shocks were delivered at increasing frequencies, and the number of attacks increased as more shocks were presented. In addition, Ulrich and Azrin (1962) found that the probability of attack for any single shock increased as the number of shocks increased. When the animals received one shock every 10 min, attacks followed approximately 50% of the shocks. When the animals received 38 shocks a min, fighting followed 85% of the shocks. The probability that a painful event will induce aggressive behavior is greater following high rates of painful stimulation.

Painful stimulation also produces attack-like responses in humans and monkeys (Azrin & Holz, 1966; Azrin, Hutchinson, & Sallery, 1964; Hutchinson, 1977). In one experiment, electric shocks were delivered to squirrel monkeys (Azrin et al., 1964). As with rats, attack was induced by shock. The animals attacked other monkeys, rats, mice, and inanimate objects, such as a stuffed doll, a round ball, or a rubber hose that they could bite. As shock intensity increased, so did the probability and duration of the attacks—a result that parallels the findings with rats.

In a review of the side effects of aversive control, Hutchinson (1977) described bite reactions by humans to aversive stimuli. Subjects were paid volunteers who were exposed to inescapable-loud noise at regular intervals. Because the noise was delivered on a predictable basis, the subjects came to discriminate the onset of the aversive stimulus. Unobtrusive measures indicated that humans would show aggressive responses (or, more precisely, bites on a rubber hose) following the presentation of loud noise. The participants' responses to noise parallels the fighting found in monkeys and other animals. Hutchinson, however, suggests that these human results should be interpreted with caution. The participants were told that they would receive aversive stimuli, but the intensity would be tolerable. Also, he noted that participants were paid to stay in the experiment, and most people would leave such a situation in everyday life if possible.

As alluded to earlier, following a period of positive reinforcement, extinction or withdrawal of reinforcement for operant behavior is capable of inducing aggressive responses. Thus, Azrin et al. (1966) were able to induce attacks on a target pigeon by alternating periods of reinforcement with periods of extinction. After continuous positive reinforcement, the onset of extinction at first increases and then decreases operant rate, which often is accompanied by an increase in aggressive behavior. Consider the vending machine that you use each morning to get a cup of coffee before class. Inserting your money is the operant, and the machine dispenses a coffee each time you insert the coins. One morning, you insert the money, but the machine does not operate (it's broken, and your behavior is on extinction). What do you do? At first, many of us would wonder what has happened, think, "Maybe I put in the wrong amount," and reinsert the required coinage (continue operant responding). But again, the machine does not operate and you now find yourself hitting the coin slot, striking other parts of the machine with your hands, kicking it, or using a magical command such as "Work, you stupid machine." As you have just taken a course in behavior analysis, you suddenly realize that it's not the stupid machine, but the period of extinction following positive reinforcement that accounts for your aggressive outburst.

Aggression also is generated by extinction following intermittent reinforcement (see Frederiksen & Peterson, 1977). After intermittent reinforcement, a period of extinction induces more aggression than after continuous reinforcement (CRF). Extinction-induced attacks occur following both FR and VR schedules, with FR schedules generating more attacks than VR contingencies. On FI schedules, in which the first response after a set period of time is reinforced, attacks directed at a restrained but protected bird (or mirror image) occur at fixed intervals ranging from 60 to 270 s. Pigeons on VI schedules of reinforcement also show induced aggression.

For humans, aggression has been induced by extinction following positive reinforcement, and by periodic reinforcement on interval and ratio schedules (Frederiksen & Peterson, 1977). The rate of attack is related to the rate of reinforcement, but the nature of this relationship in humans is not well researched. Human aggression is distributed throughout the IRI and less confined to the period just after reinforcement as in other animals. Prior history of reinforcement of the aggressor, the stimulus features of the target, availability of weapons, and the current rate of reinforcement probably combine to produce different rates of attack in humans, but again research is lacking. One possibility is that periodic reinforcement, and changes from reinforcement to extinction, provide a behavioral interpretation of the seemingly irrational mass shootings and other forms of aggression in America following loss of employment, the ending of a relationship (such as divorce), and thinning of reinforcement rates by alienated family, friends, or work associates (see Follman, Aronsen, & Pan, 2014 article on mass shootings in America in *Mother Jones* magazine).

Countercontrol
When one person punishes another's behavior, the punished individual may retaliate. This is not difficult to understand; one way to escape from punishment is to eliminate or neutralize the person (or source) who is delivering it (Azrin & Holz, 1966). This strategy is called

countercontrol, and it is shaped and maintained by negative reinforcement (i.e., removal of an aversive stimulus). Revolutions and wars are based on citizens rising up against an oppressive government. A child who is picked on by a bully may finally fight back. Adolescents with overprotective parents may rebel. All of these are examples of countercontrol.

One problem with controlling others through aversive control, especially through physical aggression, is that people who successfully use counter-aggression to stop the punishment arranged by others may be negatively reinforced for doing so. Consider a situation in which a couple's argument escalates until one of them loses their temper and strikes their partner, which suppresses their partner's arguing. This represents a series of countercontrol behaviors with one person ending the argument in a violent act. The aggressor is negatively reinforced by their partner's submission after the physical abuse. Although this does not completely explain domestic partner abuse, it does suggest that negative reinforcement plays a large role in many cases (see section "Aversive Control of Human Behavior" earlier in this chapter for more details).

Although human aggression is easily recognized, it is difficult to study in the laboratory. This is because aggressive behavior is a dangerous form of human conduct. Realizing the danger, researchers have developed procedures that protect the victim from harm. In the laboratory situation, participants are led to believe that they have an opportunity to harm another person when in reality they do not (e.g., Cherek, et al., 1997; Gustafson, 1989; Vasquez, Denson, Pedersen, Stenstrom, & Miller, 2005). In a typical experiment, participants are told that they can deliver a punisher (e.g., loud noise, electric shock, stealing points or money, or hand immersion in ice water) to another person by pressing a button on a response panel or indicating how long the hand must stay submerged in extremely cold water. The other person is in fact an accomplice or confederate of the researcher, and acts the role of victim, but does not actually receive the aversive stimulus.

There has been a debate about the reality or **external validity** of these procedures. However, evidence suggests that these methods constitute a reasonable analog of human aggression in everyday life. Participants in aggression experiments seem convinced that their actions harmed the confederate (Berkowitz & Donnerstein, 1982). Additionally, when the accomplice provokes the participants with insults or other types of aversive stimuli, they deliver greater amounts of painful stimulation than when not provoked (Baron & Richardson, 1993). Finally, people who are known to be violent usually select and deliver stronger levels of aversive stimulation than those without such a history (Cherek et al., 1997; Gully & Dengerink, 1983; Wolfe & Baron, 1971).

Aggression Breeds Aggression
The presentation of an aversive stimulus may make aggressive behavior more likely. Provocation by others is a common form of aversive stimulation that occurs in a variety of social settings. Consider a situation in which you have worked extremely hard on a term paper and you feel it is the best paper you have ever written. Your professor calls you to their office and says, "Your paper is rubbish. It lacks clarity, scholarship, and organization, and is riddled with grammatical mistakes." You protest what feels like unfair treatment, but to no avail. You storm out of the office mumbling a few choice words, and once down the hall you kick the elevator door. Later in the term you are asked to fill out a teaching evaluation and, in retaliation, you score the

professor as one of the worst teachers you have ever known. In this example, the professor's insulting remarks generated aggressive responses that ranged from kicking a door to counter-attack by negative evaluation. Generally, aggression breeds aggression (Patterson, 1976).

In terms of physical provocation, experiments show that people respond to attacks with escalating counterattacks (Borden, Bowen, & Taylor, 1971; O'Leary & Dengerink, 1973; Taylor & Pisano, 1971). In these historical experiments, participants tried to beat their opponents on a reaction-time game in which the loser received a minor electric shock. In fact, there were no actual opponents, but the participants received shocks that were programmed by the researchers. In this game, subjects were made to lose on a number of trials and the shocks from the fictitious opponent increased in magnitude. Faced with increasing physical provocation, subjects retaliated by escalating the intensity of the shocks they gave when the "opponent" lost (see Anderson, Buckley, & Carnagey, 2008 on how a history of generalized aggression or trait aggression influences aggression level in this situation).

It is important to note that these more historical experiments would not likely receive approval from an Institutional Review Board (IRB) today. An IRB is a committee at institutions that conduct human research like universities or medical hospitals; it is comprised of individuals with scientific training, non-scientific training, and community members. This committee's job is to independently review human research studies, especially regarding the ethics of the study. They protect the rights and welfare of the research participants, as well as ensure adherence to federal and state regulation and policy. IRBs were absent in the early days of aversive control research, and only began in 1974 after the heinous treatment of humans in research by the Nazi regime. Unfortunately, the United States also has not always protected humans in research, and often those who are underrepresented have received the most harm (see *Bad Blood: The Tuskegee Syphilis Experiment: A Tragedy of Race and Medicine* by James H. Jones and *The Immortal Life of Henrietta Lacks* by Rebecca Skloot for historical examples of lack of protection in research for Black Americans). As you read about some of these studies described in this textbook, pay attention to the dates when they were published. The use of shock with humans in these studies is not prevalent today because of IRBs. Even the concern of making people *think* that they shocked or harmed someone also has some ethical complexities that IRBs would consider very carefully today before allowing. The regulations of IRBs continue to change and become more protective of humans as more data and information about harm and benefit are discovered and evaluated.

The study of physical provocation's role in aggression is indeed challenging to study ethically. But it is also important to understand this type of aggression because of its high occurrence in society and in the natural world. Therefore, we rely on earlier, perhaps less ethically conducted studies as well as those that are more recent, but more ethically conducted. One older study on provocation involved physical aggression with individuals from different cultural/ethnic backgrounds: Israeli, European Russian, American, and Georgian (Jaffe, Shapir, & Yinon, 1981). These participants were asked to administer bogus shocks for "incorrect" answers to a learner, who was in fact a confederate. Although not required to do so, individuals escalated the level of shocks given for incorrect answers (punishers). When placed in groups of three, they delivered the highest levels of escalating "shocks," presumably because individual responsibility and blame is reduced in groups (in social psychology, we call this *diffusion of responsibility*). These findings were replicated across all ethnic/cultural groups, suggesting that escalation of aggression to social punishment (incorrect responses) occurs for individuals and groups regardless of cultural background.

In another study, people matched their level of aggression to the level of provocation (Juujaevari, Kooistra, Kaartinen, & Pulkkinen, 2001). Also, people retaliated more when they were provoked and subsequently presented with a minor annoyance than when they were only provoked or received no provocation. The minor annoyance became a "trigger" for retaliation when it was preceded by provocation; by itself it had no effect on aggressive behavior (Pedersen, Gonzales, & Miller, 2000). This has implications for *microaggressions*. Microaggressions are common, subtle verbal, behavioral or environmental slights (both intentional and unintentional) that communicate negative attitudes or stereotypes about culturally marginalized groups. Some examples of microaggressions include telling an articulate Black woman that she talks like a White person (being articulate is not connected to race) or asking an Asian-American student how long they have been in the United States (they were born in America, and so were their parents). These types of issues might seem like small matters to some ("I didn't mean any harm. Why so sensitive?"), but imagine having to endure these types of comments every day or have them preceded by larger aversive experiences, such as racial violence or discrimination. Sensitivity to those types of comments might increase; i.e., they could become triggers. They are called microaggressions for a reason—they are smaller than larger acts of racism (or homophobia, sexism, agism, or ableism), but their accumulated impact can feel larger to the person who is the target. This is why using care in what one says to others is a considerate practice.

Verbal insults also elicit and set the occasion for strong counterattacks. Wilson and Rogers (1975) suggest that verbal provocation can lead to physical retaliation; they noted incidents that began with verbal taunts, escalating into violent fistfights. In a laboratory study of verbal insults, Geen (1968) found that participants exposed to unprovoked, nasty comments from a confederate would retaliate with physical aggression. The participants in this study were allowed to deliver shocks to the insulting confederate (in fact, no shocks were actually given). Compared with personal frustration (a confederate preventing them from completing an assigned task) and task frustration (the task not having a solution), verbal insults produced the highest level of aggression toward the confederate.

A field study of 6th-grade children supports these findings from the laboratory. In this study, the context of the insults and reactions of other children increased perceived hostility of insults from classmates (intensity of punishment) and the amount of verbal escalation that followed. Insults related to permanent social attributes (e.g., ethnic identity) elicited humiliation and escalated the violence from verbal insults to physical aggression for both boys and girls (Geiger & Fischer, 2006).

Escalation of violence from insults to physical aggression occurs more often in males who come from a "culture of honor" (e.g., southern USA) than in those who do not (e.g., northern USA). For males who highly value honor, insults diminish the person's reputation, and retaliation involves behavior that previously restored status and respect (Cohen, Nisbett, Bowdle, & Schwarz, 1996; see McAndrew, 2009 for a discussion of challenges to status, testosterone levels, and aggression in human males). Generally, aggression breeds more aggression, and aggressive episodes may escalate toward harmful levels of physical violence, especially in cultures that propagate dignity and honor (see Skinner, 1971 on other problems of freedom and dignity).

Aggression: Response to Social Exclusion

The dictum that aggression breeds aggression can also be extended to problems of violence in American schools and other social situations. One common form of group behavior involves social exclusion of others based on their characteristics and behavior (Killen & Rutland, 2011). For example, a student who shows a high level of accomplishment in academic subjects may be excluded from the "in group," whose members call them a "nerd." Does this kind of group behavior often instigate aggression in those who receive it? An experiment has investigated this question in the laboratory (Twenge, Baumeister, Tice, & Stucke, 2001). Human participants were exposed to social exclusion by telling them that other participants had rejected them as part of the group.

The results were predictable. Social exclusion caused participants to behave more aggressively in various contexts. When insulted by another person (the target), excluded people retaliated by "blasting" the target with higher levels of aversive noise. In another experiment, the target received the same aggressive treatment even though he/she had not insulted the excluded people. This suggests that it is social exclusion itself that instigated the aggressive behavior. A further experiment showed that the effects of social exclusion on aggression could be mitigated if the target provided social praise (reinforcement) to the excluded person.

Additional studies have clarified further the broad impact of social exclusion on the learning of aggression by children. Research indicates that preschool and school-aged children that experience social exclusion show more aggressive behavior. This is generalized effects that may be maintained throughout childhood and beyond (see DeWall, Twenge, Gitter, & Baumeister, 2009; Stenseng, Belsky, Skalicka, & Wichstrom, 2014). Marginalization or social exclusion of sectors of society produces disadvantages by blocking access to certain opportunities and resources for particular groups as well as individuals—resulting sometimes in anger, alienation, and aggression (Betts & Hinsz, 2013). Behaviorally, societal restrictions imposed on marginalized groups may be analogous to prolonged timeout from reinforcement, a punishment procedure capable of inducing aggressive behavior in excluded group members.

Social Disruption as a Side Effect

When punishment is used to decrease behavior, the attempt is usually made to stop a particular response. The hope is that other unpunished behavior will not be affected. Two factors work against this—the person who delivers punishment and the setting in which punishment occurs both can become conditioned aversive stimuli (S^{ave}). Because of this conditioning, individuals often attempt to escape from or avoid the punishing person or setting. Azrin and Holz called this negative side effect of punishment **social disruption**:

> It is in the area of social disruption that punishment does appear to be capable of producing behavioral changes that are far-reaching in terms of producing an incapacity for an effective life.... For example, a teacher may punish a child for talking in class, in which case it is desired that the unauthorized vocalization of the child be eliminated but his other behaviors remain intact. We have seen previously, however, that one side effect of the punishment process was that it reinforced tendencies on the part of the individual to *escape from the punishment situation itself*. In terms of the example we are using, this means that punishment of the vocalization would not only be expected to decrease the vocalization, but also increase the likelihood of the child leaving the classroom situation. Behavior

such as tardiness, truancy, and dropping out of school would be strengthened. The end result would be termination of the social relationship, which would make any further social control of the individual's behavior impossible. This side effect of punishment appears to be one of the most undesirable aspects of having punishment delivered by one individual against another individual, since the socialization process must necessarily depend upon continued interaction with other individuals.

(Azrin & Holz, 1966, pp. 439–440, emphasis added)

It is also worth recalling the general suppressive effects of aversive stimuli. A teacher, parent, or employer (social agent) who frequently uses aversive techniques *becomes a conditioned punishing stimulus*. Once this occurs, the mere presence of the social agent can disrupt all ongoing operant behavior. This means that positive behavior falls to low levels when this person is present (see section "New Directions: Behavioral Neuroscience of Social Defeat" in this chapter).

ON THE APPLIED SIDE: COERCION AND ITS FALLOUT

In his book titled *Coercion and Its Fallout*, Murray Sidman (2001) provides a behavior analysis of coercion and its frequent use in North American society. **Coercion** is defined as the "use of punishment and the threat of punishment to get others to act as we would like, and … our practice of rewarding people just by letting them escape from our punishments and threats" (p. 1). For Sidman, coercion involves the basic contingencies of punishment and negative reinforcement. An interesting part of his book concerns escape and "dropping out" of the family, community, and society (Figure 6.11).

Dropping out—one kind of escape contingency—is a major social problem of our time. People drop out of education, family, personal and community responsibility, citizenship, society, and even life. Sidman (2001, p. 101) points out that the common element in all of these forms of conduct is negative reinforcement. Once they are involved in an aversive system, people can get out by removing themselves from the coercive situation, and this strengthens the behavior of dropping out. Sidman notes that society is the loser when people cease to participate; dropping out is nonproductive as dropouts no longer contribute to their own or society's welfare.

An unfortunate, but common, example is the school dropout. Day after day, students are sent to schools where coercion is a predominant way of teaching. Students show increasingly severe forms of dropping out. Tardiness, feigned illness, "playing hooky," and never showing up for school are common responses to the escalation of coercion in schools. Sidman summarizes the problem as follows:

> The current discipline and dropout crises are the inevitable outcome of a history of educational coercion. One may long for the days when pupils feared their teachers, spoke to them with respect, accepted extra work as punishment, submitted to being kept after school, and even resigned themselves to being beaten. But through the years, all these forms of coercive control were sowing the seeds of the system's destruction. Wherever and whenever coercion is practiced, the end result is loss of support of the system on the part of those who suffered from it. In every coercive environment, the coerced eventually find ways to turn upon the coercers. An adversarial relationship had developed between pupils and teachers, and the former victims, now parents, no longer support the system against their children.
>
> (Sidman, 2001, p. 107)

Aversive Control of Behavior **229**

Teenagers hanging out rather than going to school illustrates the problem of school coercion and students dropping out of education.
Source: Shutterstock.

Sidman goes on to note that not all teachers (or school systems) use coercion or negative reinforcement as a way to induce students to learn. Some teachers and educators are familiar with and use positive reinforcement effectively. A teacher who uses positive reinforcement looks to reward small steps of success rather than punish instances of failure. Schools that adopt positive reinforcement methods are likely to promote the enjoyment of learning as well as high levels of academic performance (Cameron & Pierce, 2002). Positive reinforcement turns dropping out into "tuning in." In this context, behavior analysts can offer new and constructive positive reinforcement techniques for teaching new behavior, and for establishing skillful academic repertoires (Sidman, 1993).

CHAPTER SUMMARY

This chapter highlighted the major contingencies of aversive control. The basic aversive contingencies were outlined in terms of positive and negative punishment and negative reinforcement. Punishment is shown to be complex in how it is conceived, analyzed, and applied. It also is shown to be an unfortunate part of life experience. Positive punishment produces or adds an event or stimulus that decreases an operant behavior's frequency. Negative punishment removes a stimulus and results in a decrease in response.

In both cases, punishment is defined by a reduced probability of response following a punishment procedure. We saw that punishment is relative and is made more effective by abrupt, intense, and immediate delivery of the punisher. The schedule of punishment (continuous is most effective), reduced effectiveness of the positive reinforcer that is maintaining the response that one wishes to decrease, and the availability of response alternatives also enhance the regulation of behavior by punishment contingencies.

Next, we turned to the control of behavior by negative reinforcement. A response that removes an aversive stimulus and results in an increase in the frequency of that response is the process of negative reinforcement. Two kinds of negative reinforcement were identified as escape and avoidance, with the only difference being the presence of the aversive in the former. The section on avoidance introduced the molecular and molar accounts of schedule performance in terms of analysis, the conditioned-suppression paradigm, and the disruption of ongoing operant behavior by periods of scheduled avoidance.

We then turned to the side effects of aversive control, and noted that avoidance behavior is persistent because operant avoidance often prevents respondent extinction of the stimulus that is conditioned to fear. A similar persistence effect was observed with exposure to inescapable punishers and learned helplessness. After a history of inescapable shocks, animals did not learn to escape subsequent shocks; people and other animals fail to emit responses that could remove the punishing events (helplessness). The implications of learned helplessness for clinical depression and its neural basis in social defeat were addressed. Displaced and countercontrol aggression were analyzed as side effects of aversive control. Analysis showed that aggression breeds aggression, and the research on human aggression supported this observation. In fact, people who control behavior by punishment often become conditioned punishers themselves (social disruption).

Finally, we looked briefly at the analysis of coercion and its fallout by Murray Sidman, emphasizing how coercive control may lead people to drop out of society and adolescents to drop out of the school system. The answer to this problem is to reduce coercive control in society and schools while increasing the regulation of behavior by positive reinforcement.

Key Words

Aversive stimulus
Avoidance
Coercion
Countercontrol
Conditioned aversive stimulus (S^{ave})
Discriminated avoidance
Displaced aggression
Escape
External validity
Learned helplessness

Microaggression
Molar account of schedule performance
Molecular account of schedule performance
Negative punishment
Negative reinforcement
Negative reinforcer
Nondiscriminated avoidance
Overcorrection
Permanence of punishment
Positive punishment

Primary aversive stimulus
Punisher
Punishment
Relativity of punishment
Response cost
Response–shock interval (RSI)
Schedule-induced aggression

Shock–shock interval (SSI)
Sidman avoidance
Social disruption
Timeout from avoidance
Timeout from positive reinforcement
Use of punishment debate

On the Web

http://morallowground.com/2011/08/16/kevin-elizabeth-schatz-christian-fundamentalists-spanked-7-year-old-daughter-to-death-because-god-wanted-them-to This is a link to the CNN story and video clip on Kevin and Elizabeth Schatz, the Christian fundamentalists who spanked their 7-year-old daughter to death. Discuss the pervasive use of corporal punishment in our society.

www.ppc.sas.upenn.edu/publications.htm Dr. Martin Seligman, who conducted the original research on learned helplessness, has turned to positive psychology, emphasizing the behaviors related to happiness and well-being. Investigate his website at the Positive Psychology Center at the University of Pennsylvania.

www.teachervision.com/lesson-plans/lesson-10155.html This site is devoted to the Art of Teaching by the Council for Exceptional Children. The webpage includes behavior management tips and advice for teachers, including classroom discipline and behavior techniques, forms, and charts. Learn how to use positive behavior management rather than punishment to manage behavior.

www.jove.com/video/4367/the-resident-intruder-paradigm-standardized-test-for-aggression The resident intruder paradigm by Dr. Koolhass and colleagues at the University of Groningen is used to study aggressive behavior in rats and may be used to induce social defeat stress for studies of behavioral neuroscience and neurobiology. This website shows the basic resident intruder protocol used for basic research. Studies of aggression or social defeat in rats must pass ethical review by a university ethics committee, and researchers must demonstrate that the scientific benefits of such studies outweigh the distress and potential injury to the animals. Before the research is carried out, criteria are set for the induction of aggression and the removal of the animals to prevent or treat injuries resulting from violent attacks.

Brief Quiz

1 In terms of aversive stimuli, attacks and foul odors are _____, while threats and failing grades are _____.
   a   potent; impotent
   b   natural; secondary
   c   primary; conditioned
   d   primitive; cultured
2 Punishment:
   a. increases the behavior on which it is contingent
   b. decreases the behavior on which it is contingent
   c. does not work
   d. should be used whenever possible

3 Research on the use of contingent electrical skin shock (CESS) punishment in the treatment of self-injurious behavior:
   a. shows that some people may develop side effects, such as emotional or mental health conditions
   b. reduces self-injurious behavior
   c. indicates that CESS treatment may eliminate the need for physical restraint or sedating drugs
   d. all of the above

4 In studies on avoidance, the time between shocks if the organism does nothing is the _____ interval and the time between an avoidance response and the next shock is the _____ interval.
   a   temporal shock; response time
   b   shock–shock; response–shock
   c   shocking; responding
   d   aversive; postponement

5 The procedure of nondiscriminated avoidance is also called:
   a   signaled avoidance
   b   sensory aversion
   c   Sidman avoidance
   d   Stevens aversion

6 In terms of operant–respondent interactions, persistence, and avoidance:
   a   operant avoidance prevents respondent extinction
   b   operant avoidance interacts with respondent aggression
   c   operant avoidance competes with respondent avoidance
   d   operant avoidance sets the occasion for respondent aversion

7 For learned helplessness, pre-exposure to escape _____ the helplessness brought on by _____ aversive stimulation.
   a   enhances; noncontingent
   b   causes; excessive
   c   augments; expected
   d   blocks; inescapable

8 With regard to displaced aggression, Ulrich and Azrin (1962) found that the probability of attack for any single shock:
   a   decreased as the number of shocks increased
   b   remained constant as the number of shocks increased
   c   increased as the number of shocks went up
   d   increased and then decreased as the number of shocks went up

9 Which are side effects of punishment?
   a. displaced aggression
   b. learned helplessness
   c. social disruption
   d. all of the above

10 In terms of dropping out, Sidman (2001) indicates that one basic element is:
   a   escape due to coercion
   b   escape due to positive reinforcement
   c   escape due to contingencies of avoidance
   d   escape due to a history of inescapable shock

Answers to Brief Quiz: 1, c (p. 187); 2, b (p. 192); 3, d (p. 201); 4, b (p. 210); 5, c (p. 210); 6, a (p. 215); 7, d (p. 216); 8, c (p. 222); 9, d (p. 215, 216, 221, 227); 10, a (p. 228).

CHAPTER 7

Operant–Respondent Interrelations: The Biological Context of Conditioning

1. Discover how operant and respondent contingencies interrelate.
2. Explore the processes of instinctive drift, sign tracking, and autoshaping.
3. Investigate the role of operant contingencies for respondent behavior.
4. Learn about the biological context of conditioning and taste aversion learning.
5. Discover how eating and physical activity contribute to activity anorexia.

So far, we have considered operant and respondent behavior as separate domains. Respondent behavior is elicited by the events (stimuli) that precede it, and operants are strengthened (or weakened) by stimulus consequences that follow them. At one time respondents were considered to involve the autonomic nervous system and smooth muscles and be involuntary whereas operants involved striated muscles and the central nervous system. Respondents were reflexive and operants were voluntary. This separation, however, is more complex and nuanced, and these specific categories are no longer tenable at both the behavioral and neural levels.

The erroneous analysis of the operating contingencies is apparent in the conditioning of sitting by a dog. Assume that you are teaching a dog to sit and you are using food reinforcement. You might start by saying, "Sit," hold a treat over the dog's head to prompt the sitting position, and follow this sitting posture with food. When training is complete, the discriminative stimulus "Sit" occasions the dog to quickly sit, which results in food. This sequence nicely fits the operant paradigm—the S^D "sit" sets the occasion for the response of sitting, and food reinforcement strengthens this behavior.

In most circumstances, however, both operant and respondent conditioning occur at the same time. If you look closely at what the dog does, it is apparent that the "Sit" command also elicits respondent behavior. Specifically, the dog salivates and looks up at you just after you say "Sit." This occurs because the "Sit" command reliably preceded, and has been correlated with, the presentation of food; "Sit" becomes a conditioned stimulus that elicits salivation and an S^D which occasions engaging in behavior that results in food (such as orienting, searching, or if trained to do so, sitting). For these reasons, the stimulus "Sit" is said to have a dual function. It

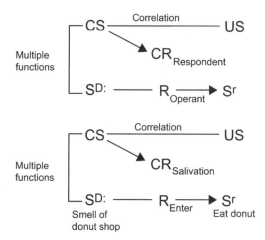

Fig. 7.1
A drawing that depicts the multiple functions of stimuli or events arranged by a contingency of reinforcement. Notice that both operant and respondent functions are interrelated in contingency procedures and respondent functions sometimes can predominate over operant functions even though the apparent procedures are operant in nature. See text for the description of orienting toward and entering the donut shop as an example of the interrelations of operant and respondent contingencies.

is an S^D in the sense that it sets the occasion for operant responses, and it is a CS that elicits respondent behavior. In an early study, Shapiro (1960) demonstrated in dogs that respondent salivation and operant lever pressing were correlated with food reinforcement.

Similar effects are seen when a warning stimulus (a tone) is turned on that signals an imminent aversive stimulus if a rat does not press a lever. The signal is a discriminative stimulus (S^D) that increases the probability of bar pressing, but it is also a CS that elicits changes in behavior such as freezing, as well as in heart rate, hormone levels, and other physiological responses (all of which can be called fear). Suppose that you are out for a before-breakfast walk and you pass a donut and coffee shop (Figure 7.1). The aroma from the shop may be a CS that elicits salivation *and* also an S^D that sets the occasion for entering the store and ordering a donut. These examples should make it clear that in many settings respondent and operant conditioning are intertwined—probably sharing common neural pathways in the brain, but modifying neuron excitability in different ways (Baxter & Byrne, 2006; Lorenzetti, Baxter, & Byrne, 2011).

ANALYSIS OF OPERANT–RESPONDENT CONTINGENCIES

When biologically relevant stimuli such as food or water are contingent on an organism's operant behavior, species-characteristic, innate (present at birth) behavior is occasionally elicited at the same time. Unconditioned reflexes are one kind of species-specific behavior often elicited during operant conditioning. This intrusion of reflexive behavior occurs because *respondent procedures are sometimes embedded in operant contingencies of reinforcement*. These respondent procedures cause species-characteristic responses that may interfere with the regulation of behavior by operant contingencies.

At one time, this intrusion of respondent behavior in operant situations was used to question the generality of operant principles and laws. The claim was that the biology of an organism overrode operant principles and behavior was said to drift toward its biological roots (Hinde & Stevenson-Hinde, 1973). Operant and respondent conditioning are, however, part of the biology of an organism. The neural capacity for operant conditioning arose on the basis of species history; organisms that changed their behavior as a result of life experience had an advantage over animals that did not do so. Behavioral flexibility and neuroplasticity (neural changes resulting from environmental contingencies) allowed for rapid adaptation to an altered environment. As a result, organisms that evolved behavioral flexibility by operant learning were more likely to survive and produce offspring.

Both evolution and learning involve *selection by consequences*. Darwinian evolution has produced both species-characteristic behavior (reflexes, fixed-action patterns, and reaction chains) and basic mechanisms for learning (operant and respondent conditioning) through natural selection. Operant conditioning during an organism's lifetime selects response topographies, rates of response, and repertoires of behavior by arranging reinforcing feedback. In this *behavior-feedback stream*, schedules of reinforcement alter neural pathways, which in turn change behavior and its reinforcing consequences, causing further adjustments to the neural pathways and behavior. One question is whether unconditioned responses (UR) in the basic reflex model (US → UR) are also selected by consequences. In this regard, it is interesting to observe that the salivary glands are activated when food is placed in the mouth; as a result, the food can be tasted, ingested (or rejected), and digested. Thus, there are notable physiological effects (consequences) following reflexive behavior. If the pupil of the eye constricts when a bright light is shown, the result is an escape from retinal pain, while at the same time avoiding retinal damage. It seems that reflexes have come to exist and operate *because these responses do something biologically important for survival*; it is the effects or consequences of these responses that maintain the operation of the reflexive behavior. One might predict that if the effects of salivating for food or blinking at a puff of air to the eye did not result in improved performance (ease of ingestion or protection of the eye), then neither response would continue, as this behavior would not add to the biological fitness of the organism (i.e., the organism would likely die before being able to reproduce).

Embedded Respondent Contingencies

The Brelands' Demonstration
Marion and Keller Breland (Figure 7.2) worked with B. F. Skinner as students, and later established a successful animal training business. They conditioned the behavior of a variety of animals for circus acts, arcade displays, advertising, and movies. In an important paper, the Brelands documented a number of instances in which species-specific behavior interfered with operant conditioning (Breland & Breland, 1961). For example, when training a raccoon to deposit coins in a box, they noted:

> The response concerned the manipulation of money by the raccoon (who has "hands" rather similar to those of primates). The contingency for reinforcement was picking up the coins and depositing them in a 5-inch metal box.
>
> Raccoons condition readily, have good appetites, and this one was quite tame and an eager subject. We anticipated no trouble. Conditioning him to pick up the first coin was simple. We started out by reinforcing him for picking up a single coin. Then the metal container was introduced, with the requirement that he drop the coin into the container. Here

Fig. 7.2
Keller and Marian Breland, who first characterized instinctive drift.
Source: Image taken from www3.uca.edu.

we ran into the first bit of difficulty: he seemed to have a great deal of trouble letting go of the coin. He would rub it up against the inside of the container, pull it back out, and clutch it firmly for several seconds. However, he would finally turn it loose and receive his food reinforcement. Then the final contingency: we put him on a ratio of 2, requiring that he pick up both coins and put them in the container.

Now the raccoon really had problems (and so did we). Not only could he not let go of the coins, but he spent seconds, even minutes rubbing them together (in a most miserly fashion), and dipping them into the container. He carried on the behavior to such an extent that the practical demonstration we had in mind—a display featuring a raccoon putting money in a piggy bank—simply was not feasible. The rubbing behavior became worse and worse as time went on, in spite of non-reinforcement.

(Breland & Breland, 1961, p. 682)

The Brelands documented similar instances of what they called instinctive drift in other species. **Instinctive drift** refers to species-characteristic behavior patterns that became progressively more invasive during training or conditioning (see Figure 7.3, which depicts a raccoon "shooting baskets" rather than depositing coins with the same hand-rubbing response). The raccoon is drifting toward the instinctive "hand rubbing" that evolved as a way of removing the exoskeleton of crayfish. The term *instinctive drift* is, however, problematic because the concept suggests a conflict between nature (biology) and nurture (environment). Behavior is said to drift toward its biological roots. However, there is no need to talk about behavior "drifting" toward some endpoint. Behavior is always appropriate to the operating environmental contingencies.

Fig. 7.3
Drawing of a raccoon performing drop shots into a basketball hoop, illustrating the animal tricks trained by the Brelands. The animals increasingly showed species-specific responses like rubbing the ball rather than the reinforced operant of dropping it into the hoop, a process called instinctive drift.

Source: Picture is taken from R. A. Dewey, *Instinctive drift. Psychology: An introduction*. Retrieved from www.intropsych.com/ch08_animals/instinctive_drift.html. Published with permission.

Recall that *respondent procedures may be embedded in an operant contingency*, and this seems to be the case for the Brelands' raccoon. Let us examine why.

In the raccoon example, the coins were presented just before the animal was reinforced with food for depositing them in the box. For raccoons, as we have seen, food elicits rubbing and manipulating of food items. Because the coins preceded food delivery, they became a CS for the respondent behavior of rubbing and manipulating (coins). This interpretation is supported by the observation that the "rubbing" behavior increased as training progressed. As more and more reinforced trials occurred, coins and food were necessarily correlated. Each conditioning trial increased the *associative strength* of the $CS_{(coin)} \rightarrow CR_{(rubbing)}$ relationship, and the behavior became more and more prominent, interfering with the operant conditioning.

Respondent processes also occur as by-products of operant procedures with rats and children. Rats hold on to marbles longer than you might expect when they receive food pellets for depositing a marble in a hole. Children manipulate tokens or coins prior to banking or exchanging them. The point is that what the Brelands found is not that unusual or challenging to an operant analysis of behavior. Today, we talk about the *interrelation of operant and respondent contingencies* rather than label these observations as a conflict between nature and nurture.

Sign Tracking
Suppose you have trained a dog to sit quietly on a mat, and reinforced the animal's behavior with food. Once this conditioning is accomplished (the dog sits quietly on the mat), you start a

238 Operant–Respondent: The Biological Context of Conditioning

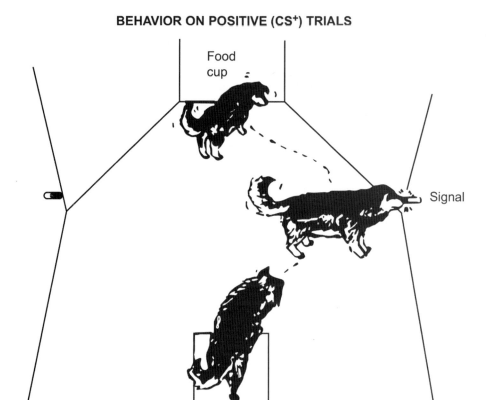

Fig. 7.4

Diagram of apparatus used in sign tracking with dogs. When the signal for food is given, the dog approaches the signal and makes "food-soliciting" responses rather than going directly to the food dish.

Source: Adapted from H. M. Jenkins, F. J. Barrera, C. Ireland, & B. Woodside (1978). Signal-centered action patterns of dogs in appetitive classical conditioning. *Learning and Motivation, 9*, pp. 272–296, 280.

second training phase. During this phase, you turn on a buzzer located on the dog's right side. A few seconds after the sound of the buzzer, a feeder delivers food to a dish that is placed 1.8 m in front of the dog. Figure 7.4 is a diagram of this sort of arrangement.

When the buzzer goes off, the dog is free to engage in any behavior it is able to emit. From the perspective of operant conditioning, it is clear what should happen. When the buzzer goes off, the dog should stand up, walk over to the dish, and eat. This is because the sound of the buzzer is an S^D that sets the occasion for the operant of going to the dish, and this response has been reinforced by food. In other words, the three-term contingency, $S^D: R \rightarrow S^r$, specifies this outcome and there is little reason to expect any other result. A careful examination of the contingency, however, suggests that the sign (sound) could be either an S^D (operant) that sets the occasion for approaching and eating the reinforcer (food), or a CS^+ (respondent) that is correlated with the US (food). In this latter case, the $CS_{(food\ signal)}$ would be expected to elicit food-related CRs.

Jenkins, Barrera, Ireland, and Woodside (1978) conducted an experiment very similar to the one described here. Dogs were required to sit on a mat and a light/tone stimulus was presented either on the left or right side of the animal. When the stimulus was presented on one side it signaled food, and when presented on the other side it signaled extinction. As expected, when the extinction stimulus came on, the dogs did not approach the food tray and for the most part ignored the signal. When the food signal occurred, however, the animals unexpectedly approached the signal (tone/light) and made what were judged by the researchers to be "food-soliciting responses" to the stimulus. Some of the dogs physically contacted the signal source, and others seemed to beg at the stimulus by barking and prancing. This behavior was called **sign tracking**, because it refers to approaching a sign (or stimulus) that has signaled a biologically relevant event (food).

The behavior of the dogs is not readily understood in terms of operant contingencies of reinforcement. As stated earlier, the animals should simply trot over to the food and eat it. Instead, the dogs' behavior appears to be elicited by the signal that precedes and is correlated with food delivery. Importantly, the ordering of stimulus → behavior resembles the CS → CR arrangement that characterizes classical conditioning. Of course, S^D: R follows the same timeline, but in this case the response should be a direct approach to the food, not to the signal. In addition, behavior in the presence of the sign stimulus appears to be food directed. When the tone/light comes on, the dog approaches, barks, begs, prances, and licks the signal. Thus, the temporal arrangement of signal followed by response, and the topography of the responses, both suggest respondent conditioning. Apparently, in this situation the pairing of the light with the unconditioned stimulus (US) properties of the food are stronger (in the sense of regulating behavior) than the operant reinforcement contingency. Because of this, the light/tone gains strength as a CS with each light/tone and food (US) conditioning trial, ensuring that sign tracking predominates in the operant situation. A caution is that one cannot entirely dismiss the occurrence of operant conditioning in this experiment. If the dog engages in a chain of responses that is followed by food, you can expect the sequence to be maintained. The sequence of going to the signal and then to the food, however, is not required by the operant contingency, and actually delays getting the food reinforcer. Overall, respondent behavior seems to confound operant control in sign-tracking situations.

Autoshaping

Shaping, the differential reinforcement of successive approximations to the final performance, is the usual way that a pigeon is taught to peck a response key (see Chapter 4). In the laboratory, a researcher operating the feeder with a hand-switch reinforces closer and closer approximations to the final performance (key pecking). Once the bird makes the first independent peck on the key, electronic programming equipment activates a food hopper and the response is reinforced. The contingency between behavior and reinforcement both during shaping and after the operant is established is clearly operant (R → S^r). This method of differential reinforcement of successive approximations requires considerable patience and a fair amount of skill on the part of the experimenter.

Brown and Jenkins (1968) reported a way to automatically teach pigeons to peck a response key. In one experiment, they first taught birds to approach and eat grain whenever a food hopper operated. After the birds were magazine trained, automatic programming turned on a key light 8 s before the grain was delivered. Next, the key light went out and the grain hopper activated. After 10–20 presentations of this key-light-followed-by-food procedure, the birds started to orient and move toward the lighted key. Eventually, all 36 pigeons in the experiment began to

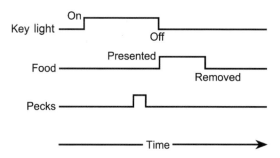

Fig. 7.5

Autoshaping procedures are based on Brown and Jenkins (1968). Notice that the onset of the light precedes the presentation of food and appears to elicit the key peck.

strike the key even though pecking never produced food. Figure 7.5 shows the arrangement between key light and food presentation. Notice that the light onset precedes the presentation of food and appears to elicit the key peck.

The researchers called this effect **autoshaping**, an automatic way to teach pigeons to key peck. Brown and Jenkins offered several explanations for their results. In their view, the most likely explanation had to do with species-characteristic behavior of pigeons. They noted that pigeons have a tendency to peck at things they look at. The bird notices the onset of the light, orients toward it, and "the species-specific look–peck coupling eventually yields a peck to the [key]" (Brown & Jenkins, 1968, p. 7). In their experiment, when the bird initiates the look–peck sequence to the key, food occurs, accidentally reinforcing the first peck.

Another possibility is that initial key pecking resulted from respondent conditioning. The researchers suggested that the lighted key had become a CS that elicited key pecks. This could occur because pigeons make unconditioned pecks (UR) when grain (US) is presented to them. In their experiment, the key light preceded grain presentation and may have elicited a conditioned peck (CR) to the lighted key (CS). Brown and Jenkins comment on this explanation and suggest that although it is possible, it "seem[s] unlikely because the peck appears to grow out of and depend[s] upon the development of other motor responses in the vicinity of the key that do not themselves resemble a peck at grain" (Brown & Jenkins, 1968, p. 7). In other words, the birds began to turn toward the key, stand close to it, and make thrusting movements with their heads, all of which led eventually to the key peck. It does not seem likely that all of these are reflexive responses. They seem more like operant approximations that form a chain culminating in pecking.

Notice that respondent behavior such as salivation, eye blinks, startles, knee jerks, pupil dilation, and other reflexes does not depend on the conditioning of additional behavior. When you touch a hot stove, you rapidly and automatically pull your hand away. This response simply occurs when a hot object is contacted. A stove does not elicit approach to it, orientation toward it, movement of the hand and arm, and other responses. All of these additional responses seem to be operant, forming a chain or sequence of behavior that includes avoiding contact with the hot stove. There also is reason to believe, however, that the orienting and movement toward the key could be part of species-typical behavior, perhaps similar to a fixed-action pattern or a reaction chain, elicited by motivationally significant events such as food (Nieuwenhuis, de Geus, & Aston-Jones, 2011).

Autoshaping has been extended to other species and other types of reinforcement and responses. Chicks have been shown to make autoshaped responses when heat was the reinforcer (Wasserman, 1973). When food delivery is signaled for rats by lighting a lever or by inserting it into the operant chamber, the animals lick and chew on the bar (Stiers & Silberberg, 1974). These animals also direct social behavior toward another rat that signals the delivery of food (Timberlake & Grant, 1975). Rachlin (1969) showed autoshaped key pecking in pigeons using electric shock as negative reinforcement. The major question that these and other experiments raise is this: What is the nature of the behavior that is observed in autoshaping and sign-tracking experiments?

In general, research has shown that autoshaped behavior is initially respondent, but when the contingency is changed so that pecks are followed by food, the pecking becomes operant. Pigeons reflexively peck (UR) at the sight of grain (US). Because the key light reliably precedes grain presentation, it acquires a CS function that elicits the CR of pecking the key. When pecking is followed by grain, however, it comes under the control of contingencies of reinforcement and it is an operant. To make this clear, *autoshaping produces respondent behavior (orienting–peck sequence directed at the lighted key) that can then be reinforced as operant behavior*. To further clarify, once respondent behavior is elicited and reinforced, it comes under the control of its consequences and is now considered to be operant behavior. A human example of respondents becoming operant might be crying. An injury may occur, causing reflexive tearing and vocalizing, and if a caretaker immediately provides comfort and a reduction in the pain, the actions of the caretaker reinforce the crying response. Subsequently, "crying" may occur simply for its consequences—comfort from caretakers. A newborn baby's transition from nursing (the reflexes of rooting, sucking, and swallowing milk) to eating solid food as they age is another example of the change from reflex to operant.

NEW DIRECTIONS: NEURAL PARTICIPATION IN AUTOSHAPING AND SIGN TRACKING

Rats often approach, contact, and attempt to consume a CS correlated with a food US, rather than going directly to the food. This behavior toward the CS is called sign tracking, which is usually viewed as a form of autoshaping (automatic shaping) in which a hungry pigeon is presented with a lighted key (CS) followed by the presentation of food (US) and the bird begins pecking at the key.

Investigators in associative learning propose that the CS in an autoshaping procedure acquires **incentive salience** or the acquisition of motivational value by the sign or cue (CS+) predicting the US. Incentive salience involves the animal attending to and approaching the CS+, and showing feeding responses to the CS+ as it would to the food US (biting and gnawing). A behavior analysis suggests that the CS+ may serve as a motivational operation (MO), increasing species-specific behavior to the sign stimulus typically directed to the US, and momentarily enhancing the effectiveness of the US in the CS–US contingency. When a drug US (heroin) is correlated with a CS+ (needles and drug preparation), the incentive salience transferred to the sign stimulus (drug paraphernalia) is said to instigate "wanting" of the drug. In humans this "wanting" effect is observed as cravings for the drug when the CS+ occurs.

To investigate the incentive salience of the sign stimulus with rats, researchers note that animals press, grasp, and bite a lever paired with food, even though there is no contingency between lever manipulation and food presentation. Furthermore, there is evidence that the CS+ (lever) acquires a conditioned reinforcement function. Rats autoshaped on a retractable lever with lever insertion as the CS+ showed sign-tracking as expected, but insertion of the lever also functioned as reinforcement for approaching the lever—much as the sound of the hopper acquires a reinforcement function in operant experiments with pigeons.

New techniques in neuroscience are available well beyond the common fMRI and PET/CAT scans to clarify the neuroanatomy and neuronal pathways of sign-tracking and incentive salience (Deisseroth, 2011). But the basic techniques of producing lesions and ablation, electro recording, direct stimulation, and chemical manipulations continue to be used in associative conditioning as functional procedures for relating behavior to neurophysiology. Studies of sign tracking in rats use these basic techniques to show how environmental contingencies (CS–US relationships) and incentives (CS+ predicting the US) together alter the brain processes and thereby regulate the behavior of organisms. A series of studies by Chang is illustrative of the advancement and integration of respondent procedures within modern neuroscience (Chang, 2013; see also Chang & Holland, 2013).

In Chang, Wheeler, and Holland (2012), for example, male rats were given ibotenic acid injections causing brain lesions. Two brain areas were of interest—the ventral striatal nucleus accumbens (NAc), an established brain center for neural reward, and the basolateral amygdala (BLA), a neural area related to emotionally arousing events. Control, sham-lesion rats received needle insertions to the same brain areas, but without infusions. After recovery from the injections, rats were autoshaped using two silent retractable levers, one on each side of a recessed liquid delivery cup. Each session involved 25 CS+ trials and 25 CS− trials. One lever (left or right) was inserted for 10 s and followed with a drop of sucrose (US) delivered upon its retraction (CS+ trial). The other lever was inserted for 10 s and retracted, but no sucrose was delivered (CS− trial). In this respondent procedure, the sucrose US is delivered regardless of the rat's behavior and food-related responses to the CS+ (sign stimulus) are used to assess the transfer of incentive salience from the US (sucrose reward) to the lever. The researchers measured autoshaping as the percentage of trials resulting in a bar press, and the rate of lever pressing on those bar-press trials. Consummatory responses directed to the CS+ lever (licking and biting) indicated sign tracking and responding directed toward the food cup indexed instrumental behavior or goal tracking. Primary results showed that sham rats pressed the CS+ bar, but not the CS− lever, indicating successful autoshaping for brain-intact animals. NAc lesions, however, impaired initiation of lever pressing while lesions to the BLA interrupted the rate of lever pressing once it occurred. Furthermore, NAc lesions impaired acquisition of sign tracking (early sessions) to the CS+ cue while lesions to the BLA impaired terminal levels of this behavior (later sessions). Lesions to both brain centers produced both of these deficits.

Overall, it seems that the NAc is involved in the acquisition of feeding-like responses to the CS+ indicative of incentive salience transferred from the US reward, while the BLA enhances incentive salience of the CS+ once it is acquired. Further research indicated that the incentive salience of the CS+ in a sign-tracking procedure does not require the integrity of the orbitofrontal cortex (OC), whereas stimulus-outcome reversal learning, prior CS− now followed by US and prior CS+ no longer followed by US, is substantially impaired (Chang, 2014). This finding suggests that complex respondent learning requires either the presence of higher-order cortical pathways, or a rerouting of critical neural circuitry when stimulus-outcome contingencies change, a form of neural plasticity.

Contingencies and Species-Specific Behavior

In discussing their 1968 experiments on autoshaping, Brown and Jenkins report that:

> Experiments in progress show that location of the key near the food tray is not a critical feature [of autoshaping], although it no doubt hastens the process. Several birds have acquired the peck to a key located on the wall opposite the tray opening or on a sidewall.
>
> (Brown & Jenkins, 1968, p. 7)

Fig. 7.6
Pigeons in a park foraging for food illustrates that pecking for food is species-specific behavior under the control of the contingencies arranged in the natural setting or the laboratory.
Source: Shutterstock.

This description of autoshaped pecking by pigeons sounds similar to sign tracking by dogs. Both autoshaping and sign tracking involve species-specific behavior elicited by food presentation. Instinctive drift also appears to be reflexive behavior elicited by food. Birds peck at grain and make similar responses to the key light. That is, birds sample or taste items in the environment by the only means available to them—beak or bill contact (see Figure 7.6 of pigeons foraging for food). In contrast, dogs make food-soliciting responses to the signal that precedes food reinforcement, behavior clearly observed in pictures of wolf pups licking the mouth of an adult returning from a hunt. Raccoons with finger-like paws rub and manipulate food items and make similar responses to coins that precede food delivery. Similarly, we have all seen humans rubbing dice together between their hands before throwing them. It is likely that autoshaping, sign tracking, and instinctive drift represent the same (or very similar) processes (see Hearst & Jenkins, 1974) and utilize common neural pathways.

One proposal is that all of these phenomena (instinctive drift, sign tracking, and autoshaping) are instances of **stimulus substitution**. Basically, when a CS (light or tone) is paired with a US (food), the conditioned stimulus is said to substitute for the unconditioned stimulus. This means that responses elicited by the CS (rubbing, barking and prancing, pecking) are similar to the ones caused by the US. Although this is a parsimonious account, there is evidence that it is wrong.

Recall from Chapter 3 that the laws of the reflex (US → UR) *do not hold* for the CS → CR relationship, suggesting that there is no universal substitution of the CS for the US. Also, in many experiments, the behavior elicited by the US is opposite in direction to the responses elicited by the conditioned stimulus, such as with conditioned compensatory responses (see "On the Applied Side: Drug Use, Abuse, and Respondent Conditioning" in Chapter 3). In addition, there are experiments conducted within the autoshaping paradigm that directly refute the stimulus substitution hypothesis.

In an experiment by Wasserman (1973), chicks were placed in a very cool enclosure. In this situation, a key light occasionally turned on closely followed by the activation of a heat lamp. All of the chicks began to peck the key light in an unusual way. The birds moved toward the key light and rubbed their beaks back and forth on it—behavior described as snuggling. These responses resemble the behavior that newborn chicks direct toward their mother when soliciting warmth. Chicks peck at their mother's feathers and rub their beaks from side to side—behavior that results in snuggling up to their mothers.

At first glance, the "snuggling to the key light" seems to be an instance of stimulus substitution. The chick behaves to the key light as it does toward its mother. The difficulty is that the chicks in Wasserman's experiment responded completely differently to the heat lamp compared to the way they responded to the key light. In response to heat from the lamp, a chick would extend its wings and stand motionless—behavior that it might direct toward intense sunlight (Wasserman, 1973). In this experiment, it is clear that the CS does not substitute for the US, because these stimuli elicit completely different responses (see also Timberlake & Grant, 1975).

Timberlake (1983) proposed an alternative to stimulus substitution. He suggested that each US (food, water, sexual stimuli, or a heat lamp) controls a distinct set of species-specific responses, a **behavior system**. Thus, for each species there is a behavior system related to procurement of food, another related to obtaining water, and another still for securing warmth and comfort. For example, the presentation of food to a raccoon activates the species-typical behavior system consisting of procurement and ingestion of food. One of these behaviors, rubbing and manipulating the food item, may be activated, depending on the CS–US (key light followed by food) contingency. Other behaviors of the system, such as bringing the food to the mouth, chewing, and swallowing may not occur. Timberlake goes on to propose that the particular responses activated by the CS or signal depend, in part, on the physical properties or features of the stimulus for that species (*incentive salience* of the CS also may be involved in activation of the behavior system for a given species, as discussed in "New Directions: Neural Participation in Autoshaping and Sign Tracking" in this chapter). Presumably, in the Wasserman experiment with chicks, properties of the key light (a visual stimulus raised above the floor) were more closely related to snuggling than to standing still and extending wings.

At the present time, it is not possible to predict which responses the CS activates in a behavior system. A researcher might predict that the CS would elicit one or more responses of the behavior system for food procurement, but cannot specify in advance which responses. One possibility is that the intensity of the reinforcer (food or heat) affects the responses activated by the CS–US contingency. For example, as the intensity of the heat source increases (approximating a hot summer day), the chick's response to the key light (sign) may change from snuggling to standing in the sun (i.e., wings open and motionless). An analysis of an animal's *ecology and evolutionary history* is necessary to predict its behavior toward environmental contingencies that are encountered during its lifetime.

Embedded Operant Contingencies

Reinforcement, Biofeedback, and Robotic Limbs

We have seen that both operant and respondent conditioning can occur at the same time. Moreover, respondent contingencies sometimes regulate responses that are usually viewed as operant behavior (pecking a lighted key), and as such these responses are respondents. There are also situations when behavior that appears to be respondent is regulated by its consequences and is therefore functionally operant behavior. Thus, operant contingencies sometimes can predominate over respondent control of behavior.

Biofeedback involves operant control of seemingly involuntary, automatic activity. Instruments are used to amplify and observe interoceptive bodily responses, which typically are regulated by respondent processes. From a clinical physiological viewpoint, over- or under-activity of the physiological system causes stress, discomfort, or illness. Providing information feedback to a patient or client, however, allows for conscious control of bodily functions, restoring homeostatic balance. Thus, temperature readings are made visually available with an electronic thermometer, muscle relaxation with electromyography, heart rate with an app on your cell phone, or carbon dioxide with a capnometry-assisted respiratory training (CART) system (Ritz, Rosenfield, Steele, Millard, & Meuret, 2014). Once these bodily responses are made observable by instrumentation, these responses may be followed by visual feedback (consequences) and take on operant functions. Patients with migraine headaches can be fitted with a thermistor to read body temperature and told to keep the meter reading in a target zone. Those with a previous history of migraines (motivational operation) usually show effective control by the reinforcing feedback of the meter readout, achieving a high rate of responding within the target range.

Cutting-edge, multidisciplinary research in brain–computer interface (BCI) also is using reinforcing feedback (visual and proprioceptive) from robotic-limb movement and relative position to train efficient use of these neuroprostheses. Thus, a modular prosthetic limb (MPL) with human-like design and appearance provides real-life movement, conveying a feeling of embodiment to users. Based on feedback training developed with monkeys in the laboratory, a user fitted with BCI is able to control the MPL with the skill and speed similar to an able-bodied person (Collinger et al., 2014). Patients with an amputated limb often show motivation (asking for and seeking) to use a robotic prosthesis to restore some of their previous capabilities, establishing feedback from the robotic limb as reinforcement for efficient movement and skilled manipulation of the appendage. The robotic limb is attached to available nerves in the remaining stump and these nerves are integrated via a computer with electrodes implanted in areas of the brain activated when a patient is instructed to move the missing appendage (completing a neural-computer-MPL circuit). After configuration of the software interface, the patient is asked to move the robotic limb with movement and positional sensory cues functioning as reinforcement for operation of the artificial appendage. Generally, from a behavioral perspective, both biofeedback and BCI illustrate the control of seemingly involuntary and/or automatic responses (respondent behavior) of the body or brain by operant reinforcement contingencies.

Reinforcement of Reflexive Behavior

In the 1960s, researchers attempted to show that involuntary reflexive or autonomic responses could be conditioned by operant procedures. Miller and Carmona (1967) deprived dogs of water and monitored their level of salivation. The dogs were separated into two groups. One group received water as reinforcement for increasing salivation, and the other group received reinforcement for decreasing salivary responses. Both groups of animals showed the expected change in amount of salivation. Thus, the dogs reinforced for increasing salivation showed more saliva flow, and the dogs reinforced for decreasing salivation showed less.

At first glance, these results seem to demonstrate the operant conditioning of salivation. However, Miller and Carmona (1967) noticed an associated change in the dogs' behavior, which could have produced the findings for salivation. Dogs that increased their saliva flow appeared to be alert, and those that decreased it were described as drowsy. Based on this possible confound, the results are suspect as salivary conditioning could be mediated by a change in the dogs' operant behavior. Perhaps drowsiness was operant behavior that resulted in decreased salivation, and being alert increased the reflex. In other words, the changes in salivation may have been part of a larger, more general behavior pattern, which was reinforced. Similar problems occurred with other related experiments. For example, Shearn (1962) showed operant conditioning of heart rate, but heart rate can be affected by a change in the pattern of breathing.

The Miller Experiments

It is difficult to rule out operant conditioning of other behavior as a mediator of reinforced reflexes. Miller and DiCara (1967), however, conducted a classic experiment in which this explanation was not possible. The researchers reasoned that operant behavior could not mediate conditioning if the animal had its skeletal muscles immobilized. To immobilize the rats, they used the drug curare. This drug paralyzes the skeletal musculature and interrupts breathing while the rats were maintained by artificial respiration. When injected with curare, the rats could not swallow food or water as reinforcement for reflexive responses. Miller and DiCara solved this problem by using electrical brain stimulation of the rats' pleasure center as reinforcement for visceral reflexes. Before the experiment was started, the rats had electrodes permanently implanted in their hypothalamus. This was done in a way that allowed the experimenters to connect and disconnect the animals from the equipment that stimulated the neural-reward center. To ensure that the stimulation was reinforcing before the experiment, the rats were trained to press a bar to turn on a brief microvolt pulse. This procedure demonstrated that the pulse functioned as reinforcement, as the animals pressed a lever for the brain stimulation.

At this point, Miller and DiCara administered curare to the rats and half received electrical brain stimulation (EBS) for decreasing their heart rate. The other animals received stimulation for an increase in heart rate. Figure 7.7 shows the results of this experiment. Both groups

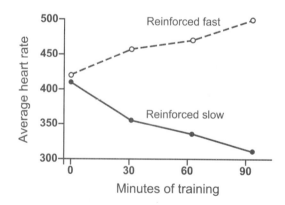

Fig. 7.7
Effects of curare immobilization of skeletal muscles and the operant conditioning of heart rate are shown (Miller & DiCara, 1967). Half the rats received electrical brain stimulation for increasing heart rate and the other half for decreasing heart rate.

started out with heart rates in the range of 400–425 beats per minute. After 90 min of contingent EBS reinforcement, the groups were widely divergent. The group reinforced for slow heart rate had rates of about 310 beats per minute, and the group reinforced for fast rate had heart rates of approximately 500 beats a minute.

Miller and Banuazizi (1968) extended this finding by monitoring intestinal contractions of rats. At the same time, the researchers measured the animals' heart rates. As in the previous experiment, the rats were injected with curare and reinforced with EBS. In different conditions, rats were required to increase or decrease intestinal contractions for EBS reinforcement. In addition, the rats were reinforced on some occasions for a decrease in heart rate, and at other times for an increase. The researchers showed that EBS reinforcement of intestinal contractions or relaxation changed these responses in the appropriate direction. In addition, the rats showed increases or decreases in heart rate depending on the contingency of EBS reinforcement. Finally, Miller and Banuazizi (1968) demonstrated that changes in intestinal contractions did not affect heart rate and, conversely, changes in heart rate did not affect contractions.

In these experiments, the contingencies of reinforcement modified behavior usually considered to be reflexive, under conditions in which skeletal responses could not affect the outcome. Furthermore, the effects were specific to the reinforced response—showing that general physiological changes related to EBS did not produce the outcomes of the experiment. Thus, autonomic responses usually elicited as URs were regulated by operant contingencies of reinforcement. Greene and Sutor (1971) extended this conclusion to humans, showing that galvanic skin responses (GSR) were controlled by contingencies of negative reinforcement.

Although autonomic responses can be regulated by operant contingencies, the operant conditioning of blood pressure, heart rate, and intestinal contractions has run into difficulties. Miller even reported problems replicating the results of his own experiments (Miller & Dworkin, 1974), concluding "that the original visceral learning experiments are not replicable and that the existence of visceral learning remains unproven" (Dworkin & Miller, 1986). The weight of the evidence does suggest that reflexive responses are, at least in some circumstances, controlled by the consequences that follow them. This behavior, however, is also controlled by predictiveness and correlation of stimuli. It is relatively easy to change heart rate by following a light (CS) with electric shock and then using the light to change heart rate (respondent contingency). It should be evident that controlling heart rate with an operant contingency is no easy task. Thus, autonomic behavior may not be exclusively tied to respondent conditioning, but respondent conditioning is particularly effective with these responses.

Clearly, the fundamental distinction between operant and respondent conditioning is operational. The distinction is operational because conditioning is defined by the operations that produce it. Operant conditioning involves a contingency between behavior and its consequences. Respondent conditioning entails the contiguity and correlation of stimuli (S–S association). Autonomic responses are usually reflexes and conditioned reflexes are best modified by respondent procedures. But when these conditioned responses are changed by the consequences that follow them, they are operant responses (an operant class). Similarly, skeletal responses are usually operant and most readily changed by contingencies of reinforcement, but when modified by contiguity and correlation of stimuli they are respondents. The whole organism is impacted by contingencies (environmental arrangement of events), whether these are designed as operant or respondent procedures. That is, most contingencies of reinforcement activate respondent processes, while Pavlovian or respondent contingencies occasionally involve the reinforcement of operant behavior.

THE BIOLOGICAL CONTEXT OF CONDITIONING

As we stated in Chapter 1, the evolutionary history, ontogenetic history, and current neurophysiological status of an organism are the **context for conditioning**. Context is a way of noting that the probability of behavior depends on certain conditions. Thus, the effective contingencies (stimuli, responses, and reinforcing events) may vary from species to species. A hungry dog is reinforced with meat for jumping a hurdle, and a pigeon flies to a particular location to get grain. These are obvious species differences, but there are other subtle effects of the **biological context**. The rate of acquisition and the level of behavior maintained by reinforcement often are influenced by an organism's neurophysiology, as determined by species history and lifetime interactions with the environment. Moreover, within a species, discriminative stimuli, responses, and reinforcing events often depend on an animal's learning history, being specific to particular situations.

Although the effects of contingencies often depend on the particular events and responses, principles of behavior such as extinction, discrimination, and spontaneous recovery show generality across species. In terms of basic principles, the behavior of schoolchildren working at math problems for teacher attention and of pigeons pecking keys for food is regulated by the principle of reinforcement even if stimuli, responses, and reinforcing events vary over species. From its early days of the 1930s, scientists have aimed to identify *general principles of behavior* using simple stimuli and responses, which were easy to execute and record, and precisely reinforce. Researchers today continue to search for general principles of learning, but remain sensitive to the interrelationships among conditioning, species history, and neurophysiology. These interrelationships are clearly present in the field of taste aversion learning.

Taste Aversion Learning

Taste aversion learning is illustrated by a study in which quail and rats were given a solution of blue salty water. After the animals drank the water, they were made sick. Following recovery, the animals were given a choice between water that was not colored but tasted salty, and plain water that was colored blue. The rats avoided the salty-flavored water, and the quail would not drink the colored solution. That is, the type of fluid avoidance depends on the species. This finding is not difficult to understand—when feeding or drinking, birds rely on visual cues, whereas rats are sensitive to taste and smell. In the natural habitat of these animals, drinking liquids that produce illness should be avoided, and has obvious survival value. Because quail typically select food based on its appearance, they avoided the colored water. In contrast, because rats usually select food by taste, they avoided the salty water associated with sickness.

Taste Aversion and Preparedness

Notice that the taste and color of the water was a compound CS ($CS_1 + CS_2$) that was paired with the US of poison, which created a UR of illness. Both species showed taste aversion learning. Thus, the animals avoided one or the other of the CS elements based on species history. In other words, the biology of the organism determined which feature became the CS, but the conditioning of the aversion by CS–US contingency (general behavior principle) was the same for both species. Of course, a bird that relied on taste for food selection would be expected to condition to the taste followed by illness contingency. This phenomenon has been called **preparedness**—quail are biologically prepared to discriminate visual features that are associated with illness, and rats respond best to a flavor–poison association. Additional experiments have shown that animals of the same species are prepared to learn some particular CS–US associations and not others.

Garcia and his colleagues conducted many important experiments concerned with the conditions that produce taste aversions in rats.[1] In a classic study, known to many as the "bright, noisy, tasty, water experiment", Garcia and Koelling (1966) were interested in determining the extent to which a taste-based CS vs. an audio-visual CS would condition to three different USs that reduce drinking water—shock, lithium chloride (a poison), or X-ray. Each of the USs was paired with either saccharin-flavored water (the tasty CS) or flashing lights and gurgling noises (bright-noisy CS)—see Figure 7.8 for these CU–US pairing conditions of the experiment. After each of the CS–US pairings, the researchers compared water intake during presentation of the CS to water intake baseline.

Figure 7.9 portrays major results of this experiment. The bright-noisy water stimulus, when paired with shock, reduced water intake. In other words, this CS–US pairing formed a conditioned aversion to water. But, when the bright-noisy CS was paired with poison and X-ray, an aversion did not form. Conversely, the tasty stimulus, when paired with poison and X-ray, resulted in a conditioned water aversion. But the tasty stimulus did not form an aversion when paired with shock. Therefore, flavor seems to condition more readily to poison and X-ray; audio-visual stimuli tend to condition more readily to shock.

These results are unusual for several reasons. During traditional respondent conditioning, the CS and US typically overlap or are separated by only a few seconds. In the experiment by Garcia and Koelling (1966), the taste CS was followed much later by the US (poison or X-ray). Also, it is often assumed that the choice of CS and US is irrelevant for respondent conditioning. Pavlov claimed that the choice of CS was arbitrary—he said that anything would do. However,

		Type of water (CS)	
		Bright-noisy	Flavored
Type of aversive stimulus (US)	Shock	Bright-noisy drinking followed by shock	Flavored drinking followed by shock
	Poison	Bright-noisy drinking followed by poison	Flavored drinking followed by poison

Fig. 7.8

Authors' rendering of conditions used to show taste aversion conditioning by rats in an experiment by Garcia and Koelling (1966).

Source: From description given in J. Garcia & R. A. Koelling (1966). Relation of cue to consequence in avoidance learning. *Psychonomic Science*, *4*, pp. 123–124. Copyright 1966 held by the Psychonomic Society, Inc.

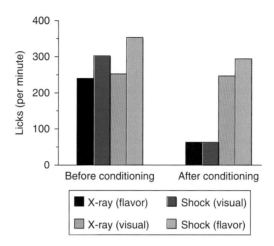

Fig. 7.9
Authors' rendering of Garcia and Koelling's (1966) major results for the taste aversion experiment are shown.
Source: Based on data from J. Garcia & R. A. Koelling (1966). Relation of cue to consequence in avoidance learning. *Psychonomic Science*, 4, pp. 123–124. Copyright 1966 held by the Psychonomic Society, Inc.

in these experiments, taste and gastrointestinal malaise produced aversion, but taste and shock did not. Thus, it appears that for some stimuli the animal is *prepared* by nature to make a connection and for others it may even be *contraprepared* (Seligman, 1970). Generally, for other kinds of respondent conditioning many CS–US pairings are required, but aversion to taste is conditioned after a single, even substantially delayed, presentation of flavor with illness. In fact, the animal need not even be conscious for an aversion to occur (Provenza, Lynch, & Nolan, 1994). Finally, the animal must experience nausea for taste aversion to condition. Strychnine is a poison that inhibits spinal neurons but does not cause sickness. It also does not cause taste aversion (Cheney, Vander Wall, & Poehlmann, 1987). The bottom line is that organisms often are prepared by evolution to condition rapidly to particular CS–US relationships (taste and illness), even when the US is considerably delayed.

The evolutionary fitness advantage of one-trial conditioning, or quickly avoiding food items that produce illness, appears obvious, but has not been directly tested. Recently, Dunlap and Stephens (2014), in the *Proceedings of the National Academy of Sciences*, provided the first experimental evidence for the evolution of preparedness for learning in *Drosophila*. The learning in this experiment concerned flies' avoidance of egg laying on agar-based media (A and B), which differed by the presence or absence of bitter, aversive quinine. To generate an experimental analog of natural selection favoring preparedness to learn, the two media types (A and B) were signaled by either a distinct odor or color (other conditions are not described here). In some situations, the odor predicted which media type to avoid (odor predicts quinine), whereas in others the color predicted best (color predicts quinine). Before flies were given a choice as to where to lay eggs, they initially experienced quinine placed on one of the media types (A or B) and signaled by distinct odor or color. Subsequently, flies chose between A and B media types where to lay eggs, using odor or color as a predictive cue. Once the flies had laid eggs, researchers artificially selected and reared the eggs deposited on the agar media *not paired* with the odor or color. If the selection reliability for the odor exceeded the reliability for color (O > C), then learning by odor was expected to have higher reproductive fitness.

Over 40 generations, populations of flies with reliable selection for odor-quinine avoidance, but unreliable selection for color-quinine, showed increased sensitivity for odor-quinine learning (preparedness) and reduced sensitivity for color-quinine. When, however, reliable selection favored the color over odor (C > O), flies show increased avoidance of quinine by color over generations. Thus, it appears that preparedness to learn in nature depends on high reproductive fitness or *reliable selection by the environment* (O > C or C > O) favoring the stimulus predicting the aversive outcome.

Taste Aversion: Novelty and Extensive CS–US Delays
Taste aversion learning has been replicated and extended in many different experiments (see Reilly & Schachtman, 2009). Revusky and Garcia (1970) showed that the interval between a flavor CS and an illness-related US could be as much as 12 h. Other findings indicate that a novel taste is more easily conditioned than one with which an animal is familiar (Revusky & Bedarf, 1967). A novel setting (as well as taste) has also been shown to increase avoidance of food when a toxin is the US. For example, Mitchell, Kirschbaum, and Perry (1975) fed rats in the same container at a particular location for 25 days. Following this, the researchers changed the food cup and made the animals ill. After this experience, the rats avoided eating from the new container. Taste aversion learning also occurs in humans, of course. Alexandra Logue at the State University of New York, Stony Brook, has concluded:

> Conditioned food aversion learning in humans appears very similar to that in other species. As in other species, aversions can be acquired with long CS–US delays, the aversion most often forms to the taste of food, the CS usually precedes the US, aversions [are] frequently generalized to foods that taste qualitatively similar, and aversions are more likely to be formed to less preferred, less familiar foods. Aversions are frequently strong. They can be acquired even though the subject is convinced that the food did not cause the subject's illness.
>
> (Logue, 1985, p. 327)

Imagine that on a special occasion you spend an evening at your favorite restaurant (Figure 7.10). Stimuli at the restaurant include your companion, waiters, candles on the table,

Fig. 7.10
An illustration of taste aversion learning in humans is shown. *Pasta primavera* is a novel taste (CS) consumed at dinner. Later that night the flu virus (US) induces sickness (UR). Subsequently, the presentation of *pasta primavera* (CS) produces avoidance of the food (CR).

Source: Shutterstock. The artwork is an author's rendering.

china, art on the wall, and many more aspects of the setting. You order several courses, most of them familiar, and "just to try it out" you have *pasta primavera* for the first time (taste-CS). What you do not know is that a flu virus (US) has invaded your body and is percolating away while you eat. Early in the morning, you wake up with a clammy feeling, a rumbling stomach, and a hot acid taste in the back of your throat. You vomit *primavera* sauce, wine, and several other ugly bits and pieces into the sink (UR).

The most salient stimulus at the restaurant was probably your date. Alas, is the relationship finished? Will you get sick at the next sight of your lost love? Is this what the experimental analysis of behavior has to do with romance novels? Of course, the answer to these questions is no. It is very likely that you will develop a strong aversion only to *pasta primavera* (taste-CS → CR-illness). Interestingly, you may clearly be aware that your illness was caused by the flu, not the new food. You may even understand taste aversion learning but, as some of the authors of this book can testify, it makes no difference. The novel taste-CS, because of its single pairing (even though delayed by several hours) with nausea, is likely to be avoided in the future.

FOCUS ON: TASTE AVERSION, NEURAL ACTIVITY, AND DRUG CRAVINGS

Behavior is intricately linked to the nervous system. Those organisms with a nervous system that allowed for behavioral adaptation survived and reproduced. To reveal the neural mechanisms that interrelate brain and behavior, neuroscientists often look for changes in the brain that underlie functional changes in behavior as the organism responds to changes and challenges in its environment. This approach has been used to understand how organisms acquire aversions to food tastes and avoidance of foods with these tastes.

One study investigated the sites in the brain that are responsive to lithium chloride (LiCl), a chemical (US) that is often used in conditioned taste aversion learning to induce nausea. The researchers used brain imaging to measure the presence of *c*-fos, a gene transcription factor and marker of neuronal activity, which is implicated in LiCl-induced illness (Andre, Albanos, & Reilly, 2007). The gustatory area of the thalamus showed elevated levels of *c*-fos following LiCl treatment, implicating this brain region as central to conditioned taste aversion. Additional research has implicated transcription factors such as *c*-fos as the biochemical substrate bridging the delay between the CS-taste and the onset of nausea (US) several hours later (Bernstein, Wilkins, & Barot, 2009). Other research by Yamamoto (2007) showed that two regions of the amygdala were also involved in taste aversion conditioning. One region is concerned with detecting the conditioned stimulus (e.g., distinctive taste) and the other is involved with the hedonic shift from positive to negative, as a result of taste aversion experience. And in a recent study, expression of *c*-fos was higher for a novel saccharin solution than for a familiar taste of saccharin, especially in two regions of the amygdala, the gustatory portions of the thalamus, and areas of insular cortex (Lin, Roman, Arthurs & Reilly, 2012). Thus, brain research is showing that several brain areas and mechanisms are involved in linking the CS-taste to the delayed nausea from toxic chemicals (US), and the subsequent aversion to tastes predictive of such nausea.

Brain sites also contribute to our urges, cravings, and excessive behavior. The cortical brain structure called the insula helps to turn physical reactions into sensations of craving. An investigation reported in *Science* showed that smokers with strokes involving damage to the insula lost their craving for cigarettes (Naqvi, Rudrauf, Damasio, & Bechara, 2007).

The insula-cortex area is also involved in behaviors whose bodily effects are experienced as pleasurable, such as cigarette smoking. Specific neural sites code for the physiological reactions to stimuli and "upgrade" the integrated neural responses into awareness, allowing the person to act on the urges of an acquired addiction (see Naqvi, Gaznick, Tranel, & Bechara, 2014 for a review of the insula's role in craving and drug use). Since humans can learn to modulate their own brain activity to reduce sensations of pain (deCharms et al., 2005), they may also be able to learn to deactivate the insula, reducing cravings associated with excessive use of drugs.

Taste Conditioning Induced by Physical Activity

Taste aversion learning is, as we have seen, a well-established conditioning process that generalizes over many species. Typically, a drug such as LiCl is used to condition taste aversion, but research on activity anorexia (Epling & Pierce, 1992) indicates that physical activity is capable of conditioning an aversion to food tastes and flavors. To directly test this hypothesis, Lett and Grant (1996) allowed hungry and thirsty rats 10 min to consume either a salt or sour solution, followed by 30 min of confinement in a running wheel. On separate unpaired trials (taste not followed by physical activity), these rats were allowed 10 min to consume the other solution and confined to 30 min in home cages (no physical activity). After three paired trials, rats avoided the solution followed by wheel running compared to the solution followed by placement in home cages. Lett and Grant (1996) concluded that rats show conditioned taste avoidance (CTA) when the flavored conditioned stimulus (CS) is repeatedly paired with wheel running. This might be because physical activity activates the sympathetic nervous system, which deactivates metabolic processes related to eating. (see Boakes & Nakajima, 2009 for CTA induced by swimming).

CTA induced by physical activity has been shown also in humans (Havermans, Salvy, & Jansen, 2009). Participants either consumed or merely tasted a flavored solution prior to 30 min of running on a treadmill at 80% of maximum heart rate. In both the consumption and the tasting-only groups, flavors followed by a single bout of intense physical activity led to a negative shift in hedonic evaluation of the paired flavor compared to another unpaired flavor not explicitly followed by treadmill running—indicating a conditioned flavor aversion. Also, participants' ratings of exercise-related gastrointestinal distress did not predict evaluation ratings of the flavor followed by the physical activity, implying that GI distress is not required to obtain CTA induced by exercising in humans. One caution for this study is that both CTA and GI distress are based on participants' ratings or evaluations rather than flavor consumption measures. Controlled research with animals, using objective tests for consumption, has implicated GI distress as a basis for exercise-induced CTA (Nakajima & Katayama, 2014).

Notice that CTA in both humans and rats involves a *forward conditioning* procedure in which the CS-taste is presented before the bouts of running (US). Sarah Salvy and her associates extended the running-wheel procedure to *backward conditioning*, in which the flavor (CS) follows the wheel running (US) (Salvy, Pierce, Heth, & Russell, 2004). Relative to unpaired controls, rats that were given backward pairings drank more of the CS-flavor, showing a *conditioned taste preference* (CTP). The observed CTP suggests that the after-effects of wheel running may act as positive reinforcement (Lett, Grant, Byrne, & Koh, 2000). If so, wheel running has bivalent properties (aversive *and* reinforcing) that produce different conditioning effects depending on the temporal placement of the CS-taste.

Subsequently, Hughes and Boakes (2008) tested the *bivalent effect of wheel running*. Rats were given access to one flavor before (CS_1) and a second flavor after (CS_2) 3 h of wheel running. Tests showed avoidance of the flavor given before running (CS_1), but preference for the flavor that followed running (CS_2). The bivalent effects, however, were found only in those rats that were given eight pre-exposure sessions of wheel running, and not in rats without this experience. That is, the bivalent effects of wheel running seemed to depend on pre-exposure of the US-activation of the sympathetic nervous system by the wheel.

Dobek, Heth, and Pierce (2012) hypothesized that the bivalent effect generalized to rats without pre-exposure to wheel running. In this study, palatable liquid food was used to encourage rats to consume equally the flavor coming before (CS_1) and after (CS_2) wheel running. Also, the wheel-running sessions lasted for 40 min, rather than 3 h, to ensure that the bivalent effect was due to equalizing exposure to the CSs and not to higher intensity of the US-wheel. Rats were given six conditioning sessions, CS_1 (flavor) → US (wheel) → CS_2 (flavor), followed by flavor consumption tests. Figure 7.11 shows that relative to a third control flavor unpaired with wheel running, a taste coming before wheel running produced less consumption (avoidance), and a taste that came after wheel running resulted in more (preference). That is, Dobek and colleagues obtained a bivalent effect (CTA and CTP) of wheel running that did not depend on pre-exposure to US-wheel.

Furthermore, it appears that the bivalent effects of wheel running do not depend on the nature of the CS-taste. Lett, Grant, Byrne, and Koh (2000) used rats to show that the after-effects of wheel running produced a *conditioned place preference (CPP)*. That is, a distinctive chamber (CS) paired with the after-effects of wheel running was preferred to an unpaired

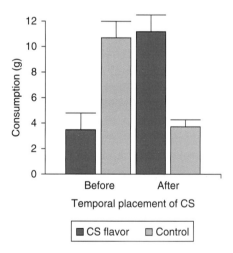

Fig. 7.11

Results are shown for the bivalent effects of wheel running. Compared to the control flavor, the flavor given before wheel running decreased on the consumption test, indicating conditioned taste aversion or CTA. The flavor given after wheel running, however, increased relative to the control flavor on the consumption test, indicating conditioned taste preference or CTP. The temporal location of the CS flavor with respect to the wheel running (US) determines its effect, CTA or CTP.

Source: From C. Dobek, C. D. Heth, & W. D. Pierce (2012). The bivalent effects of wheel running on taste conditioning. *Behavioural Processes*, *89*, pp. 36–38.

chamber. Also, Masaki and Nakajima (2008) obtained *conditioned place aversion* (*CPA*) induced by wheel running, using a forward conditioning procedure. To date, there have been no studies showing CPA to distinctive chamber preceding, and CPP to a chamber following, the same bouts of wheel running. The current evidence, however, indicates that wheel running (and vigorous physical activity) has bivalent properties—both aversive and reinforcing—resulting in aversion or preference based on the temporal location of the CS. Finally, a recent study has shown that the aversive effects of wheel running are related to gastrointestinal discomfort induced by the exercise and appear similar to other nausea-induced treatments including irradiation, motion sickness, and injection of emetic drugs (Nakajima & Katayama, 2014).

EXPERIMENTAL ANALYSIS OF ADJUNCTIVE BEHAVIOR

On time-based and interval schedules, organisms show side-effect behavior patterns not required by the contingency of reinforcement (Staddon & Simmelhag, 1971). If you received $5 for pressing a lever once every 10 min you might start to pace, twiddle your thumbs, have a sip of soda, or scratch your head between payoffs. Staddon (1977) noted that animals engage in three distinct types of behavior when food reinforcers occur on a fixed-time (FT) schedule. Immediately after food reinforcement, **interim behavior** that is independent of the schedule of reinforcement; for example, rats often groom themselves. Finally, as the time for reinforcement gets close, animals engage in food-related activities called **terminal behavior**, such as orienting toward the lever or food cup. The first of these categories, called interim or adjunctive behavior, is of most interest for the purposes of the present discussion, as it is behavior not required by the schedule but *induced by reinforcement*. Because the behavior is induced as a side effect of the reinforcement schedule, it is also referred to as **schedule-induced behavior**.

When a hungry animal is placed on an interval schedule of food reinforcement, and a water sipper, which can be sipped at any time, the animal will ingest an excessive amount of water. Falk (1961) suggested that **polydipsia** or excessive drinking is adjunctive or interim behavior induced by the time-based delivery of food. A rat that is working for food pellets on an intermittent schedule may drink as much as half its body weight during a single session (Falk, 1961). This drinking occurs even though the animal is not water deprived. The rat may turn toward the lever, press for food, obtain and eat the food pellet, drink excessively, groom itself, and then repeat the sequence. Pressing the lever is required for reinforcement, and grooming may occur in the absence of food delivery, but polydipsia is not required and appears to be induced by the schedule.

In general, **adjunctive behavior** refers to any excessive and persistent behavior pattern that occurs as a side effect of reinforcement delivery. The schedule may require a response for reinforcement (interval schedule), or it may simply be time based, as when food pellets are given every 30 s no matter what the animal is doing (FT 30 s). Additionally, the schedule may deliver food pellets on a fixed-time basis (e.g., 60 s between each pellet, FT 60 s) or it may be constructed so that the time between pellets varies (20, 75, 85, and 60 s, as in VT 60 s).

Schedules of food reinforcement have been shown to generate such adjunctive behavior as attack against other animals (Hutchinson, Azrin, & Hunt, 1968), licking at an airstream (Mendelson & Chillag, 1970), drinking water (Falk, 1961), chewing on wood blocks (Villareal, 1967), and preference for oral cocaine administration (Falk & Lau, 1997). Adjunctive behavior has been observed in pigeons, monkeys, rats, and humans; reinforcers have included water, food, shock avoidance, access to a running wheel, money, and for male pigeons the sight of a female (for reviews, see Falk, 1971, 1977; Staddon, 1977). Muller, Crow, and Cheney (1979) induced locomotor activity in college students and adolescents with developmental disabilities with fixed-interval (FI) and fixed-time (FT) token delivery. Stereotypic and self-injurious behavior of humans with developmental disabilities also has been viewed as adjunctive to the schedule of reinforcement (Lerman, Iwata, Zarcone, & Ringdahl, 1994). Thus, adjunctive behavior occurs in different species, is generated by a variety of reinforcement procedures, and extends to a number of induced responses.

A variety of conditions affect adjunctive or interim behavior, but the schedule of reinforcement delivery and the deprivation status of the organism appear to be the most important. As the time between reinforcement deliveries increases from 2 s to 180 s, adjunctive behavior increases. After 180 s, adjunctive behavior drops off, reaching low levels at approximately 300 s. For example, a rat may receive a food pellet every 10 s and drink only slightly more than a normal amount of water between pellet deliveries. When the schedule is changed to 100 s, drinking increases; and polydipsia increases again if the schedule is stretched to 180 s. As the time between pellets is further increased to 200, 250, and then 300 s, water consumption goes down. This pattern of increase, peak, and then drop in schedule-induced behavior is illustrated in Figure 7.12, and is called a bitonic function. The bitonic function has been observed in species other than the rat, and occurs for other adjunctive behavior (see Keehn & Jozsvai, 1989 for contrary evidence).

In addition to the reinforcement schedule, adjunctive behavior becomes more and more excessive as the level of deprivation increases. A rat at 80% of its normal body weight and given food pellets every 20 s drinks more water than an animal at 90% of its normal weight on the same schedule. Experiments using food reinforcement have shown that schedule-induced drinking (Falk, 1969), airstream licking (Chillag & Mendelson, 1971), and attack (Dove, 1976) increase as body weight decreases. Thus, a variety of induced activities escalate when deprivation for food is increased and food is the scheduled reinforcer. In this regard, polydipsia increases with food deprivation, but not consistently with water deprivation (Roper & Posadas-Andrews, 1981); also, preloading an animal with water does not reduce excessive drinking induced by the food reinforcement schedule (Porter, Young, & Moeschl, 1978). One possible conclusion is that

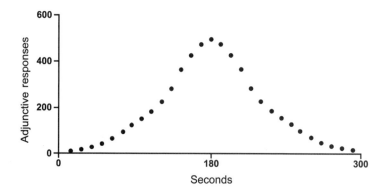

Fig. 7.12

A bitonic relationship is presented showing time between food pellets and amount of adjunctive water drinking.

schedule-induced polydipsia is increased more by motivational operations (MOs) related to the upcoming reinforcer (food pellets) than by MOs related to drinking and thirst.

Falk has noted that "on the surface" adjunctive behavior does not seem to make sense:

> [Adjunctive activities] are excessive and persistent. A behavioral phenomenon, which encompasses many kinds of activities and is widespread over species and high in predictability ordinarily, can be presumed to be a basic mechanism contributing to adaptation and survival. The puzzle of adjunctive behavior is that, while fulfilling the above criteria, its adaptive significance has escaped analysis. Indeed, adjunctive activities have appeared not only curiously exaggerated and persistent, but also energetically quite costly.
>
> (Falk, 1977, p. 326)

Falk went on to note that adjunctive behavior is similar to **displacement behavior** of animals in the wild—actions that are seemingly irrelevant, incongruous, and out of context. For example, two skylarks might stop fighting and begin to peck the ground with feeding movements. One possibility is that both adjunctive and displacement behaviors occur in situations that are generally positive for the organism, but from which the animal is likely to escape, due to the delay arranged by the schedule of reinforcement. Thus, the adaptive significance of adjunctive behavior is to maintain the animal on the schedule during a period (the delay) when it would be likely to leave or escape (Falk, 1977).

Adjunctive Behavior in Humans

Adjunctive behavior occurs in a variety of species, including humans. Excessive drug use by people may involve adjunctive behavior induced by the temporal arrangement of reinforcement. Thus, Doyle and Samson (1988) allowed human participants to drink beer while they played a gambling game that delivered monetary reinforcements. Participants on the FI 90-s payoff drank about twice as much as those assigned the FI 30-s schedule, which suggests that beer drinking was induced by the payoff schedule. In another study, Cherek (1982) allowed people to smoke while button pressing for monetary reinforcements. The average number of puffs an hour was a bitonic function of FI schedule values (30, 60, and 240 s). In both of these experiments, humans consumed higher levels of a drug (alcohol or nicotine) just after payoffs, which suggests that excessive drug taking and addiction may sometimes be schedule induced (Falk, 1998).

Another adjunctive behavior observed in humans is general restlessness, movement, and fidgeting. Lasiter (1979) monitored gross-movement activity of undergraduate students seated at a console and performing a signal-detection vigilance task. Relative to a control group (FT-control), both FT (detect-only) and FI (observation-detect) schedules increased general activity, particularly in the period immediately following the delivery and detection of the signal. These increases in movements appeared to be schedule induced, showing high rate and temporal patterning. Subsequently, Porter, Brown, and Goldsmith (1982) asked female children to press a telegraph key that delivered M&M's® candies on FR 1, FI 30-s, and FI 60-s schedules. Baselines (FR 1) were obtained before and after each FI value, and the schedules were equated for temporal lengths and the number of candies dispensed. Substantial increases in motor movements occurred after reinforcement for both FI schedules (postreinforcement vocalization and drinking were also observed). One possibility is that fidgeting and general restlessness of children with attention deficit hyperactivity disorder (ADHD) may be induced by the schedules of reinforcement arranged in the home and classroom.

Many people with developmental disabilities show a variety of stereotyped behavior that may also be schedule induced. Hollis (1973) studied stereotyped wringing of hands and head-rolling movements in institutionalized females and children with developmental delays. In the first experiment, delivery of M&M's® candies on FT 4-min and FT 30-s schedules induced the most stereotyped movements on the FT 30 s compared with baseline measures of stereotyped movements; baselines were taken when the children were fed and food deprived, but without any delivery of candies. Increases in stereotyped movement occurred in bouts following each delivery of M&M's®. Notably, FT 30-s dispenser clicks (conditioned reinforcers) temporarily induced increased stereotyped movements, but FT 30-s doorbell rings did not. In another study of adjunctive stereotyped movements, three institutionalized adult males with developmental delays served as experimental participants (Wiesler, Hanson, Chamberlain, & Thompson, 1988). Responses on a motor task were reinforced by the delivery of 8 mL of a soft drink and praise. The baseline schedule was FR 1, and three FI schedules were programmed—FI 15 s, FI 90 s, and FI 180 s. The amount of stereotyped movement on the FI 15-s schedule was about twice as much as that for FR 1 for two of the participants; the increase was less dramatic for the third participant. Also, the amount of stereotyped movement increased with the FI value. The evidence suggests that excessive stereotyped movements in humans, often described as abnormal behavior, may be induced by schedules of reinforcement. Schedules involving activities and events on the ward may induce stereotyped behavior as an unintentional by-product. [Note: feeding and caretaking schedules probably induce the stereotypic behavior of zoo animals.]

Overskeid (1992) rejected the research on adjunctive behavior in humans, pointing to the inadequate control or baseline conditions of most experiments. In contrast, Falk (1994) acknowledged that there were problems with human experiments, but concluded that the totality of the evidence supported the hypothesis that schedules of reinforcement induce adjunctive behavior in humans. Our view is that convergent evidence from experiments with a number of species, including humans and other primates, indicates that interval- and time-based schedules induce excessive behavior following reinforcement. Many seemingly abnormal, displaced, and high-rate activities of humans probably are induced inadvertently by schedules in the workplace, school, or home.

The Nature of Adjunctive Behavior

Behavioral researchers have debated the nature of adjunctive behavior, questioning whether it is operant, respondent, or a third class of biologically relevant displacement behavior. In some sense, this debate is similar to the argument about biological constraints involving instinctive drift, autoshaping, and sign tracking. Thus, with instinctive drift, pigs root the coins they are trained to deposit in a box for food reinforcement, seemingly drifting toward instinctive behavior. We noted that behavior is always appropriate to the operating contingencies and there is no need to talk about behavior drifting toward its biological roots. We also noted that respondent procedures often are embedded in operant contingencies and this appears to be the case with adjunctive behavior.

To investigate the nature of adjunctive behavior, researchers have focused on a reliable and quantitative property of adjunctive or interim behavior—the temporal distribution of licks during the inter-pellet interval. Falk (1971) referred to this feature in his analysis of schedule-induced polydipsia (SIP), concluding that excessive drinking is a form of adjunctive behavior similar to displacement activity, which occurs at a time when *probability of a pellet delivery is at its lowest* (closely following the delivery of a pellet). Staddon (1977) in his analysis of interim and terminal behavior also noted that terminal behavior (magazine entry) occurs close to pellet delivery and interim behavior occurs when delivery is unlikely (after the delivery of a pellet). And Lashley and Rossellini (1980) proposed a related interpretation where the offset of the pellet delivery becomes a Pavlovian inhibitory signal (CS−), operating to restrict SIP to periods of low-reinforcement (food pellet) probability.

An alternative account of adjunctive behavior has emphasized that food pellet delivery elicits postprandial drinking, which is adventitiously reinforced by the arrival of the next food pellet—a form of delayed reinforcement (Killeen & Pellón, 2013; Skinner, 1948b). To help account for the temporal distribution of licking to the period after pellet delivery, researchers suggested that terminal responses, such as magazine entries, compete with drinking and limit licking to an early part of the interpellet interval (Patterson & Boakes, 2012). The delayed reinforcement account has resulted in a number of experiments showing that the acquisition of adjunctive behavior is regulated in a manner similar to more common operant behaviors such as lever pressing, although adjunctive licking is more sensitive to reinforcement over longer delays than operant lever pressing (e.g., Castilla & Pellón, 2013; Pellón & Pérez-Padilla, 2013). The sensitivity to reinforcement after long delays allows adjunctive licking to be trapped by the upcoming food delivery even though there is no contingency between licking and food reinforcement—temporal proximity of response and reinforcement is enough (Killeen & Pellón, 2013).

Currently there is no resolution of the debate about the nature of adjunctive behavior, and it seems that researchers who favor a displacement account design experiments on temporal distribution gradients that support an ethological, species-specific behavior interpretation of SIP, suggesting that biologically-relevant behavior intrudes on operant schedules of reinforcement. On the other hand, behavior analysts who use general principles (delay of reinforcement) to account for adjunctive behavior continue to provide experimental evidence for SIP as operant behavior. A recent study tested the claim that magazine entries compete with adjunctive licking—displacing licking to an early part of the interpellet interval (Boakes, Patterson, Kendig, & Harris, 2015). Results based on measures of temporal distributions of licking and magazine entries, however, did not provide definitive support for response displacement or a version of delayed reinforcement, which requires the temporal distribution of magazine entries to precede that of licking. After years of research, it seems there is no definitive test yet for the nature of adjunctive behavior. One possibility is that the initial licking response is generated by phylogenetic contingencies related to evolution and natural selection, but once generated this phylogenetic behavior is malleable and subject to ontogenetic selection by contingencies of reinforcement (Skinner, 1969). At this point, biologically relevant behavior becomes operant behavior.

ON THE APPLIED SIDE: EXPERIMENTAL ANALYSIS OF ACTIVITY ANOREXIA

A paper by Routtenberg and Kuznesof (1967) reported self-starvation in laboratory rats. Cheney (one of the authors of this textbook) thought that this was an unusual effect as there are few, if any, situations in which a nonhuman animal will take its own life. Because of this, he decided to replicate the experiment, and recruited Frank Epling, who was an undergraduate student at the time, to help to run the research. The experiment was relatively simple. Cheney and Epling (1968) placed a few rats in running wheels and fed them for 1 h each day. The researchers recorded the daily number of wheel turns, the weight of the rat, and the amount of food eaten. Surprisingly, the rats increased wheel running to excessive levels, ate less and less, lost weight, and if allowed to continue in the experiment died of starvation. Importantly, the rats were not required to run and they had plenty to eat, but they stopped eating and ran as much as 10–12 miles a day.

Twelve years later, Frank Epling, at the University of Alberta, Canada, began to do collaborative research with David Pierce (author of this textbook), at the same university. They wondered if anorexic patients were hyperactive like the animals in the self-starvation experiments. If they were, it might be possible to develop an animal model of anorexia. Clinical reports indicated that many anorexic patients were indeed excessively active. For this reason, Epling and Pierce began to investigate the relations between wheel running and food intake (Epling & Pierce, 1992). The basic finding was that physical activity decreased food intake, and suppressed food intake through loss of body weight increases activity. Epling and Pierce call this feedback loop activity-based anorexia or just activity anorexia, and argue that a similar cycle occurs in some patients with anorexia (see Epling & Pierce, 1992; Epling, Pierce, & Stefan, 1983).

This analysis of eating and exercise suggests that these activities are interrelated. Depriving an animal of food should increase the reinforcing value of exercise. Rats that are required to press a lever to run on a wheel should work harder for wheel access when they are deprived of food. Additionally, engaging in exercise should reduce the reinforcing value of food. Rats that are required to press a lever for food pellets should not work as hard for food following a day of exercise. Pierce, Epling, and Boer (1986) designed two experiments to test these ideas, which we now describe.

Reinforcement Effectiveness of Physical Activity

We asked whether food deprivation increased the reinforcing effectiveness of wheel running. This is an interesting implication, because increased reinforcement effectiveness is usually achieved by withholding the reinforcing event. Thus, to increase the reinforcement effectiveness of water, a researcher typically withholds access to water, but in this case food is withheld to increase the reinforcing effectiveness of wheel access.

Nine young male and female rats were used to test the reinforcing effectiveness of wheel running as food deprivation changed. The animals were trained to press a lever to obtain 60 s of wheel running. When the rat pressed the lever, a brake was removed and the running wheel was free to turn. After 60 s, the brake was again activated and the animal had to press the lever to obtain more wheel movement for running. The apparatus that we constructed for this experiment is shown in Figure 7.13.

Once lever pressing for wheel running was stable, each animal was tested when it was food deprived and not food deprived. To measure the reinforcing effectiveness of wheel running, the animals were required to press the lever more and more for each opportunity to run—a progressive-ratio schedule (see Chapter 5 for further information about this schedule). Specifically, the rats were required to press 5 times to obtain 60 s of wheel running, then 10, 15, 20, and 25 times, and so on (you may remember this is a progressive ratio schedule of reinforcement). The point at which they gave up pressing (breakpoint) for wheel running was used as an index of the reinforcing effectiveness of exercise.

The results of this experiment are shown in Figure 7.14. Animals lever pressed for wheel running more when food deprived than when at normal weight. In other words, the animals worked harder for exercise when they were hungry. Further evidence indicated that the reinforcing effectiveness

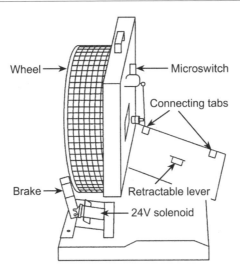

Wheel-running apparatus used in the Pierce, Epling, and Boer (1986) experiment on the reinforcing effectiveness of physical activity as a function of food deprivation.

Source: From W. D. Pierce, W. F. Epling, & D. P. Boer (1986). Deprivation and satiation: The interrelations between food and wheel running. *Journal of the Experimental Analysis of Behavior, 46*, pp. 199–210. Republished with permission.

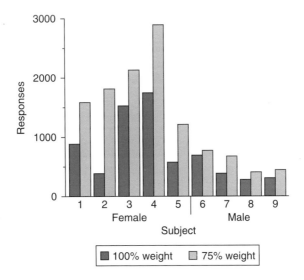

Fig. 7.14
The graph shows the number of bar presses for 60 s of wheel running as a function of food deprivation.

Source: From W. D. Pierce, W. F. Epling, & D. P. Boer (1986). Deprivation and satiation: The interrelations between food and wheel running. *Journal of the Experimental Analysis of Behavior, 46*, pp. 199–210. Republished with permission.

went up and down when an animal's weight was made to increase and decrease. For example, one rat pressed the bar 1,567 times when food deprived, 881 times when at normal weight, and 1,882 times when again food deprived. This indicated that the effect was reversible and was tied to the level of food deprivation (see Belke, Pierce, & Duncan, 2006 on substitutability of food and wheel running). Females were also more motivated to press the lever for access to the wheel.

Reinforcement Effectiveness of Food

In a second experiment, we investigated the effects of exercise on the reinforcing effectiveness of food. Four male rats were trained to press a lever for food pellets. When lever pressing occurred reliably, we tested the effects of exercise on each animal's willingness to work for food. In this case, we expected that a day of exercise would decrease the reinforcement effectiveness of food on the next day.

Test days were arranged to measure the reinforcing effects of food. One day before each test, animals were placed in their wheels without food. On some of the days before a test the wheel was free to turn, and on other days it was not. Three of the four rats ran moderately in their activity wheels on exercise days. (One rat did not run when given the opportunity. This animal was subsequently forced to exercise on a motor-driven wheel.) All of the animals were well rested (3 to 4 h of rest) before each food test. This ensured that any effects were not caused by fatigue.

Counting the number of lever presses for food, as food became more and more difficult to obtain via a progressive ratio schedule of reinforcement, allowed the reinforcement effectiveness of food to be assessed. For example, an animal had to press 5 times for the first food pellet, 10 times for the next, then 15, 20, and 25 times, and so on. As in the first experiment, the breakpoint (giving-up point) was used to measure reinforcement effectiveness. Presumably, the more effective the food reinforcer, the harder the animal would work for it.

Figure 7.15 shows that, when test days were preceded by a day of exercise, the reinforcing effectiveness of food decreased sharply. Animals pressed the lever more than 200 times when they were not allowed to run, but no more than 38 times when running preceded test sessions. Food no longer supported lever presses following a day of moderate wheel running, even though

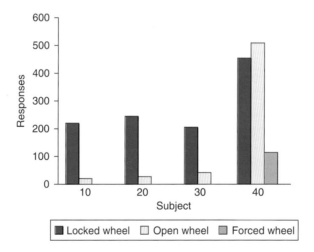

Fig. 7.15

The graph shows the number of bar presses for food when three rats (10, 20 and 30) were allowed to run on a wheel (open) as compared with no physical activity (locked wheel). (Rat 40 had a "forced wheel" condition that was shown as he was not motivated to wheel run initially.)

Source: From W. D. Pierce, W. F. Epling, & D. P. Boer (1986). Deprivation and satiation: The interrelations between food and wheel running. *Journal of the Experimental Analysis of Behavior*, 46, pp. 199–210. Republished with permission.

a lengthy rest period preceded the test. Although wheel running was moderate, it represented a large change in physical activity, since the animals were previously sedentary.

Prior to each test, the animals spent an entire day without food. Because of this, the reinforcing effectiveness of food should have increased. However, exercise seemed to override the effects of food deprivation, since responding for food decreased rather than increased. Other evidence from these experiments suggested that the effects of exercise were similar to those of feeding the animal. Although exercise reduces the reinforcing effectiveness of food, the effect is probably not because wheel running serves as a substitute for food consumption (Belke, Pierce, & Duncan, 2006).

The rat that was forced to run also showed a sharp decline in lever pressing for food (see Figure 7.15). Exercise was again moderate, but substantial relative to the animal's sedentary history. Because the reinforcement effectiveness of food decreased with forced exercise, we concluded that both forced and voluntary physical activity produce a decline in the value of food reinforcement. This finding suggests that people who increase their physical activity because of occupational requirements (e.g., ballet dancers), or just to get trim and fit, may value food less.

The Biological Context of Eating and Activity

In our view, the motivational interrelations between eating and physical activity have a basis in natural selection. Natural selection favors those animals that increase their amount of travel in times of food scarcity. During a famine, organisms can either stay where they are and conserve energy or become mobile and travel to another location. The particular strategy adopted by a species depends on natural selection. If travel led to reinstatement of food supply and staying where they were resulted in starvation, then those animals that traveled gained a reproductive advantage.

A major problem for an evolutionary analysis of activity anorexia is accounting for the decreased appetite of animals that travel to a new food patch. The fact that increasing energy expenditure is accompanied by decreasing caloric intake seems counterintuitive. From a homeostatic or energy-balance perspective, food intake and energy expenditure should be positively related. In fact, this is the case if an animal has the time to adjust to a new level of activity and food supply is not greatly reduced (Dwyer & Boakes, 1997).

When depletion of food is severe, however, travel should not stop when food is infrequently contacted. This is because stopping to eat may be negatively balanced against reaching a more abundant food patch. Frequent contact with food would signal a replenished food supply, and this should reduce the tendency to travel. Recall that a decline in the reinforcing effectiveness of food means that animals will not work hard for nourishment. When food is scarce, considerable effort may be required to obtain it. For this reason, animals ignore food and continue to travel. However, as food becomes more plentiful and the effort to acquire it decreases, the organism begins to eat (see Dixon, Ackert, & Eckel, 2003, on recovery from activity anorexia). Food consumption lowers the reinforcement effectiveness of physical activity and travel stops (see also Belke, Pierce, & Duncan, 2006 on the partial substitution of food for physical activity). On this basis, animals that expend large amounts of energy on a migration or trek become anorexic.

Behavioral Neuroscience: Activity Anorexia and Neuropeptide Y

Prominent features of activity anorexia are the suppression of food intake, escalating levels of physical activity, and a precipitous decline in body weight. These responses occur when animals face an environmental challenge of time-limited feeding (1 to 2 h a day) followed by the opportunity for food-related travel. As you can see, these behavioral observations do not suggest energy balance or homeostasis, which is the accepted model of biology.

In the homeostatic model, the hormone leptin is secreted by fat in response to weight gain, which reduces food intake. For a slimmer person with little fat, levels of leptin are lower and activate neuropeptide Y (NPY), a brain peptide that stimulates eating. Both activity-anorexic rats and humans with anorexia indeed have higher levels of NPY (Diane et al., 2011; Gendall, Kaye, Altemus, McConaha, & La Via, 1999), but they do not eat much. Finally, activity-anorexic rats and humans with anorexia are physically overactive, expending high levels of energy at a time when food is depleted (Epling & Pierce, 1992). This pattern with anorexia seems to be incompatible with the homeostatic model—at least at first glance.

But here is a possibility: NPY acts differently when animals are challenged by severe food restriction and excessive weight loss. Under these extreme conditions, NPY may no longer act on the eating centers of the brain, but now acts to stimulate a search for food (travel) and inhibit food intake (Pjetri et al., 2012). To test this possibility, Nergardh and colleagues in Sweden gave brain infusions of NPY to rats that had access to running wheels, and varied the number of hours a day for which food was available (Nergardh et al., 2007). Rats in running wheels lost more weight than sedentary controls and ran progressively more as the availability of food was reduced over experimental groups (from 24 h to 1 h per day). When food was available for only 1 h per day, body weight plummeted and the rats in this group showed the typical activity-anorexic effect. These anorexic rats also showed high levels of NPY measured by assay. But were high levels of NPY working on the eating or food search centers of the brain? The critical test occurred in rats that were given food for 2 h each day. These animals increased wheel running and decreased food intake when given brain infusions of NPY. By contrast, NPY infusions increased eating, but not wheel running, in rats that had free access to food for 24 h. The results show that NPY, a brain peptide that usually stimulates eating, acts on brain receptors to activate travel and inhibit eating during periods of food shortage or famine.

ADVANCED SECTION: AUTOSHAPING AS OPERANT–RESPONDENT INTERRELATIONSHIPS

Negative Automaintenance

When scientists are confronted with new and challenging data, they are typically hesitant about accepting the findings. This is because researchers have invested time, money, and effort in experiments that may depend on a particular view of the world. Consider a person who has made a career of investigating the free-operant behavior of pigeons, with rate of pecking a key as the major dependent variable. The suggestion that key pecking is actually respondent rather than operant behavior would not be well received by such a scientist. If key pecks are reflexive, then conclusions about operant behavior based on these responses are questionable. One possibility is to go to some effort to explain the data within the context of operant conditioning.

In fact, Brown and Jenkins (1968) suggested just this kind of explanation for their results. Recall that these experimenters pointed to the species-specific tendency of pigeons to peck at stimuli that they look at. When the light is illuminated, there is a high probability that the bird will look and peck at it. Some of these responses are followed by food, and pecking increases in frequency. Other investigators noted that when birds are magazine trained, they stand in the general area of the feeder and the response key is typically at head height just above the food tray. Anyone who has watched a pigeon knows that they have a high frequency of bobbing their heads. Since they are close to the key and are making pecking (or bobbing) motions, it is

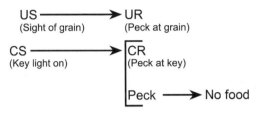

Fig. 7.16
For birds, the sight of grain (US) elicits pecking (UR) as species-specific behavior. Williams and Williams (1969) arranged omission procedures for pecking a lighted key that was usually followed by grain (autoshaping). When the key light (CS) occurred, a peck at the key (CR) caused the key light to go out and no food was presented. If the bird did not peck the lighted key, it received food. The birds pecked the key even though these responses prevented reinforcement.

possible that a strike at the key is inadvertently followed by food delivery. From this perspective, key pecks are superstitious in the sense that they are accidentally reinforced. The superstitious explanation has an advantage because it does not require postulation of a look–peck connection and it is entirely consistent with operant conditioning.

Although these explanations of pecking as an operant are plausible, the possibility remains that autoshaped pecking is at least partially respondent behavior. An ingenious experiment by Williams and Williams (1969) was designed to answer this question. In their experiment on **negative automaintenance**, pigeons were placed in an operant chamber and key illumination was repeatedly followed by food. This is, of course, the same procedure that Brown and Jenkins (1968) used to show autoshaping. The twist in the Williams and Williams procedure was that if the bird pecked the key when it was illuminated, food was not presented. This is called *omission training*, because if the pigeon pecks the key the reinforcer is omitted, or if the response is omitted the reinforcer is delivered.

The logic of this procedure is that if pecking is reflexive or has respondent properties, then it is elicited by the key light and the pigeon will reflexively strike the disk. If, on the other hand, pecking is operant, then striking the key prevents reinforcement and responses should not be maintained. Thus, the prediction is that pecking is reflexive or respondent behavior if the bird continues to peck with the **omission procedure** in place. Using the omission procedure, Williams and Williams (1969) found that pigeons frequently pecked the key even though responses did not result in reinforcement. This finding suggests that with autoshaping, when a key light stimulus precedes grain presentation, it becomes a CS that elicits pecking at the key (CR). Figure 7.16 shows this arrangement between stimulus events and responses. It is also the case that by not presenting the food (US), the key light (CS) is no longer paired with the US, and the response (CR) undergoes extinction.

The puzzling aspect of this finding is that in most cases pecking a key is regulated by reinforcement and is clearly operant. Many experiments have shown that key pecks increase or decrease in frequency depending on the consequences that follow behavior.

Autoshaping: Operant–Respondent Interrelationships

Because of this apparent contradiction, several experiments were designed to investigate the nature of autoshaped pecking as an operant or respondent. Schwartz and Williams (1972a) preceded grain reinforcement for pigeons by turning on a red or white light on two separate keys. The birds responded by pecking the illuminated disk (i.e., they were autoshaped). On some trials, the birds were presented with both the red and white keys. Pecks on the red key prevented reinforcement as in the omission procedure used by Williams and Williams (1969). Pecks on the white key, however, did not prevent reinforcement.

On these choice trials, the pigeons showed a definite preference for the white key that did not stop the delivery of grain. In other words, the birds more frequently pecked the key that was followed by the presentation of grain. Because this is a description of behavior regulated by an operant contingency (peck → food), autoshaped key pecks cannot be exclusively respondent behavior. In concluding their paper, Schwartz and Williams wrote:

> A simple application of respondent principles cannot account for the phenomenon as originally described … and it cannot account for the rate and preference results of the present study. An indication of the way operant factors can modulate the performance of automaintained behavior has been given…. The analysis suggests that while automaintained behavior departs in important ways from the familiar patterns seen with arbitrary responses, the concepts and procedures developed from the operant framework are, nevertheless, influential in the automaintenance situation.
> (Schwartz & Williams, 1972a, p. 356)

Schwartz and Williams (1972b) went on to investigate the nature of key pecking by pigeons in several other experiments. The researchers precisely measured the contact duration of each peck that birds made to a response key. When the omission procedure was in effect, pigeons produced short-duration pecks. If the birds were autoshaped, but key pecks did not prevent the delivery of grain, the duration of the pecks was long. These same long-duration pecks occurred when the pigeons responded for food on a schedule of reinforcement. Generally, it appears that there are two types of key pecks: short-duration pecks evoked (or perhaps elicited) by the presentation of grain, and long-duration pecks that occur when the bird's behavior is brought under operant control.

Ploog and Zeigler (1996) showed that the size of reinforcer that is used in autoshaping predicts the extent to which the beak is open (or gape) when a pigeon pecks a key. The delivery of larger pellets resulted in larger gapes of the beak when the pigeon pecked the key compared to smaller pellets, which resulted in smaller gapes. This aspect suggests more respondent properties of the key pecks, as the CS of the key elicits a gape that approximates the size of the US (pellet). Interestingly, larger pellets also produced higher response rates and shorter latencies to key pecking, which are more operant properties. Previous studies (e.g., Jenkins & Moore, 1973) have also shown that autoshaping with a water reinforcer results in a closed-beak peck, similar to sipping. When autoshaped with grain as a reinforcer, the gape is wide and similar to a peck that would effectively pick the grain up for eating.

Other evidence also suggests that both operant and respondent interrelationships are involved in autoshaping (see Lesaint, Sigaud, & Khamassi, 2014 for a computational model). For example, Bullock and Myers (2009) recently showed that autoshaped responding of the cynomolgus monkey (*Macaca fascicularis*) is sensitive to both negative (omission) and positive (response-dependent) contingencies, using banana pellets and stimulus-directed touch-screen responses. One possibility is that autoshaped pecking by birds is initially respondent behavior elicited by light–food pairings. Once pecking produces food, however, it comes under operant control. Even when an omission procedure is in effect both operant and respondent behavior is conditioned—suggesting that there is no uniform learning process underlying autoshaping (Lesaint, Sigaud, & Khamassi, 2014; Papachristos & Gallistel, 2006). During omission training, a response to the key turns off the key light and food is not delivered. If the bird does not peck the key, the light is eventually turned off and food is presented. Notice that light offset (dark key) is always predictive of reinforcement and becomes a conditioned reinforcer in the omission procedure. In this analysis, pecking the key is maintained by immediate reinforcement from light offset. Hursh, Navarick, and Fantino (1974) provided evidence for this conditioned reinforcement view of negative automaintenance. They showed that birds quit responding during omission training if the key light did not immediately go out when a response was made.

CHAPTER SUMMARY

This chapter has considered several areas of research on respondent–operant interrelations. Autoshaping showed that an operant response (key pecking for food) could actually be elicited by respondent procedures. Before this research, operants and respondents had been treated as separate systems subject to independent controlling procedures. The Brelands' animal training demonstrations provided a hint that the two systems were not distinct—with species-specific behavior being elicited by operant contingencies. Their work revealed the biological foundations of conditioning as well as the contributions made by biologically relevant factors. Animals are prepared by evolution to be responsive to specific events and differentially sensitive to various aspects of the environment.

Other experiments indicated that respondent behavior could be controlled by operant contingencies. The Miller studies showed that heart rate—an autonomic response—could be reinforced by electrical stimulation of the brain. The implication again is that the neural systems regulating respondent and operant behavior are interrelated, allowing for operant conditioning of behavior (heart rate) that is often considered to be hardwired.

Taste aversion is another example of biological factors underlying conditioning procedures. The findings of Garcia and Koelling indicate that some interoceptive stimuli are better paired with each other (flavor–poison) than crossing-systems stimuli (flavor–shock), illustrating how organisms are prepared for conditioning based on evolution and natural selection. Work in this area contributes to the management of toxic plant ingestion by livestock, and to the prediction and control of diet selection. Finally, we discussed activity anorexia both as a real-world human problem and as an interesting research question. What neurophysiological–behavioral mechanisms could possibly interact to drive an organism to self-starvation? It turns out that a combination of restricted access to food and the opportunity to exercise are the conditions that lead to this deadly spiral.

Key Words

Activity anorexia
Adjunctive behavior
Autoshaping
Behavior system
Biological context
Conditioned taste preference (CTP)
Context for conditioning
Displacement behavior
Incentive salience
Instinctive drift
Interim behavior
Negative automaintenance
Omission procedure (training)
Polydipsia
Preparedness
Schedule-induced behavior
Sign tracking
Stimulus substitution
Taste aversion learning
Terminal behavior

On the Web

www.youtube.com/watch?v=v6X4QJQg3cY This video discusses just one of Denver Zoo's worldwide conservation efforts, saving lions in Botswana through a method using mild condition taste aversion.

www.youtube.com/watch?v=7bD0OznhBw8 In this video, the food choices of lambs are shown to change with conditioned taste aversion induced by lithium chloride sickness. Use of plant-specific anti-toxins allows animals to consume vegetation, which contains mild toxins that would usually prevent consumption.

www.jsu.edu/depart/psychology/sebac/fac-sch/spot-peck/spot-peck.html Go to this website to read an actual scientific article on "Stereotyped adjunctive pecking by caged pigeons," by Palya and Zacny (1980), from *Animal Learning and Behavior, 8*, 293–303.

www.ctalearning.com This website provides an annotated bibliography and overview of conditioned taste aversion by Anthony Riley of the Department of Psychology, Psychopharmacology Laboratory, American University, Washington, DC.

www.youtube.com/watch?v=50EmqiYC9Xw This YouTube nicely shows the autoshaped keypeck gapes in pigeons. An example of autoshaping with water (no gape) and grain (wider gape) is shown.

Brief Quiz

1 In terms of operant contingencies and the intrusion of reflexive behavior:
   a   operant procedures elicit reflexive behavior directly by the contingencies of reinforcement
   b   reflexive behavior is elicited by respondent procedures embedded in operant contingencies
   c   respondent procedures cause species-characteristic responses
   d   both (b) and (c) are true
2 What did Brown and Jenkins (1968) conclude about autoshaping in their pigeons?
   a   the look–peck coupling is species-specific and results in pecks to the illuminated key
   b   following illumination of the key with grain eventually caused the lighted key to elicit pecking
   c   eventually an operant chain develops, culminating in pecking
   d   all of the above

3. Phenomena such as instinctive drift, sign tracking, and autoshaping have been analyzed as:
   a. stimulus substitution, where the CS substitutes for the US
   b. behavior systems activated by the US and the physical properties of the CS
   c. both (a) and (b)
   d. none of the above

4. In terms of operant conditioning of reflexive behavior, the experiment by Miller and Carmona (1967):
   a. showed conclusive results for operant conditioning of salivation
   b. showed that salivation and heart rate were both susceptible to operant conditioning
   c. showed that the increased flow of saliva was accompanied by the dogs being more alert
   d. showed all of the above

5. What does the evidence suggest about the operant conditioning of reflexive behavior?
   a. reflexes can be conditioned by operant procedures in some circumstances
   b. reflexive behavior is hardly ever controlled by respondent procedures
   c. reflexive behavior is generally controlled by operant procedures
   d. only (b) and (c) are true

6. When a CS compound (color and taste) is associated with illness, different species show avoidance to the two parts of the compound. This phenomenon is called:
   a. species readiness
   b. species set
   c. species preparedness
   d. species activation

7. What did Lett and Grant (1996) suggest with regard to activity anorexia?
   a. it could involve taste aversion induced by physical activity
   b. it probably explains taste aversion conditioning
   c. it is the first stage in taste aversion conditioning
   d. both (b) and (c)

8. Excessive drinking is technically called:
   a. polyhydration
   b. polydipsia
   c. polyfluidity
   d. polydistation

9. What is the basic finding for activity anorexia?
   a. decreased food intake increases physical activity
   b. increased food intake increases physical activity
   c. physical activity decreases food intake
   d. both (a) and (c)

Answers to Brief Quiz: 1, d (p. 234); 2, d (p. 240); 3, c (p. 236); 4, c (p. 246); 5, a (p. 247); 6, c (p. 249); 7, a (p. 253); 8, b (p. 255); 9, d (p. 260).

NOTE

1 It is worth noting that the rat is an ideal subject in these experiments for generalizing to humans. Like humans, the rat is omnivorous—it eats both meats and vegetables. Rats live wherever humans do and are said to consume 20% of the world's human food supply.

CHAPTER 8

Stimulus Control

1. Learn about stimulus control of behavior and multiple schedules of reinforcement.
2. Delve into stimulus control, behavioral neuroscience, and understanding perception.
3. Solve the problem of the "bird-brained" pigeon and implications for teaching and learning.
4. Investigate behavioral contrast and its determinants.
5. Inquire about stimulus generalization, peak shift, errorless discrimination, and fading.
6. Investigate delayed matching to sample and an experimental analysis of remembering.

In the everyday world, human behavior is changed by signs, symbols, gestures, and spoken words. Sounds, smells, sights, and other sensory stimuli that do not depend on social conditioning also regulate behavior. When social or nonsocial events precede operant behavior and affect its occurrence, they are called antecedent stimuli. An **antecedent stimulus (S)** is said to alter the probability of an operant, in the sense that the response is more (or less) likely to occur when the stimulus is present.[1]

One kind of antecedent stimulus discussed in Chapter 4 is the **S^D (esse-dee)** or **discriminative stimulus**. An S^D is an antecedent stimulus that sets the occasion for reinforcement of an operant. It predicts reinforcement will be delivered if a response occurs. In a pigeon experiment, a red light may reliably signal the presentation of food for pecking a key. After some experience, the bird will immediately peck the key when it is illuminated with the red light. Thus, the discriminative stimulus sets the occasion for a high probability of response.

The discriminative stimuli that regulate human behavior may be as simple as that in the pigeon experiment, or far more complex. A green traffic light and the word WALK sets the occasion for pedestrians to cross a street. In university libraries, the call numbers posted above the stacks and on the books are discriminative stimuli for stopping, turning corners, and other behavior, which result in finding a book. With social media and the internet, a student may find books and articles by using a search engine such as Google Scholar, Google Books, or by clicking a link that another student or professor sent them—all of this behavior involves discriminative stimuli.

DOI: 10.4324/9781003202622-8

Another kind of antecedent stimulus is called an **S^Δ** (**S-delta**). An S^Δ is a stimulus that signals extinction of an operant (see Chapter 4). For example, a rat may press a lever under a VI schedule of food reinforcement. Every now and then a tone comes on and a period of extinction is in effect in which no food is delivered. After some time, the rat will stop pressing the bar as soon as the tone is presented. Thus, the tone is defined as an S^Δ because of extinction, and lever pressing has a low probability of occurrence in its presence.

S^Δ (S-delta) stimuli that regulate human behavior also range from simple to complex. When your car is almost out of gas, a service-station sign that says CLOSED is an S^Δ for turning into that station, but also an S^D for going to your cell phone app to find the nearest alternate station since looking for fuel is reinforced with finding an open gas station. A tennis opponent who usually wins the match may become an S^Δ for playing the game. In this case, you may play tennis with others, but not with the person who always wins because that means you always lose (no reinforcement). Breakdown of communication between married couples or partners sometimes may be caused by stimuli that signal extinction for conversation. A concerned person may try to talk to their partner about a specific issue (like money), and the partner pretends to check their cell phone messages or social media. The partner's behavior is an S^Δ for conversation if the concerned person reliably stops talking when the partner looks down at their cellular device when conversation about money is initiated.

DIFFERENTIAL REINFORCEMENT AND DISCRIMINATION

Stimulus Control as a Three-Term Contingency

When an animal makes a response in one situation but not in another—that is, a **differential response**—we say that it discriminates between the situations. Alternatively, we may say that the situation exerts stimulus control over the animal's behavior. The simplest way to establish stimulus control and train a differential response or **discrimination** is to reinforce an operant in one situation and withhold reinforcement in the other. This procedure uses the basic three-term contingency involving an antecedent stimulus, the S^D, a response (R), and reinforcement (S^r). For birds in a laboratory, the contingency takes the following form:

$$S^D \text{ (red key): R (peck)} \rightarrow S^r \text{ (food)}$$

$$S^\Delta \text{ (green key): R (peck)} \rightarrow \text{extinction (no food)}$$

In the presence of the red key, pecks produce food. When the key is green, pecks are on extinction (no food). Using this basic contingency, we can establish stimulus control over the bird's behavior in the laboratory, training a red/green color discrimination, and extend the analysis to complex human behavior in everyday settings.

Figure 8.1 shows the development of a differential response (key peck) to a single key that is alternately illuminated red (predicts reinforcement) and green (predicts extinction) for 5 min. The graph shows the cumulative number of responses over a 90-min session. Pecks to the red key by a pigeon are intermittently reinforced with food. Responses emitted in the presence of the green key are extinguished and never reinforced. The procedure of alternating between periods of reinforcement and extinction is termed **differential reinforcement**.

As you can see in this idealized experiment, the pigeon begins by emitting about the same number of responses to the red and green stimuli. After about 20 min, the cumulative response curves start to separate, indicating that the bird is pecking in the presence of red

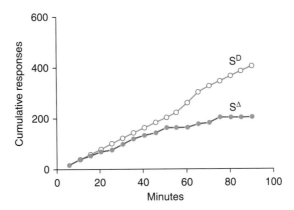

Fig. 8.1
Development of a differential response in the presence of red and green stimuli is shown. Cumulative number of responses over a 90-min session in which responses in the presence of red are reinforced and responses in the presence of green are on extinction.

more than green. At about 60 min, the pigeon seldom responds when the key is green, as shown by the leveling-off of the curve for this stimulus. Notice, however, that the cumulative curve for pecking the red key continues to rise. Because the bird pecks in the presence of a red stimulus, but does not respond when the key is green, we may say that the pigeon *discriminates* between these two colors. At this point, it is possible to label the red and green stimuli in terms of their *stimulus functions*. The red key is called a discriminative stimulus (S^D) and the green color is an S^Δ.

Suppose the bird is returned to its home cage after 90 min of such *differential reinforcement*. On the next day, the pigeon is again placed in the operant chamber and the key is illuminated with the red light. During this test session, reinforcement is not given for pecking in the presence of either red or green stimuli. Because of its previous training, the bird has a high probability of pecking the red key. Over a 60-s period, the bird may emit many responses when the S^D is present. After 60 s, the key light is changed from red to green. When the green light comes on, the probability of response is low, meaning the bird makes few pecks to the green key. By continuing to alternate between red and green, the researcher can show the stimulus control exerted by the respective colors.

Stimulus control refers to the change in behavior that occurs when either an S^D or an S^Δ is presented. When an S^D is presented, the probability of response increases; when an S^Δ is presented, the probability of response decreases. Both stimuli and their effects on behavior are important to the definition of stimulus control. For example, if you train a dog to present his paw when you say "shake" ("shake" is a verbal discriminative stimulus), that is only one-half of stimulus control. The dog must also *not* present his paw in the presence of other verbal stimuli such as "roll over" or "come here, Odee!" (conditions in which presenting a paw would not lead to reinforcement when "shake" is stated).

The stimuli that commonly control human behavior occur across all sensory dimensions. Stopping when you hear a police siren, coming to dinner when you smell food, expressing gratitude (e.g., saying "thank you") following a metaphorical or actual pat on the back, elaborating an answer because the student has a puzzled look on their face, adding salt to your soup because it tastes bland, and correctly solving a complex mathematical expression are all instances of stimulus control in human behavior.

Stimulus Control of Behavior Sequences

Suppose that you have trained a bird to reliably peck for food on an intermittent schedule in the presence of a red key. How could you use the red key to establish a simple sequence of behavior? Let us require the bird to step on a foot treadle (a lever near the floor operated by the foot) when the key is blue, and then peck the key for food when the color changes to red. To establish this two-component sequence or **response chain** you will use the red key as conditioned reinforcement (S^r) for treadle pressing as well as the S^D to peck for food. That is, because the red key has set the occasion for food reinforcement of pecking, it has acquired two *stimulus functions*, both as an S^D and as an S^r. This is because the red key not only signals the availability of food, it has also been associated with food via classical conditioning, the latter of which gives it a reinforcing function.

You begin to establish the response chain by observing the bird in the chamber pecking for food on the red key. Next, you shape the foot-treadle response to the blue light by first changing the key light to blue and look for a foot response toward the treadle. Any movement of the foot in the direction of the treadle is immediately followed by presentation of the red key as conditioned reinforcement. Also, when the key changes to red the bird pecks the key and this response is reinforced with food. You again present the blue key, look for another approximation to a foot-treadle response, and reinforce (shape) it by changing the key color to red. The bird pecks the red key and the behavior is reinforced with food. You keep following this procedure until the first definable foot-treadle response occurs, at which point the entire sequence is carried out automatically by the programmed apparatus and the bird. Once the performance is established, any person including one unfamiliar with basic behavior principles would see a bird in the chamber standing near the key and food hopper. A blue key is illuminated and the bird steps on a treadle, followed by the key color changing to red; the bird then pecks the red key and the sequence is repeated. The bird's behavior may look mysterious, but can be entirely explained by the contingencies of reinforcement.

To make the performance even more impressive, you decide to add more components to the pigeon's response chain; these include ringing a bell, climbing a perch, and pulling a string. You establish each link in the chain using the same procedures, with separate S^Ds for each component. Thus, to establish pulling a string in the presence of an orange key as the next link, you use the blue key color as conditioned reinforcement for pulling the string. You continue to add new links once each performance is well established. Now you have a bird that rings a bell when the key is green, climbs a perch when the key changes to yellow, pulls a string when it shifts to orange, steps on a foot treadle when it transitions to blue, and finally pecks the key when it turns to red. This is really impressive! These same steps can be used to train pigeons to "read" such English words as PECK or RING, by gradually fading in the letters over the colors and gradually fading out the colors (see section "Fading of Stimulus Control" later in this chapter).

Social Referencing and Behavior Chains

In early social development, infants often "look to mother" or another caretaker before interacting with unknown, ambiguous objects. Thus, infants show gaze-shift responses, looking from an object to their mothers. In this learning process, the mother's facial, vocal and gestural expressions function as discriminative stimuli (S^{D}s), which set occasion to reach for an available object that results in positive reinforcement, a sequence or chain called "social referencing." Learning of social referencing also involves discrimination of social cues, which signal that reaching for an ambiguous object may result in punishment. Figure 8.2 depicts the presumed contingencies for social referencing as a two-component response chain, resulting in reinforcement or punishment depending on the facial expression of the mother. The sequence begins with the presentation of the ambiguous stimulus (S$^D_1$, object), which occasions the gaze-shifting response (R$_1$) from the object toward the mother's face. In the presence of a joyful face (S$^D_2$) by the mother, the reaching response (R$_2$) is positively reinforced (S^R). On the other hand, in the presence of a fearful expression (S$^D_3$) by the mother, the reaching response results in punishment (S^P).

Martha Pelaez (Figure 8.3) is a behavior scientist who does work with child development and has conducted research with stimulus control in infants. One such experimental study

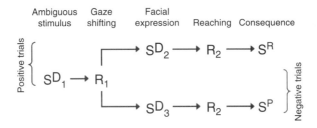

Fig. 8.2

A two-component behavior chain of social referencing is shown depicting the presumed social contingencies operating for maternal joyful faces (positive trials) and fearful faces (negative trials). S$^D_1$: ambiguous object; S$^D_2$: joyful facial expression; S$^D_3$: fearful facial expression; R$_1$: gaze shifting; R$_2$: reaching; S^P: punishing stimuli; S^R: reinforcing stimuli.

Source: Reprinted from M. Pelaez, J. Virues-Ortega, & J. Gewirtz (2012). Acquisition of social referencing via discrimination training. *Journal of Applied Behavior Analysis*, 45, pp. 23–36. Republished with permission of John Wiley & Sons, Ltd.

Fig. 8.3

Martha Pelaez, a behavior analyst who researches infant development.

Source: Published with permission from Martha Pelaez, 2021.

with mothers and their 4- and 5-month-old infants was conducted to test the two-component chain analysis of social referencing (Pelaez, Virues-Ortega, and Gewirtz, 2012), using an A-B-A-B experimental design (Chapter 2) with intervention phases staggered across infants (see "Multiple Baseline Designs" in Chapter 13). In this study, different consequences were arranged to follow reaching responses toward an ambiguous object depending on the facial expression of the infant's mother. During training, a joyful expression of the face signaled positive reinforcement and a fearful expression indicated punishment for a reaching response toward an ambiguous object (a variety of toys or puppets covered by a white cloth). Gaze shifting involved the infant turning her/his head about 90 degrees toward the mother and sighting her face; reaching for the object included the infant's movement toward the object using the upper body and arm extension, either touching it or coming within 5 cm of touching the object. On each of the randomly dispersed joyful (positive) and fearful (negative) trials in a set, an experimenter behind a puppet theater presented and shook a covered object (toy or puppet) accompanied by an unfamiliar sound. Then, the mother pointed to the covered object to prompt looking at it, and as soon as the infant looked back toward the mother's face (gaze shift), the mother displayed a previously rehearsed joyful or fearful face. Reinforcement consisted of removing the cover to reveal the hidden toy or puppet, as well as the onset of baby melodies and brightly colored lights. Punishment on negative trials resulted in the removal of the covered object (without revealing the toy or puppet) and the onset of a short unpleasant sound (buzzer, food blender, or whistle).

As shown in Figure 8.4, the experiment consisted of baseline (BL), differential reinforcement (DR_1), extinction (EXT), and return to differential reinforcement (DR_2) phases, with the number of sessions of each phase (except baseline) staggered over infants to enhance the validity of the experiment (see the solid shifting black line). Differential reinforcement involved positive reinforcement of reaching in the presence of a joyful maternal face (positive trial) and punishment of reaching in the presence of a fearful face (negative trial). During baseline and extinction phases these differential consequences were not in effect. Each panel of the figure presents the results from one of the 11 infants over sessions and is signified by "S" plus a number (S1 to S11) in the lower right portion of the panel (the graph is presented in two side-by-side sections, but is continuous over infants). The figure plots the percentage of trials for infant reaching on positive (black dots) and negative (open circles) trials, and the percentage of gaze shifting (open bars) and gaze-shifting prompts from mother (gray bars). We only highlight the major findings for reaching here; further details can be found in the original article (Pelaez, Virues-Ortega, & Gewirtz, 2012).

Notice for the baseline (BL) phase with no consequences for reaching, infants after shifting gaze show similar percentages of reaching on positive and negative facial expression trials. In the differential reinforcement (DR_1) phase, the percentage of reaching for the covered object is different for positive and negative facial expression trials, with higher percentages for positive facial expression trials for all infants. During the extinction phase (removal of differential reinforcement for reaching by maternal facial expression), all infants show a return to no differences in the percentage of reaching between positive and negative maternal facial expressions, indicating that the differential reinforcement is the causal factor in infant reaching responses. To ensure that this conclusion is correct, the researchers arranged for a return to the differential reinforcement (DR_2) phase. Notice once again that the percentage of positive (joyful faces) and negative (fearful faces) trials is different for all infants, with higher percentage of the positive trials showing more reaching by the infant than on the negative trials.

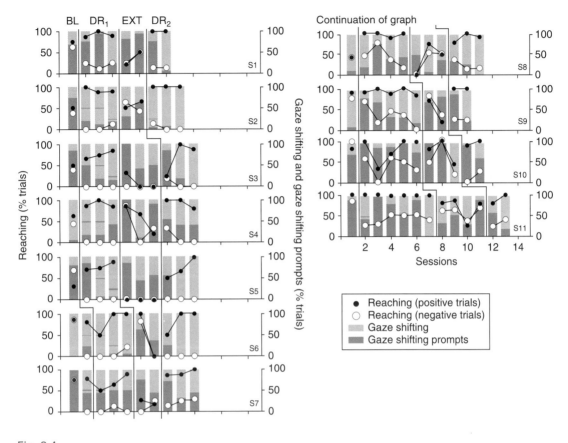

Fig. 8.4
Percentage of trials with gaze shifting and reaching by infants in the study as reported by Pelaez, Virues-Ortega, and Gewirtz (2012). The original figure has been separated into two sections for presentation in the textbook, but is continuous over the 11 infants in the original study. See text for further description.

Source: M. Pelaez, J. Virues-Ortega, & J. Gewirtz (2012). Acquisition of social referencing via discrimination training. *Journal of Applied Behavior Analysis*, 45, pp. 23–36. Republished with permission of John Wiley & Sons, Ltd.

Overall, the findings from the social referencing experiment are in accord with the behavior analysis of a two-component response chain. When an infant's gaze is shifted to a joyful maternal face, which signals positive reinforcement for reaching toward an ambiguous object or a fearful face signaling punishment, each infant learned to reach based on the observed facial expressions of the mother. Social referencing is ultimately a *generalized response chain* in which infants respond to different facial expressions of others, not only their mothers. To establish such a generalized sequence, it is likely that many repeated examples of facial expressions (multiple-exemplar training) by different human caretakers become reliable signals for a variety of responses by the infant or child resulting in differential reinforcement. Analysis of the contingencies for social referencing is necessary to teach infants and children with developmental delays this important social repertoire—using the expressions of others as signals for reinforcement of one's behavior. It should be noted, however, that while these social contingencies reliably produce social referencing in controlled settings, the everyday interactions of mothers and typically-developing infants probably activate additional processes based on our evolutionary history as a social species, involving preparedness to use facial expressions of others to guide our behavior (social referencing also has been reported for dog–human interactions, Merola, Prato-Previde, & Marshall-Pescini, 2012; and human-raised chimpanzees, Russell, Bard, & Adamson, 1997).

NEW DIRECTIONS: STIMULUS CONTROL, NEUROSCIENCE, AND WHAT BIRDS SEE

What do birds see and how do they see it? We actually do not know what humans "see," let alone birds. We do know a great deal about the structure and physiology of vision, but what is actually "seen" is only speculation. Vision and other sensations are private experiences, even though the operation of the physiological system is readily observed and analyzed.

The evolution of vision is an example of the natural selection of structures that enhance the organism's reproductive fitness. Many very primitive biological organisms have light-sensitive structures that contribute to that organism's survival even if it is only a matter of telling light from dark. Darker places are often safer than lighter places, and organisms that were sensitive to the difference produced more offspring. In terms of avian evolution both structure and function have interacted to produce the current forms of vision, making birds more viable (Goldsmith, 2006).

Vision occurs when light enters the eye and is transduced from one form of energy to another by processes in the retina—from light photons to neural impulses that travel from the eye via the optic nerve to the brain (see Donovan, 1978 on the structure and function of pigeon vision). Nothing is actually "seen" at any point. However, identifying the mechanisms in the retina and brain provides the basis for predicting how the organism might behave, as well as advancing our scientific knowledge of visual discrimination.

A direct demonstration of a bird's ability to see color, shape, and movement requires behavioral experiments wherein the bird has to discriminate aspects of these stimuli. Blough (1957) provided the first thorough behavioral assessment of color vision in pigeons, even though the anatomy of the bird eye had predicted such ability long before. Using a modified Blough procedure, Goldsmith and Butler (2005) trained parakeets to go to a yellow light for food. The researchers then presented a light composed of a mix of red and green as a comparison stimulus, and showed that with a certain mix (90% red and 10% green) the birds could not discriminate the mix from the yellow light.

These findings indicated to behavioral neuroscientists that separate retinal receptors (cones) are responsive to different hues and are used by birds to guide their behavior. Four types of cones were identified that contained different pigments and oil-droplet filters; one of these receptors is sensitive to ultraviolet (UV) wavelengths, allowing birds to discriminate colors that we cannot even imagine (see Carvalho, Knott, Berg, Bennett, & Hunt, 2010 for UV vision in long-living parrots, macaws, and cockatoos; see Lind, Mitkus, Olsson, & Kelber, 2014 for UV sensitive (UVS) and violet sensitive (VS) pigments and the transparent ocular media in 38 bird species; see Hogg et al., 2011 for UV vision in Arctic reindeer). Subsequently, behavioral experiments using operant contingencies showed that birds could make the visual discriminations predicted from the analysis of retinal receptors. The evidence suggests that UV wavelengths are seen as separate colors by birds due to the presence of UV receptors. In the everyday world of birds, these receptors allow for differential mate selection (Griggio, Hoi, & Pilastro, 2010; Hausmann, Arnold, Marshall, & Owens, 2003), improved foraging (Hastad, Ernstdotter, & Odeen, 2005), and detection of rapidly moving prey (Rubene, Hastad, Tauson, Wall, & Odeen, 2010). Overall, visual sensitivity to the UV spectrum allows for operant discriminations of the world that improve the survival and reproductive success of birds.

STIMULUS CONTROL AND MULTIPLE SCHEDULES

Behavior analysts often use multiple schedules of reinforcement to study stimulus control in the laboratory. On a **multiple schedule**, two or more simple schedules are presented one after the other and each schedule is accompanied by a distinctive antecedent stimulus—in some cases two or more reinforcement schedules are presented (neither of which is extinction), so two distinct discriminative stimuli will alternate, each signaling their respective schedules. The idealized experiment that we discussed near the beginning of this chapter is one example of a multiple schedule. Pecking was reinforced when a red light appeared on the key, and a schedule of extinction was in effect when the key color turned to green. The schedules and the associated stimuli alternated back and forth every 5 min. As indicated, these procedures result in a *differential response* to the colors—in other words, stimulus control.

In an actual experiment, presenting the component schedules for a fixed amount of time may confound the results. Without a test procedure, the researcher would not be sure that the bird discriminates on the basis of color of light or rather time. That is, time itself may have become a discriminative stimulus. Therefore, this can be controlled for using different durations of time in which each component is in effect or randomly implementing each component instead of alternating them.

A likely result of an experiment with MULT VI 2-min EXT 2-min is shown in Figure 8.5. The graph portrays the total number of responses during the RED and GREEN components for 1-h daily sessions. Notice that the bird begins by pecking equally in the presence of the RED and GREEN stimuli. Over sessions, the number of pecks to the GREEN extinction stimulus, or S^Δ, declines. By the last session, almost all of the responses occur in the presence of the RED S^D and almost none when the GREEN light is on. At this point, pecking the key can be controlled easily by presenting the RED or GREEN stimulus. When RED is presented, the bird

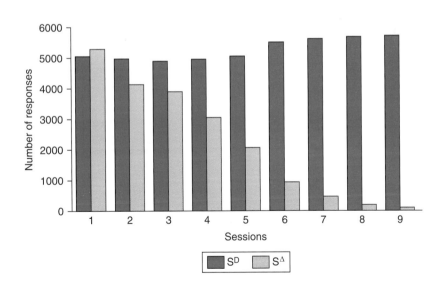

Fig. 8.5
Idealized experimental results are shown for a MULT VI 2-min EXT 2-min schedule of reinforcement. Relative to the RED VI component, pecking declines over sessions to almost zero responses per minute in the GREEN (S^Δ extinction) phase.

pecks the key at a high rate, and when the color changes to GREEN the pigeon immediately stops pecking.

Multiple Schedules: The Discrimination Index

One way to measure the stimulus control exerted by the S^D and S^Δ at any moment is to use a **discrimination index** (I_D). This index compares the rate of response in the S^D component with the sum of the rates in both S^D and S^Δ phases (Dinsmoor, 1951):

$$I_D = (S^D \text{ rate})/(S^D \text{ rate} + S^\Delta \text{ rate}).$$

The measure varies between 0.00 and 1.00. When the rates of response are the same in the S^D and S^Δ components, the value of I_D is almost 0.50, indicating no discrimination. When all of the responses occur during the S^D phase and no responses occur during the S^Δ component, the I_D is 1.00. Thus, a discrimination index of 1.00 indicates a perfect discrimination and maximum stimulus control of behavior. Intermediate values of the index signify more or less control by the discriminative stimulus.

A study by Pierrel, Sherman, Blue, and Hegge (1970) illustrates the use of the discrimination index. The experiment concerned the effects of sound intensity as a discriminative

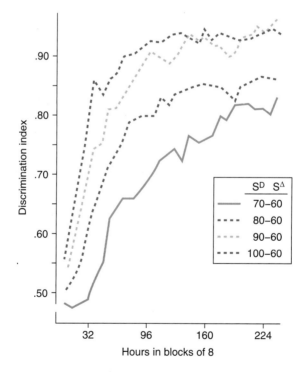

Fig. 8.6
Discrimination index (I_D) curves are shown for different values of S^D and S^Δ. Each curve is a plot of the average I_D values based on a group of four animals, repeatedly exposed to 8-h sessions of discrimination training.

Source: From R. Pierrel, G. J. Sherman, S. Blue, & F. W. Hegge (1970). Auditory discrimination: A three-variable analysis of intensity effects. *Journal of the Experimental Analysis of Behavior, 13*, pp. 17–35, Figure 1B on p. 22. Copyright 1970 held by the Society for the Experimental Analysis of Behavior, Inc. The labels for the x- and y-axes have been simplified to promote clarity.

stimulus on acquisition of a differential response. The researchers were interested in sound-intensity relationships (measured in decibels, dB) between the S^D and S^Δ. The basic idea is that the more noticeable the difference in sounds was, the better the discrimination. For example, some people have doorbells for the front and back entrances to their houses. If the chimes are very similar in sound intensity, a ring will be confusing and you may go to the wrong door. One way to correct this problem is to change the intensity of sound for one of the chimes; of course, another is to replace one chime with a buzzer.

In one of many experimental conditions, rats were trained to respond on a MULT VI 2-min, EXT schedule. The animals were separated into four equal groups, and for each group the auditory S^D for the VI component was varied while the S^Δ for the extinction phase was held constant. For each group, the S^Δ was a 60-dB tone but the S^D was different—70, 80, 90, or 100 dB. Thus, the difference in decibels or sound intensity between the S^D and S^Δ increased over groups (70–60, 80–60, 90–60, and 100 vs. 60 dB). Two 8-h sessions of the multiple schedules were presented each day, with a 4-h break between sessions.

Figure 8.6 shows the average acquisition curves for each experimental group. A mean discrimination index based on the four animals in each group was computed for each 8-h session. As you can see, all of the groups begin with I_D values of approximately 0.50 or no difference in responding between the S^D and S^Δ components. As discrimination training continues, a *differential response* develops and I_D values rise toward 1.00 or perfect discrimination. The accuracy of the discrimination, as indicated by the maximum value of I_D, is determined by the difference in sound intensity between S^D and S^Δ. In general, more rapid acquisition and a more accurate discrimination occur when the difference between S^D and S^Δ was increased.

FOCUS ON: DISCRIMINATION AND THE "BIRD-BRAINED" PIGEON

Imagine you are doing a class assignment, which involves training a pigeon to discriminate between red and green components of a multiple schedule. The assignment counts for 30% of the course grade, and you must show the final performance of the bird on the multiple schedule to your instructor. Students are given a pigeon, an operant chamber, and a microcomputer that allows them to control key color and the delivery of food from a hopper. Sessions are scheduled for 1 h each day over a 2-week period that ends with the professor's evaluation of your project. The pigeon has been food deprived, magazine trained, and taught to peck at a white-illuminated key on a VI 60-s schedule.

To create the VI schedule for the red key and extinction under the green key, you use operant-conditioning software to program your computer. The software program is set up to record the number of key pecks in each component of the multiple schedule. Your program starts a session with the key illuminated red, and the first response after an average of 60 s is reinforced with food (VI 60 s). After food is presented, the key color changes to green and extinction is in effect for an average of 60 s.

Day after day, your bird pecks at a similar rate in both the red and green components. You become more and more concerned because other students have trained their birds to peck when the key is red and stop when it is green. By the 11th session, you are in a panic because everyone else is finished, but your bird has not made much progress. You complain to your instructor that perhaps something is wrong with your bird, and it is not fair to give you a low mark because you tried your best. Your professor is a gentle behavior analyst who reminds you, "The fault is not with the pigeon; the experimental subject is always right. Let's check your contingencies in your program." You spend the night pondering the program and then, somewhat like Kohler's apes (Kohler, 1927), you "have an insight." Pecking in the extinction green key component has been reinforced with the presentation of the red key light! Voila!

You realize that the red color is always associated with food reinforcement and this suggests that the red stimulus has more than one function. It is obviously an S^D that sets the occasion for reinforced pecking. However, in addition, the stimulus itself is a conditioned reinforcer because of its association with food. Presumably, during the extinction component the bird sometimes pecked the green key, and on the basis of the computer program the color changed to red. This change in color accidentally or adventitiously reinforced pecking in the extinction component. From the bird's point of view, pecking the key during extinction turns on the red light that then allows food reinforcement. In fact, the pigeon's behavior is **superstitious** (Skinner, 1948b), because pecking in the green component does not affect the presentation of the red color.

There is an easy way to resolve this: ensure that if the bird is pecking at the end of the green extinction period, that the red reinforcement component does not begin when the pigeon is pecking. Better stated, if the pigeon is pecking during the green extinction component, and it is time for the red component to begin, a 2-s delay between the key peck and the presentation of the red-key contingency can be programmed. This contingency prevents the onset of the red stimulus (conditioned reinforcement) if responding is occurring at the moment when the extinction phase ends (or is supposed to do so). The added contingency is called **differential reinforcement of other behavior**, or **DRO**, meaning that any other behavior besides pecking will be reinforced with the red light as a conditioned reinforcer. During this 2-s delay (or DRO period), each response or peck resets the 2-s interval. If the bird does anything other than strike the key for 2 s, the red stimulus occurs. (One might say that the green stimulus sets the occasion for behavior other than key pecking, which is reinforced by the appearance of the red key.)

With this realization, you rush to the laboratory and add DRO to your computer program. At the first opportunity, you place your pigeon in the operant chamber and initiate the program. As you watch the bird's performance on the cumulative recorder, the rate of response during the S^D and S^Δ components begins to separate. After two more sessions, the discrimination index (I_D) is almost 0.90, indicating good discrimination between reinforcement and extinction components. The instructor is impressed with your analytical skills, and you get the highest mark possible for the assignment (A+). Nicely done!

This analysis has implications for teaching and learning. When most people learn from instruction but a few do not, educators, psychologists, and parents often blame the poor student, confused client, or stubborn child. They see the failure to learn as a deficiency of the person rather than a problem of contingencies of reinforcement (called "blaming the victim"; Shaver, 1985). Of course, some people and animals may have neurological and/or sensory impairment (e.g., color blindness, deafness, traumatic brain injury, sensory processing variations) that contributes to their poor performance. Nonetheless, defective contingencies of reinforcement may also contribute to, or exclusively produce, problems of discrimination and learning. In the case of the pigeon who had trouble learning, the fault was caused entirely by *adventitious reinforcement* of responding during extinction. A small change in the contingencies of reinforcement (adding DRO) made a "bird-brained" pigeon smart.

MULTIPLE SCHEDULES AND BEHAVIORAL CONTRAST

Consider an experiment by Guttman (1977) in which rats were exposed to a two-component multiple schedule with a VI 30-s reinforcement schedule in both components (MULT VI 30-s VI 30-s). A sound (white noise) signaled one component and a light signaled the other. The sound and light alternated every 3 min, and the rats made about the same number of responses in each of the components. Next, in the presence of the sound stimulus the contingencies were changed from VI to extinction (MULT VI 30-s EXT). As you might expect, the rate of response declined in the extinction component. Surprisingly, the rate of response *increased* on the VI component signaled by the light. This increase in rate occurred even though the reinforcement contingencies for the VI component remained the same. Thus, changing the contingencies of reinforcement on one schedule affected reinforced behavior on another schedule.

This effect is called **behavioral contrast** (Reynolds, 1961). Contrast refers to a negative correlation between the response rates in the two components of a multiple schedule—as one goes up, the other goes down. There are two forms of behavioral contrast, positive and negative. **Positive contrast** occurs when the rate of response *increases* in an unchanged component with a decrease in behavior in the altered or manipulated component. **Negative contrast** occurs when the rate of response *decreases* in the unchanged component with increases in response rate in the altered component.

There are several alternative interpretations of behavioral contrast. For example, when reinforcement is reduced in one component of a two-component multiple schedule, habituation to the reinforcer is less, resulting in more effective reinforcement in the unchanged component (McSweeney & Weatherly, 1998; see McSweeney & Murphy, 2014 for dynamic changes in reinforcement effectiveness). Other accounts of behavioral contrast include increased autoshaped key pecks in the unchanged component (see Chapter 7); fatigue or rest attributed to the amount of responding on the changed schedule, and compensating for response rate changes on the altered component (de Villiers, 1977; McSweeney, Ettinger, & Norman, 1981).

Behavioral Contrast: Relative Rates of Reinforcement

Although there is some dispute, one prominent account suggests that behavioral contrast results from changes in *relative rates of reinforcement*. On a two-component multiple schedule, relative rate of reinforcement for the unchanged component increases when the number of reinforcers decreases on the other schedule. Of course, relative rate of reinforcement for the unchanged component decreases when the number of reinforcers is increased on the changed schedule.

For example, on a MULT VI VI schedule, if an animal obtains 30 reinforcers each hour on the unchanged component and gets another 30 reinforcers on the other schedule, then 50% of the reinforcement occurs on both components. If the schedule is changed to MULT VI EXT, then 100% of the reinforcements occur on the unaltered component. As the *relative rate of reinforcement increases on the unchanged component, so does the rate of response*. Of course, response rate on the unchanged schedule would decrease if the relative rate of reinforcement reduces by an increase in reinforcement on the altered or manipulated component. Relative rates of reinforcement provide an account of performance on multiple schedules that is consistent with a behavioral analysis of choice and preference, using concurrent schedules of reinforcement (see Chapter 9). Thus, behavior principles of choice and preference, based on relative rates of reinforcement, may be extended to the findings on multiple schedules—extending the generality of these principles.

Experiments with food and alcohol reinforcement, however, indicate that the substitutability of the reinforcers on multiple schedules limits the impact of relative rates of reinforcement on behavioral contrast (Ettinger & McSweeney, 1981; McSweeney, Melville, & Higa, 1988). Changes in relative rates of reinforcement produced positive contrast (i.e., the rate of response increased on the unchanged schedule) when food reinforcement was continued in one component and extinction for alcohol was introduced in the other. Behavioral contrast, however, *did not occur* when alcohol reinforcement was continued and responding for food was placed on extinction. One possibility is that *alcohol is an economic substitute for food* (as rice is for potatoes), but food is not a substitute for alcohol (I'll drink to that!). That is, alcohol and food are partial substitutes. Anderson, Ferland, and Williams (1992) also reported a dramatic negative contrast wherein rats stopped responding for food and switched exclusively to responding for electrical stimulation of the brain (ESB). Relative rates of reinforcement may produce contrast only when reinforcers are economic substitutes, based on reinforcement history or biology.

Contrast effects also may be limited by the automatic reinforcement obtained by engaging in the behavior itself (Vaughan & Michael, 1982). A recent study by Belke and Pierce (2015) designed a two-component multiple schedule with wheel running as reinforcement for lever pressing in one component and wheel running as an operant for sucrose reinforcement in the other. For rats, wheel running is behavior that produces automatic reinforcement; when sucrose solution was replaced with water during extinction, operant running remained at high level even after 25 sessions of extinction; also, wheel-running and lever-pressing rates *decreased* for the other, unchanged component—results inconsistent with behavior contrast. Belke and Pierce (2015) argued that the automatic reinforcing effects of wheel running, which linked this behavior across components despite different functions (operant vs. reinforcement), provides an interpretation of the experimental results. Whether the findings are consistent with relative rates of reinforcement, with automatic reinforcement as an extraneous source of reinforcement, has not yet been determined.

After hundreds of studies of behavioral contrast, it is clear that contrast effects may occur in pigeons, rats, and even humans (Boyle, 2015; Simon, Ayllon, & Milan, 1982). In addition, contrast has been shown with various schedules of reinforcement (both ratio and interval), different kinds of responses (e.g., lever pressing, key pecking, and treadle pressing), and different types of reinforcement (e.g., food, water, and alcohol) in the component schedules. This suggests that behavioral contrast is an important process. An animal that forages successively in two patches would be expected to increase searching for food in one patch if the other patch became depleted (i.e., positive contrast). Similarly, negative contrast would occur when food in one of the patches became more abundant than that in the other. In this case, the animal would decrease foraging in the less plentiful location (Cheney, DeWulf, & Bonem, 1993). Overall, behavioral contrast in the laboratory may relate to adaptive foraging strategies of animals in the wild.

GENERALIZATION

An organism that responds in one setting, but not in another, is said to discriminate between the settings. An organism that behaves similarly in different situations is said to generalize across circumstances. **Generalization** is a common observation in everyday life. A child may call all adult males "daddy," label all small furry animals as dogs, and drink anything that looks like juice (one reason for putting child-proof caps on containers of dangerous liquids). Some students call all university teachers "profs" even though there is a variety of instructors with different credentials who do not hold the title of professor (e.g., lecturers, graduate teaching assistants). Most of us have seen an old friend at a distance only to find out that the person was not who we expected. A rude person is one who tells vulgar jokes no matter who is listening. In these and many more examples, it appears that common properties of the different stimuli set the occasion for operant behavior.

The problem is that an observer cannot be sure of the stimulus properties that regulate a common response. That is, it may not be easy for an individual to specify the geometry of dad's face, the physical characteristics that differentiate dogs from other animals, and the common aspects of different audiences for the joke teller. In the operant laboratory, however, it is usually possible to specify the exact physical dimensions of stimuli in terms of wavelength, amplitude, size, mass, and other physical properties. On the basis of experiments that use well-defined stimuli, it is possible to account for everyday examples of generalization and discrimination.

Aspects of Stimulus Generalization

Formally, **stimulus generalization** occurs when an operant that has been reinforced in the presence of a specific discriminative stimulus (S^D) is also emitted in the presence of other stimuli. The process is called stimulus generalization because the operant is emitted to new stimuli that presumably share common physical properties with the discriminative stimulus. Generalization and discrimination refer to differences in the *precision of stimulus control*. Discrimination refers to the precise control of an operant by a stimulus, and generalization involves less precise stimulus control of operant behavior.

Generalization Gradients

Guttman and Kalish (1956) conducted a classic study of stimulus generalization. Pigeons were trained to peck a key on a VI 1-min schedule of reinforcement. The key was illuminated with a green light of wavelength 550 nanometers (nm)—a wavelength of light that is approximately in the middle of the color spectrum.[2] Once the rate of key pecking for food had stabilized in the presence of the green light, the researchers tested for stimulus generalization. To do this, they exposed the pigeons to 10 additional values of wavelength (variations in color) as well as the original green light. All 11 colors were presented in a random order, and each wavelength was shown for 30 s. During these test trials, pecking the key was not reinforced (extinction); this is called a **probe trial**. Of course, as the test for generalization continued, key pecking decreased because of extinction, but the decline was equal over the range of stimuli, because different wavelengths were presented randomly. This experiment was repeated for three other groups of pigeons using three different wavelength values (colors) as discriminative stimuli—530, 580, and 600 nm.

As is shown in Figure 8.7, generalization gradients resulted from the experiment. A **generalization gradient** shows the relationship between probability of response and stimulus value. In the experiment by Guttman and Kalish (1956), probability of response is measured as the number of responses emitted by the pigeons, and stimulus value is the wavelength of

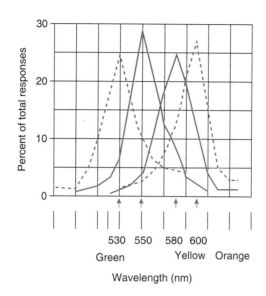

Fig. 8.7
Stimulus generalization gradients of wavelength expressed as percent of total responses by four groups of pigeons trained at different wavelengths: 530, 550, 580, and 600 nm. Notice the highest response rate (peak) of each curve is observed with the trained S^D (shown with arrows).

Source: Adapted from N. Guttman & H. I. Kalish (1956). Discriminability and stimulus generalization. *Journal of Experimental Psychology, 51*, pp. 79–88. The figure is a partial reproduction of a figure by H. S. Terrace (1966). Stimulus control. In W. K. Honig (Ed.), *Operant behavior: Areas of research and application* (p. 278). New York: Appleton, Century, Crofts.

light. To standardize the distributions, Figure 8.7 shows the percent of total responses over the range of 11 stimulus values. As you can see, a symmetrical curve with a peak at 550 nm (yellow-green training stimulus) describes stimulus generalization for pigeons trained at this wavelength (see black arrow at 550 nm). The more the new stimulus differed from the wavelength used in training, the less the percentage of total responses. Importantly, these results for the group of pigeons were typical of the curves for individual birds. In addition, similar generalization gradients were found for three other groups of pigeons using 530, 580, and 600 nm as the training stimuli (check this out). Generally, probability of response is highest for a stimulus that has signaled reinforcement (S^D), less for stimuli that are close but not identical to the S^D, and low for stimuli that substantially depart from the discriminative stimulus.

Generalization: Peak Shift

Multiple schedules may be used to study generalization gradients and often they include a component with extinction. Hanson (1959) reported an experiment with pigeons that was similar to the study by Guttman and Kalish (1956) just discussed. The procedural difference was that four groups of birds were exposed randomly to periods of signaled VI reinforcement and extinction (multiple schedule). For the experimental groups, the S^Δ component was 555, 560, 570, or 590 nm and the S^D component was always 550 nm. A control group received training only on the VI schedule, with 550 nm of light on the response key. Notice that the S^D for all groups was a key light of 550 nm, replicating one of the stimulus values used by Guttman and Kalish (1956).

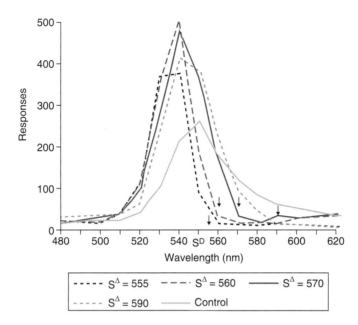

Fig. 8.8

Peak shift of a generalization gradient is portrayed. The control group shows a peak of the distribution at 550 nm that is symmetrical around this value (check this out). In contrast, the experimental groups uniformly showed a shift in the peak of the distribution that moves away from the stimulus value of the S^Δ. Arrows denote the values of the s-delta stimuli.

Source: Adapted from H. M. Hanson (1959). Effects of discrimination training on stimulus generalization. *Journal of Experimental Psychology, 58,* pp. 321–334.

Figure 8.8 shows results of the experiment by Hanson (1959). The control group that received only VI training at 550 nm produced a generalization gradient that replicates the curve of Guttman and Kalish. The peak of the distribution is at 550 nm, and is symmetrical around this value (check this out). But something interesting happened with all of the groups that received an extinction component in their training: In contrast, the experimental groups uniformly showed *a shift in the peak of the response distribution* that moved away from the stimulus value of the S^Δ. For this reason, **peak shift** refers to the change in the peak of a generalization gradient to the side of the generalization gradient that is away from the stimulus (S^Δ) that signals extinction. Peak shift effects have been observed in the "chick-a-dee" call notes of black-capped chickadees (Guillette et al., 2010), in responses to a variety of signals in the natural ecology of animals (Cate & Rowe, 2007), and in separate studies for both color and odor generalization of honeybees (Andrew et al., 2014; Martinez-Harms, Marquez, Menzel, & Vorobyev, 2014). In humans, peak shifts occur for spatial generalization, recognition of faces, and natural symmetry of faces (see Cheng & Spetch, 2002; Derenne, 2010; Spetch, Cheng, & Clifford, 2004), and peak shift gradients to complex sounds have been reported (Wisniewski, Church, & Mercado, 2009).

A final point you may also see in Figure 8.8 is that the number of responses made at the peak of each distribution is greater for the experimental groups than for the control condition. This latter finding reflects *positive behavioral contrast* that occurs on multiple schedules with S^D and S^Δ components (see the previous sections on behavioral contrast in this chapter).

ON THE APPLIED SIDE: PEAK SHIFT IN ANOREXIA NERVOSA

One aspect of body image disorder or anorexia nervosa is a private event related to distortion of body image (Gardner & Brown, 2010) and how accurately one views the size of their body. In tests for body image, an individual is given a photograph of themselves and alternative distorted photographs (Figure 8.9). Some images are manipulated to look thinner (10%, 20% thinner, etc.) and others are altered to look heavier (10%, 20% heavier, etc.) than the actual photograph. The person is asked to look at the photographs and identify the one that represents their most current body size and ideal body size. A general finding is that those with body image disorder typically choose the thinner photograph of themselves as the ideal image and a heavier image as their actual image.

A behavioral interpretation of the body image test would view the photographs as stimuli on a gradient of generalization.[3] That is, we can think of body size as a stimulus dimension that varies from extremely thin and emaciated to morbidly obese. You have probably heard about the body mass index or BMI—the ratio of weight (kg) to height squared (m^2)—as a way of characterizing body size. An extremely thin body has a BMI of less than 18.5 kg/m^2, a body that is neither thin nor fat would have a BMI of around 20–24 kg/m^2, and a severely obese body would have a BMI of around 40 kg/m^2. When viewed as a stimulus dimension, the photograph of one's own body size can be analyzed as the S^D (see 0 on Figure 8.9) and the distorted heavier photographs as the S^Δ, on a gradient of generalization. Selection of the thinner photograph as more attractive or preferred would then be interpreted as an instance of peak shift. The strongest response ("I prefer the thinner photograph") occurs to the stimulus shifted furthest from the S^Δ or heavier image on the other side of the S^D (photograph of self). From the perspective of behavior analysis, those with

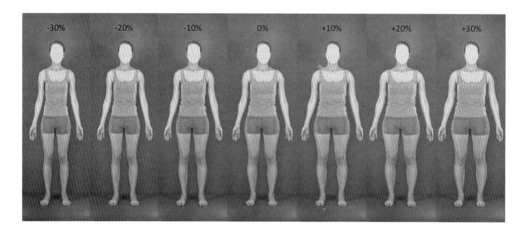

Fig. 8.9

The figure shows multiple images of a person's body as it actually is (0 = S^D). The other comparison images are manipulated to be thinner or heavier. A person with distorted body image may choose a thinner photograph of themselves as an ideal body image or perhaps select a heavier version of themselves as an actual body image, both of which may demonstrate peak shift on a gradient of generalization of body size. These distorted images may come from cultural conditioning of the "thin ideal" in which smaller body size is more strongly reinforced and larger body size is punished.

Source: Illustration of measuring body image disturbance is from the article Brooks, K. R., Mond, J. M., Stevenson, R. J., & Stephen, I. D. (2016). Body image distortion and exposure to extreme body types: contingent adaptation and cross adaptation for self and other. *Frontiers in Neuroscience*, *10*, 334.

body image disorder do not have a problem of distorted cognitions of the self, but are showing the behavioral effects of social conditioning and discrimination training from living in a culture in which thinness is viewed as an ideal body standard and larger body mass is devalued (i.e., punished) by society (Epling & Pierce, 1992). As we know, this culture can be harmful (see Hesse-Biber et al., 2006; Robles, 2009). More recently in the US, fitness, which encompasses a wider range of body types, has been more culturally supported as a beauty standard. It will be interesting to see how this affects body image stimulus control in the population.

Absolute and Relative Stimulus Control

Peak shift is an unusual effect from the point of view of absolute control by a stimulus. **Absolute stimulus control** means that the probability of response is highest in the presence of the stimulus value used in training. In fact, absolute stimulus control occurs when reinforcement is the only procedure used to establish stimulus control (no extinction training or differential reinforcement). The results of the Guttman and Kalish (1956) study and the control group of the Hanson (1959) experiment both show absolute stimulus control. In both studies, the peak of the generalization gradient is at the exact (or absolute) value of the discriminative stimulus presented during training (550 nm). When both S^D and S^Δ procedures are arranged (reinforcement and extinction), the peak of the distribution shifts away from the absolute value of the training stimulus—often analyzed as the interrelationship of excitatory and inhibitory stimulus gradients.

The shift in the peak of the generalization gradient may involve relative rather than absolute stimulus control. **Relative stimulus control** means that an organism responds to *differences* among the values of two or more stimuli. For example, a pigeon may be presented with two triangle stimuli projected on response keys; one triangle is larger and one is smaller. The pigeon is then reinforced to peck the larger of the two triangles. That is, reinforcement is contingent not on the properties of one stimulus, but that stimulus relative to the other. A correct response on one trial (a 1-in triangle when compared to a 0.5-in triangle) might be an incorrect choice on another trial (a 1-in triangle when compared to a 1.5-in triangle). Rather than responding to the absolute, or exact, size of the discriminative stimulus, the other is used as a comparison and only the one that meets the criterion of "larger" is reinforced. Similarly, the birds in the peak-shift experiments may have come under the control of the relative value of the wavelengths. Thus, the S^D was "greener" than the yellow-green S^Δ used in discrimination training. Because of this, the birds pecked most at stimuli that were relatively "greener," shifting the peak to 540 nm.

Research indicates that pigeons can readily acquire visual discriminations based on both absolute and relational features. These findings are not surprising, as pigeons have excellent color vision on the basis of their biology. But how would pigeons do when asked to respond along an auditory dimension to absolute and relational sequences of sounds? Murphy and Cook (2008) used sequences of sounds as experimental stimuli. Responses to *sequences of different sounds* using one set of pitches were reinforced, while responses to other different sound sequences made from another set of pitches, and to sequences of sounds of the same pitch repeatedly presented, were on extinction. In different experiments, the birds were asked to respond to absolute and relative features of the sound sequences, and the findings indicated that the absolute, fundamental pitch of the notes primarily controlled pigeons' behavior. Relational control occurred when the researchers increased the difficulty of the absolute discrimination, forcing the birds to rely on the relational aspects of the sounds. Pigeons naturally rely more on absolute pitch of sounds, but come under the control of relational features when this is required by the contingencies of reinforcement.

There are other ways of showing *relational control* by stimuli. To study generalization gradients and peak shift, researchers usually arrange the presentation of S^D and S^Δ so that one follows the other. This is called **successive discrimination**. An alternative procedure is labeled **simultaneous discrimination**—the S^D and S^Δ are presented at the same time, and the organism responds to one or the other. For example, a pigeon may be presented with two keys, both illuminated with white lights, but one light is brighter than the other. The bird may be reinforced for pecking the "dimmer" of the two keys. Pecks to the other key are on extinction. After training, the pigeon mostly pecks the darker of the two keys. To test that the bird's performance is caused by the difference between the two stimuli, it is necessary to present new values of luminosity and observe whether the pigeon pecks the dimmer of two keys.

Simultaneous discrimination tasks are often used in education. The television program *Sesame Street* teaches children the relations of "same" and "different" by presenting several objects or pictures at the same time. For "same", the song "One of These Things" sets the occasion for the child to identify one of several items. After the child has made a covert response, something like "It's the blue ball," the matching item is announced and shown. In this case, getting the correct answer is reinforcement for the discriminative response (see Wasserman & Young, 2010 for a discussion of same–different discriminations as the foundation for thought and reasoning).

ERRORLESS DISCRIMINATION AND FADING

Learning a discrimination usually results in mistakes. In the laboratory, when the S^D and S^Δ are presented alternately, as in successive discrimination, the organism initially makes many errors. Thus, the animal or person continues to respond in the presence of the S^Δ on the basis of generalization. As extinction and reinforcement continue, a differential response gradually occurs to the S^D and S^Δ. A pigeon is taught to peck a green key for food. Once this behavior is well established, the color on the key is changed to blue and pecking is not reinforced. The blue and green colors are alternately presented (without a DRO), and the corresponding schedules of extinction or reinforcement are in effect. During the early sessions, the onset of extinction often generates emotional behavior that interferes with ongoing operant behavior. This is also true with human learning. College students know all too well how frustrating it can be when trying to understand nuanced differences in solving algebraic problems in a math class or trying to determine the differences in negative reinforcement vs. punishment in a psychology course.

Extinction is an aversive procedure. Pigeons flap their wings in an aggressive manner and even work for an opportunity to attack another bird during the presentation of the S^Δ on a multiple schedule (Knutson, 1970). Birds will peck a key that turns off the stimulus associated with the **extinction stimulus**, implying that the stimulus is aversive. There are other problems with successive discrimination procedures. Because emotional behavior is generated, discriminative responding takes a long time to develop. In addition, spontaneous recovery of S^Δ responding from session to session interferes with the acquisition of discrimination. Finally, even after extensive training, birds and other organisms continue to make errors by responding in the presence of the signal for extinction.

Errorless Discrimination

These problems can be eliminated with a discrimination procedure described by Terrace (1963). The method is called **errorless discrimination** because the trainer or teacher does not allow the animal or person to make mistakes by responding to the extinction stimulus. As described in his 1963 paper, Terrace used early progressive training to reduce errors of discrimination. This training began when pigeons were conditioned to peck a red key at full intensity for food reinforcement. The birds were started on continuous reinforcement and moved gradually to a VI 1-min schedule. Early in this training, the key light was occasionally turned off for 5 s and extinction was in effect. Thus, a brief dark key was the S^Δ in this early phase. It is important to note that pigeons do not usually peck at a dark key, and Terrace made use of this fact.

As discrimination training continued, the dark key was gradually illuminated with a green light. The light became progressively greener and brighter and remained on for longer and longer intervals, until it stayed on for the same amount of time as the red key light. At this point, the duration of the S^D (red) was abruptly increased to 3 min while the S^Δ (green) was gradually increased to 3 min.

Now the birds were responding on a MULT VI 1-min EXT 3-min schedule. On this schedule, the red key was presented for 3 min and the pigeons pecked for food on a VI 1-min schedule during this period. After 3 min in the reinforcement component, the key color was changed from red to green, and extinction was in effect for 3 min. With these new contingencies in effect, the pigeons had sufficient time in the S^Δ component to make numerous errors, but they did not respond in the presence of the green (S^Δ) key light.

When this early progressive training was compared with standard successive discrimination procedures, there were far fewer mistakes with the errorless technique. Figure 8.10

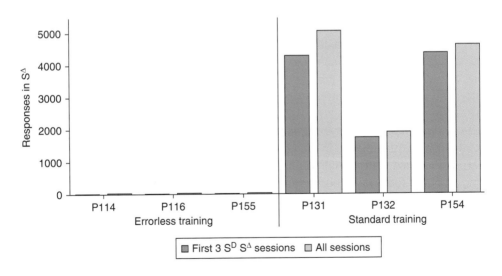

Fig. 8.10
Results are depicted of the errorless discrimination procedure used by Terrace (1963).
Source: Adapted from H. S. Terrace (1963). Discrimination learning with and without "errors." *Journal of the Experimental Analysis of Behavior, 6*, pp. 1–27, Figure 1.

shows that the three pigeons, which were trained with errorless discrimination procedures, made about 25 pecks each to the extinction stimulus (errors). Another three birds had the S^Δ introduced later in the experiment, at full intensity and for 3 min (standard method); these pigeons made between 2,000 and 5,000 pecks to the S^Δ. Compared with the errorless group, most of the pecks to the S^Δ in the standard condition occurred during the first three sessions. Overall, errorless discrimination procedures result in faster acquisition of a discrimination and substantially less incorrect responding (see Arantes & Machado, 2011 for an account of the superiority of errorless learning of a conditional temporal discrimination in pigeons).

In dolphins, errorless training methods often are used to establish discriminations on auditory and visual tasks (see Roth, 2002). Errorless methods have also been used for children and adults with developmental disabilities to teach a variety of visual, auditory, and temporal discrimination skills (Mueller, Palkovic, & Maynard, 2007), with visually impaired children to teach tactile discriminations of the Braille alphabet (Toussaint, 2011), and with typical adults to enhance learning and retention of new words (Warmington & Hitch, 2014). Additional research indicates that errorless learning methods can even be extended to complex human behavior involving flight simulation and the use of landing flares (Benbassat & Abramson, 2002). Learning to read in the English language is a difficult type of stimulus control because there are many rules and exceptions to the rules, which can be frustrating for children. There is a book that is based on errorless procedures called *Teach Your Child to Read in 100 Easy Lessons* (Engelmann, Haddox, & Brunner, 1986) that is effective at teaching the behavior of reading with few errors. (One of the authors of this textbook, Rasmussen, taught both of her sons to read at ages 4 and 3 with this book and it was not only effective, but a fun and delightful experience.)

Once discrimination has been established with errorless training, it may be difficult to reverse the roles of the S^D and S^Δ. Marsh and Johnson (1968) trained two groups of birds to discriminate between red (S^D) and green (S^Δ) stimuli. One group received errorless training and the other group received the standard discrimination procedure. After performance had stabilized, the S^D and S^Δ were reversed so that the green stimulus now signaled reinforcement and the red stimulus indicated extinction. The birds trained by the errorless method continued responding in terms of their initial training—they would not respond to the S^Δ (the new S^D from the point of view of the researcher) even when explicitly reinforced for such behavior. Birds given standard discrimination training were not as persistent, and quickly discriminated the change in contingencies.

These findings suggest that errorless procedures may be most useful in education when there is little chance of a change in the contingencies of reinforcement. For example, students may learn and retain better their multiplication tables, standard word spellings, rules for extracting a square root, and other types of rote learning with the errorless method. Students also enjoy learning more, learn very rapidly, and make few errors with errorless teaching procedures (Powers, Cheney, & Agostino, 1970). In problem-solving situations where there are many alternative solutions requiring error elimination, or where the contingencies of reinforcement change, the standard method of trial-and-error learning may produce more flexibility in responding and allow better remembering and recall, as when university students are preparing for exams (Anderson & Craik, 2006).

Fading of Stimulus Control

Errorless discrimination involves the early introduction of the S^Δ and gradual transfer of stimulus control, also called **fading**. Fading is done by gradually changing a physical aspect of an antecedent stimulus to some designated criterion. When Terrace (1963) gradually changed the dark key to the green color, this was fading in the S^Δ. Cheney and Tam (1972) used fading to transfer initial control by a color discrimination to subsequent control by line-angle tilt in pigeons; the procedure involved gradually increasing the intensity of the line segments projected on the key while decreasing the intensity of the colors. Control transferred from color to mirror-image line angles with some, but very few, errors.

Sherman (1965) gave a practical example of fading when he used the procedure to get a person diagnosed with a form of mutism to say his first words. He described the patient as:

> [A] 63-year-old man, diagnosed, in 1916, as dementia praecox, hebephrenic type. He had been in the hospital continuously for 47 years, with a history of mutism for 45 of those years. At the time of this study he was not receiving any medication or participating in psychotherapy. Periodically, when seen on the ward, ... [he] could be observed walking around mumbling softly to himself. However, all of this mumbling appeared to be nonsensical vocal behavior. In his 45-year history of mutism [he] had not exhibited any recorded instance of appropriate verbal behavior.
>
> (Sherman, 1965, p. 157)

After many sessions of reinforcement and imitation training, Sherman succeeded in getting the patient to say "food"—his first distinct utterance in 45 years. At this point, Sherman used fading to bring this response under appropriate stimulus control—responding "food" to the question "What is this?" The training was as follows:

> To obtain the word "food" from the subject when the experimenter asked, "What is this?" a fading procedure was used. With the fading procedure, the experimenter continued to hold up a bite of food each time and to deliver instructions to the subject. The behavior of the subject—that is, saying "food"—was maintained with reinforcement while the instructions to the subject were gradually changed in the following steps: (a) "Say food"; (b) "Say foo_"; (c) "Say f___"; (d) "What is this? Say f___"; (e) "What is this? Say ___"; (f) "What is this?"
>
> (Sherman, 1965, p. 158)

This example shows that the patient initially replied "food" after the experimenter said, "Say, food." The original verbal stimulus for the response "food" was gradually faded out and replaced with a new stimulus of "What is this?"

Fading procedures have been regularly used for children with autism; for example, if a child shows resistance to drinking liquids such as milk. Both for health and nutritional benefits it is sometimes necessary for children to drink (or eat) things they usually reject. In one study, the researchers treated milk avoidance by a 4-year-old girl with autism by fading out a beverage she consumed 100% of the time and fading in the milk (Luiselli, Ricciardi, & Gilligan, 2005). Using this procedure, the amount of beverage was reduced and the amount of milk was increased over training sessions. Fading allowed rapid acquisition of milk consumption without interruptions to the fading sequence by the child's usual refusal.

294 Stimulus Control

In another example, three youths with autism were taught by fading procedures to engage in conversations during shopping trips to the community store—an essential skill for living independently. The youths initially had near-zero verbal interactions with store staff, but were taught to use scripts in conversations during simulated shopping trips. The scripts were systematically faded from last word to first word, as the rates of unscripted conversations increased. Subsequently, unscripted conversations showed generalization—occurring in the presence of new store staff on actual trips to local retail stores (Brown, Krantz, McClannahan, & Poulson, 2008; see also Dotto-Fojut, Reeve, Townsend, & Progar, 2011).

In everyday life, fading is an important aspect of complex human behavior, which often goes unrecognized because of its gradual nature. Children learn to identify many objects in the world by the step-by-step transfer of stimulus control. A parent may present a glass of apple juice to a 2-year-old child and state, "Juice. Now you say juice." Eventually, after many repetitions, if the child says "juice" the glass of juice is given. Once the response "juice" has been established, stimulus control may be gradually transferred from "Now you say juice" to questions such as "What is this?" by fading. In another example, a parent may initially remain briefly at a day-care center to make the child comfortable in the new setting. Once the child starts to participate in activities, the parent discreetly leaves (fades) and stimulus control for a variety of behavior is transferred to the new situation and the teacher.

COMPLEX STIMULUS CONTROL

Up to this point, we have discussed the absolute and relative control of behavior by relatively simple configurations of stimuli, as when a red color signals reinforcement and green signals no reinforcement or when the larger of two stimuli predicts reinforcement. There are other procedures that allow for the investigation of performance regulated by more complex stimulus arrays. Complex stimulus control is the foundation of cognition.

Matching to Sample: Identity Training

One procedure often used to investigate identity ("This stimulus is a …") discriminations is called **matching to sample**. In a simple *identity matching-to-sample* (IMTS) procedure, a pigeon may be presented with three keys, as illustrated in Figure 8.11. Panel A shows a triangle projected onto the center key. The triangle is the *sample* stimulus in the sense that it is an instance of a larger set of geometric forms. To ensure that the bird attends to the sample, it is required to peck the sample key. When this *observing response* occurs, two side keys are illuminated with a triangle on one and a square on the other, which are called the *comparison* stimuli (the sample key goes off). If the bird then pecks the comparison stimulus that corresponds (is the same) to the sample (i.e., is a match), this behavior is reinforced and leads to the presentation of a new sample. Panel B shows a nonreinforced sequence (error) in which pecks to the noncorresponding or nonmatching stimulus result in extinction. Over a number of trials, the comparison stimuli appear on the left or right keys with equal probability. After some training, pigeons accurately match to sample even with new (never reinforced) samples and comparison stimuli (Blough, 1959). These generalization or transfer tests show that training resulted in a higher-order response class (generalized identity) controlled by the property of similarity or sameness of the stimulus items, a generalized stimulus class. Thus, matching to sample across numerous exemplars

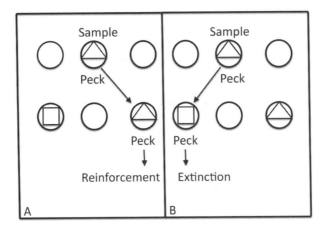

Fig. 8.11
Procedures used to train identity discrimination by a pigeon. Panel A shows that a peck to the sample key (triangle) results in two shapes on the side keys. A peck to the side key that matches the sample is reinforced. Panel B shows a sequence that is not reinforced.

(stimulus sets) leads to the formation of concepts (sameness) by pigeons. Notice how we infer that pigeons (or people) have "formed a concept" from their performance on these transfer tests. In behavior analysis, it would be incorrect to use concept as an inferred cognitive event that explains the test performance. Identity matching and multiple exemplar training (the operating contingencies) fully explain how the pigeon (or person) demonstrates the stimulus control of sameness (or any other concept) and the inferred cognitive event adds nothing to the account.

Other animals (monkeys, chimpanzees, and dolphins) have passed transfer tests on IMTS test, indicating that the same–different relationship of stimulus items controlled their behavior. In one study, researchers assessed California sea lions' (*Zalophus californianus*) ability for generalized identity matching (Kastak & Schusterman, 1994). After training two animals, Rocky and Rio, on 15 two-stimulus (visual configurations) matching-to-sample problems, the sea lions were tested for transfer of performance to 15 novel problems (Figure 8.12). The criteria for passing the transfer test included performance on the initial trial (pass/fail test item), performance on test trials compared with baseline, and performance on four-trial problem blocks. Both Rocky and Rio passed the transfer tests. Also, when given identity pairings (ring–ring; bat–bat) of stimuli previously learned as an arbitrary relation (ring–bat), both animals passed the identity generalization tests, with Rio passing on the first test and Rocky failing but passing on a second test. Rocky just needed a few more examples than Rio. The researchers concluded that the study conclusively demonstrates that California sea lions can learn and use a generalized identity-matching rule.

Generalized IMTS shows that animals can pass tests for learning of basic concepts (same/different). Reasoning by analogy is a more involved performance, which is said to show the higher cognitive capabilities of humans. Consider the analogy: fish is to water as bird is to air. To pass this item on a test (Fish is to water as bird is to ___) the student must respond to the *relational sameness* (not physical sameness) between the elements of the source domain (fish and water) and extend the sameness relation to the target domain (bird and air). To investigate

Fig. 8.12

Photograph is shown of Rio, a California sea lion, matching to sample (center key). Both Rio and another sea lion named Rocky eventually passed tests for generalized identity matching.

Source: Copyright held by Dr. Colleen Reichmuth of the Institute of Marine Sciences, University of California Santa Cruz. Published with permission.

whether nonhuman animals can learn such logical relations, researchers have developed a relational matching-to-sample (RMTS) procedure. In RMTS, the choice of AA would be correct for the sample stimulus CC, and the choice of FG would be correct for the CD sample. Notice that there is no physical similarity between the sample and the correct comparison stimulus. A correct choice of FG to the CD sample requires learning the relation of alphabetical order, not physical similarity between the stimulus items.

Wasserman and his colleagues in Russia reported that hooded crows, after IMTS training to an 80% or greater criterion, and following generalized IMTS learning, passed tests for RMTS without any further explicit training—emergent relational sameness (Smirnova, Zorina, Obozova, & Wasserman, 2015). Emergent relations have been reported for more complex types of stimulus control that use relational control, such as studies of stimulus equivalence (sameness) and relational framing (a large range of relations); these are discussed in Chapter 12. However, studies of RMTS in nonhuman animals usually have trained explicitly for relational sameness (e.g., Fagot & Maugard, 2013). In this regard, Wasserman and associates noted that the extensive IMTS experience likely contributed to the broadly applicable concept of relational sameness, but how this transfer occurred is an intriguing problem for future experiments.

BEHAVIOR ANALYSIS OF REMEMBERING AND FORGETTING

Delayed Matching to Sample and Remembering

An extension of the standard matching-to-sample task is called **delayed matching to sample (DMTS)**. This procedure was first described by Blough (1959), and involves adding a delay between the offset of the sample stimulus and the onset of the two comparison stimuli. For example, in Figure 8.13 a pigeon is presented with a center key (sample) that is illuminated with a red light. When the observing response occurs, the red sample turns off and a few seconds (e.g., 10 s) later, red and green comparison stimuli are presented on the side keys. A response to the stimulus that matches the sample is reinforced, and responses to the other stimulus are not. The basic finding is that the percentage of correct responses decreases as the delay increases (Blough, 1959; Grant, 1975).

Delayed matching to sample is a model for studying cognition and memory. For example, the time between the offset of the sample stimulus and the onset of the comparison stimuli is usually called the **retention interval**. The idea is that during this interval the organism is covertly doing something that helps to retain the information about the sample. Thus, Grant (1981) found that pigeons would "forget" the sample if they were given a sign (a vertical line on the key) that indicated that the comparison stimuli would not appear on that trial. In terms of remembering the sample, Grant reported that the pigeons performed poorly if the forget cue was presented soon after the sample went off. Performance was not as disrupted if the signal was given later in the interval. One interpretation is that the cue to forget interferes with covert rehearsal of the sample stimulus (Grant, 1981).

The cognitive metaphor of memory processes (encoding, storage, retrieval, and rehearsal) is popular in psychology. Tulving (1983) explained that **remembering** an event involves mental encoding of the event and subsequent retrieval of the information from memory due to reactivation of the encoding operations. He proposed that encoding results in a memory trace or representation of the past event. The memory trace becomes manifest when combined with retrieval processes. Thus, cognitive research into memory has emphasized how encoding produces mental representations that in turn aid in retrieval. These inferred cognitive processes and representations are not observable and are therefore speculative from a behavior analytical point of view.

Geoffrey White (2002) offers a behavioral account of memory by indicating that a behavior analysis of memory points to actions or choices (e.g., choosing between the comparison stimuli) based on the current contingencies and how those choices are in part regulated by the reinforcement history for similar choices in the past. He explains:

> Remembering is not so much a matter of looking back into the past or forward into the future as it is of *making choices at the time of remembering*. The [behavioral] approach treats remembering as a process of discriminating the relevant events from alternative possibilities. By analogy with the discrimination of objects at a physical distance, objects or events can be discriminated at a temporal distance.... That is, the discrimination is not made at the time of encoding, or learning, but at the time of remembering.
>
> (White, 2002, pp. 141–142, emphasis added)

One aspect of this behavioral approach to memory is that it challenges the well-known finding that remembering gets worse as the retention interval increases. If the discrimination

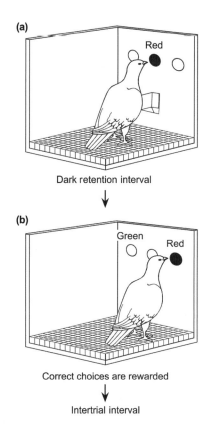

Fig. 8.13
Delayed matching to sample in a pigeon is shown. The sequence begins by the pigeon pecking a red or green sample on the center response key. A darkened chamber follows the pigeon's response to the sample during a retention interval. Next, the pigeon chooses between red and green side keys; choices that match to the sample are reinforced with food, and after a time interval another trial begins.

Source: Drawing is adapted from K. G. White (2002). Psychophysics of remembering: The discrimination hypothesis. *Current Directions in Psychological Science, 11*, pp. 141–145.

of a past event is made at the time of remembering, White suggests that it is possible to train organisms to be accurate at a specific delay. In this case, the remembering of a stimulus would be more accurate at a specific delay than with less delay or no delay. That is, remembering would not decline in accord with the retention interval.

Using the delayed matching-to-sample procedure shown in Figure 8.13, Sargisson and White (2001) compared the performance of pigeons trained with a 0-s delay and those trained with one specific delay at the outset. Typically, birds are trained to match to sample with a 0-s delay, and subsequently the delay or retention interval is gradually lengthened. In the new procedure, pigeons were trained in matching to sample at one specific delay (e.g., 4 s) and then asked to remember the sample at different retention intervals.

Figure 8.14 shows the discriminative performance for birds trained with a 0-s delay (circles). Notice that the accuracy of the discrimination decreases with the retention interval, as would be predicted by cognitive theories of memory. For pigeons trained with a 4-s delay, however, their accuracy does not systematically decrease over the retention interval (triangles). Instead,

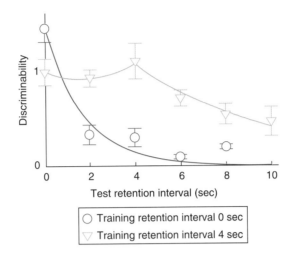

Fig. 8.14
Selected data from Sargisson and White's (2001) experiment show the accuracy of matching to sample in different groups of pigeons trained with either a 0-s retention interval or a 4-s retention interval; pigeons were tested with retention intervals that varied from 0-s to 10-s. The discriminability measure is the log of the ratio of correct to incorrect responses, and is not influenced by response bias.

Source: From K. G. White (2002). Psychophysics of remembering: The discrimination hypothesis. *Current Directions in Psychological Science*, 11, pp. 141–145. Reprinted with the permission of the American Psychological Society. Copyright 2001 held by Blackwell Publishing, Ltd.

these birds were most accurate at the training delay of 4 s, a finding that argues against the mental representation of the sample with a declining memory trace. Notice also that the birds were not trained to perform with less accuracy at brief delays. That is, the discrimination of the "to be remembered color" should have been easier at short delays (e.g., 0 s) because the sample color was observed very recently. The data show, however, that the pigeons were less accurate at delays of less than 4 s, again disconfirming a cognitive representational account. Overall, the results of the experiment support a behavioral view that remembering involves discriminative-operant behavior specific to the time interval of retrieval (see also White & Sargisson, 2011 and White, 2013 for a review).

Reverse Forgetting: Control of Retroactive Interference

The research on improved remembering at long intervals suggests that it is possible to reverse the usual forgetting which occurs with the passage of time. In a DMTS task, forgetting involves the loss of accuracy in identifying the sample stimulus over the retention interval. In the cognitive view of memory, the memory trace of the sample color weakens with time, implying that accuracy in identifying the original color should be lower with longer delays. One way to challenge the memory trace account is to use a procedure that separates accuracy in remembering the sample from the passage of time per se. White and Brown (2011) reviewed research showing that memory traces are strengthened by memory rehearsal procedures and weakened by procedures that interfere with rehearsal—typically called *retroactive interference*. The aim of their study was to reverse the forgetting of pigeons during the retention interval on a DMTS task by inserting and later removing retroactive interference.

300 Stimulus Control

Pigeons usually perform DMTS in a dark chamber. Thus, interference with remembering the sample color can be arranged by activating the houselight at the beginning of the retention interval. The houselight can be turned on throughout short retention intervals or illuminated for only a few seconds and then turned off for long retention intervals. According to memory trace and retroactive interference theories, the insertion of the houselight would be expected to increase forgetting early in the retention interval, but accuracy in remembering the sample

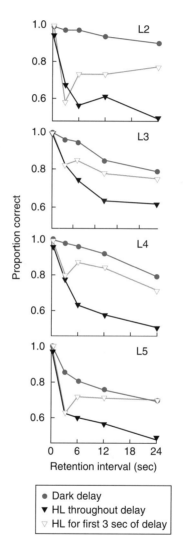

Fig. 8.15

Results are shown from study of reverse forgetting. Birds worked on a delayed matching-to-sample task involving red and green samples and retention intervals that varied from 0.2 to 24 s. The same birds made choices after the different delays under three chamber-illumination conditions: the chamber was dark throughout the retention interval (Dark delay), the chamber houselight was illuminated for the entire retention interval (HL throughout delay), or the chamber was illuminated by the houselight for the first 3 s of the retention interval and then turned off for the rest of the delay (HL for first 3 s of delay).

Source: The figure is a graph of results (Exp 2) taken from a study by K. G. White & G. S. Brown (2011). Reversing the course of forgetting. *Journal of the Experimental Analysis of Behavior, 96*, pp. 177–189. Copyright held by the Society for the Experimental Analysis of Behavior. Published with permission.

should not improve with the removal of the interference event. After removal of the interference, the memory trace would be expected to dissipate and accuracy would continue to decline.

White and Brown designed two DMTS experiments, but we shall only describe Experiment 2, which involved long retention intervals. Pigeons were trained in a dark chamber to peck a red or green sample (observing response) to initiate a retention interval that varied from 0.2 to 24 s. The retention interval ended with the red and green comparison stimuli presented on side keys. A correct choice of the comparison color resulted in food reinforcement followed by a dark period before the next trial. Incorrect choices were followed only by the dark period. The same pigeons were tested at all retention intervals with the chamber dark, with the houselight on for the first 3 s, or with the houselight on throughout the delay.

Figure 8.15 shows the results for the three experimental conditions. Notice that without any interference by the houselight (dark delay), accuracy (the proportion of correct choices) in identifying the original sample (red or green) declines over the retention interval, consistent with memory trace theory. Also, when the houselight remained on throughout the retention interval (HL throughout delay), the presumed retroactive interference (preventing rehearsal of the memorized sample) substantially reduced accuracy, as expected by a cognitive account. The critical condition involves turning on the houselight for the first 3 s of the retention interval but then turning it off, reinstating the dark chamber and removing the interference. Now accuracy drops for the 3-s delay, but recovers at the 6-s delay and then tapers off as the delay increases. This finding of a *reversal in forgetting* at a longer delay (greater accuracy at 6 s than at 3 s) is consistent with the behavioral theory of remembering and contrary to theory involving memory traces and retroactive interference.

Although further research is required, White and Brown argue that the reversal of forgetting is in accord with a discriminative-operant analysis of remembering at long intervals (Sargisson & White, 2001, as discussed previously). In the behavioral view, discriminations at short and long intervals are independent of one another and depend on the contingencies. In the case of withdrawing retroactive interference, turning on the houselight allows for extraneous sources of reinforcement of off-task behavior that compete with food reinforcement for correct choices of the comparison color. When the houselight is turned off, reinforcement of off-task behavior is withdrawn and on-task correct choices improve (greater accuracy at 6 s than at 3 s). A behavioral account in terms of extraneous sources of reinforcement fits the results without an appeal to inferred cognitive processes, and is consistent with the principles of choice discussed in Chapter 9 of this textbook.

FOCUS ON: CONCEPT FORMATION IN PIGEONS

As we have noted, principles of stimulus control are involved in many instances of so-called concept formation and abstract reasoning. People usually assume that conceptual thinking is a defining feature of humans that separates them from other animals. Although this kind of behavior is common in humans (Green, 1955), it occurs in other organisms. Herrnstein and Loveland (1964) designed an experiment to teach pigeons to distinguish humans from other objects (i.e., to learn the concept "human"). The point was to make explicit the environmental requirements for an animal, including humans, to exhibit the behavior we call "conceptual," which usually refers to some internal cognitive construct.

Consider what it means to know that this is a human being and other objects are not. Humans come in a variety of sizes, shapes, colors, postures, and many more features. Characteristics of the stimulus "human" are abstract and involve multiple-stimulus dimensions, rather than a single property such as wavelength of light. For example, human faces differ in terms of presence or absence of hair, geometric form, and several other characteristics. Defining attributes of faces include bilateral symmetry, two eyes, a nose, a mouth, and many additional features common to all people.

Although a precise physical description of humans is elusive, Herrnstein and Loveland (1964) asked whether pigeons could respond to the presence or absence of human beings in photographs. If a bird can do this, behavior is said to be controlled by the abstract property of "humanness." There is no concrete set of attributes that visually equals a human being, but there are relations among such attributes that define the stimulus class. The bird's task was to respond correctly to instances of the stimulus class, and by doing so to demonstrate concept formation. The basic procedure involved presenting many differing exemplars of a human stimulus for one group of stimuli and a second group of exemplars that was identical to the first group, except without the human. In the experiment, over 1,200 exemplar pictures were projected onto keys by a slide projector. The slides contained different photographs of natural settings, including countryside, cities, bodies of water, grass, etc. Approximately half the photographs contained at least one human being; the remainder contained no human beings. The slides did not differ in any other manner. Some slides contained human beings partly obscured by objects such as trees or automobiles. The placement of humans on the slide were varied: the center or to one side or the other, near the top or the bottom, close up or distant. Some slides contained a single person; others contained groups of various sizes. In some slides, people were clothed, partially clothed, or nude. The humans also varied in age, gender, ethnicity, and whether they were sitting, standing, or lying. There were variations in background lighting and color. The images that contained humans meant an opportunity for food (S^D) and pictures without people signaled extinction (S^Δ slides).

The results showed that the pigeons could learn to discriminate between slides with people and slides without them. Within 10 sessions of this training, every bird was responding at a higher rate to slides with humans in them. Over several months, the performance of the birds steadily improved. After extensive training, the birds were given about 80 novel slides that they had never seen before. Pigeons pecked at a high rate to new slides with people and at lower rates to slides without them. Generally, this experiment shows that pigeons can differentially respond to the abstract stimulus class of human being.

Additional experiments on teaching concept formation have been conducted with other stimulus classes and different organisms. Pigeons have discriminated trees (Herrnstein, 1979), geometric forms (Towe, 1954), fish (Herrnstein & de Villiers, 1980), one person from another (Herrnstein, Loveland, & Cable, 1976), and aerial photographs of human-made objects (Lubow, 1974). Pigeons can also discriminate abstract from impressionist art and generalize to novel examples of each category (Watanabe et al., 1995). Concept formation has also been reported for chimpanzees (Kelleher, 1958a), monkeys (Schrier & Brady, 1987), and an African gray parrot (Pepperberg, 1981). And pigeons trained to differentially respond to real objects show these responses to corresponding pictures of the objects—even when the pictures only contain novel views of the stimuli (Spetch & Friedman, 2006). Ravens have been shown to learn the concept "thief" after experience of caching food and having it removed by some human caretakers (Heinrich & Bugnyar, 2007).

Overall, this research shows that animals can learn to differentially respond to abstract properties of stimulus classes. These stimulus classes are commonly called categories when humans make similar discriminations. When people verbally describe different categories, they are said to "understand the concept." People can easily identify a computer disk and an automobile as human-made objects. When other animals show similar performances, we are reluctant to attribute the discriminative behavior to the creature's understanding of the concept, but rightfully attribute it to the learning of specific behavior.

Rather than attribute understanding to complex performances by humans or other animals, it is possible to provide an account based on evolution and the current demands of the environment. Natural selection shapes sensory, neural capacities of organisms that allow for discrimination along abstract dimensions. Birds obtain food, navigate, care for their young, and find mates largely on the basis of visual stimuli (see the section "New Directions: Stimulus Control, Neuroscience, and What Birds See" earlier in this chapter). Many of these activities require subtle adjustments to a complex and changing visual world. It is not surprising, therefore, that these creatures are readily able to learn to discriminate abstract properties of visual objects, especially when reinforcement contingencies favor such discrimination.

CONDITIONAL DISCRIMINATION

In everyday life, stimuli that regulate behavior (S^D and S^Δ) often depend on the context. Consider a matching-to-sample experiment in which a bird has been trained to match to triangles or to squares based on the sample stimulus. To turn this experiment into a conditional-discrimination task, we now add a red or green light, which illuminates the sample stimulus. The bird is required to match to the sample *only when* the background light is green, and to choose the noncorresponding stimulus only when the light is red. That is, when a green triangle is the sample, the bird must peck the comparison triangle, but when a red triangle is presented, pecks to the nonmatching circle are reinforced. Of course, if a green circle is the sample, pecks to the circle are reinforced, and when the sample turns to a red circle, pecking the triangle is the correct response. In other words, the correct (reinforced) choice is conditional upon the sample stimulus. *Conditional matching to sample* involves simultaneous discrimination of three elements in a display. The animal must respond to geometric form depending on the background color of the sample. It must also respond to the correspondence or noncorrespondence of the comparison stimuli.

Conditional discrimination is a common aspect of human behavior and is likely more common than simple three-term contingencies. A person who is hurrying to an appointment on the 15th floor of an office building will ordinarily enter the first available elevator. This same person may wait for the next lift if the elevator is full. Thus, getting on the elevator (operant) when the doors open (S^D) is conditional on the number of people in the car. In another example, you will say "8" when shown 3 + 5 and "15" if the relation is 3 × 5. Your response to the 3 and 5 is conditional on the + and × symbols. When people say and show that the spoken word "cat," the written word "cat," and a picture of a cat are equivalent, their behavior is a result of such complex discrimination training (for further information on conditional discrimination and stimulus equivalence, see Chapter 12).

ON THE APPLIED SIDE: QUALITY CONTROL AND DETECTION OF SIGNALS BY ANIMALS

In industrial settings, workers are often hired as quality-control inspectors. Quality control is usually a monotonous job that involves checking samples of a product to identify any defects. The most important skills or attributes needed for such jobs are good visual acuity and color vision. Based on these visual requirements, Thom Verhave (1966) suggested to the management of a drug company that the laboratory pigeon (*Columba livia domestica*) would be a cheap and efficient quality-control inspector. Although skeptical, the director of research for the company gave Verhave the go-ahead to train pigeons as inspectors.

The procedures were similar to a matching-to-sample (identity-matching) task. Pigeons were trained to inspect a line of drug capsules, accepting those that met a fixed standard and rejecting defective ones. In this procedure (Figure 8.16), a bird compared a drug capsule with a standard sample (a perfect one) and pecked Key 1 if it matched or Key 2 if there was a defect (a skag).

The standard capsule was fixed in position behind an inspection window. A line of capsules passed by the same window one at a time; some were perfect and others were defective. In order to initiate an inspection, the pigeon pecked at the inspection window, activating a beam of light that illuminated the sample and the comparison capsules. During training, all of the capsules on the inspection line were pre-coded by an electrical switch as either perfect or skags. If a capsule on the line was pre-coded as perfect, then the pigeon's response to Key 1 (matching response) resulted in food, turned off the beam of light behind the inspection window, and moved a new capsule into place. If a capsule was pre-coded as a skag, then a response to Key 2 (nonmatching response) turned off the illumination, moved a new capsule into the inspection window, and resulted in presentation of the food hopper. All other responses were false alarms or misses that were not

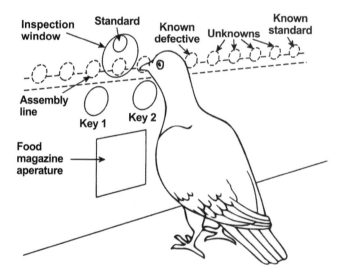

Fig. 8.16

Drawing depicts Verhave's (1966) discrimination procedures as described in the text. Pigeons were trained to inspect a line of drug capsules, accepting those that met a fixed standard and rejecting defective ones.

Source: From C. B. Ferster, S. Culbertson, & M. C. P. Boren (1975), *Behavior principles*, 2nd edition. Englewood Cliffs, NJ: Prentice-Hall, p. 558. Republished with permission. Copyright, 1975 held by Pearson Education, Inc.

reinforced and resulted in a 30-s blackout. With these contingencies in effect, the birds became about 99% accurate in identifying perfect capsules and skags.

One practical problem that Verhave faced concerned the persistence of a pigeon's performance on a real-life inspection line. In everyday life, there is no experimenter to designate perfect capsules, skags, misses, and false alarms. Without this monitoring, differential reinforcement for "hits versus misses" cannot be maintained, and a bird's performance will deteriorate over time to chance levels. A solution to this problem was to introduce capsules "known to be perfect or defective" occasionally onto the inspection line. Reinforcement or punishment was only in effect for "known" instances of matching (or nonmatching) to sample. With this procedure, sufficient differential reinforcement occurred to maintain stimulus control by the sample and comparison capsules.

In addition to the study by Verhave (1966), other researchers have attempted to use pigeons for navigation of missiles (Skinner, 1960) or to run assembly lines (Cumming, 1966). More recently, Azar (2002) reported that the US Navy in the 1970s and 1980s used pigeons to find people stranded at sea. Navy scientist Jim Simmons, PhD, trained pigeons by operant conditioning for search-and-rescue missions. The pigeons were trained to recognize objects floating in the water from an aircraft and were 93% accurate, compared with only 38% accuracy for human flight crews. When combined with human searchers, the pigeons' detection rate rose to almost 100%.

Pigeons are excellent at detecting visual signals. Other animals are highly sensitive to olfactory (smell) stimuli, making them suitable for detection of chemical odors. Animals are often "clicker trained," using a portable hand-held device that makes a "clicking" sound (Pryor, 1999; see Martin & Friedman, 2011 for correct use of the device; see also "On the Applied Side: Clicker Training" in Chapter 10 in this book). In preliminary training, the sound of the clicker is followed with food reinforcement. Once established, the sound of the clicker marks the response that matches the odor sample, and acts as conditioned reinforcement for the correct response. Withholding of clicks also acts as nonreinforcement for incorrect responses to comparison odors. The clicker allows the trainer to provide immediate reinforcement of the target behavior in a field setting where it is impossible to instantly provide food reinforcement. Clicker training with operant procedures (especially shaping and differential reinforcement) has allowed animals to perform odor-detection tasks that improve the everyday lives of people.

Landmines inflict injuries on people in more than 70 countries of the world, and more than half a million people have sustained life-changing injuries as a result of mine explosions. Allen Poling at Western Michigan University and his colleagues used African giant pouched rats (*Cricetomys gambianus*), trained by operant-conditioning procedures (clicker training), to detect the smell of the explosive chemical 2,4,6-trinitrotoluene (TNT) (Poling et al., 2011). After extensive training for accuracy in odor detection, the rats searched 93,400 m^2 of land in Gaza Province of Mozambique, finding 41 mines and 54 other explosive devices. Humans with metal detectors found no additional mines in this area. On average, the rats made 0.33 false alarms for every 100 m^2 searched, which is below the false-alarm rate standard for accrediting mine-detection animals. These findings indicate that trained pouched rats make excellent mine-detection animals, allowing human populations to reclaim their land and live free of landmine disasters (see Figure 8.17 for an image of Magawa, who holds the Guiness world record for sniffing out the most landmines). Interested readers can also see Mahoney et al., 2014 for an evaluation of landmine detection under simulated conditions.

Dogs are also highly sensitive to olfactory stimuli, and are often used for detection of illegal drugs, chemical contaminants, and explosives (Browne, Stafford, & Fordham, 2006). A recent application of operant conditioning in medicine involves odor detection by dogs of chemical markers in the urine of cancer patients. The basic idea is that urine contains volatile organic compounds

Fig. 8.17
Magawa, an African giant pouched rat holds the Guinness world record for sniffing out the most landmines (over 100), saving both human and nonhuman lives. He received a gold medal for bravery from PDSA (a British charity for animals)—a distinction that had previously only gone to dogs. He was trained by Apopo, a nonprofit organization that trains HeroRATS to detect landminds. Magawa died in retirement at the age of 8 in January 2022.

Source: British Broadcasting Company. www.bbc.com/news/world-asia-59951255

(VOCs) that mark the presence of prostate cancer (PC). These VOCs are absent from the urine of people who test negative for the disease. In one study of prostate cancer detection, a Belgian Malinois dog was clicker trained over a period of 24 months to detect the scent of urine samples from men known to have prostate cancer (Cornu, Cancel-Tassin, Ondet, Girardet, & Cussenot, 2011). After training, the dog's ability to discriminate PC samples from control urine was tested in a double-blind procedure. Samples of urine were obtained from 66 patients who had been referred to a urologist for elevated prostate-specific antigen or irregular digital rectal examination. All patients were given a prostate biopsy and divided into a group of 33 patients with cancer and a group of 33 controls with negative biopsies. For each detection test, the dog was required to identify the one cancer urine sample among six comparison samples (one sample from a person with cancer and five randomly selected control samples). The results showed that the dog correctly identified 30 of the 33 cancer cases, corresponding to an accuracy rate of 91%. Of the three incorrect identifications, one patient was given a second biopsy and found to have prostate cancer. Thus, the corrected accuracy rate for the dog was 93%. Overall, a review of the evidence suggests that dogs can discriminate VOCs in the urine and exhaled breath of cancer patients, allowing for detection of several types of cancer (Moser & McCulloch, 2010; see also Godfrey, 2014). Recently, research by Wasserman and his associates demonstrated that pigeons, birds with excellent visual ability, learned by differential reinforcement to identify with high accuracy images of benign from malignant human breast histopathology and were able to show generalization of this training to novel sets of images (Levenson, Krupinski, Navarro, & Wasserman, 2015). The birds' successful performance and training difficulties should be helpful in providing a better understanding of image perception and be useful in the development of medical imaging technology and analytical tools. Behavioral applications using dogs, pigeons, and other animals are proving to be of significant benefit to human health and wellness.

CHAPTER SUMMARY

This chapter has presented research and discussion of the stimulus conditions that set the occasion for operant behavior—changing its probability of occurrence. Influencing the probability of responding is a matter of differential reinforcement in the presence or absence of a stimulus. Such control can be produced in the absence of "errors" by the judicious use of stimulus fading. Generalization across stimuli means that there is a lack of discrimination, and responding occurs in the presence of many different stimuli. The process of remembering (memory) is treated as response probability in which a specific response is based on stimulus conditions are present in that moment as well as those from the past (a delayed conditional discrimination). The idea of a "concept" not as something inside the organism but as overt behavior under the control of a set of complex and varied stimuli is also presented. Birds were shown to learn the concept "humanness" when the contingencies supported responding to (identifying) pictures of humans and rejecting pictures without humans. Procedures such as matching to sample and training with only the S^D were discussed, and outcomes such as peak shift and behavioral contrast were highlighted. Several examples of training animals to make stimulus discriminations in everyday settings were given as evidence that operant procedures can be applied to important human problems.

Key Words

Absolute stimulus control
Antecedent stimulus (S)
Behavioral contrast
Conditional discrimination
Delayed matching to sample (DMTS)
Differential reinforcement
Differential reinforcement of other behavior (DRO)
Differential response
Discrimination
Discrimination index (I_D)
Discriminative stimulus (S^D)
Errorless discrimination
Extinction stimulus (S^Δ)
Fading
Generalization
Generalization gradient (operant)

Matching to sample
Multiple schedule
Negative contrast
Peak shift
Positive contrast
Probe trial
Relative stimulus control
Remembering
Response chain
Retention interval
S-delta (S^Δ)
Simultaneous discrimination
Stimulus control
Stimulus generalization
Successive discrimination
Superstitious behavior

On the Web

www.behaviorworks.org Susan Friedman is a psychology professor at Utah State University who has pioneered the application of applied behavior analysis (ABA) to captive and companion animals. Students from 22 different countries have participated in Susan's online courses, and she has written chapters on learning and behavior for veterinary texts as well as making frequent contributions to popular magazines. Her articles appear around the world in 11 languages. Susan has presented seminars for a wide variety of professional organizations around the world and has been nominated for the Media Award of the International Association of Behavior Analysis for her efforts to disseminate to pet owners, veterinarians, animal trainers, and zookeepers the essential tools they need to empower and enrich the lives of the animals in their care.

www.behavior.org/item.php?id=133 *Behavior Theory in Practice* (1965) by Dr. Ellen Reese is a set of four color videos (Parts I–IV) available from the Cambridge Center for Behavioral Studies, each part approximately 21 min in length. Each individual Part presents basic behavior principles in the laboratory and examples of the presence and use of those principles in a variety of everyday settings with a variety of species. For this chapter, students should order Part III: Generalization: Discrimination and Motivation.

www.youtube.com/watch?v=_eAGtAYW6mA This is a YouTube video of the HeroRAT that shows rats detecting landmines in Africa. Listen for the sound of the clicker for correct identifications and the subsequent delivery of food reinforcement by the trainers.

www.youtube.com/watch?v=h5_zJlm1B_k See if you can analyze errorless teaching in this YouTube video of the teaching of autistic children. In applied behavior analysis, the prompt is used to ensure correct responses to the S^D in early training, and high rates of reinforcement. The prompt is subsequently faded as the child comes under the control of the S^D, making few or no errors.

www.youtube.com/watch?v=yG12rqPaldc Take a look at this YouTube video on teaching a visual matching-to-sample task to a dog. See if you can identify some of the basic operant procedures used by the trainer.

www.equineresearch.org/support-files/hanggi-thinkinghorse.pdf The study of the thinking horse is the topic of this review article. See how operant discrimination training can be extended to categorization and concept formation by horses. Remember horses show such complex behavior based on the contingencies of reinforcement arranged by the trainer, not by forming mental representations in their heads.

Brief Quiz

1 An S^Δ sets the occasion upon which a response is _____ reinforced.
   a   sometimes
   b   always
   c   never
   d   maybe

2 An S^D does not cause or elicit the appearance of a response the way a _____ does.
   a   UR
   b   SR
   c   CS
   d   CR

3. In operant conditioning, what is the antecedent stimulus paired with reinforcement called?
   a. S^Δ
   b. S–R–S
   c. S^D
   d. CS

4. A two-component schedule in which both components have separate stimuli is called a:
   a. MIX
   b. CONC
   c. TAND
   d. MULT

5. To keep the onset of S^D from reinforcing responses in S^Δ, one needs to add a(n) _____ contingency.
   a. EXT
   b. IRT
   c. DRO
   d. PRP

6. If reinforcers on one schedule are depleted and responding in another schedule increases, we call this:
   a. negative contrast
   b. positive contrast
   c. substitutability
   d. anticipatory contrast

7. A change in maximal generalization responding, away from S^Δ to the other side of S^D, is called:
   a. gradient shift
   b. relative control
   c. stimulus control
   d. peak shift

8. A shaping procedure that gradually changes stimulus control from one element to another is called:
   a. approximations
   b. fading
   c. transfer
   d. conditional discrimination

9. If you trained a pigeon to turn in a circle when a TURN sign was presented, you could say that the bird was:
   a. discriminating
   b. conceptually oriented
   c. reading
   d. both (a) and (c)

10. With careful shaping and fading one might develop discrimination without:
    a. reinforcement
    b. extinction
    c. contrast
    d. errors

Answers to Brief Quiz: 1, c (p. 271); 2, c (p. 271); 3, c (p. 272); 4, d (p. 279); 5, c (p. 282); 6, b (p. 283); 7, d (p. 283); 8, b (p. 293); 9, d (p. 274); 10, d (p. 293).

NOTES

1 In this chapter, we present a classification scheme for stimuli that precede and set the occasion for reinforcement, extinction, or punishment of operant behavior. We introduce the generic term *antecedent stimulus* (S) to stand for all events that exert stimulus control over operant behavior. There are three kinds of antecedent stimuli: S^D, S^Δ, and S^{ave}. Notice that the antecedent stimulus is modified to reflect its function based on the contingencies of reinforcement that have established it (i.e., reinforcement, extinction, or punishment). The notations S^+ and S^- are also commonly used to represent the S^D and S^Δ functions of stimuli.

2 The visible color spectrum is seen when white light is projected through a prism. The spectrum ranges from violet (400 nm) at one end to red (700 nm) at the other.

3 The analysis of body image distortion as peak shift is based on an analysis by Brady Phelps in the Department of Psychology at South Dakota State University. Brady indicates he got the idea from Adam Derenne in the Department of Psychology at the University of North Dakota. Thanks to both for an interesting behavior analysis of body image disorder.

CHAPTER 9

Choice and Preference

1. Find out about how to study choice and preference in the laboratory.
2. Learn about the relative rates of reinforcement and behavioral choice.
3. Inquire about optimal foraging, behavioral economics, and self-control.
4. Investigate the matching relation on a single schedule of reinforcement.
5. Discover mathematical analysis of behavioral choice and preference.
6. Focus on behavioral neuroscience and concurrent schedules of reinforcement.

Over the course of a day, an individual makes many decisions that range from choices of great importance to ones of small consequence. Humans, for example, make over 200 decisions per day on just choices related to food (Wansink, 2007). A person makes choices when buying a new car, choosing to spend an evening with one friend rather than another, or deciding how they should study for an exam. Animals also make a variety of choices; they may choose mates with particular characteristics, select one type of food over another, or decide to leave a territory.

From a behavioral view, the analysis of **choice** is concerned with the distribution of operant behavior among alternative sources of reinforcement (options). When several options are available, one alternative may be selected more frequently than others. When this occurs, it is called a **preference** for that particular option. For example, a person may choose between two food markets, a large supermarket vs. the corner store, on the basis of price, location, and variety. Each time the individual goes to one store rather than the other, they are said to have made a choice. Eventually, the person may shop more frequently at the supermarket than the corner store, and when this occurs, the person is showing preference for the supermarket alternative.

Many people describe choosing to do something, or a preference for one activity over another, as a subjective experience. For example, you may say you like one person better than others, and based on this you feel good about spending the day with that person. From a behavioral perspective, your likes and feelings are real, but they do not provide an objective scientific account of why you decide to do what you do. To provide that account, it is necessary

DOI: 10.4324/9781003202622-9

to identify the conditions that regulate your attraction to (or preference for) the other person or friend.

EXPERIMENTAL ANALYSIS OF CHOICE AND PREFERENCE

For behavior analysts, the study of choice is based on principles of operant behavior. In previous chapters, operant behavior was analyzed in situations in which one response class was reinforced on a single schedule of reinforcement. For example, a child is reinforced with contingent attention from a teacher for correctly completing a page of arithmetic problems. The teacher provides one source of reinforcement (attention) when the child emits the target operant (math solutions). The single-operant analysis is important for the discovery of basic principles and applications. This same situation, however, may be analyzed as a choice among behavioral options. The child may choose to do math problems or emit other behavior—looking out of the window or talking to another child. This analysis of choice extends the operant paradigm or model to more complex environments in which several response and reinforcement alternatives are available.

In the everyday world, there are many alternatives that schedule reinforcement for operant behavior. A child may distribute time and behavior among parents, school, peers, and sport activities. Each alternative may require specific behavior and provide reinforcement at a particular rate and amount. To understand, predict, and change the child's behavior, all of these response–consequence relations must be taken into account. Thus, the operant analysis of choice and preference begins to contact the complexity of everyday life, offering new principles for application. Even the issue of substance abuse is viewed today as a pattern of operant choice instead of an unfortunate disease. In fact, Gene Heyman (2009) makes a compelling case for attributing substance abuse disorders to the dynamics of choice, and most effective treatments are based on an analysis of response alternatives.

The Choice Paradigm

Concurrent Schedules of Reinforcement
In the laboratory, choice and preference are modeled and investigated by arranging **concurrent schedules of reinforcement** (Catania, 1966). Figure 9.1 shows a concurrent-operant setting for a pigeon. In the laboratory, two or more simple schedules (i.e., FR, VR, FI, or VI) are simultaneously available on different response keys (Ferster & Skinner, 1957). Each key is programmed with a separate schedule of reinforcement, and the organism is free to distribute behavior between the alternative schedules. The distribution of time and behavior among the response options is the behavioral measure of choice and preference. For example, a food-deprived bird may be exposed to a situation in which the left response key is programmed to deliver 20 presentations of the food hopper each hour, while the right key delivers 60 reinforcers an hour. To obtain reinforcement from either key, the pigeon must respond according to the schedule on that key. If the bird responds exclusively to the right key (and never to the left) and meets the schedule requirement, then 60 reinforcers are delivered each hour. Because the bird could have responded to either side, we may say that it "prefers" to spend its time on the right alternative, presumably because the food payoff is three times larger.

Concurrent schedules of reinforcement have received considerable research attention because these procedures may be used as an analytical tool for understanding choice and preference. This selection of an experimental paradigm or model is based on the reasonable

Fig. 9.1
A two-key operant chamber for birds is displayed. Schedules of food reinforcement are arranged simultaneously on each key.

assumption that contingencies of reinforcement contribute substantially to choice behavior. Simply stated, all other factors being equal, the more reinforcement (higher rate) that is provided by an alternative, the more time and energy are spent on that alternative. For example, in choosing between spending an evening with either of two friends, the one who has provided the most social reinforcement is probably the one selected. Reinforcement may be social approval, affection, interesting conversation, or other aspects of the friend's behavior. The experience of deciding to spend the evening with one friend rather than the other may be something like "I just feel like spending the evening with Taylor." Of course, in everyday life choosing is seldom as uncomplicated as this, and a more common decision might be to spend the evening with both friends. To understand how reinforcement processes are working, however, it is necessary to control the other factors so that the independent effects of reinforcement, such as relative frequency, quality and duration, on choice may be observed.

Concurrent Ratio Schedules
To begin understanding choice, we start in the laboratory by controlling all variables except *the relative frequency of reinforcement*. We can use a two-key, concurrent-operant setting for humans to do this. Nowadays, concurrent schedules are arranged with computer programs (see for example Rasmussen & Newland, 2008), but Figure 9.2 shows a historical version of a concurrent schedule with distance between the two keys that may help with conceptualizing a human concurrent schedule. Consider that you are asked to participate in an experiment in which you may earn up to $50 an hour. As an experimental participant, you are taken to a room that has two response keys separated by a distance of 2.4 m (about 8 feet). Halfway between the two keys is a small opening just big enough for your hand to fit. The room is empty, except for the unusual-looking apparatus. You are told to do anything you want. What do you do? You probably walk about and inspect your surroundings and, feeling somewhat foolish, eventually press one of the response keys. Immediately following this action, $1 is dispensed by a coin machine and is held on a plate inside the small opening. The dollar remains available for about 5 s, and then the plate falls away and the dollar disappears. Assuming that you have retrieved the dollar, will you press one of the keys again? In reality, this depends on several factors: perhaps

Fig. 9.2
A classic two-key operant chamber for humans is shown. Pressing the keys results in money from a coin dispenser (middle), depending on the schedules of reinforcement. Most human operant studies are now conducted with computers.

you are wealthy and the dollar is irrelevant; perhaps you decide to "get the best of the experimenter" and show that you are not a rat; maybe you do not want to appear greedy, and so on. Assume for the moment, however, that you are a typical financially strapped student and you press the key again. After some time pressing both keys and counting the number of key presses, you discover what seems to be a rule. The left key pays a dollar for each 100 responses, while the right side pays a dollar for 250 responses. Does it make sense to spend your effort on the right key when you can make money faster and with less effort on the other alternative? Of course it does not, and you decide to spend all of your effort on the key that pays the most per unit time (it takes 2.5 times longer to press 250 times as opposed to 100 times). This same result has been found with other organisms. When two ratio schedules (in this case FR 100 and FR 250) are programmed as concurrent schedules, the alternative that produces more rapid reinforcement is chosen exclusively (Herrnstein & Loveland, 1975).

Because ratio schedules result in exclusive responding to the alternative with the highest rate of payoff, these schedules are seldom used to study choice. We have discovered something about choice: ratio schedules produce *exclusive preference* (though see McDonald, 1988 on how to program concurrent ratio schedules to produce response distributions similar to those that occur on interval schedules). Although this result is interesting, it suggests that other schedules should be used to investigate choice and preference. Once exclusive responding occurs, it is not possible to study how responses are distributed between the alternatives, which is the major objective of an experimental analysis of choice.

Concurrent-Interval Schedules
Now, consider what you might do if interval schedules were programmed on the two response keys. Remember that on an interval schedule a single response must occur after a defined or

variable amount of time. If you spend all of your time pressing the same key, you will miss reinforcement that is programmed on the other alternative. For example, if the left key is scheduled to pay a dollar on average every 2 min and the right key on average every 6 min, then a reasonable tactic is to spend most of your time responding on the left key, but every once in a while (say, once every six minutes) to check out the other alternative. This behavior will result in obtaining most of the money set up by both schedules. In fact, when exposed to concurrent-interval schedules, most animals (including humans) distribute their time and behavior between the two alternatives in such a manner (de Villiers, 1977). Thus, the first prerequisite of the choice paradigm is that *interval schedules* must be used to study the distribution of behavior.

Interval schedules are said to be independent of one another when they are presented concurrently. This is because responding on one alternative does not affect the rate of reinforcement programmed for the other schedule. For example, a fixed-interval 6-min schedule (FI 6 min) is programmed to deliver reinforcement every 6 min. Of course, a response must be made after the fixed interval has elapsed. Assume that you are faced with a situation in which the left key pays a dollar every 2 min (FI 2 min). The right key delivers a dollar when you make a response after 6 min (FI 6 min). You have 1 h a day in the experiment. If you just respond to the FI 2-min schedule, you would earn approximately $30. On the other hand, you could increase the number of payoffs an hour by occasionally pressing the FI 6-min key. This occurs because the left key pays a total of $30 each hour and the right key pays an additional $10 for a total of $40. After many hours of choosing between the alternatives, you may develop a stable pattern of responding. This *steady-state performance* is predictable. You should respond for approximately 6 min on the FI 2-min alternative and obtain three reinforcers ($3). After the third reinforcer, you may feel like switching to the FI 6-min key, on which a $1 reinforcer is immediately available. You obtain the money on this key and immediately return to the richer schedule (left key). This steady-state pattern of alternate responding may be repeated over and over with little variation.

Concurrent Variable-Interval Schedules
Recall that on variable-interval (VI) schedules, the time between each programmed reinforcer changes and the average time to reinforcement defines the specific schedule (VI 60 s). Because the organism is unable to discriminate the time to reinforcement on VI schedules, the regular switching pattern that characterizes concurrent FI performance does not occur. This is an advantage for the analysis of choice, because the organism must respond on both alternatives as switching does not always result in reinforcement. Thus, operant behavior maintained by concurrent VI VI schedules is *sensitive to the rate of reinforcement* on each alternative. For this reason, VI schedules are typically used to study choice.

Alternation: The Changeover Response
At this point, the choice paradigm is almost complete. Again, however, consider what you would do in the following situation. The two keys are separated and you cannot press both at the same time. The left key now pays a dollar on a VI 2-min schedule, while responses to the right key are reinforced on VI 6 min. The left key pays $30 each hour, and the right key delivers $10 if you respond perfectly. Assuming that you obtain all programmed reinforcers on both schedules, you earn $40 for each experimental session. What can you do to earn the most

per hour? If you stay on the VI 2-min side, you end up missing the 10 reinforcers on the other alternative. If you frequently change over from key to key, however, most of the reinforcers on both schedules are obtained. This is in fact what most animals do when faced with these contingencies (de Villiers, 1977).

Simple alternation between response alternatives prevents an analysis of choice because the distribution of behavior remains the same (approximately 50/50) no matter what the programmed rates of reinforcement. Frequent switching between alternatives may occur because of the correlation between rate of switching and overall rate of reinforcement (number of dollars per session). In other words, as the rate of switching increases, so does the hourly payoff. Another way of looking at this alternation is that organisms are accidentally reinforced for the **changeover response**. This alternation is called *concurrent superstition* (Catania, 1966), and it occurs because as time is spent on one alternative the other schedule is timing out. As the organism spends more time on the left key, the probability of a reinforcer being set up on the right key increases. This means that a changeover to the right key will be reinforced even though the contingencies do not require the changeover response. Thus, switching to the other response key is an operant that is inadvertently strengthened.

The Changeover Delay
The control procedure used to stop rapid switching between alternatives is called a **changeover delay**, or **COD** (Shull & Pliskoff, 1967). The COD contingency stipulates that responses have no effect immediately following a change from one schedule to another. After switching to the alternative, a brief time interval is required before a response can be reinforced (e.g., a 3-s delay). For example, if an organism has just changed to an alternative that is ready to deliver reinforcement, there is a 3-s delay before a response is effective. As soon as the 3-s delay has elapsed, a response is reinforced. Of course, if the schedule has not timed out, the COD is irrelevant because reinforcement is not yet available. The COD contingency operates in both directions whenever a change is made from one alternative to another. The COD prevents frequent switching between alternatives. To obtain reinforcement, an organism must spend a minimal amount of time on an alternative before switching to another schedule. For example, with a 3-s COD, changing over every 2 s will never result in reinforcement. The COD is therefore an important and necessary feature of the operant-choice procedure.

Experimental Procedures for Studying Choice
The basic paradigm for investigating choice and preference has been described. In summary, a researcher who is interested in behavioral choice should:

1 Arrange two or more concurrently available schedules of reinforcement.
2 Program interval schedules on each alternative.
3 Use variable- rather than fixed-interval schedules.
4 Require a COD to stop frequent switching between the schedules.

THE MATCHING RELATION

In 1961, an influential paper (Hernstein, 1961a) that described the distribution of behavior under concurrent schedules of positive reinforcement was published. The paper reported that pigeons matched relative rates of behavior to relative rates of reinforcement. For example, when 90% of the total reinforcement was provided by schedule A (and 10% by schedule B), approximately 90% of the bird's key pecks were allocated toward the A schedule. This matching between relative rate of reinforcement and relative rate of response is known as the **matching relation**. Today, the original mathematical statement of the matching relation and its interpretation as a simple equality of proportions (classical matching theory) is no longer tenable, being circumscribed to a subset of choice situations (McDowell, 2013). But it is essential to address proportional matching to prepare students for the *generalized matching relation* and modern matching theory as outlined in the Advanced Section of this chapter. To understand proportional matching, we turn to Herrnstein's (1961b) experiment.

Proportional Matching

The Matching Experiment
In this study (Herrnstein, 1961b) the behavior of pigeons was investigated on a two-key concurrent schedule. Concurrent VI VI schedules of food reinforcement were programmed with a 1.5-s COD. The birds were exposed to different pairs of concurrent VI VI schedules for several days. Each pair of concurrent schedules was maintained until response rates stabilized—that is, behavior on each schedule did not significantly change from session to session. After several days of stable responding, a new pair of schedule values was presented. Overall rate of reinforcement was held constant at 40 reinforcers per hour for all pairs of schedules. Thus, if the schedule on the left key was programmed to deliver 20 reinforcers an hour (VI 3 min), then the right key also provided 20 reinforcers. If the left key supplied 10 reinforcers, then the right key supplied 30 reinforcers. The schedule values that were used are presented in Figure 9.3.

The data in Figure 9.3 show the schedules operating on the two keys—A and B. As previously stated, the total number of scheduled reinforcers is held constant for each pair of VI schedules. This is indicated in the third column, in which the sum of the reinforcements per hour (Rft/h) is equal to 40 for each pair of schedules. Because the overall rate of reinforcement remains constant, changes in the distribution of behavior cannot be attributed to this factor. Note that when key A is programmed to deliver 20 reinforcers an hour, so is key B. When this occurs, the responses per hour (Rsp/h) are the same on each key. However, the responses per hour (or absolute rate) are not the critical measure of preference. Recall that choice and preference are measured as the *distribution of time or behavior* between alternatives. To express the idea of distribution, it is important to direct attention to relative measures. Because of this, Herrnstein focused on the *relative rates of response*. In Figure 9.5, the relative rate of response is expressed as a proportion. That is, the rate of response on key A is the numerator and the sum of the response rates on both keys is the denominator. The proportional rate of response on key A is shown in the final column, labeled "Relative responses."

Calculation of Proportions
To calculate the *proportional rate of response* to key A for the pair of schedules VI 4.5 min VI 2.25 min, the following simple formula is used:

$$B_a/(B_a + B_b).$$

Key	Schedule	Rft/hr	Rsp/hr	Relative reinforcement	Relative responses
A	VI 3-min	20.00	2000	0.50	0.50
B	VI 3-min	20.00	2000	0.50	0.50
A	VI 9-min	6.7	250	0.17	0.08
B	VI 1.8-min	33.30	3000	0.83	0.92
A	VI 1.5-min	40.00	4800	1.00	1.00
B	Extinction	0.00	0	0.00	0.00
A	VI 4.5-min	13.30	1750	0.33	0.31
B	VI 2.25-min	26.70	3900	0.66	0.69

Fig. 9.3

A table of schedule values and data is shown. Reinforcement per hour (Rft/h), responses per hour (Rsp/h), relative reinforcement (proportions), and relative responses (proportions) are shown.

Source: Adapted from R. J. Herrnstein (1961b). Relative and absolute strength of responses as a function of frequency of reinforcement. *Journal of the Experimental Analysis of Behavior*, 4, pp. 267–272, Figure 1 (bird 231).

The term B_a is behavior measured as the rate of response on key A, or 1,750 pecks per hour. The rate of response on key B is 3,900 pecks per hour and is represented by the B_b term. Thus, the proportional rate of response on key A is:

$$1{,}750/(1{,}750 + 3{,}900) = 0.31.$$

In a similar fashion, the *proportional rate of reinforcement* on key A may be calculated as:

$$R_a/(R_a + R_b).$$

The term R_a refers to the scheduled rate of reinforcement on key A, or 13.3 reinforcers per hour. The rate of reinforcement on key B is designated by the symbol R_b and is 26.7 reinforcers per hour. The proportional rate of reinforcement on key A is calculated as:

$$13.3/(13.3 + 26.7) = 0.33.$$

These calculations show that the relative rate of response (0.31) is very close to the relative rate of reinforcement (0.33). If you compare these values for the other pairs of schedules, you will see that the proportional rate of response approximates the proportional rate of reinforcement.

Importance of Relative Rates

Herrnstein's (1961b) paper showed that the major dependent variable in choice experiments was **relative rate of response,** because examination of behavior from one option depends on what is available from the other option. He also found that **relative rate of reinforcement** was the primary independent variable for the same reason (i.e., there are two options). Thus, in an operant-choice experiment, the researcher manipulates the relative rate of reinforcement on each key and observes the relative rate of response to the respective alternatives.

Figure 9.3 shows the independent variable, relative rate of reinforcement on key A, over a range of values. Because there are several values of the independent variable and a corresponding set of values for the dependent variable, it is possible to plot the relationship.

Figure 9.4 shows the relationship between proportional rate of reinforcement, $R_a/(R_a + R_b)$, and proportional rate of response, $B_a/(B_a + B_b)$, for pigeon 231 based on the values in Figure 9.3.

The Matching Equation for Proportional Response Rates

As the relative rate of reinforcement increases so does the relative rate of response. Furthermore, for each increase in relative reinforcement there is about the same increase in relative rate of response. This equality of relative rate of reinforcement and relative rate of response is expressed as a proportion in Equation 9.1:

$$B_a/(B_a + B_b) = R_a/(R_a + R_b). \quad \text{(Equation 9.1)}$$

Notice that we have simply taken the expressions $B_a/(B_a + B_b)$ and $R_a/(R_a + R_b)$, which give the proportion of responses and reinforcers on key A, and mathematically stated that they are equal. In verbal form, we are stating that *relative rates of response matches (or equals) relative rates of reinforcement*.

In Figure 9.4, **matching** is shown as the solid black line. Notice that this line results when the proportional rate of reinforcement exactly matches the proportional rate of response. The proportional matching equation is an ideal representation of choice behavior. The actual data from pigeon 231 approximate the matching relationship. Herrnstein (1961b) also reported the results for two other pigeons that were well described by the matching equation.

Matching Time on an Alternative

Behavioral choice can also be measured as time spent on an alternative (Baum & Rachlin, 1969; Brownstein & Pliskoff, 1968). Time spent is a useful measure of behavior when the response

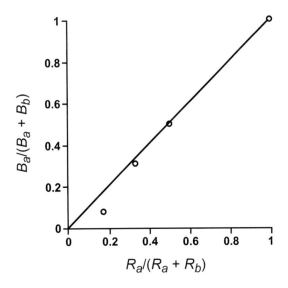

Fig. 9.4
Proportional matching of the response and reinforcement rates for bird 231.

Source: Figure is based on results from R. J. Herrnstein (1961b). Relative and absolute strength of responses as a function of frequency of reinforcement. *Journal of the Experimental Analysis of Behavior*, 4, pp. 267–272. Copyright 1961 held by the Society for the Experimental Analysis of Behavior, Inc.

is *continuous*, as in talking to another person. For the laboratory, in addition to measuring the number of responses, the time spent on an alternative may be used to describe the distribution of behavior. The proportional matching relation also can be expressed in terms of *relative time spent* on an alternative. Equation 9.2 is similar to Equation 9.1, but states the matching relationship in terms of time:

$$T_a/(T_a + T_b) = R_a/(R_a + R_b). \quad \text{(Equation 9.2)}$$

With this equation, the time spent on alternative A is represented by T_a and the time spent on alternative B is denoted by T_b. Again, R_a and R_b represent the respective rates of reinforcement for these alternatives. The equation states that the relative time spent on an alternative equals the relative rate of reinforcement from that alternative. This extension of the matching relation to continuous activities, such as standing in one place or looking at objects, is important. Most behavior outside of the laboratory does not occur as discrete responses, and time allocation has been proposed as the more fundamental measure of choice and preference (Baum, 2015). In this case, Equation 9.2, expressed as a *generalized matching equation* for time spent on alternatives, may be used to describe choice and preference (see Advanced Section of this chapter).

Matching on More Than Two Alternatives

A consideration of either Equation 9.1 or Equation 9.2 makes it evident that in order to change choice behavior, the rate of reinforcement for the target response may be adjusted; alternatively, the rate of reinforcement for other concurrent operants may be altered. Both of these procedures manipulate the relative rate of reinforcement for the specified or target behavior. Equation 9.3 represents the relative rate of response as a function of several alternative sources of reinforcement:

$$B_a/(B_a + B_b + \ldots B_n) = R_a/(R_a + R_b + \ldots R_n). \quad \text{(Equation 9.3)}$$

In the laboratory, most experiments are conducted with only two concurrent schedules of reinforcement. The proportional matching relation, however, also describes situations in which an organism may choose among several alternative sources of reinforcement, as in foraging for food (Elsmore & McBride, 1994). In Equation 9.3, behavior allocated to alternative A (B_a) is expressed relative to the sum of all behavior directed to the known alternatives ($B_a + B_b + \ldots B_n$). Reinforcement provided by alternative A (R_a) is stated relative to all known sources of reinforcement ($R_a + R_b + \ldots R_n$). Again, notice that an equality of proportions (matching) is stated.

EXTENSIONS OF THE MATCHING RELATION

The Generality of Matching

The simple equality of relative rate of response and relative rate of reinforcement describes how a variety of organisms choose among alternatives (de Villiers, 1977). McDowell (2013) has shown that the original proportion-matching equation and its interpretation as a behavioral law are not well supported by research evidence. In other words, matching is not a perfect equality; there is some error in it that makes it approximate. The *generalized matching relation* by Baum (1974b), based on response and reinforcement ratios rather than proportions, remains as a tenable law of behavior that better characterizes the error in matching (see Advanced Section of this chapter for the ratio equations; see Killeen, 2015 for a dispute about the lawfulness of generalized matching).

Generalized matching has been demonstrated in animals such as pigeons (Davison & Ferguson, 1978), wagtails (Houston, 1986), cows (Matthews & Temple, 1979), fish (Banna et al., 2011), and rats (Poling, 1978; Buckley & Rasmussen, 2012). Interestingly, matching also applies to humans across a number of different settings, including laboratory and sports (e.g., Alferink et al., 2009; Rasmussen & Newland, 2008; see Kollins et al., 1997 and Pierce & Epling, 1983 for review). Reinforcers have ranged from food (Herrnstein, 1961b) to points that are subsequently exchanged for money and course credit (Bradshaw, Ruddle, & Szabadi, 1981; Rasmussen and Newland, 2008; 2009). Behavior has been as diverse as lever pressing by rats (Norman & McSweeney, 1978), making three-point shots in basketball (Alferink et al., 2009), and conversations in humans (Conger & Killeen, 1974).

Environments in which generalized matching has been observed have included T-mazes, operant chambers, and open spaces with free-ranging flocks of birds (Baum, 1974a), as well as discrete-trial and free-operant choice by human groups (Madden, Peden, & Yamaguchi, 2002). Also, students enrolled in special education programs have been found to spend time on math problems based on the relative rates of reinforcement (e.g., Mace, Neef, Shade, & Mauro, 1994). And quantitative models of choice based on generalized matching now inform many applications of behavior analysis (Jacobs, Borrero, & Vollmer, 2013). Thus, the matching relation in its generalized form describes the distribution of individual (and group) choice behavior across species, types of response, different reinforcers, and a variety of real-world settings (see Advanced Section of this chapter).

Matching and Human Communication

An interesting test of the generalized matching relation was applied to conversation (Conger & Killeen, 1974). These researchers assessed human performance in a group discussion situation. A group was composed of three experimenters and one experimental participant. The participant was not aware that the other group members were confederates in the experiment, and was asked to discuss attitudes toward drug abuse. One of the confederates prompted the participant to talk. The other two confederates were assigned the role of an audience. Each listener reinforced the participant's talk with brief positive words or phrases such as "that's a good point" when a hidden cue light came on. The cue lights were scheduled so that the listeners gave different rates of reinforcement to the speaker. When the results for several participants were combined, relative time spent talking to the listener matched relative rate of agreement from the listener.

In a similar study, college students engaged in a 20-min discussion of youths impacted by the justice system in which they received agreement from confederates as alternative sources of reinforcement (Borrero et al., 2007). The generalized matching relation described the pooled data for relative response rates better than relative time spent talking, a finding at odds with the original experiment by Conger and Killeen. These results suggest that generalized matching operates in everyday social interaction (see Figure 9.5), but further experiments are required to clarify its application to human communication in experimentally controlled settings.

In an applied study by McDowell and Caron (2010), the generalized matching relation described the verbal behavior of boys at risk for delinquency, as they interacted with their friends.

Fig. 9.5
The matching relation can characterize conversation between two individuals.
Source: Shutterstock.

The boys' verbal responses were coded as either "rule-break talk" (speech related to the breaking of rules or social norms) or "normative talk," and positive social responses from peers were recorded as the presumed sources of reinforcement for the two verbal response classes. The generalized matching relation provided an excellent description of the boys' allocation of verbal behavior, with some deviation from exact matching and bias toward normative talk (as would be expected). Importantly, the deviation from matching was more extreme and the bias toward normative talk was lower as the risk for delinquency of the child increased. The researchers suggested that extreme deviation from matching reflects the low-reinforcement value of positive social responses for a child with delinquency, while bias away from normative talk is likely indicative of different histories of reinforcement and punishment during the upbringing of these youngsters (see Advanced Section of this chapter for more about response bias and deviations from matching).

Practical Implications of the Matching Relation

The generalized matching relation has practical implications. A few researchers have shown that the matching equations in generalized form are useful in applied settings (Borrero & Vollmer, 2002; Epling & Pierce, 1983; McDowell, 1981, 1982, 1988; Myerson & Hale, 1984). With animals, for example, generalized matching has been useful in quantifying preferences for different types of food. Bias in matching has been used to assess pigeons' preferences for wheat over buckwheat and for buckwheat over hemp (Miller, 1976). Obese and lean Zucker rats both prefer sucrose over carrot-flavored pellets (Buckley & Rasmussen, 2012), and dairy cows prefer hay over dairy meal (Matthews and Temple, 1979). One applied setting with humans where the generalized matching relation has practical importance is the classroom, where students' behavior is often maintained on concurrent schedules of social reinforcement.

Matching, Modification, and Reinforcement Schedules

In a classroom, appropriate behavior for students includes working on assignments, following instructions, and attending to the teacher. In contrast, yelling and screaming, talking out of turn, and throwing paper airplanes are usually viewed as undesirable. All of these activities, appropriate or inappropriate, are presumably maintained by teacher attention, peer approval, sensory stimulation, and other sources of reinforcement. The schedules of reinforcement that maintain behavior in complex settings such as a classroom, however, are not usually known. When the objective is to increase a specific operant and the concurrent schedules are unknown, Myerson and Hale (1984) recommend the use of VI schedules to reinforce target or problem behavior.

Recall that on concurrent ratio schedules, exclusive preference develops for the alternative with the higher rate of reinforcement (Herrnstein & Loveland, 1975). Ratio schedules are in effect when a teacher implements a grading system based on the number of correct solutions for assignments. The teacher's intervention will increase the students' on-task behavior only if the rate of reinforcement by the teacher is higher than another ratio schedule controlling inappropriate behavior, such as talking or texting to a classmate while the teacher is speaking. Basically, an intervention is either completely successful or a total failure when ratio schedules are used to modify behavior and possibly a reason that teachers sometimes say that rewards like gold stars don't work. In contrast, interval schedules of reinforcement will always redirect behavior to the desired alternative, although such a schedule may not completely eliminate inappropriate responding.

When behavior is maintained by interval contingencies, interval schedules remain the most desirable method for behavior change. Myerson and Hale used the matching equations to show that behavior-change techniques based on interval schedules are more effective than ratio interventions. They stated that:

> [I]f the behavior analyst offers a VI schedule of reinforcement for competing responses two times as rich as the VI schedule for inappropriate behavior, the result will be the same as would be obtained with a VR schedule three times as rich as the schedule for inappropriate behavior.
>
> (Myerson & Hale, 1984, pp. 373–374)

Generally, behavior change will be more predictable and successful if interval schedules are used to reinforce appropriate behavior in a classroom.

MATCHING ON SINGLE-OPERANT SCHEDULES

As with the proportional matching equation (Equation 9.1), Herrnstein's equation for *absolute rate of response* on a single-operant schedule is not considered a basic behavioral law. The response-rate equation, however, provides an excellent description of many diverse findings and is consistent with modern matching theory as stipulated by the generalized matching relation (McDowell, 2013). Here we outline Herrnstein's early analysis so that students may contact the substantial literature on this topic, follow the theoretical controversies in the coming years, and understand its behavioral applications.

The proportional matching equation (Equation 9.1) suggested that operant behavior on a single-response key is determined by rate of reinforcement for that response relative to all sources of reinforcement. Even in situations in which a contingency exists between a single response and a reinforcement schedule, organisms usually have several sources of reinforcement that are unknown to the researcher. Also, many of the activities that produce reinforcement are beyond experimental control. A rat that is lever pressing for food may gain additional reinforcement from exploring the operant chamber, scratching itself, or grooming. In a similar fashion, rather than working for teacher attention, a pupil may look out of the window, talk to a friend, or daydream. Thus, even in a single-operant setting, multiple sources of reinforcement are operating. Therefore, all operant behavior must be understood as behavior emitted in the context of other alternative sources of reinforcement (Herrnstein, 1970, 1974).

Based on these ideas, there is an equation that describes the *absolute rate of response* on a single schedule of reinforcement. This mathematical formulation is called the **quantitative law of effect**, although as we have noted its status as a perfect behavioral law is in question (McDowell, 2013). The single-operant equation states that the *absolute rate of response on a schedule of reinforcement is a hyperbolic function of rate of reinforcement on the schedule relative to the total rate of reinforcement*, both scheduled and extraneous reinforcement. Thus, as the rate of reinforcement on the schedule increases, the rate of response rapidly rises, but eventually further increases in the rate of reinforcement produce less and less of an increase in the rate of response (a hyperbolic curve; see Figure 9.6 for examples).

The rapid rise in rate of response with higher rates of reinforcement is modified by extraneous sources of reinforcement. **Extraneous sources of reinforcement** include any unknown contingencies that support the behavior of the organism. For example, a rat that is pressing a lever for food on a particular schedule of reinforcement might receive extraneous reinforcement for scratching, sniffing, and numerous other behaviors. The rate of response for food will be a function of the programmed schedule as well as the extraneous schedules controlling other behavior. In humans, a student's mathematical performance will be a function of the schedule of correct solutions as well as extraneous reinforcement for other behavior from classmates or teachers, internal neurochemical processes, and changes to the physical/chemical environment (e.g., the smell of food drifting from the cafeteria).

Extraneous reinforcement slows down the rise in rate of response with higher rates of reinforcement. One implication is that control of behavior by a schedule of reinforcement is reduced as the sources of extraneous reinforcement increase. A student who does math problems for a given rate of teacher attention would do less if extraneous reinforcement is available by looking out of the classroom window. Alternatively, the teacher would have to use higher rates of attention for problem solving when "distractions" such as smart phones are available than when there are few additional sources of reinforcement.

Experimental Evidence for the Hyperbolic Curve

The hyperbolic relation between reinforcement rate and absolute response rate has been investigated in laboratory experiments. In an early investigation, Catania and Reynolds (1968) conducted an exhaustive study of six pigeons that pecked a key for food on different variable-interval (VI) schedules. The rate of reinforcement ranged from 8 to 300 food presentations per hour. Herrnstein (1970) replotted the data on *X* and *Y* coordinates. Figure 9.6 shows the plots for the six birds, with reinforcers per hour on the X-axis and responses per minute on the Y-axis.

Figure 9.6 presents the curves that best fit these results. Notice that all of the birds produce rates of response that are described as a hyperbolic function of rate of reinforcement. Some of

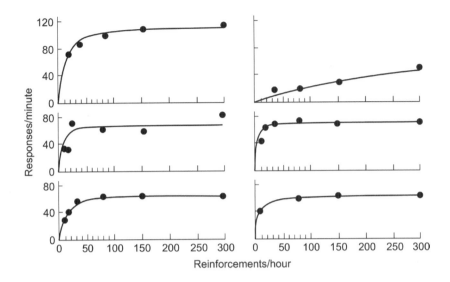

Fig. 9.6
The figure shows rate of response as a function of rate of food reinforcement for six pigeons on single VI schedules.
Source: Taken from R. J. Herrnstein (1970). On the law of effect. *Journal of the Experimental Analysis of Behavior, 13*, pp. 243–266. Copyright 1970 held by John Wiley & Sons, Ltd. Republished with permission.

the curves fit the data almost perfectly, while others are less satisfactory. Overall, Herrnstein's hyperbolic curve provides a good mathematical description of these findings and those from many other experiments.

The hyperbolic relation between reinforcement and absolute response rate, as expressed by Herrnstein's equation for the single operant, has substantial generality and is possibly based on evolution and selection for a reinforcement-learning mechanism (McDowell & Ansari, 2005). It has been extended to magnitude of food reinforcement, brain stimulation, quality of reinforcement, delay of positive reinforcement, rate of negative reinforcement, magnitude or intensity of negative reinforcement, and delay of negative reinforcement (see de Villiers, 1977 for a review). In a summary of the evidence, Peter de Villiers stated:

> The remarkable generality of Herrnstein's equation is apparent from this survey. The behavior of rats, pigeons, monkeys and ... people is equally well accounted for, whether the behavior is lever pressing, key pecking, running speed, or response latency in a variety of experimental settings. The reinforcers can be as different as food, sugar water, escape from shock or loud noise or cold water, electrical stimulation of a variety of brain loci, or turning a comedy record back on. Out of 53 tests of [the hyperbolic curve] on group data, the least-squares fit of the equation accounts for over 90% of the variance in 42 cases and for over 80% in another six cases. Out of 45 tests on individual data, the equation accounts for over 90% of the variance in 32 cases and for over 80% in another seven cases. The literature appears to contain no evidence for a substantially different equation.... This equation therefore provides a powerful but simple framework for the quantification of the relation between response strength and both positive and negative reinforcement.
>
> (de Villiers, 1977, p. 262)

According to McDowell (2013, p. 1008), however, the good fit of the hyperbolic curve is more apparent than real, as the *constant k assumption* (maximum possible rate of responding at asymptote), mathematically required to derive the hyperbolic equation from proportional matching, does not hold up to experimental tests. That is, the estimated k-value has been shown to deviate from the value expected by Herrnstein's classical matching theory as expressed by the proportion equation. An implication of this assessment is that the quantitative law of effect, and its interpretation as set forth by Herrnstein, is untenable as a behavioral law relating reinforcement to response strength. Deviations in the k-value of the absolute-rate equation, however, are consistent with the *generalized matching relation* and modern matching theory (Baum, 1974b; see deviations from matching in Advanced Section of this chapter). Thus, the generalized matching relation underpins the empirical success of the hyperbolic-rate equation, not classical theory as formulated in Herrnstein's proportional matching equation (McDowell, 2013).

ON THE APPLIED SIDE: APPLICATION OF THE SINGLE-OPERANT RATE EQUATION

Jack McDowell from Emory University was the first researcher to use Herrnstein's matching equation for a single schedule of reinforcement to describe human behavior in a natural setting. McDowell's expertise in mathematics and behavior modification spurred him to apply the matching equation for a single operant to a clinically relevant problem.

Mathematics and Behavior Modification

Carr and McDowell (1980) were involved in the treatment of a 10-year-old boy who repeatedly and severely scratched himself. Before treatment the boy had a large number of open sores on his scalp, face, back, arms, and legs. In addition, the boy's body was covered with scabs, scars, and skin discoloration, where new wounds could be produced. The boy's scratching was determined to be operant behavior as careful observation showed that the scratching occurred predominantly when he and other family members were in the living room watching television. This suggested that the self-injurious behavior was under stimulus control. In other words, the family setting made scratching more likely to occur.

Next, Carr and McDowell looked for potential reinforcing consequences maintaining the boy's self-injurious behavior. The researchers suspected that the consequences were social, because scratching appeared to be under the stimulus control of family members. In any family interaction there are many social exchanges, and the task was to identify those consequences that reliably followed the boy's scratching. Observation showed that family members reliably reprimanded the boy when he engaged in scratching. Reprimands are seemingly negative events, but the literature makes it clear that both approval and disapproval may serve as reinforcement.

Although social reinforcement by reprimands was a good guess, it was still necessary to show that these consequences in fact functioned as reinforcement. The first step was to take baseline measures of the rate of scratching and the rate of reprimands. Following this, the family members were required to withhold any verbal feedback for the boy's scratching behavior. That is, the presumed reinforcer was withdrawn (i.e., extinction), and the researchers continued to monitor the rate of scratching. Next, the potential reinforcer was reinstated; the family members again reprimanded the boy for his scratching. Relative to baseline, the scratching decreased when reprimands were withdrawn and increased when they were reinstated. This test identified the reprimands as positive reinforcement for scratching. Once the reinforcement for scratching had been identified, behavior modification was used to eliminate the self-injurious behavior.

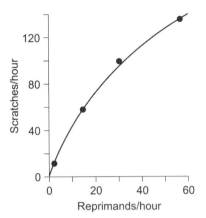

Fig. 9.7
Rate of social reinforcement and self-injurious scratching of a young boy is shown. The single-operant hyperbolic equation fit the data well.

Source: Adapted from J. J. McDowell (1981). *Quantification of steady-state operant behavior*. Amsterdam: Elsevier/North-Holland, pp. 311–324. Republished with permission of Jack McDowell.

In a subsequent report, McDowell (1981) analyzed the boy's baseline data in terms of the single-operant rate equation. He plotted the number of reprimands per hour on the X-axis and the number of scratches per hour on the Y-axis. He then fitted the matching equation for a single schedule of reinforcement to the points on the graph. Figure 9.7 shows the plot and the curve of best fit. The matching equation provides an excellent description of the boy's behavior. You will notice that most of the points are on, or very close to, the hyperbolic curve. In fact, more than 99% of the variation in rate of scratching is accounted for by the rate of reprimands. McDowell has indicated the significance of this demonstration by pointing out that the boy was in a home environment (as opposed to a laboratory environment, in which many variables would be highly controlled). Therefore, even in a more naturalistic environment, the single-operant matching equation holds.

Overall, the hyperbolic equation (single-operant rate equation) has been an important contribution to the understanding of human behavior and to the modification of human behavior in applied settings (see Fisher & Mazur, 1997; Martens, Lochner, & Kelly, 1992).

CHOICE, FORAGING, AND PREFERENCE FOR CHOICE

Optimal Foraging, Matching, and Melioration

One of the fundamental problems of evolutionary biology and behavioral ecology concerns the concept of "optimal foraging" of animals (Krebs & Davies, 1978). Foraging involves prey selection, where prey can be either animal or plant. Thus, a cow taking an occasional mouthful of grass in a field, or a redshank wading in the mud and probing with its beak for an occasional worm, are examples of foraging behavior. Because the function of foraging is to find food, foraging can be viewed as operant behavior regulated by relative rates of food reinforcement. The natural contingencies of foraging present animals with alternative sources of food called *patches*. Food patches provide items at various rates (patch density), and in this sense are similar to concurrent schedules of reinforcement arranged in the laboratory. Food patches have other properties, too, such as quality of food, effort to obtain the food, delays and barriers to food, etc. Many of these properties have been studied, but we start here with relative rates of food as a beginning, in which everything is identical between the food patches except the relative amount.

Optimal foraging is said to occur when animals obtain the *highest overall rate of reinforcement* from their foraging among alternative patches. That is, over time organisms are expected to select between patches so as to optimize (obtain the maximum possible value from) their food resources. In this view, animals are like organic computers comparing their behavioral distributions with relative outcomes and stabilizing on a response distribution that maximizes the overall rate of reinforcement, a process called **maximization**.

In contrast to the optimal foraging hypothesis, there is the process of **melioration,** which generally refers to doing the best at the moment (Herrnstein, 1997). Organisms are sensitive to *fluctuations in the momentary rates of reinforcement* rather than to long-term changes in overall rates of reinforcement. That is, an organism remains on one schedule until the local rates of reinforcement decline relative to that offered by a second schedule. Another situation involves "preference pulses", in which an organism will momentarily prefer the option that just produced a reinforcer; indeed, the larger the momentary rate of reinforcement, the longer they will stay with that option until they change to the other option (e.g., Davison & Baum, 2003).

It is not possible to examine all the evidence for melioration, matching, and maximizing in this chapter, but melioration and matching are the basic processes of choice. That is, when melioration and matching are tested in choice situations that distinguish matching from maximizing, matching theory has usually predicted the actual distributions of the behavior (Herrnstein, 1997).

One example of the application of matching theory to animal foraging has been reported by Baum (1974a; see also Baum, 1983, on foraging) for a flock of free-ranging wild pigeons. The subjects were 20 pigeons that lived in a wooden-frame house in Cambridge, Massachusetts. An opening allowed them to freely enter and leave the attic of the house. An operant apparatus with a platform was placed in the living space opposite to the outside opening to the attic. The front panel of the apparatus contained three translucent response keys and, when available, an opening allowed access to a hopper of mixed grain. The response keys were accessible by perches, so only one pigeon at a time could operate the keys and obtain food. Responses to the illuminated side keys were reinforced under concurrent VI VI schedules. Relative rates of reinforcement on the two keys were varied and the relative rate of response was measured.

Although only one bird at a time could respond on the concurrent schedules of reinforcement, Baum (1974b) treated the aggregate pecks of the group as the dependent measure. When the group of 20 pigeons chose between the two side keys, each of which occasionally produced food, the ratio of pecks to these keys was approximately equal to the ratio of grain presentations obtained from them. That is, the aggregate behavior of the flock of 20 pigeons was in accord with the *generalized matching equation*, a form of matching equation based on ratios rather than proportions (see Advanced Section at the end of this chapter). This research suggests that the matching relation in its generalized form applies to the behavior of wild pigeons in natural environments. Generally, principles of choice based on laboratory experiments can predict the foraging behavior of animals in the wild.

Preference for Choice

Animals foraging in the wild must select food items and search for food in different locations. Thus, making choices is part of the natural ecology of animals. For humans, it is notable that many democratic societies, such as America, are uniquely founded on choice, and most of the citizens of these countries value freedom of choice as a fundamental right. Throughout the day, we make repeated choices by expressing our opinion on diverse social and political issues, selecting among an enormous variety of goods and services (see Figure 9.8), and choosing to form close relationships with others in our social networks and communities. Generally, we believe in choice.

The value of choice has been investigated in the operant laboratory by having animals select between choice and no-choice options. Catania (1975, 1980) arranged for pigeons to peck an initial key to allow them to choose between two keys (choice condition) or to peck another initial key to allow them to respond on a single key (no-choice condition). The reinforcers (food) and the rates of reinforcement were identical for the choice and no-choice conditions, but the birds preferred the alternative with two keys, showing a **preference for choice** (Catania & Sagvolden, 1980). Even without differential outcomes, it seems that making choices is preferred.

Fig. 9.8
The photograph shows a busy shopping mall, illustrating the enormous variety of choice for consumer goods in America. Shoppers can select to enter different stores within the mall and choose among the goods and services offered with each store.
Source: Shutterstock.

Preference for choice has been investigated with preschool children (Tiger, Hanley, & Hernandez, 2006). Six children were presented with three colored academic worksheets that allowed access to "choice," "no-choice," and "control" arrangements. Selection of the "choice" worksheet, followed by a correct academic response, resulted in praise and presentation of a plate of five identical candies from which the child could choose only one. For selection of the "no-choice" worksheet and a correct academic response, the child received praise and presentation of a plate with one candy on it. Selection of the "control" worksheet followed by a correct academic response resulted only in praise. Modeling and prompting by the experimenter ensured that the children always made the correct academic response. The findings showed that three children consistently selected the worksheet that led to a choice of candy, and two other children showed an initial preference for choosing which was not maintained into later sessions. One child always selected the worksheet that led to the "no-choice" plate with one candy. Thus, five of the six children preferred choice—a finding that has been replicated in a subsequent study that equated the number of edible items presented on the choice and no-choice options (Schmidt, Hanley, & Layer, 2009; see Perdue, Evans, Washburn, Rumbaugh, & Beran, 2014 for choice preference in Capuchin monkeys).

Preference for choice has also been found with university students (Leotti & Delgado, 2011). On each trial, participants pressed a black key that resulted in a choice between striped and dotted keys for monetary rewards (choice condition), or pressed a white key that led to a single key, either striped or dotted, for monetary payoffs (no-choice condition). The levels and probability of monetary reward were the same for both the choice and no-choice arrangements. In this experiment, participants selected the black key leading to choice 64% of the time, which is reliably greater than would be expected by chance (50%). In another part of the study, a stimulus (circle, triangle) preceded choosing between blue and yellow keys for monetary rewards, and a second stimulus signaled pressing the key, either blue or yellow, as selected by the computer. All monetary outcomes were equated over the two signaling procedures. Participants were asked to rate the choice or no-choice cues on a scale of liking; the results showed that the stimulus associated with choice was rated higher than the one signaling no choice. The researchers also obtained brain scans showing activity in the ventral striatum, an area related to motivation and affect, when the cue signaled upcoming choice. The researchers claimed that making a choice is inherently rewarding (see Cockburn, Collins, & Frank, 2014 for a reinforcement-learning mechanism related to the value of choice).

Although preference for choice may be innately rewarding, an experimental study indicates that a history of differential reinforcement is also involved (Karsina, Thompson, & Rodriguez, 2011). After initial training about the options, participants sat facing a computer screen and used a computer mouse to make responses (Figure 9.9). On choice trials, the computer screen initially displayed two gray boxes, one containing the words "You select" and the other containing the words "Numbers generated." A click on the "You select" box (free-choice option) deactivated the "Numbers generated" box and changed the screen image to an array of eight numbered squares. Clicks by the participant on any three of the eight squares transferred the chosen numbers to a game box and allowed for a possible win. Alternatively, a click on the "Numbers generated" box (restricted-choice option) of the initial screen deactivated the "You select" option and changed the screen to the array of eight numbers but only three, chosen by the computer, were available to transfer to the game box for a possible win. Participants were told that any number in the game box that matched three randomly drawn numbers resulted

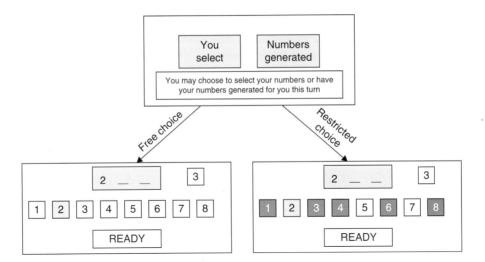

Fig. 9.9
The figure portrays the experimental arrangement for the study of reinforcement history and preference for choice. Participants selected either the "You select" or the "Numbers generated" box (top panel). Once a box was selected, the other option became dark and inoperative, and a second computer screen appeared. For example, clicking on "You select" resulted in a screen with eight numbered keys, and the participant picks any three numbers transferring them to the game box (bottom left panel). In this example, the participant has chosen the number 2 and has two more numbers to select before clicking on "Ready." The participant is told that her numbers are checked for matches and if she wins, a point is added to her total (shown as 3 points). In contrast, selection of the "Numbers generated" box deactivates the free-choice option and leads to eight keys with only three keys operative. The participant clicks on the three numbers generated for her (restricted-choice option); in this case she has already transferred the number 2 to the game box and must select 5 and 7 before clicking "Ready" and finding out if she wins and gets a point for the trial. Points are actually allocated by random-ratio (RR) schedules of reinforcement (see text for more description).

Source: Taken from A. Karsina, R. H. Thompson, & N. M. Rodriguez (2011). Effects of a history of differential reinforcement on preference for choice. *Journal of the Experimental Analysis of Behavior, 95*, pp. 189–202. Copyright 2011 held by the Society for the Experimental Analysis of Behavior. Republished with permission.

in a win, and a point registered on the screen. In fact, points for free-choice and restricted-choice arrangements were based on random-ratio (RR) schedules of reinforcement, allowing the probability of points for free-choice and restricted-choice options to be equated or varied as necessary.

Seven participants with no consistent preference, or a consistent preference for the restricted-choice option, were given differential reinforcement for selecting the free-choice option. That is, the probability of points was increased for free-choice selection and decreased for selection of the restricted-choice option. Differential reinforcement established a preference for choice in six of seven participants, a preference that lasted even when the probability of points was again equated for the free-choice and restricted-choice options. An overall conclusion is that preference for choice is a reliable finding and a history of reinforcement may establish and maintain this preference. Choice involves a history of selection between good versus better, or sooner versus later, but rarely between exactly equal alternatives. The simple option to "choose" suggests an opportunity to improve your situation, and is usually preferable (see Rost, Hemmes, & Alvero, 2014 for a concurrent-chains procedure with humans involving preference for free choice).

Preference for choice has implications for a generalized-matching analysis of behavior. The distribution of behavior on concurrent VI VI schedules of reinforcement should be biased toward a free-choice alternative, even when the rates of reinforcement on each schedule are equated. Response bias toward making a choice may help to explain opposition to government regulation of the free-market system even when government control is required. It may also relate to the continual escalation of variety and choice in all aspects of our lives. Today, we may be overwhelmed by choice, as is seen when a restaurant offers so many items that we have great difficulty choosing (Reed, Kaplan, & Brewer, 2012 reported on discounting and the value of choice, which is related to the section "Behavioral Economics, Choice, and Addiction" in this chapter). Schwartz (2004) has noted that Americans more than any other people have made a fetish out of choice, perhaps at the expense of our long-term happiness.

BEHAVIORAL ECONOMICS, CHOICE, AND ADDICTION

Choice and concurrent schedules of reinforcement have been analyzed from a microeconomic viewpoint (Rachlin, Green, Kagel, & Battalio, 1976). **Behavioral economics** involves the use of basic microeconomic concepts and principles to predict, influence, and analyze behavior in choice situations.

In economic terms, as the price of a commodity becomes more expensive, we consume less of it—our demand for it decreases. In economics, price is conceptualized as how much a commodity costs in currency (US dollars, for example). In behavior analysis, we examine price as *response cost* or how effortful it is to gain access to a reinforcer. Behaviorally, price is varied typically by a ratio schedule of reinforcement. For example, an FR 10 schedule requires 10 responses for each unit of food reinforcement (price). We could vary the price of food by changing the FR requirement. Thus, FR 20 doubles the price while FR 100 is 10 times the original price. By varying the price, it is possible to obtain a **demand curve** showing that as price increases consumption decreases (Figure 9.10). When consumption of a commodity (reinforcer) changes with price, the commodity is said to be elastic. Luxury items such as beach vacations or ski trips are highly elastic, as they are sensitive to price. Consumption of

Fig. 9.10

Illustration shows two demand curves relating price (X-axis) to consumption (Y-axis). For the inelastic demand curve, consumption does not show much decline with increases in price. In contrast, for the elastic demand curve, consumption shows an increasing decline with price.

necessities such as gasoline for cars or weekly groceries does not change much with price, and these are said to be inelastic.

The elasticity of a commodity also depends on one's income. If the cost of going to the movies increases from $10 to $20 and you live on a fixed income of $500 a week, you would probably decrease the frequency with which you go to the movies (movies are a luxury and elastic). Suppose, however, that your income increases from $500 to $1,000. You would now go to the movies as often as you did before (movies are inelastic). Therefore, elasticity and the shape of the demand curve for a given commodity (reinforcer) depends on one's income. In operant research with animals, "money" or price are operant responses and income can be created by constraining the number of responses available. In most operant experiments on choice, income is not restrained and the animal is free to vary the number of responses. If the price of a reinforcer increases (based on the schedule), the animal can adjust to the change by increasing its rate of response. If we place a rat on a budget and it only has 400 responses to allocate for a session, it is likely that the rat will be sensitive to the price. How we set up a choice situation (free response vs. budget) makes a big difference to the outcome of behavioral experiments.

Let us now turn to the classic economic problem of **substitutability**. How might a change in price of one reinforcer alter the consumption of a second reinforcer, if income is held constant? For some commodities, consumption decreases with price, but consumption of a second commodity increases. The two commodities are said to be *substitutes*. Butter and margarine can be substitutes if a shift in the price of butter results in more consumption of margarine. Beverages such as Coke and Pepsi are another example of substitutes, although conditioning of "brand loyalty" through marketing may weaken substitutability of these products. Other commodities are *independents*. As the price of one commodity increases and its consumption decreases, the consumption of a second commodity does not change. Thus, your consumption of gasoline is independent of the price of theater tickets. A third way in which commodities are related is as *complements*. As the price of one commodity increases and its consumption decreases, consumption of the other commodity also decreases. When the price of hot dogs increases and you eat less of them, your consumption of hot dog buns also decreases. Therefore, hot dogs and hot dog buns may be complements. This analysis of substitutability makes it clear that different reinforcers are not necessarily equivalent. To predict the effects of one reinforcer we often need to know its substitutability with other (alternative) reinforcers.

Substitutability, Demand, and Substance Use

Economic principles of demand and substitutability have been extended to laboratory experiments with animals working for drugs with abuse potential such as alcohol, heroin, and cocaine. For example, Nader and Woolverton (1992) showed that a monkey's choice of cocaine over food was a function of drug dose (higher doses competed more with food), but that choosing cocaine also decreased with the cocaine price (number of responses per infusion). Therefore, consumption of drugs is determined not only by the potency of the drug compared to the other options available, but also by the price of the drug. In another experiment, Carroll, Lac, and Nygaard (1989) examined the effects of a substitute commodity on the use of cocaine. Rats nearly doubled their administration of cocaine when plain water rather than a sweet solution was the option. These effects were not found in a control group that self-administered an inert saline solution—suggesting that cocaine infusion functioned as reinforcement for self-administration and the sweet solution substituted for cocaine. In this study, the presence of a substitute (sweet solution) altered the reinforcing value of the drug. One interesting observation is that at the neural level, both cocaine and the sweet solutions operate on the neural-reward centers, especially the dopamine pathway. This common neural pathway may be one of the explanations for substitutability of commodities for rats (Alsiö, et al., 2011).

In a 2005 review of behavioral economics and drug abuse, Hursh and colleagues showed that a mathematical analysis of demand curves is a way to describe the reinforcing value of different drugs (Hursh, Galuska, Winger, & Woods, 2005; see also Bentzley, Fender, & Aston-Jones, 2013 on fitting demand curves for drug self-administration). Demand curves for drugs yield quantitative measures of motivation to procure and consume drugs, and allow for comparison of the demand for similar types of drugs and between drugs that are quite different (e.g., cocaine vs. heroin). Behavioral economics also provides a way to conceptualize drug interventions, and offers new treatment strategies. For example, the analysis of substitute commodities (reinforcers) may be useful for understanding the treatment of individuals with heroin-use disorders with a drug called methadone, a less addictive opioid alternative. From an economic perspective, methadone is a partial substitute for heroin because it provides only some of the reinforcing effects of the actual drug. Also, methadone is administered in a clinical setting that is less reinforcing than the social context in which heroin is often used (Hursh, 1991). Based on this analysis, it is unlikely that availability of methadone treatment would by itself eliminate the use of heroin.

Most drugs with abuse potential cause the release of dopamine in the nucleus accumbens, which is what makes a drug reinforcing. The dopamine system, however, does not distinguish between drugs with or without abuse potential, but rather is a general locus of activation for a variety of rewarding substances. Drugs are used to excess because the reinforcing benefits are immediate and certain while the costs are uncertain and delayed. Some of the costs of excessive drug use include health problems, incarceration, interpersonal difficulties with family and friends, holding down a job or maintaining satisfactory performance in school. Thus, in a particular moment when someone is given a choice between the immediate and certain benefits of a drug versus the delayed and uncertain costs of drug use, the selection is often in favor of the now rather than later (Heyman, 2009; see also Heyman, 2014 on addiction and choice in *The New York Times*). Therefore, substance abuse can be viewed as a disorder of choice rather than a disease. This perspective is not only more optimistic than the disease model, it fits the research trends in the substance abuse literature, which shows that for most individuals, overconsumption of drugs is highly contextual.

FOCUS ON: ACTIVITY ANOREXIA AND SUBSTITUTABILITY OF FOOD AND PHYSICAL ACTIVITY

Activity anorexia is a laboratory phenomenon in which rats are placed on food restriction and provided with the opportunity to run on an activity wheel. The initial effect is that food intake is reduced, body weight declines, and wheel running increases. As running escalates, food intake drops off and body weight plummets downward, further augmenting wheel running and suppressing food intake. The result of this cycle is emaciation and, if allowed to continue, the eventual death of the animal (Epling & Pierce, 1992, 1996; Epling, Pierce, & Stefan, 1983; Routtenberg & Kuznesof, 1967). (Note that animal research ethics committees, such as Institutional Animal Care and Use Committees in the US, are in place to ensure that animals in this paradigm are carefully monitored so they do not reach extreme levels of anorexia that would place their health at risk.)

A behavioral economic model can describe the allocation of behavior between commodities such as food and physical activity (wheel running). For example, the imposed food restriction that initiates activity anorexia can be conceptualized as a substantial increase in the price of food, resulting in reduced food consumption. Low food consumption in turn increases consumption of physical activity, travel, or locomotion, which suggests that in a laboratory, food and physical activity may function as economic substitutes (see Green & Freed, 1993 for a review of these behavioral economic concepts).

In two experiments, Belke, Pierce, and Duncan (2006) investigated how animals choose between food (sucrose) and physical activity (wheel running). Rats were exposed to concurrent VI 30-s VI 30-s schedules of wheel running and sucrose reinforcement. Sucrose solutions varied in concentration (2.5, 7.5, and 12.5%). As the concentration of sucrose increased, more behavior was allocated to sucrose and more reinforcements were obtained from that alternative. Allocation of behavior to obtain wheel-running reinforcement decreased somewhat, but the rate of reinforcement did not change. The results suggested that food-deprived rats were sensitive to changes in food supply (sucrose concentration) while continuing to engage in physical activity (wheel running). In a second study, rats were exposed to concurrent variable-ratio (VR VR) schedules of sucrose vs. wheel running, wheel running vs. wheel running, and sucrose vs. sucrose reinforcement. For each pair of reinforcers, the researchers assessed substitutability by changing the prices for consumption of the commodities. The results showed that sucrose substituted for sucrose and wheel running substituted for wheel running, as would be expected. Wheel running, however, did not substitute for sucrose—the commodities were independent—but sucrose partially substituted for wheel running.

The partial substitutability of sucrose for wheel running in the experiments by Belke et al. (2006) reflects two energy-balance processes: the initiation and maintenance of travel or locomotion induced by loss of body weight and energy stores (wheel running does not substitute for food), and the termination of travel or locomotion as food supply increases (food does substitute for wheel running). In terms of activity anorexia, the fact that travel or locomotion does not substitute for food ensures that animals with low energy stores keep going on a food-related trek, even if they eat small amounts along the way. As animals contact stable food supplies, the partial substitutability of food for wheel running means that travel or locomotion would subside as food intake and body stores return to equilibrium. Behavioral economic analysis provides one way to understand the activity anorexia cycle in terms of substitutability of food and physical activity (see also Belke & Pierce, 2009, who show that demand for food (sucrose) becomes less elastic, and wheel running becomes more elastic, at low body weight; inelastic-demand functions for food by wheel-running mice also have been reported by Atalayer and Rowland, 2011).

Delay Discounting of Reinforcement Value

Demand-curve analysis of reinforcement value makes an important contribution to basic research and applications. Another problem addressed by behavioral economics involves devaluation or discounting of reinforcement value. **Delay discounting** involves choosing between small, immediate and large, delayed rewards. If you are watching your weight, you often must choose between the immediate reinforcement from a piece of chocolate cake or the long-term reinforcement of a healthy body weight and good health. If you are like most of us, you find yourself in that moment valuing the cake more and the healthy body weight less so. These kinds of decision traps are clarified when expressed in terms of immediate or delayed amounts of money.

Suppose that you have just received notification that you hit the lottery and won $100,000. You are told the full amount of your winnings will be offered to you in 10 months. However, the entity that organized the lottery may not be too excited to give you $100,000 (as they know about delay discounting), so they offer you another option: you could wait the 10 months and receive the $100,000 (called the larger, later or LL option) or you can take the money right now, but you will only receive $40,000 (called the smaller, sooner option, or SS). If you accept the SS option, you will have discounted your lottery winnings by more than 60%. That is, your winnings have lost more than 60% of their value due to the 10-month delay. Let's consider you would be happy to wait for the $100,000 in this situation, but they counter offer with this choice: would you rather have $100,000 in 10 months (LL) or $60,000 now (SS). Here, you would discount your winnings by 40%.

This higher amount of the SS option is tempting, but you decide you still want to wait for the LL. But what if the lottery makes one more counter offer: $100,000 in 10 months (LL) or $80,000 now (SS)? Here, you may decide that the $80,000 now (a 20% devaluing) is a better option than waiting for another $20,000 in 10 months; you decide to take the SS. Because the SS is now preferred over the LL option, we call this a *preference reversal*. We can use a procedure of gradually increasing the amount of the SS option offered to find the point where you switch from the LL option to the SS option. The point at which the LL and SS options are equally valued (chosen with 50% preference) is called an *indifference point* (also called the *discounted value*). For this example, the indifference point can be calculated by taking the median of the values that flank the preference reversal. In other words, when you switch from the LL when the option for $60,000 is available to the SS when $80,000 is available, the point of indifference (where the SS and LL values are equivalent) is between $60,000 and $80,000. Therefore, $70,000 is the indifference point when $100,000 is available at 10 months. In other words, $70,000 now is valued the same as $100,000 in 10 months.

We can repeat this procedure to find the indifference points when $100,000 is delayed for different amounts of time (e.g., 0 to 40 months). Usually, we obtain seven or eight indifference points over the range of the delays. At each delay, the indifference point is plotted on a graph and a mathematical curve, called the hyperbolic discounting curve, is fitted to the points. Figure 9.11 shows the hypothetical discounting curve for your lottery winnings with delay in months on the X-axis and the indifference point (also called discounted value) on the Y-axis. The indifference point or present value for $100,000 at 10 months' delay is shown as $70,000 (from previous paragraph). Six other indifference points (small circles) at different delays are also plotted. Notice that the present value drops quickly at first and then declines more slowly—the shape of the curve is *hyperbolic*.

Hyperbolic Discounting Equation

Considerable research in both human and nonhuman choice has supported a **hyperbolic discounting equation**, described by Mazur (1987).

$$V_d = \frac{A}{1+kd} \qquad \text{(Equation 9.4)}$$

In Equation 9.4, we are predicting discounted values, V_d, of the reinforcer. The amount of the reinforcer, A, is $100,000 in our example, and the value d is the delay—the variable on the X-axis. The value k is called the *discounting rate*, which must be estimated to fit a curve to the indifference points (data) obtained from the experiment. The best-fitting curve is fitted to the data by a statistical technique called nonlinear regression (available from standard statistical programs such as SPSS, R or GraphPad Prism).

To obtain the indifference points for pigeons, Mazur (1987) used an *adjusting-delay procedure*. The pigeon chooses (pecks) between left and right colored (red and green) keys that provide a small food reward delivered after a fixed delay, or a larger food reward delivered after an adjusting delay. The experiment involves forced-choice and free-choice trials. On two forced-choice trials, the bird is forced to obtain the small reward with fixed delay on one trial and the large reward with adjusting delay on the other trial. The forced choice trials are conducted to give the birds experience with each option. The forced-choice trials are followed by two free-choice trials in which the bird is free to choose between the left and right colored keys. If the small reward of fixed delay is selected on both choice trials, the indifference point has not been obtained and the delay (in 1-s steps) to the larger reward is reduced on the next block of forced- and free-choice trials. On the other hand, if the bird chooses the larger adjusting-delay reward on free-choice trials, the delay is increased in the next block of trials. Using this adjusting-delay procedure, it is possible to find the stable indifference point. The mean over several sessions of stability is used as the indifference point that is plotted on a graph. To obtain other indifference points, new fixed-delay values are set for the small reward, and the delay to the large-food reward is again adjusted over trial blocks.

Progressive-delay procedures also have been used to estimate indifference points. A *progressive-delay procedure* involves holding the delay on one alternative constant while

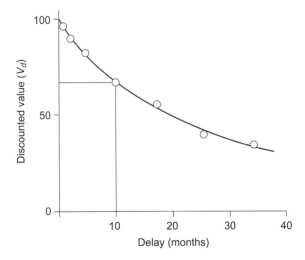

Fig. 9.11
A hypothetical discounting curve is shown based on the lottery money problem described in the text. The X-axis shows delays in months and the Y-axis is the discounted value (in thousands of dollars). The value of $100,000 plunges with delay. The intersecting point on the curve is the indifference point or discounted value of $100,000 delayed 10 months and is $70,000. In a typical experiment, a number of indifference points are obtained for $100,000 at different delays (months) and a mathematical curve is fitted to the points.

the delay on the other alternative is progressively increased across blocks of trials within an experimental session. This kind of discrete-trials experiment has been used to identify neural mechanisms related to delay discounting. In one study, rats chose between a fixed delay followed by a small drop of sucrose solution and a progressive delay leading to a large drop of sucrose (Bezzina et al., 2007). The progressive delay began equal to the fixed (8-s) delay and then increased 75% over each successive block of trials. The delay at which a rat chose the progressive alternative 50% of the time defined the indifference point. Rats with lesions to the nucleus accumbens (NAc) showed shorter indifference points than sham-operated controls, suggesting that impairment of the NAc increases the rate of discounting of the large, delayed reinforcer (see Madden & Johnson, 2011 on indifference points with nonhumans).

To establish delay discounting with humans, participants are often asked to choose between hypothetical amounts of money at given delays. In the original study by Howard Rachlin and his colleagues, participants were asked to choose between a hypothetical $1,000 right now and $1,000 after delays of 1 month to 50 years (Rachlin, Raineri, & Cross, 1991; see Rodzon, Berry, & Odum, 2011 on indifference procedures used in delay discounting with humans). At each delay value, the researchers used an *adjusting-amount procedure*, whereby the amount of money received immediately was adjusted relative to a fixed $1,000 at a set delay ($990 now vs. $1,000 in 1 month's time, adjusted in steps toward $1 now and $1,000 in 1 month's time). At each delay, for half of the participants the amount of immediate reward was adjusted downward and then repeated with an ascending progression of values. The other participants received the ascending amounts first and then the descending progression. The indifference point was calculated as the average amount of the immediate reward, based on the value at which the participant switched to the immediate reward (downward progression) and no longer preferred the delayed reward (upward progression). Seven indifference points were calculated for the $1,000 delayed for 1 month, 6 months, 1 year, 5 years, 10 years, 25 years, and 50 years. The hyperbolic equation (Equation 9.4) was calculated, and the indifference points were fitted as described previously (use of real rewards has confirmed the findings with hypothetical outcomes; see Madden & Johnson, 2011). In contrast, evolutionary models (which are beyond the scope of this textbook) account for why organisms may sometimes be short-sighted, often acting "impulsively" (choosing the SS option in the laboratory). Evolutionary models show that apparently choosing the SS option (steep delay discounting) is a special case of decision rules that increased fitness in everyday ecological settings (Fawcett, McNamara, & Houston, 2012). Other recent studies show this as well. A study by Rodriguez et al. (2021) found that present-day individuals who were economically challenged were more likely to select SS food options vs. LL food options. Therefore, under conditions of food scarcity, choosing the SS may aid in survival.

Delay Discounting Applications: Drug Abuse, Gambling, and Obesity

There are conditions, though, in which choosing the SS option may be harmful in the long run. Hyperbolic discounting has been extended to problems of "impulsive behavior", especially behavior labeled as addictive (Madden & Bickel, 2009). One trend in the literature is that people who smoke cigarettes or are diagnosed with substance use disorders with alcohol and other drugs prefer SS options over LL ones and therefore show steeper discounting of future outcomes (MacKillop et al., 2011 provides a meta-analysis of delay discounting and substance abuse). Some of this research is described here.

Using an adjusting-amount procedure, individuals who smoke cigarettes (users of nicotine) were found to have higher discounting rates (k-values of Equation 9.4) for a hypothetical delayed reward ($1,000) than non-smokers (Bickel, Odum, & Madden, 1999). This result also holds for people who drink alcohol heavily. Compared with light social-drinking controls, both heavy drinkers and problem drinkers show higher discounting of the future rewards (Vuchinich & Simpson, 1999). The difference in discounting between light-drinking controls and problem drinkers was greater than that between controls and heavy drinkers, suggesting that both alcohol consumption, and the consequences related to this consumption, affect delay discounting.

Additional studies have investigated delay discounting in those with substance use disorders. Individuals who were opioid-dependent discounted future money more than nondependent controls matched for age and gender (Madden, Petry, Badger, & Bickel, 1997). Those with opioid dependence also discounted future money ($1,000) more than they did $1,000 worth of heroin in the future. One possibility is that future money does not buy you heroin now, but future heroin ensures continued access to the drug (having a stash on hand). Individuals who use cocaine also show steeper discounting of delayed rewards than controls, but no differences in discounting rate were observed between those who use cocaine actively and those abstaining from cocaine use for at least a month (Heil, Johnson, Higgins, & Bickel, 2006). A history of cocaine use seems to be more related to discounting of delayed reinforcement value than current use of the drug.

Research has also documented high delay discounting in those who gamble compulsively, even in the absence of substance abuse. Compulsive gamblers face the choice between the behavior of gambling versus the delayed benefits of financial and interpersonal stability. As with those with substance use disorders, gamblers discount the value of future rewards more steeply than matched controls. Also, moderately high discounting rates are linked to developing problem behavior with either drugs or gambling. Even higher discounting rates predict multiple problems involving both drugs and gambling (Petry & Madden, 2011).

While DD, especially for monetary outcomes, has been studied largely as a behavioral characteristic of those with substance use disorders, it has also been applied to the study of obesity. When one considers food choices now which are especially difficult to refuse, such as those with high sugar content versus the delayed outcome of having a healthy body, this makes intuitive sense. People who are obese show greater preferences for SS money over LL money compared to those who are healthy weight (see Amlung, Petker, et al., 2016 for meta-analysis). Moreover, when faced with SS versus LL later food choices, obesity status also predicts steeper discounting with food-related outcomes, in which participants choose between SS bites of food versus LL bites of food (Hendrickson et al., 2015; Rasmussen et al., 2010). This has also been shown with obese and lean Zucker rats—obese rats prefer the SS food outcome over waiting for the LL one compared to lean controls (Boomhower, et al., 2013).

Because steeper delay discounting has been observed across a range of health behaviors, it has been referred to as a *trans-disease process* (Bickel et al., 2019), meaning that the devaluing of delayed outcomes underlies a variety of health conditions in which prevention is a LL outcome. Prevention behaviors, such as eating healthy, exercising, and managing stress compete with the immediately available effects of reinforcers like unhealthy food or drug consumption. Having good health is a life-long practice in which the immediate reinforcers of drugs and junk food, for example, must be consumed sparingly.

While it may be tempting to talk about substance abuse, obesity, and gambling as addictions, behavior analysts are careful about the use of this term for a number of reasons. One, the term addiction is a description, not an explanation. When we use it to explain someone's behavior, even when it looks irrational (e.g., someone is gambling their life savings, which compromises their future), it takes us further away from the real causes when we state, "They gamble because they have addictions." Behavior analysts seek to find the causes of behaviors and using a construct as explanation halts the search for the actual conditions that lead to strong preferences for an immediate reinforcer.

Reinforcer pathologies. Recently, in place of "addiction", behavior analysts have considered a different term for individuals who struggle with food, drug, or other reinforcer consumption to a point where it compromises their long-term health. The term **reinforcer pathology** is applied to individuals who 1) highly value consumption of a specific reinforcer (such as a drug), as evidenced by inelastic demand and 2) they also exhibit steep delay discounting for the reinforcer (Bickel et al., 2011; Jarmolowicz, et al., 2016). Research shows that those with substance abuse disorders and obesity tend to demonstrate these behavioral characteristics for drugs and food, respectively (Bickel et al., 2011; Bickel et al., 2020; DeHart et al., 2020; Lemley et al., 2016).

The term reinforcer pathology is useful for a number of reasons. It places the focus on the behavioral economic conditions of high reinforcer consumption, which assists researchers in identifying the specific causes (e.g., high availability of the drug reinforcer; fewer alterative non-drug reinforcers). The term de-emphasizes inner causes that are implicit to the use of terms such as "addiction". Because of this specificity to behavioral processes, high valuation and delay discounting of a reinforcer are more treatable than an addiction. Many assume a person with an addiction (an internal cause) will always have the addiction ("once an addict, always an addict"), even if they are abstinent for long periods of time. This seems unfairly pejorative to the person who has been drug-abstinent for a decade or longer. Indeed, the vast majority of individuals who have substance abuse problems quit on their own (called natural recovery) and for varying reasons, like they discover they are pregnant or they "age out" of the drug (Sobell, et al., 2000; Substance Abuse and Mental Health Services Administration, 2014). As to the latter, most drug use occurs during the late teens and 20s (Heyman, 2009; National Institute on Drug Abuse, 2021). With age, other life-event reinforcers (e.g., marriage, career, children) unfold and compete for behavioral allocation. As such, drug use declines with age (Substance Abuse and Mental Health Services Administration, 2014). Therefore, the term reinforcer pathology is not only a more accurate term than addiction, but it is a kinder, more hopeful term, and fits the more contextual nature of heavy reinforcer consumption.

Delay Discounting: Trait or State?

Is it possible, though, that those with substance use disorders, gambling problems, or obesity have a general tendency to value SS preferences? On the other side, are there individuals who tend to have an easier time than others waiting for LL outcomes? It appears that delay discounting has trait-like properties. One area of research that supports this includes twin studies, which accounts for genetic variation in discounting. For example, one study reported up to 62% heritability of delay discounting by monozygotic and dizygotic twins at two age points (Anokhin, Grant, Mulligan, & Heath, 2015).

Studies on test-retest reliability also support trait-like behavior. When delay discounting is tested at one time point and then again at a second time point, discounting is quite consistent and stable across time (Odum, 2011a; Odum, 2011b). Interestingly, there is also a tendency for those who steeply discount one type of delayed reward, such as money, to also steeply discount other types of rewards, such as food or drug (Odum et al., 2020), a phenomenon called *cross-commodity discounting*. Hereditary and neural factors, however, do seem to play a role in delay-discounting situations and behavioral measures of delay discounting appear to be reliable markers for substance use disorders, although further research is required to clarify these links (Bickel, Koffamus, Moody, & Wilson, 2014; Stevens & Stephens, 2011; Winstanley, 2011).

Though discounting has trait-like qualities, there is also evidence that it can be altered with different types of independent variables or interventions. In a recent review and meta-analysis, Rung and Madden (2018) showed that a number of variables, including episodic future thinking (imagining yourself in the future in a particular situation), priming, and exposure to nature scenes can switch preferences from SS money to LL money. Mindful eating is also a strategy that has been shown to switch SS food preferences to the LL (Hendrickon & Rasmussen, 2013; Hendrickson & Rasmussen, 2017).

At the present time, the evidence is mounting for trait-based variation in delay discounting; research also clearly shows that discounting patterns depend on the contingencies arranged at the moment of choice, as well as one's history of learning about immediate and delayed reinforcement. Therefore, discounting also has state-like properties.

Self-Control, Preference Reversal, and Commitment

Contingencies of immediate and delayed reinforcement are involved frequently in common problems of self-control. Students often face the choice of going out to party or staying at home and "hitting the books." Often, when given these options, students pick the immediate reward of partying with friends over the delayed benefits of studying, learning the subject matter, and achieving high grades. When a person (or other animal) selects the smaller, immediate (SS) payoff over the larger, delayed (LL) benefits, we have traditionally called the pattern **impulsive behavior**. On the other hand, a person who chooses the LL reward while rejecting the SS payoff is said to show **self-control**. In term of a student's choice of whether to party or to study, choosing to party with friends is considered impulsive behavior, while choosing to stay at home and study is self-control behavior.

It should be noted that there are some limitations with using the terms "impulsive" and "self-controlled" to describe behavior, as they have value judgments placed on them. For example, "self-controlled" seems subjectively less judgmental than "impulsive". But consider that there are situations in which an "impulsive choice" is actually a rational, reasonable choice. For example, recent research (Rodriguez, et al., 2021) shows that individuals who are food insecure (i.e., have economic or physical barriers to nutritious food) are more likely to choose SS food over LL food outcomes. In a situation in which food is uncertain and survival is at stake, it is rational for someone to take the more immediate food option. Therefore, the term "impulsive" really isn't relevant to this situation. It is also important to point out that delay

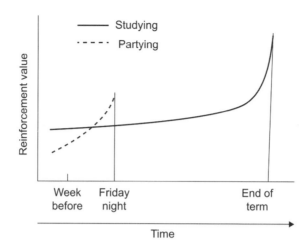

Fig. 9.12
A depiction of preference reversal for studying vs. partying is depicted. Reinforcement value of studying (solid black curve) declines rapidly and then more gradually the farther back in time the student is from the end of term. Notice, the value of studying is lower than going to the party (dashed curve) on the night of the party. However, the value of going to the party declines below the value of studying a week before the party. That is, the student's preference reverses; they prefer to study.

discounting has not historically been strongly co-related with other measures of impulsivity such as response inhibition or risk taking (see Strickland & Johnson, 2021); therefore behavior analysts are now considering delay discounting as a process separate from the construct of impulsivity.

One of the interesting things about "self-control" situations is that our preferences change over time. We may value studying over partying a week before the party, but value partying when the night of the party arrives. Based on a consideration of hyperbolic discounting, Howard Rachlin (1970, 1974) and George Ainslie (1975) independently suggested that such *preference reversals* could be analyzed as changes in reinforcement value with increasing delay. In other words, *reinforcement value decreases hyperbolically as the delay between making a choice and obtaining the reinforcer increases.*

As shown in Figure 9.12, the value of studying on the Friday night of the party (choice point) is lower than the value of having fun with friends (partying), because the payoffs for studying (learning and good grades) are delayed. At the end of term, the value of studying is high, but it declines hyperbolically—ensuring that the value of studying is less than the value of partying on the Friday night (the choice point). Delay discounting requires that we usually behave "impulsively" at the choice point by selecting the immediate, short-term payoff. In addition, the value function—value as a function of amount of reward—is steeper with long delays, perhaps explaining why studying is steeply discounted at the end of term in our example (Rachlin, Arfer, Safin, & Yen, 2015).

If we move back in time from the choice point to a week before the party, however, the value of studying relative to partying reverses. Thus, adding delay to each reinforcement option before a choice is made reverses the value of the alternative reinforcers. More generally, at some time removed from making a choice, the value of the smaller, immediate reinforcer will be less than the value of the larger, delayed reward—indicating a **preference reversal**. When

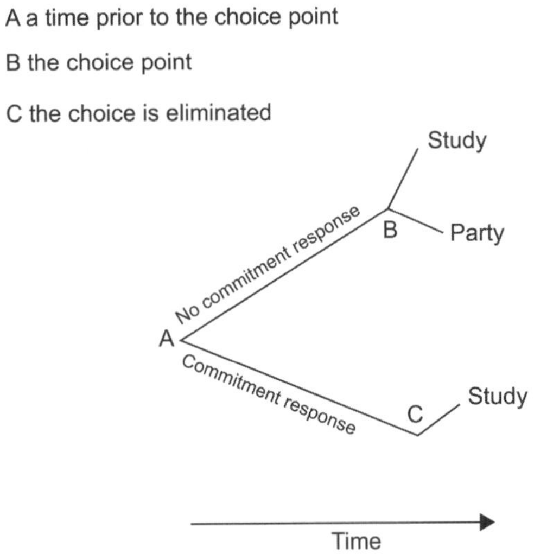

A a time prior to the choice point

B the choice point

C the choice is eliminated

Fig. 9.13
Self-control through commitment is depicted based on preference reversal (as in Figure 9.12). The student will make a commitment to study at point A (a week before the party) because the value of studying is higher than partying, but not at the choice point B (the night of the party). The commitment response removes going to the party as an option on the night of the party (C).

preference reversal occurs, people (and other animals) will make a commitment response to forgo the smaller, immediate reward and lock themselves into the larger, delayed payoff. Figure 9.13 shows the commitment procedure for eliminating the choice between studying and partying a week before the party. The **commitment response** is some behavior emitted at a time prior to the choice point that eliminates or reduces the probability of impulsive behavior. A student who has invited a classmate over to study on the Friday night of the party (commitment response) ensures that they will "hit the books" and give up the option of partying when the choice arrives (see Locey & Rachlin, 2012 for the use of commitment in a prisoner's dilemma game).

Preference reversal and commitment occur over extended periods in humans and involve many complexities. In animals, delay of reinforcement by a few seconds can change the value of the options, instill commitment, and ensure self-control over impulsiveness. As an example of preference reversal, consider an experiment by Green, Fisher, Perlow, and Sherman (1981) in which pigeons responded on two schedules of reinforcement, using a trials procedure. The birds were given numerous trials each day. On each trial a bird made a choice by pecking one of two keys. A single peck at the red key resulted in 2 s of access to grain, while a peck to the green key delivered 6 s of access to food. The intriguing aspect of the experiment involved adding a brief delay between a peck and the delivery of food. In one condition, there was a 2-s delay for the 2-s reinforcer (red key) and a 6-s delay for 6 s of access to food (green key). The data indicated that birds were impulsive, choosing the 2-s reinforcer on nearly every trial and losing about two-thirds of their potential access to food.

In another procedure, 18 additional seconds were added to the delays for each key, so that the delays were now 20 s for the 2-s reinforcer and 24 s for the 6 s of access to food. When the birds were required to choose this far in advance, they pecked the green key that delivered

6 s of access to food on more than 80% of the trials. In other words, the pigeons showed preference reversal and self-control when both reinforcers were further away.

Other research by Ainslie (1974) and Rachlin and Green (1972) shows that pigeons can learn to make a commitment response, thereby reducing the probability of impulsive behavior. Generally, animal research supports the Ainslie–Rachlin principle and its predictions. One implication is that changes in reinforcement value over extended periods also regulate delay discounting in humans (see Ainslie, 2005; Beeby & White, 2013; Rachlin, 2000; Rachlin & Laibson, 1997). In this way, behavior principles may help to explain the use of credit cards in our society (buy now, pay more later), the fact that most people have trouble saving their money, and the world financial crisis of 2008 that involved short-term financial gains and the discounting of the long-term economic costs (Lewis, 2010).

ADVANCED SECTION: QUANTIFICATION OF CHOICE AND THE GENERALIZED MATCHING RELATION

The proportional matching equations (Equations 9.1, 9.2, and 9.3) describe the distribution of behavior when alternatives differ only in rates of reinforcement. In complex environments, however, other factors also contribute to choice and preference (see Poling, Edwards, Weeden, & Foster, 2011 for different matching equations, empirical support, and a brief summary of applications).

Sources of Error in Matching Experiments

Suppose that a pigeon has been trained to peck a yellow key for food on a single VI schedule. This experience establishes the yellow key as a discriminative stimulus that controls pecking. In a subsequent experiment, the animal is presented with concurrent VI VI schedules of reinforcement. The left key is illuminated with a blue light and the right key with a yellow one. Both of the VI schedules are programmed to deliver 30 reinforcers per hour. Although the programmed rates of reinforcement are the same, the bird is likely to distribute more of its behavior to the yellow key. In this case, stimulus control exerted by yellow is an additional variable that affects behavioral choice.

In this example, the yellow key is a known source of experimental response bias that came from the bird's history of reinforcement. Many unknown variables, however, also affect choice in a concurrent-operant setting. These factors arise from both the biology and environmental history of the organism. For example, sources of error may include different amounts of effort for the responses, qualitative differences in reinforcement (e.g., food vs. water), a history of punishment, a tendency to respond to the right alternative rather than the left, and a difference in sensory capacities.

Matching of Ratios

To include these and other conditions within matching theory, it is useful to express the matching relation in terms of ratios rather than proportions. A simple algebraic transformation of Equation 9.1 yields the matching equation in terms of ratios:

1 Proportion equation: $B_a/(B_a + B_b) = R_a/(R_a + R_b)$.
2 Cross-multiplying: $B_a/(R_a + R_b) = R_a/(B_a + B_b)$.
3 Then: $(B_a \times R_a) + (B_a \times R_b) = (R_a \times B_a) + (R_a \times B_b)$.

4 Canceling: $B_a \times R_b = R_a \times B_b$.
5 Ratio equation: $B_a/B_b = R_a/R_b$.

In the ratio equation, B_a and B_b represent the rate of response or time spent on the A and B alternatives. The terms R_a and R_b express the rates of reinforcement. When relative rate of response matches relative rate of reinforcement, the ratio equation is simply a restatement of the proportional form of the matching relation.

The Power Law

A generalized form of the ratio equation may, however, be used to handle the situation in which unknown factors influence the distribution of behavior. These factors produce systematic departures from ideal matching, but may be represented as two constants (parameters) in the *generalized matching equation*, as suggested by Baum (1974b):

$$B_a/B_b = k(R_a/R_b)^a. \qquad \text{(Equation 9.5)}$$

In this form, the matching equation is represented as a **power law** in which the coefficient k and the exponent a are values that represent two potential *sources of error* for a given experiment. When these parameters are equal to 1, Equation 9.5 is the simple ratio form of the matching relation and is mathematically equivalent to the proportional matching equation.

Bias

Baum (1974b) suggested that variation in the value of k from 1 reflects preference caused by some factor that has not been identified. For example, consider a pigeon placed in a chamber in which two response keys are available. One of the keys has a small dark speck that is not known to the experimenter. Recall that pigeons have excellent visual acuity and a tendency to peck at stimuli that approximate a piece of grain. Given a choice between the two keys, a pigeon could show a systematic response *bias* for the key with a spot on it. In the generalized matching equation, the presence of such bias is indicated by a value of k different from 1. Generally, **bias** is some unknown asymmetry between the alternatives in a given experiment that affects preference over and above the relative rates of reinforcement.

Sensitivity

When the exponent a takes on a value other than 1, another source of error is present. A value of a greater than 1 indicates that changes in the response ratio (B_a/B_b) are larger than the changes in the ratio of reinforcement (R_a/R_b). Baum (1974b) called this outcome **overmatching** because relative behavior is higher than predicted from the relative rate of reinforcement. Although overmatching has been observed, it is not the most common result in behavioral-choice experiments. The typical outcome, especially for humans, is that the exponent a takes on a value of less than 1 (Baum, 1979; Davison & McCarthy, 1988; Killeen, 2015; Kollins et al., 1997; McDowell, 2013). This result is described as **undermatching** and refers to a situation in which the changes in the response ratio are less than changes in the reinforcement ratio. An interesting observation from a study by Buckley and Rasmussen (2012) found that obese Zucker rats, who find food especially reinforcing, have higher sensitivity to differences in food densities (sensitivity is above 0.8), compared to lean Zucker rats, who also undermatch, but have lower sensitivity (between 0.5 and 0.6). This difference is explained entirely by genes for obesity.

One interpretation of undermatching is that changes in relative rates of reinforcement are not well discriminated by the organism (Baum, 1974b). Sensitivity to the operating schedules is adequate when exponent *a* is close to 1 in value. Based on extensive data from 23 different studies, Baum (1979, p. 269) concluded that values of the exponent *a* between 0.90 and 1.1 are *good approximations to matching*. An organism may not detect subtle changes in the schedules, and its distribution of behavior lags behind the current distribution of reinforcement. This slower change in the distribution of behavior is reflected by a value of exponent *a* less than 1. For example, if a pigeon is exposed to concurrent VI VI schedules without a COD procedure, a likely outcome is that the bird will rapidly and repeatedly switch between alternatives. This rapid alternation usually results in the pigeon being less sensitive to changes in the reinforcement ratio, and undermatching is the outcome. A COD contingency, however, may be used to prevent the superstitious switching and increase sensitivity to the rates of reinforcement on the alternatives. Thus, the COD is a procedure that reduces undermatching as reflected by values of *a* close to 1.

Although problems of discrimination or sensitivity may account for deviations of the exponent *a* away from 1, some researchers believe that undermatching is so common that it should be regarded as an accurate description of choice and preference (Davison, 1981). In fact, the average estimate of the exponent *a* value is 0.80 across experiments with different species, which is outside the range of 0.90 to 1.1 for good matching. A generalized matching equation for time, however, produces estimates of exponent *a* closer to ideal matching, suggesting that time spent on alternatives, rather than rate of response, is the more fundamental measure of choice (Baum, 1979). On the other hand, undermatching could turn out to be the correct description of what animals do when allocating behavior to alternative sources of reinforcement—indicating that perfect matching (sensitivity = 1) is not the lawful process underlying behavioral choice and preference (Killeen, 2015). Some behavior analysts have not adopted this position, still viewing near-perfect matching as a fundamental, lawful process. Nonetheless, the origin of undermatching is still under discussion.

Estimating Bias and Sensitivity

William Baum (1974b) (Figure 9.14) formulated the **generalized matching relation** as shown in Equation 9.5. In the same article, he suggested that Equation 9.5 could be represented as a straight line when expressed in logarithmic form. In this form, it is relatively easy to portray and interpret deviations from matching (i.e., bias and sensitivity) on a line graph. Baum suggested that in linear form, the value of the slope of the line measured sensitivity to the reinforcement schedules, while the intercept reflected the amount of bias.

Algebra for a Straight Line

The algebraic equation for a straight line[1] is:

$$Y = m + n(X)$$

In this equation, n is the slope and m is the intercept. The value of X (horizontal axis) is varied, and this changes the value of Y (vertical axis). Assume that X takes on values of 1 through 10, $m = 0$, and $n = 2$. When X is 1, the simple algebraic equation is $Y = 0 + 2(1)$ or $Y = 2$. The equation can be solved for the other nine values of X and the X,Y pairs plotted on a graph. Figure 9.15 is a plot of the X,Y pairs over the range of the X values. The rate at which the line rises, or the *slope* of the line, is equal to the value of n and has a value of 2 in

Fig. 9.14
William Baum.
Reprinted with permission.

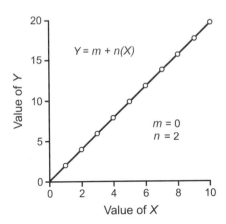

Fig. 9.15
The algebraic equation for a straight line is plotted. Slope is set at 2.0 and intercept at zero.

this example. The *intercept m* is zero in this case, and is the point at which the line crosses the *Y*-coordinate.

A Log-Linear Generalized Matching Equation

To write the matching relation as a straight line, Baum suggested that Equation 9.5 be expressed in the logarithmic form of Equation 9.6:

$$\log(B_a/B_b) = \log k + [a \times \log(R_a/R_b)]. \qquad \text{(Equation 9.6)}$$

Rft/hr A	Rft/hr B	(Ra/Rb)	X value log (Ra/Rb)	Slope (a)	Intercept (log k)	Y value log (Ba/Bb)
			MATCHING			
5	5	1	0.00	1.00	0.00	0.00
30	5	6	0.78	1.00	0.00	0.78
100	5	20	1.30	1.00	0.00	1.30
600	5	120	2.08	1.00	0.00	2.08
			UNDERMATCHING			
5	5	1	0.00	0.50	0.00	0.00
30	5	6	0.78	0.50	0.00	0.39
100	5	20	1.30	0.50	0.00	0.65
600	5	120	2.08	0.50	0.00	0.14
			BIAS			
5	5	1	0.00	1.00	1.50	1.50
30	5	6	0.78	1.00	1.50	2.28
100	5	20	1.30	1.00	1.50	2.80
600	5	120	2.08	1.00	1.50	3.58

Fig. 9.16
Application of log-linear matching equation (Equation 9.6) to idealized experimental data (scatterplot) is shown. Reinforcements per hour (Rft/h) for alternatives A and B, the ratio of the reinforcement rates (R_a/R_b), and the log ratio of the reinforcement rates (X values) are given in the table. The log ratios of the response rates (Y values) were obtained by setting the slope and intercept to values that produce matching, undermatching, or bias.

Notice that in this form, log(B_a/B_b) is the same as the Y value in the algebraic equation for a straight line. Similarly, log(R_a/R_b) is the same as the X value. The value a is the same as n and is the slope of the line. Finally, log k is the intercept, as is the term m in the algebraic equation.

The Case of Matching
Figure 9.16 shows the application of Equation 9.6 to idealized experimental data. The first and second columns give the number of reinforcements per hour delivered on the A and B alternatives. Notice that the rate of reinforcement on alternative B is held constant at 5 per hour, while the rate of reinforcement for alternative A is varied from 5 to 600 reinforcements per hour. The relative rate of reinforcement is shown in column 3, expressed as a ratio (i.e., R_a/R_b). For example, the first ratio for the data labeled "matching" is 5/5 = 1, and the other ratios may be obtained in a similar manner.

The fourth column is the logarithm of the ratio values. Logarithms are obtained from a calculator and are defined as the exponent of base 10 that yields the original number. For example, the number 2 is the logarithm of 100, since 10 raised to the second power is 100. Similarly, in Figure 9.16 the logarithm of the ratio 120 is 2.08, because 10 to the power of 2.08 equals the original 120 value.

Notice that logarithms are simply a *transformation of scale* of the original numbers. Such a transformation is suggested because logarithms of ratios plot as a straight line on X, Y coordinates, while the original ratios may not be linear. Actual experiments involve both positive and negative logarithms, since ratios may be less than 1 in value. For simplicity, the constructed examples in Figure 9.16 only use values that yield positive logarithms.

Columns 5 and 6 provide values for the slope and intercept for the log-ratio equation. When the relative rate of response is assumed to match (or equal) the relative rate of reinforcement, the slope (a) assumes a value of 1.00 and the value of the intercept (log k)

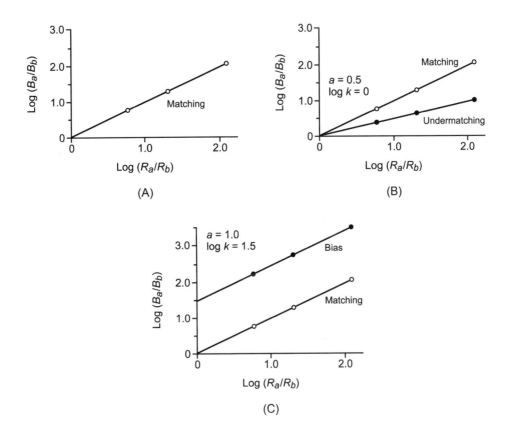

Fig. 9.17
(A) An X-Y plot of the data for "Matching" from Figure 9.16. The value of the slope is set at 1 ($a = 1$), and the intercept is set at zero ($\log k = 0$). The matching line means that a unit increase in relative rate of reinforcement [$\log(R_a/R_b)$] produces a unit increase in relative rate of response [$\log(B_a/B_b)$]. (B) An X-Y plot of the data for "Undermatching" from Figure 9.16. The value of the slope is set at less than 1 ($a = 0.5$), and the intercept is set at zero ($\log k = 0$). Undermatching with a slope of 0.5 means that a unit increase in relative rate of reinforcement [$\log(R_a/R_b)$] produces a half-unit increase in relative rate of response [$\log(B_a/B_b)$]. (C) An X-Y plot of the data for "Bias" from the data of Figure 9.16. The value of the slope is set at 1 ($a = 1$), and the intercept is more than zero ($\log k = 1.5$). A bias of this amount indicates that the new plotted data on X-Y coordinates is deflected 1.5 units from the matching line.

is zero. With slope and intercept so defined, the values of Y or $\log(B_a/B_b)$ may be obtained from the values of X or $\log(R_a/R_b)$, by solving Equation 9.5. For example, the first Y value of 0.00 for the final column is obtained by substituting the appropriate values into the log-ratio equation, $\log(B_a/B_b) = 0.00 + [1.00 \times (0.00)]$. The second value of Y is 0.78, or $\log(B_a/B_b) = 0.00 + [1.00 \times (0.78)]$, and so on.

Figure 9.17 (upper left) plots the "matching" data. The values of X or $\log(R_a/R_b)$ were set for this idealized experiment, and Y or $\log(B_a/B_b)$ values were obtained by solving Equation 9.5 when $a = 1$ and $\log k = 0$. Notice that the plot is a straight line that rises at 45 degrees. The rate of rise in the line is equal to the value of the slope (i.e., $a = 1$). This value means that a unit change in X (i.e., from 0 to 1) results in an equivalent change in the value of Y. With the intercept ($\log k$) set at 0, the line passes through the origin ($X = 0$, $Y = 0$). The result is a matching line in which log ratio of responses equals log ratio of reinforcement.

Undermatching or Sensitivity

The data in Figure 9.17 labeled "undermatching" represent the same idealized experiment. The value of the intercept remains the same (log k = 0). However, the slope now takes on a value less than 1 (a = 0.5). Based on Equation 9.6, this change in slope results in new values of Y or log(B_a/B_b). Figure 9.17 (upper right) is a graph of the line resulting from the change in slope. When compared with the matching line (a = 1), the new line rises at a slower rate (a = 0.5). This situation is known as undermatching, and implies that the subject gives less relative behavior to alternative A [log(B_a/B_b)] than expected on the basis of relative rate of reinforcement [log(R_a/R_b)]. For example, if log-ratio reinforcement changes from 0 to 1, the log ratio of behavior will change only from 0 to 0.5. This suggests poor discrimination by the animal of the operating schedules of reinforcement (low sensitivity).

Response Bias

It is also possible to have a systematic *bias* for one of the alternatives. For example, a right-handed person may prefer to press a key on the right side more than on the left. This tendency to respond to the right side may occur even though both keys schedule equal rates of reinforcement. Recall that response bias refers to any systematic preference for one alternative that is not explained by the relative rates of reinforcement. In terms of the idealized experiment, the data labeled "bias" in Figure 9.16 show that the slope of the line is 1 (matching), but the intercept (log k) now assumes a value of 1.5 rather than zero. A plot of the X or log(R_a/R_b) and Y or log(B_a/B_b) values in Figure 9.17 (lower panel) reveals a line that is systematically deflected 1.5 units from the matching line.

Experiments and Log-Linear Estimates

Setting the Values of the Independent Variable

In actual experiments on choice and preference, the values of the slope and intercept are not known until the experiment is conducted. The experimenter sets the values of the independent variable, log(R_a/R_b), by programming different schedules of reinforcement on the alternatives. For example, one alternative may be VI 30 s and the other VI 60 s. The VI 30-s schedule is set to pay off at 120 reinforcements per hour, and the VI 60-s schedule is set to pay off at 60 reinforcements per hour. The relative rate of reinforcement is expressed as the ratio 120/60 = 2. To describe the results in terms of Equation 9.6, the reinforcement ratio of 2 is transformed to a logarithm, using a calculator with logarithmic functions. Experiments are designed to span a reasonable range of log-ratio reinforcement values. The minimum number of log-ratio reinforcement values is 3, but most experiments program more than three values of the independent variable.

Each experimental subject is exposed to different pairs of concurrent schedules of reinforcement. The subject is maintained on these schedules until rates of response are stable, according to preset criteria. At this point, relative rates of response are calculated (B_a/B_b) and transformed to logarithms. For example, a subject on a concurrent VI 30-s VI 60-s schedule may generate 1,000 responses per hour on the VI 30-s alternative and 500 responses per hour on the VI 60-s schedule. Thus, the response ratio is 1,000/500 = 2, or 2 to 1. The response ratio of 2 is transformed to a logarithm. For each value of log(R_a/R_b), the observed value of the dependent variable log(B_a/B_b) is plotted on X,Y coordinates.

Choice and Preference **351**

To illustrate the application of Equation 9.6, consider an experiment by White and Davison (1973). In this experiment, several pigeons were exposed to 12 sets of concurrent schedules. Each pair of schedules programmed a different reinforcement ratio. The pigeons were maintained on the schedules until key pecking was stable from day to day. The data for pigeon 22 are plotted in Figure 9.18A on logarithmic coordinates. Notice that actual results are not as orderly as the data of the idealized experiment. This is because errors in measurement, inconsistencies of procedure, and random events operate to affect response ratios in actual experiments. The results appear to move upward to the right in a linear manner, but it is not possible to draw a simple line through the plot.

Estimates of Slope and Intercept
To find the *line that best fits* the results, a statistical technique called least-squares regression is used to estimate values for the slope and intercept of Equation 9.6. The idea is to select slope and intercept values that *minimize the errors in prediction*. For a given value of the

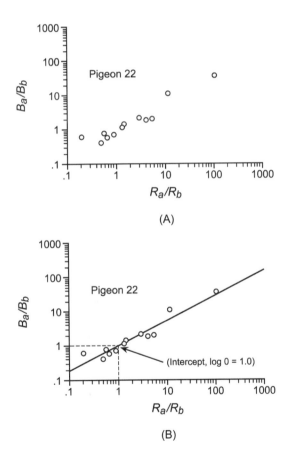

Fig. 9.18
(A) Reinforcement and response ratios for pigeon 22 plotted on logarithmic coordinates. (B) The line of best fit is added to the data from panel A.

Source: Taken from A. J. White & M. C. Davison (1973). Performance in concurrent fixed-interval schedules. *Journal of the Experimental Analysis of Behavior, 19*, pp. 147–153. Copyright 1973 held by John Wiley & Sons, Ltd. Republished with permission.

reinforcement ratio (X-axis), an error is the difference between the response-ratio value on the line (called the predicted value) and the actual or observed response ratio.

The mathematics that underlie this statistical technique are complicated and beyond the scope of this book. Most personal computers, however, have programs that will do the calculations for you. For example, you can use a program like Microsoft Excel® with a computer to obtain the best-fitting line, using linear-regression analysis. The estimate of slope was $a = 0.77$, indicating that pigeon 22 showed undermatching to the reinforcement ratios. The estimate of the intercept was zero (log $k = 0$), indicating that there was no response bias. With these estimates of slope and intercept, Equation 9.6 may be used to draw the best-fitting line.

In Figure 9.18B, the line of best fit has been drawn. You can obtain the line of best fit by substituting values for $\log(R_a/R_b)$ and finding the predicted $\log(B_a/B_b)$ values. You only need to find two points on the X,Y coordinates to draw the line. Notice that the data and best-fit line are plotted on a graph with logarithmic coordinates. Because there was no bias (log $k = 0$), the line must pass through the point $X = 1$, $Y = 1$ when R_a/R_b and B_a/B_b values are plotted on logarithmic paper.

As a final point, you may be interested in how well the generalized matching equation (or the generalized matching relation) fits the results of pigeon 22. One measure of accuracy is called explained variance. This measure varies between 0 and 1 in value. When the explained variance is 0, it is not possible to predict the response ratios from the reinforcement ratios. When the explained variance is 1, there is perfect prediction from the reinforcement ratios to the response ratios. In this instance, the explained variance is 0.92, indicating 92% accuracy. The **log-linear matching equation** is a good description of the pigeon's behavior on concurrent schedules of reinforcement.

Preference Shifts: Rapid Changes in Relative Reinforcement

In a typical matching experiment, pigeons are kept on the same concurrent VI VI schedules (reinforcement ratio) for many sessions until response rates stabilize. Once steady-state responding is obtained, the birds are presented with a new pair of VI schedules again for many sessions. This steady-state procedure is repeated until the birds have responded on all pairs of VI schedules (reinforcement ratios) planned for the experiment.

In contrast to the usual procedure, Davison and Baum (2000) used a procedure in which pigeons were given seven different pairs of VI schedules within a single session (this account is informed by Mazur & Fantino, 2014, pp. 200–202). Daily sessions presented the birds with seven component schedules each separated by a blackout period. In random order each day, the components arranged seven reinforcement ratios (R_a/R_b): 27:1, 9:1, 3:1, 1.1, 1:3, 1:9, and 1:27. The components were not signaled and the birds could not tell which pair of VI schedules was operative. This rapid change in schedules allowed the researchers to observe how the birds developed a preference for the richer schedule as successive reinforcements were delivered.

Results showed quick shifts in preference with rapid changes in the concurrent schedules of reinforcement. Pigeons started each session with almost equal responding on a randomly selected pair of VI schedules, but more responses were allocated to the richer schedule with each delivery of reinforcement. Also, the amount of shift in preference for the richer schedule depended on the reinforcement ratio—a greater shift in preference for a ratio of 27:1 than a 3:1 ratio. Detailed (fine-grain) analysis of the data showed that each delivery of reinforcement has an effect on the next choice response, a brief momentary increase in relative responding called a *preference pulse*. The effects of each reinforcement delivery, however, do not subside with the occurrence of the preference pulse, but continue to affect the response ratios after at least the last six reinforcement presentations. Not only does each delivery of reinforcement temporarily shift preference, it has *lasting short-term effects on preference*. These momentary shifts in choice responses and lasting effects for each reinforcement delivery at the molecular level provide the *underlying dynamics* for the generalized matching relation at the molar behavioral level.

NEW DIRECTIONS: BEHAVIORAL NEUROSCIENCE, MATCHING, AND SENSITIVITY

The generalized matching relation has been used to study neural events and processes using pharmacological interventions. A study by Bratcher and colleagues used the generalized matching equation (Equation 9.5) to investigate how the brain is involved in the regulation of behavior by relative rates of reinforcement (Bratcher, Farmer-Dougan, Dougan, Heidenreich, and Garris, 2005). The point was to detect the effects of specific drugs on choice behavior using the estimates of the slope *a* value (sensitivity) of the generalized matching equation.

Dopamine, a neurotransmitter in the brain, is known to be a behavioral activator, and different dopamine receptors, D1 or D2, appear to be involved in the regulation of different aspects of behavior. Thus, drug activation of D2 receptors was predicted to induce focused search for food, increased behavior directed at lever pressing, and overmatching (or matching) to relative reinforcement with estimates of the parameter *a* taking a value greater than 1. In contrast, drug activation of D1 receptors was predicted to elicit nonspecific food-related behavior, increased behavior away from lever pressing, and subsequent undermatching to relative reinforcement with estimates of *a* taking a value of less than 1.

In this study, rats were trained to press levers for food on concurrent VI VI schedules, and drug or control treatments (saline) were administered 20 min before sessions. When behavior had stabilized, response and reinforcement ratios were determined and the generalized matching relation was fit to the data, providing estimates of the *a* value or sensitivity to relative reinforcement. The results showed that the estimate of sensitivity was slightly higher than baseline with quinpirole (the D2 agonist) and substantially lower than baseline with SKF38393 (the D1 agonist). That is, as predicted, SKF38393 produced considerable *undermatching* or poor control by relative rate of reinforcement.

Analysis of video recordings of the rats' behavior indicated that quinpirole (the D2 agonist) increased chewing and sniffing of the lever and food cup—behaviors compatible with lever pressing. The D1 agonist (SKF38393) increased grooming and sniffing at some distance away from the lever—behavior incompatible with lever pressing and the reinforcement contingencies. Bratcher and colleagues suggested that "sensitivity to reward may have been due to changes in the value of the scheduled and/or any unscheduled reinforcers. That is, other behaviors elicited by D1 or D2 drug exposure ... may have taken on greater reinforcement value than operant responding" (Bratcher et al., 2005, pp. 389–390). These alternative behaviors were either compatible or incompatible with lever pressing for food, leading to the observed differences in the estimates of the slope *a* value. The researchers concluded that further study of D1 and D2 agonists, sensitivity, and the regulation of operant behavior is warranted. Furthermore, the generalized matching relation provides a *powerful analytical tool* for research on brain and behavior relationships (see Hutsell, Negus, & Banks, 2015 for cocaine vs. food choice, generalized matching, and increased sensitivity to relative price of cocaine by pharmacological and environmental treatments).

In fact, Buckley and Rasmussen (2014) used the generalized matching equation to assess the effects of the cannabinoid CB1-receptor-antagonist, rimonabant, on operant food choice with lean and obese Zucker rats. The endocannabinoids activate the CB1 receptor to modulate food intake and rimonabant is known to reduce food consumption. But less is known about the rimonabant's effect in an operant-choice situation (pairs of VI schedules) in which amount and palatability (fat/sugar composition) of food are varied. As one part of the study, the researchers assessed the effects of food amount and palatability on estimates of the bias and sensitivity parameters of the generalized matching equation. Bias for palatable food (sugar pellets vs. carrot-flavored pellets) did not increase with the administration of the CB1 antagonist. Blocking the CB1 receptor with the drug, however, increased sensitivity to food amount compared to the saline vehicle (control) in lean rats, but not obese rats. Obese Zucker rats are deficient in leptin signaling (used in energy-balance regulation), and defective leptin signaling is related to elevated levels of endocannabinoids in the hypothalamus (Marzo et al., 2001). For the two doses of rimonabant administered to the obese rats in the matching study, defects in leptin signaling would elevate levels of hypothalamic endocannabinoids, possibly obscuring the antagonistic effect of the drug on the CB1 receptor and leaving estimates of sensitivity to food amount unchanged—a result requiring clarification by further research. Pharmacological interventions on brain mechanisms together with the generalized matching relation are helping to unravel the determinants of food preferences, brain functions, and obesity.

CHAPTER SUMMARY

Why do people and other animals choose to do the things they do? Are they compelled by impulses or is their behavior random? Behavior analysts have proposed a model based on the assumption that consequences influence behavioral choices and organisms choose among many sources of reinforcement in a given moment. Choice is modeled in the laboratory by concurrent schedules of reinforcement. It begins with an understanding how the relative amount or frequency of reinforcement from two or more sources affects choice. The matching relation states that relative rates of response to reinforcement alternatives or options match the relative rates of reinforcement. This chapter describes the methods by which researchers have investigated this process. The matching relation is based on the assumption that we are always confronted with at least two alternative sources of reinforcement, and the option we choose is determined by the relative rate of reinforcement provided by that choice. This matching relation has been stated in the form of a generalized mathematical equation, and manipulation of relative rates of reinforcement, along with the bias (preference for a source of reinforcement regardless of amount) and sensitivity (how sensitive behavior is to differing densities of reinforcement) parameters, provide valuable insights into many aspects of choice behavior. Use of the matching relation has been shown to work best in the laboratory with concurrent variable-interval schedules of reinforcement, but it also applies to free-ranging wild pigeons and to social situations of humans. Other applications of the generalized matching relation have proved useful in areas such as behavioral economics. Delay discounting, which examines how an organism allocates its behavior to a smaller, immediate source of reinforcement versus a larger, delayed source, shows that the delay to reinforcement reduces it value. Delay discounting has been applied to a number of health-related behaviors, such as substance use disorders, gambling, and obesity. The generality of the generalized matching equation is remarkable, and it will continue to improve the quantification of the relationship between rates of response and rates of reinforcement.

Key Words

Behavioral economics
Bias
Changeover delay (COD)
Changeover response
Choice
Commitment response
Concurrent schedules of reinforcement
Delay discounting
Demand curve
Extraneous sources of reinforcement
Generalized matching relation
Hyperbolic discounting equation
"Impulsive" behavior
Log-linear matching equation
Matching
Matching relation

Maximization
Melioration
Overmatching
Power law for matching
Preference
Preference for choice
Preference reversal
Quantitative law of effect
Reinforcer pathology
Relative rates of reinforcement
Relative rates of response
"Self-control" behavior
Substitutability
Two-key procedure
Undermatching

On the Web

http://bio150.chass.utoronto.ca/foraging/game.html In this game of optimal foraging, students simulate a hummingbird feeding at flowers in patches arranged at random in a habitat. Each patch averages 12 flowers; each flower rewards you with 100 calories of energy on average. Travel between patches and feeding at a flower will cost you time (but not energy). Your goal is to obtain as much energy as possible in the allotted time.

www.youtube.com/watch?v=VO6XEQIsCoM Take a look at the lecture on freedom and choice by Barry Schwartz, called the *Paradox of Choice*. Schwartz states that freedom is necessary to maximize personal welfare, and that the way to freedom is to expand individual choice. In pursuing freedom of choice, however, we often find that the more choice we have the less we gain, in terms of happiness and satisfaction with life.

www.youtube.com/watch?v=xLEVTfFL7Is Warren Bickel, a leading researcher in behavioral economics, distinguishes between the impulsive and executive areas of the brain. Subsequently, he relates these areas to decisions about the future in drug addicts. The talk provides a description of behavioral treatments for addiction that help to reduce the impulsiveness of drug users, or that enhance users' consideration of future outcomes.

Brief Quiz

1. In terms of behavior, choice is concerned with:
   a. the distribution of behavior among alternative sources of reinforcement
   b. the decision-making capabilities of the organism
   c. the information processing during decision making
   d. the differential reinforcement of alternative behavior

2. Which of the following is used to investigate choice in the laboratory?
   a. an operant chamber with a single manipulandum
   b. two cumulative recorders that are running successively
   c. concurrent schedules of reinforcement
   d. both (a) and (b)

3. In order to prevent switching on concurrent schedules:
   a. program an intermittent schedule of reinforcement
   b. program a changeover delay
   c. program a multiple schedule
   d. program a DRO contingency

4. The experiment by Herrnstein (1961b) using a two-key concurrent VI VI schedule is described by:
   a. the matching relation for a single alternative
   b. the quantitative law of effect
   c. the proportional matching equation
   d. the nonmatching function for multiple alternatives

5. The generalized matching relation has described the choice behavior of:
   a. pigeons
   b. wagtails
   c. rats
   d. all of the above

6. When the response is continuous rather than discrete, use a matching equation for:
   a. time spent on each alternative
   b. rate of response on each alternative
   c. several concurrent schedules of reinforcement
   d. the single operant

7. The equation for matching of ratios of rates of response to rates of reinforcement:
   a. is stated in terms of a power law
   b. includes a value for bias
   c. includes a value for sensitivity
   d. is characterized by all of the above

8. In contrast to optimal foraging, the process of ____ has been described.
   a. maximization
   b. melioration
   c. multiple schedule inference
   d. monotonic matching

9. Behavioral economics involves the use of:
   a. economic principles to describe and analyze behavioral choice
   b. economic factors to predict animal behavior in the marketplace
   c. economic indicators when pigeons are trading goods and services
   d. economic satisfaction due to reinforcement

10 Delay discounting describes
- a how reinforcer value degrades with delay to its receipt
- b is a trait variable
- c is a state variable
- d explains why some people may struggle with preventative health behaviors, such as substance use disorders or obesity
- e all of the above

Answers to Brief Quiz: 1, a (p. 311); 2, c (p. 312); 3, b (p. 316); 4, c (p. 317); 5, d (p. 321); 6, a (p. 319); 7, d (p. 345); 8, b (p. 328); 9, a (p. 332); 10, e (p. 339)

NOTE

1 Technically, the algebraic notation for a function is $Y = f(X)$ and therefore the linear equation $Y = n(X) + m$ is correct, but here we are trying to connect a common form of the log-linear matching equation with the intercept log k added to the slope term, $a \times \log(R_a/R_b)$. This log-linear matching equation is usually estimated by the regression equation $Y = a + b(X)$. For this reason, we have written the simple linear function as $Y = m + n(X)$ in the fitting of a straight line.

CHAPTER 10

Conditioned Reinforcement

1. Inquire about conditioned reinforcement and chain schedules of reinforcement.
2. Investigate backward chaining, using it to improve your golf game.
3. Discover our preference for events linked to good news rather than bad news.
4. Learn about generalized conditioned reinforcement and the token economy.
5. Delve into the delay-reduction model of conditioned reinforcement.

Human behavior often is regulated by consequences that depend on a history of conditioning. Praise, criticism, good grades, and money are often consequences that may strengthen or weaken behavior. Such events acquire these effects because of the different experiences that people have throughout their lives. Some people have learned the value of what others say about their actions—others are indifferent to it. Henry Ford marketed and sold cars because of monetary reinforcement, status, and power, but Mother Teresa took care of the poor because she wanted to ease their suffering. In these examples, the effectiveness of a behavioral consequence may depend on a personal history of conditioning. A positive reinforcer is defined as a stimulus or event, which increases or maintains the rate of the response upon which it is contingent. The critical issue is its influence on response rate, not what exactly the stimulus or event is.

Conditioned reinforcement occurs when behavior is strengthened by events that have an effect because of a conditioning history. The important aspect of this history involves a correspondence between an arbitrary event and a currently effective reinforcer. Once the arbitrary event becomes able to increase the frequency of an operant, it is called a **conditioned reinforcer**. (It has also been called a secondary reinforcer, but conditioned reinforcer is now the accepted term.) For example, the sound of the pellet feeder operating becomes a conditioned reinforcer for a rat that presses a lever because the sound has accompanied the presentation of food. The immediate effect of lever pressing or key pecking is the *sound* of the feeder, not the consumption of food. Food is a biological or **unconditioned reinforcer** that follows the sound of the feeder. "Magazine training" is the procedure of deliberately arranging the sound of food delivery with immediate access to the food. The point in this case is to be able to deliver an auditory reinforcer, the feeder sound, wherever and whenever you wish. One way to demonstrate the conditioned-reinforcement effectiveness of the feeder sound is to arrange a contingency between a new response (e.g., pressing a spot on the wall) and the presentation of the sound, the *new-response method*. If the operant rate increases, the process is called conditioned reinforcement and the sound is a conditioned reinforcer.

In his book *The Behavior of Organisms*, Skinner (1938) described a procedure that resulted in conditioned reinforcement. Rats were exposed to a clicking sound and were given food. Later the animals were not fed, but the click was used to train lever pressing. Lever pressing increased, although it only produced the clicking sound. Because the click was no longer accompanied by food, each occurrence of the sound was also an extinction trial. For this reason, the sound declined in reinforcing effectiveness, and lever pressing for clicks decreased at the same time. It should occur to you that establishing a conditioned reinforcer (click → food) is similar to the development of a conditioned stimulus (CS) in respondent conditioning (CS → US). A previously nonfunctional stimulus is followed by a functioning reinforcer, and the nonfunctional stimulus acquires a reinforcement function, maintaining behavior upon which it is contingent. Of course, as in respondent conditioning, an extinction procedure reduces the effectiveness of the conditioned reinforcer. This reduction in reinforcement effectiveness is similar to the CS losing associative strength and control over the conditioned response (CR) when it is no longer followed by the occasional unconditioned stimulus (US).

Animals typically engage in long and complex sequences of behavior that are often far removed from unconditioned reinforcement. This is particularly true for humans. People get up in the morning, take buses to work, carry out their jobs, talk to other workers, and complete many other behavioral sequences. These operant performances occur day after day and are maintained by conditioned reinforcement. Clearly, conditioned reinforcement is a durable, long-lasting process but the new-response method does not always reveal how this occurs. Thus, behavioral researchers have turned to additional procedures, which seek to clarify the long-lasting effects of conditioned reinforcement.

ON THE APPLIED SIDE: CLICKER TRAINING

There is a major industry built around the use of conditioned reinforcement in training animals. Karen Pryor (1999; www.clickertraining.com) has applied the fact that a click from a little hand-held clicker followed often by a food treat can be used to strengthen behavior. Clicker training is like the old game of "hot and cold" where a person is searching for something and the only help given is in the form of telling the searcher whether she is "hot" (meaning close) or "cold" (meaning farther away). Clicks are used as indications of "hot" and no clicks mean "cold" for such behavior shaping (Peterson, 2004). Clicker training has been adopted by many zoo animal keepers (e.g., Lukas, Marr, & Maple, 1998; Wilkes, 1994) and is highly effective in training companion animals of all species (e.g., dogs and horses). But as a conditioned reinforcer, the trainer must remember that the backup reinforcer or treat is essential, or clicks lose their conditioned reinforcement function. Steve Martin and Susan Friedman (2011) made this point in their article on Blazing Clickers:

> Clickers, whistles and other conditioned reinforcers are valuable tools that help trainers communicate to animals the precise response they need to repeat to get a treat. When a conditioned reinforcer is reliably paired with a well-established backup reinforcer then communication is clear, motivation remains high and behaviors are learned quickly. However, when a

click isn't systematically paired with a backup reinforcer the communication becomes unclear, as evidenced by decreased motivation, increased aggression, and weak performance.... When the click begins to lose meaning because of repeated use without a treat, animals begin to search for other stimuli to predict their outcomes. They often watch for body language clues that predict the treat is imminent thereby further strengthening the behavior consequence contingency and the click is just noise. While it's true a secondary reinforcer doesn't lose its ability to strengthen behavior the first time it's used without a backup reinforcer, the number of solo clicks to extinction can't be predicted, and it can happen very quickly. So, while we may be able to get away with the occasional solo click, blazing clickers is not best training practice. When the click doesn't carry information an animal can depend on, the result is undependable behavior.

Martin and Friedman (2011, p. 4)

The principle of conditioned reinforcement used in clicker training has been extended to the teaching of medical residents by Martin Levy, an orthopedic surgeon and dog enthusiast at Montefiore Medical Center in New York, as reported in the *Scientific American* article, "Positive Reinforcement Helps Surgeons Learn" (Konkel, 2016). Levy's interest in dogs introduced him to Karen Pryor, the inventor of clicker training. Levy noticed how complex behavior could be established in dogs with clicker-training methods, and reasoned he could apply the same principles to teach his residents skills and surgical procedures—involving holding and positioning of instruments, tying surgical knots, and working with power tools.

Training surgical residents does not involve the use of a clicker and dog treat, but rather the application of the basic operant principles, especially conditioned reinforcement. Traditional surgical-training methods use demonstration and criticism of errors made by the residents, rather than acknowledging correct, skillful performance. "For a highly motivated individual [surgical resident], having a teacher acknowledge that you hit your target is in itself a pretty huge reward," says Levy. Compared to a traditional learning-by-demonstration group, the operant learning group showed more precise movements and executed the requisite tasks with greater efficiency (Levy, Pryor, & McKeon, 2016). Clearly, the operant principles underlying clicker training have wide applicability to educational and work settings where skillful and fluent performance is a requirement.

CHAIN SCHEDULES AND CONDITIONED REINFORCEMENT

The new-response method is not the only way to study conditioned reinforcement. Another approach is to construct sequences of behavior in the laboratory. A **chain schedule of reinforcement** involves two or more simple schedules (CRF, FI, VI, FR, or VR), each of which is presented sequentially and signaled by an arbitrary stimulus. Only the final or *terminal link* of the chain results in unconditioned reinforcement. Figure 10.1 shows a schematic representation of a three-component chain schedule of reinforcement. The schedule is a chain VI FR FI, and each link (or component) of the chain is signaled by a red, blue, or green light. For example, in the presence of the red light, a pigeon must emit a key peck after an average of 1 min has elapsed (VI 60 s). When the peck occurs, the light changes from red to blue and the bird must then peck the key 50 times (FR 50) to produce the green light. In the presence of the green light, a single peck after 2 min (FI 120 s) produces food and the light changes back to red (i.e., the chain starts over).

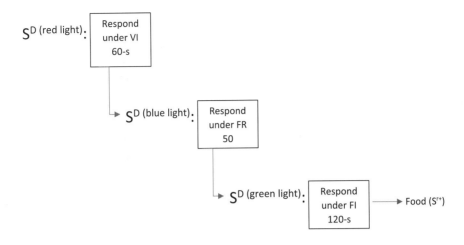

Fig. 10.1
A schematic for a three-component chain schedule of reinforcement, VI 60 s FR 50 FI 120 s. Notice that the red light only has a discriminative stimulus function, while the blue and green lights have multiple functions, including S^D and $S^{r+(cond)}$. Pecking in only the last component (green) results in the primary reinforcer of food.

When the pigeon pecks in the red component, the only consequence is that the light changes to blue. Once the blue condition is in effect, 50 responses turn on the green light. If the bird pecks for the blue and green lights in each component, the change in color is reinforcement. Recall that any stimulus that strengthens behavior is by definition a reinforcing stimulus. Thus, these lights have *multiple functions*. They are S^Ds that set the occasion for pecking the key in each link and also conditioned reinforcement, $S^{r+(cond)}$, for behavior that produces them. Figure 10.1 indicates that the red light is only a discriminative stimulus. You might suspect that it is a conditioned reinforcer, and it may have this function. The chain procedure as outlined, however, does not require a separate response to produce the red light (the last response in the chain produces food, and afterwards the red light automatically comes on), and for this reason a conditioned reinforcing function is not demonstrated.

Multiple-Stimulus Functions

Consider a sequence of two schedules, FR 50 FI 120 s, in which the components are not signaled. Formally, this is called a **tandem schedule**. A tandem is a schedule of reinforcement in which unconditioned reinforcement is programmed after completing two or more schedules, presented sequentially *without discriminative stimuli*. In other words, a tandem schedule as shown in Figure 10.2 is the same as an unsignaled chain.

Gollub (1958) compared the behavior of pigeons on similar tandem and chain schedules of reinforcement. On a tandem FI 60-s FI 60-s schedule, performance resembled the pattern observed on a simple FI 120-s schedule. The birds produced the typical scallop pattern observed on fixed-interval schedules—pausing after the presentation of food, and accelerating in response rate to the moment of reinforcement. When the tandem schedule was changed to a chain FI 60-s FI 60-s schedule by adding distinctive stimuli to the links or components, the effect of conditioned reinforcement was apparent. After some experience on the chain schedule, the birds responded faster in the initial link than they had on the tandem. In effect,

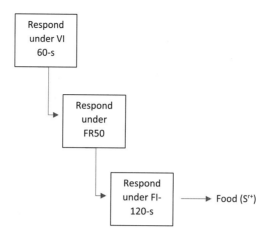

Fig. 10.2
Notice that a tandem schedule of reinforcement is the same as an unsignaled chain.

the birds produced two FI scallops rather than one during the 120-s period. This change in behavior may be attributed to the discriminative stimulus in the terminal link that also reinforced responses in the first component. In other words, the discriminative stimulus signaling the terminal link is also a conditioned reinforcer for responses in the first component of the chain—a stimulus with multiple functions (Ferster & Skinner, 1957).

Homogeneous and Heterogeneous Chains

Operant chains are classified as **homogeneous chain schedules** when the topography or form of response is similar in each component. For example, in the chain schedule discussed earlier, the bird pecks the same key in each link. Because a similar response occurs in each component, this is a homogeneous chain. In contrast, a **heterogeneous chain schedule** requires different responses for each link. Dog trainers make use of heterogeneous chains when they teach complex behavioral sequences to their animals. When going for a walk, a guide dog stops at intersections, moves forward when the traffic is clear, pauses at a curb, avoids potholes, and finds the way home. Each of these different responses is occasioned by specific stimuli and results in conditioned reinforcement. Although heterogeneous chains are common in everyday life and are created easily in the laboratory, they are usually too complex for experimental analysis. For this reason, conditioned reinforcement is typically investigated with homogeneous chains.

Chain schedules show how sequences of behavior are maintained by conditioned reinforcement in everyday life. Conditioned reinforcers in chain schedules remain effective because the terminal link leads to unconditioned reinforcement. Viewed as a heterogeneous chain schedule, going to a restaurant may involve the following links. A person calls and makes a reservation, gets dressed for the occasion, drives to the restaurant, parks the car, enters and is seated, orders dinner, and eats the meal. In this example, the S^Ds are the completion of the response requirements for each link. That is, being dressed for dinner (S^D) sets the occasion for going to the car and driving to the restaurant. Conditioned reinforcement involves the opportunity to engage in the next activity—bringing you closer to unconditioned reinforcement.

Of course, each of these components may be subdivided into finer and finer links in the chained performance. Thus, dressing for dinner is comprised of many different responses with identifiable discriminative stimuli (e.g., putting on shoes sets the occasion for tying laces). Even tying shoelaces may be separated into finer and finer links of a heterogeneous chain. The degree of detail in describing a chain performance depends on the analytical problem. An analysis of going out for dinner does not require details about how a person ties her shoes. On the other hand, a behavior analyst teaching a child with developmental disabilities to dress may focus on fine details of the chained performance.

FOCUS ON: BACKWARD CHAINING

Imagine that you have just been hired as a behavioral technician at a group home for children with developmental disabilities. One of your first assignments is to use the principle of conditioned reinforcement to teach a child to make her bed. The child has profound delays in development and has difficulties with easily following instructions or examples. She does have good motor coordination and finds potato chips to be highly reinforcing. You and the child are in one of the bedrooms, with sheets, blankets, and pillowcases stacked on the bed. You have decided to use potato chips as reinforcement for bed making.

Many people would start at the beginning of the sequence by unfolding a sheet, shaking it out, and placing it over the mattress. This tactic works for students (or children) who are easily able to follow instructions. This is not the case for this child, though, and the initial links of the chain are far removed from unconditioned reinforcement. Also, there are no conditioned reinforcers established along the way for completing the components of the chain.

The alternative way of teaching the child is to use a technique called **backward chaining**. The idea is to begin training at the end of the sequence. With backward chaining, you first teach the behavior in the terminal link of the chain. The child is reinforced with a potato chip when she places the top of the bedspread over the pillow. Once this behavior is well established, the bedspread is pulled down further. Unconditioned reinforcement now occurs when the child pulls the covers up to the pillow and then finishes making the bed. In this manner, responses that are more and more remote from the *final performance* are maintained by conditioned reinforcement (engaging in the next sequence). Of course, you often provide potato chips with social approval (i.e., "Your bed looks great!") and maintain the behavior without direct unconditioned reinforcement.

In everyday life, backward chaining has been used to train a variety of behaviors, including athletic skills. O'Brien and Simek (1983) taught golf using principles of backward chaining (see Figure 10.3). In their article they state:

> The teaching of sports has been largely unaffected by the advances in learning other operants. Golf, for example, is still routinely taught by handing the novice a driver and instructing him verbally how to get his body, arms and head to combine to hit a 250 yard drive. The usual result of such instruction is a series of swings that end in whiffs, tops and divots. This is followed by more verbal explanations, some highly complex modeling and loosely administered feedback. Endless repetitions of this chain then follow.
>
> A behavioral analysis of golf would suggest that the reinforcer for this exercise is putting the ball in the hole. The trip from tee to green represents a complex response chain in which the swing of the club up over the head and back to hit the ball is shortened as

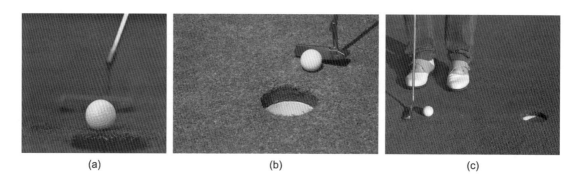

(a) (b) (c)

Fig. 10.3
Backward chaining in golf is illustrated, beginning with putting into the cup (final performance). The first step is to simply tap the ball at the lip of the cup into the hole. Once this occurs consistently, the ball is moved back from the lip and the golfer putts (small swing of club) the ball into the cup. After this performance is well established, more distance is added and the golfer must putt to the cup with more force. Once beyond the putting green, the golfer changes to other clubs appropriate to the distance and the amplitude of the swing is systematically increased. The "tee shot" begins the chain and uses a club called a driver to provide the most distance and loft toward the putting green and cup.

one gets closer to the hole. The final shot may be a putt of six inches or less leading to the reinforcement of seeing the ball disappear into the ground. This putt requires a backswing of only a few inches but involves the same basic stroke as the long backswing shot from the tee. Since the short putt seems to be the simplest response and the one closest to reinforcement, it would seem appropriate to teach the golf chain by starting with the putt and working back to the drive.

(O'Brien and Simek, 1983, pp. 175–176)

The superiority of the backward-chaining method in athletics or other areas of learning results from the principle of conditioned reinforcement. Behavior that is closest to unconditioned reinforcement is taught first. By doing this, the instructor ensures that operants in the sequence are maintained by effective consequences. With the backward-chaining method, each step in the chain may be added as the previous link is mastered.

CONDITIONED REINFORCEMENT: DETERMINANTS AND ANALYSIS

Operant chains show how complex sequences of behavior are maintained by events that have acquired a reinforcement function based on the past experience of an organism. The task for experimental analysis is to identify the critical conditions that contribute to the strength of conditioned reinforcement. It is also important to specify the factors that determine the reinforcing effectiveness of conditioned stimuli.

Effectiveness of Conditioned Reinforcement

Frequency of Unconditioned Reinforcement
The effectiveness of a conditioned reinforcer depends on the frequency of unconditioned reinforcement correlated with it. Autor (1960) found that preference for a conditioned reinforcer

increased with the frequency of unconditioned reinforcement in its presence. The effectiveness of a conditioned reinforcer increases with more and more presentations of unconditioned reinforcement, but eventually levels off. As the frequency of unconditioned reinforcement goes up, the effectiveness of a conditioned reinforcer reaches a maximum value. This relationship is strikingly similar to the increase in associative strength of a CS as described by the Rescorla–Wagner model of classical conditioning (see Chapter 3).

Variability of Unconditioned Reinforcement
Variability of unconditioned reinforcement also influences the effectiveness of a conditioned reinforcer. Fantino (1967) showed that birds preferred a conditioned reinforcer that was correlated with an alternating schedule (FR 1 half of the time and FR 99 for the other half of the trials) to one associated with a fixed schedule with the same rate of payoff (FR 50). Thus, variability of unconditioned reinforcement increases the value of a conditioned reinforcer and value is related to reinforcement effectiveness (see also Davison, 1969, 1972; Fantino, 1965; Herrnstein, 1964a). Variable schedules increase the effectiveness of conditioned reinforcement because these schedules occasionally program short intervals to unconditioned reinforcement. Compared with fixed schedules, these short intervals enhance responding and the value of stimuli correlated with them (Herrnstein, 1964b).

Establishing Operations and Effectiveness
The effectiveness of a conditioned reinforcer is enhanced by events that enhance the potency of unconditioned reinforcement. A bird responds for a light correlated with food more when it is hungry than when it is well fed. People attend to signs for bathrooms, restaurants, or hospitals when their bladders are full, when they have not eaten for some time, or when they are sick, respectively. Generally, conditioned reinforcement depends on events or stimuli that establish unconditioned reinforcement (Michael, 1982a; see Chapter 2, "Motivational Operations").

Delay to Unconditioned Reinforcement
On a chain schedule, the longer the delay between a discriminative stimulus and the delivery of the unconditioned reinforcement, the less effective the stimulus is as a conditioned reinforcer. Gollub (1958) compared the performance of pigeons on three different schedules—1) FI 5 min, 2) chain FI 1 FI 1 FI 1 FI 1 FI 1 min, and 3) tandem FI 1 FI 1 FI 1 FI 1 FI 1 min. On the simple FI 5-min schedule a blue key light occurred throughout the interval. On the chain, a different key color was associated with each of the five FI 1-min links. The components of the tandem schedule were the same as the chain schedule, but not signaled by separate colored lights, but a blue key light was on throughout the links. Birds responded to the tandem as they did to the simple FI—producing the typical FI scallop across the 5 minutes. On the chain schedule, responding was disrupted in the early components, and some of the birds stopped responding after prolonged exposure to the schedule (see also Fantino, 1969b). Disruption of responding occurs because the S^Ds in the early links (furthest from unconditioned reinforcement) signal a long delay to unconditioned reinforcement and are therefore weak conditioned reinforcers. A similar effect occurs when people give up when faced with a long delay to reinforcement on a complex, multi-component task. Students drop out of school for many reasons (see Chapter 6 on Aversive Control), but also may do so because the signs of progress are weak conditioned reinforcers—far removed from a diploma or degree.

Experimental Analysis of Conditioned Reinforcement

Many experiments use extinction procedures to investigate conditioned reinforcement. A conspicuous stimulus such as the sound of a feeder is presented just before the delivery of food. To demonstrate conditioned reinforcement, the feeder sound is subsequently used to condition a new response (e.g., pressing a spot on the wall) while food is withheld—the **new-response method for conditioned reinforcement**. If the operant rate increases, the process is conditioned reinforcement and the sound is a conditioned reinforcer. The new-response method often results in short-lived effects. Because of respondent extinction (the sound without the food), the conditioned reinforcer quickly loses its effectiveness, maintaining few responses (Kelleher & Gollub, 1962). On the other hand, considerable conditioning can occur before extinction is reached. This conclusion is in accord with Alferink, Crossman, and Cheney (1973), who found that trained pigeons continued to peck on an FR 300 schedule of hopper light presentation even with the hopper propped up so that food was always available.

Another extinction technique is called the **established-response method**. An operant that produces unconditioned reinforcement is accompanied by a distinctive stimulus, just prior to reinforcement. When responding is well established, extinction is implemented, but half of the animals continue to receive the stimulus that accompanied unconditioned reinforcement. The others undergo extinction without the distinctive stimulus. Generally, animals with the stimulus present respond more than those that do not receive the stimulus associated with unconditioned reinforcement. This result is interpreted as evidence for the effects of conditioned reinforcement.

Both extinction methods for analyzing conditioned reinforcement involve the presentation of a stimulus closely followed by unconditioned reinforcement. This procedure is similar to the CS–US pairings used in respondent conditioning. One interpretation, therefore, is that conditioned reinforcement is based on classical conditioning. This interpretation is called the stimulus–stimulus or **S–S account of conditioned reinforcement**. That is, all CSs are also conditioned reinforcers. To provide a test of the S–S account, behavior analysts devised new ways to show sustained responding on schedules of reinforcement for brief stimulus presentations where the brief stimulus is intermittently paired with food.

A Brief Stimulus Procedure Using Second-Order Schedules

We have seen that a brief stimulus such as the clicking sound of a feeder, or the presentation of a hopper light, eventually comes to support operant behavior (pecking), indicating a conditioned-reinforcement function. The extinction methods (new-response and established-response), however, do not show sustained responding for a brief stimulus only occasionally followed by food (unconditioned reinforcer). To remedy this problem, behavior analysts designed second-order schedules of brief stimulus presentations that ensured infrequent pairing of the brief stimulus with food, or unconditioned reinforcement. A **second-order schedule of reinforcement** involves two (or more) schedules of reinforcement in which completion of the requirements of one schedule is reinforced according to the requirements of a second schedule (Wing & Shoaib, 2010 described the use of second-order schedules in behavioral neuroscience).

Kelleher (1966) arranged a second-order schedule of brief stimulus presentations and infrequent delivery of food reinforcement and compared it with an identical second-order schedule (tandem) in which food never accompanied the brief stimulus. In the central experiment, pigeons responded on a second-order FR 15 (FI 4-min:W) schedule where a peck after 4 min produced a flash (0.7 s) of white light (W) on the response key for each FI 4-min component, and the birds had to complete 15 components for food reinforcement (FR15). On the 15th repetition of the FI component, the first response after 4 min produced the flash of light (W) followed immediately by presentation of the food hopper. The second-order schedule tested whether the flash of light (W) would sustain FI-component responding even though the minimum time between light flashes and food was 1 h (S-paired condition). Comparison conditions scheduled brief (0.7 s) dark (D) key, or red key light (R) stimulus changes following each of 14 FI 4-min components; on the 15th repetition, the first response after 4 min produced food reinforcement without any presentation of the brief stimulus (S-unpaired conditions).

In the S-paired condition, birds showed positively accelerated responding for the flash of white light (W) on the F1 4-min components, similar to scalloping on conventional FI schedules of food reinforcement. Omitting the white light (W) eliminated the scalloping pattern in the FI 4-min components (see Figure 10.4). Response rates increase over the fixed interval for the brief stimulus (W) paired with food, but not for the brief stimulus (D) unpaired with food. This finding indicates that the brief stimulus paired intermittently with food maintained operant behavior, functioning as a conditioned reinforcer. A complication of the findings is that response rates increased over the interval for the red light stimulus (R) that also was unpaired with food. Kelleher explained this observation by noting that the birds had extensive

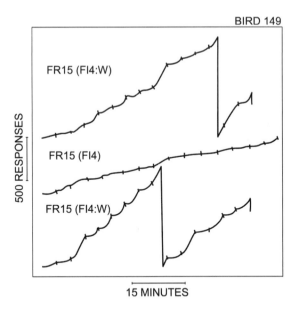

Fig. 10.4
For bird 149, effects on performance of presenting the white light (W) on the FR15(FI4:W) second-order schedule, removing the white light on the FR15(FI4) tandem schedule, and reinstating it once again, FR15(FI4:W). Notice the elimination of scalloping in each FI4 component and the reduction of response rate when the light is removed, indicating the conditioned reinforcement efficacy of the white light. The short strokes for the FR15(FI4:W) schedule indicate 0.7-s presentations of white light. For the FR15(FI4) schedule there was no stimulus change and the short strokes indicate the end of each FI4 component.

Source: From R. T. Kelleher (1966). Conditioned reinforcement in second-order schedules. Journal of the Experimental Analysis of Behavior, 9, pp. 475–485. Copyright 1966 held by John Wiley & Sons, Ltd. Republished with permission.

histories on various schedules of reinforcement that involved colored keys as discriminative stimuli. Thus, a brief stimulus paired with food functioned as conditioned reinforcement for response patterning on a schedule of reinforcement, but so did a brief stimulus with a previously established S^D function. The latter finding offers support for the **discriminative-stimulus account of conditioned reinforcement**, which states that an S^D also acquires value as a conditioned reinforcer and does not depend on being a CS associated with food.

There have been many experiments that attempted to distinguish between the S^D and S–S accounts of conditioned reinforcement (Fantino, 1977, 2008; Gollub, 1977). Thus, Schoenfeld, Antonitis, and Bersh (1950) presented a light for 1 s as an animal ate food. This procedure paired food and light, but the light could not be a discriminative stimulus, as it did not precede the food delivery. Following this training, the animals were placed on extinction and there was no effect of conditioned reinforcement. Given this finding, it seems reasonable to conclude that a stimulus must be discriminative to become a conditioned reinforcer. Current research, however, shows that *simultaneous pairing* of CS and US results in weak conditioning. For this and other reasons, it has not been possible yet to definitively test the S^D and S–S accounts of conditioned reinforcement.

On a practical level, distinguishing between these accounts of conditioned reinforcement makes little difference. In most situations, procedures that establish a stimulus as an S^D also result in that stimulus becoming a conditioned reinforcer. Similarly, when a stimulus is conditioned as a CS it almost always has an operant reinforcement function. In both cases, contemporary research (Fantino, 1977, 2008) suggests that the critical factor for conditioned-reinforcement value is the *temporal delay* between the onset of the stimulus and the later presentation of unconditioned reinforcement.

NEW DIRECTIONS: NEUROSCIENCE AND CONDITIONED REINFORCEMENT

One major issue of behavioral neuroscience is locating where in the nervous system response consequences are "evaluated" or assigned a hedonic value. That is, how does an event or stimulus such as money take on value and become an effective reinforcer? We know that conditioned reinforcers maintain behavior over long periods of time, and often in the absence of unconditioned reinforcers. These conditioned consequences also play a central role in complex social behavior.

Several brain areas (pain/pleasure centers) are known to code for hedonic value of stimuli and continuing research has refined the brain circuits involved. For example, Parkinson and colleagues reported that the amygdala is critical for the conditioned-reinforcement effects in primates (Parkinson et al., 2001). They made lesions to the amygdala of marmosets and subsequently observed insensitivity to the absence of conditioned reinforcement for pressing a touch-screen panel. In contrast, responding for unconditioned reinforcement (food) was not disrupted. Control subjects with an intact amygdala nearly ceased responding when conditioned reinforcement stopped, showing sensitivity to the contingencies. An intact and functioning amygdala seems necessary for the control of behavior by conditioned reinforcement.

The neurons of the basolateral (BL) amygdala are particularly important for encoding the value of a conditioned reinforcer (Baxter & Murray, 2002). Using second-order schedules, marmoset monkeys learned to respond on a computer screen for a tone and access to a banana milkshake (Roberts, Reekie, & Braesicke, 2007). Monkeys showed impairment of performance when given lesions to the BL amygdala. As the response requirements for access to the milkshake increased, and the frequency of the pairings between the tone (conditioned reinforcer) and the milkshake (unconditioned reinforcer) decreased, the monkeys with lesions became progressively more unable to maintain responding for contingent presentations of the tone. Control monkeys without lesions, however, continued to respond for presentations of the tone as pairings decreased. When the tone was omitted, monkeys with lesions showed insensitivity to the reinforcer efficacy—maintaining performance in the absence of the tone. Control monkeys, without impairment to the BL amygdala, were sensitive to the omission procedure and showed a marked decline in performance. The general finding is that an intact BL amygdala is required for sensitivity to contingencies of conditioned reinforcement (see also Ciano, 2008 for an account of drug seeking on second-order schedules, showing the role of the dopamine D_3 receptors situated in the BL amygdala).

Other recent evidence suggests that both aversive and rewarding stimuli, conditioned or unconditioned, affect similar brain areas. These areas include the orbitofrontal cortex, the prefrontal cortex, and the nucleus accumbens, NAc (Floresco, 2015; Roberts et al., 2007; Ventura, Morrone, & Puglisi-Allegra, 2007). The NAc septi, which is near the medial extension head of the caudate nucleus, is known to release dopamine in response to salient conditioned stimuli regardless of their hedonic valence (positive or aversive). NAc dopamine depletion slows the rate of operant responding and speeds the rate of acquisition by reducing the effectiveness of the reinforcer, not by impairing motor behavior (Salamone, Correa, Mingote, & Weber, 2003). And *in-vivo* microdialysis procedures have shown that high levels of dopamine from the NAc are present in rats that learn response–outcome relationships, but not in rats that fail to learn these relationships (Cheng & Feenstra, 2006). Clearly, the NAc and dopamine are involved in establishing and maintaining control of behavior by contingencies of reinforcement.

Behavioral neuroscience is providing a circuitry map for what goes on inside the brain when overt conditioning and learning are taking place. The interesting issue is how conditioning with arbitrary stimuli is supported by neural activity. Objects or events, which originally have no known function, can quickly become very attractive and valuable when specific brain activity occurs during conditioning. What is going on in the brain when this happens? Researchers are making progress in synthesizing how the brain and the environment work together—providing a more complete understanding of behavior and its regulation.

INFORMATION AND CONDITIONED REINFORCEMENT

Stimuli that provide information about unconditioned reinforcement may also become effective conditioned reinforcers. Egger and Miller (1962) used the extinction method to test for conditioned reinforcement. They conditioned rats by pairing two different stimuli (S_1 and S_2) with food. Figure 10.5 describes the procedures and major results. In their experiment (panel A), S_1 came on and S_2 was presented 0.5 s later. Both stimuli were turned off when the animals were given food. Both S_1 and S_2 were correlated with food, but only S_1 became an effective conditioned reinforcer (S_1^{r+}). In another condition (panel B), S_1 and S_2 were presented as before, but S_1 was occasionally presented alone. Food was never given when S_1 occurred by itself. Under these conditions, S_2 became a conditioned reinforcer (S_2^{r+}).

To understand this experiment, consider the presumed *information value* of S_2 in each situation. When S_1 and S_2 are equally correlated with food, but S_2 always follows S_1, then S_2 is *redundant*, providing no additional information about the occurrence of food. Because it is redundant, S_2 gains little conditioned-reinforcement value. In the second situation, S_1 only predicts food in the presence of S_2, and for this reason S_2 is informative and becomes a

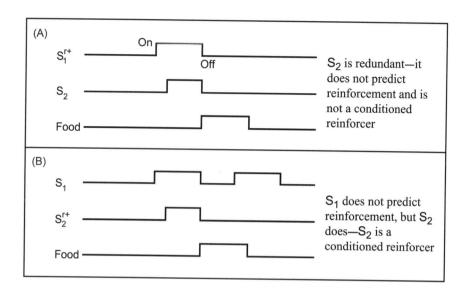

Fig. 10.5
Procedures and major results are shown for an experiment using the extinction method to test for conditioned reinforcement.

Source: Based on a description of procedures as outlined in M. D. Egger & N. E. Miller (1962). Secondary reinforcement in rats as a function of information value and reliability of the stimulus. *Journal of Experimental Psychology, 64,* pp. 97–104.

conditioned reinforcer. These results, along with later experiments (e.g., Egger & Miller, 1963), suggest that a stimulus functions as conditioned reinforcement if it provides information about the occurrence of unconditioned reinforcement.

Information Value: Good News and Bad News

The informational value of a stimulus should not depend on whether it is correlated with positive or negative events, as common sense suggests that bad news is just as informative as good news. Wyckoff (1952, 1969) designed an observing-response procedure to evaluate the strength of a conditioned reinforcer, which predicted good or bad news. In this procedure, periods of reinforcement and extinction alternate throughout a session, but stimuli (S^D and S^Δ) did not signal the shifting contingencies. This kind of alternating contingency is called a **mixed schedule of reinforcement**. A mixed schedule is the same as a multiple schedule, but without discriminative stimuli. Once the animal is responding on the mixed schedule, an observing response is added to the contingencies. The **observing response** is a topographically different operant that functions to produce an S^D or S^Δ depending on whether reinforcement or extinction is in effect. In other words, an observing response changes the mixed schedule to a multiple schedule. Figure 10.6 shows the relationships among mixed, multiple, tandem, and chain schedules of reinforcement, depending on whether the S^D is present and unconditioned reinforcement occurs in one component or all components of the schedule.

Wyckoff (1969) showed that pigeons would stand on a pedal to observe red and green colors correlated with FI 30-s reinforcement or EXT 30 s. Before the birds had an observing response available, they pecked equally in the reinforcement and extinction phases, showing

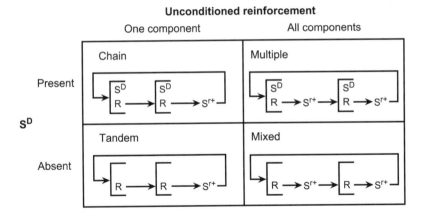

Fig. 10.6
The relationships among mixed, multiple, tandem, and chain schedules of reinforcement. These four schedules depend on whether an S^D is present or absent and whether unconditioned reinforcement occurs in one or all of the components.

failure to discriminate between the schedules. When the observing response was added, the pigeons showed a high rate of pecking in the reinforcement component and very low rates during extinction. Because the observing response was maintained, the results suggest that stimuli correlated with either reinforcement or extinction (good or bad news) acquired a conditioned reinforcement function.

Although Wyckoff's data are consistent with an information, *uncertainty reduction* view of conditioned reinforcement, it is noteworthy that his pigeons only spent about 50% of the time making the observing response. One possibility is that the birds were observing the stimulus correlated with positive reinforcement (red color) but not the stimulus that signaled extinction (green color). In other words, the birds may have only responded for good news.

In fact, subsequent experiments by Dinsmoor, Brown, and Lawrence (1972) and by Killeen, Wald, and Cheney (1980) supported the good-news interpretation of conditioned reinforcement. Dinsmoor et al. (1972) trained pigeons to peck a key on a VI 30-s schedule of food reinforcement that alternated with unpredictable periods of extinction. The birds could peck another key to turn on a green light correlated with reinforcement and a red light correlated with extinction. That is, if positive reinforcement was in effect, an observing response turned on the green light; if extinction was occurring, the response turned on the red light.

Observing responses were maintained when they produced information about both reinforcement and extinction, seemingly supporting the information hypothesis. In the next part of the experiment, observing responses only produced the green light signaling reinforcement, or the red light correlated with extinction. In this case, observing responses produced either good or bad news, but not both. When observing responses resulted in the green light correlated with reinforcement, the birds pecked at a high rate. In contrast, the pigeons would not peck a key that only produced a stimulus (red) signaling extinction. Thus, good news functions as conditioned reinforcement, but bad news does not.

The good-news conclusion is also supported by research using aversive, rather than positive, consequences. Badia, Harsh, Coker, and Abbott (1976) exposed rats to electric shocks. The shocks were delivered on several variable-time (VT) schedules, independent of the rats' behavior. During training, a light was always on and a tone occurred just before each shock. In Experiment 2 of their study, the researchers allowed the animals to press a lever that turned on the light for 1 min. During this time, if shocks were scheduled, they were signaled by a tone. In one condition, the light was never accompanied by a tone and shocks. When the light was on, the animal was completely safe from shocks. Other conditions presented more and more tones and shocks when the animal turned on the light. In these conditions, the light predicted less and less safety, and responding for the light decreased. In other words, the animals responded for a stimulus correlated with a shock-free period, but not for information about shock given by the tone signals. Once again, conditioned reinforcement is based on good news, but not bad news.

Information, Reinforcement, and Human Observing Behavior

Stimuli Linked to Bad News and No News

Many of us have been in situations like those described by the good-news theory of conditioned reinforcement. In everyday life on campus, students who usually do well on mathematics exams quickly look up their exam scores online, while those who have done poorly wait, may not check their grade at all, or will do so later (e.g., when mid-term or final grades come out). Seeing a grade is conditioned reinforcement for students who are skilled at mathematics, but not for those who find the subject difficult. In another context, investors who usually make money on the stock market keep track of their portfolio (see Figure 10.7), but those who have been losing money may seldom look at how their investments are doing. Or perhaps a person

Fig. 10.7

Illustration of a woman keeping track of her stocks on the market. According to a conditioned reinforcement account, people who have a history of observing and successfully investing (good news) in the market are likely to keep close track of their stocks and portfolio. Those who have a history of observing and losing (bad news) on the market do not keep track of their investments on a daily basis.

Source: Shutterstock.

may spend a great deal of money over the December holidays. Under normal conditions, when just a small amount of money is placed on the credit card, they may check the bill amount as soon as it is due. But after the holidays when the credit card bill comes in January, they may avoid opening the bill or email notification until they have to pay it. This also happens when receiving a bill from a physician, in which high medical costs have been charged; the bill may sit unopened on the counter for weeks. The conditioned reinforcing effects of good news maintain observing, but usually bad news does not.

Indeed, the informational and conditioned-reinforcement accounts of observing behavior have been studied in humans. In a series of experiments, Fantino and Case (1983) had human participants make observing responses, but did not require responding for reinforcement (points worth 10 cents each). Sometimes points on a counter were arranged on a VT 60-s schedule, with a point presented once per minute on average. At other times, no points were given (EXT). Thus, points were arranged on a mixed VT 60-s EXT schedule with periods of response-independent reinforcement (VT 60-s points) and periods of no reinforcement (EXT). The periods of points and no points alternated unpredictably, but participants could find out which period was in effect by responding on either of two levers. For the critical condition, observing responses on one lever produced a colored light (S^-) on a VI 60-s schedule only if the no-point (EXT) period was in effect. The light stimulus (S^-) provided information about extinction (bad news), but had no positive reinforcement value. Responses on the other lever resulted in a different colored light (S^U, uncorrelated stimulus) on a VI 60-s schedule when periods of points were scheduled (VT 60 s), but also when periods of no points were in effect (EXT). Thus, the S^U light provided no information about the schedules of points and no points. Essentially, the critical conditions of the study offered participants a choice between information about bad news and uninformative "no news." As shown in Figure 10.8 (Experiment 2), choice proportions by 4 of 6 participants strongly favored the uninformative "no news" stimulus (average choice proportion = 0.81) option. Contrary to an information (uncertainty reduction) account, no one preferred observing the stimulus (S^-) correlated with bad news. Across three experiments, 19 of 22 participants preferred (choice proportion average = 0.70) to observe the uninformative "no news" stimulus.

Over a series of studies, a stimulus correlated with bad news (high information) did not maintain human observing, but one linked to "no news" (no information) did. Preference for the bad-news stimulus over the no-news stimulus occurred only when observing the S^- permitted more efficient responding for reinforcement—a finding that is inconsistent with the information hypothesis. Thus, when observing information about bad news is linked to good news (reinforcement), people prefer to observe bad-news information to uninformative "no news" (see Case, Fantino, & Wixted, 1985). Additionally, these studies consistently showed that a stimulus correlated with good news is highly preferred to one correlated with bad news, consistent with the findings of numerous animal experiments. Overall, the conditioned reinforcement view provided a systematic and consistent account of human observing behavior.

Information and Human Observing: Supporting Evidence
In 1997, a series of experiments challenged the conditioned-reinforcement account of observing behavior, suggesting that humans actually do prefer bad-news information (Lieberman, Cathro, Nichol, & Watson, 1997). In this study, participants could respond for a bad-news message or an uninformative no-news message while playing a lottery—pressing a

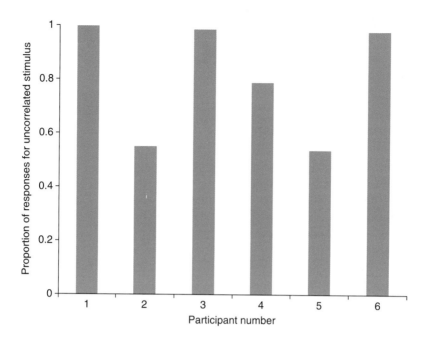

Fig. 10.8
Choice proportions are shown for each of the six participants in Experiment 2 of Fantino and Case (1983). The data indicate that 4 of 6 participants strongly preferred to observe the uninformative no-news option compared to the informative bad-news alternative. The results are evidence that the information is not the critical factor for human observing.

Source: Illustration is based on data from E. Fantino & D. A. Case (1983). Human observing: Maintained by stimuli correlated with reinforcement but no extinction. Journal of the Experimental Analysis of Behavior, 18, pp. 79–85.

button to have a computer (ERNIE) generate random numbers. After 50 button presses their best score would be saved by the computer and, after the study, the highest score of all the participants would win a monetary prize. To obtain information about how they were doing, participants could press either of two buttons. Each press (CRF) on one button produced a no-news, ambiguous message: "ERNIE says you may win, but you may lose." Each press on the other button produced the no-news message most of the time, but sometimes the bad-news, informative message, which stated, "A poor score. This one is not a winner." When each observing response was effective (CRF), participants preferred the bad-news option to the no-news alternative. A second experiment compared two groups of participants, one observing messages about periods of points and no points on CRF schedules, and another group observing messages about points on VI schedules, as in the experiment by Fantino and Case (1983). The results showed no preference for the no-news stimulus on the VI schedule, but a strong preference for the S$^-$, bad-news message in the CRF group. A third experiment included a VI group that only had a few (30) observing responses to use, and the participants had to pay for additional observing responses (VI-cost condition). The group on the standard VI observing schedule was indifferent to the no-news and bad-news options, but the VI-cost group strongly preferred the bad-news information. Overall, the results provided consistent support for the **information account** of observing behavior. People preferred the information about bad news to uninformative no news, but note that this is the case when there is no good-news option.

Conditioned Reinforcement and Useful Information

Fantino and Silberberg (2010) acknowledged that a preference for bad-news information posed an apparent problem for the conditioned-reinforcement account of human observing behavior. A series of five brief experiments investigated the role of good-news signals (S^+), no-news uninformative stimuli (S^U), and signs of bad news (S^-) for human observing behavior. Participants used a mouse to click response boxes on a computer screen to observe stimuli linked to earning occasional points worth a nickel. The points registered on the computer screen based on a mixed VT 30-s EXT schedule. In Experiment 1, the participant could click one box to sometimes see whether a point was going to register (colored pattern, S^+) or could click another box to occasionally see if no points were going to register (color pattern, S^-). The S^+ and S^- boxes were arranged to show colored patterns on two VI 30-s schedules. The participant received the good-news S^+ stimulus ("mean[ing] a nickel is coming soon") for an observing response only when the VT 30-s point schedule was actually going to pay off. They received the bad-news S^- stimulus ("mean[ing] a nickel is not coming soon") for a response only when the no-points (EXT) component of the mixed schedule was actually in effect. At other times, they received darkened boxes by responding to either the good-news or bad-news options. The participant was told that no matter which of the boxes they clicked, their responses would not affect the registration of points on the counter. The experimental results showed a consistent preference for good news over bad news, in accord with most previous studies. Also, using similar procedures, additional experiments compared a no-news (S^U) stimulus ("mean[ing] a nickel is coming soon or a nickel is not coming soon") with a signal for bad news ("mean[ing] a nickel is not coming soon") when there was virtually no correlation between observing responses and periods of winning nickels. Contrary to the findings of Lieberman and colleagues, the participants showed a preference for observing the no-news stimulus (S^U) over bad news signal (S^-) using VI schedules for observing responses, replicating the results of earlier experiments by Fantino and his colleagues.

Fantino and Silberberg (2010) conducted critical experiments to show that only when observing a bad-news stimulus or S^- correlated with good news do participants prefer the bad-news option. Recall that observing behavior in the study by Lieberman et al. (1997) occurred on CRF schedules. Assuming that points registered equally in the presence of the no-news and bad-news options, then each observing response for the bad-news stimulus (S^-) that did not produce a bad-news message was implicitly correlated with reinforcement (registration of points). Presumably, it was the implicit good news of points registering that maintained observing on the "bad news" option and the apparent preference for bad news in the experiments by Lieberman and colleagues. To test this analysis, Fantino and Silberberg arranged for each observing response to be effective (CRF or FR1) at producing stimulus patterns. For the no-news option, each observing response produced a patterned stimulus (S^U) uncorrelated with the components of the mixed schedule for winning nickels. Each response to the bad-news box, however, produced the S^- if and only if the EXT component of the mixed schedule was operative; at other times there was no stimulus change. Thus, observing no stimulus change on the bad-news option was perfectly correlated with winning nickels. Under these contingencies, participants strongly preferred the bad-news option to the no-news alternative; they showed less preference for the bad-news option when the absence of a stimulus change was imperfectly correlated with winning nickels. Fantino and Silberberg (2010) suggested that the overall results from five experiments are in accord with a conditioned-reinforcement account of human observing behavior. They also indicate that the results of their experiments are compatible with the interpretation by Lieberman and colleagues that "humans do find information reinforcing, but that this preference depends on the utility of the information" (Lieberman et al., 1997, p. 20). Thus, information about good news is useful (reinforcing), but information about bad news only has utility if it is correlated with good news, and most of the time it is not.

DELAY REDUCTION AND CONDITIONED REINFORCEMENT

Fantino and Logan reviewed the observing response studies and point out that:

> Only the more positively valued of two stimuli should maintain observing, since the less positive stimulus is correlated with an increase, not a reduction, in time to positive reinforcement (or a reduction, not an increase, in time to an aversive event).... Conditioned reinforcers are those stimuli correlated with a reduction in time to reinforcement (or an increase in time to an aversive event).
>
> (Fantino & Logan, 1979, p. 207)

This statement is based on Fantino's **delay-reduction hypothesis** (Fantino, 1969a). Stimuli closer in time to positive reinforcement, or further in time from an aversive event, are more effective conditioned reinforcers. Stimuli that signal no reduction in time to reinforcement (S^Δ) or no safety from an aversive event (S^{ave}) do not function as conditioned reinforcement. Generally, the value of a conditioned reinforcer is due to its *delay reduction*—how close it is in time to reinforcement or how far it is from punishment.

Modern views of conditioned reinforcement are largely based on the concept of delay reduction (Fantino, 1969a; Squires & Fantino, 1971). The idea is to compare the relative value of two (or more) stimuli that are correlated with different amounts of time to reinforcement. To do this, a complex-choice procedure involving **concurrent-chain schedules** is used. On these schedules, an organism may choose between alternatives that signal different amounts of time to reinforcement.

Concurrent-Chain Schedules of Reinforcement

In Chapter 9, we discussed the analysis of choice based on *concurrent schedules of reinforcement*. We also have noted the importance of chain schedules for the study of conditioned reinforcement. These schedules allow a researcher to change the temporal location of a stimulus in relation to unconditioned reinforcement. For example, the terminal-link discriminative stimulus (S^D_2) on a chain VI 20 s VI 10 s is six times closer to unconditioned reinforcement than it is on a chain VI 20 s VI 60 s. This relation is shown in Figure 10.9. In terms of time, the terminal-link that is nearer to unconditioned reinforcement (VI 10 s) should be a stronger conditioned reinforcer than one correlated with a longer delay. Thus, the terminal-link S^D_2 accompanying the VI 10-s schedule ought to be a more effective conditioned reinforcer than a discriminative stimulus correlated with VI 60 s.

For the effects of delay to be assessed, organisms must be able to choose between stimuli correlated with different reductions in time to unconditioned reinforcement. For example, using a two-key choice procedure, a chain VI 20 s VI 10 s may be programmed on the left key and a chain VI 20 s VI 60 s on the right key.

This *two-key concurrent-chains* procedure is shown diagrammatically in Figure 10.10. Consider the situation in which responses to the left key are eventually reinforced with

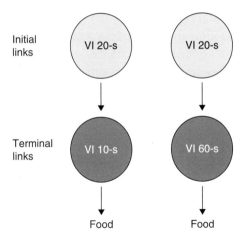

Fig. 10.9
Comparison of chain VI 20 s VI 10 s with chain VI 20 s VI 60 s. The initial links are identical, but the terminal links differ in the VI schedule value. Notice that the terminal link that has the schedule programmed that is closer in time to unconditioned reinforcement (VI 10-s) should be a more effective conditioned reinforcer.

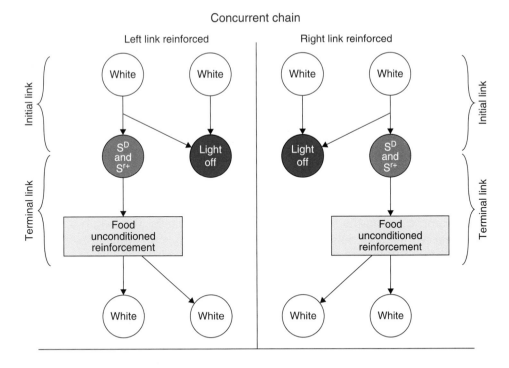

Fig. 10.10
A two-key concurrent-chains schedule of reinforcement is shown. Chain VI 20 s VI 10 s is programmed on the left key, and a chain VI 20 s VI 60 s on the right. See text for further details.

food. To start with, both left and right keys are illuminated with white lights. A bird makes left- and right-key pecks and after the left VI 20-s schedule times out, the first peck to the left key has two effects. The light on the right key goes out, and the VI 20-s schedule on that key stops timing—the key becomes dark and inoperative. At the same time, the left key changes from white to a diamond pattern. In the presence of this pattern, pecking the left key is reinforced with food on a VI 10-s schedule. After unconditioned reinforcement, both left and right keys are again illuminated with white lights and the bird chooses between the two alternatives.

A similar sequence occurs when the right key times out and the bird pecks this key. The left key becomes dark and inoperative, and the right key changes from white to a dotted pattern. In the presence of this pattern, pecking the right key is reinforced with food on a VI 60-s schedule. Following reinforcement, the discriminative stimuli in the initial links of the two chains (left and right white keys) are in effect and the bird again chooses to enter one of the terminal links (left or right).

The patterned stimuli on the left and right keys have two functions. These stimuli are S^Ds that set the occasion for pecking for food in the terminal links of the two chain schedules. In addition, the patterned stimuli function as conditioned reinforcement for pecking one or the other of the white keys in the initial-links or choice phase of the experiment. That is, reinforcement for pecking in the choice phase is the onset of the stimuli (S^D and S^r) correlated with unconditioned reinforcement in the terminal links. Because the bird is free to distribute pecks, the distribution of behavior in the initial links is a measure of the relative effectiveness of the two conditioned reinforcers.

Delay Reduction and Concurrent-Chain Schedules

Humans often respond on concurrent-chain schedules of reinforcement. A businessperson who frequently flies from Kansas City to Denver may call either Delta or American Airlines to book a ticket. Many people are trying to book flights, and the telephone lines to both companies are always busy. To contact an agent, the businessperson calls one airline and then the other. Eventually, one of the calls is successful, but both companies have recorded messages that state, "All of the lines are busy at the moment—please hold until an agent is available." After the businessperson has waited for some time, an agent answers and the ticket is booked.

In this example, calling the two airlines is the choice phase. The length of time to complete a call and get the hold message (initial-link schedules) is determined by the number of telephone lines at each airline and the number of people phoning the companies. The recorded message is conditioned reinforcement for dialing that company. The amount of time waiting on hold to book a flight (terminal-link schedule) is a function of the number of available agents. Waiting in the terminal link is reinforced by booking the flight. The sequence is repeated the next time the businessperson has a meeting in Denver.

To predict how much more (or less) reinforcing it is to be placed on hold at Delta relative to American Airlines, it is useful to consider a situation in which the initial- and terminal-link schedules are known for each company. Suppose that, on average, the telephone lines of both companies are busy for 120 s before a call is successful. In other words, the initial links for Delta and American Airlines are similar to concurrent VI 120-s schedules. The terminal-link schedules, though, are different for the two airlines. It takes an average of 30 s to talk to a Delta agent after being placed on hold. That is, the terminal link for Delta is similar to a VI 30-s schedule. After being placed on hold at American Airlines, it takes an average of 90 s to reach an agent, so that the terminal link for American Airlines is similar to a VI 90-s schedule. Thus, the sequence for booking a ticket at Delta is chain VI 120 s VI 30 s, and it is chain VI 120 s VI 90 s at American Airlines (see Advanced Section of this chapter for quantification of this example).

In this situation, Fantino's delay-reduction hypothesis predicts that the businessperson will prefer Delta to American Airlines. This is because more of the total time to reinforcement has elapsed when the person is placed on hold at Delta than with American. The conditioned reinforcement in this situation is getting the message "All of the lines are busy at the moment—please hold until an agent is available." After the message occurs, it is faster to book a ticket at Delta than at American. There has been relatively more reduction in delay to reinforcement when the Delta message occurs.

GENERALIZED CONDITIONED REINFORCEMENT

Formally, a **generalized conditioned reinforcer** is any event or stimulus that is correlated with, or exchangeable for, many sources of unconditioned reinforcement. Generalized reinforcement does not depend on deprivation or satiation for any specific reinforcer. Skinner describes its effects in the following passage:

> A conditioned reinforcer is generalized when it is paired with more than one unconditioned reinforcer. The generalized reinforcer is useful because the momentary condition of the organism is not likely to be important. The operant strength generated by a single reinforcement is observed only under an appropriate condition of deprivation—when we reinforce with food, we gain control over the hungry man. But if a conditioned reinforcer has been paired with reinforcers appropriate to many conditions, at least one appropriate state of deprivation is more likely to prevail upon a later occasion. A response is therefore more likely to occur. When we reinforce with money, for example, our subsequent control is relatively independent of momentary deprivations.
>
> (Skinner, 1953, p. 77)

Generalized Social Reinforcement

A major source of generalized conditioned reinforcement is mediated by the behavior of other people. Social consequences such as praise, attention, status, and affection are powerful reinforcers for most people. Approval, attention, affection, and praise function as **generalized social reinforcement** for human behavior (Kazdin & Klock, 1973; Kirby & Shields, 1972; Ruggles & LeBlanc, 1982; Vollmer & Hackenberg, 2001; see also Heerey, 2014 for smiles as social reinforcement). In a classroom, a child's off-task behavior may be followed regularly by attention, as when the teacher says, "What are you doing out of your seat?" The teacher may complain that the student's behavior is unmanageable. The problem, however, may concern the social-reinforcement contingency between the student's off-task behavior and the teacher's attention.

Off-task behavior usually captures the teacher's attention because it is highly intense (even aggressive) activity. Attention is reinforcing to most children because it necessarily precedes other types of reinforcement from people. When attention is contingent on off-task behavior, off-task behavior increases. The solution to the problem is not to change the child, but to alter the contingency of reinforcement. One possibility is to place the off-task behavior under extinction (i.e., ignore it) and reinforce the child's behavior with attention at any time other than when they are off-task (differential reinforcement of other behavior, or DRO). "Catch them being good" is the operative phrase.

The importance of generalized social reinforcement involving approval and affection was recognized by Skinner who pointed out that a person is likely to reinforce only the part of another's behavior of which they approve. Sign of his approval, such as a smile or verbal response such as "correct" or "that's right" can become a conditioned reinforcer. We can use this generalized reinforcer to shape the behavior of others, for example in teaching children (or adults) to speak correctly.

Skinner also discussed the submissiveness of others as generalized social reinforcement (see also Patterson, 1982, 2002). In an aggressive episode, two people use threats and possibly physical attack to control each other's behavior. Eventually, one of the combatants gives up, and this submissive behavior inadvertently serves as reinforcement for the aggressive behavior of the attacker. Giving up the argument often results in cessation of the attack by the aggressor, and removal of the attack serves as negative reinforcement for the submissive behavior of the victim. Unfortunately, the contingencies of aggression and submission arrange for an indefinite escalation of conflict, which, in more extreme forms, may result in serious harm or injury that is legally judged as assault, abuse, or murder.

The contingencies of aggression may account for many instances of abuse involving children, partners, the elderly, and individuals incarcerated in prisons and mental hospitals. To the extent that these people are dependent on the benevolence of their parents, partners, or caretakers, they must give in or submit to the demands of their keepers. Consider a person who is unemployed, has few friends, and is married to a person who physically abuses them. When the spouse becomes aggressive, the victim has little recourse other than submission. If the victim calls the police or tells a neighbor, they risk losing their home and income, and they may have learned that their spouse will only become angrier. For these reasons, the aggressor's behavior is shaped to more extreme levels.

Occasionally, victims develop an emotional attachment to the people who mistreat them, which is sometimes called Stockholm syndrome. This kind of affectionate behavior may be shaped as part of the aggressive episode. The contingencies could involve negative reinforcement, as when the aggressor's attack is reduced in intensity or removed by signs of affection from the victim. After some exposure to these contingencies, victims may even claim to love their abusers. Of course, this is an unhealthy relationship—no one should be kept in a relationship by aversive control and fear. Healthy relationships are based on mutual positive reinforcement, in which one another's emotional needs are prioritized.

There are several steps that may be taken to reduce the incidence of victim abuse in our society. One solution involves the issue of control and countercontrol. To prevent control by abuse, the victim must be able to arrange consequences that deter the actions of the aggressor. This *countercontrol* by victims is established when society provides agencies or individuals who monitor abusers and take action on behalf of the victims. Countercontrol may also involve passing laws to protect the rights of individuals who are in vulnerable situations. Another possibility is to teach alternative behavior in terms of negotiation and conflict resolution. Sociocultural environments should also support equal opportunity to resources, such as education, jobs and careers, so that individuals do not need to depend on others for resources in the event they need to escape.

Tokens, Money, and Generalized Reinforcement

Other generalized reinforcers are economic in the sense of being exchangeable for goods and services. Awards, prizes, and scholarships support an enormous range of human activity. Perhaps the most important source of economic reinforcement is money. One way to understand the reinforcing effects of money is to view it as a type of token (coins or bills) exchangeable at a later time for an almost infinite variety of goods and services.

Token reinforcement has been demonstrated in chimpanzees (Figure 10.11; see also Cowles, 1937). Chimpanzees (*Pan troglodytes*) were trained to exchange poker chips for raisins. After tokens and fruit were correlated, the animals learned to select one of several patterns to get poker chips that were later exchanged for raisins. The animals collected several tokens and then went to another room, where they inserted the chips in a vending machine for raisins. Because the discriminative operant (pattern selection) was maintained, the chips were by definition conditioned reinforcers.

Another study also showed that chimpanzees would tolerate a delay between getting a token and exchanging it for food (Wolfe, 1936). The animals earned white chips, which could

Fig. 10.11
Token reinforcement and chimpanzee behavior is depicted.
Source: Yerkes Regional Primate Research Center of Emory University. Republished with permission.

be inserted into a vending machine that immediately delivered grapes. Once the chimps were taught to insert a chip into a vending slot, the animals were taught to pull a lever to get chips. At this point, access to the vending machine was delayed, but the chimpanzees continued to work for tokens. Some animals even began to save their tokens much like people save money. When delays occurred after the chimpanzees had inserted the tokens into the vending machine, the reinforcing effectiveness of the tokens declined. [Note: the delay to reinforcement was increased, hence the delay-reduction hypothesis was supported.] This suggests that tokens bridged the interval between earning and spending, a conclusion supported by a review of token reinforcement by Hackenberg (2009, p. 262).

Experimental Analysis of Token Reinforcement

In the 1950s, Kelleher (1956, 1958b) began the experimental analysis of token reinforcement, viewing the contingencies as a sequence of interconnected schedule components similar to second-order and chain schedules. **Token schedules of reinforcement** have three distinct components involving the *token-production* schedule, the *exchange-production* schedule, and the *token-exchange* schedule (Hackenberg, 2009). Thus, when we talk about token reinforcement, we are referring to three component schedules that form a higher-order sequence. Typically, one of the component schedules is varied while the other two components remain unchanged (held constant).

An experiment by Kelleher (1958b) illustrates the experimental analysis of token reinforcement. Two chimpanzees were initially trained to press a lever on fixed-ratio (FR) and fixed-interval (FI) schedules of food reinforcement. Next, the chimps had to deposit poker chips in a receptacle for food when a window was illuminated, but not when it went dark (discriminated operant). When depositing of chips was well established, the animals were required to press a lever to obtain the poker-chip tokens. Both FR and FI schedules of token production were investigated using simple and multiple schedules. Overall, response patterns for tokens were similar to the break-and-run and scalloping patterns found on FR and FI schedules of food reinforcement.

Subsequently, Kelleher conducted a more in-depth analysis of FR schedules of token reinforcement with the same chimpanzees (No. 117 and No. 119). Responses in the presence of a white light produced tokens that could be exchanged for food at the end of the experimental session, as signaled by a red light. The chimps were required to obtain 60 tokens to produce an exchange period (exchange-production schedule FR 60), and each token could be traded for a food pellet (token exchange FR 1). The schedule of token production varied from FR 30 through to FR 125. Figure 10.12 shows the cumulative records of the final performances by the chimps on an FR 30 token-production schedule. The records show break-and-run patterns early in the session, but steady responding near the end when the chimps had collected nearly 60 tokens to be traded for food. Generally, the response patterns on FR token production showed bouts of responding and pausing typical of FR schedules—mirroring performance on FR schedules of food reinforcement (see also Smith & Jacobs, 2015 for generalized matching by rats on concurrent token-production schedules).

The findings of Kelleher with regard to FR schedules of token production have been replicated with rats and pigeons using different types of tokens. These studies indicate that tokens function as conditioned reinforcers on schedules of reinforcement, but token reinforcement is more complicated than simple schedules (Hackenberg, 2009; see Bullock & Hackenberg, 2015 for multiple-stimulus functions of tokens). Research by Bullock and Hackenberg (2006)

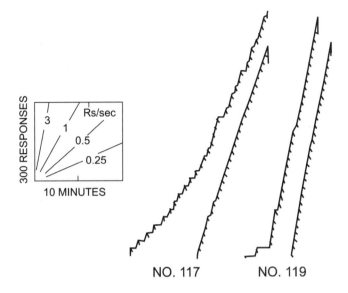

Fig. 10.12
The performance of two chimpanzees on an FR 30 schedule of token reinforcement is shown. The cumulative records indicate the characteristic break-and-run pattern of FR schedules of food reinforcement early in the session, but a high steady rate of response as the period of token exchange approaches near the end of the session. This suggests that the requirements for token exchange (collecting a specified number of tokens) modulate token production behavior on the FR 30 schedule.

Source: R. T. Kelleher (1958). Fixed-ratio schedules of conditioned reinforcement with chimpanzees. *Journal of the Experimental Analysis of Behavior, 1,* pp. 281–289. Copyright 1958 held by John Wiley & Sons, Ltd. Published with permission.

showed that steady-state responding on the FR token-production schedule is modified by the requirements set by the exchange-production schedule (obtaining 2, 4, or 8 tokens before exchanging). Response rates decreased as the FR for token production increased, an effect that was exacerbated by high exchange-production requirements (FR 8 tokens for an exchange period). Thus, response rates on token-reinforcement schedules are jointly controlled by both the token-production and exchange-production schedules.

Additional research has investigated responding for tokens when the exchange-production schedule varied and the token-production and token-exchange FR schedules remained unchanged (Foster, Hackenberg, & Vaidya, 2001). Fixed-ratio (FR) exchange-production schedules (number of tokens required for an exchange period) resulted in break-and-run patterns of responding for tokens, while VR schedules of exchange production eliminated pausing and produced higher overall response rates, similar to simple FR and VR schedules of food reinforcement. One possibility is that responding for tokens on the token-reinforcement schedule is a higher-order unit of behavior that is reinforced according to the requirements on the exchange-production schedule—as in second-order schedules of reinforcement. Hackenberg has summarized the overall findings:

> In sum, behavior under token reinforcement schedules is a joint function of the contingencies whereby tokens are produced and exchanged for other reinforcers. Other things being equal, contingencies in the later links of the chain exert disproportionate control over behavior. Token schedules are part of a family of sequence schedules that include

second-order and extended chained schedules.... Token schedules can be used to create and synthesize behavioral units that participate in larger functional units under the control of other contingencies.

(Hackenberg, 2009, p. 268)

The interdependence of the components of a token system (sequence of schedules) and the emergence of higher-order behavioral units, involving the production and exchange for other reinforcers, indicates that token research with animals may have relevance to research with humans using points and money as reinforcement.

Money and Generalized Conditioned Reinforcement

For people, money is a form of token reinforcement that maintains an enormous diversity and amount of behavior. A major difference between the chimpanzees' tokens and money is that the latter is exchangeable for an unending variety of different reinforcers. For this reason, money is a generalized conditioned reinforcer. Most behavioral experiments involving humans have used money as reinforcement. Money is relatively independent of momentary deprivation, is easily quantified, and is exchangeable for numerous goods and services outside of the laboratory.

Schedules of monetary reinforcement have been used to assess matching (see Chapter 9) and delay reduction with humans. Belke, Pierce, and Powell (1989) created a human-operant chamber, and people were required to pick up tokens from a dispenser and exchange them for 25 cents each. At first, a single token was exchanged for 25 cents, then two tokens for 50 cents, and then four tokens for $1. By extending the delay between earning and exchanging tokens, subjects learned to collect up to 40 tokens before trading them for $10.

In this experiment, there were no instructions and pressing left or right keys was shaped by monetary reinforcement. Various reinforcement schedules were then programmed to test matching, maximizing, and delay-reduction accounts of human choice and preference. Human performance on monetary schedules of reinforcement was better described by matching and maximizing models than by the delay-reduction equation. Relative rate of monetary reinforcement was the most important determinant of behavior in this situation.

The applied advantage of money and tokens traded for money is that these stimuli are tangible objects that are observed easily and the exchange value of the token can be specified precisely. For this reason, a large amount of research has been conducted on experimental communities in which economic reinforcement is scheduled for effective patterns of behavior.

ON THE APPLIED SIDE: THE TOKEN ECONOMY

One of the most important applications of behavior analysis is based on the use of tokens as generalized conditioned reinforcement. Tokens are arbitrary items such as poker chips, tickets, coins, checkmarks in a daily log, stickers, and stars or happy-face symbols given to students. To establish these objects as reinforcement, the applied researcher has a person exchange tokens for a variety of backup reinforcers. A child may exchange five stars for a period of free play, a selection of toys, access to a video game console, access to drawing materials, or an opportunity to build with a LEGO™ set.

A **token economy** is a set of contingencies or a system based on token (conditioned) reinforcement. That is, the contingencies specify when, and under what conditions, particular forms of behavior are reinforced with tokens. It is an economy in the sense that the tokens may be exchanged for goods and services much like money is in our economy. This exchange of tokens for a variety of backup reinforcers ensures that the tokens become generalized conditioned reinforcers.

Systems of token reinforcement have been used to improve the behavior of psychiatric patients (Ayllon & Azrin, 1968), children who have committed criminal acts (Fixsen, Phillips, Phillips, & Wolf, 1976), pupils in remedial classrooms (Breyer & Allen, 1975), typically-developing children in the home (Alvord & Cheney, 1994), and medical patients who follow a plan of treatment (Carton & Schweitzer, 1996; Dapcich-Miura & Hovell, 1979). Token economies have also been designed for those with alcohol and drug use disorders, incarcerated adults, nursing-home residents, and people with intellectual disabilities (Kazdin, 1977; see Dickerson, Tenhula, & Green-Paden, 2005 for token-economy studies for the treatment of schizophrenia).

One of the first token systems was designed for individuals with psychiatric disorders who lived in a large mental hospital. Schaefer and Martin (1966) attempted to modify the behavior of 40 female patients who were diagnosed with long-term schizophrenia. A general characteristic of these women was that they seemed to be disinterested in the activities and happenings on the ward. In addition, many of the women showed little interest in personal hygiene (e.g., a low probability of washing, grooming, and brushing their teeth). In general, Schaefer and Martin referred to this class of behavior as apathetic, and designed a token system to increase social and physical involvement by these patients.

The women were randomly assigned to a treatment or control group. Women in the control group received tokens no matter what they did (i.e., noncontingent reinforcement). Patients in the contingent reinforcement group obtained tokens that could be traded for a variety of privileges and luxuries. Tokens were earned for specific classes of behavior. These response classes were personal hygiene, job performance, and social interaction. For example, a patient earned tokens when she spoke pleasantly to others during group therapy. A social response such as "Good morning, how are you?" resulted in a ward attendant giving her a token and praising her effort. Other responses that were reinforced included personal hygiene, such as use of cosmetics, showering, and generally maintaining a well-groomed appearance. Finally, tokens were earned for specified jobs, such as wiping tables and vacuuming carpets and furniture.

Notice that the reinforcement system encouraged behavior that was incompatible with the label "apathetic." People who are socially responsive, groomed, and who carry out daily jobs are usually described as being more engaged with life. To implement the program, general response classes such as personal hygiene had to be specified and instances of each class, such as brushing teeth or combing hair, had to be defined. Once the behavior was well defined, the researchers trained the ward staff to identify positive instances and deliver tokens for appropriate responses.

Over a 3-month period of the study, the ward staff counted instances of involved and apathetic behavior. Responses in each class of behavior—hygiene, social interaction, and work—increased for women in the contingent-token system, but not for those who were simply given the tokens. Responses that were successful in the token economy were apparently also effective outside the hospital. Only 14% of the patients who were discharged from the token system returned to the hospital; this compared favorably with an average return rate of 28%.

Although Schaefer and Martin (1966) successfully maintained behavioral gains after the patients were discharged, not all token systems are equally effective (see Kazdin, 1983 for a review). Programs that teach social and life skills have lower return rates than those that do not. This presumably occurs because a person who is taught these skills takes better care of themselves and interacts more appropriately with others. Of course, members of the community value these social responses and reinforce and maintain the behavior (see Chapter 13 on behavior trapping).

Token economies that gradually introduce the patient to the world outside the hospital also maintain behavior better than those programs with abrupt transitions from hospital to home. A patient on a token-economy ward may successively earn day passes, overnight stays, weekend release, discharge to a group home, and eventually a return to normal living. This gradual transition to everyday life has two major effects. Contrived reinforcement on the token system is slowly reduced or faded and, at the same time, natural consequences outside of the hospital are contacted. Second, the positive responses of patients are shifted from the relatively dense schedules of reinforcement provided by the token system to the more intermittent reinforcement of the ordinary environment. Designing token systems for transitions to settings outside the institution is a topic of considerable applied importance (Paul, 2006; Wakefield, 2006).

The popularity of the token economy has waned since the 1980s, but reviews of the evidence have resulted in suggestions to reconsider its use (Dickerson et al., 2005; Matson & Boisjoli, 2009). With regard to children with intellectual disabilities and autism, there has been a preference for adopting new, untried treatments despite the evidence that token economies are highly effective at teaching a variety of intellectual and social skills. Recently, early intervention programs for autism have targeted children too young to benefit from token programs, but older children with continuing behavioral problems could still benefit from a well-designed token economy (Matson & Boisjoli, 2009; see Carnett et al., 2014 for a token economy using tokens with intrinsic interest for a youngster with autism). With more call for evidence-based treatments in the autism community, it is likely that there will be a resurgence of token-economy programs in the future.

ADVANCED SECTION: QUANTIFICATION AND DELAY REDUCTION

Consider again the example of the businessperson phoning Delta and American Airlines and how long it takes to get placed on hold at the two airlines (described in the section "Delay Reduction and Conditioned Reinforcement" earlier in this chapter). The average time to be placed on hold at both airlines is 120 s. If the person is dialing back and forth between Delta and American Airlines, the average time taken to get through is 120 s divided by the two choices, or 60 s (i.e., 120/2 = 60). This is because the initial-link schedules are simultaneously available and are both timing out.

Next, consider how long it takes to contact an agent once the businessperson has been placed on hold at one of the two airlines. In this case, the person is stuck on hold at one airline and can no longer dial the other company. The average time in the terminal links of the two chains is 30 s for Delta plus 90 s for American divided by the two links, or 60 s [i.e., (30 + 90)/2 = 60]. That is, over many bookings the person has sometimes waited 90 s for an American agent and at other times 30 s for a Delta agent. On average, the length of time spent waiting on hold is 60 s.

Based on the average times in the initial and terminal links (60 s + 60 s), the overall average total time, T, to book a flight is 120 s or 2 min. Given that it takes an average of T = 120 s to book a flight, how much will the businessperson prefer booking at Delta compared with American Airlines? Recall that it takes an average of 30 s to contact an agent at Delta and 90 s at American, after being placed on hold. This terminal-link time is represented as $t_{2\ DELTA}$ = 30 s, and $t_{2\ AMERICAN}$ = 90 s.

Of the average total time, 90 s have elapsed when the person is placed on hold at Delta ($T - t_{2\ DELTA} = 120 - 30 = 90$ s). That is, the reduction in delay to reinforcement (booking a flight) is 90 s at Delta. The delay reduction at American is 30 s ($T - t_{2\ AMERICAN} = 120 - 90 = 30$ s).

The greater the delay reduction at Delta relative to American Airlines, the greater the conditioned-reinforcement value of Delta compared with American. This relation may be expressed as follows:

$$\frac{R_{DELTA}}{R_{DELTA} + R_{AMERICAN}} = \frac{T - t_{2\ DELTA}}{(T - t_{2\ DELTA}) + (T - t_{2\ AMERICAN})}$$

$$= \frac{120 - 30}{(120 - 30) + (120 - 90)}$$

$$= \frac{90}{90 + 30}$$

$$= 0.75$$

The R values represent responses or, in this example, the number of calls to Delta (R_{DELTA}) and American ($R_{AMERICAN}$), respectively. The relative number of calls made to Delta is equal to the relative reduction in time to book a flight (reinforcement). This time is calculated as the proportion of delay reduction at Delta to the total delay reduction. According to the calculation, 0.75 or 75% of the businessperson's calls will be directed to Delta Airlines.

Experimental Test of Delay Reduction

Edmund Fantino (Figure 10.13; 1939–2015) first proposed and tested the delay-reduction analysis of conditioned reinforcement.

He proposed a general equation for preference on a concurrent-chain schedule that was based on delay reduction (1969a). Equation 10.1 is a generalized statement of the formula used to calculate preference for Delta and American Airlines.

$$\frac{R_L}{R_L + R_R} = \frac{T - t_2}{(T - t_{2L})(T - t_{2R})} \qquad \text{(Equation 10.1)}$$

In this equation, R_L and R_R represent the rate of response on the left and right initial links of a concurrent-chains schedule of reinforcement. The symbol T is the average time to reinforcement (see the airlines example for calculation). The time required in the left and right terminal links is represented by t_{2L} and t_{2R} in the equation. The equation states that relative rate of response is a function of relative reduction in time to unconditioned reinforcement.

The delay-reduction equation emphasizes conditioned reinforcement as a major determinant of choice. This is because the onset of the terminal-link S^D for each chain is correlated with a reduction in time to unconditioned reinforcement. This reduction is $T - t_{2L}$ for the left alternative and $T - t_{2R}$ for the right. Recall that the greater the reduction in time to unconditioned reinforcement signaled by a stimulus, the greater the conditioned-reinforcement value of that stimulus. The delay-reduction equation is a mathematical expression of this idea.

Fig. 10.13
Edmund Fantino.
Reprinted with permission.

Fantino (1969a) designed an experiment to test the delay-reduction equation. The subjects were six pigeons who responded for food on concurrent-chain schedules of reinforcement. In this experiment, the terminal links were always set at $t_{2L} = 30$ s and $t_{2R} = 90$ s. Notice that for the left alternative the relative rate of unconditioned reinforcement is 0.75, and according to the *proportional matching equation* the birds should spend 75% of their time on the left key. The situation is more complex when initial-link schedules are varied. Fantino's experiment involved adding initial links to the VI 30-s and VI 90-s schedules. That is, he investigated a concurrent-chains schedule with 30-s and 90-s terminal links. The schedules in the initial links were always the same for both alternatives, but the values of these schedules were varied over the course of the experiment. For example, in one condition the initial links were VI 30 s on the left and VI 30 s on the right. In another condition, the initial-link schedules were both VI 600 s. Other initial-link values between these two extremes were also investigated. The important question is what happens to the pigeons' preference for the shorter (VI 30-s) terminal link as time is added to the initial links of the chains.

Figure 10.14 shows the proportion of responses predicted by Equation 10.1 for the shorter (VI 30-s) terminal link as time is added equally to the initial links of the concurrent-chain schedule. When the schedules were chain VI 30 s VI 30 s on the left and chain VI 30 s VI 90 s on the right, the birds responded almost exclusively to the left alternative. When the chains were VI 120 s VI 30 s on the left and VI 120 s VI 90 s on the right, the pigeons showed response distributions close to proportional matching (0.75 responses on the left). Finally, when time in

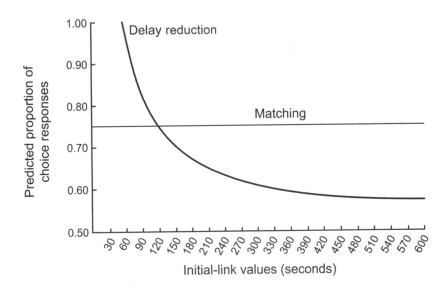

Fig. 10.14
Proportion of responses predicted by the delay-reduction equation is shown for the shorter (VI 30-s) terminal link, as time is added equally to the initial links of the concurrent-chains schedule.

Source: Adapted from E. Fantino (1969a). Choice and rate of reinforcement. *Journal of the Experimental Analysis of Behavior, 12*, pp. 723–730. Copyright 1969 held by John Wiley & Sons, Ltd. Adapted with permission.

the initial links was greatly increased to VI 600 s, the birds showed no preference for either alternative. As you can see in Figure 10.14, these results are in accord with the declining preference predicted by the delay-reduction equation.

A study by McDevitt and Williams (2010) confirmed that delay reduction determined the relative value of terminal-link stimuli (conditioned reinforcers), but also found that the *relative rate of conditioned reinforcement* influenced choice behavior. Fantino and Romanowich (2007) argued that models of choice in concurrent-chain schedules do not require a term or parameter for rate of conditioned reinforcement, but these and other recent findings suggest that conditioned-reinforcement rates may have to be included. The integration of the delay-reduction model with rate of conditioned reinforcement, however, awaits further theoretical analysis (Shahan & Cunningham, 2015 provide a possible integration of information theory, Pavlovian conditioning, and the delay-reduction model for an account of both observing behavior and choice on concurrent-chains schedules).

CHAPTER SUMMARY

In this chapter, we introduced the concept of conditioned reinforcement and research to demonstrate the variables that determine its effectiveness. There are fewer unconditioned reinforcers, but when these biologically relevant events are correlated with previously neutral stimuli, from light flashes to poker chips, these stimuli become capable of reinforcing behavior. Money is perhaps the most common and effective generalized conditioned reinforcer in human culture.

To demonstrate how conditioned reinforcement is studied, we described the use of chain schedules of reinforcement. Chain schedules involve stimuli that acquire more than one function (multiple functions). The discriminative stimulus sets the occasion for a response that is correlated with reinforcement, and a chain schedule shows that the S^D may also function as a conditioned reinforcer. The nearness in time of the conditioned reinforcer to unconditioned reinforcement is a critical factor in the delay-reduction account of conditioned reinforcement. Brain areas (the amygdala and nucleus accumbens) and neurochemicals (dopamine) participate in the regulation of behavior by contingencies of conditioned reinforcement. Organisms more often chose the terminal-link stimulus in a concurrent-chains procedure that reduces the relative delay to reinforcement. In addition, we saw that the use of backward chaining (building the chain backwards from the terminal reinforcer) to teach skills is an effective applied procedure. Finally, the text describes the systematic use of conditioned reinforcers in the form of tokens in primates, including humans. Token systems are micro-examples of money economies, and these systems have helped to manage problems of human behavior in a variety of institutional settings.

Key Words

Backward chaining
Chain schedule of reinforcement
Concurrent-chain schedule
Conditioned reinforcement
Conditioned reinforcer
Delay-reduction hypothesis
Discriminative-stimulus account of conditioned reinforcement
Established-response method
Generalized conditioned reinforcer
Generalized social reinforcement
Heterogeneous chain schedule

Homogeneous chain schedule
Information account of conditioned reinforcement
Mixed schedule of reinforcement
New-response method for conditioned reinforcement
Observing response
Second-order schedule
S–S account of conditioned reinforcement
Tandem schedule
Token economy
Token schedule of reinforcement
Unconditioned reinforcer

On the Web

www.youtube.com/watch?v=zsXP8qeFF6A Ayumu, a young chimpanzee, has learned the order of the Arabic numerals, and outperforms humans on a short-term memory task. While the performance of Ayumu is amazing and is ascribed to chimpanzee's cognitive abilities, it took several years of daily training to perfect. Try to figure out the contingencies using your basic principles of reinforcement, especially principles of matching to sample and conditioned reinforcement. Could differences in prior histories of reinforcement for recall of numbers (learning and reciting the order of numbers) account for differences between humans and chimpanzees on the memory task?

http://members.tripod.com/PoPsMin/classtokenecon.html This website focuses on the use of a token economy in classrooms of children with attention-deficit disorder (ADD).

Conditioned Reinforcement

www.youtube.com/watch?v=IC367wKGi4M This website provides a video of elementary use of clicker training with dogs. See if you can identify the basic principles of behavior and conditioned reinforcement that are used in clicker training.

www.youtube.com/watch?v=OGc8dFdQsJw A description is given on how to use a token economy for children with autism.

Brief Quiz

1. In the laboratory, when a clicking sound is followed by food, the clicking sound:
   a. takes on a conditioned reinforcement function
   b. will support an operant that produces it
   c. can become an enduring reinforcing stimulus
   d. is characterized by all of the above

2. Backward chaining involves:
   a. teaching the initial component or link first
   b. teaching the final component or link first
   c. teaching from the middle to the last component or link
   d. teaching the final component or link in random order

3. On a chain schedule of reinforcement, the longer the delay between the S^D and unconditioned reinforcement:
   a. the greater the stimulus control
   b. the less effective the S^D as a conditioned reinforcer
   c. the greater the value of the unconditioned reinforcer
   d. the less the value of the unconditioned reinforcer

4. In terms of good news and bad news, research suggests that:
   a. stimuli correlated with positive or negative reinforcement maintain an observing response
   b. stimuli correlated with punishment and extinction maintain an observing response
   c. stimuli correlated with negative reinforcement and punishment maintain an observing response
   d. stimuli correlated with positive reinforcement and extinction maintain an observing response

5. The behavior analysis of booking a flight on Delta or American Airlines illustrates:
   a. how behavior is distributed on concurrent schedules of reinforcement
   b. how behavior is distributed on a concurrent-chains schedule of reinforcement
   c. the role of delay reduction in choice situations
   d. both (b) and (c)

6. According to Skinner (1953), a generalized conditioned reinforcer:
   a. is extremely useful because it can be carried around and made contingent on behavior
   b. is not very useful because it relies on the momentary deprivation/satiation of the organism
   c. is produced by pairing a conditioned reinforcer with more than one unconditioned reinforcer
   d. is produced by backward chaining of the unconditioned and conditioned reinforcers

7. Attention from other people is usually reinforcing for children because:
   a. attention has preceded a variety of reinforcements from people
   b. attention is needed for children to develop into emotionally healthy individuals
   c. attention is around children all the time so they get used to it
   d. attention is a fundamental necessity of life that children thrive on

8 Victims sometimes become emotionally attached to people who mistreat them. This could be due to:
   a   the abuser punishing affectionate behavior of the victim
   b   the abuser negatively reinforcing affectionate behavior by the victim
   c   longing for a real emotional attachment to the parents
   d   a token economy situation

9 The research on token reinforcement and chimpanzees shows:
   a   token and food reinforcement are similar in maintaining behavior
   b   tokens can bridge the interval between earning and spending
   c   token reinforcement can maintain and train performance on discrimination tasks
   d   all of the above

10 Systems of token reinforcement in humans have been used to improve the behavior of:
   a   patients with psychiatric disorders
   b   children who engage in criminal acts
   c   typically developing children
   d   medical patients
   e   all of the above

Answers to Brief Quiz: 1, d (p. 359); 2, b (p. 363); 3, b (p. 365); 4, a (p. 370); 5, d (p. 378); 6, c (p. 379); 7, a (p. 379); 8, b (p. 380); 9, d (p. 382); 10, e (p. 384).

Correspondence Relations: Imitation and Rule-Governed Behavior

1 Inquire about contingencies of correspondence and human behavior.
2 Learn about spontaneous imitation in natural settings and the laboratory.
3 Investigate human imitation and mirror neurons.
4 Distinguish between generalized imitation and observational learning.
5 Learn about rule-governed and contingency-shaped behavior.
6 See how instructions or rules affect sensitivity to behavioral contingencies.

People often do what others do. A child who observes their brother raid the cookie jar may engage in similar behavior—at least until they are both caught by their parent. Adults sometimes watch their teenagers' dancing and repeat aspects of these performances at a neighborhood party. Both of these examples involve **correspondence relations** between the demonstrated behavior and the replicated performance. Thus, correspondence involves a special type of stimulus control where the discriminative stimulus is behavior of an individual. In the case of social modeling, we may say that the behavior of one person sets the occasion for an equivalent response by the other.

CORRESPONDENCE AND HUMAN BEHAVIOR

There are other correspondence relations established by our culture. People look for and reinforce the correspondence between *saying and doing*, or more generally between past behavior and current actions (e.g., Lovaas, 1961; Matthews, Shimoff, & Catania, 1987; Paniagua & Baer, 1982; Risley & Hart, 1968; see also Lattal & Doepke, 2001 and Silva & Lattal, 2010 on correspondence as complex conditional discrimination). When a child promises to clean their room and actually does so, their parents are pleased, whereas failure to follow through on the promise may make the parents frustrated. A large part of socialization involves arranging social reinforcement for correspondence between what is said and what is done (see Luciano, Herruzo, & Barnes-Holmes, 2001 for generalized correspondence in children; see also

DOI: 10.4324/9781003202622-11

Lima & Abreu-Rodrigues, 2010 on how "repeating what you said" contributes to generalized correspondence).

By the time a person is an adult, people expect consistency between spoken words and later performance. One kind of consistency that is upheld in social groups is between verbally expressed attitudes and behavior. A minister who preaches moral conduct and lives a moral life is valued; when moral words and moral deeds do not match, people become upset and act to correct the inconsistency. In such instances, what is said does not correspond adequately with what is done. Cognitive dissonance (Festinger, 1957; Gerard, 1994) predicts that people confronted with inconsistency (dissonance) between saying and doing would escape from it (dissonance reduction), thereby ensuring that attitudes matched behavior. Considerable research has supported this prediction. Behavior analysis helps to explain why people engage in dissonance reduction—pointing to the social contingencies that punish low correspondence between words and actions (see Egan, Santos, & Bloom, 2007 for an account of the origins of cognitive dissonance in children and monkeys).

Consistency is also important when people report on private, internal events or happenings. In these cases, the correspondence is between the internal stimulation of the body, behavior, and the verbal report. The social community tries to establish accurate descriptions of private feelings or interoreceptive stimuli (see also Chapter 1). Successful training of such reports involves reinforcing self-descriptive statements such as "I feel angry" or "I am sick" in the presence of presumed private events. Because public cues and private events usually go together, people use external, public cues from behavior to train correspondence between internal stimulation and verbal reports (see Bem, 1972 on self-perception theory; see Egan, Bloom, & Santos, 2010 for an account of how the act of choosing (public cue) leads to preference in children and monkeys; see also Chapter 12). When a child is taught how to report being hurt, parents use crying, holding the wounded area, and physical damage to infer that they are actually in pain. Because the child's behavior and the private stimulation of pain are (often) well correlated, they eventually report "I am hurt" or other internal happenings, solely on the basis of the private stimulation. The private event (painful stimulation) comes to function as a discriminative stimulus for the self-descriptive verbal response.

A problem of privacy is also faced when the community must establish consistency between private, social acts, and the report of those actions. In this case, correspondence is between *doing and saying* (Baer & Detrich, 1990; Deacon & Konarski, 1987; Lubinski & Thompson, 1987; Okouchi & Songmi, 2004). The difference between telling the truth and lying is often a difference in correspondence between doing and saying. During socialization, children are asked to report on their behavior in a variety of situations. A child who returns her empty plate to the kitchen may be asked if she ate her carrots. The response "Yes, I ate every last one" can be verified and reinforced for accuracy or correspondence (see Paniagua, 1989 on lying in children as "do-then-report" correspondence). Young children often are read traditional, moral stories ("Pinocchio"; "George Washington and the Cherry Tree") to promote honesty. Research shows that the correspondence between the moral rule of the story and the honesty of the child depends on the positive consequences for honesty emphasized in the story; stories focused on dishonesty and emphasizing negative consequences fail to promote honesty in children. Truth telling increases for children given verbal appeals to honesty emphasizing social approval, but not for verbal appeals emphasizing the negative consequences of lying (Lee et al., 2014; Talwar, Arruda, & Yachison, 2015).

This repertoire of doing and saying correspondence sometimes has serious implications in adult life. When an employee describes sexual harassment in the workplace, there is some attempt to check on the correspondence between what is reported and the actual happenings. This monitoring of doing and saying by the community is necessary to maintain the accuracy of witnesses' reports (see Critchfield & Perone, 1990 for reinforcement of accurate or truthful self-reports; see Lopez-Perez & Spiegelman, 2013 for lie-aversion presumably established by a history of social punishment). The harassed person is questioned for explicit details, the accused is asked to give their story, and accounts by other people are used to ensure exactness of the reported events. Based on this inquiry, the community ensures more reliable reports by victims, the perpetrators, and the punishment of sexual misconduct. Many aspects of legal trials involve procedures to check on and maintain correspondence between actions and recall. In the courtroom, a witness is asked to take a legal oath by swearing a solemn vow or an affirmation to tell the truth (see Figure 11.1). A witness who misleads the court by describing events and actions that did not occur is guilty of perjury and can be severely punished.

There is evidence that expressing one's feelings, saying and doing, and recalling actions and events are aspects of *verbal behavior* (Skinner, 1957). One important function of verbal behavior involves formulating and following rules, maxims, and instructions (Skinner, 1969).

Fig. 11.1

In a courtroom, a witness swears a solemn oath to tell the truth, the whole truth, and nothing but the truth. Failure to comply with the oath, involving lack of correspondence between the testimony and the facts, is the crime of perjury, punished by imprisonment.

Source: Shutterstock.

Rules may be analyzed as verbal stimuli that alter the responses of a listener. A doctor may state that "too much cholesterol increases the risk of heart attack," and the patient may then act by reducing or eliminating foods that have high cholesterol levels. Advice and other instructions regulate behavior because such rules usually have guided effective action (i.e., health has improved by following your doctor's medical recommendations).

Based on personal experiences, people often describe contingencies (formulate rules) as speakers and then follow them as listeners (rule-governed). Albert Bandura (1997) outlined how self-efficacy rules (i.e., beliefs about one's ability to cope with situations and tasks as exemplified by the children's story "The Little Engine That Could") influence performance and achievement in life. Also, social psychologists have extensively studied the impact of self-rules on thinking and actions, but have relied on social cognitive explanations of this complex behavior (Kunkel, 1997). Behavior analysts insist that following rules, even self-generated rules, is behavior maintained by contingencies of reinforcement (Galizio, 1979; Hayes, 1989b). At the end of this chapter, we analyze the listener's actions as *rule-governed behavior*. The speaker's behavior in stating rules or describing contingencies is examined as verbal behavior in Chapter 12 of this book.

Initially, we describe the process of *observational learning* as a correspondence relationship. Learning by observation involves doing what others do (imitating), in which the performance of an observer or learner is regulated by the actions of a model (correspondence). Although modeling can produce a variety of effects (e.g., social facilitation and stimulus enhancement), **imitation** requires the learner to produce a novel response that could only occur by observing a model emit a similar response (Thorpe, 1963). This kind of social learning may arise from an innate capacity for **spontaneous imitation** from an early age (see "New Directions: Imitation, Action Understanding, and Mirror Neurons" later in this chapter). More complex forms of observational learning involve contingencies that appear to build on this basic repertoire.

CORRESPONDENCE AND SPONTANEOUS IMITATION

Although doing what others do involves a large amount of social learning, this type of correspondence may have a biological basis. At the beginning of the 20th century, psychologists suggested that social organisms have an innate tendency to imitate the actions that they see or hear others perform (Baldwin, 1906; James, 1890; McDougall, 1908; Morgan, 1894). This assumption was largely based on observations that young infants seem to imitate a specific action of an adult: McDougall (1908) indicated that, as early as 4 months of age, his child would stick out his tongue when an adult did the same.

Of course, 4-month-old infants already have a considerable history of interaction with their parents, and the observed tongue protrusion may simply be attributable to social conditioning. That is, parents and other people may have smiled and laughed when the infant imitated their responses. Presumably, these social consequences strengthen imitation by the child. Another possibility is that primate infants make a lot of mouth-opening and tongue-protruding responses, especially when they are aroused by some surprising event such as an adult face. In this case, the infant's tongue protrusions merely coincide with that of the adult model and are not true imitations. Although social conditioning and arousal are plausible, research with infants and animals is providing evidence for innate or spontaneous imitation. Furthermore,

Fig. 11.2

A Japanese macaque monkey is shown washing grain to eat. Monkeys learn this behavior by observation of other monkeys in the troop.

Source: Photograph by Heather Angel of Natural Visions is published with permission.

specialized neurons in the brains of primates have been discovered that may allow for early **innate imitation** by newborns, and more complex forms of **delayed imitation** and observational learning, which involves "remembering" the modeled stimulus rather than direct stimulus control by the modeled action (see "New Directions: Imitation, Action Understanding, and Mirror Neurons" later in this chapter).

Japanese macaque monkeys seem to pass on novel behavior by observational learning. A report by Kawai (1965) describes the social transmission of an innovative method of feeding. The researchers spread grains of wheat on a sandy beach where the troop often visited. Each monkey picked the grains from the sand and ate them one at a time. Then, a young monkey began to separate the sand from the wheat more efficiently by tossing a handful of mixture into the water (see Figure 11.2). When this happened, the sand sank to the bottom and the wheat floated to the top. Using this technique, the monkey obtained more wheat with less effort. Other members of the troop observed this behavior and were soon imitating this new method of feeding and also obtaining a higher density of wheat grains with less effort than before (implicit contingency of reinforcement). Kawai indicated that observational learning transmitted many other novel behaviors, including washing the sand off sweet potatoes and swimming in the ocean (see Fuhrmann, Ravignani, Marshall-Pescini, & Whiten, 2014 for chimpanzees' observational learning—involving motor mimicking, a unidirectional-transmission process from model to observer).

Social animals have many experiences that contribute to doing what others do. It is not possible, therefore, to be sure that the initial imitation of Japanese monkeys was spontaneous or innate (based on species history) rather than acquired (based on social and operant learning). Only laboratory experiments can distinguish between acquired and spontaneous imitation.

IMITATION IN THE LABORATORY

Spontaneous and Delayed Imitation in Pigeons

Thorndike (1898) conducted the earliest experiments on imitation with cats observing the successful performances of other trained cats (model) getting out of a box. He concluded that the experiment was a dismal failure and cats (and other animals) cannot learn by mere observation. Since these early studies, it has been difficult to find an experiment that reliably demonstrates spontaneous or "innate" imitation. This is because reinforcement of the observer's behavior always confounds the results. Based on this realization, Robert Epstein (1984) designed an experiment to show spontaneous imitation with pigeons. The experimental procedures ensured that the observer bird was naive and there were no programmed (or accidental) sources of reinforcement for imitative responses.

Figure 11.3 shows the subjects and apparatus that Epstein used. Some birds served as models, and others were observers. The observers never had been in a laboratory experiment, and none of them had ever eaten from a laboratory feeder. The model and observer pigeons could see one another through a clear partition that separated the chamber into left and right compartments. Each side had exactly the same configuration. Models were always placed in the left side of the chamber, where a feeder was filled with food. Observers were placed in the right side of the chamber, where the feeder never contained food. The modeled performance in various conditions was pecking or pushing a ball, pulling on a rope, or pecking a key. All of the models were trained by operant conditioning to emit the requisite performance for food reinforcement.

There were five conditions in the first experiment. During adaptation, a naive-observer bird was placed in the right side of the chamber. One object (a ping-pong ball, rope, or key)

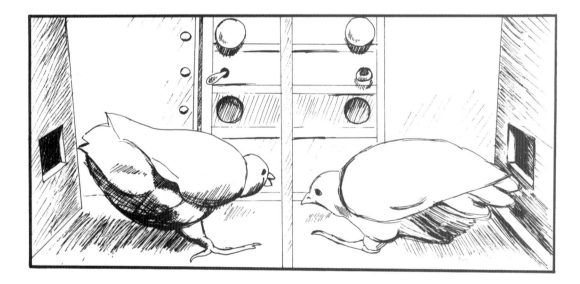

Fig. 11.3
Subjects and apparatus used to study spontaneous imitation by pigeons.
Source: Adapted from R. Epstein (1984). Spontaneous and deferred imitation in the pigeon. *Behavioral Processes*, 9, pp. 347–352.

was situated in the left compartment but not available to the bird in the right chamber. After three sessions, the same object was added to the right chamber and the naive bird was placed alone in the chamber for three sessions of baseline. Next, the object was removed from the right chamber and the model bird was added. During exposure and adaptation to the model, the model engaged in the reinforced performance of pecking the ball, pulling the rope, or pecking the key, and the observer was exposed to this performance without the object for another three sessions. Following this phase, Epstein conducted a test for model-present imitation; he added the object to the observer's chamber while the model continued to demonstrate the performance. If the observer emitted the designated response at a higher rate than during baseline, this was declared to be evidence of direct-spontaneous imitation. Finally, a test for model-absent imitation was implemented. The object remained present but the model was removed. If the observer responded to the object at a higher level than baseline, this was said to be evidence of delayed-spontaneous imitation.

Of the four observer pigeons tested, all showed more responses (key, ball, or rope) with the model present than during baseline. Two of the birds showed strong spontaneous imitation, but the effect was weaker for the other two pigeons. Birds that strongly imitated the model were found to continue this imitation even when the model was removed (i.e., model-absent imitation). The data suggested that delayed-spontaneous imitation can occur in laboratory pigeons, but the results were inconsistent over subjects.

A second experiment was conducted that specifically focused on delayed imitation. In this experiment, "peck the ball" was the imitative response. The same conditions were used as in the first experiment, but the model-present phase was omitted. Thus, the birds were never able to match their responses immediately to those of the model. The results for three new birds were clear: in each case, pecking the ball was higher after exposure to, and removal of, the model compared to baseline. Spontaneous imitation occurred even after 24 h had elapsed between watching the model and the test for imitation.

Analysis of Epstein's Experiments

These experiments on direct- and delayed-spontaneous imitation are important. Experimental procedures ensured that the occurrence of imitation could not be attributed to previous experience or current reinforcement. Thus, it appears that spontaneous imitation is a real effect and is a form of *phylogenetic behavior*. Generally speaking, then, imitative behavior occurs because it has been important for survival and reproduction of members of the species (i.e., contingencies of survival). In other words, organisms that imitated others were more likely to find food, avoid predators, and eventually produce offspring.

The phylogenetic basis of imitation is a reasonable hypothesis. As Epstein notes, however, at least three aspects of the experiments suggest that some environmental experience is also necessary. The birds were raised in a colony and may have had social encounters that contributed to imitative performance. Pigeons that are isolated from birth may show smaller effects of exposure to a model (May & Dorr, 1968). In addition, the effects of food reinforcement may have contributed to the results. Although observers were never directly reinforced with food for imitation, they did see the models eat from the feeder. In fact, Epstein remarked that occasionally the naive bird would thrust its head into the feeder hole when the model did, even though it did not receive food. Finally, only one object was present in the right and left sides of the chamber. If three objects were available, would the observer peck or pull the one the model did, without training? Each of these aspects opposes a strong conclusion about the biological basis of imitation in Epstein's experiments (see Zentall, 2006, 2011 for a discussion of a variety of alternative behavioral effects that are mistaken for imitation). Clearly, additional research is required using the controlled setting for spontaneous imitation arranged in Epstein's experiment.

The experimental research by Epstein (1984) on imitation and delayed imitation in pigeons remains controversial. Thus, one argument may be that stimulus enhancement, or pairing of a conspecific (member of a species) with the ping-pong ball, accounts for the apparent direct and delayed imitation by pigeons (recall, however, that Epstein used a ball, rope, and key). On the other hand, research using a two-action method of pecking an object or pressing a treadle supports Epstein's claim of spontaneous imitation by pigeons (Dorrance & Zentall, 2001). Although there is no further research evidence of delayed imitation by pigeons, evidence indicates that pigeons can imitate a complex, conditional discrimination (Dorrance, 2001), suggesting that delayed imitation is a possible interpretation of Epstein's results.

INFANT IMITATION RESEARCH

Spontaneous Imitation by Newborn Infants

There is evidence that spontaneous imitation occurs in human infants, almost from the moment of birth. Meltzoff and Moore (1977) were the first to report that 12- to 21-day-old infants can imitate the facial and hand movements of adult models. In these experiments, the imitative responses were tongue protrusion, mouth opening, lip protrusion, and sequential-finger movement. The infants' facial gestures and modeled stimuli are illustrated in Figure 11.4.

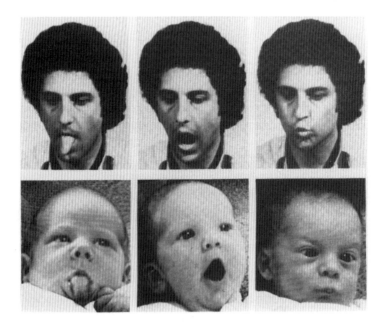

Fig. 11.4
Infants' facial gestures and modeled stimuli are shown.

Source: From A. N. Meltzoff & M. K. Moore (1977). Imitation and facial and manual gestures by human neonates. *Science*, *198*, pp. 75–78. Copyright 1977 held by the American Association for the Advancement of Science. Reprinted with permission.

Experiment 1 used three male and three female infants who ranged in age from 12 to 17 days. The experimenter presented a passive face to the infant for 90 s. Each infant was then shown four gestures or modeled stimuli in random order. Each modeled stimulus was presented four times in a 15-s presentation period. An imitation-test period followed in which the experimenter resumed presenting a passive face and the infant was monitored for imitative responses. The experimenter presented a passive face for 70 s after each new gesture.

The researchers made a videotape of the infants' behavior, and the segments were scored in random order by trained adult judges. For each segment, the judges were to order the four gestures from most likely to least likely in terms of imitation of the modeled stimulus. These judgments were collapsed to yes or no ratings of whether a particular gesture was the imitative response. In all cases, more "yes" judgments occurred when the gesture was imitative than when it was not.

Meltzoff and Moore (1977) designed a second experiment to correct some procedural problems with the first study. Six male and six female infants between 16 and 21 days old were participants. The experiment began with the researcher inserting a pacifier into the infant's mouth and presenting a passive face for 30 s. A baseline period of 150 s followed in which the pacifier was removed, but the passive face continued to be presented. Next, the pacifier was reinserted in the infant's mouth and the researcher presented one of two gestures—mouth opening or lip protrusion. The modeled stimulus was presented until the infant had watched it for 15 s. The experimenter then stopped gesturing and resumed a passive face. At this point, the pacifier was removed and a 150-s response period or imitation test began, during which time the researcher maintained a passive face. Again, the pacifier was reinserted and the second gesture was presented in the same fashion.

The videotapes were scored in random order in terms of the frequency of tongue protrusion and mouth opening. Figure 11.5 shows the frequency of response during baseline and after exposure to the experimenter's gesture. When tongue protrusions were the modeled stimulus, the infant produced this response more frequently than during baseline. On the other

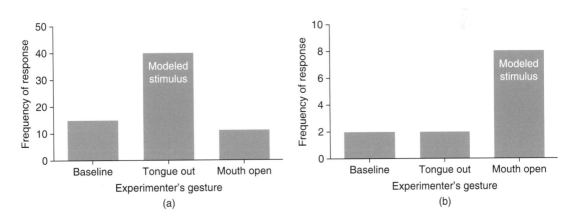

Fig. 11.5
Frequency of response during baseline and after exposure to the experimenter's gesture is shown. In panel (a) the modeled stimulus is tongue out (protrusion) and the frequency of tongue protrusions by infants increases relative to baseline and compared to the control response of mouth open. When the modeled stimulus is mouth open (panel b), the frequency of mouth opening by the infant increases relative to the control conditions.

Source: Adapted from results presented in A. N. Meltzoff & M. K. Moore (1977). Imitation and facial and manual gestures by human neonates. Science, 198, pp. 75–78.

hand, when mouth openings were the modeled stimulus, the infant frequently produced this response during the test period, but not tongue protrusions. These results suggest that newborn infants are capable of imitation of facial gestures (see Kuhl & Meltzoff, 1996 for a discussion of imitation of vocal speech sounds).

In subsequent experiments, Meltzoff and Moore (1983) showed imitation of mouth opening and tongue protrusions in newborns that were 0.7 to 71 h old. Furthermore, human neonates between a few hours and a few days old showed spontaneous imitation of finger movements (Nagy et al., 2005), and auditory–oral matching of consonant sounds (early vocal imitation), behavior thought to be beyond newborn infants' capabilities (Chen, Striano, & Rakoczy, 2004). Notably, neonatal imitation has been reported for chimpanzees (*Pan troglodytes*), suggesting that an imitative capacity at birth is not unique to *Homo sapiens* (Bard, 2007).

Jones (2009) in a review concluded that the widely accepted view of newborn infant imitation lacks supporting evidence. Newborns do match adult behaviors, but this matching is not true imitation. Tongue protrusions to adults by newborns—a well-replicated, apparently imitative behavior—are species-typical responses that occur to a range of stimuli across different sensory modalities (e.g., music). Also, early imitation studies do not rule out general arousal of species-typical responses as an alternative account for apparent newborn imitation. This conclusion, however, is still in dispute and the scientific community has not yet rejected the hypothesis of innate, neonatal imitation (e.g., Zentall, 2011).

Delayed Imitation by Human Infants

A series of studies by Meltzoff (1988a, 1988b, 1988c) indicate that infants ranging in age from 9 to 24 months imitate significantly more modeled actions than control groups over delays ranging from 24 h in the youngest infants to 4 months in the oldest. Additional research indicates that 14-month-old infants show delayed imitation of behavior modeled on television after a 24-h delay. In the same study, the researchers found delayed imitation by infants of behavior modeled by an "expert" toddler performing a novel response after a 48-h delay and a change in context from the experimental situation to the home setting (Hanna & Meltzoff, 1993). The basic findings of the Meltzoff group have been replicated with 6- to 30-month-old infants by other researchers (see Courage & Howe, 2002). The use of "expert" children as models for disabled children in classroom settings has been adopted for such behaviors as toileting, block stacking, and hand clapping (Robert Crow, personal communication).

Correspondence and Intermodal Mapping

Imitation in human newborns, if it exists, involves the infant observing a modeled gesture and responding with a set of muscle movements that correspond to the visual stimulus of modeled actions. The correspondence between the modeled stimulus and the form of response is a remarkable achievement, as the infant is unable to see its own face when she reproduces the facial gestures of the adult model (called "opaque imitation" by Zentall, 2006, 2011).

Meltzoff and Moore (1999) refer to this process as "active intermodal mapping," where infants can monitor their facial movements through proprioceptive feedback and compare this felt activity to what they see. At the present time, there is no detailed evolutionary or neuroscientific account of active intermodal mapping and spontaneous imitation in newborn infants (see Meltzoff, 1999 for a speculative account). The evidence is growing that *mirror neurons* in the brains of humans play a role in the capacity for observed goal-related actions of others, but this system does not seem to be well developed in neonates or even 6-month-old infants (Falck-Ytter, Gredeback, & van Hofsten, 2006). Thus, a definitive neuroscience account of spontaneous imitation in newborn infants is not yet available.

NEW DIRECTIONS: IMITATION, ACTION UNDERSTANDING, AND MIRROR NEURONS

Some say that imitation is the highest form of flattery. Within the last decade, neurons in the ventral premotor area of the brain have been detected that respond when a primate sees someone doing something that the animal itself has done before. These "mirror neurons" were active in macaque monkeys when the animals watched another monkey perform the same action that they had done themselves (Gallese, Fadiga, Fogassi, & Rizzolatti, 1996; Rizzolatti & Craighero, 2004). When a monkey engages in some action, neurons in its frontal lobe are active during the "doing," and a subset of these same neurons fire when the monkey just *watches* a model perform a similar response, even in newborn infant monkeys (Gross, 2006). The neurons fire as though the observer monkey was mirroring the movements of the model (see Figure 11.6). Located at the merger of the anterior dorsal visual stream into the motor cortex, mirror neurons have both motor and visual functions. Mirror-neuron cells fire during action execution and during the observation of the same action, coding both the motor act (grasping) and the action consequences (grasping for food or grasping to move an object). This core imitation circuit interacts with other neural systems related to visual, motor, and task-specific processing, but the key function is toward coding of the motor components of the model's performance, rather than visual representation of what is observed (Iacoboni, 2009).

Mirror-type cells, found in the human anterior cingulate, fire when a person is poked with a needle; surprisingly, they also fire when a patient observes someone else being poked. Some researchers have concluded that mirror neurons dissolve the distinction between "self" and "others," providing humans with the capacity for empathy, action understanding, and even

Fig. 11.6
The photograph (left) depicts modeling of tongue protrusions to a newborn rhesus monkey. The monkey's subsequent response (right) is an opaque imitation of the modeled stimulus. That is, the infant monkey is unable to see its own face when it reproduces the facial gestures of the adult human model. The mirror-neuron system has been implicated in this kind of intermodal mapping as well as other aspects of action understanding, although these claims are currently disputed.

Source: Republished under common copyright from L. Gross (2006). Evolution of neonatal imitation. *PLoS Biology, 9*, p. 1484.

language learning (claims strongly disputed by Hickok, 2014). Given the probability of mirror neurons existing in human infants, a role in action understanding would support the observations of infant-facial imitation and active intermodal mapping (Meltzoff & Moore, 1999; also see infant imitation in this chapter). The mirror circuits, however, are largely underdeveloped in early infancy, requiring more interaction with the environment to organize into a functional mirror system (Coren, Ward, & Enns, 2004, pp. 453–455). And Iriki (2006) suggested that the neural mechanisms for imitation exist in lower primates, but training (in tool use) is required for this capacity to become fully operational.

The impact of conditioning on the mirror-neuron system (MNS) is a way of showing the plasticity of the mirror circuitry (see Kysers & Gazzola, 2014 for mirror neurons and Hebbian contingency-type learning in which an increase in firing of axon neurons, through repeated and consistent activation, causes an increase in firing of nearby cells). The MNS usually supports automatic imitation of sensorimotor actions, but changes to the environmental contingencies result in adjustments by the mirror pathways. Research shows that nonmatching training can abolish or reverse the usual action-matching properties of the MNS (Ray & Heyes, 2011). In one study, participants in a reaction-time task were trained to close their hands to hand-open stimuli, and to open their hands to hand-closed images (Heyes, Bird, Johnson, & Haggard, 2005). The results showed that nonmatching training eliminated automatic imitation. That is, participants no longer made automatic, faster hand-opening responses to the open stimulus than to the closed one. Another study trained nonmatching of finger abductions or movements of a finger away from the other digits (Catmur, Walsh, & Heyes, 2007). After training, the participants showed reverse imitation, producing abductions of the little finger to observed abduction movements of the index finger. The study also showed reverse responsiveness of the MNS using motor-evoked potentials (MEPs), a common indicator of the mirror-neuron system. Thus, after training in nonmatching, observed abductions of the index finger produced larger MEPs in the little finger muscle than in the muscle of the index finger itself. In addition, using functional magnetic resonance imaging (fMRI), researchers reversed the stronger MNS response to hand movements compared with movements of the foot (Catmur et al., 2008). Participants were trained to make a hand response to a foot stimulus and a foot response to a hand stimulus. This training resulted in dominance of the foot over the hand in fMRI imaging responses of both the premotor and parietal areas of the mirror circuitry.

Together these studies show the *neural plasticity* of the MNS to modeling contingencies arranged along sensorimotor dimensions. When the contingencies require correspondence between the model stimulus (hand open) and the response of the observer (hand open), the MNS supports imitation. The MNS, however, can support reverse imitation when the modeling contingencies require noncorrespondence, as when people are given nonmatching training (hand-open model and hand-closed response). This plasticity of the MNS would allow people to use the observed behavior of others to face everyday challenges of social living. On some occasions people are reinforced for doing what others do, but sometimes doing the opposite of what others do is reinforced, as when one person pushes while another pulls. Thus, one function of the MNS may relate to social coordination of individual action.

In fact, however, the functions of mirror neurons are not entirely clear and strongly disputed (Hickok, 2014), but several possibilities have been postulated. Children with *autism* show an altered MNS that may help to explain some of their behavioral deficits (Oberman et al., 2005), such as lack of empathy (recognizing the feelings of others), age-appropriate language skills (the mirror cells are near Broca's language area), or imitation of significant others (Iacoboni et al., 1999). Such findings suggest to some researchers that the MNS could play a central role in children's socialization and language learning (see Rizzolatti, 2014 for imitation, neural mechanisms, and the rise of human culture). Furthermore, mirror neurons seem to enable observers to form action-understandings from others that do not strictly correspond to their own motor representations (Ferrari, Rozzi, & Fogassi, 2005).

As for action and empathic understanding, a study by Repacholi and Meltzoff (2007) found that 18-month-old babies could regulate their imitation of an adult model on the basis of emotions expressed by a third party toward the model (angry or neutral). This kind of emotional eavesdropping by 2-year-olds is what might be expected if the MNS allowed for action-understandings based on the emotional reactions of others, but there was no monitoring of mirror-neuron circuits or brain areas in this study. In adults learning a musical chord by observation of a guitarist (model), researchers found fMRI evidence for the involvement of the premotor cortex—a mirror-neuron circuit presumably related to action understanding during observation of the model (Buccino et al., 2004; see Kruger et al., 2014 for fMRI imaging during imitation of observed action, imitation after a delay, and observed action with no imitation). A different brain circuit, involving the middle-frontal gyrus and motor-preparation areas, became active during rest periods following observation of the model. To date, however, there has been no direct recording of mirror-neuron firings during observational learning of a complex performance in humans.

Overall, the many claims about mirror neurons, especially concerning empathy, action understanding and language learning, seem exaggerated and often extend beyond the current research evidence of behavioral neuroscience. We can view these claims, however, as tentative hypotheses, each requiring substantially more experimental testing before being accepted as scientific facts (see Hickok, 2014 for *The Myth of Mirror Neurons*; see Keysers, 2015 for a critique of Hickok's book in *Science*).

A Behavior Analysis of Spontaneous and Delayed Imitation

In the absence of a definitive account of spontaneous and delayed imitation, how should we understand this research area? One possibility is that there is no specialized capacity for innate, spontaneous imitation. Ray and Heyes (2011) suggest that imitation is entirely acquired by associative learning and increased by reinforcement. Sources of correspondence learning used in infant imitation include direct self-observation, mirrored self-observation, synchronous action, acquired-equivalence experience, and being imitated. Direct observation of one's actions is a common source of early correspondence learning as infants watch the movement of their own limbs and actively match what they see. Older infants use mirrors and reflections to engage in a range of behaviors that serve as visual stimuli to repeat the corresponding actions, such as touching a nose or making a face. Synchronous actions involve the infant and caretaker responding in a similar manner to some event, as when they both open their mouths in response to the infant being fed by a spoon. In acquired-equivalence experience, a stimulus is paired on some occasions with seeing an action, and on other occasions with the performance of the action (see Chapter 12 for a discussion of stimulus equivalence in the learning of language or verbal behavior). For example, a child hears the word "smile" on some occasions when they see someone smiling and at other times when they themselves is smiling, thereby establishing equivalence (the word "smile" = seeing a smile = smiling oneself).

Being imitated by others is the most prominent source of correspondence learning in infancy. Infants and caretakers spend much of the time in face-to-face social interaction, and 79% of the mother–infant interactions involve mothers imitating their infants' responses, increasing the opportunities for infants to do what they see and see what they do (see Chapter 12 on maternal imitation of the infant in the acquisition of language). Considering all the sources of correspondence learning and the progressive development of imitation throughout childhood, Ray and Heyes (2011) conclude that imitation is a completely ontogenetic achievement based on sensorimotor learning arising from a "richly imitogenic sociocultural environment." That is, imitation is learned by contingencies arranged by the culture, rather than activated and maintained by innate neural or cognitive mechanisms.

The claim that early imitation in humans is entirely based on infant learning established by sociocultural "imitogenic" contingencies is highly controversial. If true, it would overturn nearly 40 years of research and theory on innate neonatal imitation, beginning with the famous experiments by Meltzoff and Moore (1977) outlined earlier in this chapter. The evidence for innate imitation has been mixed, but a new study strongly indicates that neonatal imitation at birth does not occur and supports the claim that infant imitation is ontogenetic in origin, completely based on learning.

In a comprehensive longitudinal study of neonatal imitation by Virginia Slaughter and her associates, more than 100 infants were presented with actions and gestures of a social model at 1, 3, 6, and 9 weeks after birth (Oostenbroek et al., 2016). The procedure for presentation of modeled stimuli resembled the original experiment on modeling of tongue protrusions by Meltzoff and Moore (1977), with the infant responses scored as matching or not matching to the modeled stimulus. In the new procedure, however, the infant observed 11 modeled actions (tongue protrusions, mouth opening, happy face, sad face, index finger protrusion, grasping, tube protrusion, box opening, MMM sound, EEE sound and CLICK sound). At each time point, an experimenter presented the infants with each of the modeled stimuli with each stimulus presented for 15 s (five times in 3-s intervals) followed by a 15-s passive position, a trial lasting 1 min. All sessions were videotaped and infant responses were coded by two scorers, one blind to the objectives of the study; reliability of scoring was reported as good. The researchers replicated the positive results from previous studies in subsets of the data, but the overall analysis across the four time points (1, 3, 6, and 9 weeks) showed no evidence of reliable infant matching to the modeled stimuli. The authors concluded that the findings of previous research were an artifact of the restricted comparison or control conditions. It will take time for the scientific community to evaluate the general implications of the new research, but the hypothesis of innate neonatal imitation has been severely challenged.

Currently, the authors of this textbook take a compromise position that imitation by neonates is limited in range and is based on a rudimentary capacity to do what one sees (Meltzoff & Moore, 1999), but later imitation is largely due to reinforcement and other ontogenetic experiences (Ray & Heyes, 2011). That is, imitation is the phenotypic outcome of an *interaction of genes and environment*. Skinner (1984, p. 220) noted that only the first instance of any behavior is entirely attributable to genetic history. Thus, early imitation and delayed imitation by older human infants probably are less related to biology than to environmental experiences. Even so, there is no conclusive behavioral evidence showing that reinforcement history or associative sensorimotor learning accounts for delayed imitation by 6-week-old infants. A reasonable assumption is that delayed imitation involves a capacity to reproduce the modeled actions in the absence of the model, as well as a reinforcement history that substantially builds on this biological capacity.

CORRESPONDENCE, GENERALIZED IMITATION, AND OBSERVATIONAL LEARNING

Operant Imitation and Generalized Imitation

It is possible to train imitation as an operant in a social contingency of reinforcement (Miller & Dollard, 1941). The discriminative stimulus is the behavior of the model (S^D_{model}), the operant is a response that matches the modeled stimulus (R_{match}), and reinforcement is verbal praise (S^r_{social}). Matching the model is reinforced, while noncorresponding responses are extinguished. These social contingencies are similar to the discrimination experiments involving matching to sample for primary reinforcement (see Chapter 8).

Although **operant imitation** provides a straightforward account of observational learning, Albert Bandura (1969) noted that the operant account may be limited to situations in which the observer sees the model, an imitative response immediately occurs, and reinforcement follows. In everyday life, there are numerous occasions when imitation does not conform to this sequence. For example, suppose that a young child is seated in front of a television set watching *Sesame Street*, and they observe Kermit the Frog sing "It's Not Easy Being Green" for the first time. After watching Kermit's performance, the child turns off the television and goes to help their parents in the kitchen. The next day, the child begins to sing Kermit's song. The child's performance approximates the puppet's song. They may not remember every word, but they have the basic tune. Notice that the child has never before performed this sequence of responses. Because of this, directly observed reinforcement could not have strengthened their performance. Also, the child's imitative sequence occurred in the absence of the model; Kermit was not present when they imitated him. Finally, the child's imitative performance was delayed; they sang the song the next day, not immediately after Kermit's demonstration.

The Kermit-song example is typical of observational learning in everyday life, but it seems to defy an $S^D: R \to S^r$ interpretation. The imitative response is novel, and reinforcement for the song is (apparently) missing. In addition, the child sings the song one day later with the model or S^D absent. Finally, Bandura (1969) noted that there is no account of the long delay between modeled performance and later imitation.

Although Bandura (1969, 1977, 1986) has argued against an operant account based on these difficulties, Baer and his associates provided a behavior analysis of imitation that handles each of the apparent challenges to the operant paradigm (Baer & Sherman, 1964; Baer, Peterson, & Sherman, 1967). The approach is called **generalized imitation**, and is based on operant principles of discrimination and generalization (see Glossary for complete definition).

The procedures of generalized imitation begin with simple reinforcement of correspondence or matching between the modeled performance (S^D_{model}) and the imitative operant (R_{match}). The contingency requires the observer to perform the same action as the model. Reinforcement increases imitative behavior, while extinction makes it decrease. If a child is reinforced with praise for imitation of nonsense syllables by a puppet, this response will increase. When praise is withheld, imitation of the puppet declines (Baer & Sherman, 1964).

The actual discrimination procedures are shown in Figure 11.7 and involve several modeled stimuli (S^Ds) and multiple operants (R_{match}). The puppet's head nodding is an S^D for the child to

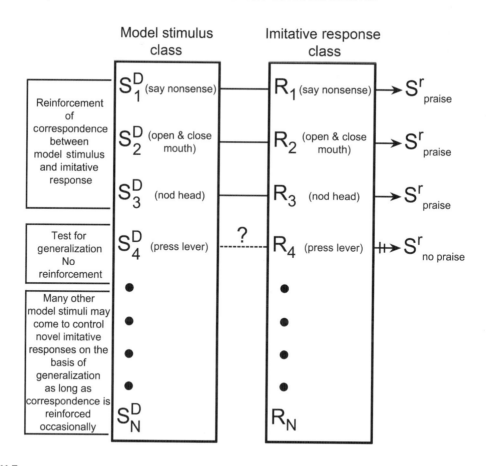

Fig. 11.7

Discrimination procedures used to establish generalized imitation are shown. After training several imitative responses, a test for generalization is given without reinforcement. Generalized stimulus (model) and response (imitation) classes eventually are formed on the basis of training the model-imitation exemplars.

Source: Based on a description of the contingencies in D. M. Baer & J. A. Sherman (1964). Reinforcement control of generalized imitation in young children. *Journal of Experimental Child Psychology, 1*, pp. 37–49.

nod their head and an extinction stimulus (S^Δ) for engaging in other behaviors, such as saying nonsense syllables or opening and closing their mouth. When the puppet opens and closes its mouth, this is an S^D for similar behavior by the child and an S^Δ for the other two responses. In each case, what the model does sets the occasion for a similar behavioral response by the child followed by reinforcement; all other responses are extinguished. This training results in a *stimulus class of modeled actions* and an *imitative response class*. The child now imitates whichever of the three responses the model performs.

The next step is to test for generalization of the stimulus and response classes. Baer and Sherman (1964) showed that a new modeled stimulus would set the occasion for a *novel imitative response*, without any further reinforcement. If the puppet began to press a lever, the child also imitated this performance, even though this response was never reinforced with praise. Thus, *generalized imitation* accounted for the appearance of novel imitative acts in children, even when these responses were never reinforced.

What about the absence of the discriminative stimulus and long delays? It is important to note that all instances of modeling and imitation involve the absence of the S^D before the imitative response occurs. That is, the model demonstrates the action (S^D presented), and after the demonstration (S^D removed) the imitative response is emitted. A contingency may be established that requires a delay of some time between the presentation of the discriminative stimulus and the imitative response (see Chapter 8 on delayed matching to sample). This is the same as when a pigeon pecks a key that matches the sample, but reinforcement depends on delaying the matching response by a few seconds. The delay between the offset of the sample stimulus and the occurrence of the matching response may be lengthened by successive approximation. Eventually, the pigeon may be accurate even after 20 s without seeing the sample.

Similarly, children may learn to delay their imitative responses. Adults may reinforce behavior of newborn infants when the baby mimics their behavior. As the child gets older, reinforcement of imitation depends on increasing delays between the modeled performance and the imitative response. If you tell a joke to someone, that person seldom repeats it in your presence. Immediate repetition of the joke does not reinforce the behavior of the listener. Later the joke is told to another audience, whose laughter reinforces the imitative performance. In this way, social contingencies generate extensive delays between the model stimulus and the imitative response.

It is important to account for the maintenance of generalized imitation. One interpretation involves conditioned reinforcement. Baer and Sherman (1964) suggest that *similarity* becomes a conditioned reinforcer. When a child is taught to imitate, reinforcement occurs only if there is correspondence between the model's actions and the learner's performance. Since reinforcement depends on similarity, imitating others becomes a conditioned reinforcer. Thus, when it occurs, imitation is automatically reinforced (see Erjavec, Lovett, & Horne, 2009 for evidence against this interpretation).

Alternatively, generalized imitation may be maintained by intermittent reinforcement. Gewirtz (1971) indicated that there was no need to postulate similarity as a conditioned reinforcer. He noted that there is no way of separating similarity from the imitative behavior it is said to explain. Intermittent reinforcement for imitation may account for the persistence of generalized imitation. Occasional reinforcement of imitation would maintain the stimulus–response relationships. That is, occasionally imitating others pays off, as when a person learns to operate a computer by watching others.

Experiments on Generalized Imitation with Human Infants

Considerable research investigating generalized imitation has been done with children (Erjavec, Lovett, & Horne, 2009). Most typically developing children, however, have extensive repertoires of imitation before they participate in an experiment. This means that the best way to analyze the determinants of generalized imitation is to use typically developing infants who have more limited imitation histories. In several studies, infants between 9 and 18 months of age were found to pass generalized-imitation tests for object-directed responses, vocalizations, and empty-handed gestures (e.g., Poulson, Kyparissos, Andreatos, Kymmissis, & Parnes, 2002). New research with added controls, however, reports no evidence for generalized imitation in infants around 1 to 2 years of age (Erjavec et al., 2009; Horne & Erjavec, 2007).

One of the most salient differences in studies that support infant generalized imitation, and those that do not, is testing for the imitative-target behaviors before modeling and correspondence training begins. Studies that use novel imitative responses which are not already present in the infants' matching repertoire fail to find generalized imitation in infants. That is, after correspondence training these infants failed to show novel imitative responses to novel modeled stimuli. One possibility is that the infants did not have the requisite skills to match their behavior to the modeled stimulus. Thus, infants were given skills training on the target behaviors before exposure to the correspondence contingencies involving matching their gesture to the model. Even with skills training, infants failed to pass tests for generalized imitation (Horne & Erjavec, 2007). In an attempt to unravel the determinants of infant generalized imitation, Erjavec et al. (2009) increased the number of exemplars for correspondence training (the number of instances of matching to the model) to establish similarity as conditioned reinforcement (Baer & Deguchi, 1985). The results showed that infants would maintain matching to the model if some of the matches were intermittently reinforced (operant imitation), but even with training in multiple exemplars, infants did not pass tests for generalized imitation.

Up to this point, contingencies of correspondence as arranged in the laboratory have not produced generalized imitation in infants or toddlers—suggesting that infants at this age may not have the ability to form higher-order (generalized) stimulus–response classes. Alternatively, training with multiple exemplars across many response topographies may be required—in other words, far more training is required than in these laboratory studies using only 12 exemplars. For example, Baer reported that it took several hundred instances of reinforced matching to establish generalized imitation in Marilla, a child with delayed development and no imitation skills (see "On the Applied Side: Training Generalized Imitation" in this chapter). Another possibility is that the emergence of generalized imitation requires verbal skills such as naming or labeling the modeled performance (as in "Do X"). If this is so, generalized imitation may increase with language acquisition and be related to the onset of observational learning in children.

More recent research indicates that generalized imitation can be established in 2- to 4-year-old children with autism using mirrors to display the modeled actions, but not with face-to-face interaction of a model and observer (Du & Greer, 2014; Miller, Rodriguez, & Rourke, 2015). Generalized-imitation training with mirrors allows for perspective taking where children learn to "see themselves as others see them." When perspective taking is incorporated into training procedures, it is likely that typically developing children also will pass tests for generalized imitation in the laboratory. Whether generalized imitation with perspective taking depicts how infants and young children learn from others' actions during everyday social development awaits further research on the experimental analysis of human development.

ON THE APPLIED SIDE: TRAINING GENERALIZED IMITATION

Donald Baer conducted the early research on generalized imitation and pioneered its application. Together with Montrose M. Wolf and Todd R. Risley at the University of Kansas (1965–2002), he founded the discipline of applied behavior analysis. He was known for his wit, intellectual brilliance, and advocacy on behalf of individuals with behavioral disabilities. In this section we learn about Baer's use of generalized imitation principles to teach imitation to a child with severe disabilities (Baer, Peterson, & Sherman, 1967).

Marilla was a child with profound developmental disabilities who had never shown signs of imitating others. At 12 years old, she had a limited repertoire of responses that included grunting sounds, following simple commands like "Sit down," dressing and feeding herself, going to the washroom, and responses such as turning a knob or opening a door. Although the staff at Firecrest School had tried their best with Marilla, they were now convinced that the child was too limited to learn anything. At this point, Baer and his associates used operant principles to teach generalized imitation to Marilla.

About an hour before lunch, Marilla was brought to a room containing a table and chairs. The training began when the teacher said, "Do this," and raised his arm (S^D). Marilla simply stared at him and did not imitate the response. The same sequence was tried several times without success. On the next attempt, the teacher raised his arm and assisted Marilla in doing the same. After this sequence, Marilla received a spoonful of her lunch and at the same time the teacher said, "Good." After several of these assisted trials, Marilla needed less and less help, and reinforcement only occurred when she lifted her arm by herself. Sometimes she raised her arm when the performance was not modeled (S^Δ); these responses were not reinforced. With this training, Marilla acquired a simple response of raising her hand when the teacher said, "Do this," and demonstrated the action (direct imitation).

Other imitative responses, such as tapping a table and parts of the body with the left hand, were established by shaping and differential reinforcement. After seven examples had been taught, the teacher said, "Do this," and tapped the arm of a chair. Marilla immediately made the same response although she had never been reinforced for doing so. This was the first instance of generalized imitation. A novel modeling stimulus (tapping the arm of the chair) resulted in a new imitative response (Marilla tapping the arm of her chair).

As more instances of reinforced imitation were added to Marilla's repertoire, the percentage of novel imitations increased. Some of the responses were more important in everyday life, such as scribbling on paper, placing geometric forms in order, crawling under a table, and burping a doll. After 120 examples of reinforced imitation, the girl would immediately imitate new examples of modeled performance.

The basic idea of generalized imitation is that reinforcement of some members of the stimulus and response classes maintains the strength of all members—including novel imitations that have never been reinforced. To show the importance of reinforcement for Marilla's novel imitations, the contingency of reinforcement was changed. The teacher continued to model various actions, but Marilla was no longer reinforced for imitating. When she did anything except imitate, however, reinforcement occurred every 30 s. This differential reinforcement of other behavior (DRO) maintains the reinforcer in the setting, places imitation on extinction, and increases behavior that is incompatible with imitation. In less than 20 sessions, both reinforced and novel imitations declined to near-zero responses for each session. Clearly, generalized imitation was maintained by reinforcement.

Next, reinforcement for imitation was reinstated and generalized imitation was acquired again. At this point, the researcher began to teach sequences or chains of imitative performance to Marilla. For example, the teacher would raise his hand and stand up; reinforcement depended on Marilla imitating this two-response sequence. With small steps, the teacher was able to add more and more responses until Marilla could follow a seven-response sequence. Many of the sequences included novel imitative responses that had never been reinforced.

In the final phase of this project, Baer and his associates decided to add vocal responses to the imitative sequences. Since Marilla made grunting sounds, the teacher said, "Do this," rose from his chair, walked to the middle of the room, and said, "Ah." Marilla followed the sequence, but when it came to the vocal response, she only made mouth movements. However, the facial expression was a good first approximation and was reinforced. Over time, closer and closer approximations occurred until Marilla completed the sequence with a well-expressed "Ah." Using fading, the teacher was able to get the girl to say "Ah" whenever he said, "Do this," and demonstrated the vocal response.

Once the imitation of various sounds was well established, the teacher combined the sounds into words and, after about 20 h of vocal imitation, Marilla could imitate words like "Hi," "Okay," "Marilla," and the names of familiar objects. When generalized imitation of motor and vocal responses was well established, new male and female experimenters were used to extend the performance to new models. Now, any teacher could work with Marilla to broaden her skills and add to her behavioral repertoire. Once a sizable imitative repertoire is available, further learning occurs much more rapidly. Rather than teaching separate responses, a person can be shown what to do. This rapid learning of complex skills is necessary for getting along in the world.

The work of Baer et al. (1967) has important practical implications for people with learning disabilities. What is less obvious is the theoretical value of this work. Baer's research shows that complex human behavior may arise from relatively simple behavior principles operating in combination. One implication is that these same principles, when added to a possible cross-modal matching capacity (Meltzoff & Moore, 1999), and perspective taking (Du & Greer, 2014) account for the development of observational learning in everyday life.

Complex Observational Learning

Albert Bandura (Figure 11.8; 1925–2021) worked on complex observational learning and self-regulatory processes for about 50 years, and is one of the most cited researchers in psychology. His work on observational learning, imitation, and aggression is discussed in the following section.

The Bobo Doll Experiment

Bandura (1965) designed an experiment to show a more complex form of learning by observation than generalized imitation. Children participated in this experiment on the imitation of aggressive behavior. As shown in Figure 11.9, each child watched a short film in which an adult demonstrated four distinctive aggressive actions toward an inflated Bobo doll (Bandura, Ross, & Ross, 1963). Every aggressive action was accompanied by a unique verbal response. While sitting on the Bobo doll, the adult punched it in the face and said, "Pow, right in the nose, boom, boom." In another sequence, the adult hit the doll with a mallet, saying, "Sockeroo, stay down." Also, the model kicked the Bobo doll and said, "Fly away," and threw rubber balls at the doll while saying, "Bang."

Some of the children saw the model rewarded by another adult, who supplied soda, a snack, and candies while saying, "Strong champion." Other children saw the model receive negative consequences. The adult scolded and spanked the model for "picking on that clown," and warned the model not to act that way again. A third group saw the modeled aggression, but no social consequences were portrayed for the aggressive behavior.

When the film ended, each child was taken to a room that contained many toys, including a Bobo doll. The child was encouraged to play with the toys and then left alone in the playroom. The researchers watched through a one-way mirror and recorded instances

Correspondence Relations: Imitation and Rule-Governed Behavior   **413**

Fig. 11.8
Albert Bandura.
Reprinted with permission.

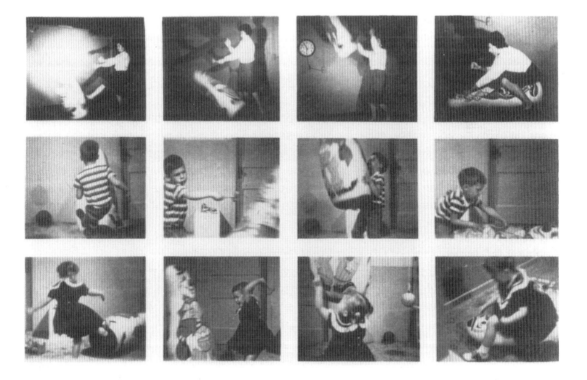

Fig. 11.9
Imitation of modeled aggression against a Bobo doll (Bandura et al., 1963). After viewing a model hit, jump on, and verbally insult a Bobo doll, children also showed these imitative aggressive responses, regardless of gender.

Source: Reprinted from A. Bandura, D. Ross, & S. A. Ross (1969). Imitation of film-mediated aggressive models. *Journal of Abnormal and Social Psychology*, *66*, pp. 3–11. Copyright 1969 held by Albert Bandura. Reprinted with permission.

of aggression and imitative aggression directed at the Bobo doll. Generally, there was a high frequency of imitative aggressive behavior toward the Bobo doll, and boys were more aggressive than girls.

Bandura (1965) also found that reward and punishment of the model's actions affected the imitation of aggression. Children who saw the model's behavior get punished were less likely to imitate aggression than those who saw the model's behavior rewarded. Children who saw the model's behavior rewarded did not differ in imitative aggression from those who watched the model perform the actions without receiving social consequences. Importantly, this means that just seeing modeled aggression (no consequences) had about as much impact on imitation as observing violence being rewarded. Finally, Bandura offered an incentive to all the children if they could remember the actions of the model in the film. With this incentive, all three groups recalled the modeled aggression at about the same level. It seemed that all of the children had learned equally from the modeled aggression, but those who witnessed punishment of the model were less inclined to perform the aggressive sequences.

Social Cognitive Interpretation of Observational Learning in Children
Bandura (1986) argued that the difference between learning and performing modeled aggression requires a cognitive theory of **observational learning**. The observer pays attention to the modeled sequence, noting the arrangement of each action. The general information in the sequence must be coded and rehearsed, as when the child says, "First sit on the Bobo, and then say the word 'pow.'" Once this abstract information is verbally coded and retained in memory, imitation is a matter of reproducing the component responses in the correct sequences. Of course, the manifestations of these cognitive processes—coded, rehearsed, and remembered—are all overt behaviors.

Complex behavior patterns, however, cannot be learned by observation until the component skills have been mastered. It is impossible to fly a plane or do an inward one-and-a-half dive by mere observation. When the separate skills have been acquired, observing others can provide information on how to sequence complex performances, especially with corrective feedback. The golf instructor may show a person how to stand, hold the golf club, and swing at the ball. This demonstration could produce a sequencing of these responses, but the person may still not hit the ball well. It takes corrective feedback from the instructor and the trajectory of the ball to improve performance. Finally, the anticipated consequences of imitation determine the likelihood of an imitative response. People who expect positive outcomes are likely to perform actions that they have witnessed; those who expect negative consequences are less likely to imitate the modeled actions.

Behavioral Interpretation of Observational Learning in Children
A behavioral interpretation of complex observational learning is that it may build on the processes of generalized imitation. As we have noted, generalized imitation provides an account of novel instances of imitation. From an operant perspective, imitation is most likely to occur in situations in which it was reinforced previously. Such behavior is unlikely to happen in situations in which it was extinguished, or in settings in which it was punished.

Suppose that Doug witnesses his brother, Barry, raiding the cookie jar before dinner. Barry is caught by his mother and sent to his room. Later, Doug steals a cookie, also is caught, and sent to his room. Over time, such experiences teach the child a rule—"what happens to others can happen to me." Based on such a learning history, children show *differential imitation* based on modeled consequences. Doug avoids activities for which Barry has been punished, and imitates the rewarded actions of his brother. This kind of conditioning history provides a plausible account of Bandura's results concerning complex observational learning.

The learning and performance differences of the Bobo doll research may also be due to previous conditioning. When Bandura offered an incentive for recalling the modeled action, he presented a discriminative stimulus that increased the probability of this verbal behavior. For most children, it is likely that being promised a reward for recalling some action is a situation that has accompanied reinforcement in the past. That is, a child may be told, "We will be proud of you if you can remember the alphabet," and the child is reinforced for reciting the ABCs. Many such instances result in a generalized tendency to recall events and actions when promised a reward. Given such a history and the incentive conditions that Bandura used, children in all three groups would show a high frequency of recalling what they have observed.

FOCUS ON: RULES, OBSERVATIONAL LEARNING, AND SELF-EFFICACY

Albert Bandura noted that observational learning in humans involves the discovery and use of abstract rules. In a dialogue with Richard Evans (1989), he stated:

> I began to develop the notion of modeling as a broad phenomenon that serves several functions. This conceptualization of modeling is concerned more with the observers' *extracting the rules* and structure of behavior, rather than copying particular examples they had observed. For example, in language learning, children are extracting the rules of how to speak grammatically rather than imitating particular sentences. Once they acquire the structure and the rules, they can use that knowledge to generate new patterns of behavior that go beyond what they've seen or heard. As they acquire the rules of language, they can generate sentences they have never heard. So modeling is a much more complex abstract process than a simple process of response mimicry.
>
> (Evans, 1989, p. 5)

From a behavioral perspective, "extracting the rules" is verbal operant behavior that describes the contingencies of reinforcement (Skinner, 1957, 1969). Both Skinner and Bandura agree about the importance of rules for human behavior, but they differ in terms of interpretation and philosophy.

Bandura (cited in Evans, 1989) talks about rules as cognitive events used to explain behavior, and Skinner (1969) views rules as *verbal descriptions of the operating contingencies*. For Skinner, following rules is behavior under the control of verbal stimuli. That is, statements of rules, advice, maxims, or laws are discriminative stimuli that set the occasion for behavior. Rules, as verbal descriptions, may affect observational learning. In this regard, Bandura's modeling experiments involve a number of distinct behavioral processes—including generalized imitation, descriptions of contingencies, and rule-governed behavior. Behavior analysts study each of these processes to understand how they may combine in complex forms of human behavior, including observational learning.

One kind of rule or description of contingency involves statements about oneself, such as, "I am a competent person who can cope with this situation." This self-description can be contrasted with statements such as, "I am an incompetent person who is unable to cope with this situation." Bandura (1997) refers to these kinds of responses as beliefs in self-efficacy, and provides evidence that these "cognitions" have a large impact on human behavior (see also Bandura & Locke, 2003 for evidence of the causal impact of self-efficacy beliefs).

From a behavior analysis view, statements of self-efficacy, as a class of verbal stimuli, can affect subsequent behavior (see the following section on "Rule-Governed Behavior"). For example, when confronted with the prospect of speaking to a large audience, John thinks (or states out loud) that he does not have the verbal skills to succeed, and estimates that his chances are only 40% for giving a well-organized, interesting, and clear presentation. Subsequently, he gives the talk and, as expected, performs at a low level. In this example, John's statement of self-efficacy describes a past history of behavior at speaking engagements or perhaps social engagements that are similar (a rule). As a rule, the verbal stimulus sets up compliance as reinforcement (e.g., establishing operation). That is, for most people, stating and following rules (compliance) have resulted in generalized social reinforcement from a verbal community. Based on social conditioning for compliance, statements of self-efficacy often predict how a person will act in subsequent (similar) situations.

RULE-GOVERNED BEHAVIOR

A large part of human behavior is regulated by verbal stimuli. Verbal stimuli are the products of speaking, writing, signing, and other forms of verbal behavior (see Chapter 12 for an analysis of verbal behavior). Rules, instructions, advice, and laws are verbal stimuli that affect a wide range of human action. The common property of these kinds of stimuli is that they describe the operating contingencies of reinforcement. The instruction "Turn on the computer and use the mouse to click the desired program in the menu" is a description of the behavior that must be executed to get a program running. Formally, rules, instructions, advice, and laws are **contingency-specifying stimuli**, describing the $S^D: R \rightarrow S^r$ relations of everyday life (Skinner, 1969).

The term **rule-governed behavior** is used when the listener's (or reader's) performance is regulated by contingency-specifying stimuli. According to this definition, a scientist shows rule-governed behavior when following specified procedures to make observations. People, as listeners, may generate their own rules when they speak. Travelers who read a map to get to their cabin might say to themselves, "Take Interstate 5 and turn left at the first exit." The self-directions are verbal rules that describe the contingencies of reinforcement that result in getting to the cabin. In an elementary school classroom, a school student may solve a set of arithmetical problems by following the square-root rule: the square root of a number, n, is the number that gives n when multiplied by itself; thus, the square root of 100 or $\sqrt{100}$ is 10 because $10 \times 10 = 100$. Rule-governed behavior is seen when a patient follows the advice of a doctor to reduce caloric intake by restriction of carbohydrates or a client implements the recommendations of an accountant to increase equity investments to reduce tax assessed by the IRS. When people obey the laws as expressed by posted speed limits, signs that say NO SMOKING, and proscriptions not to steal, the behavior is rule-governed. Following rules often depends on the explicitness of the statement, correspondence to actual contingencies or accuracy, complexity as discriminative stimuli, source of the verbal statement (self or other), and temporal relations (immediate, delayed, or remote). In turn, these features of effective rules control rule-following behavior depending on a listener's history of reinforcement as arranged by the verbal community (Pelaez, 2013).

When solving a problem, people often make up or construct their own discriminative stimuli (response-produced stimuli). A person who has an important early morning appointment may set an alarm clock for 6 a.m. Technically, setting the alarm is **precurrent behavior**, or an operant that precedes some other response. This behavior produces a discriminative stimulus that sets the occasion for getting up and going to the meeting. Thus, a major function of *precurrent behavior* is the **construction of S^Ds** that regulate subsequent action.

People may also construct discriminative stimuli through written words or spoken sounds (verbal stimuli). For example, a person may make a shopping list before going to the supermarket. Making a list is precurrent behavior, and the list is a discriminative stimulus for choosing groceries. Similarly, economical shoppers may say to themselves, "Only buy products that are on sale." This verbal stimulus acts as something like the grocery list in the previous example. As a rule, the verbal expression points to the relation between the stimuli, behavior, and reinforcement in the marketplace (see Taylor & O'Reilly, 1997 for a description of the use of self-instruction in shopping by people with mild learning disabilities). The words *on sale* identify a property of products that is correlated with saving money (reinforcement). The rule makes it easier to discriminate a good deal from a bad one, is easily recalled, and may be executed in any relevant situation (e.g., "buy low and sell high" or "there is no free lunch").

RULE-GOVERNED AND CONTINGENCY-SHAPED BEHAVIOR

People are said to solve problems either by discovery or by instruction. From a behavioral perspective, the difference is between the direct effects of contingencies (discovery) and the indirect effects of rules (instruction). When performance is attributed to direct exposure to reinforcement contingencies, behavior is said to be **contingency-shaped**. As previously noted, performance set up by constructing and following instructions (and other verbal stimuli) is termed *rule-governed behavior* (Catania, Matthews, & Shimoff, 1990; Hayes, 1989b).

Skinner (1969) illustrated the differences between contingency-shaped and rule-governed behavior in his analysis of a baseball player "catching the ball" and a naval commander "catching a satellite":

> The behavior of a baseball outfielder catching a fly ball bears certain resemblances to the behavior of the commander of a ship taking part in the recovery of a re-entering satellite. Both (the outfielder and commander) move about on a surface in a direction and with a speed designed to bring them, if possible, under a falling object at the moment it reaches the surface. Both respond to recent stimulation from the position, direction, and speed of the object, and they both take into account effects of gravity and friction. The behavior of the baseball player, however, has been almost entirely shaped by contingencies of reinforcement, whereas the commander is simply obeying rules derived from the available information and from analogous situations.
>
> (Skinner, 1969, p. 146)

Although behavior attributed to rules and that attributed to contingencies may occasionally look the same, the variables that affect performance are in fact quite different. One difference is motivational—reinforcement determines the rate of response (probability) for a given setting, while rules only affect how the response is executed (topography). Recall that a rule is a special kind of discriminative stimulus and that S^Ds affect behavior because they set the occasion for reinforcement. This means that *rule following itself must arise from contingencies of reinforcement*. The advice of a friend is taken only because such directions have been useful in the past. For example, a friend may have recommended a certain restaurant and you found it enjoyable. Based on these consequences, you are now more likely to follow your friend's advice, especially for dining.

Reinforcement for following the advice of others in various situations may establish a general tendency to do what others recommend. This kind of reinforcement history may underlie a *generalized susceptibility to social influence* (Orne & Evans, 1965). You probably know someone who is a sucker for a sales pitch. Many sales pitches are presented as advice, in the sense that a salesperson describes the benefits (reinforcement) of owning a product. Often, however, the purchase results in more benefits to the seller than to the buyer. The television evangelist does not have a material product, but uses advice, promises, and threats of retribution to get people to send in money.

When directions have been backed up with social punishment rather than natural consequences, they are called orders and commands. Individuals follow orders, regardless of the particular commands, because they have been punished for disobedience (blind obedience, Milgram, 1974). Figure 11.10 shows a famous, though controversial, experiment on obedience to authority by Dr. Stanley Milgram (left) with a shock panel that human participants from New Haven, Connecticut, were ordered to use. The right-hand photograph shows an elderly man (learner) who was given the supposed shocks. Participants in the experiment delivered bogus

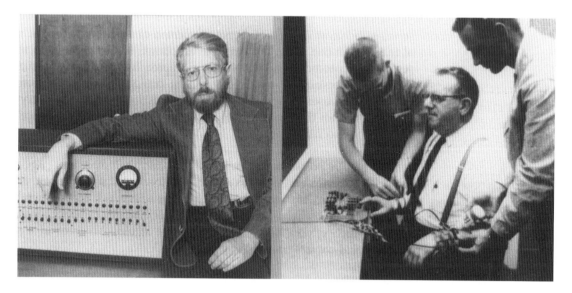

Fig. 11.10

Stanley Milgram's (1974) famous study of obedience to authority illustrates the impact of orders and commands on human behavior. Based on the experimenter's orders, subjects administered what they thought were increasingly severe electric shocks to a 59-year-old man who complained of a heart condition. Stanley Milgram (left) with the shock generator used in the obedience experiment.

Source: Photograph of victim (right) being strapped into the chair is from the film *Obedience* © 1965 by Stanley Milgram and distributed by Penn State Media Sales. Both photographs are reproduced with the permission of Alexandra Milgram.

shocks that they believed to be real. Many participants delivered the shocks even though the older man complained of a heart problem. The tendency to obey the commands of the authority (the experimenter) outweighed the signs and sounds of distress from the elderly victim (see Burger, 2009 for a replication of Milgram's basic findings; see also Reicher, Haslam, and Smith, 2012 for obedience and participants' identification with science versus the moral values of the general community; see Perry, 2012 for a methodological and ethical critique of Milgram's obedience experiments; but see Travis, 2013 for a review of Perry's book and a defense of Milgram's research).

The importance of reinforcement contingencies in establishing and maintaining rule-following behavior are clearly seen with ineffective rules and instructions. One kind of rule that is likely to be weak is based on statistical analysis of contingencies. For example, it is unlikely that a person will give up smoking merely on the basis of the directive "Stop smoking—smoking causes cancer." The actual health consequences are too remote and the statistical chances of getting cancer too unlikely. Of course, smoking usually declines when a person gets cancer, but at this point it is too late. When rules describe delayed and improbable events, it is necessary to find other reasons to follow them, as when the community establishes immediate consequences for smoking by ostracizing smokers to special restricted areas.

Government reports about second-hand smoke and its effects have led some communities to classify public smoking as illegal. Towns and cities arrange fines and other penalties for failure to obey the no-smoking law. In this case, smokers follow the anti-smoking rule for reasons unrelated to smoking itself (i.e., social punishment). A similar effect is obtained when smoking is described as sinful or shameful and religious sanctions are used to promote compliance. Generally, social contingencies may be used to establish rule-following behavior when natural contingencies are too remote or improbable to be effective.

FOCUS ON: INSTRUCTIONS AND CONTINGENCIES

In his discussion of rule-governed and contingency-shaped behavior, Skinner (1969) speculated that instructions might affect performance differently than the actual contingencies of reinforcement. One way to test this idea is to expose humans to reinforcement procedures that are accurately or inaccurately described by the experimenter's instructions. If behavior varies with the instructions while the actual contingencies remain the same, this would be evidence for Skinner's assertion (see Hackenberg & Joker, 1994 on correspondence between instructions and contingencies).

An early study by Lippman and Meyer (1967) showed that human performance on a fixed-interval (FI) schedule varied with instructions. When subjects were told that points (exchanged for money) would become available after a specific amount of time, their performance was characterized by a low rate of response, appropriate to the fixed interval. In contrast, subjects who were told that points depended on a certain number of responses produced a high and steady rate of response. In a similar kind of study, Kaufman, Baron, and Kopp (1966) placed subjects on a variable-interval (VI) schedule of reinforcement and told them that points were available on either a fixed-interval or a variable-ratio (VR) basis. Performance was more in accord with the experimental instructions than with the actual VI contingencies.

The fact that instructions, in these experiments, seem to override the actual contingencies has been used to argue against a reinforcement analysis of human behavior. Bandura

Fig. 11.11
Mark Galizio.
Reprinted with permission.

(1971, 1974) linked instructions to modeling. He argued that both of these procedures activate subjects' expectations, which in turn affect subsequent behavior. This means that expected reinforcement, rather than actual contingencies, is the stronger determinant of human behavior. In addition, Dulany (1968) disputed the claim that instructions were complex discriminative stimuli. He argued that there was no evidence to show that instructions gain (or lose) control over behavior because of selective reinforcement.

Mark Galizio (1979) (Figure 11.11) addressed both objections when he showed that following instructions is in fact a discriminative operant. In a series of important experiments, human subjects responded to avoid the loss of money. Subjects received a payment to attend experimental sessions, and they could turn a handle to avoid a loss of 5 cents from their earnings. When they turned the handle to the right, the onset of a red light and loss of money were postponed. Subjects were exposed to four different contingencies during a session. A change in the contingency was signaled by one of four amber lights. One condition had no losses, but the other three had costs scheduled every 10 s. For the conditions in which costs occurred, each response delayed the next loss for either 10, 30, or 60 s.

To vary instructional control, labels were placed above the amber lights that signaled each condition. When instructions were accurate, there were no discrepancies between the labels and the contingencies. Thus, the component in which each response postponed the loss for 10 s was labeled correctly as "10 s," as were the "30-s," "60-s," and "no-loss" components.

Galizio (1979) also created conditions of inaccurate instructions in which the labels did not match the actual contingencies. In a no-contact condition, all of the components were changed to no losses, but the labels incorrectly described different response requirements. If subjects behaved in accord with the instructions, they made unnecessary responses, but there was no monetary loss. As you might expect, people followed the rules. For example, subjects turned the handle more when the label said "10 s" than when it said "60 s."

At this point, a contact condition was implemented in which losses occurred every 10 s in all components. The signs still read "10 s," "20 s," "60 s," and "no loss." In this situation, responding to the instructions produced considerable loss of earnings. Consider a person who turned the handle every 60 s but lost money every 10 s. Subjects quickly stopped following the instructions and responded in terms of the actual contingencies of reinforcement.

Overall, the results of Galizio's experiments provide strong support for the view that instructional control is a form of rule-governed behavior (see Buskist & Miller, 1986; Hayes, Brownstein, Haas, & Greenway, 1986; Horne & Lowe, 1993; Ribes & Martinez, 1998; Ribes & Rodriguez, 2001; see Hayes & Ju, 1997 and Plaud & Newberry, 1996 for analysis and applications of rule-governed behavior). In accord with numerous other experiments, subjects were found to rapidly acquire appropriate responses to the contingencies when instructed about how to behave.

Importantly, the influence of instructions depended on the consequences of following these rules (see Svartdal, 1992). When the costs of rule-following behavior increased, people no longer followed the rules. Additional evidence showed that following instructions could be brought under stimulus control (O'Hora, Barnes-Holmes, and Stewart, 2014 found evidence that understanding instructions and following them involve different sources of stimulus control). Thus, people follow instructions in situations that have signaled reinforcement, but not in situations signaling extinction or aversive consequences. Finally, Galizio showed that accurate instructions have reinforcing properties, a characteristic shared by simple discriminative stimuli. People not only respond to instructions but also seek out reliable descriptions of the contingencies (see Pelaez, 2013).

The degree to which consequences play a role in rule-following can also be applied to the extent to which people around the world followed guidelines for reduction of risk of the COVID-19 virus during the pandemic that began in 2019. When this chapter was written, there were over 1 million COVID-related deaths in America and over 6 million COVID-related deaths worldwide within a two-year period since the pandemic began (Centers for Disease Control and Prevention, 2022a). Americans were offered suggestions (and in some cases, mandates) on how to reduce the spread and severity of COVID-19 from the Centers for Disease Control and Prevention and World Heath Organization—federal and global agencies, respectively, that pool epidemiological data, analyze trends, and make recommendations for reducing risk. There were two especially effective behaviors for reducing risk—wearing masks and getting the COVID-19 vaccine. One would imagine that contingencies such as the number of deaths would influence the correspondence between the recommendations (or rules) from government agencies and the behavior that would reduce the spread of a highly contagious virus. However, in America (and all over the world, for that matter), many refused these tactics. Indeed, some even protested them (see Figure 11.12). Why would people willfully not follow a set of rules that could save their lives?

The answer is complex, but likely has to do with consequences of rule following. Part of the issue was that only a small percent of individuals who contracted COVID-19 died (often less than 2% of the cases)—that is, there was a relatively low likelihood of death even if a person contracted COVID-19. Moreover, the symptoms varied in intensity from one individual to another. Some who contracted COVID-19 were asymptomatic or had mild symptoms; others had very severe symptoms that required hospitalization. Therefore, to many who directly experienced COVID-19, the virus did not appear deadly at all. In addition, many of those who died from COVID-19 were older or had risk factors, making it difficult for younger, healthier people to identify the contingencies of severe illness or death with themselves. Further

Fig. 11.12
A vaccine protester in Malta who, despite the scientific evidence that supports that COVID-19 vaccines are safe and effective, holds a sign that says that the vaccine is dangerous.
Source: www.reuters.com/world/europe/malta-sees-biggest-protest-yet-against-covid-measures-2022-01-16/

complicating the issues was misinformation that was spread about the effectiveness, side effects, and safety of masks and vaccines (see Figure 11.12). *Requiring* people to wear masks or get vaccinated for a virus that, to them, did not seem deadly, did not help with compliance. Indeed, it occasioned countercontrol. Therefore, the correspondence between the rules and consequences of COVID-19 mitigation was not clear.

Rules as Function-Altering Events

Although the discriminative function (S^D) of rules is well established, several researchers (Malott, 1988; Michael, 1982a, 1983; Schlinger & Blakely, 1987) argued that contingency-specifying stimuli have additional, and perhaps even more crucial, effects. Rules can act as **function-altering events**, altering the function of other stimuli and thereby the strength of relations among these stimuli and behavior (Schlinger & Blakely, 1987). A passenger on an airplane is instructed to respond to a drop in cabin pressure by "placing the yellow oxygen mask over your mouth and breathing normally." The instruction is a *function-altering event* that sets up the "dangling yellow mask" as a discriminative stimulus for placing the mask on the face. In the absence of the rule, the dangling mask might occasion looking at it or asking for the airline attendant (see Schmitt, 2001 for a discussion of rule following after a delay). The function-altering effect of the airline rule is shown when the passengers put on their masks only at the appropriate moment. Also, the probability of placing masks on faces is higher for those who are given the instruction.

Rules may alter the discriminative functions of stimuli in more complex ways, as when a person is given detailed instructions. An individual may be told, "See George about buying the car, but if Sherry is there don't make an offer." As a result of this verbal description, the listener emits a conditional discrimination: George is S^D for making an offer, and Sherry is an S^Δ for this behavior. Notice that without the detailed instruction or rule, George and Sherry may have no discriminative functions when buying a car.

Following Rules and Joint Control

Lowenkron (1999) discussed how rules control behavior. That is, how do verbal statements of contingencies (rules) emitted at one time regulate behavior at a later time? Barry presented three problems for a behavior analysis of rules: first, *memory function* or how rules have effects after a delay; secondly, *recognition function* or how the event specified by a rule is known; and thirdly, *response function* or how the specified event occasions the specific response. To answer these questions without using the language of cognition and mental events, Lowenkron introduced the notion of **joint control** where two verbal stimuli exert stimulus control over a common verbal topography (Lowenkron, 2006; see Fields & Spear, 2012 for complex joint control involving graph-to-text correspondence, as in "paying attention" to features of the graph, the description of the graph, and the correspondence between the two).

Figure 11.13 depicts a task that involves joint control. The problem is for you to locate the number 103020 in the array at the top of the figure. This is a type of matching-to-sample problem with the array as the comparison. Try to do it now. Finding the correct sequence required joint control by verbal stimuli over the terminal verbal response "103020, I found it." Given the statement of the problem, you probably rehearsed or repeated the sequence to yourself (memory function) as you looked at the array of numbers to verbally identify the correct sequence (recognition function). The two verbal stimuli, repeating the required number and identifying that number in the array, jointly controlled the terminal verbal response "103020, I found it" (response function) (see Chapter 12 for formal analysis of the verbal responses by the speaker; see Lowenkron, 2004 for an extension of this analysis to word–object meaning).

In another common example, you are getting ready for a birthday party and there is a cake in the oven. Your friend has to go to the store to buy some soft drinks and says to you, "When

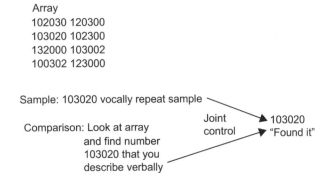

Fig. 11.13

Joint control by verbal stimuli is illustrated. The two verbal stimuli, repeating the required number and identifying that number in the array, jointly control the terminal verbal response "103020, I found it" (response function).

Source: Based on B. Lowenkron (1999). Joint control of rule following: An analysis of purpose. *Annual Meetings of the Association for Behavior Analysis in Chicago*. Retrieved from www.calstatela.edu/faculty/zlowenk/toc.html.

the cake has risen, take it out of the oven" (rule). You are likely to repeat the rule ("better see if the cake is done") and check the cake as it is baking. Notice that the memory function of the rule is fulfilled by repetition or verbal rehearsal of the rule statement over time, not by a mental event or cognition. At some point, you repeat the rule to "check the cake," look in the oven, and verbally identify that "the cake has risen." The verbal stimulus "the cake has risen" fulfills the recognition function without reference to cognitive events. Rehearsal of the rule statement and verbally identifying that the cake has risen exert joint control over the terminal verbal response "the cake is done; take it out of the oven" and removal of the cake from the oven.

Notice that the form or topography of the terminal response is completely specified by your friend's statement of the rule—you say that the cake is done and take it out of the oven to comply with your friend's request (rule). Failure to follow the rules often results in social punishment, as people get upset when their instructions are not reinforced with compliance. Also, rules that are ignored usually have additional aversive consequences, such as ruined cakes and spoiled birthday parties (see Cerutti, 1989 on collateral consequences of rule following). As we have seen in this chapter, the contingencies of reinforcement ensure that we often follow the rules of others and the rules that we give to ourselves (self-generated rules). Thus, rule-governed behavior is operant behavior regulated by contingencies of reinforcement.

CHAPTER SUMMARY

In this chapter, we learned about correspondence relations, focusing on imitation and rule-governed behavior. Spontaneous imitation occurs in several species, including human infants, but the evidence is still controversial. The discovery of mirror neurons suggests that early imitation in infants and other primates is based on neural mechanisms related to our evolutionary history, but again there are disputes about the claims and evidence. We propose that operant imitation, generalized imitation, and complex observational learning build on a basic capacity for imitation. Observational learning seems to integrate imitation with rule-following behavior to produce behavior that is transmitted from one person to another and from one generation to the next. Rule-governed behavior concerns the effects of verbal stimuli on the behavior of the listener. That is, instructions and other rules (other verbal stimuli) are products of the behavior of the speaker that regulate the behavior of the listener. We discovered that rules play a large and important role in the regulation of human behavior, not as mental events but as verbal descriptions of the contingencies. Rule-following behavior is maintained by social and collateral contingencies, but instructed behavior often appears to be insensitive to reinforcement contingencies. When inaccurate rules generate behavior with high costs, however, people give up following the rules and respond to the actual contingencies. One way to understand rule-governed behavior is based on joint control where two verbal stimuli combine to control a common form of verbal response. In the next chapter, we examine contingencies of reinforcement that regulate the behavior of the speaker, or what Skinner (1957) called the analysis of verbal behavior.

Key Words

Construction of S^Ds
Contingency-shaped behavior
Contingency-specifying stimuli
Correspondence relations
Delayed imitation
Function-altering event
Generalized imitation
Imitation
Innate imitation
Joint control
Observational learning
Operant imitation
Precurrent behavior
Rule-governed behavior
Spontaneous imitation

On the Web

www.pigeon.psy.tufts.edu/avc/zentall/default.htm On this website, Zentall and Akins address the issue of imitation in animals, including the evidence, function, and mechanisms. The distinction between apparent imitation and "true" imitation is one topic of interest.

www.ted.com/talks/vs_ramachandran_the_neurons_that_shaped_civilization Neuroscientist Vilayanur Ramachandran outlines the fascinating functions of mirror neurons. Only recently discovered, these neurons allow us to learn complex social behaviors, some of which formed the foundations of human civilization as we know it. As we noted in the chapter, the presumed multiple functions of mirror neurons are very much disputed and require more research evidence before being accepted as scientific facts.

www.youtube.com/watch?v=8aBbnz7hZsM The research on infant imitation by Andrew Meltzoff and his associates was outlined in this chapter. Go to this YouTube webpage to find out more about babies being born to learn, about the study of early imitation and other related research by Dr. Meltzoff.

www.youtube.com/watch?v=C6ju2-ljWhs&list=PL-WYHqmfGbt9-fSHaBdudnhRCrJpNpWTH Following directions and instructions is an important part of the rule-governed behavior of children, especially before going to elementary school. In this *Sesame Street* episode, "Furry Potter and the Goblet of Cookies," observational learning is used to teach children to follow directions of placing cookies correctly in two containers, using different features of the cookies. Try to analyze the episode in terms of the correspondence relations used in this chapter. Note the consequences of following the instructions correctly, such as getting into the castle and getting untangled from the spider web.

Brief Quiz

1 The relation between saying and doing is formally a _____ relation.
    a   cognitive consistency
    b   correspondence
    c   synergistic
    d   dose–response

2 Imitation requires that the learner emits a _____ response that could only occur by observing a(n) _____ emit a similar response.
    a   significant; peer
    b   operant; organism
    c   novel; model
    d   similar; conspecific

3   What did Thorndike (1898) conclude from imitation experiments with animals?
    a   animals show amazing intelligence
    b   animals can easily imitate another member of their species
    c   animals do not show a capacity for intelligence
    d   animals cannot learn by observation

4   The second experiment by Epstein (1984) concerned _____ _____ and showed that spontaneous imitation in pigeons occurred after _____ had elapsed.
    a   delayed imitation; 24 h
    b   deferred matching; 48 h
    c   delayed sampling; 24 h
    d   deferred equivalence; 48 h

5   In the study (Experiment 1) of infant imitation by Meltzoff and Moore (1977), the researchers:
    a   used 12- to 21-day-old infants
    b   presented a passive face for 90 s
    c   presented four gestures in random order
    d   did all of the above

6   Both humans and pigeons seem to engage in spontaneous imitation. The appearance of similar behavior:
    a   shows that humans and pigeons share much in common
    b   shows identical functions of the behavior
    c   shows that structure and function go together
    d   shows none of the above

7   With regard to generalized imitation, which of the following statements are true?
    a   generalized imitation is based on principles of social cognition
    b   generalized imitation provides an account of imitation after a delay
    c   generalized imitation is part of Bandura's cognitive theory
    d   all of the above

8   To show the importance of _____ for Marilla's _____ imitations, Don Baer changed the _____ of reinforcement.
    a   stimulus control; immediate; contingency
    b   stimulus control; spontaneous; quality
    c   reinforcement; novel; contingency
    d   reinforcement; immediate; quality

9   Rule-governed behavior involves:
    a   control by contingency-specifying stimuli (verbal stimuli)
    b   the effects of instructions on the behavior of the listener
    c   the effects of advice given by a counselor
    d   all of the above

10 Rule following is _____ regulated by _____ of reinforcement.
    a   behavior; contingencies
    b   cognitively; expectancies
    c   mentally; contingencies
    d   socially; expectancies

Answers to Brief Quiz: 1, b (p. 394); 2, c (p. 396); 3, d (p. 398); 4, a (p. 400); 5, d (p. 401); 6, d (pp. 399–402); 7, b (p. 409); 8, c (p. 411); 9, d (p. 415); 10, a (p. 419).

CHAPTER 12

Verbal Behavior

1. Identify the difference between language and verbal behavior.
2. Discover the operant functions and basic units of verbal behavior.
3. Investigate the emergence of naming as a higher-order verbal operant class.
4. Learn about three equivalence relations: reflexivity, symmetry, and transitivity.
5. Delve into the behavioral neuroscience of derived stimulus relations.
6. Inquire about three-term contingencies and the natural speech of infants.

Humans are social animals. Most of the daily life of people takes place in the company of others. An important aspect of human social behavior involves what we do with words, as in speaking, writing, signing, and gesturing. Behavior analysts use the term *verbal behavior* to refer to this kind of human activity. In this chapter, verbal behavior is analyzed according to the same principles of behavior that have been used throughout this book. The analysis explores the role of contingencies of reinforcement in the regulation of verbal behavior.

In terms of behavior analysis, Lee (1981a) notes that the concept of language tends to obscure environment–behavior relationships. Language usually directs research attention to grammar, syntax, and unobservable mental representations and processes (structure), rather than to the objective conditions that influence the behavior of a speaker or writer (function). Catania (1998) has also noted that the "language of reference" implicitly proceeds from words to objects in the world. The possibility that environmental contingencies regulate our speaking and writing is not usually considered.

LANGUAGE AND VERBAL BEHAVIOR

People usually use the term "language" when they talk about speaking and other forms of communication. Although some researchers argue that language is behavior (Baer & Guess, 1971), others use the term to refer to a set of linguistic habits (Hockett, 1958, 1968), while still others point to the underlying innate universal grammar that is presumed to organize spoken and written words (e.g., Chomsky, 1957; but see Lieberman, 2015 for evolutionary

DOI: 10.4324/9781003202622-12

and neurobiological evidence against universal grammar). Some view language as a cultural phenomenon that does not depend on individual behavior or mental structures (Sanders, 1974). Finally, language is said to consist of three main features involving vocabulary, syntax, and meaning (Erlich, 2000, p. 140). As you can see, there is little agreement on the definition of language. The most important implication of this confusion is that language may not be a useful concept for a natural-science approach to speaking, communicating, and other forms of verbal behavior.

To rectify these problems, Skinner (1957) introduced the term *verbal behavior*. This term helps to redirect attention to the operating contingencies controlling the speaker's behavior. In contrast to the term *language*, **verbal behavior** deals with the performance of a speaker and the environmental conditions that establish and maintain such performance. That is, verbal behavior concerns the *function* of what we do with words that are spoken, written, or signed. Some of the functions of verbal behavior that have been researched include how we learn to talk about things and events in the world, how we learn to communicate our feelings and emotions, and how the listener's response to what we say shapes what we talk about (see Chapter 11 for an analysis of the listener's rule-governed behavior).

FOCUS ON: SPEAKING AND EVOLUTION OF THE VOCAL APPARATUS

The way in which people of a culture reinforce the verbal behavior of others changes over time, and modification of the social contingencies alters what people say. If biological evolution did not lead to the mental rules or universal grammar of language, as is usually assumed (Dessalles, 2007, pp. 153–164), it is useful to ask what role evolution and biology played in human speech and communication. In his 1986 paper on evolution and verbal behavior, Skinner speculated about the role of natural selection for vocal behavior or speaking. He stated:

> The human species took a crucial step forward when its vocal musculature came under operant control in the production of speech sounds. Indeed, it is possible that all the distinctive achievements of the species can be traced to that one genetic change. Other species behave vocally, of course, and the behavior is sometimes modified slightly during the lifetime of the individual … but … the principal contingencies have remained phylogenetic.… Some of the organs in the production of speech sounds were already subject to operant conditioning. The diaphragm must have participated in controlled breathing, the tongue and jaw in chewing and swallowing, the jaw and teeth in biting and tearing, and the lips in sipping and sucking, all of which could be changed by operant conditioning. Only the vocal cords and pharynx seem to have served no prior operant function. The crucial step in the evolution of verbal behavior appears, then, to have been the genetic change that brought them under the control of operant conditioning and made possible the coordination of all of these systems in the production of speech sounds.
>
> (Skinner, 1986, p. 117)

Skinner's evolutionary analysis of the human anatomy allowing for vocal speech has been confirmed subsequently by research in acoustics, physiology, and anatomy. Although there is much controversy about the evolution of human speech, most scientists now agree on the "principles of physiology and the 'laws' of physical acoustics that determine the capabilities of the anatomy involved in the production of human speech" (Lieberman, Laitman, Reidenberg, & Gannon, 1992, p. 447).

The production and variation of audible speech sounds are severely limited in mammals, including the great apes. Human infants retain the nonhuman primate and mammalian location of the larynx (upper neck, C3), and cannot make the vowel sounds [i], [u], and [a] that are present in adult speech. Over the first 2 to 3 years, the larynx of the child shows a developmental descent to the adult position (C6) in the neck. This descent is accompanied by other anatomical changes that dramatically alter the way children breathe and swallow. During this developmental period, neuromuscular control is extended within the larynx and pharynx, with changes beginning even before the descent of the larynx has occurred.

Phonation, or the production of audible sources of speech sound, is made possible by the activity of the larynx—involving the movement of the vocal folds or cords. Complex neuromuscular control ensures that the vocal folds are able to move inward before the onset of speech, and rapidly in and out for continued speech production (Lieberman et al., 1992). Furthermore, the supralaryngeal-vocal tract consists of both the oral and nasal airways that provide the phonetic quality of the sound (Figure 12.1). This tract serves as an acoustic filter (in much the same way as a pair of sunglasses filters light), suppressing the passage of sounds at particular frequencies, but allowing passage of those at other frequencies (Lieberman, 2007).

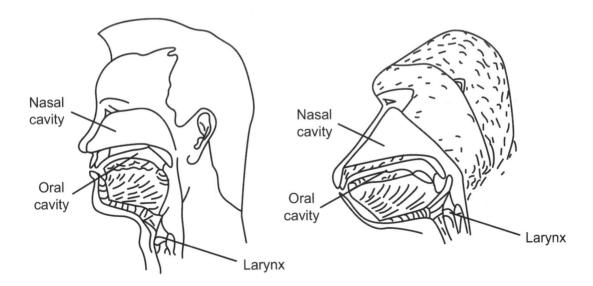

Fig. 12.1
Voiced sounds of human and nonhuman primates involve vibration of the vocal folds of the larynx. Sound travels up the vocal tract where the oral and nasal cavities of the vocal tract act as a filter. This filter allows passage of acoustic energy at some frequencies and attenuates energy at other frequencies according to their size- and shape-specific transfer function. The sound radiated at the lips thus reflects the combined action of an acoustic source and a filter. The larynx lies lower in the vocal tract of the adult human compared with nonhuman or young children, creating a second large cavity (the pharynx) at the back of the mouth.

Source: A. A. Ghazanfar & D. Rendall (2008). Evolution of human vocal reproduction. *Current Biology, 18*(11), pp. R457–R460. Artwork by Michael Graham. Adapted with permission of the American Psychological Association from D. Rendall, J. R. Vokey, & C. Nemeth (2007). Lifting the curtain on the Wizard of Oz: Biased voiced-based impressions of speaker size. *Journal of Experimental Psychology and Human Perception Performance, 33*, pp. 1208–1219.

The human tongue, originally adapted for swallowing, has been modified by natural selection to allow for the formation of the basic vowel sounds [i], [u], and [a], which are fundamental to adult speaking (Lieberman, 2012). The shape and position of the tongue gradually change with the developing infant. The newborn tongue is flat and mostly resides in the oral cavity, limiting production of the basic vowels related to complex speech. As the tongue gradually descends into the pharynx (throat) during development, it assumes a posterior rounded contour, carrying the larynx down with it. By the age of 6 to 8 years, the tongue's oral and pharyngeal proportions are equal (1:1) and its shape is fully formed, allowing for control of the vocal musculature and tongue required for adult speaking and communication. In humans, the FOXP2 transcription factor (encoded by the FOXP2 gene) is implicated in enhanced motor control and learning of speech by increasing synaptic plasticity and dendrite connectivity in the basal ganglia and other neural circuits (Lieberman, 2012).

Clearly, the human vocal apparatus and neural capacity for speech has allowed for fine-grain operant control of verbal behavior (Skinner, 1957). During the course of human evolution, the breathing and digestive regions were modified from the two-tube system of our earliest hominid ancestors to the intersecting upper tracts that we have today. These changes would have involved substantial "restructuring in the respiratory, digestive and vocalizing patterns and would have occurred contemporaneously with parallel changes in central and peripheral neural control" (Lieberman et al., 1992, p. 324; see also Lieberman, 2012). It is likely that many minor modifications of morphology had already appeared in the genus *Homo* more than one million years ago (Lieberman, 2015). These modifications were further elaborated by natural selection and appeared as the integrated respiratory, upper digestive, and vocal tract of *Homo sapiens* approximately 500,000 years ago (see MacLarnon & Hewitt, 2004 on the evolution of human breath control; see also Lieberman, 2012 and Takemoto, 2008 on the evolution of the human tongue). Specification of the selection pressures that favored such changes currently is a primary focus of research on the evolution of human speech (de Boer, 2005; Lieberman, 2007, 2014).

VERBAL BEHAVIOR: SOME BASIC DISTINCTIONS

Verbal behavior refers to the vocal, written, and signed behavior of a speaker, writer, or communicator. This behavior operates on the listener, reader, or observer, who arranges for reinforcement of the verbal performance in a particular setting. A person may ask a server for "a demitasse of vichyssoise." The speaker's behavior affects the listener, who in turn supplies reinforcement—serving the soup. A similar effect is produced if the person writes their order on a piece of paper. In this case, the written words function like the spoken ones; the server reads the order and brings the meal. Verbal behavior therefore substantially expands the ways that humans can produce effects on the world.

Verbal behavior allows us to affect the environment *indirectly* (Vargas, 1998). This contrasts with nonverbal behavior, which often results in direct and automatic consequences. When you walk toward an object, you come closer to it. If you lift a glass, there is a direct and automatic change in its position. Verbal behavior, on the other hand, only works through its effects on other people. To change the position of a lamp, the speaker says "Hold up the blue lamp" to a listener who is inclined to respond. Notice that reinforcement of the verbal response is not automatic, as many conditions may affect what the listener does. The listener may not hear you, may be distracted, or may not understand you—picking up the red lamp rather than the blue one. Generally, the social contingencies that regulate verbal behavior are complex, subtle, and highly flexible.

The Range of Verbal Behavior

Although verbal behavior is usually equated with speaking, vocal responses are only one of its forms. In addition to talking, a person emits gestures and body movements (non-verbals) that indirectly operate on the environment through their effects on others. In most cultures, a frown sets the occasion for others to remove some aversive event while a smile may signal the observer to behave in ways that produce positive reinforcement. In fact, frowns and smiles have such consistent and pervasive effects on others that some researchers have considered these gestures to be universal symbols (Rosenberg & Ekman, 1995).

Another kind of verbal behavior involves manual signing rather than speech sounds. In American Sign Language (ASL), the speaker produces arm and hand movements that are functionally similar to speech sounds. In this case, regulation of the listener's behavior is along a visual dimension. Deaf speakers may also acquire complex finger movements known as "finger spelling" that function like letters in the English alphabet. Figure 12.2 illustrates some of the basic manual movements of ASL and digital positions for finger spelling.

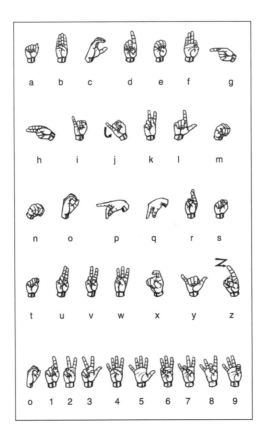

Fig. 12.2
Examples of American Sign Language (ASL) and finger-spelled letters.
Source: From T. J. O'Rourke (1978). *A basic vocabulary: American Sign Language for parents and children*. Silver Spring, MD: T. J. Publishers. Copyright 1978 T. J. Publishers, Inc.

In the behavioral view, writing is verbal behavior that functions to regulate the behavior of a reader. Although written words and sentences have little formal similarity to spoken ones, the two modes of communication have equivalent functions. Recall that behavior analysts classify behavior in terms of its functions, and for this reason both writing and speaking are commonly categorized as *verbal operants*.

Speaking, Listening, and the Verbal Community

The behavior of the speaker (or writer) is functionally different from the behavior of the listener (or reader). That is, the conditions that regulate speaking are distinct from those that affect listening. In the field of psycholinguistics, the distinction between speaking and listening is often blurred by talking about language encoding and decoding. Since both are treated as aspects of language (i.e., the transmission of meaning), there is little attempt to analyze the separate functions of such behavior. In fact, Skinner (1969) used the term *rule-governed behavior* to describe the behavior of the listener and *verbal behavior* to specify the performance of the speaker. Of course, in any actual communication between two people, each person alternates as speaker and listener, or occasionally a person even talks to herself—acting as both speaker and listener in the same body (see Silbert, Honey, Simony, Poeppel, & Hasson, 2014 for a possible common neural mechanism supporting everyday speaking and listening, usually denoted as speech production and comprehension). The repertoires of behavior of speakers and listeners are, however, analytically distinct from Skinner's perspective.

Rule-governed behavior refers to the effects of words in the forms of instructions, advice, maxims, and laws on the listener's behavior (see Chapter 11). In this view, rules are seen as complex discriminative stimuli, and the principles that govern stimulus control also regulate the behavior of the listener. Although many behavior analysts have accepted this perspective, others have suggested that rule-governed behavior involves additional processes (Hayes, 1989b).

Regardless of one's view about the behavior of the listener, verbal behavior requires special attention because the consequences of verbal behavior are *mediated by the actions of others* (Vargas, 1998). The way a person speaks is shaped by the consequences supplied by the listener. A busy parent may not respond to the polite response of "Milk, please" by their child. However, a change in form to "Give me milk!" may induce compliance. Inadvertently, the parent is teaching their child to give commands in a loud voice. Subtle contingencies of reinforcement shape the style, dialect, tonal quality, and other properties of speaking.

The contingencies that regulate verbal behavior arise from the practices of people in the **verbal community**. It is the verbal community (community of listeners) that arranges the social contingencies to establish and maintain the practices of listeners with respect to the verbal behavior of speakers. These practices are part of the culture of the group, which have evolved over generations (Skinner, 1953). The practices of the verbal community therefore refer to the customary ways in which people of the culture reinforce the behavior of a speaker. When linguists analyze the grammar of a language, they extract rules that describe the reinforcing practices of the verbal community. For example, the grammatical rule "[i] before [e] except after [c]" describes a requirement for reinforcement set by the community; the written spelling *received* is reinforced while *recieved* is not. Thus, verbal behavior is established and maintained by the reinforcing practices of the community, and these practices change based on cultural evolution (see Mesoudi, 2016, who emphasizes imitation as a mechanism for high-fidelity transmission of cultural practices, while underplaying the critical importance of contingencies of reinforcement arranged by the verbal community in establishing and maintaining customary practices of the group). The analysis of cultural change in terms of verbal practices requires the integration of several fields of study including anthropology, archeology, and linguistics, and is beyond the scope of this textbook (but see Chapter 14 for a brief analysis of cultural transmission and change).

Social Use of Words in a Verbal Community

In a mentalistic view of language, words are said to refer to things in the world. That is, words somehow have the power to represent, communicate, and express the world as perceived by the speaker. The speaker is said to encode by syntax (grammar) and semantics (meaning) the message or information that is transmitted by speech and decoded or comprehended by the listener. This information-transmission view of language has gained considerable popularity, especially in the age of computers and cellular communication.

Behavior analysts offer a different viewpoint, proposing that the social environment shapes the way in which we use words (Guerin, 2003; Skinner, 1957). That is, the way we talk and what we say are a function of social contingencies, involving effects or consequences arranged by members of a verbal community. If you yell at a rock or at the moon, it has no effect. On the other hand, when you yell "You're not listening to me!" at your partner there may be usually observable social consequences, such as the partner now listening to the speaker. There could also be verbal counter-aggression and disruption of the social relationship, depending on how the verbal operant was delivered. Generally, we say things in ways that previously induced others to do something for us, compelled others to say things, elicited or occasioned people's attention, and influenced others to remain in social relationships with us. The long-range consequences of speech and word-use reside in gaining access to the social and economic resources mediated by members of diverse verbal communities (Guerin, 2003), from close relationships in dating, intimacy, and families to friendship and collegial relationships of school, community, and workplace.

One of the ways in which words are used is to "establish facts," in the sense that our words correspond to events and objects in the everyday world. When words about the world correspond to actual happenings, people (or oneself) are likely to adopt and repeat those words, usually with conviction (response strength), and to act in ways consistent with the verbal description. In everyday language, a person who "believes" some statement of fact is likely to be persuaded by it. For example, if you can persuade a neighbor that you have a good reputation, this person may tell others about your standing in the community, help you when you need it, and provide you with a variety of social and economic resources.

To establish facts and persuade others (or themselves), speakers use words to give accounts of their own or others' actions. A major emphasis of social psychology is attribution, or giving accounts and explanations of the actions of others and ourselves (Kelley, 1987). In the traditional view, attributions are cognitions generated by the perceiver to make sense of the world and predict what others will do in social interactions. Guerin (2003, p. 260) argues, however, that attributions are not used to understand the social world, but are, in fact, verbal strategies or ways to persuade others (and oneself) with words, thereby gaining and maintaining access to social and economic resources. For example, an explanation or attribution of why Taylor and Ollie are fighting may attribute the cause to Taylor's tendency to "sit on the couch" rather than look for employment ("Taylor and Ollie are fighting because Taylor is lazy and a good-for-nothing")—an internal attribution to Taylor's disposition. In contrast, an explanation of the conflict might involve giving an external attribution, describing the cause of fighting as the "lack of employment opportunities" in the job market and "the loss of income." A behavior analysis suggests that speakers will use a form of attribution (internal or external) to prevent challenges to their "factual" accounts and to maintain social credibility, rather than just to explain and understand the actions of others. Thus, a speaker who is talking to an audience predisposed against Taylor (Ollie's sibling) is likely to provide an internal (dispositional) account of their actions; an audience predisposed toward Taylor would occasion an external (marketplace) attribution for the pair's dispute.

Control by the audience over the speaker's causal accounts of actions (internal vs. external) is clearly seen in the so-called self-serving bias, which attributes personal success to internal factors and failures to external sources. Research shows that people provide internal accounts of their own actions when they expect positive consequences from others. A pianist who gives an outstanding gala performance may claim to television viewers that it was the outcome of her perseverance and dedication to music over a lifetime. Another pianist who gives a mediocre or dismal rendition of the musical work might point to aspects of the situation (an inattentive and restless audience) or context that disrupted her performance (external attribution). The internal attribution claims responsibility for the successful performance and positive social consequences from listeners; external attribution for failure weakens social responsibility, reducing blame and other negative social consequences. Overall, accounts or explanations of actions may be analyzed as the use of words by speakers (verbal behavior) shaped by the social consequences of listeners, or members of a verbal community (for a more complete analysis of speaker–listener conversation, see Guerin, 2003).

OPERANT FUNCTIONS OF VERBAL BEHAVIOR

In his book *Verbal Behavior*, Skinner (1957) presented a preliminary analysis of this kind of human activity. Although some linguists have treated Skinner's work as a behavioral theory of language, it is more likely that the book represents a set of testable hypotheses about verbal behavior (MacCorquodale, 1970). Skinner described verbal behavior in terms of the principles found in the operant laboratory. Such an analysis must ultimately be judged in terms of its adequacy. That is, it must deal with the facts of the speaker's behavior in natural settings and the experimental and observational evidence that supports or refutes such an account (Sautter & LeBlanc, 2006). In this section, the basic verbal classes are outlined using the distinctions made by Skinner (1957) as well as clarifications made by others (e.g., Michael, 1982b; Oah & Dickinson, 1989). Remember that the basic units or elements of verbal behavior would be combined, elaborated, and extended in any comprehensive account of speaking, talking, and communicating. How these basic units are integrated into actual complex speech requires continuing analysis and research, as found in the journal *The Analysis of Verbal Behavior*, published by the Association for Behavior Analysis International.

Functional Operant Units: Manding and Tacting

Verbal behavior may be separated into two broad operant classes, namely manding and tacting, based on the regulating conditions. These two operant classes involve the functions of getting what you want from others (manding) and making contact or reference to things and happenings in the world (tacting).

When you say "Give me the book," "Don't do that," "Stop," and so on, your words are regulated by motivational conditions—deprivation for the book or another person doing something unpleasant. In behavior analysis, this verbal behavior is called manding. **Manding** refers to a class of verbal operants whose form is regulated by establishing operations (e.g., deprivation, aversive stimulation, etc.). The word *manding* comes from the common English word *commanding*, but commanding is only a small part of this operant class.

Everyday examples of manding include asking someone for a glass of water when you are thirsty, or requesting directions from a stranger when you are lost. Notice that specific reinforcement is made effective for manding by some *establishing operation*. A glass of water reinforces asking for it when you are deprived of water, and directions are reinforcement for requesting them when you are lost. Common forms of manding include speaking or writing orders, asking questions, requesting objects or things, giving flattering comments to others, and promoting commercial products (e.g., "Buy this detergent").

There is another major class of verbal operants. **Tacting** is defined as a class of verbal operants whose form is regulated by nonverbal discriminative stimuli (nonverbal S^D), which is shaped and maintained by generalized conditioned reinforcement from the verbal community. A child is tacting when she says "The sun is orange" in the presence of the midday sun on a beach with their parent. In this example, the presence of the sun in the sky (and the relevant property of color) is a nonverbal S^D for tacting by the child. The operant class of tacting is maintained by generalized conditioned reinforcement from the verbal community (e.g., mother, father, teacher, friends, and others), usually in the form of corrective feedback such as "Yes," "Right," and so on. The word *tacting* comes from the more familiar term *contacting*, and refers to verbal behavior that makes contact with events in the world (nonverbal S^Ds). Everyday examples of tacting include describing a scene, identifying objects, providing information about things or issues, and reporting on your own behavior and feelings and that of others.

Occasionally, it is difficult to distinguish between manding and tacting. A child who says "juice" in the presence of a glass of apple juice could mean "give juice" or "that is a glass of juice." If the response is equivalent to "give juice," it is functioning as manding—controlled by deprivation and the specific reinforcement of "getting juice"—but if the response is controlled by the nonverbal stimulus of the glass of juice, it is tacting. In another example, a person who says "I believe you have the sports page" may be tacting the nonverbal stimulus (the sports page), or manding specific reinforcement (getting the sports page). The issue is often resolved by the listener saying "Yes, I do" and continuing to read the paper. If the original response was manding, the listener's reply will not function as reinforcement (the sports page is not given). In this case, the speaker is likely to clarify the disguised manding by asking, "May I please have the sports page?"

In a further example, a person who picks up their significant other for a date may say, "You look attractive tonight" (Figure 12.3). Again, *the form or topography of response cannot distinguish manding from tacting*. If the speaker's verbal response is regulated by abstract

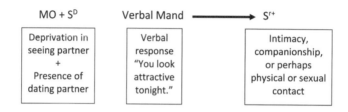

Fig. 12.3
An analytical diagram of manding sexual reinforcement on a date. The MO or motivational operation of deprivation of not having seen the dating partner in a while plus the presence of the date as a discriminative stimulus (S^D) sets up the conditions for the speaker's response, "You look attractive tonight." The verbal mand is maintained by a previous history of specific reinforcement (S^r_{sexual}). Under different controlling conditions the same response could be a verbal tact of the listener's "attraction" (a complex nonverbal discriminative stimulus). Notice that the form of response (what the speaker says) does not tell us whether they are manding or tacting. Only an analysis of the controlling conditions for the verbal response can clarify its function—if the MO is withdrawn (chances of physical of sexual contact) but the speaker still says their date is "attractive," the verbal response is likely a tact.

properties of "attractive" (nonverbal S^D) of the listener, then the speaker is tacting. On the other hand, the verbal response could be regulated by motivation for intimacy or physical or sexual reinforcement and, if so, the speaker is manding. Only an analysis and test of the relevant contingencies can distinguish between manding and tacting behavior. Thus, the listener could test the controlling contingencies by withholding physical or sexual reinforcement—testing the veracity of their date's flattering remarks.

Many advertisements and television commercials are disguised manding, in the sense that the verbal responses of an announcer seem to describe the benefits of the product (tacting), but are in fact requests to buy it (manding). A television actor dressed in a doctor's white coat states that "Xprin relieves pain and muscle ache and is available at your local drug store." The verbal description of the product (relieves pain) suggests tacting by the speaker (actor), but there are manding (profit) aspects to the verbal behavior. Given these conflicting contingencies, listeners learn how to reveal the disguised manding of a speaker (testing the controlling variables), and speakers learn to conceal their obvious manding of a listener (making the verbal description of the product appear as tacting its benefits). Persuasion and attitude change may therefore be analyzed in terms of manding, tacting, and the interaction of speakers and listeners (Bem, 1965).

RESEARCH ON VERBAL BEHAVIOR: MANDING AND TACTING

Training of Verbal Operants

According to Skinner (1957), the basic classes of verbal behavior are functionally independent in the sense that the relations involved in manding are distinct from those that define tacting. This **functional independence** means that it is possible to teach manding and tacting as separate, unrelated operant classes. It also implies that there is no basic ordering of the verbal repertoire; that is, it is not necessary to train manding before tacting, or vice versa. The hypothesis of functional independence of the basic verbal response classes is one feature that distinguishes Skinner's functional analysis of verbal behavior from other approaches to language development (Carr & Miguel, 2013; Gamba, Goyos, & Petursdottir, 2014). In this section, research on basic verbal relations is outlined and assessed in terms of functional independence of the response classes.

Mand Relations

Recall that the mand relation is defined by an establishing operation (EO) and specific reinforcement. An establishing procedure regulates the topography or form of manding behavior and sets up a specific consequence as reinforcement. To train manding, the most direct procedure is to manipulate an EO and reinforce the verbal response with the specified consequence. In the animal laboratory, EOs usually involve a history of deprivation for some event that functions as primary reinforcement, such as food.

Most human behavior in everyday settings, however, is regulated by conditioned and generalized conditioned reinforcement. To investigate the manding of conditioned reinforcement, Michael (1988) suggested the use of a **conditioned establishing operation (CEO)**. The procedure is called the blocked-response CEO, in which a response that usually occurs is blocked because of the temporary absence of a specific condition, stimulus, or event. For example, you may leave your seminar notes at home as you rush to the university. Because you cannot complete the behavioral sequence or chain of giving a seminar presentation, obtaining the notes would function as reinforcement for sending a text message to your roommate or running back to your dorm room to get them. The notes would not have a reinforcement function during a casual lunch with an old friend, for example, because they are not necessary to this behavioral sequence. Whenever an event or stimulus is required to complete a behavior chain, withholding the event should establish it as reinforcement for operant behavior (see Michael, 2000 for a more extensive analysis).

Hall and Sundberg (1987) used the blocked-response CEO to train manding by students who were deaf with developmental delays. The first step was to teach a sequence or chain of responses. For example, a student was taught to open a can of fruit with a can opener, to pour the contents into a bowl, and to eat it with a spoon. When the sequence was trained, the student was given the items to complete the chain, except that one was missing. In this situation, the teacher reinforced a previously trained verbal response that specified the missing item (manding) by supplying the object. Since students came to emit such verbal responses, it appears that CEO and specific reinforcement are regulating conditions for manding behavior (see also Carroll & Hesse, 1987).

There are other studies of mand training that did not manipulate an establishing operation (Savage-Rumbaugh, 1984; Simic & Bucher, 1980; Sundberg, 1985). In these studies, humans, apes, and pigeons were required to produce a response that specified a particular object (food items or toys). The objects were shown to the subject to evoke an appropriate mand response (EO). When the verbal response occurred, the object was given, which functioned as specific reinforcement.

For example, in the study by Savage-Rumbaugh (1984), chimpanzees were shown a number of food items. If the animal pointed to the corresponding symbol on a communication panel, the item was given as reinforcement. Chimpanzees readily acquired this kind of verbal response and even more complex symbolic communication (Savage-Rumbaugh, 1986; Savage-Rumbaugh, Shanker, & Taylor, 1998). However, there is some question as to the exact controlling conditions. The food items may have functioned as discriminative stimuli that set the occasion for selecting the corresponding symbol key, in which case the chimpanzee's behavior involved tacting rather than manding. Because the sources of control were complex, the behavior is best described as *impure manding*, being controlled by food items as discriminative stimuli and specific reinforcement when food deprived (EO).

In chimpanzee studies, pointing to a food symbol is manding when deprived, as it results in getting the item. Pointing at something is a type of mand in which the response topography or form (index finger extended) remains constant, but the response is directed at different stimuli (banana or apple). This contrasts with human speech, in which the topography of the vocal response varies with the establishing operation (EO) and specific reinforcement (e.g., "give food" vs. "give water"). Vocal manding facilitates discrimination by a listener as the form of response varies, perhaps resulting in more rapid and precise compliance (reinforcement). Although pointing to what you want is formally manding, saying what you want is much more effective—especially if the listener is in another room, or the object is out of sight.

Finally, manding can involve control of verbal behavior by contingencies of negative reinforcement. In one study, applied behavior analysts developed a program to increase the quality of life for three children with autism. The researchers taught the youngsters manding for the removal of nonpreferred items. This negatively reinforced manding generalized to other untrained items as well (Yi, Christian, Vittimberga, & Lowenkron, 2006). Additional studies have shown that individuals, trained in manding using one motivational operation (MO), show manding under a different MO without further training (Lechago, Carr, Grow, Love, & Almason, 2010).

Tact Relations
To train tacting responses, a speaker must come to emit a verbal operant whose form depends on a nonverbal discriminative stimulus. A second requirement is that the operant class should be acquired and maintained by *nonspecific reinforcement*. Reinforcement is nonspecific if the reinforcer for one response exerts no stimulus control over the form of the next response. In animal studies, a response may qualify as tacting even if it is reinforced with food, as long as food reinforcement does not set the occasion (S^D) for a subsequent verbal response or the selection of the next symbol. For example, a chimpanzee may be offered an apple, and when it selects the symbol key for apple it is given a piece of banana. The presentation of the banana cannot set the occasion for pressing the symbol for apple on the next trial.

Tact relations have been investigated with chimpanzees, in particular Kanzi the bonobo chimpanzee (Figure 12.4). Savage-Rumbaugh (1984) used pointing to symbol keys as the verbal response. When the experimenter displayed an item of food (apple), a response to the corresponding symbol on the lexigram board resulted in praise and the delivery of a different item of food (banana). Thus, the item of food used as reinforcement always differed from the one on display.

In this situation, the display of an item of food was a nonverbal S^D that set the occasion for a response to the appropriate symbol key (tacting). Since reinforcement was nonspecific, the consequences of behavior could not regulate pointing to a particular symbol. Because the chimpanzee points to the apple symbol (in the presence of an apple) and is reinforced with a banana, we can be sure that the verbal response is tacting rather than manding.

Note that chimpanzees' symbol pointing came under the control of the displayed food items and therefore qualified as tacting. Thus, in this experiment, the *topography of the tact was the same* (finger pointing), but its location changed. In contrast, vocal tacting of humans

Fig. 12.4
A photograph of Kanzi the bonobo chimpanzee with his trainer Sue Savage-Rumbaugh, emeritus scientist formerly at the Great Ape Trust, Des Moines, Iowa. Kanzi is shown with portable, lexigram boards used to train tacting and other complex forms of verbal behavior.

Published with permission of Sue Savage-Rumbaugh. Copyright held by Bonobo Hope.

by speaking involves *changes in topography depending on the nonverbal stimulus* (e.g, "That's a chair" or "There's a table"). Finally, the delivery of a food item is probably not necessary, and generalized conditioned reinforcement (e.g., praise, acceptance, or attention) alone could be used to train tacting in both apes and human children (see Savage-Rumbaugh et al., 1993; Savage-Rumbaugh, Shanker, & Taylor, 1998; see also Carr & Miguel, 2013 and Sundberg, 1996 for a behavioral analysis).

Researchers have also used pigeons to investigate tact relations. Michael, Whitley, and Hesse (1983) trained tacting based on changes in response topography. Pigeons received nonspecific reinforcement (food) that depended on a bird emitting a particular form of response in the presence of a nonverbal discriminative stimulus. For example, a thrust of the head was reinforced when a red ball was presented, and turning in a circle produced reinforcement when a blue ball was the discriminative stimulus. Functionally, this is equivalent to a child who says "That's a red coat" and "This is a brown coat" and is reinforced by acceptance of the description by the listener. Tacting in the pigeons was successfully established even though the contingencies required correspondence between the nonverbal stimulus and the form of the bird's response. An unanswered question of this research is whether pigeons (or chimpanzees) can show generalization of a tact relation. That is, without further training, would the respective responses for blue and red occur when the objects were triangles or squares rather than balls?

In terms of application, there are behavioral experiments of humans with language delays, which trained tacting as part of a more general program of language acquisition (Carroll & Hesse, 1987; Guess, 1969; Lamarre & Holland, 1985; Lee, 1981a); also, children with autism have learned tacting to the actions of others (Williams, Carnerero, & Perez-Gonzalez, 2006). In one study with preschool children, Carroll and Hesse (1987) investigated the effects of alternating between training of mand and tact responses. During training of mand responses, manding an object produced the item. For training of tact responses, the experimenter presented the objects as discriminative stimuli and provided praise as reinforcement for correct responses. Results indicated that the children responded appropriately to the verbal contingencies, and that *mand training facilitated the acquisition of tacting.* Thus, manding "Give cup" increased the acquisition of tacting "That's a cup." This latter finding is interesting because it suggests that under some conditions, manding and tacting are not independent classes of behavior (e.g., Sigafoos, Doss, & Reichle, 1989; Sigafoos, Reichle, Doss, Hall, & Pettitt, 1990). Apparently, these verbal operant relations may interrelate as when parts of the response forms are shared—both involve the word "cup."

Experiments by Lamarre and Holland (1985) with typically-developing children, and Lee (1981b) for humans with language delays, also concerned the acquisition of tacting (see Partington, Sundberg, Newhouse, & Spengler, 1994 for tact training of a child with autism). In these experiments, one object was placed on the left and another on the right. The tact response was saying "On the right" or "On the left" depending on the position of the object. For example, the experimenter would prompt, "Where is the dog?" The subject who answered "On the right" when the dog was on the right side of a flower received social praise as reinforcement. This type of training successfully established verbal responses that contacted the position of an object. In another version of tact training, Guess (1969) trained verbal responses that contacted the quantity of an object. Speakers with language deficits were taught to emit the singular form of a noun when a single object was shown, and to emit the plural form if two identical items were presented.

In these experiments, correct responses produced food, rather than praise. Thus, the subject was presented with a single cup and saying "Cup" rather than "Cups" was reinforced with food. Food may be defined as nonspecific reinforcement in such studies, as it does not exert any stimulus control over the next verbal response "Cup." In humans, both generalized conditioned reinforcement (praise, approval, or attention) and nonspecific reinforcement (food in the preceding example) may be used to establish tacting to various features of the nonverbal environment (position or quantity of objects).

Overall, Skinner's description of the controlling variables for manding and tacting (Skinner, 1957) has been verified by research on a variety of animals, including primates, young children, and humans with language deficits. This research shows that manding is verbal behavior under the control of an establishing operation (EO) and specific reinforcement. In contrast, tacting is verbal behavior controlled by nonverbal discriminative stimuli and generalized conditioned reinforcement (or nonspecific reinforcement). The experimental analysis of manding and tacting has resulted in a technology of training verbal behavior in humans who do not show basic verbal skills (Carr & Miguel, 2013; Sundberg & Michael, 2001). A basic question is whether these verbal classes (manding and tacting) are learned together or are acquired separately.

Functional Independence of Basic Verbal Classes

Skinner (1957) proposed the functional independence of mand and tact response classes. Because the contingencies controlling these verbal responses are distinct, it should be possible to establish manding of an object or action without the speaker tacting it, and vice versa. Traditional cognitive or linguistic analysis does not make such distinctions, and would not predict that asking for something and identifying it by words arise from different contingencies of reinforcement set by the verbal community. For this reason, it is important to assess the research, testing Skinner's *functional independence hypothesis* (Gamba, Goyos, & Petursdottir, 2014).

We have already seen that manding can emerge with tact training, and tacting sometimes appears along with the acquisition of manding—evidence seemingly contrary to Skinner's functional independence hypothesis. Also, in the everyday development of language, there is no indication that asking for things and identifying them develop separately in children. In fact, some behavior analysts argue that manding and tacting are necessarily interrelated by higher-order verbal skills involving derived stimulus relations (relational frames), or by higher-order naming relations (Barnes-Holmes, Barnes-Holmes, & Cullinan, 2000; Hayes, Barnes-Holmes, & Roche, 2001; Horne & Lowe, 1996).

Nonetheless, Lamarre and Holland (1985) demonstrated the functional independence of manding and tacting in children between 3 and 5 years old. As previously noted, the children were trained to respond "On the left" or "On the right" to identify the location of an object (tacting) or to ask the experimenter to place an object on the left or right of another one (manding). When tested under extinction conditions, children who were given tact training did not show reliable evidence of manding, and those trained in manding failed to demonstrate tacting. Thus, the training of one verbal operant did not result in the emergence of the other, demonstrating the functional independence of these verbal response classes. After finding this result for initial training, all of the children were trained to emit the verbal operant (manding or tacting) that had not been trained in the first phase of the study. Once the children had learned both manding and tacting, they were tested again for independence of function. Mand responses to left or right locations were now reversed, and tests were made for reversal of tacting; also, tacting to objects at these locations was reversed and tests were made for reversal of mand responses. Again, the results demonstrated functional independence of these verbal classes, even when the verbal response forms for manding and tacting were identical. The results clearly supported Skinner's functional independence hypothesis.

Subsequently, Petursdottir, Carr, and Michael (2005) used preschool children to systematically replicate and extend the findings of Lamarre and Holland (1985). Using stickers and praise as reinforcement, the researchers trained the children initially to complete two four-piece assembly tasks—constructing a cube and solving a puzzle. Next, they were trained to emit tact responses to each of the four pieces of one task and to emit mand responses to the separate pieces of the other task, using arbitrary word forms. Probe trials (tests) without reinforcement were used to assess the effects of training on the untrained response class (manding or tacting) for each child. Following mand training, 4 out of 4 children reliably emitted tact responses on probe trials, but tact training produced unreliable effects on tests for mand responses.

These findings differ from those of Lamarre and Holland (1985) in that the children showed transfer between verbal response classes, rather than complete functional independence. One possibility is that the contingencies differed between the two studies. The earlier study by Lamarre and Holland (1985) required the children to emit verbal responses to an abstract stimulus property (location on the left or on the right), whereas the more recent study by Petursdottir et al. (2005) used a concrete stimulus (pieces of puzzles or cubes) to establish the requisite verbal behavior. For young children, transfer between manding and tacting might increase if the training stimuli were more concrete. In fact, when the stimulus properties were abstract, Twyman (1996) found functional independence of new mand and tact responses to an abstract property (whole crayon) in educationally disabled children with existing mand and tact responses.

Another way to account for the differences in functional independence over studies relates to establishing operations. The earlier study by Lamarre and Holland did not use an explicit establishing operation (EO) for mand training, but the assembly-task study did—using concealed pieces of the task as a specific EO. Differences in explicitness of the EO between studies would mean that children in the earlier location study (on the left of or on the right) were not trained in mand responses, whereas those in the assembly task were. If mand responses were not trained, there could be no transfer between mand and tact response classes, leading Lamarre and Holland to the incorrect conclusion that there was functional independence.

Notably, the reinforcement histories of the children also differed between the two studies. When children have a relevant history of manding and tacting outside the laboratory, as would be the case for verbal responses involving objects (pieces of puzzle), these response classes would show more cross-transfer in an experimental setting. Without such a history, as with verbal responses related to abstract locations or other abstract stimulus properties, mand training would not result in transfer to tact responses, and vice versa; the response classes would be independent.

A recent study on manual signing by typically developing infants suggests that, when reinforcement histories are controlled, transfer between mand and tact responses does not occur (Normand, Machado, Hustyi, & Morely, 2011). The participants were three children between 8 and 15 months old, none of whom exhibited vocal language or manual signing before the study (one infant did occasionally gesture). To establish an explicit EO, mand training occurred at times when infants were usually hungry (after a nap or before dinner), or when an object (a rattle) was out of sight. Training for one infant involved emitting an informal sign for "apple sauce," bringing the open palm of one hand to the crown of the head (not formal ASL). Another infant was trained to emit a sign for "pears," bringing the palm of one hand to the nose. The target response for the third infant was an informal sign for "rattle," defined as one hand hitting the thigh with a double movement. Mand training involved social modeling of the sign and specific reinforcement of the infant's verbal response (sign) by delivery of the food or object. Test sessions were arranged to assess the acquisition of manding and the possible transfer between mand and tact responses (echoic responses were also assessed). Tests for manding involved withholding the stimulus item (food or object) prior to a session, and following each target response (sign) with the presentation of the stimulus item (no model prompting). To assess tact responses, the stimulus items were freely available to the infants for a period prior to and throughout the test session. In accord with the functional definition of tacting, brief verbal praise followed each signing response for these tests (other control conditions were also arranged, but are not discussed here).

Compared with baseline sessions without modeling and reinforcement, all of the infants emitted the target sign more during sign-training conditions. For test sessions, mand responses for the stimulus items were reliably observed for all infants, but none of them showed evidence of tacting the items. In this study, the acquisition of mand responses (informal signing of items controlled by an explicit EO and specific reinforcement) did not increase the acquisition of tacting (informal signing of items and generalized reinforcement by praise). The infants had no history of sign training before the study, and the verbal repertoires did not show transfer within the experiment, a result consistent with the findings of Lamarre and Holland (see also Nuzzolo-Gomez & Greer, 2004 on functional independence in children with autism/developmental disabilities who did not emit the target mand or tact responses before training). At the present time, the functional independence of mand and tact response classes is not completely resolved, but the weight of the evidence is tipping toward Skinner's hypothesis (Skinner, 1957) when reinforcement histories are controlled within the study (a review by Gamba, Goyos, & Petursdottir, 2014, found at least weak support for functional independence of mand and tact responses).

ADDITIONAL VERBAL RELATIONS: INTRAVERBALS, ECHOICS, TEXTUALS, AND AUTOCLITICS

Intraverbal Relations

Other verbal responses also depend on discriminative stimuli. **Intraverbal behavior** is a class of verbal operants regulated by *verbal* discriminative stimuli. Verbal stimuli arise from verbal behavior; a verbal response by a speaker ("one, two, three …") may be a stimulus for a subsequent verbal operant by the same speaker ("four"). When a verbal response ("Mary") exactly replicates the verbal stimulus ("Say Mary"), we may say that there is *point-to-point correspondence* between them. In this case, the verbal behavior is defined as echoic (see p. 445). Intraverbal behavior, however, has no point-to-point correspondence between the verbal stimulus ("jack, queen …") and the response ("king").

In a conversation by cell phone texting, one person "texts" and the other responds to the message. Katherine writes, "I went shopping today" and Jason responds, "What did you get?" Notice that there is no exact correspondence between the two texting responses, meeting the criterion of an intraverbal exchange. Our ability to answer questions, tell stories, describe events, solve problems, recall the past, and talk about the future depends on an extensive repertoire of intraverbal behavior.

In everyday language, thematically related words (or sentences) are examples of intraverbal behavior. For example, the verbal response "Fish" to the spoken words "Rod and reel" is an intraverbal operant; saying "Water" in response to the written word LAKE is also intraverbal behavior. On the other hand, the person who says "water" to the spoken sound "water" is not showing intraverbal regulation; in this case, there is exact correspondence between the response and the stimulus, and the response is echoic.

Intraverbal learning is part of our educational curriculum (see Figure 12.5). In elementary school, students are initially taught addition of numbers, as when a child says or writes the number "3" in response to the problem "1 + 2 equals …" An older child may recite the multiplication table, as in "5 × 5" equals "25." In this example, the verbal stimulus "5 × 5" exerts direct stimulus control over the response "25," and the relation is intraverbal. In contrast, a student who derives the answer "25" by adding five 5s, counting by 5s, or counting the cells in a 5 × 5 matrix is tacting the number or set of elements rather than emitting intraverbal behavior. As you can see, the training of academic behavior in young children

Fig. 12.5
The photograph depicts learning the addition of numbers by an advanced student. This kind of instruction is involved in the acquisition of the intraverbal response "824" when the student is asked "What's 213 + 227 + 384" by her teacher.
Source: Shutterstock.

involves several verbal relations, including tacting and intraverbal behavior (see Partington & Bailey, 1993).

Children and adults with language deficits may also benefit from intraverbal training (Carr & Miguel, 2013; Sundberg & Michael, 2001; Sundberg, Endicott, & Eigenheer, 2000), and empirical studies of intraverbal behavior steadily have increased over the past few years, but the range of intraverbal response forms investigated in these studies is somewhat limited (Sautter & LeBlanc, 2006). In one study, Sundberg and Sundberg (2011) described the age-related changes in intraverbal behavior of 39 typically developing children (aged 23 to 61 months) and 71 children with autism (aged 35 months to 15 years), using an 80-item inventory. Here we focus on the age-related changes in intraverbal behavior of typically developing children.

The inventory contains eight sets or groups of intraverbal questions that are posed to the child and increase in difficulty, involving multiple sources of stimulus control within the question. For example, a Group 1 level question requires the child to emit a simple "fill in" response to "Twinkle, twinkle, little____," while a Group 8 level question asks "How is a car different from a bike?", requiring a complex-verbal conditional discrimination. The child's exact response was recorded next to each question and subsequently scored as correct or erroneous by two individuals (over 90% reliable scoring). For typically developing children, the data showed that the *intraverbal scores were correlated with age*, with inventory scores generally increasing as the child became older. Also, the intraverbal scores of these typically developing children substantially improved after 3 years of age, and then rose toward the maximum score of 80 at 5 years of age.

Fig. 12.6
The photograph illustrates a young mother interacting with her baby. An echoic response is trained when the mother says "Mama" and the baby, intently looking at her lips, makes the approximation "Ma" that is reinforced with smiles from the mother. Eventually over time, the mother says "Mama" and the child emits "Mama" as an echoic response.
Source: Shutterstock.

Echoic Relations

When there is *point-to-point correspondence* between the stimulus and the response, verbal behavior may be classified as either echoic or textual, depending on the criterion of *formal similarity*. The contingencies of echoic behavior require formal similarity whereas the contingencies of textual behavior do not (see p. 414). **Formal similarity** requires that the verbal stimulus and the product of the response be in the same mode (auditory or visual) and have exact physical resemblance (same sound pattern). **Echoic responses** are a class of verbal operants regulated by a verbal stimulus in which there is *correspondence and formal similarity* between the stimulus and the response. Saying "This is a dog" to the spoken stimulus "This is a dog" is an example of an echoic response in human speech. Generally, *echoic behavior is generalized imitation along a vocal dimension* (see Chapter 11 and Poulson et al., 1991).

Echoic behavior occurs at an early age in an infant's acquisition of speech. The child who repeats "Dada" or "Mama" to the same words uttered by a parent is showing echoic operant behavior. In this situation, any product of behavior (sound pattern of child) that closely replicates the verbal stimulus (modeled sound pattern) is reinforced (see Figure 12.6).

Although reinforcement appears to occur along a dimension of acoustical correspondence, the contingencies of echoic behavior are probably based more on the *matching of phonetic units*. The learning of echoic behavior likely begins with the basic units of speech called phonemes. Coordinated movements of the child's larynx, tongue, and lips result in phonemes (e.g., "ma"), which replicate parts of adult speech ("ma … ma"). When articulations by the child correspond to those of the adult, the acoustical patterns also overlap. Adults who hear speech-relevant sounds ("ma") often provide social consequences (e.g., smiling, praise, or cuddling) that are paired with these acoustical patterns. On this basis, the duplication of speech sounds itself comes to function as automatic reinforcement for speech-relevant articulations by the child (Yoon & Bennett, 2000).

It is important to emphasize that echoic behavior is not simply the duplication of sounds. As a verbal operant, echoic performance is regulated by specific reinforcement contingencies based on articulation. Echoic contingencies in humans involve reinforcement by listeners of correspondence of basic speech units rather than the mere reproduction of sounds. These units begin as phonemes (i.e., the smallest sound units to which listeners react), expand to words, and eventually may include full phrases and sentences. In contrast, parrots and other birds duplicate the sounds that they hear (Pepperberg, 1981, 2000), but their behavior is not necessarily verbal in Skinner's sense (Skinner, 1957). Parrots reproduce sounds or noises even when these responses produce no change in the behavior of the listener. For this reason, an infant's speech is echoic behavior, but a parrot's "speech" is not.

Echoic contingencies are most prevalent during language acquisition. This means that an infant's vocalizations will have more echoic components than the speech of an adult. It also implies that adult speech may become more echoic when a person is learning to speak a second language. Thus, a Spanish teacher may demonstrate word pronunciation to a student who initially makes many errors in articulation. The teacher gives repeated examples, and the student is reinforced for correct pronunciation. After some practice and correction, the student's pronunciation becomes close to that of the teacher. Only when the speech units correspond is the student said to show competence in pronunciation of Spanish.

Textual Relations

Verbal behavior is textual when there is *no formal similarity* between the stimulus and the response. **Textual behavior** is defined as a class of verbal operants regulated by verbal stimuli where there is correspondence between the stimulus and the response, but no formal similarity. The most common example of textual behavior is reading out loud. The child looks at the text, which reads, "Good night, moon," and emits the spoken words, "Good ... night ... moon ... ." In adult reading, the behavior is also textual but the "out loud" aspect is no longer emitted—the person reads silently so that the response is now a private event. Textual behavior is also observed when an administrative assistant *takes dictation* from their supervisor. In this case, hearing the words "Dear Professor Smith ..." spoken by the supervisor sets the occasion for writing these words by the administrative assistant. Again, correspondence between the stimulus and the response occurs, but there is no formal similarity.

Autoclitic Relations

The **autoclitic** is a form of verbal behavior that modifies the consequences produced by other verbal responses. It is verbal behavior used in conjunction with, and controlled by, primary verbal units, such as mand, tact, and intraverbal responses. Skinner (1957) described five categories of autoclitic relations: descriptive, qualifying, quantifying, manipulative, and relational (for a summary, see Howard & Rice, 1988). For example, descriptive autoclitic responses (e.g., I think, I doubt, I see, I hear) are used to further specify the control by a nonverbal stimulus over the speaker's tacting. At a railway station, Paul says to Stephanie, "*I see* the train is coming," adding "I see" to inform her (the listener) that stimulus control over his tacting "the train is coming" is visual rather than auditory. By adding the autoclitic response to his tacting of the train's arrival, Paul provides more precise stimulus control over Stephanie's behavior (looking for the train), increasing the probability of generalized conditioned reinforcement ("Okay") from her. An in-depth coverage of autoclitic relations is beyond the scope of this textbook, and it is convenient for further analysis to collapse Skinner's five categories into two—autoclitic tacting and autoclitic manding (Peterson, 1978).

The defining property of autoclitic manding is that these verbal units are controlled by motivational operations (MO) that make it reinforcing for the speaker to modify the mand, thereby increasing stimulus control over the listener's behavior. If deprived of water, Kelly may say "Get me some water" to their sibling, Terry. Alternatively, Kelly may say "*Could you please* get me some water," with the first three words having no meaning (stimulus control) on their own, but functioning with the mand response to increase the likelihood that Terry ("you") will comply. Thus, the autoclitic unit "Could you please …" is added to manding of the water, sharpening stimulus control over the listener's behavior (getting water) and increasing the probability of reinforcement for the speaker (receiving water). Many parts of speech that have no meaning on their own are added to the speaker's mand responses based on the practices of the verbal community.

As we mentioned earlier, autoclitic responses are used to modify the stimulus control exerted by the speaker's tacting over the listener's response. Looking toward the gas gauge, Thelma says to her buddy Louise (the driver), "*I doubt* we are out of gas," as opposed to the response "*I'm sure* we are out of gas." For the listener, the words "I doubt" versus "I'm sure" control a conditional discrimination with respect to stopping for gas at the next station. You stop for gas if your buddy is "sure" that there is nothing left in the tank. For the speaker, the autoclitic is added as a verbal response under the stimulus control of the primary tact "We are out of gas," as indicated by the gauge pointer and the time since the last fill-up. That is, the speaker discriminates the strength (or weakness) of his own verbal behavior when saying "I'm sure" versus "I doubt." In a situation where you may be out of gas, the speaker must be able to tell whether saying "We are out of gas" is appropriate. The verbal community sets up the repertoire of "responding to one's own behavior," presumably because of the benefits that accrue to listeners (e.g., avoiding an empty gas tank) when speakers make such complex-verbal conditional discriminations.

A study by Howard and Rice (1988) is informative about the training involved in the early acquisition of autoclitic responses regulated by weak stimulus control of primary tacting. Four preschool children were initially trained to tact colors (red, yellow, and blue), geometric shapes (square, circle, and triangle), and alphabet letters (L, H, and M), using praise and "happy faces" exchanged for time with preferred activities (e.g., drawing a picture) as generalized reinforcement. Tact training involved the experimenter asking, "What is this?," followed by presentation of a stimulus from one of the stimulus sets (colors, shapes, or letters) and the child identifying it with a verbal response (e.g., "red," "square," or "M"). Once 90% or more of the children's tact responses were correct, they were trained to emit the autoclitic response "like" to weak stimulus control by the primary tact response. The experimenter now presented the child with a distorted example of a stimulus from one set (e.g., red-like color) and asked, "What is this?" If no autoclitic response occurred, the trainer modeled the correct response ("like red," "like square," or "like M"), withdrew the stimulus, and presented it again a few seconds later. Each stimulus set was trained separately until the children met a criterion of 90% or more correct. At this point autoclitic training continued, but probe trials were used to assess generalization of autoclitic responses (generalized response class). For probe trials, new distorted examples of stimuli from the three stimulus sets (never trained) were presented without reinforcement.

Although only one child completed all of the phases of the study, the overall results from probe trials showed that the children acquired the generalized autoclitic response "like X," occasioned by weak stimulus control of the primary tact. Other findings showed that the ability to tact a stimulus is not sufficient to obtain autoclitic responses to distorted examples, and that training across multiple stimuli within sets must occur before a generalized response class or autoclitic frame ("like X") is formed. Howard and Rice (1988, pp. 57–58) noted that the "exact nature of the stimulus that evokes autoclitics of weakness" is difficult to specify. Thus, "there is no particular characteristic of the nonverbal stimulus [that controls the tact response] one can point to and say, 'It is this dimension [of the stimulus] which is controlling the autoclitic *like*.'" Given this problem, Howard and Rice argue that "it seems reasonable to assume that the speaker is reacting to some private stimulation related to the reduced tendency [low probability of response] to make the primary tact response."

Without a public stimulus to accompany the private stimulation of "response tendency," difficult to train descriptive autoclitic behavior. In applied settings where the focus is on developing verba_____ in children with autism, the complexity of training autoclitics is recognized Behavior analysts recommend training a wide variety of primary verbal operants (mand, tac echoic, and intraverbal responses) to allow for "communication effectiveness" without direct training of "grammatical correctness" (Sundberg & Michael, 2001, p. 13). A broad range of primary verbal units would allow the child with autism to contact the natural contingencies of verbal community—adding grammatical correctness (autoclitics) to the child's verbal repertoire without direct instruction. An unfortunate implication of this view is that basic research on autoclitic behavior has stalled, and applied studies have not clarified the contingencies for training each of Skinner's five classes of autoclitics. Overall, more basic and applied research is needed on contingencies of reinforcement and autoclitic behavior.

FOCUS ON: HIGHER-ORDER VERBAL CLASSES AND THE NAMING RELATION

The basic units of verbal behavior are combined to form higher-order classes, allowing for the greater novelty and spontaneity in a child's language production. Previously, we described the emergence of *generalized imitation* as a higher-order operant class arising from reinforcement of multiple exemplars of modeling and imitation (see Chapter 11). Once a child has acquired generalized imitation, any new modeled stimulus presented by an adult occasions a corresponding novel imitation by the child. Once established, generalized imitation allows children to rapidly learn new forms of behavior from models—adding observational learning to the basic process of learning by successive approximation (see Chapter 4 on shaping).

In a seminal article, Horne and Lowe (1996) proposed that naming something (object, place, or action) involves a generalized operant class that substantially expands the verbal repertoire of the child. Analytically, the **naming relation** or *the generalized class of naming* arises from verbal contingencies that integrate the echoic and tact response classes of the child as speaker with the conditional discrimination behavior of the child as listener.

Seated on the floor, a parent may ask or instruct their child to point to the ball among the other toys, model and prompt pointing to the ball, and differentially reinforce pointing at the ball with praise and approval. Thus, the parent in this episode is teaching the child to follow instructions—the basis of comprehension or listening. Within the same episode, the parent is likely to say "Ball" while pointing to the ball and provide reinforcement for the child's echoing or repeating the mother's word. Notice that the social contingencies arranged by the parent require both listening and speaking by the child. The child as listener follows the parent's instruction, pointing to the ball. As the speaker, the child emits the word "Ball" as an echoic response to a similar utterance by their parent. The naming relation also involves tact training, as when the parent points to the ball and asks "What is this?", adding the prompt "Say, ball," and the child responds by saying the word "Ball."

After extensive name training across various objects, we may say that the child knows the names of toys, such as ball, doll, and truck. That is, the child says "Ball" (and not "Doll" or "Truck") when we hold up a ball and ask, "What is this?" Also, we expect the child to look in the direction of the ball when we say, "Where's the ball?", and to point at it, go to it, and pick it up if it is in sight. And when a new toy car is introduced, the parent may say, "This is a car" and ask the child, "What is this?" without further training. A child who answers "Car" to this question shows acquisition of the generalized verbal relation—a naming response has emerged without explicit reinforcement.

A variety of research studies of children with language delays indicate that the naming relation arises from multiple-exemplar instructions (MEI) (Greer, Stolfi, & Pistoljevic, 2007), involving rotation of the child's listener and speaker responses during training. After training to establish basic verbal units and listening, MEI alternates among instructions to match, instructions to point, and instructions to tact arranged in different sequences. For example, in the stimulus set of birds (parrot, canary, ostrich, cardinal, hummingbird), a child without naming capability may be shown pictures of birds and asked to "Match ostrich" by placing a target picture on top of the duplicate picture in the set. The child is then asked to "Point to parrot" among the set of five birds. This is followed by an opportunity to say "Hummingbird" (tact without prompt) in response to the presentation of a picture of a hummingbird, or tacting the bird with the addition of a vocal antecedent "What is this?" For the set of gems (diamond, sapphire, ruby, amethyst, and emerald), the child may be asked to "Point to diamond," and then asked to "Match the amethyst," followed by an opportunity to tact a picture of the ruby, saying "Ruby" with or without the vocal prompt "What is this?"

One research question is whether this kind of MEI is required to establish the generalized class of naming, or whether single exemplar instruction (SEI) would also result in the higher-order verbal class (Greer et al., 2007). For SEI training, the child's listener and speaker responses are not rotated during instruction, but each learning unit (instructed matching to sample, instructed pointing to objects, and tacting of objects with and without vocal prompts) is trained separately in massed practice sessions. During training by either MEI or SEI, correct responses are reinforced (with tokens and praise) while incorrect responses by the child result in the trainer demonstrating the correct response (correction), repeating the vocal antecedent, and the child providing the accurate response (no tokens and praise).

The results showed that MEI training increased novel unreinforced listener and speaker components of naming. None of the children who were given SEI training showed emergent or novel naming, even though they received the same amount of training on the learning units.

The data suggest that rotation of the speaker–listener components found in MEI training is required for the acquisition of naming in children who lack a naming repertoire. (The interested reader should see also Barnes-Holmes, Hayes, Barnes-Holmes, & Roche, 2001 for the interpretation of naming as a **relational frame** involving derived stimulus relations, bidirectionality, and combinatorial entailment.) Thus, through MEI the child learns that listener and speaker responses go together, establishing the generalized verbal class of naming. The speaker–listener repertoires remain independent with SEI training, and there is no evidence of generalized naming with these procedures. Notably, children who did not initially learn generalized naming with SEI training subsequently showed naming of novel objects when retrained with the MSI protocol (see also Mahoney, Miguel, Ahearn, & Bell, 2011 for the use of motor responses (as opposed to vocal responses) in tact training and the emergence of naming and stimulus categorization in preschool children).

One possibility is that parents use unplanned MEI to establish naming in typically developing children, as part of everyday language learning. Further research is required on the natural contingencies involved in the development of naming in children without language deficits.

SYMBOLIC BEHAVIOR AND STIMULUS EQUIVALENCE

For most people, the flag of their nation is a significant symbol. When some Americans see the flag, they may think of the United States, freedom, and apple pie. Other Americans may see the symbol of a flag as a symbol of oppression, hate, or inequality. This suggests that symbolic behavior involves the training of **stimulus equivalence**. The presentation of one class of stimuli (e.g., flags) sets the occasion for responses made to other stimulus classes (e.g., countries). This seems to be what we mean when we say that the flag stands for, represents, or signifies a country or values of a country. Equivalence relations such as these are an important aspect of human behavior. For example, when teaching a child to read, spoken words (names of animals) are trained to visual stimuli (pictures of animals) and then to written symbols (written words for animals). Eventually, the written word is said to stand for (or mean) the actual object, in the same sense that a flag stands for a country or its values. In this section, we shall examine the behavior analysis of equivalence relations as a scientific account of symbolic activity and meaning (see Linehan, Roche, & Stewart, 2010 for an experimental analysis of equivalence relations in the design of challenging computer games that result in a high level of game enjoyment).

Basic Equivalence Relations

When stimulus class A is shown to be interchangeable with stimulus class B (if A = B then B = A), we may say that the organism shows *symmetry* between the stimulus classes. Symmetry is only one form of equivalence relation. A more elementary form of equivalence is called *reflexivity*, as noted in Chapter 8. In this case, an A to A relation (A = A) is established so that, given the color red on a sample key, the organism responds to the comparison key with the identical color (red). A child who is given a picture of a cat and then finds a similar picture of a cat in a set of photographs is showing reflexivity or identity matching.

Reflexivity and symmetry are basic logical relations of the mathematics of sets and beyond the scope of this book (see www.math.uah.edu/stat/foundations/Equivalence.html for a formal presentation). Here we give examples of basic equivalence relations. A child who is presented with the number 1 shows reflexivity when she points to 1 in an array of numbers 2, 3, 1, 4, 5. The same child shows symmetry if, when given the number 2, she selects the display XX rather than X or XXX and when given XX she selects 2 from the array 3, 2, 1, 5, 4.

There is one other equivalence relation in mathematics. This is the relation of transitivity. If the written numbers *one*, *two*, and *three* are equivalent to the arithmetic numbers 1, 2, and 3 and these arithmetic numbers are equivalent to displays X and XX and XXX it logically follows that *one*, *two*, and *three* are equivalent to displays X and XX and XXX. That is, if A = B and B = C, then A = C (transitivity).

Experimental Analysis of Equivalence Relations

Although equivalences are logically required by mathematics, it is another thing to show that such relations govern the behavior of organisms. In terms of behavior, three stimulus classes (A, B, and C) are called equivalent when an organism has passed tests for reflexivity, symmetry, and transitivity (Sidman & Tailby, 1982). Thus, after equivalence training in early reading, the spoken sound "Say cat," the word CAT, and the picture of a cat acquire equivalence of function for a young child (Figure 12.7).

A complete experiment for stimulus equivalence consists of both identity and symbolic matching procedures. In **identity matching**, the researcher presents a sample stimulus (e.g., a triangle) and two options (e.g., triangle or circle). The procedure is repeated over multiple examples of sample and comparison options. The organism is reinforced for choosing the

Fig. 12.7

An illustration is shown of an equivalence relation in early reading. After training by many examples through parent and child interaction, the sound of the spoken word, "Say, cat," the word CAT, and the picture of a cat acquire equivalence in function. This is the beginning of early reading. Stimulus equivalence (reflexivity, symmetry, and transitivity) seems to underlie much of human language and complex symbolic behavior.

Source: Photo: iStockphoto.

option that corresponds to the sample, establishing *generalized matching-to-sample* or identity matching. **Symbolic matching** involves presenting one class of stimuli (geometrical forms) as the samples and another set of stimuli (different line angles) as the options. Reinforcement depends on an arbitrary relation triangle = horizontal in the laboratory, or flag = country in American culture. After the reinforced relations are trained, tests are made for each kind of equivalence relation. *The question is whether reflexivity, symmetry, and transitivity occur without further training.* To make this clear, identity and symbolic matching are training procedures that allow for stimulus equivalence, but the procedures do not guarantee it. We shall describe such an experiment in a step-by-step manner.

Figure 12.8 shows the identity-matching procedures used to show reflexivity. The training involves identity matching for line angles or geometric forms by a pigeon. The bird is presented with three keys that may be illuminated as shown in the two displays (Display A or B). For each display, two sets alternate on the three keys. A set includes a sample key and two option keys. For the sake of clarity, in our example the option (side key) that matches the sample is always shown on the left of the displays, and the nonmatching option is on the right. *In real experiments, of course, the position of the matching stimulus varies from trial to trial, eliminating any left or right bias.* A peck on the sample key illuminates the option keys, and pecks to the matching key produce food and the next sample. Pecks to the nonmatching key are not reinforced and lead to the next trial (the next sample).

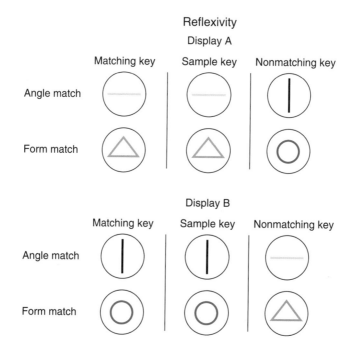

Fig. 12.8

Identity (reflexivity) matching procedures used to establish reflexivity in pigeons as described by Catania (1984). First, train matching to angles using the *angle match* arrangements in Displays A and B (top). Next, train matching to form using the *form match* arrangements in Displays A and B (bottom). Finally, test for reflexivity or generalized identity matching using color matching displays (not shown). See text for a description of the contingencies.

Reflexivity

Reflexivity involves showing an equivalence relation for a stimulus class (A = A), using a procedure of identity matching. In Display A (angle match) of Figure 12.8, the sample key presents a line angle (horizontal). When the pigeon pecks the sample key, horizontal and vertical stimuli are presented on the side keys (matching and nonmatching). Pecks to the horizontal matching key are reinforced with food, while pecks to the vertical nonmatching key are not. The next trial may present Display B (angle match). Now the sample is a vertical line. If the bird pecks the vertical-line matching key, it receives food, but pecks to the horizontal-line nonmatching key are extinguished. Based on this training and many more matching-to-sample trials, the bird learns to identify line angles (identity matching).

Similar procedures may be used to train identity matching based on geometric form. In Display A (form match) of Figure 12.8, the form display is based on triangles and circles. When Display A is in effect, the sample key is illuminated with a triangle. Pecks to the sample produce the two options—triangle and circle. Pecks to the key that matches the sample are reinforced, while pecks to the nonmatching geometric form are placed on extinction. A new trial may result in Display B (form match). In this case, the sample is a circle. When the bird pecks the sample, two options are presented on the side keys (circle and triangle). Pecks to the key with a circle produce food, but pecks to the key with a triangle are extinguished. Using these procedures, the pigeon learns to identify geometric forms.

Reflexivity includes *a test for generalization of identity matching*. A test for reflexivity would involve testing for generalization of identity matching based on the training of matching to sample by angle and form. For example, a bird trained to identity match to angle and form could be tested with colors (green or red) as the sample and comparison stimuli. A bird that pecks to the color that matches the sample, *without specific training* on colors, shows reflexivity or generalized identity matching. Lowenkron (1998) showed that many instances of generalized identity matching involved the training of joint stimulus control in the sense that two stimuli come to regulate a common response topography.

Symmetry

Recall that **symmetry** occurs when stimulus class A is shown to be interchangeable with stimulus class B (if A = B then B = A). Figure 12.9 shows the procedures used to train symbolic matching and the tests for symmetry. These procedures are implemented only after a bird has shown identity matching. For example, symbolic matching occurs if the bird is trained to discriminate geometric shapes on the basis of angles (angle-to-form discrimination). Symmetry occurs if the bird can pass *a test for reversal* (form-to-angle discrimination) *without further training*.

This procedure is shown by the angle-to-form display in Figure 12.9 (Display A). Pecks to the horizontal sample illuminate the side options—a triangle or a circle. In the presence of the horizontal-line sample, pecks to the triangle are reinforced while pecks to the circle are not. When Display B is presented, the sample is the vertical line and pecks to the circle are reinforced while pecking the triangle is on extinction.

Once the matching of angle to geometric form is well established, a **reversal test** (form to angle) is conducted without further reinforcement. In a reversal test of Display A, the bird is presented with a triangle as the sample and the question is whether it pecks the side key

454 Verbal Behavior

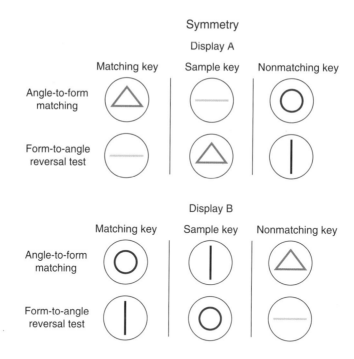

Fig. 12.9
Symbolic matching procedures used to train and test for symmetry in pigeons, as described by Catania (1984). First, train angle-to-form matching (Angle = Form) using the arrangements of Displays A and B (top). Next, test for reversal or symmetry (Form = Angle) using the form-to-angle arrangements of Displays A and B (bottom). See text for a description of the contingencies.

with the horizontal line. Because horizontal = triangle was trained, the bird shows symmetry if it pecks the horizontal comparison key when presented with a triangle sample (triangle = horizontal). Similarly, because vertical = circle was trained, symmetry is shown if the bird pecks the vertical side key of Display B when the circle is presented as the sample (circle = vertical). In everyday language, the bird responds as if the horizontal line stands for triangle and as if the vertical line means circle. The percentage of "correct" responses during the test (without reinforcement) is the usual measure of symbolic performance on this symmetry task.

Transitivity

The relation of **transitivity** occurs when stimulus classes A, B, and C are shown to be equivalent (if A = B and B = C then A = C). Figure 12.10 illustrates the procedures that may be used to train and test a pigeon for transitivity. These procedures would be used only if a bird had passed the tests for reflexivity and symmetry. Rows 1 and 5 of Displays A and B of the figure present the angle-to-form (symbolic matching) procedures for symmetry that were described earlier (horizontal = triangle; vertical = circle). To test for transitivity, the pigeon is trained to produce an additional discrimination. Rows 2 and 6 of Displays A and B illustrate this training. The pigeon is reinforced for matching a geometric form to intensity of illumination on the option keys—darker or lighter key. For example, in row 2 of Display A, pecking the lighter option key is reinforced when a triangle is the sample (triangle = lighter) and pecking the darker key

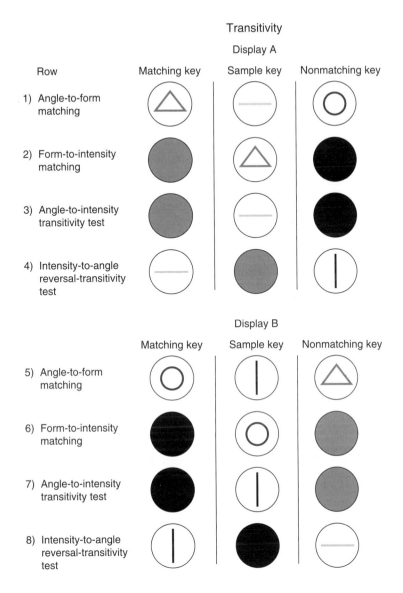

Fig. 12.10
Symbolic matching procedures used to establish and test for transitivity in pigeons, as described by Catania (1984). First, train angle-to-form matching (Angle = Form) using the arrangements in rows 1 and 5 of Displays A and B. Next, train form-to-intensity matching (Form = Intensity) using the arrangements in rows 2 and 6 of Displays A and B. Following training, conduct a transitivity test (Angle = Intensity) using the arrangements in rows 3 and 7 of Displays A and B. Finally, conduct reversal-transitivity tests (Intensity = Angle) using the arrangements in rows 4 and 8 of Displays A and B. See text for a description of the contingencies.

is not reinforced; also, row 6 of Display B shows that pecking the darker key produces food when a circle is the sample (circle = darker) while pecking the lighter option is on extinction.

Notice that the bird is trained such that horizontal = triangle and vertical = circle (rows 1 and 5) and has shown reversal on tests of symmetry. Given this performance, if triangle = lighter and circle = darker (rows 2 and 6), then the following relations could occur without explicit training on transitivity tests: horizontal = lighter, and lighter = horizontal (rows 3 and 4); also, vertical = darker, and darker = vertical (rows 7 and 8). These tests would establish transitivity in the pigeon, showing that the bird responds to the set of line angles as it does to the set of geometric forms, and responds to the set of geometric forms as it does to the set of light intensities (A = B = C). This performance would be similar to that of a person who responds to the written word DOG in the same way as they do to a picture of a dog or the spoken word "dog." The stimuli are said to be equivalent because they regulate the same operant class. Lowenkron (1998) proposed that many instances of *joint stimulus control* (equivalence between stimuli) of response forms underlie human language-related performances involving both logical (relative size and distance) and semantic (word meaning) relations.

NEW DIRECTIONS: BEHAVIORAL NEUROSCIENCE AND DERIVED STIMULUS RELATIONS

Neuroscientists are searching for the brain mechanisms that support the behavior of language use. For example, when stimulus class A is shown to be interchangeable with stimulus class B we say that the two classes show equivalence or perhaps that the person has formed a concept. Reinforcement of behavior under the control of distinct sets of arbitrary conditional-stimulus relations (A = B; B = C) often results in the appearance or emergence of new *derived stimulus relations* (A = C or C = A, equivalence classes), which might represent the underlying foundation of semantic (meaning) relations of human language or, alternatively, semantic meaning could aid in the emergence of derived relations (Amtzen, Nartey, & Fields, 2015). In applied settings, clinical problems often are treated by cognitive-behavioral therapy; such therapy assumes that disturbances of emotion and cognition cause behavior problems. Behavior analysts, however, suggest that these complex derived stimulus relations are regulated by the emergent functions of verbal stimuli (Guinther & Dougher, 2015).

Much is known behaviorally about the formation of stimulus equivalence and other derived stimulus relations, but the brain mechanisms related to this ability are not well described. Researchers have used functional magnetic resonance imaging (fMRI) of the brain of conscious humans to reveal the neural areas linked to both trained and derived stimulus relations (Dickins, 2005; Schlund, Cataldo, & Hoehn-Saric, 2008; Schlund, Hoehn-Saric, & Cataldo, 2007). Research of this nature promises to clarify the neurobiological substrates of symbolic verbal behavior usually taken to indicate conceptual and other higher-order cognitive processes.

Using a matching-to-sample task (MTS), human participants were trained to form conditional relations within stimulus classes (Schlund et al., 2007). That is, the participants were asked to respond to sample and comparison symbols and learn the symbolic equivalences. This type of conditional responding is suspected to recruit specific frontal-parietal and frontal-subcortical brain areas central to higher cognitive functioning. A neuroimaging procedure using fMRI allowed the researchers to correlate blood-oxygen-level activation in areas of the brain with the behavioral discriminations for individual participants.

Responding to both conditional and derived stimulus relations activated similar regions of the brain, but the magnitude was greater for trained relations in frontal areas of the brain (see Dickins et al., 2001 on involvement of the dorsolateral prefrontal cortex, a part of the brain related to semantic processing in language). The researchers also observed predominantly right-hemisphere activation, suggesting that the complex conditional responding in this type of task (MTS) is mediated more by nonverbal than by verbal processes (see also O'Regan, Farina, Hussey, & Roche, 2015 for event-related brain potentials (ERPs), showing larger right-posterior P3a activation for derived equivalence relations than for directly trained relations). A subsequent study showed that transitive-relational responses activate the anterior hippocampus in the medial temporal lobe (memory areas), but symmetrical relations increase blood-oxygen flow to the parahippocampus (memory encoding area), as well as the frontal and parietal lobe regions (Schlund et al., 2008). The common activation of these frontal and medial temporal regions appears to support conditional responding to both directly trained relations and derived or emergent relations.

Notably, the frontal area of the brain is often where injuries occur as a result of automobile accidents. This brain region is not critical for life, and victims often survive their injuries, but are frequently diagnosed as cognitively impaired. Rehabilitation efforts for these victims might be improved by the combined results from neuroimaging and studies of derived stimulus relations. In this regard, more research is required to better isolate the motor and neural functions of equivalence class formation and to specify more precisely the sensory and motivational aspects of derived or emergent relations.

Research on Equivalence Relations

Murray Sidman and his colleagues established the basic procedures to assess equivalence relations in both human participants and nonhuman subjects (Sidman & Tailby, 1982; Sidman, Rauzin, Lazar, Cunningham, Tailby, & Carrigan, 1982). These methods help to eliminate alternative hypotheses for failure to observe emergent equivalence relations (symmetry and transitivity) in different species. To ensure that animals have the prerequisite repertoires for equivalence class formation, baseline training is given for A to A (color to color) and B to B (geometric shape to geometric shape) identity discriminations and for A to B (color to shape) discriminations, involving matching of a sample to a set of comparisons. An evaluation is then made of the animal's ability to make successive discriminations of all samples and simultaneous discriminations among all comparisons (Saunders & Spradlin, 1989). The method also includes probe tests for reflexivity, symmetry, and transitivity, as well as unreinforced probe trials within an ongoing baseline of reinforced trials. If probes do not show emergent or untrained relations (e.g., symmetry), then performance on baseline trials is used to assess any decline in the basic (A–A, B–B, or A–B) relations necessary for equivalence class formation. Finally, the rate of reinforcement on baseline trials is thinned to reduce the discrimination of nonreinforcement on probe trials. With this procedure, a failure to find emergent relations is unlikely due to discriminated extinction on probe trials.

Using these standardized methods, Sidman and colleagues (1982) reported that monkeys and baboons could not pass tests for equivalence that were easily passed by children, even children with developmental delays (see Carr, Wilkinson, Blackman, & McIlvane, 2000). It seems that nonhuman subjects lacked critical experiences (e.g., multiple-exemplar training, control of location variables, and generalized identity matching) that are naturally arranged by the environment of humans. Subsequently, a limited number of nonhuman studies claimed that animals could pass tests for reflexivity, symmetry, and transitivity (McIntire, Cleary, & Thompson, 1987; Vaughn, 1988) or symmetry and transitivity (D'Amato, Salmon, Loukas, & Tomie, 1985; Richards, 1988). These studies are controversial, as some researchers asserted that the animals did not demonstrate generalized relations—all of the relations were directly trained (e.g., Hayes, 1989a; Saunders, 1989). Also, in nonhuman research there is some question as to whether the pigeon (or ape) is picking out the key that matches the sample or is merely doing *exclusion*, rejecting the nonmatching option (Carrigan & Sidman, 1992).

A review of symmetry in nonhuman animals identified 24 studies in species ranging from rats to chimpanzees, but the overall results were equivocal, with about 55% of the studies showing mixed or strong evidence of emergent relations (Lionello-DeNolf, 2009). Two studies with sea lions (Rocky and Rio, described in Chapter 8 in the section "Matching to Sample: Identity Training") showed clear evidence of symmetry and transitivity, a result attributed to the animals' history of multiple-exemplar training and the use of multiple S⁻ comparison stimuli during training (Kastak, Schusterman, & Kastak, 2001; Schusterman & Kastak, 1993). Figure 12.11 shows Rio during training of the A–B relation (given CRAB choose TULIP) and the B–C relation (given TULIP choose RADIO). The third photograph in this sequence shows the test for one of the four C–A, emergent relationships, illustrating both symmetry and transitivity (given RADIO choose CRAB). Another study reported unambiguous evidence of symmetry in pigeons. These experiments used successive matching-to-sample to control for variation in stimulus location and intermixed sessions of A–B with A–A and B–B identity training to control for temporal placement of the stimulus as either a sample or a comparison (Frank & Wasserman, 2005). Follow-up experiments showed that pigeons given similar training, but without the intermixed sessions of identity training, did not show emergent symmetry on probe trials. Also, pigeons that were initially trained only with arbitrary matching trials did not show symmetry on probe trials, but subsequently demonstrated emergent symmetry when intermixed identity training was added to the baseline-training trials.

We noted the role of multiple-exemplar instruction in the naming relation, and the evidence suggests that multiple-exemplar training (MET) is a critical feature for the emergence of symbolic equivalence classes in nonhuman animals. Exactly why training with multiple exemplars works is not well understood, but it probably relates to reduced control by *where and when* a stimulus appears (Swisher & Urcuioli, 2015; Zentall, Wasserman, & Urcuioli, 2014). MET ensures that samples and comparisons appear in multiple locations, allowing for the generalization among stimuli over locations. In addition, MET reduces control by the temporal location of the sample and comparison stimuli. That is, on training trials, samples always appear first and comparisons second, a temporal feature that gains control of behavior and interferes with equivalence learning. One possibility is that the intermixed sessions of identity training that were used for pigeons showing emergent symmetry also reduce control by irrelevant temporal and spatial features of sample and comparison stimuli. Further research using MET-type procedures with numerous training trials should eventually reveal how to reliably establish equivalence classes and emergent relations in nonhuman animals.

In fact, Zentall and his colleagues (2014) indicated that reliable associative learning of the symmetry relations already has been established in nonhuman animals. And Smirnova, Zorina, Obozova, and Wasserman (2015) recently observed derived stimulus relations indicative of analogical reasoning in crows trained by relational matching-to-sample (see also Miguel et al. 2015 for a relational tacting procedure that results in the emergence of analogical reasoning in humans). The evidence is mounting that stimulus equivalence and other derived stimulus relations occur across species and this behavior is more pronounced in humans.

Stimulus Equivalence and Application

At the applied level, stimulus equivalence training has been helpful to those who lack reading skills, and in the development of educational curricula based on derived stimulus relations (for a review, see Rehfeldt, 2011). Researchers have used people with developmental delays

Fig. 12.11
Photograph of Rio the California sea lion is shown. Colleen Kastak and Ronald J. Schusterman raised and trained Rio at the Pinniped Cognition & Sensory Systems Laboratory in Santa Cruz, CA. Rio is shown during training of the A-B relation (given CRAB choose TULIP) and the B-C relation (given TULIP choose RADIO). The third photograph in this sequence shows a test for one of the four C-A emergent relations, illustrating both symmetry and transitivity (given RADIO choose CRAB).

Source: Copyright held by Dr. Colleen Reichmuth of the Institute of Marine Sciences, University of California Santa Cruz. Republished with permission.

who could pass a reflexivity test (identity matching) but, before training, failed to show symmetry or transitivity (Sidman & Cresson, 1973; Sidman, Cresson, & Wilson-Morris, 1974; see also Lazar, 1977). Study participants were given training in symbolic matching. They were presented with one of 20 spoken names and asked to select the corresponding picture from a comparison set (A = B training). Next, the participants were trained to select printed words from a set when given one of the 20 names (A = C training). After both training procedures, participants displayed four untrained relations without further training—two symmetry and two transitivity relations. Subjects showed B to A, and C to A, reversals—when given a picture they emitted the corresponding name and when given a printed word they said it. In addition, the participants showed two transitivity relations. When given a picture (car, boy, or dog), the participants selected the corresponding printed word (B = C), and when given the printed word, they selected the corresponding picture (C = B).

During training the participants were presented with three stimulus classes that contained 20 elements in each class (spoken words, pictures, and written words). Forty instances of symbolic matching were reinforced (spoken words = pictures, and spoken words = written words). Tests revealed that 80 new instances of correspondence were established indirectly from training (B = A, C = A, B = C, and C = B). Notice that equivalence training generated many new derived or untrained stimulus relations, which may relate to the generative nature of language, where people generate many novel sentences never before heard or said (Chase, Ellenwood, & Madden, 2008; Stewart, McElwee, & Ming, 2013).

As you can see, the reinforcement of symbolic matching resulted in a preliminary form of reading by people with developmental disabilities. The limits on this training have not been established, but it seems obvious that equivalence relations make up a large part of human education (mathematics, science, reading, etc.). Equivalence classes are not the same as discriminative stimuli, because S^Ds cannot be exchanged for the responses they occasion. Clearly, equivalence relations define symbolic performance and are an important part of the experimental analysis of verbal behavior (see Sidman, 1994).

One problem with this conclusion is a growing literature which shows that reinforcement is not necessary for the emergence of derived stimulus functions (Minster, Elliffe, & Muthukumaraswamy, 2011; Zentall et al., 2014). The idea is that stimulus–stimulus (S–S) correlation, the degree to which one stimulus goes together with another (e.g., sight of dog with word "dog"), is the basic determinant of emergent stimulus relations, not reinforcement contingencies as required by Sidman (2000). Several reports support this conclusion (e.g., Smeets, Barnes-Holmes, & Nagle, 2000). In fact, Minster et al. (2011) designed an experiment with humans to include conditions where stimulus relations were correlated during training but, at the same time, explicitly placed on operant extinction. They found that emergent stimulus relations were established by the stimulus–stimulus correlations alone. This means that reinforcement contingencies and S–S correlations each produce similar outcomes. Since reinforcement contingencies always contain S–S correlations, it may be that S–S correlation is the driving force behind equivalence and derived-stimulus learning. For training symbolic behavior, explicit reinforcement may not be as critical as arranging correlations among common stimulus features (e.g., a real dog, a picture of a dog, and the word "dog").

ON THE APPLIED SIDE: THREE-TERM CONTINGENCIES AND NATURAL SPEECH

At the most basic level, behavior analysts suggest that the acquisition of verbal behavior is governed by contingencies of reinforcement. An important question is whether humans arrange verbal contingencies in their everyday interactions. Evidence of operant contingencies in casual speech is important for a comprehensive account of verbal behavior. When observational research shows natural dependencies between speakers and listeners, we can be more confident that our understanding of speaking (and writing) is not an artifact of laboratory procedures. Also, evidence of verbal contingencies without explicit control by an experimenter suggests that laboratory findings may eventually have general applicability. For both of these reasons, the studies by Moerk (1990) of contingency patterns in mother–child verbal episodes are an important contribution to the analysis of verbal behavior.

Data, Transcripts, and Findings

The data are based on a reanalysis of the verbal interactions between a child named Eve and her mother. The original observations were collected by Brown (1973) as part of a larger study of mother–child interaction. Eve and her mother were observed in their home during everyday activities. When the study began, Eve was 18 months old, and she was 28 months old at the end of the research. Brown collected numerous samples of verbal interaction between Eve and her mother over this 10-month period. Moerk selected all odd-numbered samples and analyzed 2 h of transcribed audio recording for each of these samples.

Moerk and two trained research assistants coded the transcripts. Observational categories included verbal behavior emitted by both mother and child (Eve). For example, sentence expansion involved the mother adding syntactic elements to her child's utterance (e.g., Eve says, "See boy" and her mother says, "You see the boy"), while sentence reduction occurred when Eve omitted elements that were originally present in her mother's speech (e.g., mother says, "Give the toy to mommy" and Eve says, "Give toy mum"). The research focuses on the arrangement of such verbal utterances in mother–child–mother interactions.

Moerk (1990) found that many different mother–child–mother verbal sequences ended with maternal reinforcement. Reinforcement was defined as feedback from the mother that confirmed that Eve's utterance was linguistically acceptable (e.g., "Yes," "Right," "OK," and so on). A sequence that often occurred was the mother saying a new or rare word (model) that was repeated by Eve (imitation) and followed by her acceptance by the mother (reinforcement). Another three-term pattern involved the mother repeating what she had just said, Eve emitting an approximation to this utterance, and her mother ending the sequence with words of acceptance.

The findings indicate that three-term contingencies (maternal verbal stimulus, child verbal imitation, and maternal reinforcement) characterized many of the verbal episodes of early language learning, and are compatible with Skinner's functional analysis of verbal behavior (see also Moerk, 2000; see also Golinkoff, Can, Soderstrom, & Hirsh-Pasek, 2015, p. 341, for evidence that adult contingent responsiveness to infants' pointing to objects (tacting) is central to early language learning).

Relational Frames

Equivalence among stimuli is not the only derived relation that occurs with verbal stimuli. Other relations, such as larger, smaller, happier, greater, less than, etc. can also transfer function without directly experiencing associated contingencies. For example, Dougher et al. (2007)

first trained three abstract visual symbols (we will call them A, B, and C) to three sizes (small, medium and large, respectively) with matching-to-sample procedures. For example, when stimulus A and B are presented, choosing stimulus A is reinforced, because it is the smallest. This is relational stimulus control (see chapter 8). Then, participants were trained to press a key on a keyboard at whatever rate they desired in the presence of only the medium stimulus (B). When A and C stimuli were presented as probe trials, participants clicked slower when A was presented and faster when C was presented. Therefore, the relations of smaller and larger transferred to other stimuli without the participant being directly reinforced for these relations; this is called a relational frame. Then, the researchers wanted to determine the extent to which the function of a stimulus would transfer to the stimuli in the frame. A mild, painless shock was paired only with stimulus B (the medium one) and skin conductance was measured. Sure enough, when skin conductance was tested to stimuli A and C, participants showed a smaller amount of skin conductance to A compared to B and the largest amount of skin conductance to C. Importantly, all that was trained was 1) the ABC relational frame 2) skin conductance to stimulus B. What was *learned*, however, was differential relations to A and C without A and C ever having direct experience with shock. The function of shock transferred to the other stimuli in a manner that represented the ABC relational frame. Relational frame theory, then, can account for the rapid acquisition of relations to verbal stimuli and their transfer of function; this indeed is foundational to a behavioral account of language and cognition (see Barnes-Holmes & Harte, 2022; Barnes-Holmes & Roche, 2001; Hughes & Barnes-Holmes, 2016 for reviews).

Advanced Section: A Formal Analysis of Mand and Tact Relations

In his book *Verbal Behavior*, Skinner (1957) discusses the formal differences in behavior regulation between manding and tacting. In this section on advanced issues, we explore the social contingencies that establish and maintain these two classes of verbal behavior. The contingencies are somewhat complex, and diagrams of the interrelation of the speaker and listener help to depict the controlling variables.

The Mand Relation

A formal analysis of the mand relation is depicted in Figure 12.13. A **social episode** involves the social interaction of speaker and listener. The line through the middle of the diagram separates the speaker's events and actions from those of the listener. Each person completes a behavioral sequence or chain ($S^D: R \rightarrow S^r + S^D: R \rightarrow S^r \ldots$), and social interaction involves the intermingling of these chains or the **interlocking contingencies** (examine this in Figure 12.13). In the diagram of a social episode, an arrow ($\rightarrow$ horizontal or vertical) means "produces or causes"; thus, a verbal response by one person may cause an event, condition, or stimulus for the behavior of the other person (vertical arrow). That is, the verbal behavior of one person functions as a stimulus and/or consequence in the behavior chain of the other individual. Also, within the behavioral sequences of each individual, the verbal operants produce effects or consequences (horizontal arrow) supplied by the behavior of the other person (check this out).

In the example shown in Figure 12.12, we assume that two people are seated at a counter in a cafeteria. Dinner is placed in front of the speaker, but the ketchup is out of reach and situated near the other person or listener. In this context, the presence of food on the table is an *establishing operation* (EO) for behavior that has produced ketchup in the past (see Michael, 1982a, 1993; see also Chapter 2). The EO also makes getting ketchup a reinforcing event in this situation.

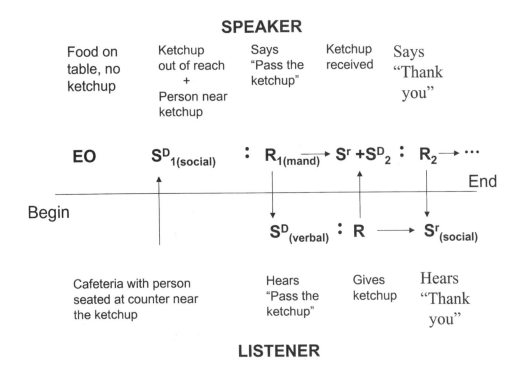

Fig. 12.12
Analysis of the mand relation between speaker and listener based on Skinner (1957). The person who needs ketchup is analyzed as the speaker and another customer is the listener. EO = establishing operation; SD = discriminative stimulus; R = operant; S^r = reinforcement. See text for a description of the verbal contingencies.

In addition to the EO, the speaker's mand response ("Pass the ketchup") in Figure 12.12 is regulated by the presence of ketchup near the listener ($S^D_{1\text{ speaker}}$). The first vertical arrow, passing from the listener's side of the interaction (operant chain) to the speaker's side, shows the causal effect of the listener. If there were no other people in the restaurant, it is likely that the speaker would get out of the seat and get the ketchup themself. The presence of a listener increases the probability that the speaker will say "Pass the ketchup" rather than get it. This means that the listener functions as part of the discriminative stimulus (S^D_1) in this social episode. Together, the out-of-reach ketchup and the presence of the listener (S^D_1) set the occasion for (:) a verbal response (R_1) by the speaker.

The speaker's verbal response (R_1) of "Pass the ketchup" affects the listener as a stimulus. The causal effect of the speaker's behavior on the listener is shown as a vertical downward arrow from R_1 (speaker) to the listener's side of the interaction (operant chain). The words "Pass the ketchup" uttered by the speaker are a verbal stimulus for the listener (S^D_1) that sets the occasion for (:) the listener to pass the ketchup ($R_{1\text{ listener}}$). In this social episode, the listener's response of passing the ketchup ($R_{1\text{ listener}}$) is reinforcement for the speaker's verbal operant ($S^r_{1\text{ speaker}}$). Because the speaker's verbal response ("Pass the ketchup") produces specific reinforcement (getting the ketchup) from the listener, the verbal operant is formally manding. As previously stated, *manding* is verbal behavior set up by an EO (out-of-reach ketchup) and maintained by specific reinforcement (getting the ketchup) mediated by the listener's behavior.

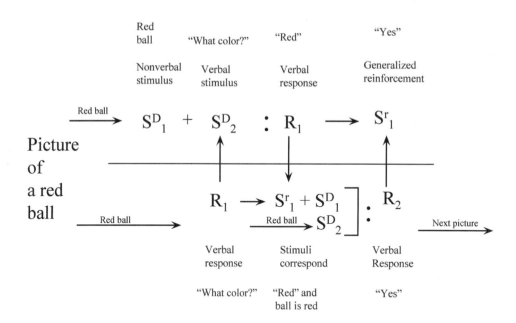

Fig. 12.13
Analysis of the tact relation between speaker and listener is depicted, based on Skinner (1957). Both the student and teacher emit verbal behavior during the social episode, but tacting by student (speaker) is the focus of the analysis. S^D = discriminative stimulus; R = operant; S^r = reinforcement. See text for a description of the verbal contingencies.

In this situation, the listener's response of passing the ketchup has **multiple functions** for the speaker's behavior ($S^r_1 + S^D_{2\ speaker}$). Passing the ketchup not only functions as reinforcement for the mand response, but it also functions as a discriminative stimulus for the next response by the speaker. Thus, the same event (the listener giving the ketchup) can have several causal effects on the speaker's behavior. Based on the discriminative function of the listener's response ($S^D_{2\ speaker}$), getting the ketchup sets the occasion for (:) the speaker saying "Thank you," a verbal response ($R_{2\ speaker}$) that serves as generalized conditioned reinforcement for the listener's behavior ($S^r_{1\ listener}$). The "Thank you" response also serves as the ending point for this social episode, releasing the listener from obligations with respect to the speaker.

The Tact Relation
Figure 12.13 depicts a formal analysis of the tact relation. As with manding, the verbal episode involves the *interlocking contingencies* of a speaker and a listener. In this example, the speaker is a student and the listener is a teacher. The social episode begins in a classroom with the teacher showing pictures of objects to a young student.

When a picture of a red ball is displayed, this event causes (horizontal arrow) the teacher to say, "What color?" (Figure 12.13). The teacher's question (R_1) produces a verbal stimulus to the student (vertical arrow upward). In this situation, the student's answer depends on *both* the nonverbal stimulus of the red ball ($S^D_{1\ speaker}$) and the teacher's question ($S^D_{2\ speaker}$). Notice that the student will give a different answer if the question is "What shape?"

The student's answer of "Red" is formally tacting ($R_{1\ speaker}$) because the operant is regulated by the nonverbal stimulus (redness of ball). In this example, the student's tacting produces a verbal stimulus ($S^D_{1\ listener}$) for the teacher that may or may not *correspond* to the specified physical property of the ball ($S^D_{2\ listener}$). If the student's answer of "Red" corresponds to the color of the ball, the teacher's question "What color?" is reinforced ($S^r_{1\ listener}$). Notice how the speaker and listener complete individual operant chains (e.g., $S^D: R \rightarrow S^r + S^D: R \rightarrow S^r$ ...) that are *interlocking*, in the sense that the behavior of each person causes stimulation and reinforcement for the behavior of the other.

In terms of analysis, the teacher's question "What color is the ball?" is manding. This verbal response is reinforced by correspondence between the student's tacting and the actual color of the object. When correspondence occurs, this condition sets the occasion for the teacher saying ($R_{2\ listener}$) "Yes" and turning to the next picture (noncorrespondence may lead to repeating the question, perhaps in a different way). The teacher's verbal response ("Yes") produces generalized conditioned reinforcement ($S^r_{1\ speaker}$) for the student's tact response and functions to maintain the verbal operant class.

Finally, it is useful to compare the controlling variables for the mand and tact relations. As we have seen in the teacher–student example, the form or topography of the tact response depends on an appropriate nonverbal stimulus. The redness of the ball regulated the student's verbal response. In contrast, mand responses depend on an establishing operation (EO) such as deprivation. The dinner without ketchup regulated asking for it. Generalized conditioned reinforcement (acceptance, praise, and attention) serves to maintain the verbal operant class of tacting. In contrast, specific reinforcement related to an EO (getting the ketchup) maintains the operant class of mand responses.

CHAPTER SUMMARY

People talk to each other, and in different countries they use different languages. It seems clear that this ability is usually learned in childhood, and according to the principles that govern other operant behavior. Evidence suggests, however, that the vocal apparatus and its neuromuscular features may have evolutionary origins allowing for extensive and complex production and control of speech sounds. Skinner defined speaking, writing, and gesturing as verbal behavior, and proposed an analysis based on basic operant principles. This analysis begins with a description of the function of language as established by the reinforcing practices of the verbal community. Manding and tacting are two broad classes of verbal operant behavior. Manding is a verbal form regulated by establishing operations and specific reinforcement. Tacting is a form regulated by nonverbal discriminative stimuli and maintained by generalized conditioned reinforcement. A verbal interaction between two people may involve manding, tacting, and many other verbal response classes (e.g., intraverbals) regulated by verbal stimuli. Finally, symbolic behavior and stimulus equivalence were discussed as examples of "higher-order" activities involving verbal operants. The equivalence relations of reflexivity, symmetry, and transitivity were shown to extend from formal mathematics to the control of behavior. Stimulus classes exist when

> organisms have passed tests for these three relationships. Relational frames can also be learned, which form the basis of more complex emergent stimulus relations. Verbal behavior is possibly the most complex of human activities, and its intricacies continue to engender much research and behavior analysis.

Key Words

Autoclitic (verbal behavior)
Conditioned establishing operation (CEO)
Echoic responses
Formal similarity
Functional independence
Identity matching
Interlocking contingencies
Intraverbal behavior
Manding
Multiple functions (of stimuli)
Naming relation
Reflexivity

Relational frame
Reversal test
Social episode
Stimulus equivalence
Symbolic matching
Symmetry
Tacting
Textual behavior
Transitivity
Verbal behavior
Verbal community

On the Web

http://now.uiowa.edu/2014/12/crows-are-smarter-you-think This webpage describes the research by Ed Wasserman and his colleagues in Russia on derived relations (analogical reasoning) in crows arising from training birds on a relational matching-to-sample task. You can get the original article by Smirnova, Zorina, Obozova, and Wasserman published in *Current Biology* (2015) and found in the references for this textbook.

www.ted.com/talks/deb_roy_the_birth_of_a_word.html MIT researcher Deb Roy wanted to understand how his infant son learned language—so he wired up his house with video cameras to catch every moment (with exceptions) of his son's life, then parsed 90,000 hours of home video to watch "gaaaa" slowly turn into "water." This is astonishing, data-rich research with deep implications for how we learn verbal behavior.

www.youtube.com/watch?v=QKSvu3mj-14 This is Part 1 from *Cognition and Creativity*, a video by Robert Epstein and B. F. Skinner. The clip introduces basic conditioning and then goes on to present the Jack and Jill study of communication in pigeons. See if you can analyze how Epstein and Skinner got pigeons to show such complex "verbal" behavior. After you finish Part 1, proceed to www.youtube.com/watch?v=erhmslcHvaw for Part 2 of this video, showing how to analyze other forms of behavior usually attributed to cognition and higher mental processes.

www.youtube.com/watch?v=dBUHWoFnuB4 This video features research and life with Kanzi, a bonobo chimpanzee, and Dr. E. Sue Savage-Rumbaugh at the Georgia State University Language Research Center in Atlanta, GA.

Brief Quiz

1. _____ introduced the term *verbal behavior* to deal with the _____ of the speaker.
   a. Chomsky; transformational grammar
   b. Skinner; performance
   c. Crick; conscience
   d. Pavlov; conditioned responses

2. According to the behavioral or functional account, sign language, gestures, and body movements are instances of:
   a. nonverbal communication
   b. message transmission
   c. verbal behavior
   d. culture and tradition

3. One kind of conditioned establishing operation (CEO) called the _____ involves withholding an object or item necessary to complete a behavioral sequence.
   a. no item method
   b. absent object technique
   c. interrupted item method
   d. blocked response

4. When a verbal response depends on a verbal discriminative stimulus, the verbal relations are:
   a. manding
   b. tacting
   c. intraverbal
   d. textual

5. In echoic behavior, when _____ by the child correspond to those of the adult, the _____ patterns also overlap.
   a. sounds; temporal
   b. phonemes; reverberation
   c. speech; phoneme
   d. articulations; acoustical

6. When John says to his sister Katherine "*Could you please* get me a glass of water" rather than "Get me a glass of water," the basic unit added to his mand is called a (an):
   a. textual editing
   b. autoclitic response
   c. extended tacting
   d. none of these

7. A response such as "I have butterflies in my stomach" can be analyzed as: _____.
   a. generalized tacting
   b. generalized manding
   c. formal manding
   d. formal tacting

8. When reinforcement is based on matching of geometric forms to different line angles, the procedure is called:
   a. identity matching
   b. matching to sample
   c. transitivity matching
   d. symbolic matching

9   If a picture of a dog, the spoken word "dog," and the written word DOG all regulate the same behavior, we say that the stimulus classes are:
    a   overlapping
    b   the same
    c   equivalent
    d   confounded

10 **ADVANCED SECTION:** In terms of interlocking contingencies, a vertical arrow (downward) from the speaker's operant chain to that of the listener's indicates that:
    a   the speaker's behavior causes stimulation and/or reinforcement for the listener's behavior
    b   the listener is causally motivated to behave as the speaker requests
    c   the speaker is causally motivated to produce a condition for the listener
    d   the interaction between the speaker and the listener is mutually anticipated

Answers to Brief Quiz: 1, b (p. 428); 2, c (p. 430); 3, d (p. 437); 4, c (p. 443); 5, d (p. 445); 6, b (p. 446); 7, a (p. 465); 8, d (p. 452); 9, c (p. 457); 10, a (p. 463).

CHAPTER 13

Applied Behavior Analysis

1. Learn about applied behavior analysis, its methods, and data recording.
2. Investigate contingency management of drug abstinence.
3. Explore the impact of behavior analysis in teaching and education.
4. Discover a common structure of programming for the treatment of autistic behavior.
5. Focus on the obesity crisis and behavior management of eating and body weight.

The experimental analysis of behavior is a science that easily lends itself to application. This is because the focus of the discipline is on those environmental events or causes that directly alter the behavior of individual organisms. Almost half a century ago, behavior analysts proposed that operant and respondent principles controlling behavior in the laboratory probably regulate human and nonhuman behavior in the everyday world (Dollard & Miller, 1950; Skinner, 1953). Thus, principles of behavior allow us to change socially significant human conduct. Based on this assumption, Skinner (1948a) wrote his second book, *Walden Two*, as a novelized description of a utopian society based on behavior principles. At least two experimental communities developed utilizing many of the principles of *Walden Two*: Twin Oaks in the Eastern USA and Los Horcones near Hermosillo, Mexico (Fishman, 1991; McCarty, 2012).

Principles of behavior change have been used to improve the performance of university students (Moran & Malott, 2004; Pear, Schnerch, Silva, Svenningsen, & Lambert, 2011), increase academic skills (Alberto & Troutman, 2013), teach children with developmental delays and autism self-care and communication skills (Boutot & Hume, 2012; Kodak & Bergmann, 2020; McGaha-Mays & Heflin, 2011), reduce phobic reactions (Jones & Friman, 1999; Shabani & Fisher, 2006), persuade children and adult drivers to wear seat belts (Van Houten et al., 2010; Van Houten, Reagan, & Hilton, 2014), encourage drivers to obey stop signs and increase driving distance when following other cars during cell phone use (Arnold & Van Houten, 2013; Austin, Hackett, Gravina, & Lebbon, 2006), prevent occupational injuries (Geller, 2006, 2011), preventing gun play and promoting safety skills in children (Miltenberger & Gross, 2011; Miltenberger et al., 2005), increase drug abstinence and other healthy behavior (Donohue, Karmely, & Strada, 2006; Hand, Heil, Sigmon, & Higgins, 2014; Higgins, Silverman, & Heil,

DOI: 10.4324/9781003202622-13

2007). Behavioral interventions have had an impact on clinical psychology, medicine, counseling, job effectiveness, sports training, and environmental protection. Applied experiments have ranged from investigating the behavior of psychotic individuals to analyzing (and altering) contingencies of entire institutions (see Kazdin, 2013). Thus, principles of behavior derived from experimental and applied research have wide-scale applicability because the world actually operates according to these principles.

CHARACTERISTICS OF APPLIED BEHAVIOR ANALYSIS

Behavioral principles, research designs, observational techniques, and methods of analysis transfer readily to an applied science. When this is done to improve performance or solve social problems, the technology is called applied behavior analysis (Baer, Wolf, & Risley, 1968). Thus, **applied behavior analysis (ABA)** or behavioral engineering is a field of study that focuses on the application of the principles, methods, and procedures of the science of behavior (Ayllon & Michael, 1959). Because applied behavior analysis is a wide field of study, it cannot be characterized by a single definition. Nonetheless, several features in combination distinguish applied behavior analysis as a unique discipline.

Concentration on Research

Behavior therapists and applied researchers are committed to a scientific analysis of human behavior. What a person does and the events that regulate behavior are objectively identified. In this regard, operant and respondent conditioning are assumed to control most human activity regardless of how verbal behavior, generalized imitation, equivalence relationships, and neurophysiology complicate the analysis. Although behavior principles are widely applied to human behavior, these principles also are used broadly in the management of nonhuman behavior, such as in pets (Friedman, Edling, & Cheney, 2006; www.Behaviorworks.org), zoo animals (Maple & Perdue, 2013; Markowitz, 1981), and livestock (Foster, Temple, & Poling, 1997; Provenza, Gregorini, & Carvalho, 2015; Provenza, Villalba, Cheney, & Werner, 1998; www.Behave.net).

Applied behavior analysis involves two major areas of research which entail the application of operant and respondent principles to improve human behavior. A good deal of literature has documented the success of this enterprise (see the *Journal of Applied Behavior Analysis* from its beginning in 1968 for many examples). Thousands of experiments and applications have shown how basic conditioning principles can be used in a variety of complex settings. Problems unique to the applied context have been addressed and treatment packages designed for the modification of behavior have been described and evaluated (see Martin & Pear, 2015).

Another set of studies have not focused directly on behavior change, but are a part of applied behavior analysis and currently discussed as *translational research* (Mace & Critchfield, 2010; Critchfield, 2011 also proposed that basic experimental analysis furthers translational research). Translational investigations often involve an analysis of everyday human behavior in a social context and the implications for improving the human condition. For example, studies of the environmental contingencies related to human cooperation, competition, successful teaching practices, and coercive family dynamics often identify basic principles of complex human interaction (Dymond & Critchfield, 2002; Epling & Pierce, 1986). Researchers in this area of applied behavior analysis are attempting to specify the operating contingencies, which produce or modify a variety of social problems (Lamal, 1997).

Behavior Is the Primary Focus

Applied behavior analysts focus on the observable behavior of people and other animals in non-laboratory settings. Behavior is not considered to be an expression of inner agencies or causes such as personality, cognition, and attitude. Thus, marital difficulties, out-of-control children, public littering, phobic reactions, poor performance in exams, excessive energy use, negative self-descriptions, and many other social and personal difficulties are analyzed as problems of behavior. Interventions for these and other human problems are directed at changing environmental contingencies (stimulus-control variables and behavior consequences) to improve behavior.

Of course, people think, feel, and believe a variety of things associated with what they do. Individuals who are experiencing difficulty in life may have unusual thoughts and feelings. A depressed person may feel worthless and think that nobody likes them. The same person does not spend much time visiting friends, going to social events, or engaging in the usual activities of life. A behavioral intervention for this problem would probably focus on increasing the person's activity, especially social interaction, which maximizes contact with reinforcement. The individual may be asked to set goals for completing various tasks and reinforcement is arranged when these activities are accomplished. When people become more socially involved, physically active, and complete daily tasks, they usually do not describe themselves as depressed. In this and many more cases, a change in reinforcement or the density of reinforcement of daily activities produces a change in feeling and thinking (Cautela, 1984, 1994).

The Importance of Conditioning

Our discussion should make it clear that problem behavior may, in most cases, be understood in the same fashion as any other behavior. Principles of conditioning are neutral with respect to the form and frequency of behavior. Maladaptive, annoying, and dangerous conduct may be produced inadvertently by environmental contingencies, just like more positive responses to life events.

Consider an institutional setting in which three staff nurses are in charge of 20 children with disabilities. The nurses are busy and as long as the children behave well, they are left alone. This natural response to a strenuous work schedule may, for some children, result in deprivation for adult attention. When one of the children accidentally hits her head and is hurt, very likely a staff member rushes over to comfort the child. It is possible that head hitting would increase in frequency because it has been reinforced by contingent attention (e.g., Lovaas & Simmons, 1969). Of course, when people are injured, they cannot be ignored. One way to deal with such a conundrum would be to provide plenty of social reinforcement for appropriate play, academic activities, ward chores, self-hygiene, and other daily activities. This tactic is called **differential reinforcement of other behavior**, or **DRO** (e.g., Hammond, Iwata, Fritz, & Dempsey, 2011; Lindberg, Iwata, Kahng, & DeLeon, 1999). In the preceding example, the procedure would strengthen responses that are incompatible with self-injury and reduce deprivation for adult attention. Another useful strategy is **differential reinforcement of alternative behavior**, or **DRA**, which targets a specific alternative response for reinforcement (see Petscher, Rey, & Bailey, 2009 for a review of DRA in the context of developmental disabilities).

Although much human behavior is a function of contingencies of reinforcement, biological factors also produce behavior change. A person who has experienced a stroke, a child with fetal alcohol syndrome, an individual in the later stages of Alzheimer's Disease, or an adult suffering from Huntington's chorea may emit responses that are a function of brain damage, toxic agents, disease, and genetics. Even when this is the case, however, principles of conditioning often can be used to improve behavior (Epling & Pierce, 1990, pp. 452–453).

ABA in Reducing Challenging Behavior

In some circumstances, access to non-contingent reinforcement (NCR) in general in a person's environment can reduce challenging behavior. This is typically thought to decrease the motivation (establishing operation) to engage in challenging or maladaptive behavior. Ingvarsson, Kahng, and Hausman (2008) conducted a study with an 8-year-old girl diagnosed with autism spectrum disorder, mild cerebral palsy, moderate cognitive impairment, and obsessive-compulsive disorder. The girl, Manuela, had been admitted to an inpatient behavioral unit for the assessment of dangerous challenging behavior. She engaged in aggression, disruptions, and self-injurious behavior. Manuela's parents reported she engaged in challenging behavior when denied access to edible and other favored items.

The researchers first conducted an assessment to understand why Manuela was engaging in the challenging behavior. They found consequences such as providing escape from a task and getting access to snacks were maintaining the challenging behavior. For intervention the researchers compared three conditions: (1) Manuela was given a preferred edible item before each time she was asked to do a task (high density non-contingent reinforcement), (2) Manuela was given a preferred edible before every fourth task she was asked to do (low density non-contingent reinforcement), and (3) Manuela was only given a preferred edible contingent on compliance (differential reinforcement of alternative behavior; DRA). The researchers found decreases in challenging behavior relative to baseline, the same amount of decrease in challenging behavior across high and low density of reinforcement, and similar decreases between high density NCR and DRA. This information suggested non-contingent reinforcement was effective in reducing challenging behavior and high and low densities of reinforcement had the same effect on challenging behavior. And, furthermore, the non-contingent reinforcement of challenging behavior was as effective at reducing the challenging behavior as contingent reinforcement of an alternative behavior. What is also interesting is the reinforcer used during NCR was shown to be functionally relevant with respect to the problem behavior as predicted by the researchers' assessment. A benefit of this study was the demonstration that extinction was not necessary to reduce rates of challenging behavior. Simply increasing non-contingent reinforcement for Manuela was effective in reducing her challenging behavior and helping her live a healthier life.

Direct Behavior Change

Applied behavior analysts usually focus directly on the environmental events that generate and maintain behavior. Typically, target behavior and the events that precede and follow this behavior are counted for several days. During this baseline, treatment is withheld so that a later change in behavior can be evaluated. This *behavioral assessment* also provides information about stimulus control (events that precede the behavior) and contingencies of reinforcement (events that follow behavior) that maintain responses.

A general example of a direct behavior change procedure is **behavioral contracting**. Following a baseline period of assessment, a behavioral plan of action may be negotiated between the behavior therapist, the client, and concerned others. This plan usually includes a statement of target responses, consequences that follow different actions, and long-term

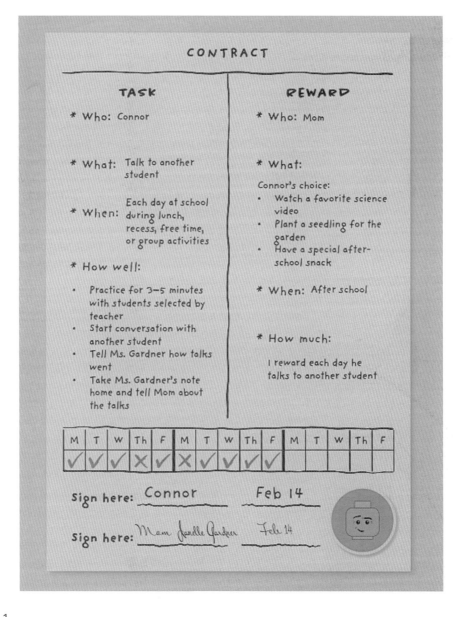

Fig. 13.1
A home/school behavior contract for a student with autism who wanted to make friends with classmates. (From Dardig, J. C., & Heward, W. L. [2022]. *Let's Make a Contract: A Positive Way to Change Your Child's Behavior*. The Collective Book Studio. Reprinted by permission of the authors.)

goals (final performance). Then, a detailed **behavioral contract** is drawn up that objectively specifies what is expected of the client and the consequences that follow behavior (Dardig & Heward, 2022). Figure 13.1 is a home/school contract created for Connor, a 10-year-old boy with autism, who wanted to make friends with his classmates. The task, "Talk to another student," is done in school. The reward, "Connor's choice" from a menu of activities, is received at home. The contract includes a chart for Connor's teacher to record his task completion each day. In addition to the contract pictured here, Connor's teacher created a simple script he could follow when starting a conversation and arranged for him to practice the script with several classmates. Successful completion of this contract relies on daily communication among the teacher, the child, and his mother. When Connor successfully interacts with classmates, the natural reinforcers of positive attention from his peers will maintain the behavior of talking to other students, and the contract can be faded out.

Applied behavior analysts do not typically focus on what has been called the "therapeutic process" for behavioral change. While psychotherapy indeed helps many people with a variety of problems, including identifying problems in relationships, it often does not include specific behavioral targets for change. For example, a person may discover the reasons why they drink too much alcohol, and that it might be harming the relationship with their spouse, but putting together a treatment to *change* this behavior may be missing. Behavior analysts have traditionally preferred to arrange contingencies of reinforcement to alter behavior problems.

Some behavior analysts, though, see psychotherapy and behavior analysis as compatible. Steven Hayes at the University of Nevada in Reno is a clinical behavior analyst who has emphasized the importance of rule-governed and derived stimulus relations in a therapeutic setting. From Hayes's perspective of acceptance and commitment therapy (ACT) and relational frame theory (RFT), talking is a form of social influence that may be used to change the client's actions. That is, instructions and relational operant learning allow people to verbally frame life events and this framing may be influenced in a therapeutic setting to alter the client's probability of behavior (see Hayes, 1987; Hayes, Strosahl, & Wilson, 2012; Hughes & Barnes-Holmes, 2016; Zettle & Hayes, 1982).

While ACT is one of the most used clinical treatments today, most applied behavior analysts are trained to treat behavior with direct contingency management. Others, however, are investigating the practical importance of instructions, rules, and therapeutic advice—verbal stimuli maintained by remote contingencies of social reinforcement (Hayes, 1989b; see also De Houwer, Barnes-Holmes, & Barnes-Holmes, 2015, who propose that functional and cognitive therapies are not necessarily incompatible).

Programming for Generality

For direct treatment of challenging behavior, applied behavior analysts focus on the generality of behavior change (Baer, 1982b; Stokes & Baer, 1977). That is, researchers attempt to ensure that their interventions produce *lasting changes in behavior* that occur in all relevant settings. As noted in Chapter 8 on stimulus control, when behavior is reinforced in the presence of a particular stimulus, a gradient of generalization is produced that falls on both sides of the discriminative stimulus (S^D). Rather than rely on the organism to generalize automatically in an appropriate manner, the applied behavior analyst often attempts to *program for generality* (i.e., teaching generalization directly).

Generality of behavior change involves three distinct processes: stimulus generalization, response generalization, and behavior maintenance (Martin & Pear, 2015, pp. 150–157).

Behavior change has generality if the target response(s) occurs in a variety of situations, spreads to other related responses, and persists over time. Stimulus generalization occurs when the person responds similarly to different situations (e.g., a person greets one friend as she does another). **Response generalization** occurs when a target response is strengthened and other similar responses increase in frequency (e.g., a child who is reinforced for building a house out of LEGO™ may subsequently arrange the pieces in many different ways). **Behavior maintenance** refers to how long a new behavior persists after the original contingencies are removed (e.g., a person with anorexia when taught to eat properly shows long-lasting effects of treatment if they maintain adequate weight for many years).

Donald Baer (see section "On the Applied Side: Training Generalized Imitation" in Chapter 11) emphasized the importance of training behavioral generality, and provided the following illustration:

> Suppose that a client characterized by hypertension has been taught systematic progressive relaxation techniques on the logic that the practice of relaxation lowers blood pressure a clinically significant amount, at least during the time of relaxation, and that the technique is such that relaxation can be practiced during all sorts of everyday situations in which the client encounters the kinds of stress that would raise blood pressure if self-relaxation did not pre-empt that outcome. Suppose that the relaxation technique has been taught in the clinician's office, but is to be used by the client not only there, but in the home, at work, and recreation settings in which stress occurs. Thus, generalization of the technique across settings, as well as its maintenance after clinical treatment stops, is required.
>
> (Baer, 1982b, p. 207)

To program for generality of behavior change, Baer (1982b) suggests a variety of procedures that affect stimulus and response generalization, and behavior maintenance. First, stimulus generalization of relaxation (or any other behavior) is promoted when the last few training sessions are given in situations that are as similar as possible to everyday settings. Second, when relaxation training is done in a variety of different contexts, such as different rooms with different therapists and at different times of day, stimulus generalization increases. Finally, a therapist who trains relaxation in the presence of stimuli that elicit hypertension in everyday life is programming for stimulus generalization.

Response generalization is increased when the client is taught a variety of ways to obtain the same effect. For example, to relax and reduce her blood pressure, the client may be taught meditation skills or mindfulness strategies, progressive muscle relaxation, and controlled breathing. In addition, a person may be taught to produce new forms of response, as when the therapist says, "Try to find new ways of relaxing and reducing your blood pressure" and reinforces novel responses by the client.

Behavior change may be programmed to last for many years if operant responses contact sources of reinforcement outside of the therapeutic setting. Applied behavior analysts who teach their clients skills that are reinforced by members of the social community are programming for behavior maintenance. This sort of programming has been called **behavior trapping** because, once learned, the new behavior is "trapped" by natural, everyday contingencies of reinforcement (e.g., Durand, 1999; Hansen & Lignugaris/Kraft, 2005; Stokes, Fowler, & Baer, 1978).

Focus on the Social Environment

From a behavioral point of view, it is the physical environment and social system that require change, not the person. James Holland stated:

> Our contingencies are largely programmed in our social institutions and it is these systems of contingencies that determine our behavior. If the people of a society are unhappy, if they are poor, if they are deprived, then it is the contingencies embodied in institutions in the economic system, and in the government, which must change. It takes changed contingencies to change behavior.
>
> (Holland, 1978, p. 170)

Behavior-change programs are usually more circumscribed in their focus than Holland recommends (but see section "Level 3: Selection and Evolution of Culture" in Chapter 14 for a discussion of cultural design). Applied behavior analysts have seldom been in a position to change, or recommend changes in, institutional contingencies. They have targeted more local contingencies involving family and community.

Most behavior-change programs attempt to identify and alter significant variables that maintain target responses. As we have said, these variables are usually in the person's social environment. For this reason, treatment programs are often conducted in schools, hospitals, homes, prisons, and the community at large (see Glenwick & Jason, 1980; Jason, 1998; Lamal, 1997; Mattaini & McGuire, 2006). Parents, teachers, friends, coworkers, bosses, and partners typically control significant sources of reinforcement that maintain another person's behavior. These individuals are often involved and instructed in how to change contingencies of reinforcement to alter a client's behavior. This is especially relevant with interactions in a family where the "problem" is the child's behavior, but the solution is changing the parents' contingency management (Latham, 1994).

RESEARCH STRATEGIES IN APPLIED BEHAVIOR ANALYSIS

In Chapter 2 we discussed A-B-A-B reversal designs for operant research. For single-subject research, basic or applied, the A-B-A-B reversal design has the highest level of **internal validity**—ruling out most extraneous factors. While a reversal design is always preferred, there are practical and ethical difficulties that restrict its use in applied settings.

In natural settings, behavior is often resistant to a reversal procedure. For example, using contingencies of reinforcement to increase socially acceptable playing may alter a child's shyness. If the reinforcement procedure is now withdrawn, the child will probably continue playing with other children (the point of the intervention). This may occur because the shy child's behavior is maintained by the social reinforcement they now receive from playmates. In other words, the child's *behavior is trapped* by other sources of reinforcement. While this is a good result for the child, it is not a useful outcome in terms of inference about causation and research design. This is because the applied analyst cannot be absolutely sure that the original improvement in behavior was caused by the intervention.

Another difficulty with the reversal design in applied settings is that it requires the withdrawal of a reinforcement procedure that was probably maintaining improved behavior. For example, a psychiatric patient may bite their arms, a form of self-injury, which may lead to the use of arm restraints. An applied behavior analyst instead might implement a DRO procedure, in which any behavior except arm biting is reinforced. Under DRO, the arm biting is substantially reduced to a point at which the restraints are no longer necessary. Although we cannot be sure that the DRO contingency caused the reduction in self-injury, it would be inadvisable to remove the contingency during a reversal only to show that it was effective. Thus, the A-B-A-B reversal design is sometimes inappropriate for ethical reasons.

Multiple Baseline Designs

To solve the problems raised by the A-B-A-B reversal design, applied behavior analysts have developed other single-subject designs. **Multiple baseline designs** demonstrate experimental control and help to eliminate alternative explanations for behavior change (Christ, 2007). There are three major types of multiple baseline designs as first described by Hall, Cristler, Cranston, and Tucker (1970). These designs are a multiple baseline across settings, a multiple baseline across participants, and a multiple baseline across behaviors.

Multiple Baseline Across Settings

In the **multiple baseline across settings** design, a reinforcement procedure is applied in one situation, but is withheld in other settings. When behavior only changes in the situation where it is reinforced, then the contingency is applied to the same response in another setting. Hall and his associates used this design in a modification of children's tardiness in getting to class after recess or lunch (Hall et al., 1970).

Figure 13.2 shows the multiple baseline across settings used by Hall et al. (1970). The researchers used what they called a "patriots chart" to modify lateness after lunch, and after morning and afternoon recesses. Children in the fifth grade who were on time for class had their names posted on the chart—an intervention that was easy and low cost to the teacher. As you can see, punctuality improved when the chart was posted. Notice that the chart was first posted after lunchtime, but it was not introduced following morning or afternoon recesses. The number of students who were late for class after lunch declined from about eight to less than two. This was not the case for the recess periods; the number of students who were tardy after recess remained at four or five. Next, the researchers continued to post the patriots chart after lunch, but they added the chart following the morning recess. When this occurred, all of the students were on time for class following both lunch and morning recess. Finally, when the chart was also posted following the afternoon recess, all of the students were on time for all of the class periods. The multiple baseline across settings design demonstrates an effect of the intervention by staggering the introduction of the independent variable over time and settings (see also Alberto, Heflin, & Andrews, 2002).

Multiple Baseline Across Participants

A similar design logic is used in the **multiple baseline across participants** design, when an intervention is progressively introduced to different participants who exhibit similar target behavior (see Iwata, Pace, Kalsher, Cowdery, & Cataldo, 1990, Study 2, on the use of this design to assess modification of self-injurious behavior). In Experiment 2, Hall et al. (1970) attempted to improve three students' scores on French quizzes. Modification involved a requirement to stay after school for tutoring if the student scored below a C grade on a quiz. The contingency was first introduced to Dave, then to Roy, and finally to Debbie. Figure 13.3 indicates that

478 Applied Behavior Analysis

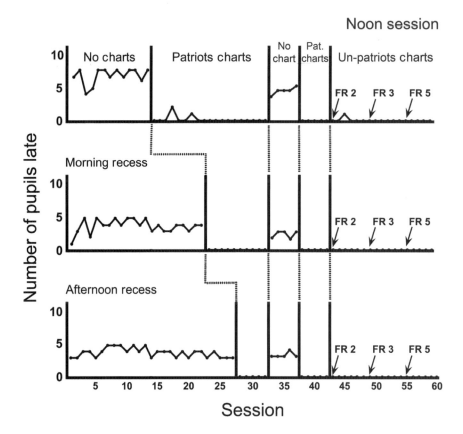

Fig. 13.2
The multiple baseline design across settings is depicted.

Source: The figure is from R. V. Hall, C. Cristler, S. S. Cranston, & B. Tucker (1970). Teachers and parents as researchers using multiple baseline designs. *Journal of Applied Behavior Analysis*, *3*, pp. 247–255. Copyright 1970 John Wiley & Sons, Ltd. Reprinted with permission.

Dave's quiz performance dramatically improved when the contingency was applied. The other students also showed improvement only after the contingency went into effect. All of the students received grades of C or better when contingency management was used to improve their performance in the French class.

Multiple Baseline Across Behaviors
A **multiple baseline across behaviors** design is used when a reinforcement procedure is applied progressively to several operant behaviors. In this case, the participant, setting, and consequences remain the same, but different responses are sequentially modified. Hall et al. (1970) provided an example of this design with a 10-year-old girl when they modified her after-school reading, working on a Campfire honors project, and practicing of the clarinet. The girl had to spend at least 30 min on an activity or else she had to go to bed early. She had to go to bed 1 min earlier for every min less than 30 min she spent on an activity. As you can see from Figure 13.4, practicing the clarinet was modified first and time spent playing the

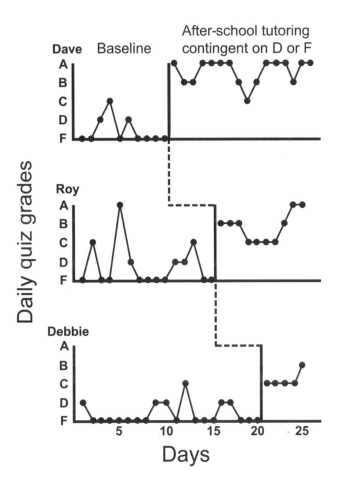

Fig. 13.3
The multiple baseline design across participants is depicted.

Source: The figure is from R. V. Hall, C. Cristler, S. S. Cranston, & B. Tucker (1970). Teachers and parents as researchers using multiple baseline designs. *Journal of Applied Behavior Analysis*, *3*, pp. 247–255. Copyright 1970 John Wiley & Sons, Ltd. Reprinted with permission.

instrument increased from about 15 to 30 min. Next, both practicing the clarinet and working on the Campfire project were targeted, and both performances were at about 30 min. Finally, reading for book reports was modified and all three target responses occurred for 30 min. The avoidance contingency seems to be effective because each behavior changes when the contingency is introduced, but not before.

Multiple baseline and A-B-A-B reversal designs are the most frequently used research methods in applied behavior analysis. There are, however, many variations of these basic designs that may be used to increase internal validity or to deal with specific problems in the applied setting (e.g., Carr & Burkholder, 1998). Often the basic designs are combined in various ways to be certain that the effects are due to the independent variable. In fact, Hall et al. (1970) used a reversal phase in their experiment on tardiness and the patriots chart, but for reasons of clarity this was not shown in Figure 13.2. There are many other designs that

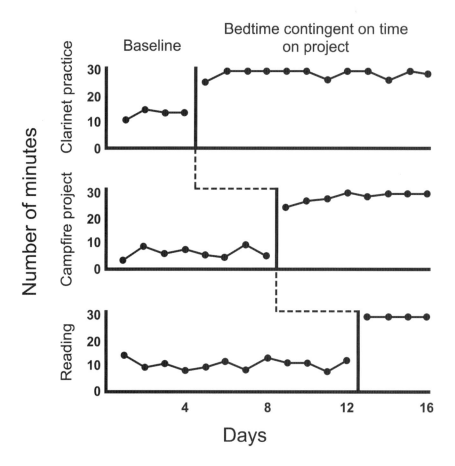

Fig. 13.4
The multiple baseline design across behaviors is depicted.

Source: The figure is from R. V. Hall, C. Cristler, S. S. Cranston, & B. Tucker (1970). Teachers and parents as researchers using multiple baseline designs. *Journal of Applied Behavior Analysis*, *3*, pp. 247–255. Copyright 1970 John Wiley & Sons, Ltd. Reprinted with permission.

are useful in a given situation. A *changing criterion* design involves progressive increases (or decreases) in the performance criterion for reinforcement. For example, a hyperactive child is reinforced for spending progressively more time on academic work. At first the child may be required to spend 3 min working quietly, then 5 min, then 10 min, and so on. The child's behavior is measured at each level of the criterion. A research example of this design is given in the section on self-control in this chapter (see also Belles & Bradlyn, 1987).

Issues of Measurement in Applied Behavior Analysis

It is relatively easy to objectively define an operant in the laboratory. Responses are often defined by electrical switch closures and there is no dispute about the occurrence. When responses occur, computers and other electronic equipment record them. In the applied setting, definition and measurement of the behavior are much more difficult, especially when parents, teachers, and psychologists are used to identify challenging behavior. In this regard, Kazdin has made the point that:

> Identification of the target behavior may appear to be a relatively simple task. In a given setting (e.g., the home, school, or work place), there is general agreement as to the "problems" of the clients whose behaviors need to be changed and as to the general goals of the program. Global or general statements of behavioral problems are usually inadequate for actually beginning a behavior modification program. For example, it is insufficient to select as the goal alteration of aggressiveness, learning deficits, speech, social skills, depression, psychotic symptoms, self-esteem, and similar concepts. Traits, summary labels, and personality characteristics are too general to be of much use. Moreover, definitions of the behaviors that make up such general labels may be idiosyncratic among different behavior change agents (parents, teachers, or hospital staff). The target behaviors have to be defined explicitly so that they can actually be observed, measured, and agreed upon by individuals administering the program.
>
> (Kazdin, 1989, p. 54)

Kazdin goes on to discuss three criteria for an adequate response definition (see also Johnston & Pennypacker, 1993). The first criterion is *objectivity*. This means that the response definition should refer to observable features of behavior in clearly specified situations. *Clarity* of definition is another requirement. This means that the description of the response can be read and then clearly restated by a trained research assistant or observer. Finally, the definition should be *complete* in the sense that all instances of the behavior are distinguished from all non-occurrences. Thus, a troublesome student may be objectively defined as one who talks without permission when the teacher is talking and who is out of their seat without permission during a lesson. The definition is clear in that it is easily understood and may serve as a basis for actual observation. Completeness is also shown, as only these two responses are instances of the troublesome behavior class, and any other responses are not.

This definition of response assumes that there is a problem with the student's performance, not the teacher's judgment. The applied behavior analyst must be sensitive to the possibility that the teacher is too critical of the student. It is possible that many students talk without permission and leave their seats during lessons. The teacher, however, only gets upset when Anna, for example, is running about or talking during instruction. In this case, response definition may be accurate and modification successful, but the intervention is unfair. Applied behavior analysts must constantly be aware of whether they are *part of the solution or part of the problem* (Holland, 1978). If the problem lies with the teacher, it is his or her behavior that requires change.

Recording Behavior

Once a suitable response is defined, the next step is to record the behavior when it occurs. The simplest tactic is to record every instance of the response. Practically, this strategy may be very time-consuming and beyond the resources of most applied behavior analysts. One alternative is to count each instance of behavior only during a certain period of the day (e.g., lunch, recess, or first class in the morning). This method of observation is called event recording for specified periods.

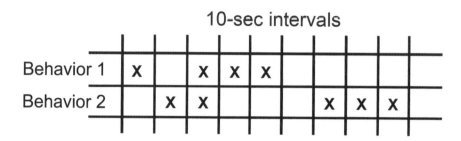

Fig. 13.5
The interval recording method is shown as used in behavioral observation and measurement.

Another strategy is to select a block of time and divide the block into short, equal intervals. This is called **interval recording**. For example, a 30-min segment of a mathematics class may be divided into 10-s segments. Regardless of the number of responses, if the behavior occurs in a given 10-s segment, then the observer records it as a single event. One way in which this could be done is to have an observer wear a headset connected to a cassette tape recorder that plays a tape that beeps every 10 s. When the target behavior occurs, the observer records it on a piece of paper divided into segments that represent the 10-s intervals (see Figure 13.5). After each beep, the observer moves to the next interval.

Time sampling is another method of recording used in applied behavior analysis. This technique samples behavior over a long time scale, with observations made at specified times throughout the day. For example, a patient from a psychiatric ward may be observed every 30 min, as a nurse does the rounds, and instances of delusional talk are recorded. Again, the issue is whether the target behavior is occurring at the time of the observation, not how many responses are made.

When behavior is continuous, **duration recording** is the preferred method of observation. Continuous behavior involves responses such as watching television, riding a bicycle, or sitting in a chair. When behavior is continuous rather than discrete, an observer may use a stopwatch to record the duration of occurrence. When the person is sitting in a chair the watch is timing, and when the person does something else the watch is stopped.

Reliability of Observations

No matter what method of recording behavior is used, **reliability of observation** is a critical issue. Briefly, reliability of observation involves the amount of agreement among observers who independently record the same response. For example, two observers may sit at the back of a classroom and use 10-s intervals to record the occurrence of Jessica's out-of-seat behavior. After 30 min of observation, each researcher has recorded 180 intervals of 10 s. One way to assess reliability is to count the number of times both observers agree that the behavior did or did not occur within an interval. This can be accomplished by video recording the participant during an observation period and then having two observers score the responses later from the tape. Reliability is usually calculated as the percentage agreement between observers, ranging from zero to 100%. Generally, applied behavior analysts strive for reliability of greater than 80%.

CONTINGENCY MANAGEMENT AND SUBSTANCE USE DISORDERS

Substance use disorder is a major health problem in America. Estimates by the National Institutes of Health indicate that substance abuse costs our society over half a trillion dollars each year, exceeding the costs of all other major medical illnesses (Silverman, Roll, & Higgins, 2008). The indirect costs in terms of early deaths and destroyed lives are immeasurable. Given the devastating impact of substance use disorders, the federal government has invested large sums of money to find effective treatment programs. One highly effective intervention, called **contingency management** (**CM**), uses operant principles to arrange contingencies to *increase abstinence* from drug use, promote adherence to taking prescribed medication, and increase retention in treatment programs (Burch, Morasco, & Petry, 2015; Higgins & Katz, 1998; Higgins & Petry, 1999; Higgins et al., 1991).

Reinforcement of Abstinence

Contingency management is a treatment that has been widely applied to promote drug abstinence (for reviews see Bolivar et al., 2021; Higgins & Petry, 1999; Higgins et al., 2007). Typically, individuals with substance use disorders (SUD) are required to provide urine specimens several times a week, and the samples are screened for presence of a drug. Notice that the use of urine specimens allows for an objective assessment of drug use outside the clinic; verbal reports of use may be unreliable. When urine samples are negative for drug use (drug free), clients receive reinforcement consisting of vouchers for personally valued goods and services. In voucher-based CM programs, the value of the vouchers increases as the client shows longer and longer periods of abstinence. Presentation of urine samples that indicate drug use results in withholding of vouchers (no reinforcement) and, in some programs, the purchasing value of vouchers is reset to the original value or reduced.

One way to assess the effectiveness of the voucher-exchange contingency of CM is to compare it with the standard *community reinforcement approach* (*CRA*), which has been shown to be one of the most effective programs in the treatment of SUDs (Hendrik et al., 2004), but an approach without specific contingencies for abstinence. CRA is based on the assumption that environmental contingencies, especially social and economic ones, play a central role in the regulation of drug and alcohol use (Hunt & Azrin, 1973). Consequently, CRA uses social, recreational, familial, and vocational reinforcement contingencies to support a drug-free lifestyle.

In a treatment study of cocaine dependence in Spain (Garcia-Fernandez et al., 2011), researchers randomly assigned cocaine users to CRA and CRA-plus-vouchers conditions. The CRA involved five training components: drug avoidance skills, lifestyle change, relationship counseling, other-substance abuse, and other-psychiatric problems. The components of the CRA were applied individually to vocational counseling, lifestyle change, relationship counseling, and other-psychiatric problems, while the rest of the components were applied in group-therapy sessions (Budney & Higgins, 1998). In the CRA condition, urine specimens were collected twice a week and participants were informed of their urinalysis results immediately after submitting their specimens. Participants, however, did not receive vouchers for remaining drug free. Cocaine patients in the CRA-plus-vouchers condition also were informed of their urinalysis results, but urine samples were taken three times a week and the program added vouchers contingent on abstinence.

The voucher program used points valued at 35 cents each; the first drug-free test earned 10 points, with a 5-point increase for each subsequent and consecutive negative sample. Patients also earned a 40-point bonus for three consecutive drug-free tests. Failure to submit a sample or a positive test for cocaine set the value back to the initial 10 points, but submission of three consecutive negative tests returned the value to its level before the reset. Points could not be lost once they had been earned. Participants on this point system earned an average of about $900 over a 12-week period. After this initial period, on a random basis, only half of the specimens collected twice a week earned points, and patients now earned about $750 on average. Points were exchangeable for vouchers that allowed patients to purchase a variety of goods and services appropriate to a drug-free lifestyle.

The results of the study are shown in Figure 13.6. The percentage of drug-free tests was 80% on average for the CRA condition, but 97% for the CRA with added vouchers (panel A)—a 17% improvement over the standard behavioral program. Panel B shows that 93% of the patients in the CRA-plus-vouchers group remained abstinent for 4 weeks, compared with 53% in the CRA condition. At 12 weeks, 72% of the patients in the CRA-plus-vouchers program remained drug free, compared with 38% in the CRA condition (Petry, Martin, Cooney, and Kranzler, 2000 reported similar percentages of abstinence from alcohol at 8 weeks, using

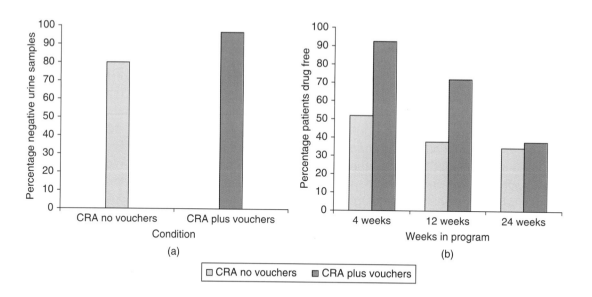

Fig. 13.6

The graph portrays the major findings of a study of contingency management (CM) of cocaine substance use disorder by Garcia-Fernandez, et al. (2011). Panel A shows that the percentage of negative urine samples (abstinence) is greater with the Community Reinforcement Approach (CRA) with added voucher-based CM than without the vouchers. Panel B shows that the percentage of patients drug free after 4 and 12 weeks is higher with CRA plus vouchers than CRA alone. After 24 weeks, CRA and CRA plus vouchers have a similar percentage of patients drug free, indicating the voucher program did not maintain abstinence in the long run.

Source: Graphs are based on findings reported by G. Garcia-Fernandez, R. Secades-Villa, O. Garcia-Rodriguez, E. Sanchez-Hervas, J. Fernandez-Hermida, & S. T. Higgins (2011). Adding voucher-based incentives to community reinforcement approach improves outcomes during treatment for cocaine dependence. *American Journal on Addictions*, 20, pp. 456–461.

opportunities to win prizes as reinforcement). Approximately 38% of the patients in the CRA-plus-vouchers group succeeded in remaining drug free throughout the entire 24 weeks of the study, compared with 35% of the patients in CRA condition. Notably, both behavioral programs (CRA and CRA plus vouchers) after 6 months attained substantially higher percentages of abstinence than usual-care and 12-step programs (AA philosophy) for cocaine abuse (0–20%; Hendrik et al., 2004). In one study, a 12-step program achieved only 5% abstinence at the end of 4 months of treatment (Higgins et al., 1993).

Figure 13.6 (panel B) also indicates that the initially high effectiveness of the voucher-exchange contingency diminished over the 24-week period of treatment. One possibility is that the switch from 3-weekly urine tests to 2-weekly specimens with random testing weakened control of abstinence by the voucher program. Participants made less money ($900 vs. $750) when urine samples were reduced and random testing began (weeks 13 to 24), even though the maximum amount of possible earnings increased (from $1,400 to $2,200). These results indicate that it is difficult to maintain the large treatment gains offered by voucher-based CM and to maintain abstinence once the CM procedure is removed. In fact, in one study, researchers found very low abstinence rates (6.6% on average) at follow-up for high-use cigarette smokers in a residential drug treatment program, after using contingent vouchers and motivational interviewing to reduce sustained smoking (Rohsenow et al., 2015). New research on choice as melioration (Heyman, 2009; Kurti & Dallery, 2012) and self-control as discounting (Madden & Bickel, 2009; see Chapter 9) may be helpful in designing CRA and CM programs to maintain long-term abstinence, as well as arranging environmental contingencies for a drug-free lifestyle.

Cessation of Smoking During Pregnancy

Contingency management has been extended to the treatment of other socially important problems. Smoking during pregnancy is the leading preventable cause of poor pregnancy outcomes in the USA, causing immediate and longer-term negative effects for mothers and their children. CM with financial incentives is an evidence-based method for promoting abstinence from cigarette smoking during pregnancy—especially for economically disadvantaged mothers who have the highest smoking rates (Burch et al., 2015 also reported that CM works well for those with SUDs who receive financial assistance for a physical disability). Higgins and colleagues conducted a review of the studies in which pregnant women earned vouchers exchangeable for retail items contingent on biochemically verified abstinence from recent smoking (Higgins et al., 2012). Results from six controlled trials with economically disadvantaged pregnant smokers supported the efficacy of a voucher-based CM program for increasing smoking abstinence rates before the birth of the infant and early postpartum. In addition, results from three randomized trials provided evidence that the CM intervention improves fetal growth, mean birth weight, percentage of low-birth-weight deliveries, and breastfeeding duration. The systematic use of CM shows promise as a way of promoting smoking cessation among economically disadvantaged pregnant and recently postpartum women, and as a treatment to improve childbirth outcomes.

Evaluation of Contingency Management and Substance Use Disorders

Quantitative reviews of SUDs show that CM outperforms other interventions, including cognitive-behavioral therapy (Dutra et al., 2008; Lussier, Heil, Mongeon, Badger, & Higgins,

2006; Prendergast, Podus, Finney, Greenwell, & Roll, 2006). Given the strong evidence of its effectiveness, it is noteworthy that community based treatment providers are not using the "better mousetrap" (Roll, Madden, Rawson, & Petry, 2009). It might be easy to blame the providers of SUD treatment for their ignorance of CM or judgmental bias, but as behavior analysts, we would be wise to analyze the behavior of the providers (those who adopt the CM technology) (Roll et al., 2009).

One possibility is that providers of treatment are not familiar with behavioral principles, CM technology, and behavioral language such as "contingency" and "management." Most treatment providers are trained with a clinical focus, using predominantly cognitive theories. These non-specialists may be "turned off" when a component of the treatment package that makes sense in terms of behavior principles seems to be impractical for a clinical setting (who has time to monitor the target behavior?). Decisions to adopt a new technology instead of an alternative treatment also involve assessment of economic costs and benefits. When there are substantial costs to implementing a treatment program, the benefits of adopting the new technology must be considerable and immediate. However, the research on better outcomes at lower economic cost (cost-effectiveness) of contingency management relative to other types of therapy for SUD remains inconclusive (Shearer, Tie, & Byford, 2015). Without evidence of substantial cost-effectiveness, government funding may stall and agencies may resist adoption of CM treatments.

Furthermore, the implementation of a CM treatment often results in immediate costs for the adopters, and delayed signs of benefit for drug-using clients. In terms of immediate costs, clinical staff are required to conduct regular drug tests that involve wage and material costs for collecting samples, testing specimens, and recording results. Most providers indicate that they could afford to conduct tests once a week, but effective CM programs require repeated drug testing several times a week, a component of treatment that makes CM unacceptable to many agencies and clinicians. Voucher-based interventions also have costs related to the use of monetary reinforcement. Providers indicate that $50 a month per client is a feasible cost, but research shows that effective CM interventions spend about $50 a week, a cost that is impractical for many clinical settings. Adoption of CM treatment thus depends on identifying high-valued reinforcements at low cost that can effectively compete with drug reinforcement. Few, if any, have been identified to date—suggesting that vouchers for abstinence are a necessary expense in the treatment of SUD.

Ultimately, the amount of money available for treatment of SUD (and CM interventions) depends on cultural values and social contingencies (Lamal, 1997). Evidence indicates that effective drug-abuse treatments provide positive economic returns to society (Cartwright, 2000), and are far less expensive in the long run than placing those with SUDs in prisons (Chandler, Fletcher, & Volkow, 2009). Recently, the state of California became the first state to receive federal approval to cover contingency management through Medicaid and they are piloting a statewide program to target the reduction of cocaine and methamphetamine use.

A large sector of the American public, however, condemns the use of drugs as immoral and criminal conduct, deserving of punishment by imprisonment. Based on this public sentiment, prisons in the USA are more overcrowded with drug offenders than those in any other nation (Bewley-Taylor, Hallam, & Allen, 2009). Cultural values on race complicates the issues further. For example, as Black Americans are disproportionately arrested and incarcerated for drug-related crimes compared to White Americans (see Gaston, 2019; Pettit & Western, 2004). Until there is a shift toward equal valuation of American citizens and a focus on rehabilitation relative to deterrence by punishment, government funding of CM drug treatment is likely to remain inadequate—subverting widespread clinical adoption of this effective behavioral technology.

Contingency Management and Online Resources

In this chapter, we have emphasized the evidenced-based research on contingency management and substance use disorders. There are, however, other important advances in the use of contingency management, especially in the use of online help resources. These resources have not been evaluated for effectiveness by controlled trials, but are noted here as interesting advances in the use of new technologies to assist in the delivery of contingency management (CM) interventions.

> If you ever had a New Year's resolution you didn't keep, there's help for you. An online contingency management program, *Stickk.com*, is an example of a website designed to help people achieve their personal behavioral goals—goals that seldom are realized. Behavioral economists at Yale University developed *Stickk.com* to allow you to define your behavioral objectives, set up a contract, and have the option of naming a referee to enforce it. As part of the contract, you can choose to set a penalty for failing to reach your target. The penalty could merely be emailing your failure to a list of friends (or enemies), or paying some of your hard-earned cash as a backup. You can make a commitment to paying the money to anyone you want, including an "anti-charity," which for a liberal could be giving the money to the Republican Party. The more money you put on the line, the better you do, according to an analysis by *Stickk.com* of 125,000 contracts over the past three years. The likelihood of success for people who don't make a commitment of some kind is only 29%, but it rises to 59% if you designate a referee, and to 71.5% if you make a monetary commitment. A contract that includes both a referee and a financial investment has about an 80% success rate.
>
> (John Tierney, "Be It Resolved," *The New York Times Sunday Review*, January 5, 2012)

One problem with the *Stickk.com* program and other "self-help" websites is the use of punishment. If you fail to reach your goal, you lose money and it goes to an organization you cannot stand. This can work to generate avoidance, but it does not generate a pleasant outcome for goal attainment. Commitment of money is a good strategy, as money is usually an effective consequence. We would recommend, however, refunding the deposit for achieving one's goal, as well as recruiting friends and colleagues to provide positive social support. Another strategy is to reinforce alternative behavior by altering one's lifestyle—a procedure often overlooked by self-help websites.

Those who have body-weight and obesity problems might want to try another online website. One online treatment program offers the opportunity to sign up for weight-loss challenges. The website *HealthyWage.com* arranges challenges and contests that allow participants to win money for losing body weight (see also *Gym-pact.com* and *Dietbet.com*). The company indicates that it receives financial support from insurers, health services and hospitals, food companies, and the government. These agencies donate money in the belief that paying incentives for people to lose weight costs less than paying for the healthcare of an obese population.

The *HealthyWage.com* website offers three challenges that involve different requirements, betting, and payoffs. Here we describe the most basic challenge. The 10% challenge allows you to bet money on losing weight; you double your money for losing 10% of your weight in 6 months. At registration, you pay $100 and if you lose 10% of your weight in 6 months, the company pays you $200. Your starting and ending weights must be verified (by a health club or physician) and you are required to give weekly reports of your weight on the website. We should note that betting on your ability to lose weight is not the best way to manage body weight over a lifetime. Also, the effectiveness of this and other weight-loss challenges has not been established yet by independent clinical trials, but online CM is becoming a more popular method of delivering individual treatment for those who are overweight or obese (see also "Behavior Management of the Obesity Crisis" in this chapter).

BEHAVIOR ANALYSIS IN EDUCATION

Behavior principles have been applied in a wide variety of educational settings (Sulzer-Azaroff, 1986; Twyman, 2014; West & Young, 1992). University students have shown better academic performance after being taught with a personalized system of instruction (PSI), either in the classroom (Cook, 1996; Keller, 1968; Kulik, Kulik, & Cohen, 1980) or with computers (Pear, Schnerch, Silva, Svenningsen, & Lambert, 2011; Springer & Pear, 2008). In addition, learning has been accelerated, both for elementary-school children and for university students learning English with computers, by precision teaching (Cuzzocrea, Murdaca, & Oliva, 2011; Lindsley, 1972). Athletic performance has been improved by applying behavior principles to physical education (Colquitt, Pritchard, & McCollum, 2011; Martin & Hrycaiko, 1983; Pocock, Foster, & McEwan, 2010). And children with autism have benefited from the teaching of social and living skills (Flynn & Healy, 2012; Lovaas, 1987; Maurice, 1993). These are just a few of the many applications of behavior principles to education (Twyman, 2014). In this section we focus on two examples, but there are many more educational applications than are reported here, including programmed learning (Skinner, 1968), direct instruction (Engelmann & Carnine, 1982), and interteaching (Boyce & Hineline, 2002).

From a behavioral viewpoint, the ideal "teacher" is a personal tutor who knows all the topic information, all the appropriate behavioral science (i.e., shaping, chaining, reinforcement schedules, establishing operations, and other principles), and who keeps accurate track of student performance, delivering immediate reinforcement for maximum effectiveness. Our culture is getting close to creating such a "teacher" in the form of computer-based, online-technology. Khan Academy (khanacadem.org), which effectively teaches mathematics (among other topics) at all grade levels is an excellent example. Artificial intelligence and robotics may soon produce personal tutors, perhaps in the near future (Barreto & Benitti, 2012).

For the moment, Headsprout™ is an online computer-based, early-reading program that demonstrates the use of new technology to deliver educational services to parents and children. This program effectively teaches young children the basic skills and strategies for successful reading. In the Headsprout system, children learn to read with positive reinforcement, discovering that letters and sounds go together to make words, the words make sentences, and sentences make stories. The children go on to learn that stories convey meaning and can be read either to extract information or just for fun and pleasure. Basic skills and strategies for reading are acquired in an exciting "learn-by-doing" manner, involving the instruction and mastery of skills such as phonemic awareness (sounds go with words), print awareness, phonics, sounding out, segmenting, and blending (using sound elements to decode words). Additional programmed instruction teaches vocabulary development, reading fluency, and reading comprehension.

The success of Headsprout is based on four teaching fundamentals. First, the instructional program arranges the reading material so that children succeed (high rate of reinforcement). Second, children practice and learn the material until they master the skill or strategy. Third, the skill is practiced and reinforced until the child is able to consistently, quickly, and accurately perform it (fluency). Finally, the child is given cumulative reviews of skills and strategies to improve retention and transfer their reading skills to material outside of the program (Huffstetter, King et al., 2010; Layng, Twyman, & Stikeleather, 2004). When parents purchase the *Headsprout Early Reading* package (40 episodes), they are sent a package containing a progress map to track their child's progress, stickers to place on the map after the child has mastered an episode, and story books to read at designated points in the program. Overall, Headsprout is an example of effective computer-based behavioral technology that is currently being used to teach basic reading to many children.

A Personalized System of Instruction

The traditional lecture method used to instruct college and university students has remained largely unchanged for the last thousand years. A teacher stands in front of a number of students and talks about their area of expertise. There are variations on this theme—students are encouraged to participate in discussion, discover new facts on their own, reach conclusions through a series of planned questions, and become active rather than passive learners. During lectures, various forms of logic are used to arrive at conclusions and classroom demonstrations are arranged. Basically, the lecture method of instruction is the same as it has always been; however, presenting material is not equivalent to teaching the subject matter.

Fred Keller recognized that the lecture method of college teaching was inefficient for some. He reasoned that anyone who had acquired the skills needed to attend college was capable of successfully mastering most or all college courses. Some students might take longer than others to acquire expertise in a course, but the overwhelming majority of students would be able to do so eventually. If behavior principles were to be taken seriously, there were no bad students, only bad teachers.

In a seminal article, titled "Good-Bye, Teacher ...," Keller (1968) outlined a college teaching method based on principles of operant conditioning. He called his teaching method a **personalized system of instruction (PSI)** (also called the "Keller Plan"; Sherman, Ruskin, & Semb, 1982). Basically, PSI courses are organized so that students move through the course at their own pace. Some students may finish the course within a few weeks, whereas others require a semester or longer.

Course material is broken down into many small units of reading and, if required, laboratory assignments. Students earn points (conditioned reinforcement) for completing unit tests and lab assignments. Mastery of the lab assignments and unit tests is required. If test scores are not close to perfect, the test (usually in a different form) is taken again. The assignments and tests build on one another, so they must be completed in order. Undergraduate proctors are recruited to assist with running the course. These individuals tutor students and mark unit tests and laboratory assignments. Proctors are "chosen for [their] mastery of the course content and orientation, for [their] maturity of judgment, for [their] understanding of the special problems that confront ... beginner[s], and for [their] willingness to assist [with the course]" (Keller, 1968, p. 81). Lectures and class demonstrations are an optional privilege; students may or may not attend them. Lectures are scheduled once the majority of students in the class have passed a sufficient number of unit tests to indicate that they are ready to appreciate the lectures; no exams are based on these lectures. The course instructor designs the course, makes up the tests, delivers the optional lectures, adjudicates disputes, and oversees the course.

Comparison studies have evaluated student performance on PSI courses against performance of students given non-PSI computer-based instruction, audio-tutorials, traditional lecture-based teaching, visual-based instruction, and other programmed-instruction methods. College students instructed by PSI outperformed students taught by these other methods when given a common final exam (for a review, see Lloyd & Lloyd, 1992; see Pear et al., 2011 for PSI computer-based instruction). Despite this positive outcome, logistical problems in organizing PSI courses, teaching to mastery level (most students get an A for the course), and allowing students more time than the allotted semester to complete the course have worked against the wide adoption of PSI in universities and colleges (Binder & Watkins, 1989).

Precision Teaching

Ogden Lindsley extended the method of free-operant conditioning to humans, emphasizing Skinner's dictum to focus on rate of response (Lindsley, 1972, 1991). In what became known as **precision teaching**, Lindsley (1990a, 1990b) devised a method of systematic instruction that encouraged students and teachers to target specific behaviors; to count, time, and graph the responses; and to revise instructional procedures based on the charted data (see also Binder, 2010; Binder & Watkins, 2013). As an instructional system, precision teaching has four guiding principles: a focus on directly observable behavior, rate as the basic behavioral measure, the charting of behavior on a Standard Celeration Chart, and "the learner knows best."

To *focus on behavior*, precision teaching translates learning tasks into concrete, directly observable behaviors that can be counted, timed, and recorded. Private behavior such as silent reading must be made public. A child who is poor at silent reading might be asked to read out loud so that counts of the number of correct words can be obtained (a measure of "decoding" skills). To assess their comprehension skills, the teacher might provide a list of questions to the student after they have read silently a passage from a book. Following this, the teacher would count the number of correct answers that the student made on the quiz.

Once behavior is defined, the rate of response is used as the basic measure of learning (or performance). The rate is the average number of correct responses during the period of assessment, or *counts per minute*. The use of rate of correct responses (frequency/time) focuses instruction on **fluency** involving accuracy *and* high frequency. When a performance becomes fluent, the behavior is retained for longer, persists during long periods on the task, is less affected by distractions, and is more likely to be available in new learning situations—to combine with other well-learned behaviors (see Binder, 1996; West & Young, 1992).

Next, *the rate is plotted* on a Standard Celeration Chart, allowing the teacher and student to observe improvement in the target behavior for each week (Figure 13.7). The degree

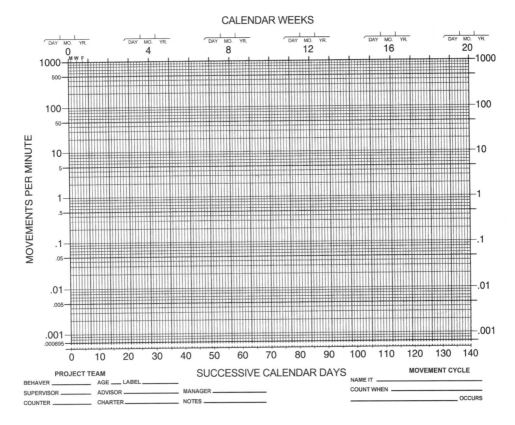

Fig. 13.7
A Standard Celeration Chart using 6-cycle semilogarithmic coordinates, with counts per minute (rate) on the Y-axis and calendar days on the X-axis. Using this 6-cycle scale, a student whose rate of spelling targeted words increases from 2 to 4 per minute would appear the same as another student whose rate increased from 20 to 40 a minute. The students show the same proportional amount of improvement. When the charted rate doubles from one week to the next, we say it is accelerating at "times 2." When the charted rate is cut in half from one week to the next, we say it is decelerating at "divided by 2." A straight line from the lower left corner to the upper right has an angle of 30 degrees and represents a x2 change in behavior per week, the objective of accelerating precision teaching interventions.

of acceleration (or deceleration) is a useful measure of learning in academic settings. In this regard,

> When data are plotted on the standard celeration chart, learning is generally represented by a straight or nearly straight line. The value of the slope of the line which best fits the distribution of values [plotted rates over days] on a logarithmic scale is thought of as an "index of learning." The steeper the slope, the faster the learning is; the flatter the slope, the slower the learning is.
>
> (West & Young, 1992, p. 132)

The whole idea of precision teaching is to improve learning in a way that is objective and quantifiable. Teachers and students work out plans for improvement, implement the instructional procedures, and assess the effects of the interventions.

The general rule is that *the learner knows best*. That is, if a student is progressing according to the instructional plan, then the program is appropriate for that student. In contrast, if the targeted behavior for a student shows low **celeration**, the program needs to be changed. In other words, precision teaching requires that we alter the teaching strategy rather than blame the student (e.g., by assuming that "John is unteachable"). Thus, the student is always "right" and, in the context of low improvement, new instructional procedures are required to improve learning and performance (Carnine, 1995).

Precision teaching is a cost-effective technology (according to a survey by Albrecht, cited in Lindsley, 1991) that has been successfully applied to teach learners ranging from the young student with developmental disabilities to university graduate students (White, 1986). Binder and Watkins (1990) reported on a precision teaching program conducted in Great Falls, Montana, in the early 1970s. Over a 4-year period, teachers at Sacajawea elementary school added 20–30 min of precision teaching to their regular curriculum. On the *Iowa Test of Basic Skills*, the students who were given precision teaching improved between 19 and 40 percentile points compared with other students in the district. More generally, improvements of two or more grade levels per year of instruction are commonly observed in precision teaching classrooms (e.g., Binder & Watkins, 2013; West, Young, & Spooner, 1990), and significant learning gains are achieved for students with developmental delays or autism (Fischer, Howard, Sparkman, & Moore, 2010; Holding, Bray, & Kehle, 2010; Johnson and Layng, 1994). These instructional gains have been attributed to the effects of rate building on fluent performance, a conclusion that has been challenged. Experimental studies of the learning gains from precision teaching have not ruled out the effects of practice and high rates of reinforcement as the critical factors, rather than the arrangement of instructional material designed to alter response rate and fluency (Chase, Doughty, & O'Shields, 2005; Doughty, Chase, & O'Shields, 2004; see Binder, 2004 and Kubina, 2005 for counterarguments).

Although highly successful in promoting rapid and fluent learning, precision teaching remains only a small part of mainstream education, as do most behavioral programs applied to education (Twyman, 2014). Skinner (1984) pointed to several steps that are in line with precision teaching that needed to be taken to advance educational systems: (1) be clear about what is to be taught, (2) teach first things first, in an ordered sequence of progression, (3) stop making all students advance at essentially the same pace, and (4) program the subject matter—a good program of instruction guarantees a great deal of successful action.

In the end, the problem of education rests within the culture. Skinner stated that "a culture that is not willing to accept scientific advances in the understanding of human behavior, together with the technology that emerges from these advances, will eventually be replaced by a culture that is" (Skinner, 1984, p. 953). The survival of a culture depends on education of the young. The question is, who is willing to adopt and promote a scientific approach to effective education?

APPLICATIONS OF BEHAVIOR PRINCIPLES: SELF-CONTROL AND AUTISM

Training Self-Control

In applied behavior analysis, **self-control** techniques may be taught to clients who are then better able to manage their own behavior. One common technique for self-control is called self-reinforcement. Belles and Bradlyn (1987) conducted an interesting study to modify the behavior of heavy smoking by arranging self-reinforcement and self-punishment remotely (by phone). The client was a 65-year-old man who lived 200 miles away from the clinic. The researchers arranged a treatment program with the client and his wife. For each day that he smoked less than a specified number of cigarettes, he added $3 to a savings fund that was used to buy items that he wanted. When he exceeded the agreed number of cigarettes, he had to send a $25 check to the therapist, who donated the money to a charity that was unacceptable to the client. His wife verified the number of cigarettes he smoked each day by unobtrusively monitoring his behavior.

A **changing criterion design** was used to evaluate the effectiveness of the self-control procedure (see Gast, 2010, pp. 383–389, for a complete description of the advantages and limitations of this design). In this design, the criterion for the number of cigarettes smoked each day was progressively lowered over 95 days. The effects of self-reinforcement are shown if the subject meets or falls below the criterion set by the researchers. Figure 13.8 shows the effects of the treatment. A horizontal line shows the target level for each period, indicating that the client generally matched his behavior to this criterion. Notice that although the criterion generally decreased, the researchers occasionally set a value higher than a previous phase and the client's behavior changed in accord with the contingencies. After 81 days on the program, the client's cigarette consumption had declined from about 85 to 5 cigarettes each day. At this point, he was satisfied with his progress and said that he wanted to remain at this level. Follow-up reports on his smoking over 18 months showed that he continued to smoke only 5 cigarettes a day.

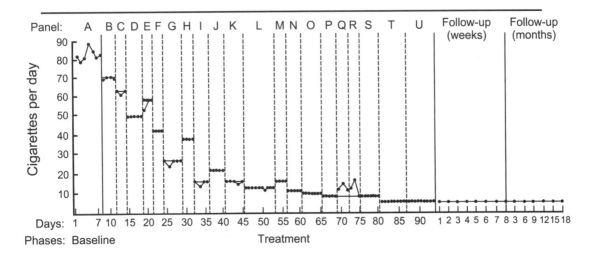

Fig. 13.8
Graph of the changing criterion design used in the modification of excessive smoking.

Source: Figure is from D. Belles & A. S. Bradlyn (1987). The use of the changing criterion design in achieving controlled smoking in a heavy smoker. *Journal of Behavior Therapy and Experimental Psychiatry, 18*, pp. 77–82. Copyright 1987 held by Elsevier, Ltd. Published with permission.

Behavior Analysis in the Treatment of Autism

Children who are diagnosed with autism spectrum disorder (ASD) show an early lack of social interaction with parents, other family members, and peers. For example, these children often resist being held and may show resistance if picked up or hugged. As children with autism age, they may be mistakenly thought to be deaf, as they may not talk or even establish eye contact when spoken to. Many show repeated stereotyped patterns of behavior such as rocking back and forth, spinning a top, or wiggling their fingers in front of their eyes. More than 85% of children exhibit language delays or are not verbal at all. The prognosis for alleviation of these symptoms depends on severity, but often the overwhelming majority of such children require extended care and supervision (Ghezzi, Williams, & Carr, 1999).

Ivar Lovaas (1927–2010) researched and developed an applied behavior analytic (ABA) treatment of autism from the 1960s onward (Lovaas, 1966, 1977, 1987; McEachin, Smith, & Lovaas, 1993). Lovaas (1977, 1987) described an **early intensive behavioral intervention (EIBI)** in which youngsters are given 40 or more hours each week of behavioral intervention designed to increase social behavior, teach speaking and communication, and reduce or eliminate self-stimulation and aggressive behavior. Most of the treated children with autism showed significant improvement in their daily functioning. Incredibly, after 2 years of EIBI treatment for children with autism less than 30 months old, 50% of these youngsters were later indistinguishable from typically developing schoolchildren. No other treatment of children

Effective Educational Programs for Autism
US National Research Council

- Early entry into intervention
- Intensive instructional programming (25 or more hours a week, 5 or more days a week, 12 months a year)
- One-to-one or small-group instruction to achieve clearly defined individualized goals
- Specialized training by discrete trials and incidental teaching
- Systematic and individualized instruction
- Focus on development of spontaneous social communication, adaptive skills, appropriate behaviors, play skills, and cognitive and academic skills

Fig. 13.9

Summary of the guidelines for an effective program in the treatment of autism as set by the National Research Council of the United States.

Source: Extracted from R. M. Foxx (2008). Applied behavior analysis treatment of autism: The state of the art. *Child and Adolescent Psychiatric Clinics of North America, 17*, pp. 821–834.

with autism has produced such dramatic improvement (Foxx, 2008; Lovaas, 1993; Maurice, 1993; Schopler & Mesibov, 1994).

A meta-analysis of studies has supported the effectiveness of ABA-based early intervention programs (Peters-Scheffer, Didden, Korzilius, & Sturmey, 2011; but see Warren et al., 2011 for a discussion of reservations and methodological concerns). An evidence-based review panel of the New York Department of Health concluded that of 18 types of interventions for autism, only ABA combined with EIBI was effective, whereas auditory integration therapy, facilitated communication, floor time, sensory integration therapy, touch therapy, music therapy, hormones, vitamin therapies, and special diets were not recommended (Foxx, 2008). Furthermore, the National Research Council of the United States set out guidelines for effective educational programs for autism, and only ABA with EIBI met or exceeded these standards (Figure 13.9). A review of autism spectrum disorders concluded that next to genetic studies, ABA is one of most researched areas and provides the most promising treatment results of any research area to date (Matson et al., 2012). Most attention has focused on EIBI, parent training, functional assessment, challenging behaviors, independent living, and social skills training.

NEW DIRECTIONS: AUTISM, MIRROR NEURONS, AND APPLIED BEHAVIOR ANALYSIS

Autism is a multiply-determined and multiply-expressed disorder that has become alarmingly prevalent. The source of the disordered behaviors labeled as autism is not clear, but the treatment of choice is EIBI by skilled behavior analysts (Charlop-Christy & Kelso, 1997; Foxx, 2008; Ghezzi, Williams, & Carr, 1999). Because of its unknown etiology and often extraordinary behavioral effects, people are desperate to try different treatments and to promote a variety of possible causes, some of which are clearly fads based on pseudoscientific beliefs (Todd, 2015).

Autism spectrum disorders (ASD) are almost exclusively defined by behavior, primarily in terms of social deficits such as poor eye contact, low communication skills, noncompliance, and lack of social play (Klin, Chawarska, Rubin, & Volkmar, 2004; Volkmar, Carter, Grossman, & Klin, 1997). The child with autism looks like a typically developing child, but does not show age-appropriate behavior—especially social behavior. A set of molecular pathways is abnormally expressed in the brains of people with autism (Voineagu et al., 2011) and several genetic mutations are associated with autism (Sanders et al., 2012), but at present the causal evidence for a genetic factor is not definitive.

The discovery of the "mirror-neuron" system (see Chapter 11), or the action-observation network (AON) in humans, has provided hope for improved behavioral treatment of autistic children. Youngsters with ASD are said to lack development in the AON system, suggesting they would benefit from early intensive behavioral intervention (EIBI) focused on imitation (Kana, Wadsworth, & Travers, 2011). Behavioral interventions targeting response components of visual discrimination, generalized imitation (Baer, Peterson, & Sherman, 1967), and observational learning (Bandura, 1977) should be especially effective in engaging and organizing the AON of youngsters with autism (Vivanti & Rogers, 2014). A recent study, however, did not find fMRI activation differences in the AON between typically developing and ASD adolescents during passive video-viewing of hand actions such as grasping an object (Pokorny et al., 2015). That is, for observing hand actions of others, there was no fMRI evidence of global impairment of the mirror network in the ASD sample compared with typically developing adolescents (see also Enticott et al., 2013 for motor-evoked potentials during action observation). The AON, however, consists of at least 10 brain sub-areas (Pokorny et al., 2015) and observing hand actions must be integrated with other neural sub-systems to execute the complex behavioral sequences involved in imitation and observational learning (Kana et al., 2011). Thus, early deficits in AON sub-areas related to imitation and observational learning have not been ruled out.

The mirror-neuron system allows the observer to process information about self-performed actions and consequences (goal-directed behavior) and, using parts of the same sensorimotor system, respond to actions, emotions, and goal-directed behaviors of others (Oberman & Ramachandran, 2007; see Hickok, 2014 for a critique of the evidence for mirror neurons; see Keysers, 2015 for a critique of Hickok's, 2014 analysis). Neural assessment of the AON during intensive behavioral training could greatly enhance our understanding and treatment of childhood autism (Thompson, 2007, pp. 429–430; see also McGill & Langthorne, 2011 for developmental disabilities). One possibility is that early intensive behavioral intervention (EIBI) is effective for up to 50% of children with autism with intact neural systems because EIBI helps to organize and activate the mirror system in the developing child, as Iriki (2006) suggests is the case for other primates.

Obviously, neural mechanisms are altered during early experience, and the immature nervous system is especially sensitive to environmental influences during maturation and development. Thompson (2007) suggested that "early differential reinforcement of discriminative responding [by children with autism] to visual images of hand and arm movements may promote gene expression in the mirror neuron system" (p. 437), as potential epigenetic effect on neural plasticity and learning. EIBI has produced lasting modifications of autistic behavior in some children, and it is possible that interventions aimed at priming the AON, such as discrimination training of the actions and emotions modeled by others, could be even more effective.

Unfortunately, there is a subgroup of children with autism who do not benefit from 30 or more hours of weekly behavioral treatment (Thompson, 2007, p. 429). Modern neural imaging could help to demarcate those children who are unable to form synapses or specific neural pathways, especially in the mirror-neuron network, and profit from EIBI. These individuals' behavior, however, could be treated by alternative behavioral procedures designed for their specific deficits and excesses, which might include additional interventions by specialists in fields such as speech pathology and physical therapy.

THE NEURODIVERSITY MOVEMENT AND APPLIED BEHAVIOR ANALYSIS

Traditionally, in behavior analysis, the target goals for behavioral change have been toward "normal" or "typical" development. One can understand this approach, as parents of children, especially those with developmental disabilities, often have the desire for their children to be socially accepted and treated consistent with the status quo. One problem with this, however, is that behavioral variations that deviate from "normal" are referred to as deficits or excesses—language that indirectly pathologizes behavior. In some instances, this seems appropriate, such as with self-injurious behavior. The goal to decrease self-injury because it is harmful to the individual is a laudable endeavor. However, not all behavioral variations are harmful. For example, we have mentioned variability as an important behavior that enhances social interaction, and invariability (e.g., restrictive and repetitive behavior or stereotypy) as interfering and limiting in some situations. Reducing stereotypy may help a person function better in society, but it may also be a behavior that makes those with ASD very uncomfortable and even cause anxiety and feel they have to hide it. Therefore, for some, overall reduction of the target of stereotypy may not be ideal or even necessary. It may be that waiting for the appropriate time and place to engage in stereotypy (as opposed to completely eliminating it) is an agreeable solution for a target behavior (see Potter et al., 2013).

Consider this from the standpoint of more socially acceptable behavior: one can imagine a fun behavior such as dancing and singing might interfere with daily activities if it was happening too frequently and in inappropriate contexts, however, scheduling time to dance and sing throughout the day or at the end of the day, or only in specific contexts, might be more beneficial for the individual.

In the 1990s, the Neurodiversity Movement began. This paradigm reconceptualizes differences in brain and the resulting behavior that people (such as those with ASD) have as variation in the brain and behavior, as opposed to pathologies. Through this lens, there is acceptance of a wider range of behavior and the goal is to find ways to let individuals with variations in brain and behavior to live in harmony with the world around them, as opposed to applying the neurotypical standard. In the present day, more applied behavior analysts are examining behavioral differences through this lens. Perhaps we may see this movement gain even greater momentum in time.

Behavior Analytic Early Intervention Approaches to Autism

A recent review by Smith and Iadarola (2015) classified Individual, Comprehensive ABA as a well-established intervention for children with autism disorder. This update built on a review by Rogers and Vismara (2008) that concluded Lovaas (1987) early intensive behavior intervention (EIBI) model as a well-established approach to treating ASD. Smith and Iadarola list commonalities of empirically supported interventions for ASD in line with ABA principles (for discussion and review see Iovannone, Dunlap, Huber, & Kincaid 2003):

1. Individualized services and supports: making use of the particular interests and individual learning style of each child with ASD to increase engagement in activities through interventions such as reinforcement systems and incorporation of preferred activities into intervention sessions.
2. Systematic intervention planning: selecting goals and instructional procedures based on a data-based assessment of each child, monitoring progress, and troubleshooting as needed.
3. Comprehensible, structured environments such as using visual schedules to help children with ASD anticipate transitions between activities and organized work spaces to facilitate task completion.
4. Specific intervention content to address the impairments in social communication and restricted, repetitive behaviors that define ASD.
5. Functional approach to problem behavior: assessing the function or purpose of the behavior and selecting intervention strategies based on this assessment.
6. Family involvement to promote consistency between home and the intervention setting, take advantage of the family's knowledge of the child with ASD, and overcome difficulties that children with ASD are likely to have in conveying information from one setting to another.

Based on the success and effectiveness of the Lovaas treatment approach, a number of institutes, organizations, and companies that use early-intensive behavioral intervention (EIBI) for autism have been developed. Typical institutes and organizations might use the treatment package described by Lovaas as a foundation, but many programs are supplemented by recent evidence-based work in the field. For example, the *Picture Exchange Communication System* (PECS; Bondy & Frost, 1994; Cummings, Carr, & LeBlanc, 2012) is used for children who have problems with vocalization.

Another included approach is the Verbal Behavior Approach by Mark Sundberg to establish functional verbal responses, which are necessary for communication (Barbera, 2007).

ABA programs for treating autism may vary in terms of the components included in the intervention. Organizations and ABA companies have included programmed generalization aids in the transfer and maintenance of learned skills from a highly structured training situation to the child's everyday environment (i.e., home, school, and community) (see Neely et al., 2016 for review). In comprehensive ABA-based programs, parents are taught the necessary behavioral skills for training their child. The treatment package includes specific intervention strategies for accomplishing behavioral outcomes. For example, the children's behavior is reinforced for reciprocating vocal interaction when the teacher talks to them. Appropriate life skills, such as eating meals with utensils, dressing oneself, and personal hygiene (i.e., brushing teeth and combing hair) are reinforced with tokens and social approval. (see Ivy et al., 2017 for review). Verbal skills, including manding and tacting, also are targets for behavior change. Autism service staff members monitor progress and, if necessary, provide advice for program changes.

EIBI programming typically involves discrete-trial training (DTT) procedures in which the teacher presents a stimulus and the response by the child is reinforced (or corrected) (Smith, 2001). Teaching should also include generalization of subskills to new situations, trainers, and response variations within the standard-teaching setting. Best practice teaching is focused on training stimulus generalization. Once the child masters an appropriate response in the presence of a specific S^D, the teacher varies properties of the S^D while maintaining the appropriate response. The S^D may be "What am I doing?" along with hand waving by the teacher, and the child is reinforced for saying "You're waving your hand." Next, the teacher may say "I'm doing what?" or "Hey, what's happening here?" (varying the S^D), and reinforcement remains contingent on the response "You're waving your hand." Finally, the training emphasizes maintaining learned concepts and skills. The training ensures that the child demonstrates generalization of skills when lessons are changed from one location or time to another. Early intervention should also involve programmed environmental distractions, similar to everyday interruptions in a classroom (building *behavioral momentum*, as described by Nevin and Grace, 2000). The child is taught to maintain accurate responding in the face of these random interruptions.

Research on daily living skills and hygiene has been conducted with children with ASD. A child may be able to use the toilet and to dress themselves, but may be unable to select the clothes to wear. In the classroom, a child may be able to write on a piece of paper when instructed by the teacher, but may be unable to get a piece of paper on their own. Formally, the training at this level is focused on completion of extended *behavior chains* or sequences. Other research that builds upon and uses the practices in DTT involves training and generalization of social skills (greetings, reciprocity, and empathy) that will be necessary to interact with others in everyday settings (e.g., classroom, playground, and home). For example, the child is taught to discriminate between greeting their parents and saying hello to a playmate (Figure 13.10; see also Chapter 12 on manding; see Ganz, Davis, Lund, Goodwyn, & Simpson, 2012 for a supportive meta-analysis of PECS on targeted and non-targeted behavioral objectives).

What about transition from the behavioral program to regular classrooms? The first thing to note is that almost no children with autism move from traditional, nonbehavioral treatment programs to public-school classrooms (Lovaas, 1987). Comprehensive intervention programs with children with ASD were given by Maurice, Green, and Luce (1996) concerning intensive behavioral treatment of autism. Their manual indicates that intensive behavioral programs work

Fig. 13.10
Photograph shows Brenda Terzich, an applied behavior analyst, and a child using the Picture Exchange Communication System (PECS) at the ABC school for autism in Sacramento, CA. PECS is a behavioral technology developed by Bondy that allows the nonvocal child to communicate with others and acquire a basic verbal repertoire of manding and tacting. Published with permission.

best when children are less than 5 years of age (the younger the better), the program includes at least 30 h of treatment per week, and children continue in the program for at least 2 years. Under these conditions, even very low-functioning children make large behavioral gains. Due to the success and subsequent demand for in-home and center-based behavioral treatment, demand for ABA practitioners has exponentially increased (BACB, 2022). Over 70% of ABA practitioners work in the professional emphasis of Autism Spectrum Disorder (BACB, n.d.).

> **OVERVIEW OF THE BEHAVIOR ANALYST CERTIFICATION BOARD**
>
> The Behavior Analyst Certification Board (BACB) is a non-profit organization established in 1998, which evolved out of a program of credentialing and continuing education for behavioral service providers in the state of Florida. The BACB is accredited by the National Commission for Certifying Agencies (NCCA), which developed the first standards for professional certification programs to help ensure health, welfare, and safety of the public. Today the BACB's roles are to operate certification programs, establish practice standards, administer examinations, and provide ethic requirements and a disciplinary system for each of its certification programs. To date, the BACB has over 176,000 members (BACB, n.d.).

Video Modeling, Mobile Technology, and Autism

In Chapter 11, we saw that modeling, imitation, and observational learning were critical to the rapid acquisition of human behavior, especially for those with developmental disabilities and autism (see "On the Applied Side: Training Generalized Imitation" in that chapter). People with autism and other learning difficulties are now able to access video modeling (VM), which is a video demonstration by a model of an integrated sequence of skills to perform some task, and video prompting (VP). VP is a video of a model showing the separate steps or components of a task, using new mobile technology (Ayres & Langone, 2005). This technology includes portable DVD players, laptop computers, and personal digital assistants (PDA), as well as the popular iPod, iPad, and iPhone. For many people with autism, using these devices is highly reinforcing, as it allows them to obtain self-generated information about living and life skills, and allows them fit into the wider community that endorses the use of such technology (Bereznak, Ayres, Mechling, & Alexander, 2012).

In a study of video self-prompting and mobile technology, Bereznak et al. (2012) used iPhones to teach vocational and daily-living skills to three male high school students with autism spectrum disorder (ASD). The students were taught to use the devices to access video prompting of the steps to use a washing machine, a microwave to make noodles, and a photocopying machine. For example, to use the washing machine, a task analysis defined the following steps or components: turn dial to setting for regular wash, pull dial to start running water, open the door, take cap off detergent, pour detergent into washer, put cap back on detergent, put detergent on counter, put clothes in the washer, and close the door. After instruction in how to use an iPhone to watch a training video, the student was situated before a washing machine, given the iPhone showing play, forward, and rewind buttons, and told to "touch play to start doing the wash." The student now touched the play feature and a single word appeared on the screen describing the step and followed by the video presentation with an audio description for the first step. At the end of each step, the video displayed a stop sign, which set the occasion for the student to press the pause button. After hitting the pause button, the student was asked to imitate the behavior shown on the video clip. Incorrect responses (not matching to the video model) were blocked and corrected, while the correct response received a schedule of verbal praise, such as "Nice job turning the dial."

The results indicated an immediate effect of video prompting on each targeted skill (using a washing machine, making noodles, and photocopying) across the three students, indicating that an iPhone can serve as an effective self-prompting device to teach daily-living and vocational skills to adolescent students with ASD (see Domire & Wolfe, 2014 for a review—concluding that VP is more effective than VM in training a broad array of targeted skills). One hidden benefit is that self-prompting on an iPhone is a valued behavior in America, indicating to others some degree of autonomy and self-determination. Thus, behavior principles along with mobile technology may generate self-regulation of actions (not just rote learning)—giving the person with ASD greater independence, self-determination, and intrinsic motivation.

BEHAVIORAL TREATMENT AND PREVENTION: THE PROBLEM OF OBESITY

In recent years, behavior analysts have focused attention on the factors that produce behavior problems (Bellack, Hersen, & Kazdin, 2011). Animal models of disordered behavior have been developed that provide insight into the causes of problem behavior involving depression, anorexia, and other psychiatric disorders (see Epling & Pierce, 1992; Keehn, 1986; Seligman & Maier, 1967). Other researchers have been concerned with promoting behavior related to physical health. The area of **behavioral medicine** is a multidisciplinary field that includes behavior-change programs targeted at health-related activities such as following special diets, self-examination for early symptoms of disease, exercising, taking medicine, stopping smoking, and other health-related issues (Friman, Finney, Glasscock, Weigel, & Christophersen, 1986; Gellman & Turner, 2012; Pomerleau, 1979). The idea is that many health problems such as obesity and diabetes may be prevented or treated by integrating knowledge about biological, genetic, and behavioral risk factors.

The Obesity Crisis

Obesity is usually viewed clinically as a body-weight status that elevates the risk of health problems, such as type-2 diabetes, metabolic syndrome, and heart disease. In contrast to this medical view, a biobehavioral analysis indicates that genotype (genetic makeup) may be related to an animal's survival of food-related challenges (a version of "thrifty gene theory" by Neel, 1962). From a biobehavioral perspective, obesity results from variation in genotype (obese-prone or lean-prone) in combination with an environment offering free access to high-calorie, energy-dense foods. Indeed, like many animals, humans evolved the ability to eat and store excessive amount of food as energy, since food scarcity was part of the evolutionary milieu; there is genetic variation in this ability. It has only been in the last 50–100 years that mass food production has led to an abundance of food availability in industrialized nations. Not surprisingly, obesity rates, especially in the last 40 years have risen. For example, over 40% of Americans are obese (Centers for Disease Control, 2022b) and an additional 30% are overweight. This trend suggests that there is a combination of genetic (i.e., an obese-prone genotype that selects for greater fat storage) and environmental factors (e.g., high production of processed foods that are high in refined carbohydrates) that combine to create variation in obesity.

An obese-prone genotype confers an advantage under conditions of unpredictable changes in food supply—allowing for storage and conservation of energy and increased foraging during periods of food depletion (Diane et al., 2011; Pierce, Diane, Heth, Russell, & Proctor, 2010). In the laboratory, rats on food restriction with the opportunity to (wheel) run show activity anorexia (AA), a cycle of suppressed food intake, declining body weight, and escalating wheel activity that may lead to starvation and death (Epling, Pierce, and Stefan, 1983 referred to this cycle as activity-based anorexia). One test of the biobehavioral hypothesis of obesity is to expose obese-prone rats to the *AA challenge* to see if this genotype would survive under famine-like conditions.

Obese-prone (cp/cp) JCR:LA-cp rats lack the ObR-leptin receptor, eliminating control by leptin, a major hormone for regulation of energy intake and energy expenditure. In a free-feeding environment, cp/cp (obese-prone) rats show overeating, excessive body-weight gain, and pronounced inactivity—leading to obesity, metabolic syndrome, and cardiac failure in adult animals. Thus, when obese-prone rats are exposed to an environment of unlimited, high-energy food supply, their risk for poor health increases. In contrast, JCR:LA-cp lean-prone rats with an intact ObR-leptin receptor eat moderately when food is freely available, maintain a healthy body weight, and remain physically active into adulthood. These rats adapt well to a food-abundant environment. Studies of the AA challenge, however, indicate that lean-prone rats do not survive the test, reaching the starvation criterion (75% of body weight) in 3 to 5 days. Would juvenile obese-prone rats outlast their lean-prone counterparts in the AA situation?

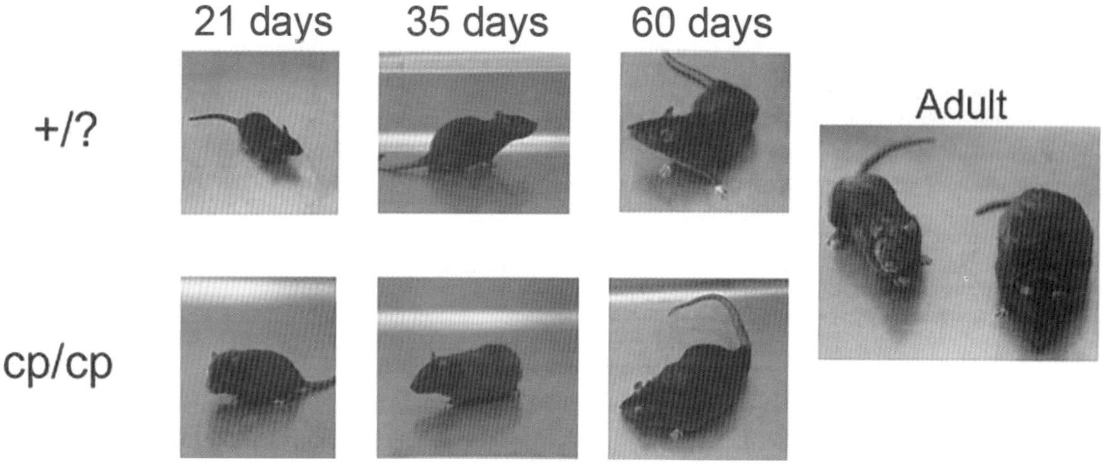

Fig. 13.11
The phenotypic development of JCR:LA-cp male rats from weanling pups to adulthood in a free-feeding environment. Lean-prone rats of this strain eat moderately, remain physically active and keep a trim body weight. Obese-prone (cp/cp) rats are characterized by overeating, pronounced inactivity, and excessive gains in body weight. In the study, both lean-prone and obese-prone rats (35–40 days old) were exposed to the AA challenge of 1.5 h access to food and 22.5 h access to running wheels—testing the evolutionary hypothesis that an obese-prone genotype allows for survival during periods of unpredictable food shortages and food-related travel.

Source: Photographs provided by Dr. Spencer Proctor, Director of the Cardiovascular Diseases Laboratory, The Alberta Institute of Diabetes, University of Alberta, Edmonton, Alberta, Canada. Dr. James C. Russell of the Faculty of Medicine (retired) at the University of Alberta developed the JCR:LA-cp strain as an animal model for obesity, heart disease, and the metabolic syndrome.

To answer this question, adolescent obese-prone and lean-prone rats aged 35–40 days (see Figure 13.11) were exposed to the AA challenge (1.5 h of food access and 22.5 h of access to running wheels), and measures of behavior and metabolism were obtained. Notably, at this age, obese-prone rats were similar in body weight to lean littermates.

The findings supported the *adaptive hypothesis of obesity*. Juvenile obese-prone rats gained a survival advantage over lean-prone animals when confronted with the food restriction and wheel-running AA challenge. Young obese-prone rats survived approximately twice as long as lean-prone juveniles, even though their initial body weight did not differ from lean littermates before the challenge (Figure 13.12, left). Food intake by obese-prone rats was similar to that of lean-prone animals for the AA-challenge period, but body composition measures indicated that juvenile obese-prone rats conserved fat mass while lean-prone rats had depleted fat reserves, a finding that was confirmed by metabolic measures.

During food restriction, the daily wheel running of juvenile obese-prone (cp/cp) rats was similar to that of lean-prone juveniles, but the young obese-prone rats maintained this pace for more days and covered three times the distance (Figure 13.12, right). Other evidence showed less physiological stress in the obese-prone rats compared with the lean-prone animals, allowing for extensive daily travel. In the natural world, the obese-prone genotype would have adaptive value, allowing animals to search for food over extended distances and increasing the likelihood of contact with a stable source of food.

The evidence suggests that having an obese-prone genotype may be deleterious when food is abundant, but is adaptive when food is scarce and food-related travel is initiated.

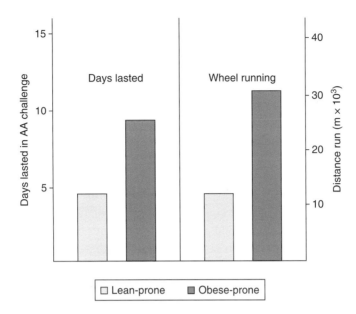

Fig. 13.12
Results of the experiment exposing lean-prone and obese-prone rats to the AA challenge (1.5 h access to food and 22.5 h access to running wheels). The left side of the graph shows the days lasted in the AA challenge, indicating that obese-prone rats lasted about twice as long as their lean-prone counterparts. The right side of the graph shows the results for wheel running in the AA challenge. Obese-prone rats traveled about three times the distance as lean-prone animals during the AA challenge, behavior that would increase the likelihood of contact with a food source in the natural environment. The overall results indicate that an obese-prone genotype had survival value when animals were faced with unpredictable food shortages and famines.

Source: W. D. Pierce, A. Diane, C. D. Heth, J. C. Russell, & S. D. Proctor (2010). Evolution and obesity: Resistance of obese-prone rats to a challenge of food restriction and wheel running. *International Journal of Obesity*, *34*, pp. 589–592.

Additional research (Diane et al., 2011) shows that experience with food restriction and an obese-prone genotype combine to further increase survival in the AA challenge. Thus, both genotype and food environment contributed to survival during unpredictable food shortages, which require extended food-related travel. In our current food-rich environment, however, an obese-prone genotype that favors overconsumption and conservation of fat reserves is not an advantage—causing many health-related problems.

Behavior Management of the Obesity Crisis

One implication of the research on obese-prone rats and the AA challenge is that modification of the food environment, especially for obese-prone children, is a necessary step toward solving the obesity crisis in America. As the rising medical costs of obesity have effects on the public purse, one would expect political and social policies to regulate the food industry and advertising of food products (see, for example, Centre for Science in the Public Interest news release on mandating nutrition information on menus of chain-restaurants, Ottawa, Canada, May 8, 2012). While these large-scale changes could alter community and family contingencies currently supporting children's high consumption of energy-dense, fattening foods, they have not yet materialized in ways that have altered society in profound ways.

Childhood obesity is also being tackled by behavioral interventions targeted at nursery and school-age children. Pauline Horne and Fergus Lowe (1946–2014), behavior analysts at Bangor University in the UK, developed the *Food Dudes Healthy Eating Program* (http://food-dudes.co.uk/prog_over.htm) to instill healthy food choices in the nation's children. The program is based on social modeling and reinforcement principles (Horne et al., 2011). One part of the program involves schoolchildren watching DVDs of the adventures of Food Dudes, who are "cool" superhero children battling the Junk Punks to save the Life Force. The Junk Punks weaken the Life Force by depriving the world of fruit and vegetables. The Food Dudes gain special powers by eating fruit and vegetables, and are able to thwart the scheming of the Junk Punks. The purpose of the Food Dude series of DVDs is to use principles of modeling and imitation to increase the schoolchildren's consumption of fruit and vegetables at meals and snack times, both at school and at home. Food Dude Rewards (stickers and Food Dude prizes) are also part of the program. When children initially select and eat fruits and vegetables, they receive Food Dude Rewards and soon come to enjoy the taste of these foods, allowing the rewards to be gradually phased out. In a recent study with preschool children, the researchers concluded that:

> The modeling and rewards intervention proved to be a powerful means of producing lasting increases in young children's snack-time consumption of fruit and vegetables … [and] despite the fact that there were never any reward contingencies [at lunchtime] … the effects of the snack-time intervention generalized strongly to a different time of day and a different meal context…. [At follow-up] there were large and significant increases in lunchtime consumption [of fruits and vegetables].
>
> (Horne et al., 2011, pp. 382–383)

The Food Dudes program for prevention of childhood obesity has spread throughout the schools of the UK and Ireland, and there are indications that other countries, including Canada and the USA, are interested in developing programs based on the success of the Bangor model. Although the Food Dudes program trains youngsters to select and prefer a variety of fruits and vegetables, at the present time evidence is lacking to show that these behavior modifications actually result in lower rates of childhood obesity, which are maintained into adulthood. Hopefully longitudinal follow-up studies, relating selection of fruits and vegetables by children to prevention of body-weight gain and obesity, are planned for the future.

Another necessary step toward solving the obesity crisis is to design and implement behavior management programs for overweight adults in North America. Research shows that behavior-focused techniques that are effective in treating autism, stuttering, and substance abuse may also be effective in managing weight and preventing weight gain (Freedman, 2011). Studies show that recording calories, eating patterns, exercise, and body weight is essential. In addition, setting modest, achievable, step-by-step goals helps to shape an active lifestyle, and joining a support group—such as a runners' club or a virtual group of dieters—increases the chances of success. Overall, there are a variety of behavioral strategies to help solve the obesity crisis in North America, even as we await advances in the control of metabolic processes related to body weight (e.g., brown fat and making white fat cells burn rather than store energy, Spiegelman, 2008; Zielinska, 2012).

FOCUS ON: CONDITIONED OVEREATING AND CHILDHOOD OBESITY

Taste conditioning plays a role in the development of obesity. Families that provide children with sweetened foods and drinks may inadvertently contribute to overeating and childhood obesity, according to research using respondent conditioning to induced overeating (Davidson & Swithers, 2004; Pierce, Heth, Owczarczyk, Russell, & Proctor, 2007; Swithers & Davidson, 2008; Swithers, Doerflinger, & Davidson, 2006). The researchers found that sweet food—even those with non-caloric artificial sweeteners—disrupted the body's capacity to regulate energy intake, resulting in overeating and weight gain. The findings help to explain why increasing numbers of children and adults in North America lack the ability to regulate energy intake.

Pierce et al. (2007) suggested that being able to match calorie intake with the body's needs involves the ability to learn that sweet food tastes predict the amount of calories ingested (see also Swithers, 2015; Swithers & Davidson, 2008). Thus, both obese-prone and lean-prone juvenile rats are prepared to learn that particular tastes signal caloric energy. Based on this associative learning, the use of calorie-wise foods may undermine the body's natural ability to regulate energy intake and body weight. In this way, diet foods and drinks could lead to **conditioned overeating** and obesity in children, which is especially a problem for obese-prone individuals.

Early experiences can teach young rats that specific food flavors are useful for predicting the energy content of foods. Subsequently, the animals use food tastes to determine the body's need for calories. When food flavors have been associated with low caloric energy, as with diet foods and drinks, juvenile rats eat more than their bodily need after ingesting a nutritious snack or pre-meal containing that flavor. Overall, the research shows that young rats overeat and gain weight when the taste of food has been predictive of low energy content.

Obesity is a significant risk factor for both type-2 diabetes and cardiovascular disease, and is an increasing major health problem in North America and Europe. In this regard, it is important to note that the food industry generates an extensive variety of products, some of which offer attractive tastes but have little or no caloric energy. In fact, it has become commonplace for people to eat calorie-wise foods and drinks or add artificial sweeteners rather than consume less of high-energy foods (i.e., small servings). Given the ubiquity of the "diet craze," many children may learn that the taste of food often predicts low energy value. Children with such a dietary history may not effectively regulate their caloric intake over the course of a day—overeating at dinner by failing to compensate for intake of palatable high-calorie snack foods during and after school. The best strategy, according to the researchers, is to keep diet foods away from youngsters and give them small portions of regular foods to avoid weight gain and obesity (Pierce et al., 2007; Swithers, 2015).

CHAPTER SUMMARY

In this chapter, we have presented many examples of applied behavior analysis. Issues of observer reliability, irreversibility of treatment, multiple baseline designs, and fluency and rate as dependent measures were highlighted. We emphasize a focus on behavior and its functions, such as gaining attention for self-abusive behavior. The behavioral approach to therapy involves direct interventions on the problem behavior by manipulating the environment, as illustrated by voucher-based contingency management of substance use disorders. Applied behavior analysts usually reject appeals to disturbance of cognitive or mental processes in favor of changing the world in which people live.

Several systematic educational programs were described that have shown major improvements in student achievement. For decades, precision teaching and PSI have produced superior pupil progress, but these approaches are largely ignored in mainstream education. Applications of behavior principles have also been used to train behavior and educate children with autism, helping many of them to enter mainstream educational systems.

We also reviewed research on the obesity crisis, behavior management of obesity, and conditioned overeating with implications for childhood obesity. This area of research illustrates the use of behavior principles in health and medicine, indicating that applied behavior analysts are making important contributions to human welfare. The value of behavior analysis as a natural science is that it derives from, and applies to, all behaving organisms including humans. The principles of behavior were discovered by scientific experiments, which were replicated by countless studies and numerous real-world applications using a great variety of procedures, subjects, and settings. We have seen that the basic behavioral model is compatible with the findings of neuroscience and neurobiology and is sufficient to account for whatever behavior is under study, while also allowing for controlled interventions in applied settings. Wide-scale adoption of behavioral technology has been slow in education and other applied areas, perhaps because Western values of freedom and dignity often oppose the effective use of reinforcement contingencies in everyday life.

Key Words

Applied behavior analysis
Behavior maintenance
Behavior trapping
Behavioral contract
Behavioral medicine
Celeration
Changing criterion design
Conditioned overeating
Contingency management (CM)
Differential reinforcement of alternative behavior (DRA)
Differential reinforcement of other behavior (DRO)
Duration recording
Early intensive behavioral intervention (EIBI)
Fluency

Intensive behavioral intervention
Internal validity
Interval recording
Multiple baseline across behaviors
Multiple baseline across participants
Multiple baseline across settings
Multiple baseline designs
Personalized system of instruction (PSI)
Precision teaching
Reliability of observation
Response generalization
Self-control
Time sampling

On the Web

www.bacb.com Do you want to become a certified behavior analyst? Go to this website to get more information on the steps and examinations involved. The Behavior Analyst Certification Board is a nonprofit corporation established as a result of credentialing needs identified by behavior analysts, state governments, and consumers of behavior analysis services.

www.contractingwithkids.com Companion website for Dardig and Heward's "Let's Make a Contract." Contract forms and other resources available here.

www.nsf.gov/news/news_summ.jsp?cntn_id=122294 The National Science Foundation (NSF) funded the P.A.D. simulator-trainer for clinical breast examination by clinicians. This video describes the development of this behavioral technology by Mark Goldstein and the scientists at MammaCare, and how it is being used to save lives by providing training in palpation and search of breast tissue.

http://rsaffran.tripod.com/aba.html This website contains a collection of Internet resources for parents of children with autism and related disorders based on applied behavior analysis. It provides help in finding service providers and private schools, and information about parental experiences, intervention principles, and much more.

www.abcreal.com Find out more information about the ABC school of autism and Applied Behavioral Consultants, Inc. A video presentation is also available of the treatment program based on early intensive behavioral intervention, PECS, and the verbal behavior approach.

www.iaba.com Go to the website of the Institute for Applied Behavior Analysis in southern California, which offers behavioral management services as well as educational and employment services to people with developmental disabilities.

Brief Quiz

1. Applied behavior analysis is a field of study that focuses on the application of the _____, methods, and procedures of the science of behavior.
   a. equations
   b. principles
   c. research
   d. findings
2. In terms of a behavioral contract, the details usually specify:
   a. what is expected of the client
   b. the level of attention required
   c. the consequences that follow behavior
   d. both (a) and (c)
3. What is behavior trapping?
   a. the fact that animal behavior leads to trapping by hunters
   b. Don Baer got trapped by a behavioral contingency
   c. new behavior is trapped by the natural contingencies
   d. an attempt to overcome the traps that our behavior causes

4 Which of the following are multiple baseline designs?
   a multiple baseline across settings
   b multiple baseline across subjects
   c multiple baseline across behaviors
   d all of the above
5 In the treatment of substance use disorders, contingency management involves:
   a the use of fading medical testing
   b the use of vouchers contingent on abstinence
   c the use of contingent attention
   d the use of reinforcement of alternative behavior
6 Fred Keller wrote a seminal article on college teaching called _____.
   a "Farewell to College Education"
   b "Good-Bye, Teacher …"
   c "Keller on Teaching"
   d "So Long to Higher Education"
7 Belles and Bradlyn (1987) conducted a smoking cessation study using a _____ design.
   a multiple baseline across subjects
   b A-B-A-B reversal
   c changing criterion
   d factorial
8 What is the first step in early intervention programs for children with autism?
   a teaching of "splinter skills" to able students
   b teaching of stimulus generalization to those who are advanced
   c training and generalization of social skills, especially with parents
   d discrete trials of stimulus, response, and reinforcement
9 In the adaptive hypothesis of obesity, a(n) _____ confers an advantage under _____ food supply.
   a obese-prone genotype; unpredictable
   b lean-prone genotype; diminishing
   c biological predisposition; varying
   d evolutionary mutation; reduced
10 Which of the following is a component of the Food Dudes program?
   a use of stickers and Food Dudes prizes as reinforcement
   b use of differential reinforcement of other behavior (DRO)
   c use of social modeling by superheroes eating fruits and vegetables
   d both (a) and (c)

Answers to Brief Quiz: 1, b (p. 470); 2, d (p. 473); 3, c (p. 475); 4, d (p. 477); 5, b (p. 483); 6, b (p. 489); 7, c (p. 493); 8, d (p. 494); 9, a (p. 502); 10, d (p. 504).

Three Levels of Selection: Evolution, Behavior, and Culture

1. Explore three levels of selection by consequences: evolution, behavior, and culture.
2. Investigate genetic and operant control of behavior in the marine snail, *Aplysia*.
3. Learn about epigenetic modifications and the innate social behavior of carpenter ants.
4. Discover how verbal behavior contributes to the transmission of cultural practices.
5. Inquire about cultural evolution at the behavioral and cultural levels.

Behavioral researchers suggest that **selection by consequences** is the operating principle for biology, behavior, and culture (e.g., McDowell, 2004, 2013; Pierce & Epling, 1997; Skinner, 1981; Wilson, Hayes, Biglan, & Embry, 2014; see Santana, 2015 for challenges to behavioral-level selection). It is a general form of causation that goes beyond the push–pull mechanistic Newtonian model of physics (Hull, Langman, & Glen, 2001). In terms of evolution (level 1), selection by consequences involves natural selection or the selection of genes based upon an organism's reproductive success. At the level of behavior (level 2), selection by consequences is described by the principle of reinforcement—the selection and changes in operant behavior by the effects it produces. A third level of selection occurs in terms of culture (level 3). Cultural selection involves the change of practices (common ways of doing things) based on large-scale consequences for the group—involving greater efficiency, lower costs, and higher likelihood of survival.

In this chapter, selection by consequences is examined at the genetic, behavioral, and cultural levels (see Figure 14.1). In showing the parallels among these different levels, behavior analysts seek to integrate the study of behavior with biology on the one hand and the social sciences on the other (Wilson, Hayes, Biglan, & Embry, 2014). The attempt is not to reduce behavior to biology, or culture to behavior. Rather, it is to show the common underpinnings of all life science in terms of the extension and elaboration of basic principles.

DOI: 10.4324/9781003202622-14

CONTINGENCY OF SURVIVAL

Ecological environment : Genotype ⟶ Benefits/costs for reproduction

[In a specific habitat, species characteristics resulting from differences in genotype produce more (or less) reproductive success—process is natural selection]

CONTINGENCY OF REINFORCEMENT

Situation : Operant ⟶ Reinforcement/punishment

[In a specific situation, a particular response from an operant class produces reinforcement (or extinction) that increases (or decreases) the rate of occurrence of the operant—the process is reinforcement]

METACONTINGENCY

Technological environment : Cultural practice ⟶ Benefits/costs for survival of group

[In a specific technological environment, a particular kind of cultural practice produces outcomes for the group that increase (or decrease) the practice—process of cultural selection]

Fig. 14.1
Selection by consequences operates at three levels: evolution, behavior, and culture.

LEVEL 1: EVOLUTION AND NATURAL SELECTION

The evolutionary history of a species, or **phylogeny**, is the outcome of natural selection. Darwin (1859) showed how organisms change or evolve in accord with this principle (Figure 14.2). Based on a thorough analysis of life forms, Darwin concluded that reproductive success was the underlying basis of evolution. That is, individuals with more children passed on a greater number of their characteristics to the next generation.

Darwin noticed structural differences among members of sexually-reproducing species. Except for identical (monozygotic) twins, individuals in the population vary in their physical features. Thus, birds like the thrush show variation in color of plumage, length of wings, and

Fig. 14.2
Charles Darwin (1809–1882) discovered the principle of natural selection.
Republished with permission of the Archives of the History of American Psychology, Center for the History of Psychology, The University of Akron.

thickness of beak. Based on differences in their features, some individuals in a population are more successful than others at surviving and producing offspring. Differences in reproductive success occur when certain members of a species possess attributes and behavior that make them more likely to survive and reproduce in a given environment. Generally, individuals with features that meet the survival requirements of a habitat produce more offspring than others. As the number of descendants with those features increases, the genetic traits of these individuals are more frequently represented in the population. If there is a fecundity (surplus) of individuals produced and there exists even small variability between individuals, those with the most fit characteristics will be selected and hence multiply. This process of differential reproduction is called **natural selection**, and the change in the genetic makeup of the species is **evolution**.

Contingencies of Survival

From a behavioral viewpoint, natural selection involves **contingencies of survival** (Skinner, 1984). The habitat, niche, or environment inadvertently sets requirements for survival of individuals. Members of a species who exhibit features and behavior appropriate to the contingencies survive and reproduce. Those with less appropriate characteristics have fewer offspring and their genetic line may become extinct. Natural selection therefore occurs as particular organisms satisfy (or fail to satisfy) the contingencies of survival.

An important implication of a contingency analysis of evolution is that the requirements for survival and reproductive success may change gradually or suddenly. For example, during the time of the dinosaurs, the collision of a large asteroid with the earth drastically changed the climate, fauna, and temperature of the planet in a very brief time (e.g., Alvarez, 1982; Alvarez, Asaro, & Michel, 1980). Given these changes in environmental contingencies, dinosaurs could not survive and reproduce. The small mammals, which possessed features and behavior more appropriate to the new prevailing contingencies, however, increased their reproductive success. Changes in the contingencies due to large-scale disasters may, therefore, occasionally favor characteristics and behavior that have advantages in a changed environment. This would occur even though these characteristics may have been a disadvantage in the past (see Gould, 2002 for a punctuated-equilibrium view of evolution; see Dawkins, 1976, 2004 for a discussion of gradual genetic selection).

Phenotype, Genotype, and Environment

Evolutionary biologists distinguish between phenotype and genotype. An organism's **phenotype** refers to all the characteristics and behavior observed during the lifetime of an individual. For example, an individual's size, color, and shape are anatomical features of phenotype. Behavioral features include taste preferences, aggressiveness, shyness, and many others. Different phenotypic attributes of individuals may or may not reflect underlying genetic variation.

The **genotype** refers to the actual genetic makeup of the organism. Some observable characteristics are largely determined by genotype, while other features are strongly influenced by experience (see Chapters 1 and 6). However, as shown in Figure 14.3, much of the phenotypic variation results from an interaction of genes and environment. Thus, the height of a person is attributable to both genes and nutrition working together. Evolution only occurs when the phenotypic variation among individuals is based on differences in genotype. If differences in height or other features did not result from genetic differences, natural selection for tallness (or shortness) could not occur. This is because there would be no genes for height to pass on to the next generation. People who engage in bodybuilding by lifting weights and taking steroids may substantially increase their muscle size (phenotype), but this characteristic will not be passed on to their children; it is not heritable. Natural selection can only work when there are genes that underlie differences in physical features.

Sources of Genetic Variation

There are two major sources of heritable genetic variation: recombination of existing genes, and mutation. Genetic differences from recombination arise from sexual reproduction. This is because the blending of male and female genes produces an enormous number of random

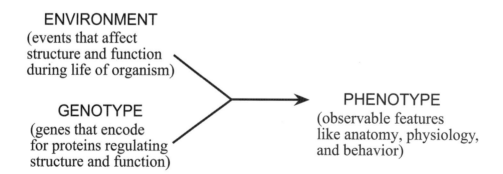

Fig. 14.3
Phenotype is a product of genotype and environment.

combinations in offspring. Although sexual recombination produces variation, the number of genetic combinations is constrained by the existing pool of genes. In other words, there is a finite number of genes in a population, and this determines the amount of variation caused by sexual reproduction.

Mutation occurs when the genetic material (e.g., genes or chromosomes) is altered. These changes are accidents that affect the genetic code or instructions carried by an ovum or sperm. For example, naturally occurring background radiation may alter a gene site, or a chromosome may break during the formation of sex cells or gametes. Such mutations are passed on to offspring, who display new characteristics. In most instances, mutations produce physical features that work against an organism's survival and reproductive success. However, on rare occasions mutations produce traits that improve reproductive success. The importance of mutation is that it is the source of new genetic variation. All novel genetic differences are ultimately based on mutation.

Natural selection depends on genetic variation arising from recombination and mutation. Genes code for proteins, which in turn regulate embryonic development and structural form (Mukherjee, 2016). This means that differences in genes result in phenotypic differences in the structure (e.g., size and form of the brain) and physiology (e.g., release of hormones) of organisms. Selection occurs when specific genes underlying these phenotypic features contribute to reproductive fitness. Individuals with such characteristics have more offspring, ensuring that their genes occur at a higher frequency in the next generation. (Note: epigenetic regulation of gene expression during the organism's lifetime, in response to internal or external environmental changes, contributes to phenotypic variability; this source of variation may allow for evolution by natural selection, Jablonka & Lamb, 2002, pp. 93–94; see also in this chapter, "New Directions: Epigenetic Reprogramming of Social Behavior in Carpenter Ants.")

GENETIC REGULATION OF BEHAVIOR

Behavioral Rigidity

As we have noted, the behavior of organisms is always a phenotypic expression of genes and environment. Genes closely regulate some behavioral characteristics, and in such instances the environment plays a subsidiary role. For example, in some species, defense of territory occurs as a ritualized sequence of behavior called a fixed-action pattern (e.g., Tinbergen, 1951). The sequence or chain is set off by a specific stimulus, and the component responses are repeated almost identically with each presentation of the stimulus (see Chapter 3). The behavior pattern is based on a "genetic blueprint," and the environment simply initiates the sequence.

For example, the male stickleback fish will aggressively defend its territory from male intruders during the mating season. The fish shows a fixed sequence of threatening actions that are elicited by the red underbelly of an intruding male. Tinbergen (1951) showed that this fixed-action pattern occurred even to cigar-shaped pieces of wood that had a red patch painted on the bottom (see Chapter 3). In addition, he showed that a male intruder with its red patch hidden did not evoke the threatening sequence. Generally, the male stickleback is genetically programmed to carry out the attack sequence given a specific stimulus at a particular moment in time. Presumably, in the evolutionary history of sticklebacks, those males that threatened or attacked an intruder gained a reproductive advantage (driving off competing males), accounting for the occurrence of the fixed-action pattern in this species.

FOCUS ON: GENETIC CONTROL OF A FIXED-ACTION PATTERN

Scheller and Axel (1984) reported on the genetic control of a complex behavioral sequence. They used recombinant DNA technology to isolate a subset of gene locations that control the egg-laying sequence of the marine snail *Aplysia*.

Recombinant DNA technology is beyond the scope of this book, but the important thing is that these procedures can be used to identify gene sites that encode for specific neuropeptides (see Taghert & Nitabach, 2012 for an overview of neuropeptide modulation of invertebrate neural systems). In the experiment by Scheller and Axel (1984), the researchers isolated a set of gene sites that coordinated the release of several peptides. These chemicals caused neurological changes that invariably produced the egg-laying sequence.

Using techniques of genetic manipulation, Scheller and Axel were able to "turn on" the gene sites that controlled a complex and integrated sequence of behavior. In this sequence, the snail first contracts the muscles of the reproductive duct and expels a string of egg cases. Next, the animal grasps the egg string in its mouth and waves its head, behavior that typically functions to remove eggs from the duct. It then attaches the tangle of string to a solid surface. This behavioral sequence is shown in Figure 14.4. The fixed-action pattern was activated in an unmated snail by direct manipulation of the egg-laying hormone (ELH) gene.

The DNA sequences that control egg laying may play an important role in other aspects of this animal's behavior. For example, the genetic material that encodes for head-waving behavior may be duplicated and appear in other genes that regulate feeding (Sossin, Kirk, & Scheller, 1987; Taghert & Nitabach, 2012). In this regard, Scheller and Axel suggested:

> The same peptide may be incorporated in several different precursors encoded by different genes. Consider head waving in *Aplysia*. A characteristic waving of the snail's head takes place during feeding as well as during egg laying. The same peptide or peptides could elicit the same behavioral component (head waving) in two very different contexts. To this end the head-waving peptide (or peptides) may be encoded in some other gene—one implicated in feeding behavior—as well as the ELH gene. In this way complex behaviors could be assembled by the combination of simple units of behavior, each unit mediated by one peptide or a small number of peptides.
>
> (Scheller & Axel, 1984, p. 62)

When environments were stable and predictable, the replication of the same DNA sequence in a new genetic context may be one way in which organisms evolved complex behavior (see also the role of gonadotropin-releasing hormone (GnRH) in the control of complex behavior and locomotion of *Aplysia*, but not in its reproduction, in Sun & Tsai, 2011). This solution involves using the same genetic code in different combinations. Although a high level of behavioral complexity may be achieved in this manner, the resulting behavior is tightly controlled by the underlying genetic context.

Some forms of animal communication are strongly influenced by genotype. For example, the waggle dance of the honeybee (Figure 14.5) is a highly ritualized sequence of behavior,

Fig. 14.4
The egg-laying sequence of the marine snail (*Aplysia*) is shown. The sequence involves (1) expelling a string of egg cases, (2) grasping the egg string by the mouth, (3) waving the head to draw the string out of the duct, and (4) affixing a triangle of string to a solid substance. This behavior was elicited by genetic procedures that activated the gene coding for egg-laying hormone (ELH) and other peptides associated with egg-laying behavior.

Source: Reproduced from R. H. Scheller & R. Axel (1984). How genes control innate behavior. *Scientific American, 250*, pp. 54–62. Copyright 1984 held by the Estate of Ikuyo Tagawa Garbar. Reprinted with permission.

under genetic control (Johnson, Oldroyd, Barron, & Crozier, 2002). The dance guides the travel of other bees (Frisch, 1967) and varies with the particular habitats in which honeybees evolved (Dornhaus & Chittka, 2004). After abundant foraging, a bee returns to the hive and begins to dance while other bees observe the performance in the dark. Subsequently, bees that observed the dance fly directly to the foraging area in a so-called beeline. However, stimulus control by the waggle dance is not entirely rigid, as experienced and successful foragers only briefly observe the dance before departing, perhaps to confirm that the flowers are still yielding forage (Biesmeijer & Seeley, 2005).

The position of the sun with respect to food plays an important role in determining the initial dance. A bee may dance for several hours and during this time the dance changes. These behavioral adjustments occur as the position of the sun with respect to food is altered by the

516 Three Levels of Selection: Evolution, Behavior, and Culture

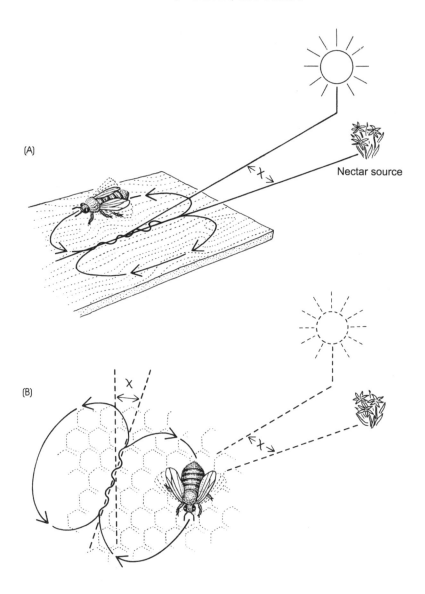

Fig. 14.5
The dance of a honeybee illustrates a phylogenetic form of communication in animals. When the bee returns from a nectar source, the dance begins with the insect waggling its abdomen. The number of waggles and direction of movement control the flight pattern of other bees that observe the performance. The orientation of the food source, relative to the current position of the sun, also is indicated by the waggle dance.

Source: Figure is taken from J. Alcock (1989). *Animal behavior: An evolutionary approach*. Sunderland, MA: Sinauer Associates, p. 207. Copyright 1989 held by Sinauer Associates, Inc. Reprinted with permission.

rotation of the earth. Thus, the bee's dancing corrects for the fact that the sun rises and falls over the course of a day.

The survival and reproductive value of the dance relates to increased food supply for the hive. One problem is accounting for the occurrence of the dance before other bees responded to it—that is, before the dance had survival value. Presumably, the distance and direction that

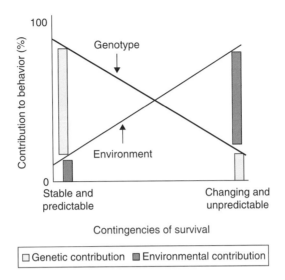

Fig. 14.6
When contingencies of survival are relatively stable and predictable, genetic regulation of behavior is predominant (e.g., fixed-action patterns) and the environment plays a subsidiary role. As contingencies of survival become more uncertain, the role played by the environment and conditioning increases, while direct genetic regulation of behavior declines.

bees traveled had some effect on their behavior. Signs of fatigue and phototropic movements may have varied with distance and the position of the sun when they returned. Bees that evolved sensitivities to what other foragers did could respond to these aspects of behavior—relying on genes that coded for specific neurochemicals. Over time, natural selection favored variations in phototropic (and other) movements that made honeybee dancing more effective. Foraging bees would dance in conspicuous ways that allowed other bees to travel more accurately to the food source (for a similar analysis, see Skinner, 1984, p. 116).

Fixed-action patterns and the communication of bees are examples of behavior that predominantly is regulated by genes and is usually termed species-specific. In both instances, complex sequences of behavior are activated by specific stimuli and carried out in a highly ritualized manner. As shown in Figure 14.6, this form of behavior regulation was selected when the habitat or ecological niche of an animal was relatively stable and predictable.

Behavioral Flexibility (Learning)

When organisms were faced with unpredictable and changing environments, natural selection favored **behavioral flexibility**—adjusting one's behavior on the basis of past experience (alternatively, behavioral flexibility may have allowed for exploitation of novel environments, Leal & Powell, 2012). In this case, genes played a subsidiary role, primarily coding for *general processes of learning*. These processes allowed an organism to adjust to changing environmental requirements throughout its lifespan (see Davies, Krebs, & West, 2012, pp. 18–20, on phenotypic or behavioral plasticity related to climate change and breeding times; epigenetic control of gene expression could also be involved in phenotypic plasticity as described in

Chapters 1 and 6). Flexibility of behavior in turn contributed to the reproductive success of the organism. Skinner noted the reproductive advantage of behavioral flexibility:

> Reproduction under a much wider range of conditions became possible with the evolution of two processes through which individual organisms acquired behavior appropriate to novel environments. Through respondent (Pavlovian) conditioning, responses paired in advance by natural selection could come under the control of new stimuli. Through operant conditioning, new responses could be strengthened (reinforced) by events which immediately followed them.
>
> (Skinner, 1984, p. 477)

In other words, respondent and operant conditioning are general learning processes that are themselves genetically determined.

There is evidence for the selection of conditioning (e.g., Hirsch & McCauley, 1977; Lofdahl, Holliday, & Hirsch, 1992). In a classic experiment, Hirsch and McCauley (1977) showed that the blowfly, *Phormia regina*, could be classically conditioned and that the process of conditioning was heritable. Blowflies can be trained to extend their proboscis (or snout) whenever water is applied to their feet, if they are given sugar that is paired with foot wetting. Even though this conditioned reflex is learned, the process of establishing the reflex can be modified dramatically by artificial selection. Flies varied in the number of elicited responses to the conditioned stimulus on trials 8–15, and were assigned a conditioning score between 0 and 8. Subjects with higher conditioning scores were selected and mated with each other, as were subjects with lower scores. A control group of flies was mated independent of their conditioning scores.

As shown in Figure 14.7, over seven generations, flies selected for conditioning showed increasingly more conditioned responses on test trials than their ancestors. When conditioning was selected against, each generation of flies showed less conditioned responses than the previous population. Flies that were mated regardless of conditioning scores (control) did not show a change over generations. At the end of seven generations, there was no overlap in the distribution of conditioning scores for the three groups—indicating that selection resulted in three separate populations of flies.

The experiment by Hirsch and McCauley (1977) demonstrates that conditioning of a specific reflex has a range of variability. Based on this variation, selection can enhance the process of conditioning or eliminate it for distinct behavioral units. From a behavioral view, contingencies of survival continually mold the degree of behavioral flexibility of organisms—extending (or removing) the process of conditioning to a wide range of responses (see also Lofdahl, Holliday, & Hirsch, 1992 on selective breeding of excitatory conditionability in the fruit fly, *Drosophila melanogaster*).

The presence of genetic variation for learning ability in animals opens the way for experiments asking how, and under what ecological conditions, improved conditioning should evolve. Mery and Kawecki (2002) investigated the experimental evolution of learning ability in *Drosophila melanogaster*. Over 51 generations, experimental populations of flies were exposed to conditions favoring associative conditioning of oviposition substrate choice (choice of a medium—pineapple vs. orange—on which to lay eggs). Flies that learned to associate a chemical stimulus (quinine) with a particular substrate, and still avoided this medium several hours after the stimulus had been removed, were selected for breeding. After 15 generations of selection, the experimental populations showed a marked ability to avoid laying eggs

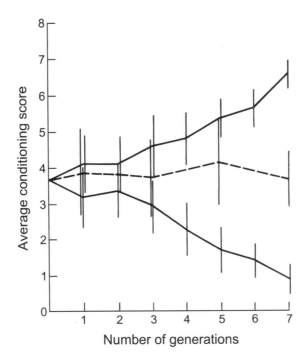

Fig. 14.7
Artificial selection for respondent conditioning in the blowfly, *Phormia regina*, reported by Hirsch and McCauley (1977). Flies mated for high-conditioning scores (solid line upward) showed more conditioned responses over generations than flies that were randomly paired (dashed line). Other flies that were mated on the basis of low-conditioning scores (solid line downward) did progressively worse than their ancestors.

Source: The graph is from J. Hirsch & L. McCauley (1977). Successful replication of, and selective breeding for, classical conditioning in the blowfly (*Phormia regina*). *Animal Behaviour*, 25, pp. 784–785. Copyright held by *Animal Behaviour*. Reprinted with permission.

on substrates that had contained the quinine stimulus several hours earlier. The improved conditioned avoidance was also observed when the flies were faced with a choice between novel substrates. Generally, the study demonstrates that these behavioral changes are caused by artificial selection for rate of learning and retention of the avoidance response following removal of the aversive stimulus (better remembering). One implication is that natural selection would produce similar changes in conditionability and response retention in the everyday world of these organisms (see also Schlichting & Wund, 2014 on the role of phenotypic plasticity and epigenetics in evolution).

Evolution, Behavioral Flexibility, and Body Form

The Cambrian explosion of animal life around 530 million years ago is probably the most spectacular diversification in evolutionary history (Figure 14.8), and there is evidence for the hypothesis that one of the key factors driving this great diversification in life forms was associative learning (Dukas, 2013; Ginsburg & Jablonka, 2010). According to the learning hypothesis, the evolution of associative conditioning required only small modifications in already evolved neural mechanisms. Once the basic conditioning mechanisms appeared on the evolutionary scene, associative learning enabled animals to exploit new niches, promoted new types of behavior

Fig. 14.8

A depiction of the Cambrian explosion around 530 million years ago, involving the relatively rapid appearance of the major animal phyla (divisions) as found in the fossil record. Evidence indicates that before 580 million years ago, most organisms were simple cells living in colonies. Then, over about 70 to 80 million years the rate of evolution exponentially increased and the diversity of life forms began to approximate those of today.

Source: Photograph by D. W. Miller published with permission.

related to predation, and led to adaptive responses fixed through genetic accommodation processes. Ginsburg and Jablonka explained that:

> Organisms with … associative learning had an enormous selective advantage: they were able to adapt ontogenetically to a variety of biotic and abiotic environments and to use new resources. Their learnt behaviors guided where and how they looked for food and protection, how they sought mates, how they reacted to predators and to competitors, and were fundamental to the construction of the niches that they and their offspring inhabited…. For example, if an animal learnt that food is usually available in a particular area and consequently it tended to stay and reproduce there, its offspring would have the same learning environment and learning opportunities. Natural selection would then favor any physiological or morphological features that improved adaptation to this learning environment. The explosion of new behaviors and new ecological opportunities that followed the evolution of associative learning would have been accompanied by an explosion of new, matching, morphological adaptations.
>
> (Ginsburg & Jablonka, 2010, p. 15)

Thus, one possibility is that natural selection for associative learning drove the evolution of animal morphology or body forms, a proposition contrary to the traditional biological view that form determines function (learning).

NEW DIRECTIONS: EPIGENETIC REPROGRAMMING OF SOCIAL BEHAVIOR IN CARPENTER ANTS

Recall that epigenetics is a biological science concerned with changes in *gene expression* during an organism's lifetime without alterations to the gene or DNA sequence. In Chapter 1, we described DNA methylation and histone acetylation as mechanisms at the cellular level that tighten and loosen the chromatin structure surrounding genes, allowing for differences in gene expression (transcription and translation). And we discovered that histone acetylation, involving histone acetyltransferase enzymes or HATs, makes DNA more accessible for transcription—allowing for enhanced retention of learning following fear conditioning (e.g., freezing to a CS previously associated with foot shock, see Chapter 6). Histone deacetylation, on the other hand, involves histone deacetylase (HDAC) enzymes, which keep DNA more tightly wrapped around the histone cores (tight chromatin structure)—making it harder for DNA transcription. Here we describe research published in *Science* by Shelley Berger and her associates on epigenetic reprogramming of social behavior in ants, using laboratory techniques of histone modification with HAT and HDAC *inhibitors* (Simola et al., 2015). The study shows that seemingly innate, caste-specific behavior is actually under control of the ant's environment and epigenetic makeup—changing the individual's epigenome by environmental manipulations (feeding/injecting chemical inhibitors) reprograms its behavior to that of ants from a different caste. (Note: the hypothesized role of DNA methylation in caste-specific behavior differences is currently disputed; see Libbrecht, Oxley, Keller, & Kronauer, 2015.)

For the epigenetic study, researchers used carpenter ants (*C. floridanus*) of two distinct castes called minors and majors (Figure 14.9), which show striking differences in social behavior. Notably, all ants in a colony are sisters, sharing nearly identical genetic makeup—yet ants markedly differ in physical appearance and behavior, depending on caste. Minors are small foragers, searching for and gathering food. In contrast, majors are soldiers with large heads and mandibles used to defeat enemies and transport large pieces of food. Importantly, genes related to brain development and neurotransmission (neuroplasticity) are expressed more in minors than majors—suggesting differential gene expression is involved in the different behaviors of majors and minors.

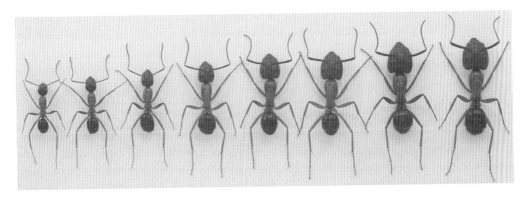

Fig. 14.9

Photograph of Florida carpenter ant workers. The workers vary in form and size from petite minors (far left) to rugged majors (far right). Minor and major ants were used in the epigenetic study described in the textbook.

Source: Photograph by Melanie Couture and Dominic Ouellette from *Science* magazine, "Researchers nearly double the size of workers" by Mitch Leslie, March 11, 2012. Copyright held by the American Association for Advancement of Science (AAAS). Published with permission.

In the *Science* article, the researchers first established that compared to majors, minor ants did most of the foraging and scouting for the colony. Next, minors and majors fed HDAC *inhibitors* (HDACi, class I and II) showed enhanced foraging compared to colony- and age-matched untreated controls; HDACi-treated minors were the first to find food (scouting) whereas majors never scouted. These findings suggested that HDACi treatment allows for greater intrinsic histone acetylation (HAT activation), increasing the foraging-related behavior of ants from both castes, but more so in minors. These behavioral gains in minors subsequently were suppressed by feeding them a CBP HAT inhibitor (CBP HATi), thereby inhibiting histone acetylation of genes having predominantly neuronal functions. [Note: CBP is CREB-binding protein, an intrinsic transcriptional HAT coactivator, having acetyltransferase and transcriptional functions.] As noted, majors seldom engaged in foraging and almost never scouted; however, micro-injections of HDACi into the brains of new adult majors (just emerged from pupal stage) increased the number of scouts from this caste and immediately enhanced foraging to levels typically observed only in minors, with effects of a single injection lasting up to 50 days. Notably, co-injection of CBP HATi suppressed the foraging of majors induced by HDACi treatment—suggesting that CBP HAT activation is critical to the reprogramming of caste-specific behavior in major ants. Whether these histone modifications of majors' behavior are confined to a "developmental window" or can be extended throughout the lifetime is not resolved by this research.

Epigenetic regulation of caste-specific behavior, and the resulting division of labor within the colony, probably had an evolutionary advantage—allowing ants to adapt to sudden environmental shifts within their lifetime. When faced with extended famine or enhanced predation, ants that evolved epigenetic control of neuroplasticity and behavior could adjust the caste ratios (majors to minors) as a rapid survival strategy. For example, under conditions of food scarcity, epigenetic changes during development, involving cellular responses to low food supply, would result in more adult minors that scout and forage for food—increasing the probability of contacting a reliable food source for the colony. Another implication of Simola et al.'s (2015) study is that CBP HAT and HDACs may help to regulate the organized social activities of many species (invertebrates, vertebrates, and mammals) as these enzymes are highly conserved and known to control behavioral plasticity, learning, and memory. Additionally, differences in CBP HAT activity and gene expression by caste may fine-tune the neural pathways in minors, allowing for enhanced foraging as learned behavior. If this is the case, carpenter ants may serve as a viable animal model to further investigate the epigenetics of learning and memory in a relatively simple organism with well-defined behavioral differences (see also Bonasio, 2012).

LEVEL 2: BEHAVIORAL SELECTION BY REINFORCEMENT

The evolution of operant behavior allowed variation and selection to work throughout the lifetime of an organism. Many organisms evolved genetic programs that coded for operant regulation, perhaps building on capacities for associative learning. For some species, natural selection ensured that operant control extended to more and more aspects of behavior. Individuals who inherited an extensive capacity for operant conditioning could adjust to complex and changing, often social, situations on the basis of behavioral consequences. Thus, selection by reinforcement became a major mode of ontogenetic adaptation (Glenn & Field, 1994).

Evolution and Behavioral Dynamics

McDowell (2010; see also McDowell, 2013) outlined an evolutionary theory of behavioral dynamics built on the idea that *behavior is selected by its consequences*. The theory uses three rules—a selection rule, a reproduction rule, and a mutation rule—to generate expected adaptive behavior in single and concurrent-operant situations. Predictions of the theory are then compared with the distribution of responses by live animals (e.g., pigeons). To implement the theory, each behavior in the population of potential behaviors is represented by a 10-bit string of 0s and 1s, extending from [0000000000] to [1111111111]. The 10-bit string is the behavior's *genotype* that can be decoded into a decimal integer between 0 (all 0s) and 1023 (all 1s) that identifies the particular behavior and is conceptually its *phenotype*. In a random draw of 100 potential behaviors from the population of permissible phenotypes (0 to 1023), only a particular range or set of values would succeed in generating reinforcement. For example, all behaviors in the range 513 to 641 (operant class) could potentially operate a lever to produce reinforcement, although each behavior (e.g., phenotype 523) is unique in its topography or some other property.

Once a randomly selected behavior is emitted, a new population of potential behaviors is generated using 100 pairs of parent behaviors. For example, one parent behavior might be the integer 235 [0011101011] and the other parent behavior could be the integer 115 [0001110011]. The method of choosing parents depends on whether the emitted behavior was reinforced. If the emitted behavior is reinforced, the parent behaviors are chosen in accord with the theory's *selection rule* (not outlined here). If the emitted behavior is not reinforced, parent behaviors are chosen at random. After the parent behaviors have been chosen, each mother–father pair is mated in accord with a *reproduction rule* yielding a child behavior 123 [**000**1**11**0**11**] with a 10-bit string composed of bits (0s and 1s) taken from parent #1 behavior (in bold type) and other bits taken from parent #2 behavior. *Mutation* may also occur, as when the child behavior phenotype 123 is changed to 115 by a random change in the seventh bit from 1 to 0 yielding the string [0001110011]. Using the three rules of selection, reproduction, and mutation, the evolutionary theory is used to generate the adaptive behavior of a virtual (computer-generated) organism that can be compared with the actual behavior of organisms in operant settings. Other aspects of the theory and its implementation are beyond the scope of this book.

McDowell's evolutionary theory of behavioral dynamics fits quantitatively with the behavior of live organisms adapting to operant contingencies. The rules of the evolutionary theory when implemented on a computer generate behavioral output consistent with modern matching theory and preference pulses in rapidly changing environments (see Chapter 9; see also Kulubekova, 2012), suggesting that steady-state matching may be a higher-level outcome of Darwinian rules operating in dynamic behavioral systems (McDowell & Popa, 2010). One issue is how to tie the theory of adaptive behavior to Darwinian principles at the level of neural mechanisms. Referring to Edelman's theory of selection of neuronal groups (Edelman, 1987, 2007), McDowell noted that:

> The bit, the bit sequence, the bit string, and the bit string class [of my theory] are realized in [Edelman's theory] as a neuron, a synapse, a neuronal group, and a collection of degenerate neuronal groups [all having the same function]. In addition, the action of selection, which is carried out formally by the parental selection function in the [evolutionary theory of behavior dynamics] is realized in [Edelman's theory] by the operation of diffuse value

systems [such as the dopamine pathways] in the brain that alter synaptic strengths or thresholds.

(McDowell, 2010, p. 364)

One future possibility is that Darwinian principles may allow for a more comprehensive theory of brain function and behavior—a unified theory based on Skinner's notion of selection by consequences.

Selection for Operant Processes

Sigrid Glenn noted the biological advantage of operant processes and selection by behavioral consequences:

> The instability of local environments and what might be considered a limit on behavioral complexity in genetic programs appears to have given rise to a less cumbersome and more rapid sort of variation and selection. Instead of building thousands of complex behavioral relations into DNA, evolution built a few programs for behavioral processes that allowed changing environments to build behavior repertoires "as needed" during the lifetime of individuals. A relatively small change in a bit of DNA could result in profound changes in the possibility for ontogenetic adaptability if that change involved a gene for a behavioral process. All that was required as a first step was genetically uncommitted activity and susceptibility of that activity to selection by behavioral consequences.

(Glenn, 1991, p. 43)

The evolution of operant conditioning, a range of uncommitted behavior, and susceptibility to certain kinds of reinforcement resulted in a second level of selection. Behavioral selection supplemented and extended selection by consequences at the evolutionary level (i.e., natural selection).

FOCUS ON: OPERANT REGULATION IN THE MARINE SNAIL, *APLYSIA*

An example of how behavioral selection supplements the genetic control of behavior is seen in the marine snail, *Aplysia* (Figure 14.10) (see "Focus On: Genetic Control of a Fixed-Action Pattern" in this chapter). This simple organism serves as a model system to investigate the regulation of operant feeding (Brembs, Lorenzetti, Reyes, Baxter, & Byrne, 2002; see also operant conditioning by heat in the fruit fly, *Drosophila melanogaster*, related to the *ignorant* gene; the role of the ancestral form of the FoxP gene in fruit flies' motor learning, related to vocal learning in songbirds and language in humans; evidence in the *Drosophila* that operant and respondent conditioning are distinct, involving two largely separate molecular processes; and the role of the behavior-initiating neuron in operant conditioning of the great pond snail, *Lymnaea stagnalis*; Brembs, 2003, 2008, 2011; Mendoza et al., 2014).

Intact, freely behaving snails with extracellular electrodes in the buccal ganglia were observed during food ingestion. Researchers recorded the neural activity that previously accompanied eating and delivered it to the buccal ganglia contingent on spontaneous biting (no food present). This procedure increased biting during the session and also when tested 24 h later. These observations showed operant learning supported by neural activity (biting reinforced by

Fig. 14.10
Photograph of the marine snail, *Aplysia*, used to study the role of operant processes in the regulation of feeding. Intact, freely behaving snails with extracellular electrodes in the buccal ganglia were observed during food ingestion. Recorded neural activity that accompanied eating was delivered contingent on spontaneous biting (no food present), and biting increased during the session, and even occurred at some level when tested 24 h later.

Source: Common access provided by *Wikimedia*.

eating-related neural activity), and "memory" or retention of that learning (reoccurrence sometime later) in the intact animal.

Subsequently, the researchers removed the buccal ganglia from trained and untrained (yoked control) snails and assessed sensitivity to depolarizing current. Results showed that buccal-motor patterns became ingestion-like as a result of training. Finally, one cell called B51, which is active during feeding (and could be the site of operant "memory"), received a brief puff of dopamine contingent upon depolarization and membrane changes. A significant decrease occurred in the B51's threshold compared with control cells, indicating an operant reinforcement effect at the cellular level. That is, the contingent dopamine served to differentially alter the cell's activity so that it produced ingestion-like motor patterns.

The research on operant conditioning of *Aplysia* indicates that *operant selection by consequences* is highly conserved over species at neural and behavioral levels. Clearly, the neural network and single-cell activity are modifiable by the effects of what happens after activation. As in any operant, the consequences at the neural and cellular levels affect the reoccurrence of the neuronal activity, and thereby contribute to the regulation of feeding behavior in *Aplysia*. In addition, the work on *Aplysia* illustrates the continuum of behavioral regulation based on both genetic and environmental influences. That is, analysis of egg laying and feeding by *Aplysia* suggests regulation of behavior by gene-activated hormones and neurotransmitters combined with operant conditioning at the neurocellular level.

Operant Selection and Extinction

The unit of selection at the behavioral level is the operant. The operant is a functional unit of behavior. At the neurocellular level, operants involve increased activation, integration, and consolidation of neurons. Thus, the neuron is the physical unit on which behavioral selection works (see Chapter 4 for conditioning the neuron; see Donahoe, 2002 and Guerra & Silva, 2010 for neural-network analysis of conditioning; see Stein & Belluzzi, 2014 for operant conditioning of individual neurons; see Brembs et al., 2002 for operant conditioning of feeding in *Aplysia* at the neurocellular level). An operant is composed of variations in response forms that make contact with the environment, an operant class. Response forms vary from moment to moment, and some variations change the environment in ways that increase those forms (selection by consequences). A child who manipulates a rubber duck in the bathtub may inadvertently squeeze it in ways that produce a squeaking sound. If the sound functions as reinforcement, those ways of squeezing that produce squeaking increases over time. Recall that reinforcers are defined not by what they are (squeaking rubber duck), but by their effects on behavior (selection of response variants by consequences).

If few (or no) response variations are reinforced, the operant decreases and the process is extinction. Thus, all members of an operant class cease to exist when variants no longer result in reinforcement. The sound device in the toy duckling may break, and squeezing it in different ways no longer has the characteristic effect. Over time, the child would squeeze the rubber duck less and less as the operant class undergoes extinction.

Extinction, as we have seen in Chapter 4, not only eliminates operants but also generates behavioral variation. Greater variation in the behavior increases an individual's chances of contacting the prevailing contingencies—often critical to survival and reproductive success. In the bathtub, the child may push the broken rubber duck under the water, emitting a response that has never occurred before. The effect of this behavior may be to generate bubbles on the surface that, in turn, reinforce the child's novel behavior.

A more profound example of extinction and behavioral variation concerns people trying new ways of doing things when old ways no longer work (or do not work well). Thomas Edison's invention of the electric light bulb involved behavioral variation and selection. To generate electric light, Edison collected and tested a variety of materials to produce an effective lamp filament (Etzkowitz, 1992, p. 1005). He was known as the trial-and-error inventor, but a better description of his performance is "trial and success." Invention occurs when novel forms of response (trying different filaments) are generated by extinction and the appropriate response (using a tungsten filament) has been selected by the prevailing contingencies of reinforcement (effective and efficient light). Creativity, originality, or invention can be defined as non-repetitive perseverance, and control by operant contingencies contributes to such behavior.

Susceptibility to Reinforcement

Contingencies of reinforcement resemble contingencies of survival (Skinner, 1984). Many animals eat and copulate simply because these responses have contributed to survival and reproduction. Male black-widow spiders copulate and are then eaten by their mates. For these animals, copulating only had survival value—passing on the genetic code even though the individual dies following the act. Other organisms evolved sensory systems that allowed food and sexual contact to reinforce behavior. That is, animals whose actions resulted in sexual contact were more likely to act that way again. At this point, organisms had two redundant reasons for eating and copulating—genetic fitness and reinforcement.

When food and sexual contact became reinforcing, new forms of behavior often tangentially related to eating and copulating could be established. Animals could acquire new ways of finding, gathering, and processing foods based on reinforcement. Similarly, sexual reinforcement could establish and maintain a diversity of actions. These include looking at erotic objects, seeking out a diversity of sexual partners, attracting a desirable mate, and performing a variety of sexual responses (e.g., genital contact with different parts of body or position of intercourse).

Susceptibility to reinforcement may sometimes depend on the species and the particular behavior. Species-specific reinforcement is shown by the fact that chaffinches (*Fringilla coelebs*) will peck a disk for food, but not for contingent presentation of birdsong. However, the same birds will step on a perch for species-specific song, suggesting the biological preparedness of the response–reinforcer relationship (Hinde & Stevenson-Hinde, 1973; see also Chapter 7).

Primates also may be susceptible to species-specific reinforcement. The work of Harlow and Zimmerman (1959) on mother–infant attachment suggests that "contact comfort" may function as reinforcement for infants staying close to and preferring their mothers. Infants who only received food reinforcement from their mothers did not show strong attachment behavior. These findings again suggest that the response–reinforcer relations is biologically prepared, especially in many social species.

Organisms that are susceptible to reinforcement may acquire behavior that is not adaptive (Pierce & Epling, 1988). One paradoxical by-product of natural selection for operant conditioning is that people sometimes behave in ways that have immediate benefit for survival, but may be harmful in the long run of today's world. Humans choose foods that are not healthful and engage in sexual behavior that may lead to an unplanned pregnancy or a sexually transmitted disease. While conditioned reinforcement helped our species survive and reproduce, one can see that in today's world when food is plentiful and survival in most parts of the world is highly likely, conditioned reinforcement is no longer necessarily linked to reproductive success. People learn to use birth control, love adopted children, and risk their lives to help others. The point is that susceptibility to reinforcement has been evolutionarily adaptive, but this sensitivity may generate behavior that serves other cultural purposes, such as recreation and entertainment.

EVOLUTION, REINFORCEMENT, AND VERBAL BEHAVIOR

Social Signals

As we noted earlier, a honeybee signals the location of food by dancing in ways that affect the travel of other bees. This form of communication involves a high degree of genetic regulation. Genes, as we have seen, also code for the general behavioral processes known as respondent and operant conditioning. Once these learning capacities evolved, signaling and responses to signals could be acquired on the basis of an organism's interaction with the environment.

The acquisition of human gestures in terms of selection by consequence are good examples. Gesturing is behavior that results from social contingencies of reinforcement.

Fig. 14.11

Photograph shows a gesture for "come here and see this."
Source: iStock.

A social contingency involves the behavior of two (or more) people who arrange stimuli and reinforcement for each other's actions (Skinner, 1986). The person who sees a surprising sight may physically pull a companion toward the view and be reinforced by their friend's reactions to the sight. On later occasions, a pulling motion with the arm or a smaller hand gesture (see Figure 14.11) may occur before the companion is within reach. The friend may avoid being pulled to the sight by coming when the gesture is first made. The reinforcement contingencies composed of each person's behavior establish and maintain this social episode.

Although social contingencies are clearly involved in human signs and gestures, other processes may play an important role. The research on stimulus equivalence discussed in Chapter 12 is relevant to signs and gestures (Dickins & Dickins, 2001). Humans easily distinguish equivalent stimulus classes, but other organisms do not. Gestures and signs may stand for or be equivalent to other stimulus classes. A smile, the spoken words "good job," and a thumbs-up gesture become equivalent when they have a similar effect on behavior. Equivalence relations depend on discrimination of reflexivity (A = A and B = B), symmetry (if A = B then B = A), and transitivity (if A = B and A = C, then B = C). Complex transitivity or other derived stimulus relations seem to involve evolution of species-specific capacities for discrimination as well as general behavioral processes such as operant conditioning.

Humans readily generate and respond to iconic or representational signs when communication is required but speaking is not possible. For example, Brown (1986) recounts a study that examined children with deafness whose parents refused to use American Sign Language (they believed that signing would reduce the odds of acquiring vocal speech) (Goldin-Meadow & Morford, 1985). Each of the 10 children with deafness independently developed on their own a similar repertoire of gesture-based iconic signs to communicate with their parents. Presumably, some particular ways of signing were more effective than others in altering the behavior of the parents. These children were also compared to three younger children (not deaf), who were learning to speak. The children could hear, and they also showed iconic signing that gradually diminished as their own vocal speech increased. This later finding suggests that speech has some advantages over gestures and iconic signs when speakers and listeners have normal-hearing abilities (Goldin-Meadow & Wagner, 2005).

Corballis (1999, 2003) reviewed several perspectives dealing with the probable evolution of language from manual gestures, hand to mouth (see also Gillespie-Lynch, Greenfield, Lyn, & Savage-Rumbaugh, 2014 on gestures, symbols, and evolution of language in apes and humans). He indicated that with the emergence of the genus *Homo*, which involved increased brain size and adequate vocal apparatus, contingencies could have further promoted the development of speech and language. In this regard, it is useful to provide a behavior analysis of the evolution of speech sounds.

Vocal and Speech Sounds

Natural selection must have been important in the evolution of vocalization and speech sounds. Compared with gestures and iconic signs, sounds can affect a listener's behavior when it is too dark to see, others are out of sight, or no one is looking at the speaker. Spoken sounds also have an advantage for speakers whose hands are full—whether they are warding off prey or holding weapons to attack an enemy.

Most of the organs that allowed for speech sounds probably evolved for other reasons (see Chapter 12). The diaphragm was used in breathing, the tongue and jaws were involved in eating, and the lips could take in water by sucking and sipping. The vocal cords and pharynx did not play a direct role in survival, but may have evolved in social species that could benefit from the calls and cries of others (see Blumstein & Armitage, 1997 on social alarms in marmots; see also Pollard & Blumstein, 2012 for a review of evolution and communication complexity in ground squirrels, prairie dogs, and marmots).

There were probably several other important steps in the evolution of human speech. One involved the extension of operant processes to a range of speech-relevant behavior. Each organ that contributed to speech was initially reflexive—the organism responding to specific stimulation. Survival must have been better served when operant processes supplemented reflexive behavior. An organism could breathe as a reflex elicited by high levels of circulating carbon dioxide, or it could hold its breath to avoid a predator. Based on natural selection, more and more speech-relevant behavior came under the control of its consequences. Compared with the great apes, humans made an evolutionary leap when the vocal apparatus was supplied with nerves (i.e., innervated) for operant regulation, as described in Chapter 12.

The step to operant regulation of the vocal musculature is not sufficient to account for speech. Evolution must also have resulted in the coordination of all the systems involved in the production of speech. The great apes have complete operant control of their hands, but have not developed a sophisticated system of signs, gestures, or symbols. Children show early iconic signing that shifts toward spoken words as more and more speech is acquired. Both iconic signing and spoken words require that the speaker and listener respond to abstract stimulus relations along several dimensions. Thus, neural coordination of speech probably built on, and added to, specialized capacities for discrimination involving the visual, auditory, and motor systems. In less technical terms, humans evolved systems for symbolic behavior and these systems were eventually integrated with those of speech (Pierce & Epling, 1988).

Speech sounds constitute a limitless pool of uncommitted behavior. This behavior is spontaneously emitted at high frequency, but plays no direct role in survival (Skinner, 1986). From a behavioral view, wide variation in spontaneous speech sounds allowed for selection of vocal operants by reinforcement supplied by listeners. Thus, Osgood (1953) found that an infant's babbling included all of the speech sounds that make up the different languages of the world. Vocal responses similar to the verbal community increase, while dissimilar speech drops out of the repertoire. One possibility is that a child's speech sounds are shaped toward adult forms by listeners' reinforcement of successive approximations, as well as by repetition and modeling of adult speech by listeners (Risley & Hart, 2006).

Goldstein, King, and West (2003) manipulated mothers' reactions to infants' vocalizations and found that phonological aspects of babbling increase with contingent social stimulation from mothers (reinforcement), but not with noncontingent maternal behavior. Social shaping by reinforcement creates rapid shifts in infant vocalization to more advanced speech sounds. Interestingly, birdsong of juvenile male cowbirds is shaped toward mature forms used in mating by subtle reactions of females called wing strokes (West & King, 1988). Thus, social shaping by reinforcement is probably a general mechanism for the development of adult vocal forms of communication across diverse vocal species.

Verbal Behavior

The evolution of operant processes, the coordination of speech systems, and a large variety of uncommitted speech sounds allowed for the regulation of vocal operants by others. A person in an English-speaking community learns to speak in accord with the verbal practices of the community. Thus, the way a person speaks is shaped by the reinforcement practices of others. On a specific occasion, the community provides reinforcement for certain ways of speaking and withholds reinforcement or supplies aversive stimulation for other unacceptable responses. In this manner, the individual eventually conforms to the customary practices of the community and, in so doing, contributes to the perpetuation of the culture.

Verbal behavior (see Chapter 12) allows people to coordinate their actions. When people observe rules, take advice, heed warnings, and follow instructions, their behavior is rule governed. Rule-governed behavior (see Chapter 11) allows people to profit from what others say. If a fellow camper reports that a bear is near your tent, you can move the tent to a different camping site. A student looking for a good course may benefit from the advice of another student. In these examples, the listener or person who responds to the verbal report avoids an aversive event (the bear) or contacts positive reinforcement (a good course). Children are taught to follow advice and instructions. Parents and others provide simple verbal stimuli that set the occasion for reinforcement of the child's compliance. In this way, the child is taught to listen to what others say.

As we have noted, listeners benefit from the verbal reports of others. For this reason, listeners are inclined to reinforce the person who provides useful instructions. In a verbal community, people are taught to express their appreciation for the advice received from others. For example, in an English-speaking community, people say "Thank you" and other variations of this response when given directions, advice, or instructions. These verbal responses by the listener reinforce the behavior of the speaker.

Verbal behavior evolved (level 2) in the sense that particular ways of speaking were more or less effective in regulating the behavior of listeners within a verbal community. Response variation and selection ensured that many ways of speaking were tried and more and more people adopted successful combinations of speech sounds.

Speaking by Whistling: Ecology and Verbal Contingencies

How people speak depends on the verbal contingencies of reinforcement. Ecological and geographic aspects of the environment (topography of the terrain) alter verbal contingencies, in the sense that listeners cannot reinforce the verbal responses of speakers when they cannot hear and understand what is being said. Under these conditions, new ways of speaking and listening should evolve, and this is what happened on La Gomera, one of the Canary Islands in the Atlantic off the coast of North Africa (Classe, 1957; see also Meyer, 2015 on whistle speech throughout the world).

The island has a harsh terrain consisting of high mountain peaks and low ravines, gorges, and valleys, which separate the hamlets of the Gomeros, a Spanish-speaking people

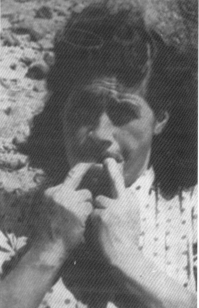

Fig. 14.12
Photograph of the rugged landscape (left) of La Gomera that favored the use of whistle speech or *silbo* by the islanders. A Gomera woman is shown (right) using her fingers and articulation of the tongue to whistle in Spanish. The major function of whistled speech is that it can be clearly heard and understood by villagers several miles away—overcoming the communication obstacles set by the disjointed terrain of the island.

Source: A. Classe (1957). The whistled language of La Gomera. *Scientific American*, *196*, pp. 111–120. Published with permission of Nature Publishing Group.

(Figure 14.12). A short distance of 500 yards may take an hour or so to walk on the rough paths, which are not much more than goat trails. To say the least, communication over a distance has been a central problem to people on the island. Long ago the Gomeros began to use a whistled speech, which allowed them to speak to each other over extended distances across the disjoined landscape. The language is called *silbo* (Spanish for "whistle"), and it is not just a signal system, but is considered to be a version of the Spanish language. Whistled Spanish uses fingers positioned in the mouth to facilitate the whistle and articulation by the tongue to produce verbal responses, as in ordinary speech. Its main utility is that the whistles can be heard and clearly understood over great distances—much farther than shouting in everyday Spanish. On a good day without wind, a listener more than a mile away can understand the *silbo* Spanish of the speaker. A skillful *silbador* can whistle messages that travel 3 miles or more, and the record is said to be up to 8 miles (see Meyer, 2007 for further analysis of whistled languages).

Speaking by whistling shows that verbal behavior evolves and adapts to the changing contingencies of reinforcement (level 2 selection). In the everyday world of Spanish speakers, talking with words allows for reinforcement by listeners. For the Gomeros, the rugged landscape has altered the usual verbal contingencies, and whistling Spanish has evolved as a way of solving the problem. The customary way of speaking, by means of words or whistles, refers to the common linguistic practices of a verbal community—the common standards for speech (a grammar). These verbal practices served as the underlying basis for a third level of selection (level 3 selection), namely the selection and evolution of cultural practices.

LEVEL 3: SELECTION AND EVOLUTION OF CULTURE

The evolution of operant processes and verbal behavior allowed for the emergence of human culture. Sigrid Glenn (1988, 1989, 2004) proposed a behavior analysis of culture, building on the works of Skinner (1953) and anthropologist Marvin Harris (1979). Social scientists often talk about culture as the ideas and values of a group. In contrast, a behavior analysis suggests that a culture involves the usual ways of acting and speaking in a community. These customary forms of behavior (customs and mores) are the cultural practices of the group.

Cultural Practice

From a behavioral perspective, cultural practices involve the interlocking operant behavior of many people—the members of a culture. Each person's behavior provides stimulation and reinforcement for the actions of others. A **cultural practice** is therefore defined in terms of interlocking social contingencies—where the behavior of each person supports the behavior of other members of the community. The pattern of behavior that arises from the *interlocking contingencies* is the type of practice (e.g., what people do in that culture).

This view of culture suggests that what people do in a particular community is determined by the function of a practice. The ancient Romans adopted military tactics that were highly effective in most battles. For example, Roman soldiers would form a close body of men, called a *phalanx*, and interlock their shields as a common barrier against the enemy. Although there are many ways to conduct a battle, this military maneuver became popular because of its effectiveness. In other words, what people in a particular culture do is a function of the previous benefits and costs of that practice. With changes in technology (the products of a culture), the phalanx and the interlocking of shields became obsolete—the costs in terms of casualties and lost battles increased relative to the benefits of this military maneuver.

Cultural practices for the group are functionally similar to operants for individuals. Both operants and cultural practices are selected by consequences, but at different levels of selection (Lamal, 1997). Thus, a cultural practice increases when people have benefited from it. The practice of making water jars involves alternative sets of interlocking operants that resulted in a common outcome. One person gathers clay, another person makes the pot, and a consumer trades something for the jar. The common outcome of such a practice is greater efficiency in transporting and storing water. There are many ways of storing and transporting water, including shells, hollow leaves, woven baskets, clay pots, and indoor plumbing. The cultural form that predominates (e.g., plumbing in North America) reflects the basic processes of selection by consequences. In terms of selection, operants are selected by contingencies of reinforcement and cultural practices are selected by metacontingencies (see Norton, 1997 on geographic practices, rule-governed behavior, and metacontingencies).

Metacontingencies and Cultural Practices

Glenn (Figure 14.13) made important contributions to the behavior analysis of culture when she first described the metacontingencies of cultural practices. **Metacontingencies** refer to contingent relations between practices, involving *interlocking behavioral contingencies* (IBCs), and the effects or consequences of those practices for the group (Glenn, 1988, 2004). For example, to free a car stuck in a snow drift, the driver steers and accelerates while two passengers push from behind (IBC). The IBC, or the division of cooperative behavior between driver and passengers, satisfies the metacontingency (everyone gets to work on time).

In one study of metacontingencies, participants played a game where the group had to pick from an 8 × 8 matrix with plus (+) or minus (−) signs equally and randomly distributed in the cells (Vichi, Andery, & Glenn, 2009). Participants received chips eventually exchanged for

Fig. 14.13
Sigrid Glenn.
Published with permission.

money at the end of a session. Each round of the game involved participants betting chips by placing them in a box. The group then agreed on a row number (1–8) and the experimenter selected a column by a criterion unrevealed to the participants. An intersection of the row and column (cell) with a plus sign (+) resulted in the experimenter doubling the chips in a betting box; a minus sign (–) resulted in subtraction of half the chips in the box. Next, the participants had to deposit some chips to a common pool or "savings" box (collective decision) and divide the rest among the individual members. The chips in the common pool were distributed by the group members based on a collective agreement and exchanged for money at the end of the session.

The choice of a column by the experimenter was based on how the chips were divided among the members on the previous round—either equally or unequally. The division of chips (equal or unequal) by the group members, or the *interlocking behavioral contingency*, served as the dependent variable. Two independent contingencies were arranged by the experimenter. In condition A, the experimenter arranged plus signs (+) when chips were *equally* divided in the previous round. For condition B, plus signs were given when chips were *unequally* divided among participants. Conditions A and B were reversed after 10 consecutive winning rounds (A-B-A-B design). Overall, the results showed that the collective interlocking behavioral contingency (equal or unequal distribution of resources) increased with the contingent outcomes (wins or losses) arranged for these practices—the metacontingencies (see also Ortu, Becker, Woelz, & Glen, 2012 for regulation of interlocking behavioral contingencies (practices) by metacontingencies in an iterated prisoner's dilemma game—even when the group practices reduce individual gains).

In the context of American education, the concept of metacontingency implies selection of instructional strategies that result in more and better trained students of science, but this may not occur. In complex cultures such as the USA, competing (or concurrent) metacontingencies often mean that the "best" educational practice is not selected. A less than optimal form of scientific education may prevail for some time because teaching science is only part of the function of education. For example, the manifest function of education is to teach reading, writing, and arithmetic. The hidden or latent function of schooling includes keeping people out of the workforce and categorizing them into high-, medium-, and low-status groups based on educational attainment. Thus, the form of education that predominates is one that has produced the most overall benefit to the community, group, or society. If the relative outcomes of an educational practice resulting in low scientific competence exceed those of a system that yields high scientific achievement, then the less adequate educational practice would predominate in the culture (see Lamal & Greenspoon, 1992 on metacontingencies and the US Congress; see also Lamal, 1997). In fact, in Chapter 13, we saw that behavioral methods for teaching (PSI and precision teaching) were more effective than traditional methods, but the latter were retained by the American educational system.

FOCUS ON: MACROCONTINGENCY, DEPLETING RESOURCES, AND COSTLY USE OF PUNISHMENT

Metacontingencies involve selection by consequences of the interlocking behavior of group members—the selection of cultural practices. When the operant behavior of multiple individuals generates a *cumulative effect* (sum of the individual effects) for the group, without an explicit interlocking behavioral contingency, we may describe this relation as a **macrocontingency** (Borba, Tourinho, & Glenn, 2014; Malott & Glenn, 2006). One type of macrocontingency concerns the ecological costs arising from the overuse of limited resources—involving widespread consumption of forests, pastures, air, energy, fisheries, and water systems (see Platt, 1973 for a discussion of other kinds of macrocontingencies or social traps). In his classic article on "The Tragedy of the Commons," Hardin (1968) described a macrocontingency involving the use of a public-grazing area for cows called the "commons." Farmers are free to use the commons to feed cows and increase profit, but the more cows each farmer feeds the faster the grazing area is depleted (given a fixed rate of resource repletion), until there is no grass left and no more grazing. Notice that this kind of social problem involves the behavior of multiple individuals each making repeated choices between short-term and long-term contingencies of reinforcement (see Chapter 9 on choice and self-control for similar contingencies). Hardin's analysis suggested that individuals acting independently and pursuing their own self-interest (immediate reinforcement) would eventually deplete the limited resource (cumulative effect), even when it is not in anyone's long-term interest to do so. In America today, burning of fossil fuels and pollution is an example of the limited-resource macrocontingency. Each user (or company) can dump the waste products from burning oil and gas from cars and households into the atmosphere for short-term individual gains, but the behavior of multiple users destroys the air quality and raises the global temperature—long-term cumulative effects that are in no one's best interest.

Experimental Analysis of Resource Management

In the laboratory, it is possible to arrange contingencies for resource management and experimentally analyze the conditions leading to overuse rather than conservation of limited resources. In an early study, schedules of reinforcement with immediate and delayed consequences were arranged for resource use (Brechner, 1977). Student participants in groups of three responded individually for points exchanged toward class credit. Every 10 responses (FR 10) added one point to each student's total, but subtracted a point from a *common pool* of points (limited resource) that replenished at a fixed rate. If the three participants responded for points faster than the replenishment rate, the pool depleted and the experiment ended before the students could obtain the maximum class credit. The researcher also varied the maximum size of the pool and the opportunity for the students to communicate. The results indicated that participants depleted the common resource when the pool was small and when users could not agree through communication to reduce the rate of resource use (this research on resource use is related to experiments on social cooperation and exchange; Fehr & Gintis, 2007).

Effects of Communication and Costly Use of Punishment

The effects of communication and costly use of punishment (using punishment at a personal cost) on resource management were investigated with a computer simulation, as reported in the journal *Science* (Janssen, Holahan, Lee, & Ostrom, 2010). The experiment involved five participants (resource users) gathering or harvesting tokens from a *common pool* of tokens (stars) distributed over a computer-simulated grid of cells. Participants collected tokens from the pool by pressing the arrow keys to move their avatars around the screen, and pressed the space bar to collect each token (worth 2 cents) during each decision period. The resource rapidly depleted when everyone gathered tokens for themselves as quickly as possible without considering the long-term cumulative effects; participants who pursued only immediate reinforcement sat looking at a dark screen not making any money. On the other hand, if participants restrain themselves by maintaining at least 50% of the tokens on the screen, they could triple their earnings. In periods when use of punishment was allowed, participants could subtract two tokens from another user at the cost of one of their own tokens, as long as they had funds in their account (costly punishment). Written communication, when allowed, involved text messages in a "chat room" before a decision period. Each experimental condition consisted of three consecutive 4-min periods of costly punishment (P), communication (C), or a combination of both (CP), and three consecutive 4-min periods when neither communication nor punishment (NCP) was allowed.

When the experiment started with NCP, the participants depleted the resource (pool of tokens) within about 90 s, showing that without punishment and communication the participants responded in terms of immediate reinforcement, as expected by Hardin's analysis of the "tragedy of the commons." In contrast, an opportunity to communicate reduced the rate of collecting tokens by each group member and increased the pool of tokens (and profit) for all. Participants' messages in C periods focused on the timing and location of the tokens. A common message was to refrain from gathering tokens for a set length of time, to allow the resource to regenerate. When participants started the game with C periods, the amount of earnings did not drop off when participants were no longer able to communicate, upholding their prior agreements.

When P without communication was introduced after NCP (or at the start of the experiment), gross earnings dropped, as the contingency of punishment was not specific without a preceding verbal stimulus. That is, removal of tokens (punishment) without prior messages did not stipulate what behavior produced the loss—collecting too many tokens, gathering tokens in a particular location, collecting tokens in a specific pattern, or moving the avatar too quickly on the screen. For P phases, participants did not use punishment in half of the periods when it was allowed. Without communication, they were reluctant to punish token gathering by others based on the threat of retaliation. When participants in P conditions did use punishment to control token collecting, going too fast or having too many tokens were given as reasons, which matched with the data record of punishment events actually delivered.

Interestingly, communication with punishment (CP) did not lead to as long-lasting cooperative behavior as communication without punishment (C). It appears that *the use of punishment erodes prior cooperative agreements*, which are then less likely to persist when punishment and communication are withdrawn. For CP periods, the main reason participants gave for punishing others, when they did so, was for not following the agreements. In addition to using costly punishment, participants in CP periods would use messages to scold others who were viewed as free riders, taking advantage of the increased tokens in the pool, but not contributing themselves.

These experiments confirm that people reluctantly use punishment to manage the uncooperative behavior of others, even when it is costly for them to do so (costly punishment). The use of punishment, however, does not solve the problem of resource overuse, and often results in high collective losses. When communication is allowed, users make agreements to manage overuse, increase gains, and continue to manage resources even when communication is no longer possible. These gains are not sustained when punishment is used and communication is withdrawn, presumably because punishment erodes interpersonal trust (i.e., counting on others to do their part or contribute their fair share; Molm, Takahashi, & Peterson, 2000). At the practical level, conservation of limited resources should not improve when users impose fines and taxes (costly punishment) for overuse or misuse of resources, even when accompanied by reasonable pleas and agreements for careful resource management. Negotiated agreements to limit resource use without costly punishment, however, should be helpful in overcoming the contingencies for resource use, at least until individual users experience the long-term cumulative benefits of sound resource management.

ORIGIN, TRANSMISSION, AND EVOLUTION OF CULTURAL PRACTICES

Cultural practices arise and are transmitted over generations. Galef and Allen (1995) showed how diet preferences could be established, diffused, and socially transmitted by rats. Founder colonies of four rats were taught an arbitrary food preference. Subsequently, members of the colonies were slowly replaced with naive rats that had no food preferences. The food preference was still maintained in the third generation of replacements, although none of these animals had received the original training. Thus, a food practice established by arbitrary nonsocial contingencies was transmitted and maintained by implicit social contingencies of the group.

Another example comes from a story about animals in a zoo enclosure. A troop of baboons was provided with a choice between a preferred food (bananas) and a less appetizing laboratory chow (Pierce, 1991). As expected, the baboons consistently chose to eat bananas. Following a baseline period, the researchers established a negative reinforcement contingency for eating the less preferred food. Whenever any animal approached the bananas, the entire colony was drenched with water from a hose that was used to clean the enclosure. After exposure to this contingency, the troop attacked any member that approached the bananas. Eventually, all members of the troop were exclusively eating the less preferred chow and avoiding cold showers.

The researchers then removed the reinforcement contingency—approaching and eating the bananas no longer resulted in being soaked with water. As you might expect, the group did not test the operating contingencies and continued to attack any member that went toward the preferred food. At this point, the contingencies had established a *cultural taboo* that was highly resistant to change. Thus, social contingencies and observational learning contributed to the maintenance of the food taboo even though the original negative reinforcement contingencies had long since been removed.

Harris (1974) provided a functional analysis of the origin and transmission of many human cultural practices. To illustrate, in India the cow is deified and many Hindus do not eat beef. This was not always the case—when the last ice age ended the peoples of Northern India raised and ate cattle, sheep, goats, and many agricultural products. Cattle, however, have some advantages other than just providing a source of meat; they may be easily herded and trained to pull plows or carts.

The population density increased greatly in the Ganges River valley, and by 300 BC the people of the valley had destroyed the trees surrounding the river. As a result, the risk of drought increased and farms decreased in size. Small farms have little space for animals, but draft animals were essential for working the land and transporting agricultural products. Cows provided traction, milk, and meat, but the farmer who ate his cow lost milk production and a working animal. Thus, the people of India faced a social trap involving the immediate benefit of eating beef and the long-term loss of the cows' other advantages.

A cost–benefit analysis suggests that it was better to keep a cow than to eat it. To avoid this social trap, the cow was deified and eating beef became a cultural taboo. The Hindu community has maintained this practice right up to modern times. Other cultures have food taboos that may be analyzed in terms of the function of cultural practices. Until very recently, Catholics did not eat meat on Fridays, many Islamic and Jewish people will not eat pork, and Chinese people do not drink cow milk (Harris, 1974; see Kangas, 2007 on Marvin Harris's cultural materialism and its ties to behavior analysis; see also Ward, Eastman, & Ninness, 2009 for an experimental analysis of cultural materialism).

Cultural Evolution

Cultural evolution presumably begins at the level of the individual where technological effects reinforce variation in individual behavior. An inventor may discover a new way of making a wheel, a farmer finds a food crop that produces higher yields, and a teacher may find a novel way to teach reading. A culture is said to evolve when the community adopts these innovative practices.

Adoption of innovations depends on the metacontingencies facing the group. For example, a new food crop with higher yield is selected when the metacontingencies favor increased grain production. This could occur when a community is impoverished or when higher-yielding crops support the feeding of domestic animals used for work or consumption. Higher-yield crops may not be selected when food is overly abundant, when increased grain supply leads to problems of storage, or when a new crop attracts pests that spread disease.

Adoption of Novel Behavior
A troop of Japanese macaques on Koshima Island is well known for its innovations and traditions (Kawamura, 1959). In one example, an infant female called Imo began to wash sweet potatoes to remove the grit. This behavior was later observed in Imo's playmates and her mother, who taught it to another offspring. Imo was also among the first to take up swimming and to throw wheat kernels on the water. Throwing wheat on the water removed the sand that was mixed with the kernels, because the sand sank and the kernels floated. Both of these practices were eventually adopted by the entire troop (Tomasello, 2009; Whiten & Boesch, 2001; see Whiten, McGuigan, Marshall-Pescini, & Hopper, 2009 for social learning, imitation, and chimpanzee culture; see also Whiten, Hinde, Laland, & Stringer, 2011 for evidence of continuity between nonhuman and human culture).

Transfer of Cultural Practices to a New Situation
How would an American eat seaweed salad if they had never observed anyone use chopsticks before? They would probably follow the customary practice of North American culture and use a fork (Call & Tennie, 2009). Gruber and colleagues presented two adjacent troops of chimpanzees with a problem equivalent to eating seaweed salad. The researchers drilled

Fig. 14.14
Photograph of chimpanzees from a troop that uses sticks to extract substances from cracks and holes. In the study, the chimps were given logs with two holes filled with honey, one shallow and the other deeper. The animals could extract the honey from a shallow hole with their fingers but had to use a tool to get the honey in a deeper hole. One group traditionally used leaves to remove liquid from holes while the other commonly used sticks. The two chimpanzee groups applied their customary "table manners" to the problem of eating the honey: one group used leaves to get at the honey, while the other group used sticks.

Source: Published originally by J. Call & C. Tennie (2009). Animal culture: Chimpanzee table manners? *Current Biology*, *19*, pp. R981–R983. Republished with permission of the Max Planck Institute for Evolutionary Anthropology.

holes at two depths into wooden logs, and filled them with honey (Gruber, Muller, Strimling, Wrangham, & Zuberbuhler, 2009). The chimps could extract the honey from the shallow hole with their fingers, but had to use a tool to remove the honey from the deeper hole. One group traditionally used leaves to remove liquid from holes, while the other group commonly used sticks. The two chimpanzee groups applied their customary "table manners" to the problem of eating the honey—one group used leaves to get at the honey, while the other group used sticks (Figure 14.14). Preference for sticks or leaves remained consistent within each group, matching the chimpanzees' natural foraging techniques in other situations. The evidence indicates that chimpanzee groups use their traditional cultural practice to solve a new feeding problem, an example of transfer of a practice from one situation to another (but see Gruber, Reynolds, & Zuberbuhler, 2010 for a discussion of the continuing debate on chimpanzee culture).

Persistence of Cultural Practices
A common observation is that cultural practices in humans often remain unchanged over many generations, even when these customs make no sense. A practice may persist for many years because the members of the group who engage in it fail to contact a change in the metacontingencies. At the individual level, a person may conform to a meat-eating taboo

because the social contingencies arranged by the group (religious proclamations and sanctions) avert contact with the positive reinforcement contingency (i.e., eating beef tastes good and adds nutritional protein to one's diet).

Cultural practices may also persist when the metacontingencies are stable and the current practices are sufficient to meet the contingencies. For centuries, the only way that books were manufactured was by having scribes make written copies. As a cultural practice, copying books by hand allowed for more standardized transmission of knowledge than that by word of mouth. Better methods of food preparation, house construction, agriculture, waste disposal, and transportation could be described in a common manner and passed from one generation to another. Thus, written transcription satisfied the metacontingencies for passing on social knowledge, but it was not the only way to reproduce books. The advent of the printing press allowed for an alternative form of book duplication that was less costly in time and effort, more productive, and much faster. Given these benefits, transcription by hand was eventually made obsolete by the invention of the printing press, but for a while both forms of the practice were used for duplication of social knowledge. In fact, transcription by hand is still used today, although it is done infrequently and for specialized purposes—involving translation of a work from one language to another. In unstable and rapidly changing environments, people learn many ways of doing things that are retained in the culture (persistence of social knowledge), but they often select among a few, less persistent forms of behavior that have resulted in high payoffs in the recent past (Rendell et al., 2011).

A final point is that an innovation such as the printing press often determines whether a culture remains static or dynamic. Individual inventions produce variation in cultural practices in much the same way that genetic mutations produce changes in species characteristics. That is, new forms of individual behavior (e.g., designing and making can openers) are generated by aversive contingencies or extinction (satisfying a need). These novel behaviors and products (can openers) are occasionally adopted by others and propagated as ways of meeting the metacontingencies (efficient storage and access to food with low risk of disease). Generally, *variation in form and selection by consequences* operate at different levels—producing both individual innovation and cultural evolution.

CHAPTER SUMMARY

This chapter provides many examples of how *selection by consequences* can operate at three levels of analysis: evolution, behavioral, and cultural. Our point is that adaptive behavior that provides benefits can lead to changes at the genetic, behavioral, and cultural levels. Selection is a universal process and, although we for the most part have restricted our analysis to the behavior of organisms, such a process has much greater range. The application of contingency analysis has proved useful in accounting for many diverse observations, from changes in conditioning ability in flies, to acquisition of verbal behavior in humans, to the creation and maintenance of cultural dietary taboos. Alternative forms of explanation for these observations have not proved as helpful as assuming and applying the unifying concept of selection by consequences. A scientific approach to behavior based on selection at three levels is not only possible, but is ongoing and substantiated by thousands of research studies. The findings from these explorations have led to further questions about the regulation of behavior in more complex and widespread areas. The next step is the full integration of behavior analysis, neuroscience, biology, and social science within a common framework based on evolution and selection. We hope that this textbook is a beginning to such an integration.

Key Words

Behavioral flexibility
Contingencies of survival
Cultural evolution
Cultural practice
Evolution
Genotype
Macrocontingency

Metacontingency
Mutation
Natural selection
Phenotype
Phylogeny
Selection by consequences

On the Web

www.ucmp.berkeley.edu/history/evolution.html This website explores the theory of evolution and the history of evolutionary thought.

http://en.wikipedia.org/wiki/Richard_Dawkins Go to Wikipedia and explore the contributions and controversies of Richard Dawkins on evolution, especially his dispute with religion.

http://videolectures.net/yaleeeb122f07_stearns_lec33/ Evolution, behavior, and game theory have implications for solving problems such as the tragedy of the commons discussed in this chapter. Take a look at the lecture by Dr. Stephen Steams at Yale University to learn more about this subject.

www.youtube.com/watch?v=aUCoLel5Qxg Observe a video clip of the waggle dance of bees and explore new research on dialects of bees' communication in the hive at YouTube for NOVA PBS. Although the dance involves genetic control of behavior, learning also plays a role in this fascinating ritual of bees.

www.youtube.com/watch?v=PgEmSb0cKBg The whistle speech of La Gomera Island, *silbo*, is a form of Spanish that has been passed down over the centuries to handle the problem of communication at a distance in the rugged, disjointed landscape of the island. Take a look!

Brief Quiz

1 The single common principle operating at the level of evolution, behavior, and culture is:
    a   selection by design
    b   survival of the fittest
    c   phylogenetic contingencies
    d   selection by consequences

2 Natural selection involves:
    a   reproductive fitness
    b   operant variation
    c   reproductive diversity
    d   ontogenetic adaptation

3. At what levels does selection by consequences occur?
   a. cultural
   b. biological
   c. behavioral
   d. all of the above
4. Two major sources of genetic variation are mutations and:
   a. phenotypic genes
   b. sexual recombination
   c. random novelty
   d. social pressure
5. Egg laying in *Aplysia* is an example of:
   a. genetic control of behavior
   b. environmental control of behavior
   c. basic instincts
   d. released action patterns
6. The behavior of invertebrates such as *Aplysia* is totally:
   a. controlled by genes
   b. learned
   c. cultural
   d. a and b
7. Operant behaviors are selected by:
   a. chromosomes
   b. consequences
   c. choice
   d. cognitions
8. The ability to have one's behavior strengthened by reinforcement is:
   a. learned
   b. heritable
   c. determined
   d. disadvantageous
9. Social signaling:
   a. is mostly genetically regulated in bees
   b. is mostly due to reinforcement contingencies in humans
   c. may involve stimulus equivalence in humans
   d. all of the above
10. What allowed for the emergence of human culture?
    a. evolution of operant processes
    b. evolution of verbal behavior
    c. social signaling by bees
    d. both (a) and (b)

Answers to Brief Quiz: 1, d (p. 509); 2, a (p. 511); 3, d (p. 510); 4, b (p. 512); 5, a (p. 514); 6, d (p. 525); 7, b (p. 525); 8, b (p. 522); 9, d (p. 528); 10, d (p. 532).

GLOSSARY

A-B-A-B design. This is the most basic single-subject research design. Also called a *reversal design*, it is ideally suited to show that specific features of the environment regulate an organism's behavior. The A phase, or baseline, is used to measure behavior before the researcher introduces an environmental change. During baseline, the experimenter takes repeated measures of the behavior under study, and this establishes a criterion against which any changes (attributed to the independent variable) may be assessed. Following the baseline phase, an environmental condition is changed (B phase) and behavior is measured repeatedly. If the independent variable, or environmental condition, has an effect, then the behavioral measure (dependent variable) will change—increase or decrease. Next, the baseline phase is reintroduced (A) and behavior is again measured. Since the treatment is removed, behavior should return to baseline levels. Finally, the independent variable is introduced again and behavior is reassessed (B). According to the logic of the design, behavior should return to a level observed in the initial B phase of the experiment. This second application of the independent variable helps ensure that the behavioral effect is caused by the manipulated condition.

Abolishing operation (AO). In contrast to the establishing operation, an abolishing operation (AO) decreases the effectiveness of behavioral consequences, and momentarily reduces behavior that has resulted in those consequences in the past. Thus, providing frequent social attention for a period (*noncontingent* attention) functions as an abolishing operation. That is, a period of noncontingent attention subsequently decreases the effectiveness of attention as a social reinforcer, and reduces self-injurious behavior maintained by adult attention.

Absolute stimulus control. When operants are regulated by the physical properties of one stimulus (color or hue), this is called absolute stimulus control. See also relative stimulus control.

Acceptance and Commitment Therapy (ACT). A third-wave behavioral therapy intervention that uses acceptance and mindfulness strategies, as well as commitment and behavior change strategies, to increase the range of behaviors that are acceptable to a person in a given situation (as opposed to only avoidance, for example).

Activity anorexia (AA). Following a period of food restriction, physical activity suppresses food intake and declining body weight increases activity. This negative feedback loop is called activity anorexia in rats and a similar cycle occurs in many anorexic patients.

Adjunctive behavior. Also called interim behavior. On interval schedules of reinforcement, or time-based delivery of food, organisms often show excessive behavior within the interreinforcement interval (IRI). For example, rats may drink up to three times their usual daily water intake (polydipsia) over a 1-h session. This behavior immediately follows reinforcement and is a side effect of periodic food delivery.

Ad libitum weight. The body weight of an organism that has free access to food 24 h a day.

Antecedent stimulus. An antecedent stimulus is any stimulus or event that changes the probability of operant behavior. There are three kinds of antecedent stimuli: S^D, S^Δ, and S^{ave}. An S^D increases the probability of response by predicting reinforcement, and an S^Δ makes responding less likely by predicting extinction. A S^{ave} may increase or decrease the likelihood of operant behavior, depending on the operating contingency (negative reinforcement or punishment).

Applied behavior analysis. This is a branch of behavior analysis that uses behavior principles to solve practical problems such as the treatment of autism or improved teaching methods. Applied behavior analysis is also referred to as behavioral engineering.

Associative strength. During respondent conditioning, the term *associative strength* is used to describe the relation between the conditioned stimulus (CS) and the magnitude of the conditioned response (CR). In general, associative strength increases over conditioning trials and reaches some maximum level.

Assumption of generality. The assumption of generality implies that the effects of contingencies of reinforcement extend over species, reinforcement, and behavior. For example, a fixed-interval (FI) schedule is expected to produce the scalloping pattern for a pigeon pecking a key for food and a child who is solving mathematics problems for teacher approval, all other things being equal.

Autism spectrum disorder. A broad range of conditions characterized by challenges with social interaction, repetitive behaviors, and non-verbal communication.

Autoclitic (verbal behavior). The autoclitic is a form of verbal behavior that modifies the consequences produced by other verbal responses. It is verbal behavior used in conjunction with, and controlled by, primary verbal units such as mands, tacts, and intraverbals. Skinner (1957) described five categories of autoclitic relations: descriptive, qualifying, quantifying, manipulative, and relational.

Autoshaping. Refers to a respondent conditioning procedure that generates skeletal responses. For example, a key light is turned on a few seconds before grain is presented to a pigeon. After several pairings of key light and grain, the bird begins to peck the key. This effect was first reported as autoshaping, an automatic way to teach pigeons to key peck.

Aversive stimulus. Refers to an event or stimulus that an organism escapes or avoids.

Avoidance. See negative reinforcement.

Backward chaining. A procedure used to train a chained performance. The basic idea is to first train behavior that is closest to primary reinforcement; once responding is established, links in the chain are added that are farther and farther from primary reinforcement. Each link in the chain is reinforced by the S^D which is also a conditioned reinforcer signaling the next component in the sequence.

Backward conditioning. In the respondent procedure of backward conditioning, the unconditioned stimulus (US) comes on before the conditioned stimulus (CS). The general consensus has been that backward conditioning is unreliable, and many researchers question whether it occurs at all. There is evidence that backward conditioning can occur when the CS has biological significance (e.g., the sight of a predator).

Baseline. The term refers to the base rate of behavior against which an experimental manipulation is measured. An uncontrolled baseline is the rate of an operant before any known conditioning; a controlled baseline (e.g., the rate of response on a variable-interval 60-s schedule) may be established to assess the effects of an experimental manipulation (e.g., presentation of intermittent shocks).

Baseline sensitivity. The term means that a low dose of a drug can cause substantial changes in baseline behavior. More generally, a behavioral baseline that varies with small increases in the independent variable is said to show sensitivity.

Behavior. Refers to everything that an organism does, including covert actions like thinking.

Behavioral contract. A behavioral plan of action that is negotiated between a client, child, or spouse and concerned others is a behavioral contract. The plan usually includes a statement of target responses, consequences that follow different actions, and long-term goals. The contract objectively specifies what is expected of the person in terms of behavior and the consequences that follow.

Behavioral contrast. Contrast refers to an inverse relationship between the response rates for two components of a multiple schedule—as one goes up the other goes down. There are two forms of contrast, positive and negative. Positive contrast occurs when rate of response in an *unchanged* component of a multiple schedule *increases* with a decline in behavior in the other schedule. Negative contrast occurs when rate of response in an *unchanged* component schedule *decreases* and an increase in behavior occurs in the other component of a multiple schedule.

Behavioral economics. The use of economic concepts (price, substitute commodity, etc.) and principles (e.g., marginal utility) to predict, control, and analyze the behavior of organisms in choice situations.

Behavioral flexibility. When organisms were faced with unpredictable and changing environments, natural selection favored those individuals whose behavior was flexible—adjusting on the basis of past experience. In this case, genes played a subsidiary role coding for general processes of learning. These processes allowed an organism to adjust to changing environmental requirements throughout its lifespan. Flexibility of behavior in turn contributed to the reproductive success of the organism.

Behavioral medicine. Behavior-change programs that target health-related activities such as following special diets, self-examination for early symptoms of disease, exercising, taking medicine, and so on. In many instances, the idea is that problems of behavior that affect health may be prevented before treatment is necessary.

Behavioral neuroscience. Refers to a scientific area that integrates the science of behavior (behavior analysis) with the science of the brain (neuroscience). Areas of interest include the effects of drugs on behavior (behavioral pharmacology), neural imaging and complex stimulus relations, choice and neural activity, and the brain circuitry of learning and addiction.

Behavioral variability. Refers to the animal's tendency to emit variations in response form in a given situation. The range of behavioral variation is related to an animal's capabilities based on genetic endowment, degree of neuroplasticity, and previous interactions with the environment. Behavioral variability in a shaping procedure allows for selection by reinforcing consequences and is analogous to the role of genetic variability in natural selection.

Behavior analysis. Behavior analysis is a comprehensive experimental approach to the study of the behavior of organisms. Primary objectives are the discovery of principles and laws that govern behavior, the extension of these principles over species, and the development of an applied technology.

Behavior analysts. These people are researchers and practitioners of behavior analysis.

Behaviorism. A term that refers to the scientific philosophy of behavior analysis.

Behavior maintenance. Refers to how long a new behavior persists after the original contingencies are removed (e.g., an anorexic patient who is taught to eat properly shows long-lasting effects of treatment if he maintains adequate weight for many years).

Behavior system. A species-specific set of responses elicited by a particular unconditioned stimulus (US). That is, for each species there is a behavior system related to procurement of food, another related to obtaining water, and still another for securing warmth.

Behavior trapping. Refers to the teaching of new behavior that, once established, is "trapped" by natural contingencies of reinforcement—the contingencies of everyday life.

Bias. In the generalized matching equation, response bias (k) refers to some unknown asymmetry between the alternatives in a given experiment that affects preference over and above the relative rates of reinforcement; can also mean a preference (Baum, 1974b).

Biological context. The evolutionary history and biological status of an organism are part of the context for specific environment–behavior interactions.

Blocking. In respondent compound conditioning, a conditioned stimulus (CS) that has been associated with an unconditioned stimulus (US) blocks a subsequent CS–US association. A CS_1 is paired with a US until the conditioned response reaches maximum strength. Following this conditioning, a second stimulus or CS_2 is presented at the same time as the original CS_1, and both are paired with the US. On test trials, the original CS_1 elicits the conditioned response (CR) but the second stimulus or CS_2 does not.

Break and run. Refers to a pattern of response, seen on a cumulative record, that occasionally develops on fixed-interval (FI) schedules. There is a long postreinforcement pause (PRP) followed by a brief burst of responses that result in reinforcement.

Breakpoint. The highest ratio value completed on a progressive-ratio (PR) schedule of reinforcement.

Celeration. The word *celeration* is used in precision teaching to denote two kinds of behavior change, acceleration and deceleration. Acceleration occurs when the rate of target behavior (frequency/time) is increasing over days, while deceleration involves decreasing rate over this period. A graph of the rates over days allows for evaluation of behavior change and revision of the instructional components based on the observed celeration (change in rate over days).

Chain schedule of reinforcement. A chain schedule of reinforcement refers to two or more simple schedules (CRF, FI, VI, FR, or VR), each of which is presented sequentially and signaled by an S^D. Only the final or terminal link of the chain results in primary reinforcement. See also heterogeneous and homogeneous chain schedules.

Change in associative strength. A factor that affects the increment in associative strength on any one trial is the change in associative strength which is the difference between the present strength of the conditioned stimulus (CS) and its maximum possible value.

Change in level (baseline to treatment). One of the inspection criteria for visual assessment of behavior change is the change in level or average (response rate and percentage) from baseline to treatment.

Changeover delay (COD). A changeover delay is a control procedure that is used to stop rapid switching between alternatives on concurrent schedules of reinforcement. The COD contingency stipulates that responses do not have an effect immediately following a change from one schedule to another. After switching to a new alternative, a brief time is required before a response is reinforced. For example, if an organism has just changed to an alternative schedule that is ready to deliver reinforcement, there is a brief delay before a response is effective. As soon as the delay has elapsed, a response is reinforced. The COD contingency operates in both directions whenever a change is made from one alternative to another.

Changeover response. On a concurrent schedule, a changeover is a response that an organism emits when it switches from one alternative to another.

Changing criterion design. A research design primarily used in applied behavior analysis. The rate of target behavior is progressively changed to some new criterion (up or down). For example, the criterion for the number of cigarettes a person smokes each day could be progressively lowered over several months. The effects of the independent variable are shown if the subject meets or falls below the criterion for any set of days (e.g., the criterion is 20 cigarettes for week 3, but changes to 10 by week 6).

Choice. From a behavioral view, choice is the distribution of operant behavior among alternative sources of reinforcement (e.g., concurrent schedules of reinforcement).

Coercion. Coercion is defined as the "use of punishment and the threat of punishment to get others to act as we would like, and to our practice of rewarding people just by letting them escape from our punishments and threats" (Sidman, 2001, p. 1). That is, coercion involves the basic contingencies of punishment and negative reinforcement.

Commitment response. The commitment response is some behavior emitted at a time prior to the choice point that eliminates or reduces the probability of impulsive behavior. A student who invites a classmate over to study on Friday night (commitment response) ensures that she will "hit the books" and give up partying when the choice arrives.

Compound stimuli. In respondent conditioning, two (or more) conditioned stimuli (e.g., tone and light) called a compound are presented together and acquire the capacity to evoke a single conditioned response (e.g., salivation).

Concurrent-chain schedule. Refers to two or more chain schedules that are simultaneously available. See also chain schedule of reinforcement and concurrent schedules of reinforcement.

Concurrent schedules of reinforcement. Involves two or more schedules of reinforcement (e.g., FR, VR, FI, VI) that are simultaneously available. Each alternative is associated with a separate schedule of reinforcement and the organism is free to distribute behavior to the schedules.

Conditional discrimination. A conditional discrimination is a differential response to stimuli that depends on the stimulus context (a four-term contingency of reinforcement). Consider a matching-to-sample experiment where a bird has been trained to match to triangles and squares based on the sample stimulus. To turn this experiment into a conditional-discrimination task, a houselight is inserted that may be turned on or off. The bird is required to match to the sample when the houselight is on and to choose the noncorresponding stimulus when the houselight is off. Conditional matching to sample involves simultaneous discrimination of three elements in a display. The animal must respond to geometric form depending on the sample, to the correspondence or noncorrespondence of the comparison stimuli, and to the condition of the houselight (on/off). See also matching to sample.

Conditioned aversive stimulus (Save). An aversive stimulus based on a history of conditioning. See aversive stimulus.

Conditioned establishing operation (CEO). Involves an establishing operation that depends on a history of reinforcement for completing a behavioral sequence or chain. One procedure is called the blocked-response CEO, in which a response that usually occurs is blocked because of the temporary absence of a specific condition, stimulus, or event. For example, you may leave your seminar notes at home as you rush to the university. Because you cannot complete the behavioral sequence of giving a seminar presentation, obtaining the notes would function as reinforcement for making a telephone call to get them. The notes would not have a reinforcement function during a casual lunch with an old friend because they are not necessary to this behavioral sequence. Whenever an event or stimulus is required to complete a behavior chain, withholding the event will establish it as reinforcement for operant behavior.

Conditioned overeating. Refers to a procedure of pairing a food taste (salt or sweet) with low caloric energy. When high-energy foods are consumed with tastes that have predicted low calorie content, juvenile rats overeat at their regular meals. The basic effect is that diet foods can cause overeating in children as an unintended side effect of taste conditioning.

Conditioned place preference (CPP). A procedure where the conditioned stimulus (CS) is a particular place or location and the sweet-flavored solution is the unconditioned stimulus (US). The solution is given in one distinct chamber (stripes) but not in another (white) and the animal shows a preference by a choice test for the location paired with the solution.

Conditioned reflex. See conditioned response and conditioned stimulus.

Conditioned reinforcement. Refers to the presentation of a conditioned reinforcer and the subsequent increase in rate of the operant that produced it.

Conditioned reinforcer. A conditioned reinforcer is an event or stimulus that has acquired its effectiveness to increase operant rate on the basis of an organism's life or ontogenetic history.

Conditioned response (CR). An arbitrary stimulus, such as a tone, is associated with an unconditioned stimulus (US) that elicits reflexive behavior (e.g., food elicits salivation). After several pairings, the stimulus is presented alone. If the stimulus now elicits a response (tone now evokes salivation), the response to the tone is called a conditioned response (CR).

Conditioned stimulus (CS). An arbitrary stimulus, such as a tone, is associated with an unconditioned stimulus (US) that elicits reflexive behavior (e.g., food elicits salivation). After several pairings, the stimulus is presented alone. If the stimulus now elicits a response (tone evokes salivation), it is called a conditioned stimulus (CS).

Conditioned-stimulus function. An event or stimulus that has acquired its function to elicit a response on the basis of respondent conditioning. When a tone is followed by food in the mouth, the tone becomes a conditioned stimulus (CS) for salivation.

Conditioned suppression. In conditioned suppression, a previously CS (e.g., tone, light, etc.) is paired with an aversive US such as an electric shock. After several pairings, the originally CS becomes a conditioned aversive stimulus (CS^{ave}). Once the CS^{ave} has been conditioned, its onset suppresses ongoing operant behavior. A rat may be trained to press a lever for food. After a stable rate of response is established, the CS^{ave} is introduced. When this occurs, the animal's lever pressing is suppressed.

Conditioned taste aversion (CTA). A sweet-flavored liquid may function as a conditioned stimulus (CS) in taste aversion conditioning and drug-induced sickness (lithium chloride) may serve as the unconditioned stimulus (US). After repeated pairings of the flavor or taste with the drug, the animal shows avoidance of the sweet-flavored solution.

Conditioned withdrawal. When a conditioned stimulus (CS) that accompanies drug use is presented, people are said to have "cravings" and this respondent process is called conditioned withdrawal. The CS elicits reactions that are ordinarily countered by the unconditioned stimulus (US). However, when the US is not delivered and the conditioned response (CR) reactions occur, people experience withdrawal. A heroin addict can have their withdrawal symptoms immediately terminated by a heroin injection. If you are accustomed to having a cigarette after a meal, the craving you experience can be alleviated with a smoke.

Confirmation bias. The tendency to interpret new information as one confirming one's already existent beliefs or verbal statements.

Construction of SDs. In solving problems, people make up or construct their own discriminative stimuli. A person who has an important early morning appointment may set an alarm clock for 6:00 a.m. Technically, setting the alarm is precurrent behavior, or an operant that precedes some other response or performance. That is, setting the alarm is behavior that results in the alarm ringing at 6:00 a.m., setting the occasion for getting up and going to the meeting. A major function of precurrent behavior is the construction of S^Ds that regulate subsequent action. See also precurrent behavior.

Context for conditioning. Refers to the ontogenetic and phylogenetic histories of an organism, including its current physiological status as well as contextual events or stimuli that are present when conditioning occurs.

Context of behavior. Refers to the fact that environment–behavior relationships are always conditional—depending on other circumstances.

Contextual stimuli. In terms of operant and respondent conditioning, contextual stimuli are uncontrolled sights, sounds, smells, and so on that are the background for conditioning. These stimuli are conditioned at the same time that behavior is strengthened.

Contingencies of survival. Refers to the contingencies (in the sense of "if–then" requirements) that result in differential reproduction or natural selection. The habitat or ecological environment sets requirements for the survival of individuals and their genes. Members of a species who exhibit features and behavior appropriate to the contingencies survive and reproduce, and those with less appropriate characteristics have fewer offspring. Natural selection (differential reproduction) therefore occurs as particular organisms satisfy (or fail to satisfy) the contingencies of survival.

Contingency (respondent). In respondent conditioning, contingency refers to a correlation between conditioned stimulus (CS) and unconditioned stimulus (US). Rescorla (1988) suggested that a positive correlation between CS and US, rather than the mere pairing of these stimuli, is necessary for conditioning. For operant conditioning, see contingency of reinforcement.

Contingency management. One highly effective behavioral intervention is called contingency management (CM). This intervention uses operant principles to arrange contingencies to promote desired behavior. In the context of drug abuse, contingency management is used to increase abstinence from drug use, to promote adherence to taking prescribed medication, and to increase retention in treatment programs.

Contingency of reinforcement. A contingency of reinforcement defines the relationship between the occasion, the operant class, and the consequences that follow the behavior (e.g., $S^D: R \rightarrow S^r$). We change the contingencies by altering one of the components and observing the effect on behavior. For example, a researcher may change the rate of reinforcement for an operant in a given situation. In this case, the $R \rightarrow S^r$ component is manipulated while the $S^D: R$ component is held constant. Contingencies of reinforcement can include more than three terms as in conditional discrimination (e.g., four-term relations); also, the effectiveness of reinforcement contingencies depends on motivational events called establishing operations (e.g., deprivation and satiation).

Contingency-shaped behavior. Refers to operant behavior that is directly under the control of contingencies of reinforcement, as opposed to rule-governed behavior.

Contingency-specifying stimuli. Refers to a technical term for verbal stimuli that regulate the behavior of listeners. Rules, instructions, advice, maxims, and laws are contingency-specifying stimuli in the sense that the verbal stimulus describes an actual contingency of reinforcement of everyday life. See rule-governed behavior.

Contingent response. In the response deprivation hypothesis, the contingent response is the activity obtained by making the instrumental response, as in the contingency if activity A occurs (instrumental response), then the opportunity to engage in activity B (contingent response) occurs.

Continuous reinforcement (CRF). When each response produces reinforcement (e.g., each lever press produces food), the schedule is called CRF or continuous reinforcement.

Correlation. As used in respondent conditioning, the percentage of conditioning trials in which the conditioned stimulus (CS) is followed by the unconditioned stimulus (US), and the percentage of trials in which the CS is not followed by the US. See also contingency.

Correspondence relations. Survival or reinforcement contingencies that select for equivalence, matching, or similarity between (a) the behavior of a model and observer, as in imitation; (b) what a person says and what is done (say–do correspondence); (c) what is done and what is said (do–say correspondence); (d) private stimulation and the verbal report (describing emotions); and (e) an instruction or rule and what is done (rule-governed behavior).

CS-pre-exposure effect. An alternative term for latent inhibition.

Countercontrol. Behavior that occurs in response to controlling an aversive stimulus, often in social situations. One type of countercontrol is an aggressive counterattack.

Cultural evolution. Cultural evolution begins at the level of the individual, when its technological effects reinforce behavior. An inventor may discover a new way of making a wheel; a farmer finds a food crop that produces higher yields; and a teacher may find a novel way to teach reading. A culture is said to evolve when the community adopts these innovations and the practice (e.g., using a higher-yield type of wheat) is passed on from one generation to the next.

Cultural practice. A cultural practice is defined in terms of interlocking social contingencies—where the behavior of each person supports the behavior of other members of the community. The pattern of behavior that arises from the interlocking contingencies is the type of practice (i.e., what people do in that culture).

Culture. Culture is usually defined in terms of the ideas and values of a society. However, behavior analysts define culture as all the conditions, events, and stimuli arranged by other people that regulate human action.

Cumulative record. A cumulative record is a real-time graphical representation of operant rate. Each response produces a constant upward increment on the Y-axis, and time is indexed on the X-axis. The faster the rate of response is, the steeper the slope or rise of the cumulative record. See also cumulative recorder.

Cumulative recorder. Refers to a laboratory instrument that is used to record the frequency of operant behavior in real time (rate of response). For example, paper is drawn across a roller at a constant speed, and each time a lever press occurs a pen steps up one increment. When reinforcement occurs, this same pen makes a downward deflection. Once the pen reaches the top of the paper, it resets to the bottom and starts to step up again. See also cumulative record.

Delay discounting. Delay discounting involves decisions between small, immediate and large, delayed rewards. If you are watching your weight, you often must choose between the immediate reinforcement from a piece of chocolate cake or the long-term reinforcement of body-weight loss and improved health. If you are like most of us, you find yourself eating the cake and forgoing the weight loss. That is, the large benefit in the future is devalued or discounted.

Delayed conditioning. A respondent conditioning procedure in which the conditioned stimulus (CS) is presented a few seconds before the unconditioned stimulus (US) occurs.

Delayed imitation. Refers to imitation of the modeled stimulus after a delay and in the absence of the model or modeled stimulus. Delayed imitation is considered to require more cognitive abilities than direct imitation (i.e., delayed imitation involves remembering the modeled stimulus).

Delayed matching to sample (DMTS). On a matching-to-sample task, the comparison stimuli are presented sometime after the sample stimuli are turned off. See also matching to sample.

Delay-reduction hypothesis. Stimuli that signal a decrease in time to positive reinforcement, or an increase in time to an aversive event, are more effective conditioned reinforcers. Generally, the value of a conditioned reinforcer is attributed to its delay reduction—how close it is to reinforcement or how far it is from punishment.

Demand curve. A demand curve is a mathematical curve showing how consumption decreases with price. When consumption of a commodity (reinforcer) rapidly decreases with price, the commodity is said to be *elastic*. Luxury items (European vacations) are highly elastic, being sensitive to price. Consumption of necessities (groceries) does not change much with price and are said to be *inelastic*.

Dependent variable. The variable that is measured in an experiment, commonly called an effect. In behavior analysis, the dependent variable is a measure of the behavior of an organism. One common dependent variable is the rate of occurrence of an operant (e.g., the rate of lever pressing for food).

Deprivation operation. Refers to the procedure of restricting access to a reinforcing event. Withholding an event or stimulus increases its effectiveness as a reinforcer.

Differential reinforcement. In discrimination procedures, differential reinforcement involves reinforcement in the presence of one stimulus (S^D) but not in other settings (S^Δ). The result is that the organism comes to respond when the S^D is presented and to show a low probability of response in settings that have not resulted in reinforcement (S^D). A differential response in S^D and S^Δ situations is called discrimination and an organism that shows this differential response is said to discriminate the occasion for reinforcement.

Differential reinforcement may be based on a property of operant behavior and in this case results in *response differentiation*. For example, when reinforcement is based on short interresponse times (IRT, 2–5 s), the distribution of IRTs becomes centered around short intervals. A change in the contingencies to reinforce longer IRTs (20–25 s) produces a new distribution centered around long intervals. See response differentiation.

Differential reinforcement of alternative behavior (DRA). In applied behavior analysis, the undesirable behavior is placed on extinction while alternative behavior, incompatible with the undesirable response, is reinforced. This differential reinforcement procedure often results in an increase in alternative desirable behavior and a decrease in the undesirable response.

Differential reinforcement of other behavior (DRO). Refers to reinforcement for any behavior other than a target operant. For example, after a period of time the applied behavior analyst delivers reinforcement for any behavior other than "getting out of seat" in a classroom. The target behavior is on extinction and any other behavior is reinforced.

Differential response. When an organism makes a response in one situation but not in another, we say that the animal discriminates between the situations or makes a differential response.

Direct replication. Repeating the procedures and measures of an experiment with several subjects of the same species (e.g., pigeons) is called direct replication. If each pigeon is exposed to a fixed-interval 30-s schedule of food reinforcement and each bird shows a scalloping pattern of pecking the key (i.e., a low rate of response following reinforcement that increases to a high rate at the moment of reinforcement), then the experimental procedures show direct replication.

Discriminated avoidance. Refers to avoidance behavior emitted to a warning stimulus. For example, a dog stops barking when its owner shouts "Stop!"

Discriminated extinction. Refers to a low rate of operant behavior that occurs as a function of an S^Δ. For example, the probability of putting coins in a vending machine with an "out of order" sign on it is very low.

Discrimination. When an organism makes a differential response to two or more stimuli (or events), we can say the animal discriminates between them. This process is called discrimination.

Discrimination index (ID). This index compares the rate of response in the S^D component to the sum of the rates in both S^D and S^Δ phases.

ID = (SD rate)/(SD rate + SΔ rate). The measure is a proportion that varies between 0.00 and 1.00. Using the I_D measure, when the rates of response are the same in the S^D and S^Δ components, the value of I_D is 0.50, indicating no discrimination. When all responses occur during the S^D phase, the S^Δ rate is zero and I_D is 1. Thus, a discrimination index of 1 indicates a perfect discrimination and maximum stimulus control of behavior. Intermediate values of the index signify more or less control by the discriminative stimulus.

Discriminative function. When an organism's behavior is reinforced, those events that reliably precede responses come to have a discriminative function. These events are said to *set the occasion* for behavior and are called discriminative stimuli. Discriminative stimuli acquire this function because they predict (have been followed by) reinforcement.

Discriminative stimulus (SD). Refers to an event or stimulus that precedes an operant and sets the occasion for operant behavior (antecedent stimulus).

Discriminative-stimulus account of conditioned reinforcement. Refers to the hypothesis that it is necessary for a stimulus to be a discriminative stimulus (S^D) in order for it to be a conditioned reinforcer. The hypothesis has been largely discounted, and the weight of the evidence supports Fantino's (1969b) delay-reduction hypothesis. See delay-reduction hypothesis.

Displaced aggression. When an aversive stimulus is delivered, an organism may display aggressive behavior towards another organism or object, even if that individual or object is not the source of the aversive.

Displacement behavior. Displacement behavior is observed in the natural environment and is characterized as irrelevant, incongruous, or out of context. That is, the behavior of the animal does not make sense given the situation, and the displaced responses do not appear to follow from immediately preceding behavior. Like adjunctive behavior (see definition in this Glossary), displacement responses arise when consummatory activities like eating are interrupted or prevented.

Duration recording. When behavior is continuous, duration recording is a method of observation. An observer may use a stopwatch, or other timing device, to record the duration of behavior. When a person is sitting in a chair, the watch is timing; and when the person leaves the chair, the watch is stopped.

Early intensive behavioral intervention (EIBI). Lovaas (1977, 1987) described an early intensive behavioral intervention (EIBI) where children were given 40 or more hours each week of behavioral intervention designed to increase social behavior, teach speaking and communication, and eliminate self-stimulation and aggressive behavior. Most children with autism showed significant improvement in their daily functioning.

Echoic responses. When there is point-to-point correspondence between the stimulus and response, verbal behavior may be classified as echoic. A further requirement is that the verbal stimulus and the echoic response must be in the same mode (auditory, visual, etc.) and have exact physical resemblance (e.g., same sound pattern). An echoic is a class of verbal operants regulated by a verbal stimulus in which there is correspondence and topographic similarity between the stimulus and response. Saying "This is a dog" to the spoken stimulus "This is a dog" is an example of an echoic response in human speech.

Elicited (behavior). Respondent (CR) and reflexive (UR) behavior are elicited in the sense that the behavior is made to occur by the presentation of a stimulus (CS or US).

Emitted (behavior). Operant behavior is emitted in the sense that it occurs at some probability in the presence of a discriminative stimulus (S^D), but the S^D does not force its occurrence.

Emotional response. Refers to a response such as "wing flapping" in birds that occurs with the change in contingencies from reinforcement to extinction. A common emotional response is called aggression (attacking another organism or target).

Environment. The functional environment is all the events and stimuli that affect the behavior of an organism. The environment includes events "inside the skin" like thinking, hormonal changes, and pain stimulation.

Errorless discrimination. In errorless discrimination, the trainer does not allow the organism to make mistakes by responding to the extinction stimulus. Initially S^D and S^Δ are very different, but differences between the stimuli are gradually reduced as training progresses. The procedure eliminates the emotional behavior generated by extinction with other discrimination-training methods. For example, pigeons flap their wings in an aggressive manner and work for an opportunity to attack another bird during the presentation of the S^ on a multiple schedule. This behavior does not occur when errorless discrimination is used in training.

Escape. See negative reinforcement.

Established-response method. In terms of conditioned reinforcement, an operant that produces unconditioned reinforcement is accompanied by a distinctive stimulus just prior to reinforcement. When responding is well established, extinction is implemented but half of the subjects continue to get the stimulus that accompanied unconditioned reinforcement. The other subjects undergo extinction without the distinctive stimulus. Generally, subjects with the stimulus present respond more than the subjects who do not get the stimulus associated with unconditioned reinforcement. This result is interpreted as evidence for the effects of conditioned reinforcement.

Establishing operation (EO). Formally, an establishing operation is defined as any change in the environment that alters the effectiveness of some stimulus or event as reinforcement and simultaneously alters the momentary frequency of the behavior that has been followed by that reinforcement. Thus, an establishing operation has two major effects: (a) it increases the momentary effectiveness of reinforcers supporting operant behavior, and (b) it increases the momentary probability of operants that have produced such reinforcement. For example, the most common establishing operation is deprivation for primary reinforcement. This procedure has two effects. First, food becomes an effective reinforcer for any operant that produces it. Second, behavior that has previously resulted in getting food becomes more likely.

Evolution. In terms of biology, the change in the genetic makeup of the species as observed in the expressed characteristics of its members.

Experimental analysis of behavior. The method of investigation most commonly used in behavior analysis. The method involves breaking down complex environment–behavior relations into component principles of behavior. The analysis is verified by arranging experimental procedures that reveal the underlying basic principles and controlling variables. This involves intensive experimentation with a single organism over an extended period, rather than statistical assessment of groups exposed to experimental treatments.

External validity. External validity refers to the extent that experimental findings generalize to other behaviors, settings, reinforcers, and populations. That is, does the cause-and-effect relationship found in an experiment occur at different times and places when the original conditions are in effect?

Extinction. The procedure of extinction involves the breaking of the contingency between an operant and its consequence. For example, bar pressing followed by food reinforcement no longer produces food. As a behavioral process, extinction refers to a decline in the frequency of the operant when an extinction procedure is in effect. In both instances, the term *extinction* is used correctly.

Extinction burst. A rapid burst of responses when an extinction procedure is first implemented.

Extinction stimulus (S^Δ). An S^Δ (pronounced S-delta) is a stimulus that sets the occasion for a decrease in operant responses. For example, an "out of order" sign on a vending machine decreases the probability of putting money in the machine. See S-delta.

Extraneous sources of reinforcement (Re). Involves all non-programmed sources of reinforcement that regulate alternative behavior—reducing the control of behavior on a specified schedule of reinforcement. Extraneous sources of reinforcement include any unknown contingencies that support the behavior of the organism. For example, a rat that is pressing a lever for food on a particular schedule of reinforcement could receive extraneous reinforcement for scratching, sniffing, and numerous other behaviors. The rate of response for food will be a function of the programmed schedule as well as the extraneous schedules controlling other behavior. In humans, a student's mathematical performance will be a function of the schedule of correct solutions as well as extraneous reinforcement for other behavior from classmates or teachers, internal neurochemical processes, and changes to the physical/chemical environment (e.g., smell of food drifting from the cafeteria). See also quantitative law of effect.

Fading. The procedure involves transferring stimulus control from one value of a stimulus to another. This is done by gradually changing a controlling stimulus from an initial value to some designated criterion.

Falsifiability. The capacity for a statement or hypothesis to be proven wrong or incorrect

First-order conditioning. In first-order respondent conditioning, an apparently neutral stimulus is paired with an unconditioned stimulus (US). When this occurs, the control of the response to the US is transferred to the neutral stimulus, which is now called a conditioned stimulus (CS).

Fixed-action pattern (FAP). A sequence or chain of behavior set off by a specific stimulus. The component responses are repeated almost identically with each presentation of the stimulus. Fixed-action patterns are based on a "genetic blueprint," and the environment simply initiates the sequence. For example, the male stickleback fish will aggressively defend its territory from male intruders during mating season. The fish shows a fixed sequence of threatening actions that are elicited by the red underbelly of an intruding male.

Fixed interval (FI). The fixed interval is a schedule of reinforcement in which an operant is reinforced after a fixed amount of time has passed. For example, on a fixed-interval 90-s schedule (FI 90), one bar press after 90 s results in reinforcement. Following reinforcement, another 90-s period goes into effect; and after this time has passed, another response will produce reinforcement.

Fixed ratio (FR). The fixed ratio is a response-based schedule of reinforcement that delivers reinforcement after a fixed number of responses are made. For example, on a fixed ratio 10 (FR 10), the organism must make 10 responses per reinforcement.

Fluency. In precision teaching, the use of rate (frequency/time) focuses instruction on *fluency* or accuracy *and* high frequency. When a performance becomes fluent, the behavior is retained longer, persists during long periods on the task, is less affected by distractions, and is more likely to be available in new learning situations (i.e., to combine with other well-learned behaviors).

Force of response. Reinforcement can be made contingent on the force or magnitude of response. Force or magnitude is a property or dimension of behavior.

Formal similarity. A term used in verbal behavior to define echoic behavior. Formal similarity requires that the verbal stimulus and the product of the response be in the same mode (auditory, visual, etc.) and have exact physical resemblance (e.g., same sound pattern).

Free-operant method. In the free-operant method, an organism may repeatedly respond over an extensive period of time. The organism is "free" to emit many responses or none at all. More accurately, responses can be made without interference from the experimenter (as in a trials procedure).

Functional analysis. An analysis of behavior in terms of its products or consequences. Functionally, there are two basic types of behavior, operant and respondent. The term *respondent* defines behavior that increases or decreases because of the presentation of a stimulus (or event) that precedes the response. Such behavior is said to be elicited, in the sense that it reliably occurs when the stimulus is presented. There is a large class of behavior that does not depend on an eliciting stimulus. This behavior is called emitted and spontaneously occurs at some frequency. When emitted behavior is strengthened or weakened by the events that follow the response, it is called operant behavior. Thus, operants are emitted responses that increase or decrease depending on the consequences they produce.

Functional independence. A term used in verbal behavior to describe the independence of the operant classes of manding and tacting. Formally, each operant class is controlled by separate contingencies of reinforcement; training mand relations would not necessarily affect the training of tact relations or vice versa.

Function-altering event. Verbal stimuli such as rules and instructions can alter the function of other stimuli and, thereby, the strength of relations among stimuli and behavior. For example, an instruction about what to do in an airline emergency can establish stimulus control by a "dangling yellow mask" over the behavior of "placing the mask over your face and breathing normally."

Generality. An experimental result has generality when it is observed in different environments, organisms, and so on. For example, the principle of reinforcement generalizes over species, settings, responses, and reinforcers. In a pigeon, the peck-for-food relationship depends on the establishing operation of deprivation for food in the immediate past. For humans, who have an extensive capacity for operant conditioning, going to a soda machine to get a cold drink on a hot afternoon is an effective contingency. In both examples, establishing operations and reinforcement are the operating principles.

Generalization. Emitting similar behavior in different situations. An organism is said to show generalization if it fails to discriminate between one situation and another.

Generalization gradient (operant). Generalization occurs when an organism responds to values of the S^D (or fewer responses to the S^Δ) that were not trained during acquisition. A generalization gradient is the function (graph) that relates values of the S^D (intensity of light) to a measure of response strength (operant rate).

Generalization gradient (respondent). Generalization occurs when an organism shows a conditioned response (CR) to values of the conditioned stimulus (CS) that were not trained during acquisition. A generalization gradient is the function (graph) that relates values of the CS (loudness of tone) to a measure of response strength (amount of CR).

Generalized conditioned reinforcer. A conditioned reinforcer that is backed up by many other sources of reinforcement is a generalized conditioned reinforcer. Money is a good example of a generalized conditioned reinforcer. Cash may be exchanged for a large variety of goods and services. Human behavior is regulated by generalized reinforcement, involving social attention, approval, and affection.

Generalized imitation. A reinforcement procedure used to teach the generalized response and stimulus classes "do as I do." The procedure involves reinforcement of correspondence between modeled performance and imitative operants. After training a number of exemplars, a novel modeled stimulus is presented without reinforcement and a new imitative response occurs that matches the modeled performance. Generalized imitation involves both stimulus generalization of the class of modeled stimuli and response generalization of the class of imitative responses.

Generalized matching relation (generalized matching law). Proportion equations like $B_a/(B_a + B_b) = R_a/(R_a + R_b)$ describe concurrent performance when alternatives differ only in rate of reinforcement. However, in complex environments, other factors also contribute to choice and preference. These factors arise from the biology and environmental history of the organism. For example, sources of error may include different amounts of effort for the responses, qualitative differences in reinforcement such as food versus water, a history of punishment, a tendency to respond to the right alternative rather than the left, and sensory capacities.

To include these and other conditions within the matching law, it is useful to express the law in terms of ratios rather than proportions (i.e., $B_a/B_b = R_a/R_b$). When relative rate of response matches relative rate of reinforcement, the ratio equation is simply a restatement of the proportional form of the matching law. A generalized form of the ratio equation may, however, be used to handle the situation in which unknown factors influence the distribution of behavior. These factors produce systematic departures from ideal matching but may be represented as two constants (parameters) in the generalized matching equation: $B_a/B_b = k(R_a/R_b)^a$. In this form, the matching equation is known as the generalized matching law. The coefficient k and the exponent a are values that represent two sources of error for a given experiment. When these parameters are equal to 1, the equation is the simple ratio form of the matching law.

Generalized social reinforcement. A generalized conditioned reinforcer that is also a social reinforcer increases or maintains operant behavior. Praise is a social reinforcer backed up by many sources of reinforcement. See also generalized conditioned reinforcer.

Genotype. Genotype refers to the genetic makeup of the organism. Some observable characteristics are largely determined by genotype, other features are strongly influenced by experience, but most result from an interaction of genes and environment. Thus, the height of a person is attributable to both genes and nutrition.

Habituation. Habituation occurs when an unconditioned stimulus (US) repeatedly elicits an unconditioned response (UR). The frequent presentation of the US produces a gradual decline in the magnitude of the UR. When the UR is repeatedly elicited it may eventually fail to occur at all.

Heterogeneous chain schedule. A heterogeneous chain requires different responses for each link of the chain schedule. Dog trainers make use of heterogeneous chains when they teach complex behavioral sequences to their animals. In going for a walk, a seeing-eye dog stops at intersections, moves forward when the traffic is clear, pauses at a curb, avoids potholes, and finds the way home. Each of these different responses is occasioned by specific stimuli and results in conditioned reinforcement. See also chain schedule.

History of reinforcement. Refers to the reinforcement contingencies that an organism has been exposed to during its lifetime including the changes in behavior due to such exposure.

Homeostasis. Walter Cannon coined the word in 1932 as the tendency of a system to remain stable and to resist change. In terms of a biological system, homeostasis refers to the regulation of the system by negative feedback loops. For example, the body maintains a temperature within a very fine tolerance. If the environment warms up or cools down, physiological mechanisms (sweating or shivering) involving the sympathetic and parasympathetic nervous systems are activated to reduce the drift from normal body temperature. Homeostasis involves self-regulation to maintain an internal environment in a stable or constant condition by means of multiple dynamic equilibrium adjustments.

Homogeneous chain schedule. Operant chains are classified as homogeneous when the topography or form of response is similar in each link of the schedule. For example, a bird pecks the same key in each component of the chain. Each link in the schedule produces a discriminative stimulus for the next link, and the S^D is also a conditioned reinforcer for the behavior that produces it. See also chain schedule.

Hyperbolic discounting equation. Denotes a delay-discounting equation by Mazur (1987) that shows a hyperbolic decay: $V_d = \frac{A}{1+kd}$. In the equation, we are predicting discounted values, V_d, of the reinforcer. The amount of the reinforcer, A, is $100,000 in our example, and the value, d, is the delay—the variable on the X-axis. The value, k, is called the *discounting rate*, which must be estimated to fit a curve to the indifference points (data) obtained from the experiment. The value of a delayed reinforcer plunges with time in a hyperbolic manner.

Hypothetical construct. Unobservable events or processes that are postulated to occur and that are said to explain behavior are called hypothetical constructs. For example, Freud's mental device "ego" is a hypothetical construct that is used to explain self-gratifying behavior. In cognitive psychology, terms like "cognitive representation" or "mental imagery" are hypothetical terms that are said to explain the behavior of knowing and observing the world. From a behavioral perspective, the difficulty is that the mental constructs are easily invented, are inferred from the behavior they are said to explain, and are inherently unobservable with direct observation. That is, there is no objective way of getting information about such events except by observing the behavior of people or other organisms.

Identity matching. In identity matching, the researcher presents a sample stimulus (e.g., a triangle) and two options (e.g., triangle or circle). The procedure is repeated over multiple examples of sample and comparison options. The organism is reinforced for choosing the option that corresponds to the sample, establishing *generalized matching to sample* or identity matching. See also reflexivity.

Imitation. True imitation requires that the learner emits a novel response that could only occur by observing a model emit a similar response.

Immediacy of change (baseline to treatment). In visual inspection of behavioral data, we assume that the cause of a change in behavior must immediately precede the change. In behavior analysis, immediacy is assessed using the last three data points of the baselines and the first three data points for the treatment phases. Immediacy of change also is assessed from the treatment phase to the return to baseline. For the high-impact results, the change in the dependent variable is almost immediate with the changes in the independent variable (baseline to treatment or treatment to baseline).

Immediate causation. Refers to the kind of mechanism studied by physics and chemistry—the "billiard ball" type of process where we try to isolate a chain of events that directly result in some effect. Thus, in physiology the bar pressing of a rat for food or a gambler playing roulette could each involve the release of endogenous opiates and dopamine in the hypothalamus.

Incentive salience. Involves the acquisition of motivational value by the sign or cue (CS+) predicting the unconditioned stimulus (US) in an autoshaping or sign-tracking procedure.

Independent variable. The variable that is manipulated, changed, or controlled in an experiment, commonly called a cause. In behavior analysis, a change in the contingencies of reinforcement, the arrangement of events that precede and follow the behavior of an organism (e.g., changing the rate of reinforcement).

Information account of conditioned reinforcement. A hypothesis suggesting that a stimulus becomes a conditioned reinforcer if it provides information about the occurrence of primary reinforcement. This notion has been largely discounted and replaced by Fantino's (1969b) delay-reduction hypothesis. See also delay-reduction hypothesis.

Instinctive drift. Species-characteristic behavior that becomes more and more invasive during operant training is called instinctive drift.

Instrumental response. In the response deprivation hypothesis, the instrumental response is the behavior that produces the opportunity to engage in some activity.

Intensive behavioral intervention. Refers to a term used in the treatment of autistic behavior where the child is given targeted or planned interventions for behavioral excesses and deficits of 30 or more hours each week. This kind of programmed behavioral intervention is most effective with youngsters under 4 years of age.

Interim behavior. See adjunctive behavior.

Interlocking contingencies. In social episodes involving manding and tacting, each person (speaker and listener) completes a behavioral sequence or chain ($S^D: R \rightarrow S^r + S^D: R \rightarrow S^r$ ...), and the verbal relations involve the intermingling of these chains or the interlocking contingencies. In an interlocking contingency, the behavior of one person causes stimulation and reinforcement for the behavior of the other, and vice versa.

Intermittent reinforcement effect. Intermittent reinforcement schedules generate greater resistance to extinction than continuous reinforcement (CRF). The higher the rate of reinforcement, the greater the resistance to change; however, the change from CRF to extinction is discriminated more rapidly than between intermittent reinforcement and extinction.

Intermittent schedule of reinforcement. A schedule programmed so that some rather than all operants are reinforced. In other words, an intermittent schedule is any schedule of reinforcement other than continuous (CRF).

Internal validity. When many extraneous variables are ruled out by an experimental design, the research has high internal validity. That is, changes in the dependent variable may be reasonably attributed to changes in the independent variable (cause → effect). Internal validity is the minimum requirement for all experiments.

Interreinforcement interval (IRI). The interreinforcement interval (IRI) is time between any two reinforcers. Research shows that the postreinforcement pause (PRP) is a function of the IRI. As the time between reinforcements becomes longer, the PRP increases. On fixed-interval (FI) schedules the PRP is approximately one-half the IRI. For example, on a FI 300-s schedule (in which the time between reinforcements is 300 s), the average PRP will be 150 s. On fixed ratio (FR), the evidence indicates similar control by the IRI, as the ratio requirement increases the PRP becomes longer. See postreinforcement pause (PRP).

Interresponse time (IRT). The time between any two responses is called the interresponse time (IRT). The IRT may be treated as a conditionable property of operant behavior; for example, the IRTs on a variable-interval (VI) schedule of reinforcement are much longer than on a variable-ratio (VR) schedule. VI schedules are said to differentially reinforce long IRTs while VR schedules differentially reinforce short IRTs.

Interval recording. Refers to a measurement strategy used in applied behavior analysis to assess the rate of target behavior. A block of time is selected and divided into short, equal intervals, and if the target behavior occurs it is recorded once in an appropriate time bin. For example, a 30-min segment of mathematics class may be divided into 10-s bins. Regardless of the number of responses, if the behavior occurs in a given 10-s segment, then the observer records it as a single event.

Interval schedules. These are schedules of reinforcement based on the passage of time and one response after that time has elapsed.

Intraverbal behavior. Intraverbal behavior involves a class of verbal operants regulated by verbal discriminative stimuli. In everyday language, thematically related words (or sentences) are examples of intraverbal relations. For example, the verbal response "Fish" to the spoken words "Rod and reel" is an intraverbal response; saying "Water" to the written word LAKE is also intraverbal behavior. Thus, intraverbal relations arise from verbal behavior itself. A previous verbal response by a speaker is a stimulus for a subsequent verbal operant.

***In-vitro* reinforcement.** A method used to investigate reinforcement in the neuron, increasing calcium bursts or firings by injection of dopamine agonists or other agents.

Joint control. Refers to the notion that two verbal stimuli exert stimulus control over a common verbal topography. In finding the correct sequence of numbers in an array, repeating the required number and identifying that number in the array jointly control the terminal verbal response "[number] I found it."

Latency. Refers to the time from the onset of one event to the onset of another. For example, the time it takes a rat to reach a goal box after it has been released in a maze.

Latent inhibition. A term used to denote that the animal's learning of the CS–US relation is reduced or inhibited by pre-exposure of the CS, revealed by an acquisition test following the conditioning phase.

Law of effect. As originally stated by Thorndike, the law refers to stamping in (or out) some response. A cat opened a puzzle-box door more rapidly over repeated trials. Currently the law is stated as the principle of reinforcement: operants may be followed by consequences that increase (or decrease) the probability or rate of response.

Law of intensity–magnitude. As the intensity of an unconditioned stimulus (US) increases, so does the magnitude or size of the unconditioned response (UR).

Law of the latency. As the intensity of the unconditioned stimulus (US) increases, the latency (time to onset) of the unconditioned response (UR) decreases.

Law of the threshold. At very weak intensities a stimulus will not elicit a response, but as the intensity of the eliciting stimulus increases there is a point at which the response is evoked. That is, there is a point below which no response is elicited and above which a response always occurs.

Learned helplessness. Learned helplessness involves exposing an animal to inescapable and severe aversive stimulation (shocks). Eventually the animal gives up and stops attempting to avoid or escape the situation. Next, an escape response that under ordinary circumstances would be acquired easily is made available, but the animal does not make the response. The organism seems to give up and become helpless when presented with inescapable aversive stimulation.

Learning. Refers to the acquisition, maintenance, and change of an organism's behavior as a result of lifetime events (the ontogeny of behavior). In everyday language, learning often is used to refer to transitional changes in behavior (e.g., from not knowing to knowing one's ABCs) but conditions that maintain behavior in a steady-state are also part of what we mean by learning (e.g., continuing to recite the alphabet).

Limited hold. A limited hold is a contingency where the reinforcer is available for a set time after an interval schedule has timed out. Adding a limited hold to a variable-interval (VI) schedule increases the rate of responding by reinforcing short interresponse times (IRTs).

Log-linear matching equation. To write the matching law as a straight line, we may write the log-linear equation:

$\log(B_a/B_b) = \log k + [a \times \log(R_a/R_b)]$. Notice that in this form, $\log(B_a/B_b)$ is the Y variable and $\log(R_a/R_b)$ is the X variate. The constants a and $\log k$ are the slope and intercept, respectively. See generalized matching law.

Macrocontingency. When the operant behavior of multiple individuals generates *a cumulative effect* for the group, without an explicit interlocking behavior contingency, we may describe this relation as a macrocontingency. A cumulative effect for the group is sometimes generated by many individuals emitting operant behavior for immediate reinforcement—each person pursuing her personal interests (immediate reinforcement), as exemplified by the use of limited resources. Compare to metacontingency.

Magazine training. Refers to following the click of the feeder (stimulus) with the presentation of food (reinforcement). For example, a rat is placed in an operant chamber and a microcomputer periodically turns on the feeder. When the feeder is turned on, it makes a click and a food pellet falls into a cup. Because the click and the appearance of food are associated in time you would, after training, observe a typical rat staying close to the food magazine, and quickly moving toward it when the feeder is operated (see conditioned reinforcer).

Manding. The word *manding* comes from the common English word *commanding*, but commanding is only part of this operant class. Manding is a class of verbal operants whose form is regulated by establishing operations (e.g., deprivation, aversive stimulation, etc.) and specific reinforcement. When you say "Give me the book," "Don't do that," "Stop," and so on, your words are regulated by motivational conditions or establishing operations (e.g., deprivation for the book, or by another person doing something unpleasant). The establishing operation (no ketchup) regulates the topography of manding ("give ketchup") and ensures that a particular event functions as specific reinforcement (getting ketchup).

Matching (relation). When the relative rate of response matches (or equals) the relative rate of reinforcement. In proportional matching, the proportional rate of response on alternatives A and B equals the proportional rate of reinforcement on the two alternatives. The matching relation is also called the matching law. Matching also has been expressed in an equation using ratios with sources of error and in this form is called the generalized matching law. See relative rate of response and relative rate of reinforcement.

Matching to sample. A procedure used to investigate recognition of stimuli is called matching to sample. For example, a pigeon may be presented with three keys. A triangle or sample stimulus is projected onto the center key. To ensure that the bird attends to the sample, the pigeon is required to peck the sample key. When this happens, two side keys are illuminated with a triangle on one and a square on the other, called the comparison stimuli. If the bird pecks the comparison stimulus that corresponds to the sample, this behavior is reinforced and leads to the presentation of a new sample. Pecks to the noncorresponding stimulus result in extinction and the next trial. See identity matching.

Maximization. In this economic view of behavior, humans and other animals are like organic computers that compare their behavioral distributions with overall outcomes and eventually stabilize on a response distribution that maximizes overall rate of reinforcement. See melioration as an alternative view.

Maximum associative strength. In the Rescorla–Wagner model, a conditioned stimulus (CS) can acquire only so much control over a conditioned response (CR). This is the maximum associative strength for the CS. Thus, a tone (CS) that is paired with 1 g of food will have maximum associative strength when conditioned salivation (CR) to the tone is about the same amount as the unconditioned salivation (UR) elicited by the food (US). That is, an unconditioned stimulus (US) elicits a given magnitude of the unconditioned response (UR). This magnitude sets the upper limit for the CR. The CS cannot elicit a greater response than the one produced by the US.

Melioration. An explanation of how organisms come to produce matching on concurrent schedules of reinforcement. In contrast to overall maximizing of reinforcement, Herrnstein (1997) proposed a process of melioration (doing the best at the moment). Organisms, he argued, are sensitive to fluctuations in the momentary rates of reinforcement rather than to long-term changes in overall rates of reinforcement.

Metacontingency. A metacontingency refers to contingent relations between practices, as parts of *interlocking behavioral contingencies* (IBCs), and the effects or consequences of those practices for the group or culture. For example, to free a car stuck in a snow drift, the driver steers and accelerates while the two passengers push from behind. The IBC (stuck car) sets up the division of cooperative behavior (the practice) as a way to get the car moving—satisfying the metacontingency (everyone gets to destination). Compare to macrocontingency.

Mixed schedule of reinforcement. A mixed schedule is two or more basic schedules (CRF, FR, FI, VI, VR) presented sequentially in which each link ends with primary reinforcement (or in some cases extinction) and the component schedules are not signaled by discriminative stimuli. In other words, a mixed schedule is the same as an unsignaled multiple schedule. See multiple schedule of reinforcement.

Modal action pattern (MAP). The term denotes the behavioral flexibility of seemingly fixed-action patterns. The major topographic features of these reflex combinations may appear similar across individuals and situations, but there are numerous idiosyncratic differences. For example, all robins (*Turdus migratorius*) build nests that appear very similar in construction. But, it is clear they do not all build in the same location, or use the same materials.

There is great individual variation in all phases of nest construction, suggesting modification by the environment (ontogeny).

Molar account of schedule performance. Molar accounts of behavior on schedules of reinforcement or punishment are concerned with large-scale factors that regulate responding over a long period of time. For example, the average time between reinforcers for an entire session and the overall reduction in shock frequency are molar-level variables.

Molecular account of schedule performance. Molecular accounts of behavior on schedules of reinforcement or punishment focus on small moment-to-moment relationships between behavior and its consequences. For example, the time between any two responses (IRT) and the response–shock interval (R–S) are molecular-level variables.

Motivational operation (MO). To capture both the establishing and abolishing effects of events that precede reinforced behavior (or punishment), it is useful to introduce a more inclusive concept. The motivational operation (MO) refers to any event that alters the reinforcement effectiveness of behavioral consequences and changes the frequency of behavior maintained by those consequences.

Multiple baseline across behaviors. A multiple baseline research design across behaviors is used when a reinforcement procedure is applied progressively to several operants. In this case, the subject, setting, and consequences remain the same, but different responses are modified sequentially.

Multiple baseline across participants. A research design in which an intervention is introduced progressively for different subjects who exhibit similar target behavior. The same behavior (e.g., stealing) is first modified for subject 1, and baselines are collected for subjects 2 and 3. Next, the behavior of subject 2 is changed while the rate of target behavior for subjects 1 and 3 continues to be assessed. Finally, the treatment procedure is applied to subject 3.

Multiple baseline across settings. In this research design, a reinforcement procedure is applied in one situation but is withheld in other settings. When behavior changes in the situation where it is reinforced, the contingency is applied to the same response in another setting.

Multiple baseline designs. A class of research designs used primarily in applied behavior analysis. See multiple baseline across behaviors, multiple baseline across participants, and multiple baseline across settings.

Multiple functions (of stimuli). A given event or stimulus, such as a student saying "The ball is red," can have several functions in the control of behavior (e.g., $S^r + S^D$). For example, the response can function as reinforcement for the teacher's question "What color is the ball?" and at the same time function as a discriminative stimulus for the teacher saying "Yes."

Multiple schedule. A multiple schedule is two or more basic schedules (CRF, FR, FI, VI, VR) presented sequentially, each link ending with primary reinforcement (or in some cases extinction); the component schedules are signaled by discriminative stimuli. In other words, a multiple schedule is the same as a chain schedule, but each link produces primary reinforcement. See chain schedule of reinforcement.

Mutation. Mutation occurs when the genetic material (e.g., genes or chromosomes) of an individual changes. These changes are accidents that affect the genetic code carried by ova or sperm. For example, naturally occurring background radiation may alter a gene site or a chromosome may break during the formation of sex cells or gametes. Such mutations are passed on to offspring, who display new characteristics.

Naming relation. Horne and Lowe (1996) proposed that naming something (object, place, or action) involves a generalized operant class that substantially expands the verbal repertoire of the child. Analytically, the naming relation or *the generalized class of naming* arises from verbal contingencies that integrate the echoic and tact response classes of the child as speaker with the conditional-discrimination behavior of the child as listener.

Natural selection. Refers to the differential reproduction of the members of a species and their genetic endowment. Based on a thorough analysis of life forms, Darwin concluded that reproductive success was the underlying basis of evolution. That is, individuals with more offspring pass on a greater number of their characteristics (genes) to the next generation.

Negative automaintenance. Birds are autoshaped to peck a key, but in negative automaintenance food is not presented if the bird pecks the key. This is also called an omission procedure or training because food reinforcement is omitted if key pecking occurs.

Negative contrast. See behavioral contrast.

Negative punishment. Negative punishment is contingency that involves the removal of an event or stimulus following behavior and decreasing the rate of response. The negative punishment procedure requires that behavior (watching television) is maintained by positive reinforcement (entertaining programs) and the reinforcer is removed (TV turned off) if a specified response occurs (yelling and screaming). The probability of response is reduced by the procedure.

Negative reinforcement. Negative reinforcement is a contingency where an ongoing stimulus or event is removed (or prevented) by some response (operant) and the rate of response increases. If it is raining, opening and standing under an umbrella removes the rain and maintains the use of the umbrella on rainy days. When operant behavior increases by removing an ongoing event or stimulus the contingency is called *escape*. The contingency is called *avoidance* when the operant increases by preventing the onset of the event or stimulus. Both escape and avoidance involve negative reinforcement.

Negative reinforcer. A negative reinforcer is any event or stimulus that increases the probability (rate of occurrence) of an operant that removes or prevents it. See also negative reinforcement.

Neuroplasticity. Refers to alterations of neurons and neural interconnections during a lifetime by changes in environmental contingencies.

New-response method for conditioned reinforcement. First, a neutral stimulus is associated with a reinforcing event (sound of feeder is followed by food), and after this procedure the stimulus (sound of feeder) is shown to increase the frequency of some operant behavior.

Nondiscriminated avoidance. A procedure used to train avoidance responding in which no warning stimulus is presented is called nondiscriminated or Sidman avoidance. See also negative reinforcement.

Observational learning. From a social cognitive viewpoint, the observer pays attention to the modeled sequence, noting the arrangement of each action. The general information in the sequence must be coded and rehearsed. Once this abstract information is retained in memory, imitation is a matter of reproducing the component responses in the correct sequences. From a behavioral perspective, observational learning involves the integration of generalized imitation, rule-governed behavior, and verbal behavior. Each of these components is addressed separately in behavior analysis.

Observing response. The observing response is a topographically different operant that functions to produce a discriminative stimulus (S^D) or extinction stimulus (S^Δ) depending on whether reinforcement or extinction is in effect. In other words, an observing response changes a mixed schedule of reinforcement to a multiple schedule. See mixed and multiple schedules.

Omission procedure (training). See negative automaintenance.

Ontogenetic. Each organism has a unique life history (ontogeny) that contributes to its behavior. Ontogenetic changes in behavior are caused by events that occur over the lifetime of an individual. Ontogenetic history builds on species history (phylogeny) to determine when, where, and what kind of behavior will occur at a given moment. See also phylogenetic.

Ontogenetic selection. The selection of operant behavior during the lifetime of an organism is ontogenetic selection. The process involves operant variability during periods of extinction and selection by contingencies of reinforcement. An organism that alters its behavior (adaptation) on the basis of changing life experiences is showing ontogenetic selection. In this ontogenetic form of adaptation, the topography and frequency of behavior increase when reinforcement is withheld (increase in operant variability). These behavioral changes during extinction allow for the selection of behavior by new contingencies of reinforcement. Thus, a wild rat that has been exploiting a compost heap may find that the homeowner has covered it. In this case, the rat emits various operants that may eventually uncover the food. The animal may dig under the cover, gnaw a hole in the sheathing, or search for some other means of entry. A similar effect occurs when food in the compost heap is depleted and the animal emits behavior that results in getting to a new food patch. In the laboratory, this behavior is measured as an increase in the topography and frequency of bar pressing as the schedules of reinforcement change.

Operant. An operant is behavior that operates on the environment to produce a change, effect, or consequence. These environmental changes select the operant appropriate to a given setting or circumstance. That is, particular responses increase or decrease in a situation as a function of the consequences they produced in the past. Operant behavior is emitted (rather than elicited) in the sense that the behavior may occur at some frequency before any known conditioning.

Operant aggression. Refers to aggressive behavior that is reinforced (increased) by the removal of an aversive event arranged by another member of the species. See also negative reinforcement.

Operant chamber. A laboratory enclosure or box used to investigate operant conditioning. An operant chamber for a rat is a small, enclosed box that typically contains a lever with a light above it and a food magazine or cup connected to an external feeder. The feeder delivers a small food pellet when electronically activated.

Operant class. Refers to a class or set of responses that vary in topography but produce a common environmental consequence or effect. The response class of turning on the light has many variations in form (turn on light with left index finger, or right one, or side of the hand, or saying to someone "Please turn on the light").

Operant conditioning. An increase or decrease in operant responses as a function of the consequences that have followed these responses.

Operant imitation. Operant imitation is imitative behavior controlled by its consequences. See imitation.

Operant level. Refers to the rate of an operant before any known conditioning. For example, the rate of key pecking before a peck–food contingency has been established.

Operant rate. See rate of response.

Operant variability. Operant behavior becomes increasingly more variable as extinction proceeds. From an evolutionary view, it makes sense to try different ways of acting when something no longer works. That is, behavioral variation increases the chances that the organisms will reinstate reinforcement or contact other sources of reinforcement, increasing the likelihood of survival and reproduction of the organism.

Overcorrection. Overcorrection is a positive punishment procedure that uses "restitution" to reduce or eliminate destructive or aggressive behavior. Overcorrection may also involve *positive practice*, requiring the violator to intensively practice an overly correct form of the action.

Overmatching. In the generalized matching equation, a value of *a* greater than 1 indicates that changes in the response ratio (B_a/B_b) are larger than changes in the ratio of reinforcement (R_a/R_b). This outcome occurs because relative behavior increases faster than predicted from relative rate of reinforcement. See also generalized matching law.

Overshadowing. This effect occurs when a compound stimulus is used as the conditioned stimulus (CS) in a respondent conditioning experiment. For example, a light + tone (CS) may be presented at the same time and be associated with an unconditioned stimulus (US) such as food. The most *salient* property of the compound stimulus comes to regulate exclusively the conditioned response. Thus, if the tone is more salient than the light, only the tone will elicit salivation.

Paradoxical effects of punishment. Refers to the evidence that response-produced shock resembles some of the effects of positive reinforcement (FI scalloping). The shocks, however, do not actually function as positive reinforcement.

Partial reinforcement effect (PRE). See intermittent reinforcement effect.

Peak shift. A shift that occurs in the peak of a generalization gradient away from an extinction stimulus (S^Δ) is called peak shift. See generalization gradient.

Permanence of punishment. Refers to a debate as to whether punishment by itself, without additional procedures like extinction or reinforcement of alternative behavior, can permanently eliminate undesirable behavior.

Personalized system of instruction (PSI). A college teaching method based on principles of operant conditioning and designed by Fred Keller (1968). Keller called his teaching method a personalized system of instruction or PSI. Basically, PSI courses are organized such that students move through the course at their own pace and they are reinforced for completing small course units.

Phenotype. An organism's phenotype refers to anatomical and behavioral characteristics observed during the lifetime of the individual. For example, an individual's size, color, and shape are anatomical aspects of phenotype. Behavioral features include taste preferences, aggressiveness, and shyness. Different phenotypic attributes of individuals may or may not reflect underlying genetic variation.

Phylogenetic. Behavior relations that are based on the genetic endowment of an organism are called phylogenetic and are present on the basis of species history. Behavior that aids survival or procreation is often (but not always) unlearned. This is because past generations of organisms that engaged in such behavior survived and reproduced. These animals passed on (to the next generation) the characteristics (via genes) that allowed similar behavior. Thus, species history provides the organism with a basic repertoire of responses that are evoked by environmental conditions. See also ontogenetic.

Phylogeny. Phylogeny is the species history of an organism.

Placebo effect. Concerns the effect of an inert substance such as a sugar pill on the "physiological well-being" of a patient. That is, patients treated with sugar pills show improvements relative to a no-treatment control group.

Polydipsia. Polydipsia or excessive drinking is adjunctive behavior induced by the time-based delivery of food. For example, a rat that is working for food on an intermittent schedule may drink as much as half its body weight during a single session. This drinking occurs even though the animal is not water deprived. See also adjunctive behavior.

Positive contrast. See behavioral contrast.

Positive punishment. Refers to a procedure that involves the presentation of an event or stimulus following behavior that has the effect of decreasing the rate of response. A child is given a spanking for running into the street and the probability of the behavior is decreased.

Positive reinforcement. Positive reinforcement is a contingency that involves the presentation of an event or stimulus following an operant that increases the rate of response.

Positive reinforcer. A positive reinforcer is any stimulus or event that increases the probability (rate of response) of an operant when presented.

Postreinforcement pause (PRP). The pause in responding that occurs after reinforcement on some intermittent schedules (e.g., FR, FI) is called the postreinforcement pause.

Power law for matching. See generalized matching.

Precision teaching. In what became known as precision teaching, Ogden Lindsley devised a method of systematic instruction that encouraged students and teachers to target specific behaviors; count, time, and graph them; and revise instructional procedures based on the charted data. The use of the Standard Celeration Chart for graphing change in response rate over days is a prominent feature of this teaching method.

Precurrent behavior. Refers to operant behavior that precedes a current response. Precurrent behavior often functions to establish stimulus control over subsequent operant behavior, as when a person sets the alarm for 6:00 a.m. (precurrent behavior) to ensure stimulus control by the clock over waking up and going to an appointment or job (current behavior). In this example, both the precurrent and current behavior are maintained by the reinforcement contingency (e.g., avoiding the consequences of being late). When precurrent behavior is private, as in thinking about chess moves, the behavior provides S^D control over the actual movement of the chess pieces. Thinking about chess moves and actual moves are maintained by the contingency of reinforcement involving getting a momentary advantage and ultimately winning the game. See construction of S^Ds.

Preference. When several schedules of reinforcement are available concurrently, one alternative may be chosen more frequently than others. When this occurs, we say that the organism shows a preference for that alternative.

Preference for choice. When equated for differential outcomes, humans and other animals show a preference for options that allow them to make choices compared with options that limit or restrict the opportunity to choose.

Preference reversal. The term refers to the change in value of a reinforcer as a function of time to the choice point (as in self-control). For example, people make a commitment to save their money (monthly deduction at the bank) rather than spend it because the value of saving is greater than spending when far from the choice point (getting paid). At the choice point, spending is always higher in value than saving the money.

Premack principle. A higher-frequency behavior will function as reinforcement for a lower-frequency behavior.

Preparedness. Some relations between stimuli, and between stimuli and responses, are more likely because of phylogenetic history. This phenomenon has been called preparedness. For example, a bird that relies on sight for food selection would be expected to associate the appearance of a food item and illness, but rats that select food on the basis of taste quickly make a flavor–illness association.

Preratio pause. The number of responses (ratio size) required and the magnitude of the reinforcer have both been shown to influence postreinforcement pause (PRP). Calling this pause a "post" reinforcement event accurately locates the pause but the ratio size is what actually controls it. Hence, many researchers refer to the PRP as a preratio pause. See postreinforcement pause (PRP).

Primary aversive stimulus. Refers to an aversive stimulus that has acquired its properties as a function of species history. See aversive stimulus.

Primary laws of the reflex. The primary laws of the reflex include (1) the law of the threshold, (2) the law of intensity–magnitude, and (3) the law of the latency. These laws govern the US → UR relationship.

Private behavior. Behavior that is only accessible to the person who emits it (e.g., thinking).

Probability of response. The probability that an operant will occur on a given occasion (measured as rate of response).

Progressive-ratio (PR) schedule. Refers to a schedule where the number of responses (ratio) increases (or decreases) after reinforcement. For example, a pigeon on an increasing progressive ratio may be required to make 2 responses for access to food, then 4, 8, 16, 32, and higher ratios. In a foraging model, the increasing progressive-ratio schedule simulates a depleting patch of food.

Punisher. A stimulus that decreases the frequency of an operant that produces it.

Punishment. As a procedure, punishment involves following an operant with a punisher. Usually, the operant is maintained by positive reinforcement so that punishment is superimposed on a baseline of positive reinforcement. Punishment also refers to a decrease in operant behavior when followed by a punisher or when reinforcement is withdrawn contingent on responding. See positive and negative punishment.

Quantitative law of effect. The law states that the absolute rate of response on a schedule of reinforcement is a hyperbolic function of rate of reinforcement on the schedule relative to the total rate of reinforcement (both scheduled and extraneous reinforcement).

That is, as the rate of reinforcement on the schedule increases, the rate of response also rises, but eventually further increases in rate of reinforcement produce less and less of an increase in rate of response (hyperbolic). Also, the rise in rate of response with increasing rate of reinforcement is modified by extraneous sources of reinforcement (R_e). Extraneous reinforcement reduces the rate of response on the reinforcement schedule. One implication is that control of behavior by a schedule of reinforcement is weakened by sources of extraneous reinforcement.

A proportional matching equation is one mathematical expression of the quantitative law of effect. The equation relates absolute response and reinforcement rates, using alternative sources of reinforcement as the context. The equation may be derived from a restatement of the proportional matching law and is written as $B_a/(B_a + B_e) = R_a/(R_a + R_e)$. In this equation, B_e refers to all behavior directed to extraneous sources of reinforcement, and R_e represents these sources. The term B_a represents rate of response on the programmed schedule, and R_a is the rate of scheduled reinforcement.

Range of variability (in assessment). Changes in level produced by the treatment must be assessed in relation to the visual inspection of the range of variability of the dependent variable. The range of variability is the difference between highest and lowest values of the dependent measures in baseline and treatment phases of the experiment.

Rate of response (operant rate). Refers to the number of responses that occur in a given interval. For example, a bird may peck a key for food two times per second. A student may do math problems at the rate of 10 problems per hour.

Ratio schedules. Response-based schedules of reinforcement are ratio schedules; these schedules are set to deliver reinforcement following a prescribed number of responses. The ratio specifies the number of responses for each reinforcer.

Ratio strain. A disruption of responding that occurs when a ratio schedule is increased rapidly. For example, faced with a change in the schedule from continuous reinforcement (CRF) to the large fixed-ratio (FR) value, an animal will probably show ratio strain in the sense that it pauses longer and longer after reinforcement. This occurs because the time between successive reinforcements contributes to the postreinforcement pause (PRP). The pause gets longer as the interreinforcement interval (IRI) increases. Because the PRP makes up part of the interval between reinforcements and is controlled by it, the animal eventually stops responding. Thus, there is a negative feedback loop between increasing PRP length and the time between reinforcements (IRI). See postreinforcement pause (PRP) and interreinforcement interval (IRI).

Reaction chain. Reaction chains are phylogenetic sequences of behavior. An environmental stimulus sets off behavior that produces stimuli that set off the next set of responses in the sequence; these behaviors produce the next set of stimuli and so on. Presenting stimuli that prompt responses ordinarily occurring in the middle part of the sequence will start the chain at that point rather than at the beginning. Reaction chains are like consecutive sets of reflexes where the stimulus that elicits the next response in the sequence is produced by the previous reflex.

Reflex. When an unconditioned stimulus (US) elicits an unconditioned response (US → UR), the relationship is called a reflex.

Reflexivity. Involves showing an equivalence relation for a stimulus class (A = A), using a procedure of identity matching. A pigeon shows reflexivity when the bird repeatedly matches samples of line angles to identical line angles in the comparison displays. Also, on a generalization test, the bird matches color samples to color comparisons without any specific training on colors. The bird has learned to find the "same" dimension (angle or color) in the comparisons as portrayed in the samples.

Reinforcement. Involves an increase in the rate of operant behavior as a function of its consequences. Also, refers to the procedure of presenting a reinforcing event when a response occurs.

Reinforcement efficacy. Most of the applied research on progressive-ratio (PR) schedules uses the giving-up or breakpoint as a way of measuring reinforcement efficacy or effectiveness, especially of drugs like cocaine. The breakpoint for a drug indicates how much operant behavior the drug will sustain at a given dose. If breakpoints for two drugs are different, we can say that the drug with the higher breakpoint has greater reinforcement efficacy.

Reinforcement function. Any event (or stimulus) that follows a response and increases its frequency is said to have a reinforcement function. If an infant's babbling increases due to touching by the mother, we can say that maternal touching has a reinforcement function.

Reinforcer pathology. The presence of two distinct but likely interacting repertoires involving consumption of a reinforcer that tend to be at the extremes of the distribution of behaviors. The two processes are 1) high valuation of a reinforcer, as defined by inelastic demand and 2) the excessive preference for the immediate consumption of a commodity despite long-term negative outcomes (delay discounting).

Reinstatement (of response). The recovery of behavior when the reinforcer is presented alone (response independent) after a period of extinction. In an operant procedure, reinstatement involves reinforcement of a response followed by extinction. After extinction, response-independent reinforcement is arranged and the opportunity to respond is removed (using retractable levers). This is followed by tests that reinstate the opportunity to respond (response levers available).

Relative rate of reinforcement. When two or more sources of reinforcement are available (as on a concurrent schedule), relative rate of reinforcement refers to the rate of reinforcement delivered on one alternative divided by the sum of the rates of reinforcement from all sources of reinforcement. Relative rate of reinforcement is a measure of the distribution of reinforcement between or among alternatives.

Relative rate of response. When two or more sources of reinforcement are available (as on a concurrent schedule), relative rate of response refers to rate of response on one alternative divided by the sum of the response rates on all alternatives. Relative rate of response is a measure of the distribution of behavior between or among alternative sources of reinforcement.

Relative stimulus control. Relative stimulus control involves the organism responding to differences between two or more stimuli. For example, a pigeon may be trained to peck in the presence of the larger of two triangles rather than to the absolute size of a triangle. See also absolute stimulus control.

Relational frame theory. A behavioral theory of human cognition and communication that is based on relations between stimuli (e.g., A < B < C). Transfer of function occurs within the frame when one of those stimuli is associated with a function or consequence, such as an aversive stimulus. For example, if B is associated with an aversive that causes fear, a smaller amount of fear will transfer to A and a greater amount to stimulus B. Relational frames explain the rapid and generative manner of learning without direct experience with contingencies of reinforcement or punishment.

Reliability of observation. In applied behavior analysis, reliability of observation involves the amount of agreement among observers who independently record the same behavior. One way to assess reliability is to count the number of times two observers agree that a target behavior did (or did not) occur. This can be expressed as a percentage agreement that varies from 0 to 100%. Generally, applied behavior analysts strive for reliability of greater than 80% agreement.

Remembering. The verb *remembering* (or *forgetting*) is used to refer to the effect of some event on behavior after the passage of time (as opposed to the noun *memory*, which seems to refer to a mental representation stored in the brain). According to White (2002), remembering is not so much a matter of looking back into the past or forward into the future as it is of *making choices at the time of remembering*.

Remote causation. Typical of sciences like evolutionary biology, geology, and astronomy. In this case, we explain some phenomenon by pointing to remote events that made it likely. Thus, natural selection for coloration explains the current frequency of the characteristic in the population.

Renewal (of responding). One type of post-extinction effect is called renewal, involving the recovery of responding when the animal is removed from the extinction context. In respondent extinction, such recovery of responding is well established and is thought to occur because of inhibitory learning to the extinction context (Bouton, 2004). Once the animal is removed from the extinction setting, the contextual cues for inhibition no longer occur and responding recovers. A similar effect is observed with operant behavior after extinction, but the evidence is not as extensive.

Repertoire (of behavior). All the behavior an organism is capable of emitting on the basis of species and environmental history.

Replication (of results). Replication of results is used to enhance both internal and external validity of an experiment. If results replicate over time and place, it is likely that the original findings were due to the experimental variable and not due to extraneous conditions (internal validity). Replication also establishes that the findings have generality in the sense that the effects are not limited to specific procedures, behaviors, or species (external validity). See also direct and systematic replication.

Rescorla–Wagner model. The basic idea of the Rescorla–Wagner model of respondent conditioning is that a conditioned stimulus (CS) acquires a limited amount of associative strength on any one trial. The term *associative strength* describes the relation between the CS and the magnitude of the conditioned response (CR). In general, associative strength increases over conditioning trials and reaches some maximum level. A given CS can acquire only so much control over a CR. This is the maximum associative strength for the CS. Thus, a tone (CS) paired with 1 g of food will have maximum associative strength when conditioned salivation (CR) has the same strength as unconditioned salivation (UR) elicited by the gram of food (US). The magnitude of the UR to the US sets the upper limit for the CR. The CS cannot elicit a greater response (CR) than the one produced by the US.

Resistance to extinction. Refers to the perseverance of operant behavior when it is placed on extinction. Resistance to extinction is substantially increased when an intermittent schedule of reinforcement has been used to maintain behavior. See intermittent reinforcement effect.

Respondent. Respondent is behavior that increases or decreases by the presentation of a conditioned stimulus (CS) that *precedes* the conditioned response (CR). We say that the presentation of the CS regulates or controls the respondent (CR). Respondent behavior is elicited, in the sense that it reliably occurs when the CS is presented. The notation system used with elicited behavior is CS → CR. The CS causes (arrow) the CR.

Respondent acquisition. Refers to the procedure of pairing the conditioned stimulus (CS) with the unconditioned stimulus (US) over trials when respondent level for the CS is near zero. Also, refers to the increase in magnitude of the conditioned response (CR) when respondent level for the CS is near zero.

Respondent conditioning. Respondent conditioning occurs when an organism responds to a new event based on a history of pairing with a biologically important stimulus. The Russian physiologist Ivan Pavlov discovered this form of conditioning at the turn of the 20th century. He showed that dogs salivated when food was placed in their mouths. This relation between the food stimulus and salivation is called a reflex and occurs because of the animal's biological history. When Pavlov rang a bell just before feeding the dog, it began to salivate at the sound of the bell. In this way, new features (sound of bell) controlled the dog's respondent behavior (salivation). Thus, presenting stimuli together in time (typically CS then US) is the procedure for respondent conditioning. If a conditioned stimulus (CS) comes to regulate the occurrence of a conditioned response (CR), respondent conditioning has occurred.

Respondent discrimination. Respondent discrimination occurs when an organism shows a conditioned response to one stimulus but not to other similar events. A discrimination procedure involves positive and negative conditioning trials. For example, a positive trial occurs when a CS+ such as a 60-dB tone is followed by an unconditioned stimulus like food. On negative trials, a 40-dB tone is presented (CS−) but not followed by food. Once a differential response occurs (salivation to 60 dB but not to 40 dB), we may say that the organism discriminates between the tones.

Respondent extinction. The procedure of respondent extinction involves the presentation of the conditioned stimulus (CS) without the unconditioned stimulus (US) after acquisition has occurred. As a behavioral process, extinction refers to a decline in the strength of the conditioned response (CR) when an extinction procedure is in effect. In both instances, the term *extinction* is used correctly.

Respondent generalization. Respondent generalization occurs when an organism shows a conditioned response (CR) to values of the conditioned stimulus (CS) that have not been trained. For example, if a tone of 375 Hz is followed by food, a dog will salivate at maximum level when this tone is presented. The animal, however, may salivate to other values of the tone. As the tone differs more and more from 375 Hz, the CR decreases in magnitude.

Response chain. Refers to a sequence of discriminative stimuli and responses where each response produces a change in the stimulus controlling behavior. Once established, each discriminative stimulus (S^D) in the chain has two functions—acting as a conditioned reinforcer for the response that produced it and as a discriminative stimulus for the next response in the sequence.

Response class. A response class refers to all the forms of the performance that have a similar function (e.g., putting on a coat to keep warm). In some cases, the responses in a class have close physical resemblance, but this is not always the case. For example, saying "Please open the door" and physically opening the door are members of the same response class if both result in an open door.

Response cost. Refers to a negative punishment procedure in which conditioned reinforcers (tokens) are removed contingent on behavior, and the behavior decreases.

Response deprivation. Occurs when access to the contingent behavior is restricted and falls below its baseline (or free-choice) level of occurrence.

Response deprivation hypothesis. The principle that organisms work to gain access to activities that are restricted or withheld (deprivation), presumably to reinstate equilibrium or free-choice levels of behavior. This principle is more general than the Premack principle, predicting when any activity (high or low in rate) will function as reinforcement.

Response differentiation. When reinforcement is contingent on some difference in response properties, that form of response will increase. For example, the force or magnitude of response can be differentiated; if the contingencies of reinforcement require a forceful or vigorous response in a particular situation, then that form of response will predominate. In another example, when reinforcement is based on short interresponse times (IRT, 2–5 s), the distribution of IRTs becomes centered on short intervals. Changing the contingencies to reinforce longer IRTs (20–25 s) produces a new distribution centered on long intervals. See differential reinforcement.

Response generalization. Response generalization occurs when a target response is strengthened and other similar responses increase in frequency (e.g., a child reinforced for building a house out of LEGO™ subsequently may arrange the pieces in many different ways).

Response hierarchy. With regard to responses *within a response class*, a response hierarchy refers to the order or likelihood of the response forms in the class based on response properties (effort) or probability of reinforcement in a given situation. For a child, the parents may have differentially reinforced shouting rather than quiet conversation at the dinner table and loud talk has a higher probability of occurrence at dinner than talk at less volume. For a *free-choice or baseline assessment* (Premack, 1962), the responses in different classes for a situation are arranged in a hierarchy (*between response classes*) by relative frequency or probability of occurrence. For a rat the probability of eating, drinking, and wheel running might form a hierarchy with eating occurring most often and wheel running least.

Response–shock interval (R–S). On an avoidance schedule, the time from a response that postpones shock to the onset of the aversive stimulus, assuming another response does not occur. See also the shock–shock interval (S–S).

Resurgence. After a period of reinforcement, the increase in behavioral variability or topography during extinction is called resurgence.

Retention interval. The time between the offset of the sample stimulus and the onset of the comparison stimuli is the retention interval.

Reversal test. Once the matching of angle to geometric form is well established, a reversal test (form to angle) is conducted without any further reinforcement. In a reversal test, the bird is presented with a triangle as the sample and the question is whether it pecks the side key with the horizontal line. Because horizontal = triangle was trained, the bird shows symmetry if it pecks the horizontal comparison key when presented with a triangle sample (triangle = horizontal). Similarly, because vertical = circle was trained, symmetry is shown if the bird pecks the vertical side key when the circle is presented as the sample (circle = vertical). In everyday language, the bird responds as if the horizontal line stands for triangle and as if the vertical line means circle. The percentage of "correct" responses during the test (without reinforcement) is the usual measure of symbolic performance on this reversal test.

Rule-governed behavior. Denotes the effects of contingency-specifying stimuli on the listener's behavior. When instructions, rules, advice, maxims, and laws regulate operant behavior, the behavior is said to be rule governed. Control by instructions can make operant behavior insensitive to the operating contingencies of reinforcement.

Run of responses. A fast burst of responding is called a run. For example, after the postreinforcement pause (PRP) on a fixed-ratio (FI) schedule, an organism will rapidly emit the responses required by the ratio.

Salience. The symbol S in the Rescorla–Wagner equation is a constant that varies between 0 and 1, and may be interpreted as the salience (e.g., dim light versus bright light) of the conditioned stimulus (CS) based on the sensory capacities of the organism. The constant S (salience) is estimated after conditioning and determines how quickly the associative strength of the CS rises to maximum. That is, a larger salience coefficient makes the associative strength of the CS rise more quickly to its maximum.

Satiation. Repeated presentations of a reinforcer weaken its effectiveness, and for this reason rate of response declines. Satiation refers to this effect, and the repeated presentation of a reinforcer is called a satiation operation.

Scalloping. Refers to the characteristic pattern of response seen on a cumulative record produced by a fixed-interval (FI) schedule. There is a pause after reinforcement, then a few probe responses, and finally an increasingly accelerated rate of response to the moment of reinforcement.

Schedule-induced behavior. See adjunctive behavior.

Schedule of reinforcement. In relation to responses, a schedule of reinforcement is the arrangement of the environment in terms of discriminative stimuli and behavioral consequences. Mechner notation describes these behavioral contingencies.

Science of behavior. See behavior analysis.

S-delta (S^Δ). When an operant does not produce reinforcement, the stimulus that precedes the operant is called an S-delta (S^Δ). In the presence of an S-delta, the probability of emitting an operant declines. See extinction stimulus.

Second-order conditioning. Second-order conditioning involves pairing two CSs (CS_1 + CS_2), rather than a CS and US (CS + US). Pavlov (1927/1960) conducted the early experiments on second-order conditioning. The tick of a metronome was paired with food. The sound of the metronome came to elicit salivation. Once the ticking sound reliably elicited salivation, Pavlov paired it with the sight of a black square (CS_1 + CS_2). Following several pairings of the metronome beat with the black square, the sight of the black square elicited salivation.

Second-order schedule of reinforcement. A second-order schedule involves two (or more) schedules of reinforcement in which completion of the requirements of one schedule is reinforced according to the requirements of a second schedule.

Selection by consequences. From a behavioral viewpoint, the principle of causation for biology, behavior, and culture is selection by consequences. With regard to biology, mutation and sexual reproduction ensure a range of variation in genes that code for the features of organisms. Some physical attributes of the organisms, coded by genes, meet the requirements of the environment. Organisms with these adaptive features survive and reproduce, passing their genes to the next generation (phylogenetic). Organisms without these characteristics do not reproduce as well and their genes are less represented in the subsequent generations. Natural selection is a form of selection by consequences that occurs at the biological level.

Selection by consequences has been extended to the level of behavior as the principle of reinforcement. Operant behavior is an expressed characteristic of many organisms, including humans. Organisms with an extensive range of operant behavior adjust to new environmental situations on the basis of the consequences that follow behavior. This kind of selection occurs over the lifetime of the individual (ontogenetic) and behavior change is a form of evolution. Brain neurons are probably the units selected at the behavioral level; the interplay of neurons allows for behavior to be passed on from one moment to the next (transmitted). The process of the selection and change of operant behavior is analogous to evolution and natural selection at the genetic level. Reinforcement is therefore an ontogenetic process that extends selection by consequences to the level of behavior.

A third level of evolution and selection occurs at the cultural level (cultural selection). The unit of selection at this level is the cultural practice or meme. Cultural practice involves the interlocking behavior of many people. As with operant behavior itself, cultural practices vary in form and frequency. Different ways of doing things are more or less successful in terms of efficiency, productivity, and survival of group members. Generally, group-level outcomes or effects (metacontingencies) increase or decrease the rate of adoption and transmission of practices in the population. The fit between current practices and new ways of doing things (e.g., technology) plays a role in adoption and transmission of innovations by the group. Although an innovative technology or method may be more efficient, it may also be more costly to change from traditional to new ways of doing things.

Self-control. From a behavioral perspective, self-control occurs when a person emits a response that affects the probability of subsequent behavior—giving up immediate gains for greater long-term benefits or accepting immediate costs for later rewards. When people (and other organisms) manage their behavior in such a way that they choose the more beneficial long-range consequences, they are said to show self-control.

Sensory preconditioning. In respondent compound conditioning, two stimuli such as light and tone are repeatedly presented together (light + tone) without the occurrence of a US (preconditioning). Later, one of these stimuli (CS_1) is paired with an unconditioned stimulus (US) and the other stimulus (CS_2) is tested for conditioning. Even though the second stimulus (CS_2) has never been directly associated with the US, it comes to elicit the conditioned response (CR).

Sexual selection. Refers to the increased reproductive success of genes that code for attributes or behavior attractive (having a stimulus function) to the opposite sex. Individuals with these features and underlying genes have increased chances of copulation and more offspring compared with those who lack such attractiveness.

Shaping. The method of successive approximation or shaping may be used to establish a response. This method involves the reinforcement of closer and closer approximations to the final performance. For example, a rat may be reinforced for standing in the vicinity of a lever. Once the animal is reliably facing the lever, a movement of the head toward the bar is reinforced. Next, closer and closer approximations to pressing the lever are reinforced. Each step of the procedure involves reinforcement of closer approximations and nonreinforcement of more distant responses. Many novel forms of behavior may be shaped by the method of successive approximation.

Shock–shock interval (S–S). The shock–shock interval is the scheduled time between shocks using an avoidance procedure. The S–S interval is the time from one shock to the next if the avoidance response does not occur. See also the response–shock interval (R–S).

Side effects of schedules of reinforcement. Also referred to as schedule patterns, a side effect of a schedule of reinforcement is a predictable behavioral pattern that is not required of the contingency of the schedule, but occurs with remarkable replicability. Examples include the fixed-ratio break-and-run pattern or the fixed-interval scallop.

Sidman avoidance. See nondiscriminated avoidance.

Sign tracking. Sign tracking refers to approaching a sign (or stimulus) that signals a biologically relevant event. For example, dogs are required to sit on a mat and a stimulus that signals food is presented to the animal. When the food signal is presented, the dogs approach the stimulus and make food-soliciting responses to it.

Simultaneous conditioning. A respondent conditioning procedure in which the conditioned stimulus (CS) and unconditioned stimulus (US) are presented at the same moment. Compared with delayed conditioning, simultaneous conditioning produces a weaker conditioned response (CR).

Simultaneous discrimination. In simultaneous discrimination, the S^D and S^Δ are presented at the same time and the organism is reinforced for responding to the relative properties of one or the other. For example, a pigeon may be presented with two keys, both illuminated with white lights, but one light is brighter than the other. The bird is reinforced for pecking the dimmer of the two keys. Pecks to the other key are placed on extinction. After training, the pigeon will peck the darker of any two keys. See also relative stimulus control.

Single-subject research. Experimental research that is concerned with discovering principles and conditions that govern the behavior of single or individual organisms. Each individual's behavior is studied to assess the impact of a given experimental variable. In behavioral research, a change in the contingencies of reinforcement is assessed for each bird, rat, or human (e.g., changing the schedule of reinforcement, the operant, or the discriminative stimuli).

Social disruption. Refers to a negative side effect of punishment in which the person who delivers punishment and the context become conditioned aversive stimuli. Individuals will attempt to escape from or avoid the punishing person or setting.

Social episode. A social episode involves the interlocking contingencies between speaker and listener, as when a customer asks the waiter for a napkin and gets it (manding). The episode begins with the customer spilling her coffee (establishing the napkin as reinforcement) and ends when the waiter provides the napkin (reinforcement) and the customer says, "Thank you."

Spontaneous imitation. Refers to innate imitation based on evolution and natural selection (a characteristic of the species) rather than experiences during the lifetime of the individual. See imitation and generalized imitation.

Spontaneous recovery (operant). After a period of extinction, an organism's rate of response may be close to operant level. After some time, the organism is again placed in the setting and extinction is continued. Responding initially recovers, but over repeated sessions of extinction the amount of recovery decreases. Repeated sessions of extinction eliminate stimulus control by extraneous features of the situation and eventually "being placed in the setting" no longer occasions the operant.

Spontaneous recovery (respondent). An increase in the magnitude of the conditioned response (CR) after respondent extinction has occurred and time has passed. A behavioral analysis of spontaneous recovery suggests that the CS–CR relation is weakened by extinction, but the context or features of the situation elicit some level of the CR. During respondent conditioning, many stimuli not specified by the researcher as the conditioned stimulus (CS), but present in the experimental situation, come to regulate behavior.

S–S account of conditioned reinforcement. Refers to the hypothesis that it is necessary for a stimulus to be paired with primary reinforcement to become a conditioned reinforcer. The hypothesis has been largely discounted, and the weight of the evidence supports Fantino's (1969b) delay-reduction hypothesis. See delay-reduction hypothesis.

Steady-state performance. Schedule-controlled behavior that is stable and does not change over time is called steady-state performance. For example, after an extensive history on VI 30 s, a rat may press a lever at approximately the same rate day after day.

Stimulus class. Stimuli that vary across physical dimensions but have a common effect on behavior belong to the same stimulus class.

Stimulus control. A change in operant behavior that occurs when either an S^D or S^Δ is presented is called stimulus control. When an S^D is presented, the probability of response increases; and when an S^Δ is given, operant behavior has a low probability of occurrence.

Stimulus equivalence. Involves the presentation of one class of stimuli (e.g., flags) that occasions responses to other stimulus classes (e.g., countries). This seems to be what we mean when we say that the flag stands for, represents, or signifies our country. Equivalence relations such as these are an important aspect of human behavior. For example, in teaching a child to read, spoken words (names of animals) are trained to visual stimuli (pictures of animals) and then to written symbols (written words for animals). Eventually, the written word is said to stand for the actual object, in the same sense that a flag stands for a country.

Stimulus function. When the occurrence of an event changes the behavior of an organism, we may say that the event has a stimulus function. Both respondent and operant conditioning are ways to create stimulus functions. During respondent conditioning, an arbitrary event like a tone comes to elicit a particular response, like salivation. Once the tone is effective, it is said to have a conditioned-stimulus function for salivation. In the absence of a conditioning history, the tone may have no specified function and does not affect the specified behavior.

Stimulus generalization. Stimulus generalization occurs when an operant reinforced in the presence of a specific discriminative stimulus also is emitted in the presence of other stimuli. The process is called stimulus generalization because the operant is emitted to new stimuli that presumably share common properties with the discriminative stimulus.

Stimulus substitution. When a CS (e.g., light) is paired with a US (e.g., food) the CS is said to substitute for the US. That is, food evokes salivation and by conditioning the light elicits similar behavior.

Structural approach. In the structural approach, behavior is classified in terms of its form or topography. For example, many developmental psychologists are interested in the intellectual growth of children. These researchers often investigate what a person does at a given stage of development. The structure of behavior is emphasized because it is said to reveal the underlying stage of intellectual development. See also functional analysis.

Substitutability. This term is used to denote that a change in price of one reinforcer alters the consumption of a second reinforcer, holding income constant. For some commodities, consumption decreases with price, but consumption of a second commodity increases. The two commodities are said to be *substitutes*. Butter and margarine are substitutes if a shift in the price of butter results in more consumption of margarine. Beverages like Coke and Pepsi are another example of substitutes. Other commodities are *independents*. As the price of one commodity increases and its consumption decreases, the consumption of a second commodity does not change. Thus, your consumption of gasoline is independent of the price of theater tickets. A third way that commodities are related is as *complements*. As the price of one commodity increases and its consumption decreases, consumption of the other commodity also decreases. When the price of hot dogs increases and you eat fewer of them, your consumption of hot dog buns, relish, etc. also decreases.

Successive approximation. See shaping.

Successive discrimination. A procedure used to train differential responding is called successive discrimination. The researcher arranges the presentation of S^D and S^Δ so that one follows the other. For example, a multiple schedule is programmed so that a red light signals variable-interval (VI) food reinforcement; this is followed by a green light that indicates extinction is in effect.

Superstitious behavior. Behavior that is accidentally reinforced is called superstitious. For example, a parent may inadvertently strengthen aggressive behavior when a child is given his or her allowance just after fighting with a playmate. Switching from one alternative to another may be accidentally reinforced on a concurrent schedule if the alternative schedule has reinforcement setup. In this case, the organism is accidentally reinforced for a change from one schedule to another.

Symbolic matching. In a matching-to-sample task, symbolic matching involves the presentation of one class of stimuli as the sample (geometrical forms) and another set of stimuli (different line angles) as the comparisons. Reinforcement depends on an arbitrary relation (triangle = vertical).

Symmetry. When stimulus class A is shown to be interchangeable with stimulus class B (if A = B, then B = A), we may say that the organism shows symmetry between the stimulus classes. After training a form-to-angle discrimination (triangle = vertical), a reversal test is conducted without reinforcement using line angles as the sample and geometric shapes as the comparisons (vertical = triangle). An organism that passes the reversal test is said to demonstrate symmetry of angles and forms. See also reversal test.

Systematic replication. Refers to increasing the generality of an experimental finding by conducting other experiments in which the procedures are different but are logically related to the original research. An experiment is conducted with rats to find out what happens when food pellets are presented contingent on lever pressing. The observation is that lever pressing increases when followed by food pellets. In a systematic replication, elephants step on a treadle to produce peanuts. The observation is that treadle pressing increases. Both experiments are said to show the effects of positive reinforcement contingencies on operant behavior. See also direct replication.

Tacting. Denotes a class of verbal operants whose form is regulated by specific nonverbal discriminative stimuli. For example, a child may see a cat and say "Kitty." The word *tact* comes from the more familiar term *contact*. Tacting is verbal behavior that makes contact with the environment. In common parlance we say the people make reference to the world (language of reference), but in behavior analysis the world (stimuli) controls the verbal response class of tacting.

Tandem schedule. A tandem schedule is two or more basic schedules (CRF, FR, FI, VI, VR) presented sequentially in which only the final link ends with primary reinforcement (or in some cases extinction) and the component schedules are not signaled by discriminative stimuli. In other words, a tandem schedule is the same as an unsignaled chain schedule.

Taste aversion learning. When a distinctive taste (e.g., flavored liquid) is paired with nausea or sickness induced by a drug, X-ray, or even physical activity, the organism shows suppression of intake of the paired flavor.

Temporal pairing. In respondent conditioning, the pairing of the conditioned stimulus (CS) and unconditioned stimulus (US) in time. Temporal pairing is technically called CS–US contiguity.

Terminal behavior. On a schedule of reinforcement, as the time for reinforcement gets close, animals engage in activities related to the presentation of the reinforcer. For example, a rat will orient toward the food cup.

Textual behavior. Denotes a class of verbal operants regulated by verbal stimuli where there is correspondence between the stimulus and response, but no topographical similarity. The most common example of textual behavior is reading out loud. The child looks at the text, "See Dick, see Jane," and emits the spoken words, "See Dick, see Jane." The stimulus and response correspond, but the stimulus is visual and the response is vocal.

Timeout from avoidance. Refers to negative reinforcement of behavior that terminates, prevents, or postpones the avoidance contingencies of work or life. We value holidays, leaves of absence, and other periods that temporarily suspend or remove the everyday "shocks" and behavioral requirements that pervade our lives.

Timeout from positive reinforcement. This is a negative punishment procedure where the wrongdoer loses access to positive reinforcement for a specified period of time for engaging in the undesirable behavior.

Time sampling. A method of recording used mostly in applied behavior analysis. Behavior is sampled over a long time scale. The idea is to make observations at specified times throughout the day. For example, a patient on a psychiatric ward may be observed every 30 min, as a nurse does the rounds, and instances of psychotic talk are recorded.

Token economy. A reinforcement system based on token reinforcement; the contingencies specify when, and under what conditions, particular forms of behavior are reinforced. The system is an economy in the sense that tokens may be exchanged for goods and services, much like money is in our economy. This exchange of tokens for a variety of backup reinforcers ensures that the tokens are conditioned reinforcers. Token economies have been used to improve the behavior of psychiatric patients, juvenile delinquents, pupils in remedial classrooms, medical patients, those with alcohol and substance use disorders, prisoners, nursing-home residents, and persons with developmental disability.

Token schedule of reinforcement. Token schedules of reinforcement have three distinct components involving the token-production schedule, the exchange-production schedule, and the token-exchange schedule (Hackenberg, 2009). When we talk about token reinforcement, then, we are referring to three-component schedules that compose a higher-order sequence. Typically, one of the component schedules is varied while the other two components remain unchanged (held constant).

Tolerance (to a drug). When more of a drug (US) is needed to obtain the same drug effects (UR), we talk about drug tolerance. In respondent conditioning, the counteractive effects to CSs are major components of drug tolerance.

Topography. Refers to the physical form or characteristics of the response. For example, the way that a rat presses a lever with the left paw, the hind right foot, and so on. The topography of response is related to the contingencies of reinforcement in the sense that the form of response can be broadened or restricted by the contingencies. The contingency of reinforcement may require only responses with the left paw rather than any response that activates the microswitch—under these conditions left paw responses will predominate. Generally, topography or form is a function of the contingencies of reinforcement.

Trace conditioning. A respondent conditioning procedure in which the conditioned stimulus (CS) is presented for a brief period and after some time passes the unconditioned stimulus (US) occurs. Generally, as the time between the CS presentation and the occurrence of the US increases, the conditioned response (CR) becomes weaker. When compared to delayed conditioning, trace conditioning is not as effective.

Transition state. Refers to the instability of behavior generated by a change in contingencies of reinforcement. For example, when continuous reinforcement (CRF) contingencies are changed to FR 10, the pattern of response is unstable during the transition. After prolonged exposure to the FR contingency, the performance eventually stabilizes into a regular or characteristic pattern. See also steady-state performance.

Transitivity. An organism shows transitivity when it responds to stimulus class A as it does to stimulus class C or A = C after training that A = B and B = C. For example, if the written words *one, two, three* are equivalent to the arithmetic numbers 1, 2, and 3 and the words and these arithmetic numbers are equivalent to X and X, X and X, X, X then it logically follows that the words *one, two, and three* are equivalent to X and X, X and X, X, X—the relationship is transitive. An organism is said to show transitivity when it passes tests for transitivity after training for symbolic matching of stimulus class A (angles) to stimulus class B (geometric forms) and B (geometric forms) to C (intensity of illumination).

Trend (in baseline). A trend is a systematic decline or rise in the baseline values of the dependent variable. A drift in baseline measures can be problematic when the treatment is expected to produce a change in the same direction as the trend.

Trial-and-error learning. A term coined by Thorndike (1898) that he used to describe results from his puzzle-box and maze-learning experiments. Animals were said to make fewer and fewer errors over repeated trials, learning by trial and error.

Two-key procedure. On a concurrent schedule of reinforcement, the alternative schedules are presented on separate response keys.

Unconditioned reinforcer. Denotes a reinforcing stimulus that has acquired its properties as a function of species history. Although many reinforcers such as food and sex are general over species, other reinforcers such as the song of a bird or the scent of a mate are particular to a species. Behavior analysis, evolutionary biology, and neuroscience are necessary to describe, predict, and control the behavior regulated by unconditioned reinforcement.

Unconditioned response (UR). All organisms are born with a set of reflexes (US → UR). These relationships are invariant and biologically based. The behavior elicited by the unconditioned stimulus (US) is called the unconditioned response (UR).

Unconditioned stimulus (US). All organisms are born with a set of reflexes (US → UR). These relationships are invariant and biologically based. The eliciting event for the reflex is called the unconditioned stimulus (US).

Undermatching. In the generalized matching equation, the exponent *a* takes on a value less than 1. This result is described as undermatching and occurs when changes in the response ratio are less than changes in the reinforcement ratio. The effect is interpreted as low sensitivity to the programmed schedules of reinforcement. See also generalized matching law.

Use of punishment debate. Concerns the arguments and evidence for and against the use of punishment to control self-injurious and aggressive behavior in positive behavioral support programs.

US-pre-exposure effect. A procedure where animals are first given repeated exposures to the US by itself and then a series of CS → US pairings (conditioning). Compared to animals given pairings with a novel US, those familiar with the US (pre-exposed) show weaker and slower conditioning on the acquisition test.

Variable interval (VI). Refers to a schedule of reinforcement in which one response is reinforced after a variable amount of time has passed. For example, on a VI 30-s schedule, the time to each reinforcement changes but the average time is 30 s.

Variable ratio (VR). Refers to a response-based schedule of reinforcement in which the number of responses required for reinforcement changes after each reinforcer. The average number of responses is used to index the schedule. For example, a rat may press a lever for reinforcement 50 times, then 150, 70, 30, and 200. Adding these response requirements for a total of 500, then dividing by the number of separate response runs (5), yields the schedule value, VR 100.

Verbal behavior. Verbal behavior refers to the vocal, written, and gestural performances of a speaker, writer, or communicator. This behavior operates on the listener, reader, or observer, who arranges for reinforcement of the verbal performance. Verbal behavior only has indirect effects on the environment. This contrasts with nonverbal behavior, which usually results in direct and automatic consequences. When you walk toward an object, you come closer to it. Verbal behavior, on the other hand, works through its effects on other people. To change the position of a lamp, the speaker states "Lift the lamp at the back of the room" to a listener who is inclined to respond. Although verbal behavior is usually equated with speaking, vocal responses are only one of its forms. For example, a person may emit gestures and body movements that indirectly operate on the environment through their effects on others. A frown sets the occasion for others to remove some aversive event, while a smile may signal the observer to behave in ways that produce positive reinforcement.

Verbal community. The contingencies that regulate verbal behavior arise from the practices of people in the verbal community. The verbal community refers to the customary ways that people reinforce the behavior of the speaker. These customary ways or practices have evolved as part of cultural evolution. The study of the semantics and syntax of words and sentences (linguistics) describes the universal and specific contingencies arranged by the verbal community. In the behavioral view, language does not reside in the mind but in the social environment of the speaker.

REFERENCES

Abramson, L. Y., Seligman, M. E. P., & Teasdale, J. D. (1978). Learned helplessness in humans: Critique and reformulation. *Journal of Abnormal Psychology, 87*, 49–74.

Ader, R., & Cohen, N. (1981). Conditioned immunopharmacologic responses. In R. Ader (Ed.), *Psychoneuroimmunology* (pp. 281–319). New York: Academic Press.

Afifi, T. O., Mota, N. P., Dasiewicz, P., MacMillan, H. L., & Sareen, J. (2012). Physical punishment and mental disorders: Results from a nationally representative sample. *Pediatrics, 130*, 184–192. doi:10.1542/peds.2011–2947.

Ainslie, G. W. (1974). Impulse control in pigeons. *Journal of the Experimental Analysis of Behavior, 21*, 485–489.

Ainslie, G. W. (1975). Specious reward: A behavioral theory of impulsiveness and impulse control. *Psychological Bulletin, 82*, 463–496.

Ainslie, G. W. (2005). Précis of breakdown of will. *Behavioral and Brain Sciences, 28*, 635–673.

Alberto, P. A., & Troutman, A. C. (2013). *Applied behavior analysis for teachers* (9th ed.). Upper Saddle River, NJ: Pearson.

Alberto, P. A., Heflin, L. J., & Andrews, D. (2002). Use of the timeout ribbon procedure during community-based instruction. *Behavior Modification, 26*, 297–311.

Alferink, L. A., Critchfield, T. S., Hitt, J. L., & Higgins, W. J. (2009). Generality of the matching law as a descriptor of shot selection in basketball. *Journal of Applied Behavior Analysis, 42*(3), 595–608. https://doi.org/10.1901/jaba.2009.42-595

Alferink, L. A., Crossman, E. K., & Cheney, C. D. (1973). Control of responding by a conditioned reinforcer in the presence of free food. *Animal Learning and Behavior, 1*, 38–40.

Alsiö, J., Nordenankar, K., Arvidsson, E., Birgner, C., Mahmoudi, S., Halbout, B., et al. (2011). Enhanced sucrose and cocaine self-administration and cue-induced drug seeking after loss of VGLUT2 in midbrain dopamine neurons in mice. *Journal of Neuroscience, 31*, 12593–12603.

Alvarez, L. W. (1982). Experimental evidence that an asteroid impact led to the extinction of many species 65 million years ago. *Proceedings of the National Academy of Sciences, 80*, 627–642.

Alvarez, L. W., Asaro, F., & Michel, H. V. (1980). Extraterrestrial cause for the cretaceous–tertiary extinction—Experimental results and theoretical interpretation. *Science, 206*, 1095–1108.

Alvord, J. R., & Cheney, C. D. (1994). *The home token economy.* Cambridge, MA: Cambridge Center for Behavioral Studies.

Amat, J., Aleksejev, R. M., Paul, E., Watkins, L. R., & Maier, S. F. (2010). Behavioral control over shock blocks behavioral and neurochemical effects of later social defeat. *Behavioural Neuroscience, 165*, 1031–1038.

Amat, J., Baratta, M. V., Paul, E., Bland, S. T., Watkins, L. R., & Maier, S. F. (2005). Medial prefrontal cortex determines how stressor controllability affects behavior and dorsal raphe nucleus. *Nature Neuroscience, 8*, 365–371.

Amlung, M., Petker, T., Jackson, J., Balodis, I., & MacKillop, J. (2016). Steep discounting of delayed monetary and food rewards in obesity: A meta-analysis. *Psychological Medicine, 46*(11), 2423–2434. https://doi.org/10.1017/S0033291716000866

Amtzen, E., Nartey, R. K., & Fields, L. (2015). Enhanced equivalence class formation by the delay and relational functions of meaningful stimuli. *Journal of the Experimental Analysis of Behavior, 103*, 524–541.

Anderson, C. A., Buckley, K. E., & Carnagey, N. L. (2008). Creating your own hostile environment: A laboratory examination of trait aggressiveness and the violence escalation cycle. *Personality and Social Psychology Bulletin, 34*, 462–473.

Anderson, C. D., Ferland, R. J., & Williams, M. D. (1992). Negative contrast associated with reinforcing stimulation of the brain. *Society for Neuroscience Abstracts, 18*, 874.

Anderson, N. D., & Craik, F. I. (2006). The mnemonic mechanisms of errorless learning. *Neuropsychologica, 44*, 2806–2813.

Andre, J., Albanos, K., & Reilly, S. (2007). C-fos expression in the rat brain following lithium chloride-induced illness. *Brain Research, 1135*, 122–128.

Andrew, S. C., Perry, C. J., Barron, A. B., Berthon, K., Peralta, V., & Cheng, K. (2014). Peak shift in honey bee olfactory learning. *Animal Cognition, 17*, 1177–1186.

Anger, D. (1956). The dependence of interresponse times upon the relative reinforcement of different interresponse times. *Journal of Experimental Psychology, 52*, 145–161.

Anokhin, A. P., Grant, J. D., Mulligan, R. C., & Heath, A. C. (2015). The genetics of impulsivity: Evidence for the heritability of delay discounting. *Biological Psychiatry, 77*, 887–894.

Anrep, G. V. (1920). Pitch discrimination in a dog. *Journal of Physiology, 53*, 367–385.

Antonitis, J. J. (1951). Response variability in the white rat during conditioning, extinction, and reconditioning. *Journal of Experimental Psychology, 42*, 273–281.

Aoyama, K., & McSweeney, F. K. (2001). Habituation may contribute to within-session decreases in responding under high-rate schedules of reinforcement. *Animal Learning & Behavior, 29*(1), 79–91. https://doi.org/10.3758/BF03192817

Appel, J. B. (1961). Punishment in the squirrel monkey *Saimiri sciurea*. *Science, 133*, 36.

Appel, J. B., & Peterson, N. J. (1965). Punishment: Effects of shock intensity on response suppression. *Psychological Reports, 16*, 721–730.

Arantes, J., & Machado, A. (2011). Errorless learning of a conditional temporal discrimination. *Journal of the Experimental Analysis of Behavior, 95*, 1–20.

Arcediano, F., & Miller, R. R. (2002). Some constraints for models of timing: A temporal coding hypothesis perspective. *Learning and Motivation, 33*, 105–123.

Arnold, M. L., & Van Houten, R. (2013). Increasing following headway with prompts, goal setting, and feedback in a driving simulator. *Journal of Applied Behavior Analysis, 44*, 245–254.

Atalayer, D., & Rowland, N. E. (2011). Comparison of voluntary and foraging wheel activity on food demand in mice. *Physiology & Behavior, 102*, 22–29.

Austin, J., & Delaney, P. F. (1998). Protocol analysis as a tool for behavior analysis. *The Analysis of Verbal Behavior, 15*, 41–56.

Austin, J., Hackett, S., Gravina, N., & Lebbon, A. (2006). The effects of prompting and feedback on drivers' stopping at stop signs. *Journal of Applied Behavior Analysis, 39*, 117–121.

Austin, J., Hatfield, D. B., Grindle, A. C., & Bailey, J. S. (1993). Increasing recycling in office environments: The effects of specific, informative cues. *Journal of Applied Behavior Analysis, 26*(2), 247–253. https://doi.org/10.1901/jaba.1993.26-247

Autor, S. M. (1960). The strength of conditioned reinforcers as a function of frequency and probability of reinforcement. Unpublished doctoral dissertation, Harvard University, Cambridge, MA. Retrieved from http://krypton.mnsu.edu/%7Epkbrando/CommentaryP_C.htm.

Avargues-Weber, A., & Giurfa, M. (2013). Conceptual learning by miniature brains. *Proceedings of the Royal Society: B, 280*. doi:10.1098/rspb.2013.1907.

Ayllon, T., & Azrin, N. H. (1968). *The token economy: A motivational system for therapy and rehabilitation*. New York: Appleton-Century-Crofts.

Ayllon, T., & Michael, J. (1959). The psychiatric nurse as a behavioral engineer. *Journal of the Experimental Analysis of Behavior*, *2*, 323–334. https://doi.org/10.1901/jeab.1959.2-323

Ayres, K. M., & Langone, J. (2005). Intervention and instruction with video for students with autism: A review of the literature. *Education and Training in Developmental Disabilities*, *40*, 183–196.

Azar, B. (2002). Pigeons as baggage screeners, rats as rescuers. *Monitor on Psychology*, *33*, 42–44.

Azrin, N. H. (1956). Effects of two intermittent schedules of immediate and nonimmediate punishment. *Journal of Psychology*, *42*, 3–21.

Azrin, N. H., & Holz, W. C. (1966). Punishment. In W. K. Honig (Ed.), *Operant behavior: Areas of research and application* (pp. 380–447). New York: Appleton-Century-Crofts.

Azrin, N. H., Besalel-Azrin, V. A., & Azrin, R. D. (1999). *How to use positive practice, self-correction, and overcorrection*. Austin, TX: Pro-Ed.

Azrin, N. H., Hake, D. F., & Hutchinson, R. R. (1965). Elicitation of aggression by a physical blow. *Journal of the Experimental Analysis of Behavior*, *8*, 55–57.

Azrin, N. H., Holz, W. C., & Hake, D. F. (1963). Fixed-ratio punishment. *Journal of the Experimental Analysis of Behavior*, *6*(2), 141–148. https://doi.org/10.1901/jeab.1963.6-141

Azrin, N. H., Hutchinson, R. R., & Hake, D. F. (1966). Extinction-induced aggression. *Journal of the Experimental Analysis of Behavior*, *9*, 191–204.

Azrin, N. H., Hutchinson, R. R., & Sallery, R. D. (1964). Pain aggression toward inanimate objects. *Journal of the Experimental Analysis of Behavior*, *7*, 223–228.

Badia, P., Harsh, J., Coker, C. C., & Abbott, B. (1976). Choice and the dependability of stimuli that predict shock and safety. *Journal of the Experimental Analysis of Behavior*, *26*, 95–111.

Baer, D. M. (1982a). The imposition of structure on behavior and the demolition of behavioral structures. In D. J. Bernstein (Ed.), *Response structure and organization: The 1981 Nebraska symposium on motivation* (pp. 217–254). Lincoln: University of Nebraska Press.

Baer, D. M. (1982b). The role of current pragmatics in the future analysis of generalization technology. In R. B. Stuart (Ed.), *Adherence, compliance and generalization in behavioral medicine* (pp. 192–212). New York: Brunner/Mazel.

Baer, D. M., & Deguchi, H. (1985). Generalized imitation from a radical-behavioral viewpoint. In S. Reiss & R. Bootzin (Eds.), *Theoretical issues in behavior therapy* (pp. 179–217). New York: Academic Press.

Baer, D. M., & Detrich, R. (1990). Tacting and manding in correspondence training: Effects of child selection of verbalization. *Journal of Experimental Analysis of Behavior*, *54*, 23–30.

Baer, D. M., & Guess, D. (1971). Receptive training of adjectival inflections in mental retardates. *Journal of Applied Behavior Analysis*, *4*, 129–139.

Baer, D. M., & Sherman, J. A. (1964). Reinforcement control of generalized imitation in young children. *Journal of Experimental Child Psychology*, *1*, 37–49.

Baer, D. M., Peterson, R. F., & Sherman, J. A. (1967). The development of imitation by reinforcing behavioral similarity to a model. *Journal of the Experimental Analysis of Behavior*, *10*, 405–416.

Baker, A. G., Steinwald, H., & Bouton, M. E. (1991). Contextual conditioning and reinstatement of extinguished instrumental responding. *The Quarterly Journal of Experimental Psychology Section B: Comparative and Physiological Psychology*, *43*, 199–218.

Baer, D. M., Wolf, M. M., & Risley, T. R. (1968). Some current dimensions of applied behavior analysis. *Journal of Applied Behavior Analysis*, *1*(1), 91.

Baker, T. B., & Tiffany, S. T. (1985). Morphine tolerance as habituation. *Psychological Review*, *92*, 78–108.

Baldwin, J. M. (1906). *Mental development, methods, and processes*. New York: Macmillan.

Bandura, A. (1965). Influence of models' reinforcement contingencies on the acquisition of imitative responses. *Journal of Personality and Social Psychology*, *1*, 589–595.

Bandura, A. (1969). *Principles of behavior modification*. New York: Holt, Rinehart, & Winston.

Bandura, A. (1971). Vicarious and self-reinforcement processes. In R. Glaser (Ed.), *The nature of reinforcement* (pp. 228–278). New York: Academic Press.

Bandura, A. (1974). Behavior theory and the models of man. *American Psychologist, 29*, 859–869.

Bandura, A. (1977). *Social learning theory*. Englewood Cliffs, NJ: Prentice-Hall.

Bandura, A. (1986). *Social foundations of thought and action: A social cognitive theory*. Englewood Cliffs, NJ: Prentice-Hall.

Bandura, A. (1997). *Self-efficacy: The exercise of control*. New York: Freeman.

Bandura, A., & Locke, E. (2003). Negative self-efficacy and goal effects revisited. *Journal of Applied Psychology, 88*, 87–99.

Bandura, A., Ross, D., & Ross, S. A. (1963). Imitation of film-mediated aggressive models. *Journal of Abnormal and Social Psychology, 66*, 3–11. https://doi.org/10.1037/h0048687

Banna, K. M., DeVries, D., & Newland, M. C. (2011). Choice in the bluegill (Lepomis macrochirus). *Behavioural processes, 88*(1), 33–43.

Barbera, M. L. (2007). *The verbal behavior approach: How to teach children with autism and related disorders*. London, UK: Jessica Kingsley Publishers.

Bard, K. A. (2007). Neonatal imitation in chimpanzees (Pan troglodytes) tested with two paradigms. *Animal Cognition, 10*, 233–242.

Barnes-Holmes, D., & Harte, C. (2022). Relational frame theory 20 years on: The Odysseus voyage and beyond. *Journal of the Experimental Analysis of Behavior*.

Barnes-Holmes, D., Barnes-Holmes, Y., & Cullinan, V. (2000). Relational frame theory and Skinner's verbal behavior: A possible synthesis. *The Behavior Analyst, 23*, 69–84.

Barnes-Holmes, Y., Hayes, S., Barnes-Holmes, D., & Roche, B. (2001). Relational frame theory: A post-Skinnerian account of human language and cognition. *Advances in Child Development and Behavior, 28*, 101–138.

Baron, A., & Galizio, M. (1983). Instructional control of human operant behavior. *The Psychological Record, 33*, 495–520.

Baron, A., & Galizio, M. (2005). Positive and negative reinforcement: Should the distinction be preserved? *The Behavior Analyst, 28*, 85–95.

Baron, A., & Galizio, M. (2006). The distinction between positive and negative reinforcement: Use with care. *The Behavior Analyst, 29*, 141–151.

Baron, R. A., & Richardson, D. R. (1993). *Human aggression*. New York: Plenum Press.

Barreto, F., & Benitti, V. (2012). Exploring the educational potential of robotics in schools: A systematic review. *Computers & Education, 58*, 978–988.

Bartlett, S. M., Rapp, J. T., Krueger, T. K., & Henrickson, M. L. (2011). The use of response cost to treat spitting by a child with autism. *Behavioral Interventions, 26*(1), 76–83.

Baum, M. (1965). An automated apparatus for the avoidance training of rats. *Psychological Reports, 16*, 1205–1211.

Baum, M. (1969). Paradoxical effect of alcohol on the resistance to extinction of an avoidance response in rats. *Journal of Comparative and Physiological Psychology, 69*, 238–240.

Baum, W. M. (1974a). Choice in free-ranging wild pigeons. *Science, 185*, 78–79.

Baum, W. M. (1974b). On two types of deviation from the matching law: Bias and undermatching. *Journal of the Experimental Analysis of Behavior, 22*, 231–242.

Baum, W. M. (1979). Matching, undermatching, and overmatching in studies of choice. *Journal of the Experimental Analysis of Behavior, 32*, 269–281.

Baum, W. M. (1983). Studying foraging in the psychological laboratory. In R. L. Mellgren (Ed.), *Animal cognition and behavior* (pp. 253–278). New York: North-Holland.

Baum, W. M. (1993). Performance on ratio and interval schedules of reinforcement: Data and theory. *Journal of the Experimental Analysis of Behavior*, *59*, 245–264.

Baum, W. M. (2001). Molar versus molecular as a paradigm clash. *Journal of the Experimental Analysis of Behavior*, *75*, 338–341.

Baum, W. M. (2002). From molecular to molar: A paradigm shift in behavior analysis. *Journal of the Experimental Analysis of Behavior*, *78*, 95–116.

Baum, W. M. (2015). Driven by consequences: The multiscale molar view of choice. *Managerial and Decision Economics*, May 18. doi:10.1002/mde.2713.

Baum, W. M., & Rachlin, H. C. (1969). Choice as time allocation. *Journal of the Experimental Analysis of Behavior*, *12*, 861–874.

Baxter, D. A., & Byrne, J. H. (2006). Feeding behavior of Aplasia: A model system for comparing cellular mechanisms of classical and operant conditioning. *Learning & Memory*, *13*, 669–680.

Baxter, M. G., & Murray, E. A. (2002). The amygdala and reward. *Nature Reviews Neuroscience*, *3*, 563–573.

Beavers, G. A., Iwata, B. A., & Gregory, M. K. (2014). Parameters of reinforcement and response-class hierarchies. *Journal of Applied Behavior Analysis*, *47*, 70–82.

Beavers, G. A., Iwata, B. A., & Lerman, D. (2013). Thirty years of research on the functional analysis of problem behavior. *Journal of Applied Behavior Analysis*, *46*, 1–21.

Beck, H. P., Levinson, S., & Irons, G. (2009). Finding Little Albert: A journey to John B. Watson's infant laboratory. *American Psychologist*, *64*, 605–614.

Beeby, E., & White, K. G. (2013). Preference reversal between impulsive and self-control choice. *Journal of the Experimental Analysis of Behavior*, *99*, 260–276.

Behavior Analyst Certification Board. (2022). US employment demand for behavior analysts: 2010–2021. Littleton, CO.

Behavior Analyst Certification Board. (n.d). *BACB certificant data*. Retrieved from www.bacb.com/BACB-certificant-data.

Belke, T. W., & Pierce, W. D. (2009). Body weight manipulation, reinforcement value and choice between sucrose and wheel running: A behavioral economic analysis. *Behavioural Processes*, *80*, 147–156.

Belke, T. W., & Pierce, W. D. (2015). Effects of sucrose availability on wheel running as an operant and as a reinforcing consequence on a multiple schedule: Additive effects of extrinsic and automatic reinforcement. *Behavioural Processes*, *116*, 1–7.

Belke, T. W., Pierce, W. D., & Duncan, I. D. (2006). Reinforcement value and substitutability of sucrose and wheel running: Implications for activity anorexia. *Journal of the Experimental Analysis of Behavior*, *86*, 97–109.

Belke, T. W., Pierce, W. D., & Powell, R. A. (1989). Determinants of choice for pigeons and humans on concurrent-chains schedules of reinforcement. *Journal of the Experimental Analysis of Behavior*, *52*, 97–109.

Bellack, A. S., Hersen, M., & Kazdin, A. E. (2011). *International handbook of behavior modification and therapy* (2nd ed., paperback reprint of original 1990 ed.). New York: Springer.

Belles, D., & Bradlyn, A. S. (1987). The use of the changing criterion design in achieving controlled smoking in a heavy smoker: A controlled case study. *Journal of Behavior Therapy and Experimental Psychiatry*, *18*, 77–82.

Bem, D. J. (1965). An experimental analysis of self-persuasion. *Journal of Experimental Social Psychology*, *1*, 199–218.

Bem, D. J. (1972). Self-perception theory. In L. Berkowitz (Ed.), *Advances in experimental social psychology*: 6 (pp. 1–62). New York: Academic Press

Benbassat, D., & Abramson, C. I. (2002). Errorless discrimination learning in simulated landing flares. *Human Factors and Aerospace Safety*, 2, 319–338.

Bentzley, B. S., Fender, K. M., & Aston-Jones, G. (2013). The behavioral economics of drug self-administration: A review and new analytical approach for within-session procedures. *Psychopharmacology*, 226, 113–125.

Bereznak, S., Ayres, K. M., Mechling, L. C., & Alexander, J. L. (2012). Video self-prompting and mobile technology to increase daily living and vocational independence for students with autism spectrum disorders. *Journal of Developmental and Physical Disabilities*, 24, 269–285.

Berkowitz, L., & Donnerstein, E. (1982). External validity is more than skin deep: Some answers to criticism of laboratory experiments. *American Psychologist*, 37, 245–257.

Bernard, C. (1927). *An introduction to the study of experimental medicine*. New York: Macmillan (original work published in 1865).

Bernstein, I. L., Wilkins, E. E., & Barot, S. K. (2009). Mapping conditioned taste aversion associations through patterns of c-fos expression. In S. Reilly & T. R. Schachtman (Eds.), *Conditioned taste aversion* (pp. 328–340). New York: Oxford University Press.

Bertaina-Anglade, V., La Rochelle, C. D., & Scheller, D. K. (2006). Antidepressant properties of rotigotine in experimental models of depression. *European Journal of Pharmacology*, 548, 106–114.

Berton, O., McClung, C. A., DiLeone, R. J., Krishnan, V., Renthal, W., Russo, S. J., et al. (2006). Essential role of BDNF in the mesolimbic dopamine pathway in social defeat stress. *Science*, 311, 864–868.

Betts, K. R., & Hinsz, V. B. (2013). Group marginalization: Extending research on interpersonal rejection to small groups. *Personality and Social Psychology Review*, 17, 355–370.

Bewley-Taylor, D., Hallam, C., & Allen, R. (2009). *The incarceration of drug offenders: An overview*. Oxford, UK: The Beckley Foundation Drug Policy Programme. Retrieved from http://idpc.net/publications/2009/06/incarceration-drug-offenders-overview-beckley-briefing-16

Bezzina, C. W., Chung, T. C., Asgari, K. K., Hampson, C. L., Brody, S. S., Bradshaw, C. M., et al. (2007). Effects of quinolinic acid-induced lesions of the nucleus accumbens core on inter-temporal choice: A quantitative analysis. *Psychopharmacology*, 195, 71–84.

Bickel, W. K., Athamneh, L. N., Basso, J. C., Mellis, A. M., DeHart, W. B., Craft, W. H., & Pope, D. (2019). Excessive discounting of delayed reinforcers as a trans-disease process: Update on the state of the science. *Current Opinion in Psychology*, 30, 59–64.

Bickel, W. K., Athamneh, L. N., Snider, S. E., Craft, W. H., DeHart, W. B., Kaplan, B. A., & Basso, J. C. (2020). Reinforcer pathology: implications for substance abuse intervention. *Recent Advances in Research on Impulsivity and Impulsive Behaviors*, 139–162.

Bickel, W. K., Jarmolowicz, D. P., Mueller, E. T., & Gatchalian, K. M. (2011). The behavioral economics and neuroeconomics of reinforcer pathologies: implications for etiology and treatment of addiction. *Current Psychiatry Reports*, 13(5), 406–415. https://doi.org/10.1007/s11920-011-0215-1

Bickel, W. K., Jarmolowicz, D. P., Mueller, E. T., & Gatchalian, K. M. (2011). The behavioral economics and neuroeconomics of reinforcer pathologies: implications for etiology and treatment of addiction. *Current Psychiatry Reports*, 13(5), 406–415.

Bickel, W. K., Koffamus, M. N., Moody, L., & Wilson, A. G. (2014). The behavioral- and neuro-economic process of temporal discounting: A candidate behavioral marker of addiction. *Neuropharmacology: Part B*, 76, 518–527.

Bickel, W. K., Odum, A. L., & Madden, G. J. (1999). Impulsivity and cigarette smoking: Delay discounting in current, never, and ex-smokers. *Psychopharmacology*, 146, 447–454.

Bierley, C., McSweeney, F. K., & Vannieuwkerk, R. (1985). Classical conditioning of preferences for stimuli. *Journal of Consumer Research, 12*, 316–323.

Biesmeijer, J. C., & Seeley, T. D. (2005). The use of waggle dance information by honey bees throughout their foraging careers. *Behavioral Ecology and Sociobiology, 59*, 133–142.

Biglan, A. (1995). Translating what we know about the context of antisocial behavior into a lower prevalence of such behavior. *Journal of Applied Behavior Analysis, 28*, 479–492. https://doi.org/10.1901/jaba.1995.28-479

Biglan, A., (2015). *The Nurture Effect: How the Science of Human Behavior Can Improve Our Lives & Our World*. Oakland, CA: New Harbinger Publications.

Bijou, S., & Baer, D. M. (1978). *Behavior analysis of child development*. Englewood Cliffs, NJ: Prentice-Hall.

Binder, C. (1996). Behavioral fluency: Evolution of a new paradigm. *The Behavior Analyst, 19*, 163–197.

Binder, C. (2004). A refocus on response-rate measurement: Comment on Doughty, Chase, and O'Shields. *The Behavior Analyst, 27*, 281–286.

Binder, C. (2010). Building fluent performance: Measuring response rate and multiplying response opportunities. *Behavior Analyst Today, 11*, 214–225.

Binder, C., & Watkins, C. L. (1989). Promoting effective instructional methods: Solutions to America's educational crisis. *Future Choices, 1*, 33–39.

Binder, C., & Watkins, C. L. (1990). Precision teaching and direct instruction: Measurably superior instructional technology in schools. *Performance Improvement Quarterly, 3*, 74–96.

Binder, C., & Watkins, C. L. (2013). Precision teaching and direct instruction: Measurably superior instructional technology in schools. *Performance Improvement Quarterly, 26*, 73–115.

Bitterman, M. E., Menzel, R., Fietz, A., & Schafer, S. (1983). Classical conditioning of the proboscis extension in honeybees (*Apis mellifera*). *Journal of Comparative Psychology, 97*, 107–119.

Bjork, D. W. (1993). *B. F. Skinner: A life*. New York: Basic Books.

Blass, E. M., Ganchrow, J. R., & Steiner, J. E. (1984). Classical conditioning in newborn humans 2–48 hours of age. *Infant Behavior and Development, 7*, 223–235.

Blough, D. S. (1957). Spectral sensitivity in the pigeon. *Journal of the Optical Society of America, 47*, 827–833.

Blough, D. S. (1959). Delayed matching in the pigeon. *Journal of the Experimental Analysis of Behavior, 2*, 151–160.

Blough, D. S. (1966). The reinforcement of least-frequent interresponse times. *Journal of the Experimental Analysis of Behavior, 9*, 581–591.

Blumstein, D. T., & Armitage, K. B. (1997). Alarm calling in yellow-bellied marmots: 1. The meaning of situationally variable alarm calls. *Animal Behaviour, 53*, 143–171.

Boakes, R. A., & Nakajima, S. (2009). Conditioned taste aversions based on running or swimming. In S. Reilly & T. R. Schachtman (Eds.), *Conditioned taste aversion: Behavioral and neural processes* (pp. 159–178). New York: Oxford University Press.

Boakes, R. A., Patterson, A. E., Kendig, M. D., & Harris, J. A. (2015). Temporal distributions of schedule-induced licks, magazine entries, and lever presses on fixed- and variable-time schedules. *Journal of Experimental Psychology: Animal Learning and Cognition, 41*, 52–68.

Boisseau, R. P., Vogel, D., & Dussutour, A. (2016). Habituation in non-neural organisms: evidence from slime moulds. *Proceedings of the Royal Society B: Biological Sciences, 283* (1829), 20160446.

Bolin, B. L., Reynolds, A. R., Stoops, W. W., & Rush, C. R. (2013). Relationship between oral D-amphetamine self-administration and ratings of subjective effects: Do subjective-effects ratings correspond with a progressive-ratio measure of drug-taking behavior? *Behavioural Pharmacology, 24*, 533–542.

Bolívar, H. A., Klemperer, E. M., Coleman, S. R., DeSarno, M., Skelly, J. M., & Higgins, S. T. (2021). Contingency management for patients receiving medication for opioid use disorder: a systematic review and meta-analysis. *JAMA psychiatry*, *78*(10), 1092–1102. Retreived in 2022 from www.cdc.gov/obesity/data/adult.html

Bolles, R. C. (1970). Species-specific defense reactions and avoidance learning. *Psychological Review*, *77*, 32–48.

Bonasio, R. (2012). Emerging topics in epigenetics: Ants, brains, and noncoding RNAs. *Annals of the New York Academy of Sciences*, *1260*, 14–23.

Bonasio, R., Tu, S., & Reinberg, D. (2010). Molecular signals of epigenetic states. *Science*, *330*, 612–616.

Bondy, A., & Frost, L. (1994). The picture exchange communication system. *Focus on Autistic Behavior*, *9*, 1–19.

Boomhower, S., & Rasmussen, E.B., and Doherty, T. (2013). Impulsive choice in the obese Zucker rat. *Behavioural Brain Research*, *241*, 214–221.

Borba, A., Tourinho, E. Z., & Glenn, S. S. (2014). Establishing the macrobehavior of ethical self-control in an arrangement of macrocontingencies in two microcultures. *Behavior and Social Issues*, *23*, 68–86. doi:10.5210/bsi.v.23i0.5354.

Borden, R. J., Bowen, R., & Taylor, S. P. (1971). Shock setting behavior as a function of physical attack and extrinsic reward. *Perceptual and Motor Skills*, *33*, 563–568.

Boren, J. J. (1961). Resistance to extinction as a function of the fixed ratio. *Journal of Experimental Psychology*, *4*, 304–308.

Borrero, J. C., & Vollmer, T. R. (2002). An application of the matching law to severe problem behavior. *Journal of Applied Behavior Analysis*, *35*, 13–27.

Borrero, J. C., Crisolo, S. S., Tu, Q., Rieland, W. A., Ross, N. A., Francisco, M. T., et al. (2007). An application of the matching law to social dynamics. *Journal of Applied Behavior Analysis*, *40*, 589–601.

Bostow, D. E. (2011). The personal life of the behavior analyst. *The Behavior Analyst*, *34*, 267–282.

Bouton, M. E. (2004). Context and behavioral processes in extinction. *Learning & Memory*, *11*(5), 485–494. www.learnmem.org/cgi/doi/10.1101/lm.78804

Bouton, M. E. (2014). Why behavior change is difficult to sustain. *Preventive Medicine*, *68*, 29–36. doi:10.1016/j.ypmed.2014.06.010.

Boutot, E. A., & Hume, K. (2012). Beyond timeout and table time: Today's applied behavior analysis for students with autism. *Education and Training in Autism and Developmental Disabilities*, *47*(1), 23–38. www.jstor.org/stable/23880559

Bower, G. H., & Hilgard, E. R. (1981). *Theories of learning*. Englewood Cliffs, NJ: Prentice-Hall.

Boyce, T. E., & Hineline, P. N. (2002). Interteaching: A strategy for enhancing the user-friendliness of behavioral arrangements in the college classroom. *The Behavior Analyst*, *25*, 215–226.

Boyle, M. (2015). A translational investigation of positive and negative behavioral contrast. PhD dissertation, Utah State University. Retrieved from http://digitalcommons.usu.edu/etd/4234

Bradshaw, C. A., & Reed, P. (2012). Relationship between contingency awareness and human performance on random ratio and random interval schedules. *Learning and Motivation*, *43*, 55–65.

Bradshaw, C. M., Ruddle, H. V., & Szabadi, E. (1981). Studies of concurrent performance in humans. In C. M. Bradshaw, E. Szabadi, & C. F. Lowe (Eds.), *Quantification of steady-state operant behaviour* (pp. 79–90). Amsterdam: Elsevier/North-Holland.

Brainard, M. S., & Doupe, A. J. (2002). What songbirds teach us about learning. *Nature*, *417*, 351–358.

Brainard, M. S., & Doupe, A. J. (2013). Translating birdsong: Songbirds as a model for basic and applied medical research. *Annual Review of Neuroscience*, *36*, 489–517.

Bratcher, N. A., Farmer-Dougan, V., Dougan, J. D., Heidenreich, B. A., & Garris, P. A. (2005). The role of dopamine in reinforcement: Changes in reinforcement sensitivity induced by D1-type, D2-type, and non-selective dopamine receptor agonists. *Journal of the Experimental Analysis of Behavior, 84*, 371–399.

Bray, S., & O'Doherty, J. (2007). Neural coding of reward-prediction error signals during classical conditioning with attractive faces. *Journal of Neurophysiology, 97*, 3036–3045.

Brechner, K. C. (1977). An experimental analysis of social traps. *Journal of Experimental Social Psychology, 13*, 552–564.

Breland, K., & Breland, M. (1961). The misbehavior of organisms. *American Psychologist, 16*, 681–684.

Brembs, B. (2003). Operant conditioning in invertebrates. *Current Opinion in Neurology, 13*, 710–717.

Brembs, B. (2008). Operant learning of *Drosophila* at the torque meter. *Journal of Visualized Experiments, 16*, 731.

Brembs, B. (2011). Spontaneous decisions and operant conditioning in fruit flies. *Behavioural Processes, 87*, 157–164.

Brembs, B., Lorenzetti, F. D., Reyes, F. D., Baxter, D. A., & Byrne, J. H. (2002). Operant reward learning in *Aplysia*: Neuronal correlates and mechanisms. *Science, 296*, 1706–1708.

Brettell, L., & Martin, S. (2017). Oldest *Varroa* tolerant honey bee population provides insight into the origins of the global decline of honey bees. *Sci Rep* **7**, 45953.

Breyer, N. L., & Allen, G. L. (1975). Effects of implementing a token economy on teacher attending behavior. *Journal of Applied Behavior Analysis, 8*, 373–380.

Brody, H. (2000). *The placebo response*. New York: Harper Collins.

Brooks, D. C., & Bouton, M. E. (1993). A retrieval cue for extinction attenuates spontaneous recovery. *Journal of Experimental Psychology: Animal Behavior Processes, 19*, 77–89.

Brooks, K. R., Mond, J. M., Stevenson, R. J., & Stephen, I. D. (2016). Body image distortion and exposure to extreme body types: contingent adaptation and cross adaptation for self and other. *Frontiers in Neuroscience, 10*, 334. https://doi.org/10.3389/fnins.2016.00334

Brown, J. L., Krantz, P. J., McClannahan, L. E., & Poulson, C. L. (2008). Using script fading to promote natural environment stimulus control of verbal interactions among youths with autism. *Research in Autism Spectrum Disorders, 2*, 480–497.

Brown, P. L., & Jenkins, H. M. (1968). Auto-shaping of the pigeon's key-peck. *Journal of the Experimental Analysis of Behavior, 11*, 1–8.

Brown, R. (1973). *A first language: The early stages*. Cambridge, MA: Harvard University Press.

Brown, R. (1986). *Social psychology: The second edition*. New York: Free Press.

Browne, C., Stafford, K., & Fordham, R. (2006). The use of scent-detection dogs. *Irish Veterinary Journal, 59*, 97–104.

Brownstein, A. J., & Pliskoff, S. S. (1968). Some effects of relative reinforcement rate and changeover delay in response-independent concurrent schedules of reinforcement. *Journal of the Experimental Analysis of Behavior, 11*, 683–688.

Bruce, S., & Muhammad, Z. (2009). The development of object permanence in children with intellectual disability, physical disability, autism and blindness. *International Journal of Disability, Development and Education, 56*, 229–246.

Bruzek, J. L., Thompson, R. H., & Peters, L. C. (2009). Resurgence of infant caregiving responses. *Journal of the Experimental Analysis of Behavior, 92*, 327–343.

Buccino, G., Vogt, S., Ritzl, A., Fink, G. R., Zilles, K., Freund, H. J., et al. (2004). Neural circuits underlying imitation learning of hand actions: An event-related fMRI study. *Neuron, 42*, 323–334.

Buckley, J. L., & Rasmussen, E. B. (2012). Obese and lean Zucker rats demonstrate differential sensitivity to rates of food reinforcement in a choice procedure. *Physiology & Behavior, 108*, 19–27.

Buckley, J. L., & Rasmussen, E. B. (2014). Rimonabant's reductive effects on high densities of food reinforcement, but not palatability, in lean and obese Zucker rats. *Psychopharmacology, 231*, 2159–2170.

Buckley, K. B. (1989). *Mechanical man: John Broadus Watson and the beginnings of behaviorism.* New York: The Guilford Press.

Budney, A. J., & Higgins, S. T. (1998). *National institute on drug abuse therapy manuals for drug addiction: Manual 2. A community reinforcement approach: Treating cocaine addiction (NIH Publication No. 98–4309).* Rockville, MD: US Department of Health and Human Services.

Bullock, C. E., & Hackenberg, T. D. (2006). Second-order schedules of token reinforcement with pigeons: Implications for unit price. *Journal of the Experimental Analysis of Behavior, 85*, 95–106.

Bullock, C. E., & Hackenberg, T. D. (2015). The several roles of stimuli in token reinforcement. *Journal of the Experimental Analysis of Behavior, 103*, 269–287.

Bullock, C. E., & Myers, T. W. (2009). Stimulus-food pairings produce stimulus-directed touch-screen responding in cynomolgus monkeys (*Macaca fascicularis*) with or without a positive response contingency. *Journal of the Experimental Analysis of Behavior, 92*, 41–55.

Burch, A. E., Morasco, B. J., & Petry, N. M. (2015). Patients undergoing substance abuse treatment and receiving financial assistance for a physical disability respond well to contingency management treatment. *Journal of Substance Abuse Treatment, 58*, 67–71.

Burger, J. M. (2009). Replicating Milgram: Would people still obey today? *American Psychologist, 64*, 1–11.

Bushell, D., Jr., & Burgess, R. L. (1969). Characteristics of the experimental analysis. In R. L. Burgess & D. Bushell, Jr. (Eds.), *Behavioral sociology: The experimental analysis of social processes* (pp. 145–174). New York: Columbia University Press.

Buske-Kirschbaum, A., Kirschbaum, C., Stierle, H., Jabaij, L., & Hellhammer, D. (1994). Conditioned manipulation of natural killer (NK) cells in humans using a discriminative learning protocol. *Biological Psychology, 38*, 143–155.

Buskist, W. F., & Miller, H. L. (1986). Interaction between rules and contingencies in the control of fixed-interval performance. *The Psychological Record, 36*, 109–116.

Call, J., & Tennie, C. (2009). Animal culture: Chimpanzee table manners? *Current Biology, 19*, R981–R983.

Cameron, J., & Pierce, W. D. (2002). *Rewards and intrinsic motivation: Resolving the controversy.* Westport, CT: Bergin & Garvey.

Cameron, J., Banko, K. M., & Pierce, W. D. (2001). Pervasive negative effects of rewards on intrinsic motivation: The myth continues. *The Behavior Analyst, 24*, 1–44.

Cameron, J., Pierce, W. D., Banko, K. M., & Gear, A. (2005). Achievement-based rewards and intrinsic motivation: A test of cognitive mediators. *Journal of Educational Psychology, 97*, 641–655.

Carnett, A., Raulston, T., Lang, R., Tostanoski, A., Lee, A., Sigafoos, J., et al. (2014). Effects of a perseverative interest-based token economy on challenging and on-task behavior in a child with autism. *Journal of Behavioral Education, 23*, 368–377.

Carnine, D. (1995). Rational schools: The role of science in helping education become a profession. *Behavior and Social Issues, 5*, 5–19.

Caroni, P., Donato, F., & Muller, D. (2012). Structural plasticity upon learning: Regulation and functions. *Nature Reviews Neuroscience, 13*, 478–489.

Carr, D., Wilkinson, K. M., Blackman, D., & McIlvane, W. J. (2000). Equivalence classes in individuals with minimal verbal repertoires. *Journal of the Experimental Analysis of Behavior, 74*, 101–114.

Carr, E. G., & McDowell, J. J. (1980). Social control of self-injurious behavior of organic etiology. *Behavior Therapy, 11*, 402–409.

Carr, J. E., & Burkholder, E. O. (1998). Creating single-subject design graphs with Microsoft Excel™. *Journal of Applied Behavior Analysis, 31*, 245–251.

Carr, J. E., & Miguel, C. F. (2013). The analysis of verbal behavior and its therapeutic applications. In G. J. Madden (Ed.), *APA handbook of behavior analysis: Vol. 2. Translating principles into practice* (pp. 329–352). Washington, DC: American Psychological Association.

Carrigan, P. F., Jr., & Sidman, M. (1992). Conditional discrimination and equivalence relations: A theoretical analysis of control by negative stimuli. *Journal of the Experimental Analysis of Behavior, 58*, 183–204.

Carroll, M. E., Lac, S. T., & Nygaard, S. L. (1989). A concurrently available nondrug reinforcer prevents the acquisition of decreases in the maintenance of cocaine-reinforced behavior. *Psychopharmacology, 97*, 23–29.

Carroll, R. J., & Hesse, B. E. (1987). The effects of alternating mand and tact training on the acquisition of tacts. *The Analysis of Verbal Behavior, 5*, 55–65.

Carton, J. S., & Schweitzer, J. B. (1996). Use of a token economy to increase compliance during hemodialysis. *Journal of Applied Behavior Analysis, 29*, 111–113.

Cartwright, W. S. (2000). Cost-benefit analysis of drug treatment services: Review of the literature. *Journal of Mental Health Policy and Economics, 3*, 11–26.

Carvalho, L. S., Knott, B., Berg, M. L., Bennett, A. T. D., & Hunt, D. M. (2010). Ultraviolet-sensitive vision in long-lived birds. *Proceedings of the Royal Society B: Biological Sciences, 278*, 107–114.

Case, D. A., Fantino, E., & Wixted, J. (1985). Human observing: Maintained by negative information stimuli only if correlated with improved response efficiency. *Journal of the Experimental Analysis of Behavior, 54*, 185–199.

Castilla, J. L., & Pellón, R. (2013). Combined effects of food deprivation and food frequency on the amount and temporal distribution of schedule-induced drinking. *Journal of the Experimental Analysis of Behavior, 100*, 396–407.

Catania, A. C. (1966). Concurrent operants. In W. K. Honig (Ed.), *Operant behavior: Areas of research and application* (pp. 213–270). Englewood Cliffs, NJ: Prentice-Hall.

Catania, A. C. (1975). Freedom and knowledge: An experimental analysis of preference in pigeons. *Journal of the Experimental Analysis of Behavior, 24*, 89–106.

Catania, A. C. (1980). Freedom of choice: A behavioral analysis. In G. H. Bower (Ed.), *The psychology of learning and motivation, 14* (pp. 97–145). New York: Academic Press.

Catania, A. C. (1984). *Learning*. Englewood Cliffs, NJ: Prentice-Hall.

Catania, A. C. (1998). *Learning*. Englewood Cliffs, NJ: Prentice-Hall.

Catania, A. C. (2008). The *Journal of the Experimental Analysis of Behavior* at zero, fifty, and one hundred. *Journal of the Experimental Analysis of Behavior, 89*, 111–118.

Catania, A. C., & Harnard, S. (1988). *The selection of behavior*. New York: Cambridge University Press.

Catania, A. C., & Reynolds, G. S. A. (1968). A quantitative analysis of the responding maintained by interval schedules of reinforcement. *Journal of the Experimental Analysis of Behavior, 11*, 327–383.

Catania, A. C., & Sagvolden, T. (1980). Preference for free choice over forced choice in pigeons. *Journal of the Experimental Analysis of Behavior, 34*, 77–86.

Catania, A. C., Matthews, B. A., & Shimoff, E. H. (1990). Properties of rule-governed behaviour and their implications. In D. E. Blackman & H. Lejeune (Eds.), *Behaviour analysis in theory and practice: Contributions and controversies* (pp. 215–230). Hillsdale, NJ: Lawrence Erlbaum Associates.

Cate, C. T., & Rowe, C. (2007). Biases in signal evolution: Learning makes a difference. *Trends in Ecology and Evolution, 22*, 380–387.

Catmur, C., Gillmeister, H., Bird, G., Liepelt, R., Brass, M., & Heyes, C. (2008). Through the looking glass: Counter-mirror activation following incompatible sensorimotor learning. *European Journal of Neuroscience, 28*, 1208–1215.

Catmur, C., Walsh, V., & Heyes, C. (2007). Sensorimotor learning configures the human mirror system. *Current Biology*, *17*, 1527–1531.

Cautela, J. R. (1984). General level of reinforcement. *Journal of Behavior Therapy and Experimental Psychiatry*, *15*, 109–114.

Cautela, J. R. (1994). General level of reinforcement II. Further elaborations. *Behaviorology*, *2*, 1–16.

Centers for Disease Control and Prevention (2022a). COVID data tracker. Retrieved September 26, 2022 from https://covid.cdc.gov/covid-data-tracker/#datatracker-home

Centers for Disease Control and Prevention (2022b). Overweight and Obesity: Adult Obesity Facts. Retrieved September 26, 2022 from www.cdc.gov/obesity/data/adult.html

Cerutti, D. (1989). Discrimination theory of rule-governed behavior. *Journal of the Experimental Analysis of Behavior*, *51*, 251–259.

Chance, P. (1999). Thorndike's puzzle boxes and the origins of the experimental analysis of behavior. *Journal of the Experimental Analysis of Behavior*, *72*, 433–440.

Chandler, R. K., Fletcher, B. W., & Volkow, N. D. (2009). Treating drug abuse and addiction in the criminal justice system. *Journal of the American Medical Association*, *301*, 183–190.

Chang, S. (2013). Neural basis of autoshaped lever pressing. Dissertation, Johns Hopkins University.

Chang, S. (2014). Effects of orbitofrontal cortex lesions on autoshaped lever pressing and reversal learning. *Brain Research*, *273*, 52–56.

Chang, S., & Holland, P. C. (2013). Effects of nucleus accumbens core and shell lesions on autoshaped lever-pressing. *Behavioral Brain Research*, *256*, 36–42.

Chang, S., Wheeler, D. S., & Holland, P. C. (2012). Roles of n. accumbens and basolateral amygdala in autoshaped lever pressing. *Neurobiology of Learning and Memory*, *97*, 441–451.

Charlop-Christy, M. H., & Kelso, S. E. (1997). *How to treat the child with autism: A guide to treatment at the Claremont Autism Center*. Claremont, CA: Marjorie H. Charlop-Christy.

Chase, P. N., Doughty, S. S., & O'Shields, E. (2005). Focus on response rate is important but not sufficient: A reply. *The Behavior Analyst*, *28*, 163–168.

Chase, P. N., Ellenwood, D. W., & Madden, G. J. (2008). A behavior analytic analogue of learning to use synonyms, syntax, and parts of speech. *The Analysis of Verbal Behavior*, *24*, 31–54.

Chen, G., & Steinmetz, J. E. (1998). A general-purpose computer system for behavioral conditioning and neural recording experiments. *Behavioral Research Methods, Instruments, & Computers*, *30*, 384–391.

Chen, X., Striano, T., & Rakoczy, H. (2004). Auditory-oral matching behavior in newborns. *Developmental Science*, *7*, 42–47.

Cheney, C. D. (1996). Medical nonadherence. In J. R. Cautela & Waris Ishaq (Eds.), *Contemporary issues in behavior therapy* (pp. 9–21). New York: Springer.

Cheney, C. D., & Epling, W. F. (1968). *Running wheel activity and self-starvation in the white rat*. Unpublished manuscript Cheney, WA: Department of Psychology, Eastern Washington State University,.

Cheney, C. D., & Tam, V. (1972). Interocular transfer of a line tilt discrimination without mirror-image reversal using fading in pigeons. *Journal of Biological Psychology*, *14*, 17–20.

Cheney, C. D., Bonem, E., & Bonem, M. (1985). Changeover cost and switching between concurrent adjusting schedules. *Behavioural Processes*, *10*, 145–155.

Cheney, C. D., DeWulf, M. J., & Bonem, E. J. (1993). Prey vulnerability effects in an operant simulation of foraging. *Behaviorology*, *1*, 23–30.

Cheney, C. D., vanderWall, S. B., & Poehlmann, R. J. (1987). Effects of strychnine on the behavior of Great Horned Owls and Red-Tailed Hawks. *Journal of Raptor Research*, *21*, 103–110.

Cheng, J., & Feenstra, M. G. P. (2006). Individual differences in dopamine efflux in nucleus accumbens shell and core during instrumental learning. *Learning and Memory*, *13*, 168–177.

Cheng, K., & Spetch, M. L. (2002). Spatial generalization and peak shift in humans. *Learning and Motivation*, *33*, 358–389.

Cheng, S. C., Quintin, J., Cramer, R. A., Shepardson, K. M., Saeed, S., Kumar, V., et al. (2014). mTOR- and HIF-1-mediated aerobic glycolysis as metabolic basis for trained immunity. *Science*, *345*, *6204*. doi:10.1126/science.1250684.

Cheng, T. D., Disterhoft, J. F., Power, J. M., Ellis, D. A., & Desmond, J. E. (2008). Neural substrates underlying human delay and trace eyeblink conditioning. *Proceedings of the National Academy of Sciences*, *105*, 8108–8113.

Cherek, D. R. (1982). Schedule-induced cigarette self-administration. *Pharmacology, Biochemistry, and Behavior*, *17*, 523–527.

Cherek, D.R., Moeller, F.G., Schnapp, W., & Dougherty, D.M. (1997). Studies of violent and nonviolent male parolees: I. Laboratory and psychometric measurements of aggression, *Biological Psychiatry*, *41*, 514–522.

Chillag, D., & Mendelson, J. (1971). Schedule-induced airlicking as a function of body-weight in rats. *Physiology and Behavior*, *6*, 603–605.

Chittka A, & Chittka L (2010) Epigenetics of royalty. *PLOS Biology 8*(11): e1000532. https://doi.org/10.1371/journal.pbio.1000532

Chomsky, N. (1957). *Syntactic structures*. The Hague: Mouton.

Christ, T. J. (2007). Experimental control and threats to internal validity of concurrent and nonconcurrent multiple baseline designs. *Psychology in the Schools*, *44*, 451–459.

Ciano, P. D. (2008). Drug seeking under a second-order schedule of reinforcement depends on dopamine D_3 receptors in the basolateral amygdala. *Behavioral Neuroscience*, *122*, 129–139.

Classe, A. (1957). The whistled language of La Gomera. *Scientific American*, *196*, 111–120.

Cockburn, J., Collins, A. G. E., & Frank, M. J. (2014). A reinforcement learning mechanism responsible for the valuation of free choice. *Neuron*, *83*, 551–556.

Cohen, D., Nisbett, R. E., Bowdle, B. F., & Schwarz, N. (1996). Insult, aggression, and the southern culture of honor: An "experimental ethnography." *Journal of Personality and Social Psychology*, *70*, 945–960.

Cohen, P. S. (1968). Punishment: The interactive effects of delay and intensity of shock. *Journal of the Experimental Analysis of Behavior*, *11*, 789–799.

Cohen, S. L., Richardson, J., Klebez, J., Febbo, S., & Tucker, D. (2001). EMG Biofeedback: The effects of CRF, FR, VR, FI and VI schedules of reinforcement on the acquisition and extinction of increases in forearm muscle tension. *Applied Psychophysiology and Biofeedback*, *26*, 179–194.

Cohn, S. L. (1998). Behavioral momentum: The effects of the temporal separation of rates of reinforcement. *Journal of the Experimental Analysis of Behavior*, *69*, 29–47.

Collinger, J. L., Kryger, M. A., Barbara, R., Betler, T., Bowsher, K., Brown, E. H. P., et al. (2014). Collaborative approach in the development of high-performance brain-computer interfaces for a neuroprosthetic arm: Translation from animal models to human control. *Clinical Translation Science*, *7*, 52–59.

Colquitt, G., Pritchard, T., & McCollum, S. (2011). The personalized system of instruction in fitness education. *Journal of Physical Education, Recreation and Dance*, *82*, 1–58.

Conger, R., & Killeen, P. (1974). Use of concurrent operants in small group research. *Pacific Sociological Review*, *17*, 399–416.

Conyers, C., Miltenberger, R., Maki, A., Barenz, R., Jurgens, M., Sailer, A., & Kopp, B. (2004). A comparison of response cost and differential reinforcement of other behavior to reduce disruptive behavior in a preschool classroom. *Journal of Applied Behavior Analysis*, *37*, 411–415.

Cook, D. (1996). Reminiscences: Fred S. Keller: An appreciation. *Behavior and Social Issues, 6*, 61–71.

Cooper, J. O., Heron, T. E., & Heward, W. L. (2007). *Applied behavior analysis* (2nd ed.). Upper Saddle River, NJ: Pearson.

Corballis, M. C. (1999). The gestural origins of language. *American Scientist, 87*, 138–145.

Corballis, M. C. (2003). From mouth to hand: Gesture, speech, and the evolution of right-handedness. *Behavioral and Brain Sciences, 26*, 199–208.

Coren, S., Ward, L. M., & Enns, J. T. (2004). *Sensation and perception* (6th ed.). Hoboken, NJ: John Wiley & Sons.

Cornu, J. N., Cancel-Tassin, G., Ondet, V., Girardet, C., & Cussenot, O. (2011). Olfactory detection of prostate cancer by dogs sniffing urine: A step forward in early diagnosis. *European Urology, 59*, 197–201.

Courage, M. L., & Howe, M. L. (2002). From infant to child: The dynamics of cognitive change in the second year of life. *Psychological Bulletin, 129*, 250–277.

Courtney, K., & Perone, M. (1992). Reduction in shock frequency and response effort as factors in reinforcement by timeout from avoidance. *Journal of the Experimental Analysis of Behavior, 58*, 485–496.

Cowles, J. T. (1937). Food-tokens as incentive for learning by chimpanzees. *Comparative Psychology Monographs, 14*, 1–96.

Critchfield, T. S. (2011). Translational contributions of the experimental analysis of behavior. *The Behavior Analyst, 34*(1), 3–17. https://doi.org/10.1007/BF03392245

Critchfield, T. S., & Perone, M. (1990). Verbal self-reports as a function of speed, accuracy, and reinforcement of the reported performance. *The Psychological Record, 40*, 541–554.

Crossman, E. K., Trapp, N. L., Bonem, E. J., & Bonem, M. K. (1985). Temporal patterns of responding in small fixed-ratio schedules. *Journal of the Experimental Analysis of Behavior, 43*, 115–130.

Cumming, W. W. (1966). A bird's eye glimpse of men and machines. In R. Ulrich, T. Stachnik, & J. Mabry (Eds.), *Control of human behavior* (pp. 246–256). Glenview, IL: Scott Foresman & Co.

Cummings, A. R., Carr, J. E., & LeBlanc, L. A. (2012). Experimental evaluation of the training structure of the Picture Exchange Communication System (PECS). *Research in Autism Spectrum Disorders, 6*, 32–45.

Cuzzocrea, F., Murdaca, A. M., & Oliva, P. (2011). Using precision teaching method to improve foreign language and cognitive skills in university students. *International Journal of Digital Literacy and Digital Competence, 2*, 50–60.

D'Amato, M. R., Salmon, D. P., Loukas, E., & Tomie, A. (1985). Symmetry and transitivity of conditional relations in monkeys (*Cebus apella*) and pigeons (*Columba livia*). *Journal of the Experimental Analysis of Behavior, 44*, 35–47.

Dale, R. H. I. (2008). The spatial memory of African elephants (*Loxodonta africana*): Durability, interference, and response biases. In N. K. Innis (Ed.), *Reflections on adaptive behavior: Essays in honor of J. E. R. Staddon* (pp. 143–170). Cambridge, MA: MIT Press.

Dapcich-Miura, E., & Hovell, M. F. (1979). Contingency management of adherence to a complex medical regimen in elderly heart patients. *Behavior Therapy, 10*, 193–201.

Dardig, J. & Heward, B. (2022). *Let's Make a Contract*. Oakland, CA: The Collective Book Studio.

Darley, J. M., Glucksberg, S., & Kinchla, R. A. (1991). *Psychology*. Englewood Cliffs, NJ: Prentice-Hall.

Darwin, C. (1859). *On the origin of species by means of natural selection*. London: John Murray.

Davidson, T. L., & Swithers, S. E. (2004). A Pavlovian approach to the problem of overeating. *International Journal of Obesity, 28*, 933–935.

Davies, N. B., Krebs, J. R., & West, S. A. (2012). *An introduction to behavioral ecology*. West Sussex, UK: Wiley-Blackwell.

Davis, D. R., Kurti, A. N., Skelly, J. M., Redner, R., White, T. J., & Higgins, S. T. (2016). A review of the literature on contingency management in the treatment of substance use disorders, 2009–2014. *Preventive Medicine, 92*, 36–46. https://doi.org/10.1016/j.ypmed.2016.08.008

Davison, M. C. (1969). Preference for mixed-interval versus fixed-interval schedules. *Journal of the Experimental Analysis of Behavior, 12*, 247–252.

Davison, M. C. (1972). Preference for mixed-interval versus fixed-interval schedules: Number of component intervals. *Journal of the Experimental Analysis of Behavior, 17*, 169–176.

Davison, M. C. (1981). Choice between concurrent variable-interval and fixed-ratio schedules: A failure of the generalized matching law. In C. M. Bradshaw, E. Szabadi, & C. F. Lowe (Eds.), *Quantification of steady-state operant behaviour* (pp. 91–100). Amsterdam: Elsevier/North-Holland.

Davison, M. C., & Baum, W. M. (2000). Choice in a variable environment: Every reinforcer counts. *Journal of the Experimental Analysis of Behavior, 74*, 1–24.

Davison, M., & Baum, W. M. (2003). Every reinforcer counts: Reinforcer magnitude and local preference. *Journal of the Experimental Analysis of Behavior, 80*(1), 95–129. https://doi.org/10.1901/jeab.2003.80-95

Davison, M. C., & Ferguson, A. (1978). The effect of different component response requirements in multiple and concurrent schedules. *Journal of the Experimental Analysis of Behavior, 29*, 283–295.

Davison, M. C., & McCarthy, D. (1988). *The matching law: A research review*. Hillsdale, NJ: Lawrence Erlbaum Associates.

Dawkins, R. (1976). *The selfish gene*. London: Oxford University Press.

Dawkins, R. (2004). *The ancestor's tale: A pilgrimage to the dawn of evolution*. Boston, MA: Houghton Mifflin.

Day, J. J., & Sweatt, J. D. (2011). Cognitive neuroepigenetics: A role for epigenetic mechanisms in learning and memory. *Neurobiology of Learning and Memory, 96*, 2–12.

De Boer, B. (2005). Evolution of speech and its acquisition. *Adaptive Behavior, 13*, 281–292.

De Brugada, I., Hall, G., & Symonds, M. (2004). The US-preexposure effect in lithium-induced flavor-aversion conditioning is a consequence of blocking by injection cues. *Journal of Experimental Psychology: Animal Behavior Processes, 20*, 58–66.

De Houwer, J., Barnes-Holmes, Y., & Barnes-Holmes, D. (2016). Riding the waves: A functional-cognitive perspective on the relations among behaviour therapy, cognitive behaviour therapy and acceptance and commitment therapy. *International Journal of Psychology, 51*(1), 40–44.

De Houwer, J., Thomas, S., & Baeyens, F. (2001). Associative learning of likes and dislikes: A review of 25 years of research on human evaluative conditioning. *Psychological Bulletin, 127*, 853–869.

De Villiers, P. (1977). Choice in concurrent schedules and a quantitative formulation of the law of effect. In W. K. Honig & J. E. R. Staddon (Eds.), *Handbook of operant behavior* (pp. 233–287). Englewood Cliffs, NJ: Prentice-Hall.

Deacon, J. R., & Konarski, E. A., Jr. (1987). Correspondence training: An example of rule-governed behavior? *Journal of Applied Behavior Analysis, 20*, 391–400.

deCharms, R. C., Maeda, F., Glover, G., Ludlow, D., Pauly, J. M., Soneji, D., et al. (2005). Control over brain activation and pain learned by using real-time functional MRI. *Proceedings of the National Academy of Sciences, 102*, 18626–18631.

Deci, E. L., Koestner, R., & Ryan, R. M. (1999). A meta-analytic review of experiments examining the effects of extrinsic rewards on intrinsic motivation. *Psychological Bulletin, 125*, 627–668.

DeFulio, A., & Hackenberg, T. D. (2007). Discriminated timeout avoidance in pigeons: The roles of added stimuli. *Journal of the Experimental Analysis of Behavior, 88*, 51–71.

DeHart, W. B., Snider, S. E., Pope, D. A., & Bickel, W. K. (2020). A reinforcer pathology model of health behaviors in individuals with obesity. *Health Psychology, 39*(11), 966–974. https://doi.org/10.1037/hea0000995

Deisseroth, K. (2011). Optogenetics. *Nature Methods, 8*, 26–29.

Demuru, E., & Palagi, E. (2012). In bonobos yawn contagion is higher among kin and friends. *PLoS One, 7*, e49613. doi:10.1371/journal.pone.0049613.

Derenne, A. (2010). Shifts in postdiscrimination gradients within a stimulus dimension based on bilateral facial symmetry. *Journal of the Experimental Analysis of Behavior, 93*, 485–494.

Derenne, A., & Baron, A. (2002). Preratio pausing: effects of an alternative reinforcer on fixed- and variable-ratio responding. *Journal of the Experimental Analysis of Behavior, 77*, 273–282.

Dessalles, J. L. (2007). *Why we talk: The evolutionary origins of language.* New York: Oxford University Press.

DeWall, C. N., Twenge, J. M., Gitter, S. A., & Baumeister, R. F. (2009). It's the thought that counts: The role of hostile cognition in shaping aggressive responses to social exclusion. *Journal of Personality and Social Psychology, 96*, 45–59.

Diane, A., Pierce, W. D., Heth, C. D., Russell, J. C., Richard, D., & Proctor, S. D. (2011). Feeding history and obese-prone genotype increase survival of rats exposed to a challenge of food restriction and wheel running. *Obesity, 20*, 1787–1795.

Dias, B. G., & Ressler, K. (2014). Parental olfactory experience influences behavior and neural structure in subsequent generations. *Nature Neuroscience, 17*, 89–96,

Dickerson, F. B., Tenhula, W. N., & Green-Paden, L. D. (2005). The token economy for schizophrenia: Review of the literature and recommendations for future research. *Schizophrenia Research, 75*, 405–416.

Dickins, D. W. (2005). On the aims and methods in the neuroimaging of derived relations. *Journal of the Experimental Analysis of Behavior, 84*, 453–483.

Dickins, D. W., Singh, K. D., Roberts, N., Burns, P., Downes, J. J., Jimmieson, P., et al. (2001). An fMRI study of stimulus equivalence. *NeuroReport: Brain Imaging, 12*, 405–411.

Dickins, T. E., & Dickins, D. W. (2001). Symbols, stimulus equivalence and the origins of language. *Behavior and Philosophy, 29*, 221–244.

DiMarzo, V., Goparaju S., Wang, L., Liu, J., Batkal, S., Jaral, Z., et al. (2001). Leptin regulated endocannabinoids are involved in maintaining food intake. *Nature, 410*, 822–825.

Dinsmoor, J. A. (1951). The effect of periodic reinforcement of bar-pressing in the presence of a discriminative stimulus. *Journal of Comparative and Physiological Psychology, 44*, 354–361.

Dinsmoor, J. A. (1977). Escape, avoidance, punishment: Where do we stand? *Journal of the Experimental Analysis of Behavior, 28*, 83–95.

Dinsmoor, J. A. (2001a). Stimuli inevitably generated by behavior that avoids electric shock are inherently reinforcing. *Journal of the Experimental Analysis of Behavior, 75*, 311–333.

Dinsmoor, J. A. (2001b). Still no evidence for temporally extended shock-frequency reduction as a reinforcer. *Journal of the Experimental Analysis of Behavior, 75*, 367–378.

Dinsmoor, J. A., Brown, M. P., & Lawrence, C. E. (1972). A test of the negative discriminative stimulus as a reinforcer of observing. *Journal of the Experimental Analysis of Behavior, 18*, 79–85.

Dixon, D. R., Vogel, T., & Tarbox, J. (2012). A brief history of functional analysis and applied behavior analysis. In J. L. Matson (Ed.), *Functional assessment of challenging behaviors* (pp. 3–24). Autism and Child Psychopathology Series. New York: Springer.

Dixon, P. D., Ackert, A. M., & Eckel, L. A. (2003). Development of, and recovery from, activity-based anorexia in female rats. *Physiology and Behavior, 80*, 273–279.

Dobek, C., Heth, C. D., & Pierce, W. D. (2012). Bivalent effects of wheel running on taste conditioning. *Behavioral Processes, 89*, 36–38. https://doi.org/10.1016/j.beproc.2011.10.009.

Dollard, J., & Miller, N. E. (1950). *Personality and psychotherapy.* New York: McGraw-Hill.

Domire, S. C., Wolfe, P. (2014). Effects of video prompting techniques on teaching daily living skills to children with autism spectrum disorders: A review. *Research and Practice for Persons with Severe Disabilities, 39*, 211–226.

Domjan, M. (2016). Elicited versus emitted behavior: Time to abandon the distinction. *Journal of the Experimental Analysis of Behavior, 105*, 231–245.

Donahoe, J. W. (2002). Behavior analysis and neuroscience. *Behavioural Processes, 57*, 241–259.

Donohue, B. C., Karmely, J., & Strada, M. J. (2006). Alcohol and drug abuse. In M. Hersen (Ed.), *Clinician's handbook of child behavioral assessment* (pp. 337–375). San Diego, CA: Elsevier Academic Press.

Donovan, W. I. (1981). Maternal learned helplessness and physiologic response to infant crying. *Journal of Personality and Social Psychology, 40*, 919–926.

Donovan, W. J. (1978). Structure and function of the pigeon visual system. *Physiological Psychology, 6*, 403–437.

Dornhaus, A., & Chittka, L. (2004). Why do bees dance? *Behavioral Ecology and Sociobiology, 55*, 395–401.

Dorrance, B. R. (2001). Imitative learning of conditional discriminations in pigeons. *Dissertation Abstracts International: Section B: The Sciences & Engineering, 61* (11-B) 6169.

Dorrance, B. R., & Zentall, T. R. (2001). Imitative learning in Japanese quail depends on the motivational state of the observer at the time of observation. *Journal of Comparative Psychology, 115*, 62–67.

Dos Santos, C. V., Gehm, T., & Hunziker, M. H. L. (2010). Learned helplessness in the rat: Effect of response topography in a within-subject design. *Behavioural Processes, 86*, 178–183.

Dotto-Fojut, K. M., Reeve, K. F., Townsend, D. B., & Progar, P. R. (2011). Teaching adolescents with autism to describe a problem and request assistance during simulated vocational tasks. *Research in Autism Spectrum Disorders, 5*, 826–833.

Dougher, M. J., Hamilton, D. A., Fink, B. C., & Harrington, J. (2007). Transformation of the discriminative and eliciting functions of generalized relational stimuli. *Journal of the experimental analysis of behavior, 88*(2), 179–197.

Doughty, A. H., Giorno, K. G., & Miller, H. L. (2013). Effects of reinforcer magnitude on reinforced behavioral variability. *Journal of the Experimental Analysis of Behavior, 100*, 355–369.

Doughty, S. S., Chase, P. N., & O'Shields, E. (2004). Effects of rate building on fluent performance: A review and commentary. *The Behavior Analyst, 27*, 7–23.

Dove, L. D. (1976). Relation between level of food deprivation and rate of schedule-induced attack. *Journal of the Experimental Analysis of Behavior, 25*, 63–68.

Doyle, T. A., & Samson, H. H. (1988). Adjunctive alcohol drinking in humans. *Physiology and Behavior, 44*, 775–779.

Draganski, B., Gaser, C., Busch, V., Schuierer, G., Bogdahn, U., & May, A. (2004). Neuroplasticity: Changes in gray matter induced by training. *Nature, 427*, 311–312.

Du, L., & Greer, R. D. (2014). Validation of adult generalized imitation topographies and the emergence of generalized imitation in young children with autism as a function of mirror training. *The Psychological Record, 64*, 161–177.

Dube, W. V., & McIlvane, W. J. (2001). Behavioral momentum in computer-presented discriminations in individuals with severe mental retardation. *Journal of the Experimental Analysis of Behavior, 75*, 15–23.

Duhigg, C. (2012). *The power of habit: Why we do what we do in life and business*. Toronto: Doubleday Canada.

Dukas, R. (2013). Effects of learning on evolution: Robustness, innovation and speciation. *Animal Behaviour, 85*, 1023–1030.

Dulany, D. E. (1968). Awareness, rules, and propositional control: A confrontation with S–R behavior theory. In T. Dixon & D. Horton (Eds.), *Verbal behavior and behavior theory* (pp. 340–387). New York: Prentice Hall.

Dunlap, A. S., & Stephens, D. W. (2014). Experimental evolution of prepared learning. *Proceedings of the National Academy of Sciences, 111*, 11750–11755.

Durand, V. M. (1999). Functional communication training using assistive devices: Recruiting natural communities of reinforcement. *Journal of Applied Behavior Analysis, 32*, 247–267.

Dussutour, A. (2021). Learning in single cell organisms. *Biochemical and Biophysical Research Communications, 564*, 92–102.

Dutra, L., Stathopoulou, G., Basden, S. L., Leyro, T. M., Powers, M. B., & Otto, M. W. (2008). A meta-analytic review of psychosocial interventions for substance use disorders. *American Journal of Psychiatry, 165*, 179–187.

Dworkin, B. R., & Miller, N. (1986). Failure to replicate visceral learning in the acute curarized rat preparation. *Behavioral Neuroscience, 100*, 299–314.

Dwyer, D. M., & Boakes, R. A. (1997). Activity-based anorexia in rats as failure to adapt to a feeding schedule. *Behavioral Neuroscience, 111*, 195–205.

Dymond, D., & Critchfield, T. S. (2002). A legacy of growth: Human operant research in *The Psychological Record, 1980–1999. The Psychological Record, 52*, 99–108.

Eckerman, D. A., & Lanson, R. N. (1969). Variability of response location for pigeons responding under continuous reinforcement, intermittent reinforcement, and extinction. *Journal of the Experimental Analysis of Behavior, 12*, 73–80.

Edelman, G. M. (1987). *Neural Darwinism: The theory of neuronal group selection.* New York: Basic Books.

Edelman, G. M. (2007). Learning in and from brain-based devices. *Science, 318*, 1103–1105.

Egan, L. C., Bloom, P., & Santos, L. R. (2010). Choice-induced preferences in the absence of choice: Evidence from a blind two-choice paradigm with young children and capuchin monkeys. *Journal of Experimental Social Psychology, 46*, 204–207.

Egan, L. C., Santos, L. R., & Bloom, P. (2007). The origins of cognitive dissonance: Evidence from children and monkeys. *Psychological Science, 18*, 978–983.

Egger, M. D., & Miller, N. E. (1962). Secondary reinforcement in rats as a function of information value and reliability of the stimulus. *Journal of Experimental Psychology, 64*, 97–104.

Egger, M. D., & Miller, N. E. (1963). When is reward reinforcing? An experimental study of the information hypothesis. *Journal of Comparative and Physiological Psychology, 56*, 132–137.

Eibl-Eibesfeldt, I. (1975). *Ethology: The biology of behavior.* New York: Holt, Rinehart and Winston.

Eisenberger, R., & Cameron, J. (1996). The detrimental effects of reward: Myth or reality? *American Psychologist, 51*, 1153–1166.

Eisenberger, R., & Shanock, L. (2003). Rewards, intrinsic motivation, and creativity: A case study of conceptual and methodological isolation. *Creativity Research Journal, 15*, 121–130.

Elsmore, T. F., & McBride, S. A. (1994). An eight-alternative concurrent schedule: Foraging in a radial maze. *Journal of the Experimental Analysis of Behavior, 61*, 331–348.

Engelmann, S., & Carnine, D. (1982). *Theory of instruction: Principles and application.* New York: Irvington.

Engelmann, S., Haddox, P., & Bruner, E. (1986). *Teach your child to read in 100 easy lessons.* Simon and Schuster.

Enticott, P. G., Kennedy, H. A., Rinehart, N. J., Bradshaw, J. L., Tonge, B. J., Daskalakis, Z. J., et al. (2013). Interpersonal motor resonance in autism spectrum disorder: Evidence against a global "mirror system" deficit. *Frontiers in Human Neuroscience,* May 23. doi:10.3389/fnhum.2013.00218.

Epling, W. F., & Pierce, W. D. (1983). Applied behavior analysis: New directions from the laboratory. *The Behavior Analyst, 6*, 27–37.

Epling, W. F., & Pierce, W. D. (1986). The basic importance of applied behavior analysis. *The Behavior Analyst, 9*, 89–99.

Epling, W. F., & Pierce, W. D. (1990). Laboratory to application: An experimental analysis of severe problem behaviors. In A. C. Repp & N. N. Singh (Eds.), *Perspectives on the use of nonaversive and aversive interventions for persons with developmental disabilities* (pp. 451–464). Sycamore, IL: Sycamore Publishing Co.

Epling, W. F., & Pierce, W. D. (1992). *Solving the anorexia puzzle: A scientific approach*. Toronto: Hogrefe & Huber.

Epling, W. F., & Pierce, W. D. (Eds.) (1996). *Activity anorexia: Theory, research, and treatment*. Mahwah, NJ: Lawrence Erlbaum Associates, Inc.

Epling, W. F., Pierce, W. D., & Stefan, L. (1983). A theory of activity-based anorexia. *International Journal of Eating Disorders, 3*, 27–46.

Epstein, L. H., Saad, F. G., Handley, E. A., Roemmich, J. N., Hawk, L. W., & McSweeney, F. K. (2003). Habituation of salivation and motivated responding for food in children. *Appetite, 41*(3), 283–289. https://doi.org/10.1016/S0195-6663(03)00106-5

Epstein, R. (1984). Spontaneous and deferred imitation in the pigeon. *Behavioral Processes, 9*, 347–354.

Epstein, R. (1985). Extinction-induced resurgence: Preliminary investigation and possible application. *Psychological Record, 35*, 143–153.

Erjavec, M., Lovett, V. E., & Horne, P. J. (2009). Do infants show generalized imitation of gestures? II. The effects of skills training and multiple exemplar matching training. *Journal of the Experimental Analysis of Behavior, 91*, 355–376.

Erlich, P. R. (2000). *Human natures: Genes, cultures, and the human prospect*. New York: Penguin.

Estes, W. K., & Skinner, B. F. (1941). Some quantitative properties of anxiety. *Journal of Experimental Psychology, 29*, 390–400.

Ettinger, R. H., & McSweeney, F. K. (1981). Behavioral contrast and responding during multiple food–food, food–water, and water–water schedules. *Animal Learning and Behavior, 9*, 216–222.

Etzkowitz, H. (1992). Inventions. In E. F. Borgatta & M. L. Borgatta (Eds.), *Encyclopedia of sociology*, 2 (pp. 1004–1005). New York: Macmillan.

Evans, R. I. (1989). *Albert Bandura, the man and his ideas—a dialogue*. New York: Praeger.

Fagot, J., & Maugard, A. (2013). Analogical reasoning in baboons (*Pappio papio*): Flexible reencoding of the source relation depending on the target relation. *Learning & Behavior, 41*, 229–237.

Falck-Ytter, T., Gredeback, G., & von Hofsten, C. (2006). Infants predict other people's action goals. *Nature Neuroscience, 9*, 878–879.

Falk, J. L. (1961). Production of polydipsia in normal rats by an intermittent food schedule. *Science, 133*, 195–196.

Falk, J. L. (1964). Studies on schedule-induced polydipsia. In M. J. Wayner (Ed.), *Thirst: First international symposium on thirst in the regulation of body water* (pp. 95–116). New York: Pergamon Press.

Falk, J. L. (1969). Schedule-induced polydipsia as a function of fixed interval length. *Journal of the Experimental Analysis of Behavior, 9*, 37–39.

Falk, J. L. (1971). The nature and determinants of adjunctive behavior. *Physiology and Behavior, 6*, 577–588.

Falk, J. L. (1977). The origin and functions of adjunctive behavior. *Animal Learning and Behavior, 5*, 325–335.

Falk, J. L. (1994). Schedule-induced behavior occurs in humans: A reply to Overskeid. *Psychological Record, 44*, 45–63.

Falk, J. L. (1998). Drug abuse as adjunctive behavior. *Drug and Alcohol Dependence, 52*, 91–98.

Falk, J. L., & Lau, C. E. (1997). Establishing preference for oral cocaine without an associative history with a reinforcer. *Drug and Alcohol Dependence, 46*, 159–166.

Fantino, E. (1965). Some data on the discriminative stimulus hypothesis of secondary reinforcement. *Psychological Record, 15*, 409–414.

Fantino, E. (1967). Preference for mixed-versus fixed-ratio schedules. *Journal of the Experimental Analysis of Behavior, 10*, 35–43.

Fantino, E. (1969a). Choice and rate of reinforcement. *Journal of the Experimental Analysis of Behavior, 12*, 723–730.

Fantino, E. (1969b). Conditioned reinforcement, choice, and the psychological distance to reward. In D. P. Hendry (Ed.), *Conditioned reinforcement* (pp. 163–191). Homewood, IL: Dorsey Press.

Fantino, E. (1977). Conditioned reinforcement: Choice and information. In W. K. Honig & J. E. R. Staddon (Eds.), *Handbook of operant behavior* (pp. 313–339). Englewood Cliffs, NJ: Prentice-Hall.

Fantino, E. (2008). Choice, conditioned reinforcement, and the Prius effect. *The Behavior Analyst, 31*, 95–111.

Fantino, E., & Case, D. A. (1983). Human observing: Maintained by stimuli correlated with reinforcement but no extinction. *Journal of the Experimental Analysis of Behavior, 18*, 79–85.

Fantino, E., & Logan, C. A. (1979). *The experimental analysis of behavior: A biological perspective*. San Francisco, CA: W. H. Freeman.

Fantino, E., & Romanowich, P. (2007). The effect of conditioned reinforcement rate on choice: A review. *Journal of the Experimental Analysis of Behavior, 87*, 409–421.

Fantino, E., & Silberberg, A. (2010). Revisiting the role of bad news in maintaining human observing behavior. *Journal of the Experimental Analysis of Behavior, 93*, 157–170.

Fawcett, T. W., McNamara, J. M., & Houston, A. (2012). When is it adaptive to be patient? A general framework for evaluating delayed rewards. *Behavioural Processes, 89*, 128–136.

Fehr, E., & Gintis, H. (2007). Human motivation and social cooperation: Experimental and analytical foundations. *Annual Review of Sociology, 33*, 43–64.

Feldman, M. A. (1990). Balancing freedom from harm and right to treatment for persons with developmental disabilities. In A. C. Repp & N. N. Singh (Eds.), *Perspectives on the use of nonaversive and aversive interventions for persons with developmental disabilities* (pp. 261–271). Sycamore, IL: Sycamore Publishing Co.

Ferrari, P. F., Rozzi, S., & Fogassi, L. (2005). Mirror neurons responding to observation of actions made with tools in monkey ventral premotor cortex. *Journal of Cognitive Neuroscience, 17*, 212–226.

Ferster, C. B., & Skinner, B. F. (1957). *Schedules of reinforcement*. New York: Appleton-Century-Crofts.

Ferster, C. B., Culbertson, S., & Boren, M. C. P. (1975). *Behavior principles*. Englewood Cliffs, NJ: Prentice-Hall.

Festinger, L. (1957). *A theory of cognitive dissonance*. Stanford, CA: Stanford University Press.

Fields, L., & Spear, J. (2012). Measuring joint stimulus control by complex graph/description correspondences. *The Psychological Record, 62*, 279–294.

Fields, R. D. (2009). *The other brain*. New York: Simon & Schuster.

Figlewicz, D. P., & Sipols, A. (2010). Energy regulatory signals and food reward. *Pharmacology Biochemistry and Behavior, 97*, 15–24.

Filby, Y., & Appel, J. B. (1966). Variable-interval punishment during variable-interval reinforcement. *Journal of the Experimental Analysis of Behavior, 9*, 521–527.

Fiorillo, C. D., Tobler, P. N., & Schultz, W. (2003). Discrete coding of reward probability and uncertainty by dopamine neurons. *Science, 299*, 1898–1902.

Fischer, J. L., Howard, J. S., Sparkman, C. R., & Moore, A. G. (2010). Establishing generalized syntactical responding in young children with autism. *Research in Autism Spectrum Disorders*, *4*, 76–88.

Fisher, W. W., & Mazur, J. E. (1997). Basic and applied research on choice responding. *Journal of Applied Behavior Analysis*, *30*(3), 387–410. https://doi.org/10.1901/jaba.1997.30-387

Fishman, S. (1991). The town B. F. Skinner boxed. *Health*, *5*, 50–60.

Fixsen, D. L., Phillips, E. L., Phillips, E. A., & Wolf, M. M. (1976). The teaching-family model of group home treatment. In W. E. Craighead, A. E. Kazdin, & M. J. Mahoney (Eds.), *Behavior modification: Principles, issues, and applications* (pp. 310–320). Boston, MA: Houghton Mifflin.

Floresco, S. B. (2015). The nucleus accumbens: An interface between cognition, emotion and action. *Annual Review of Psychology*, *66*, 25–52.

Flynn, L., & Healy, O. (2012). A review of treatments for deficits in social skills and self-help skills in autism spectrum disorder. *Research in Autism Spectrum Disorders*, *6*, 431–441.

Follman, M., Aronsen, G., & Pan, D. (2014). A guide to mass shootings in America. *Mother Jones*. First published 2012 and updated May 24, 2022. Retrieved from www.motherjones.com/politics/2012/07/mass-shootings-map?page=1.

Foreman, A. M. (2009). Negative reinforcement by timeout from avoidance: The roles of shock-frequency reduction and response-effort reduction. MSc. Thesis. Morgantown, WV: Department of Psychology, Eberly College of Arts and Sciences, West Virginia University.

Foster, T. A., Hackenberg, T. D., & Vaidya, M. (2001). Second-order schedules of token reinforcement with pigeons: Effects of fixed- and variable-ratio exchange schedules. *Journal of the Experimental Analysis of Behavior*, *76*, 159–178.

Foster, T. M., Temple, W., & Poling, A. (1997). Behavior analysis and farm animal welfare. *The Behavior Analyst*, *20*, 87–95.

Foxx, R. M. (2008). Applied behavior analysis treatment of autism: The state of the art. *Child and Adolescent Psychiatric Clinics of North America*, *17*, 821–834.

Foxx, R. M., & Azrin, N. H. (1973). The elimination of autistic self-stimulatory behavior by overcorrection. *Journal of Applied Behavior Analysis*, *6*, 1–14.

Francis, G. (2014). Too much success for recent groundbreaking epigenetic experiments. *Genetics*, *198*, 449–451.

Frank, A. J., & Wasserman, E. A. (2005). Associate symmetry in the pigeon after successive matching-to-sample training. *Journal of the Experimental Analysis of Behavior*, *84*, 147–165.

Frederiksen, L. W., & Peterson, G. L. (1977). Schedule-induced aggression in humans and animals: A comparative parametric review. *Aggressive Behavior*, *3*(1), 57–75. https://doi.org/10.1002/1098-2337(1977)3:1<57::AID-AB2480030106>3.0.CO;2-D

Freedman, D. H. (2011). How to fix the obesity crisis. *Scientific American*, *304*, 40–47.

Freeman, J. H., & Steinmetz, A. B. (2011). Neural circuitry and plasticity mechanisms underlying delay eyeblink conditioning. *Learning and Memory*, *19*, 666–677.

Fridlund, A. J., Beck, H. P., Goldie, W. D., & Irons, G. (2012). Little Albert: A neurologically impaired child. *History of Psychology*, *15*, 1–34.

Friedman, S. G., Edling, T., & Cheney, C. D. (2006). The natural science of behavior. In G. J. Harrison & T. L. Lightfoot (Eds.), *Clinical avian medicine* (pp. 46–59). Palm Beach, FL: Spix.

Friman, P. C., Finney, J. W., Glasscock, S. T., Weigel, J. W., & Christophersen, E. R. (1986). Testicular self-examination: Validation of a training strategy for early cancer detection. *Journal of Applied Behavior Analysis*, *19*, 87–92.

Frisch, K. von (1967). *The dance language and orientation of bees*. Cambridge, MA: Harvard University Press.

Fritz, J. N., Iwata, B. A., Hammond, J. L., & Bloom, S. E. (2013). Experimental analysis of precursors to severe problem behavior. *Journal of Applied Behavior Analysis*, *46*, 101–129.

Fryer, R. G., Jr. (2010). Financial incentives and student achievement: Evidence from randomized trials. *Working Paper 15898*. Retrieved from www.nber.org/papers/w15898

Fuhrmann, D., Ravignani, A., Marshall-Pescini, S., & Whiten, A. (2014). Synchrony and motor mimicking in chimpanzee observational learning. *Scientific Reports, 4*, 5283. doi:10.1038/srep05283.

Galef, B. G. Jr., & Allen, C. (1995). A new model system for studying behavioural traditions in animals. *Animal Behaviour, 50*, 705–717.

Galizio, M. (1979). Contingency-shaped and rule-governed behavior: Instructional control of human loss avoidance. *Journal of the Experimental Analysis of Behavior, 31*, 53–70.

Gallese, V., Fadiga, L., Fogassi, L., & Rizzolatti, G. (1996). Action recognition in the premotor cortex. *Brain, 119*, 593–609.

Gallup, A. C. (2011). Why do we yawn? Primitive versus derived features. *Neuroscience & Biobehavioral Reviews, 35*, 765–769.

Gallup, A. C., Swartwood, L., Militello, J., & Sacket, S. (2015). Experimental evidence of contagious yawning in budgerigars (*Melopsittacus undulates*). *Animal Cognition, 18*, 1051–1058. doi:10.1007/s10071-015-0873-1.

Gamba, J., Goyos, C., & Petursdottir, A. I. (2014). The functional independence of mands and tacts: Has it been demonstrated empirically? *Analysis of Verbal Behavior, 31*, 10–38. doi:10.1007/s40616-014-0026-7.

Gamble, E. H., & Elder, S. T. (1990). Conditioned diastolic blood pressure: The effects of magnitude of acquired response and feedback schedules on resistance to extinction. *International Journal of Psychophysiology, 9*, 13–20.

Ganz, J. B., Davis, J. L., Lund, E. M., Goodwyn, F. D., & Simpson, R. L. (2012). Meta-analysis of PECS with individuals with ASD: Investigation of targeted versus non-targeted outcome, participant characteristics, and implementation phase. *Research in Developmental Disabilities, 33*, 406–418.

Gapp, K., Jawaid, A., Sarkies, P., Bohacek, J., Pelczar, P., Prados, J., et al. (2014). Implication of sperm RNAs in transgenerational inheritance of the effects of early trauma in mice. *Nature Neuroscience, 17*, 667–669. doi:10.1038/nn.3695.

Garcia, J., & Koelling, R. A. (1966). Relation of cue to consequence in avoidance learning. *Psychonomic Science, 4*, 123–124.

Garcia-Fernandez, G., Secades-Villa, R., Garcia-Rodriguez, O., Sanchez-Hervas, E., Fernandez-Hermida, J., & Higgins, S. T. (2011). Adding voucher-based incentives to community reinforcement approach improves outcomes during treatment for cocaine dependence. *American Journal on Addictions, 20*, 456–461.

Gardner, E. T., & Lewis, P. (1976). Negative reinforcement with shock-frequency increase. *Journal of the Experimental Analysis of Behavior, 25*, 3–14.

Gardner, R. M., & Brown, D. L. (2010). Body image assessment: A review of figural drawing scales. *Personality and Individual Differences, 48*(2), 107–111.

Gast, D. L. (2010). *Single subject research methodology in behavioral sciences*. New York: Routledge.

Gaston, S. (2019). Enforcing race: A neighborhood-level explanation of Black–White differences in drug arrests. *Crime & Delinquency, 65*(4), 499–526. https://doi.org/10.1177/0011128718798566

Geen, R. G. (1968). Effects of frustration, attack, and prior training in aggressiveness upon aggressive behavior. *Journal of Personality and Social Psychology, 9*, 316–321.

Geiger, B., & Fischer, M. (2006). Will words ever harm me? Escalation from verbal to physical abuse in sixth-grade classrooms. *Journal of Interpersonal Violence, 21*, 337–357.

Geller, E. S. (2006). Occupational injury prevention and applied behavior analysis. In A. C. Gielen, D. A. Sleet, & R. J. DiClemente (Eds.), *Injury and violence prevention: Behavioral science theories, methods, and applications* (pp. 297–322). San Francisco, CA: Jossey-Bass.

Geller, E. S. (2011). Psychological science and safety: Large-scale success at preventing occupational injuries and fatalities. *Current Directions in Psychological Science, 20*, 109–114.

Geller, I., & Seifter, J. (1960). The effects of meprobamate, barbiturates, d-amphetamine and promazine on experimentally induced conflict in the rat. *Psychopharmacologia, 1*, 482–492.

Geller, I., Kulak, J. T., Jr., & Seifter, J. (1962). The effects of chlordiazepoxide and chlorpromazine on punished discrimination. *Psychopharmacologia, 3*, 374–385.

Gellman, M., & Turner, J. R. (2012). *Encyclopedia of behavioral medicine*. New York: Springer Publishing.

Gendall, K. A., Kaye, W. H., Altemus, M., McConaha, C. W., & La Via, M. C. (1999). Leptin, neuropeptide Y, and peptide YY in long-term recovered eating disorder patients. *Biological Psychiatry, 46*, 292–299.

Gerard, H. B. (1994). A retrospective review of Festinger's a theory of cognitive dissonance. *Psychological Critiques, 39*, 1013–1017.

Gershoff, E. T. (2002). Corporal punishment by parents and associated child behaviors and experiences: A meta-analytic and theoretical review. *Psychological Bulletin, 128*, 539–579.

Gewirtz, J. L. (1971). The roles of overt responding and extrinsic reinforcement in "self-" and "vicarious-reinforcement" phenomena and in "observational learning" and imitation. In R. Glaser (Ed.), *The nature of reinforcement* (pp. 279–309). New York: Academic Press.

Ghazanfar, A. A., & Rendall, D. (2008). Evolution of human vocal reproduction. *Current Biology, 18*(11), R457–R460. http://dx.doi.org/10.1016/j.cub.2008.03.030.

Ghezzi, P., Williams, W. L., & Carr, J. (1999). *Autism: Behavior analytic perspectives*. Reno, NV: Context Press.

Gillespie-Lynch, K., Greenfield, P. M., Lyn, H., & Savage-Rumbaugh, S. (2014). Gestural and symbolic development among apes and humans: Support for a multimodal theory of language evolution. *Frontiers in Psychology, 5*, 01228. doi:10.3389/fpsyg.2014.01228.

Ginsburg, S., & Jablonka, E. (2010). The evolution of associative learning: A factor in the Cambrian explosion. *Journal of Theoretical Biology, 266*, 11–20.

Giurfa, M. (2007). Behavioral and neural analysis of associative learning in the honeybee: A taste from the magic well. *Journal of Comparative Physiology A, 193*, 801–824.

Glenn, S. S. (1988). Contingencies and metacontingencies: Toward a synthesis of behavior analysis and cultural materialism. *The Behavior Analyst, 11*, 161–179.

Glenn, S. S. (1989). Verbal behavior and cultural practices. *Behavior Analysis and Social Action, 7*, 10–14.

Glenn, S. S. (1991). Contingencies and metacontingencies: Relations among behavioral, cultural, and biological evolution. In P. A. Lamal (Ed.), *Behavioral analysis of societies and cultural practices* (pp. 39–73). New York: Hemisphere.

Glenn, S. S. (2004). Individual behavior, culture, and social change. *The Behavior Analyst, 27*, 133–151.

Glenn, S. S., & Field, D. P. (1994). Functions of the environment in behavioral evolution. *The Behavior Analyst, 17*, 241–259.

Glenwick, D., & Jason, L. A. (1980). *Behavioral community psychology: Progress and prospects*. New York: Praeger.

Godfrey, A. (2014). Canine scent detection of human cancers: Is this a viable technique for detection? *Veterinary Nursing Journal, 29*, 392–394.

Goetz, E. M., & Baer, D. M. (1973). Social control of form diversity and the emergence of new forms in children's blockbuilding. *Journal of Applied Behavior Analysis, 6*, 209–217.

Goldiamond, I. (1962). Perception. In Arthur J. Bachrach (Ed.), *Experimental foundations of clinical psychology* (pp. 280–340). New York: Basic Books.

Goldin-Meadow, S., & Morford, M. (1985). Gesture in early child language: Studies of deaf and hearing children. *Merrill-Palmer Quarterly, 31*, 145–176.

Goldin-Meadow, S., & Wagner, S. M. (2005). How our hands help us learn. *Trends in Cognitive Sciences*, *9*, 234–241.

Goldsmith, T. H. (2006). What birds see. *Scientific American*, *295*, 68–75.

Goldsmith, T. H., & Butler, B. K. (2005). Color vision of the budgerigar (*Melopsittacus undulates*): Hue matches, tetrachromacy, and intensity discrimination. *Journal of Comparative Physiology A: Neuroethology, Sensory, Neural, and Behavioral Physiology*, *191*, 933–951.

Goldstein, M. H., King, A. P., & West, M. J. (2003). Social interaction shapes babbling: Testing parallels between birdsong and speech. *Proceedings of the National Academy of Sciences*, *100*, 8030–8035.

Golinkoff, R. M., Can, D. D., Soderstrom, M., & Hirsh-Pasek, K. (2015). (Baby) talk to me: The social context of infant-directed speech and its effects on early language acquisition. *Current Directions in Psychological Science*, *24*, 339–344.

Gollub, L. R. (1958). *The chaining of fixed-interval schedules*. Unpublished doctoral dissertation, Harvard University, Cambridge, MA.

Gollub, L. R. (1977). Conditioned reinforcement: Schedule effects. In W. K. Honig & J. E. R. Staddon (Eds.), *Handbook of operant behavior* (pp. 288–312). Englewood Cliffs, NJ: Prentice-Hall.

Gollub, L. R. (1991). The use of computers in the control and recording of behavior. In I. H. Iverson & K. A. Lattal (Eds.), *Experimental analysis of behavior: Part 2* (pp. 155–192). New York: Elsevier.

Gott, C. T., & Weiss, B. (1972). The development of fixed-ratio performance under the influence of ribonucleic acid. *Journal of the Experimental Analysis of Behavior*, *18*, 481–497.

Gould, S. J. (2002). *The structure of evolutionary theory*. Cambridge, MA: Harvard University Press.

Grant, D. S. (1975). Proactive interference in pigeon short-term memory. *Journal of Experimental Psychology: Animal Behavior Processes*, *1*, 207–220.

Grant, D. S. (1981). Short-term memory in the pigeon. In N. E. Spear & R. R. Miller (Eds.), *Information processing in animals: Memory mechanisms* (pp. 227–256). Hillsdale, NJ: Lawrence Erlbaum Associates.

Green, D. M., & Swets, J. A. (1966). *Signal detection theory and psychophysics*. New York: John Wiley & Sons.

Green, E. J. (1955). Concept formation: A problem in human operant conditioning. *Journal of Experimental Psychology*, *49*, 175–180.

Green, L., & Freed, D. E. (1993). The substitutability of reinforcers. *Journal of the Experimental Analysis of Behavior*, *60*, 141–158.

Green, L., Fisher, E. B., Perlow, S., & Sherman, L. (1981). Preference reversal and self control: Choice as a function of reward amount and delay. *Behavior Analysis Letters*, *1*, 43–51.

Greene, W. A., & Sutor, L. T. (1971). Stimulus control of skin resistance responses on an escape-avoidance schedule. *Journal of the Experimental Analysis of Behavior*, *16*, 269–274.

Greer, R. D., Stolfi, L., & Pistoljevic, N. (2007). Emergence of naming in preschoolers: A comparison of multiple and single exemplar instruction. *European Journal of Behavior Analysis*, *8*, 109–131.

Griffiths, R. R., & Thompson, T. (1973). The post-reinforcement pause: A misnomer. *The Psychological Record*, *23*(2), 229–235.

Griggio, M., Hoi, H., & Pilastro, A. (2010). Plumage maintenance affects ultraviolet colour and female preference in the budgerigar. *Behavioural Processes*, *84*, 739–744.

Grissom, N., & Bhatnagar, S. (2009). Habituation to repeated stress: Get used to it. *Neurobiology of Learning and Memory*, *92*, 215–224.

Groopman, J. (2015). Oliver Sacks, the doctor. *The New Yorker*. Retrieved August 30, 2022 from www.newyorker.com/news/news-desk/oliver-sacks-the-doctor.

Gross, L. (2006). Evolution of neonatal imitation. *PLOS Biology*, *4*, e311. doi:10.1371/journal.pbio.0040311.

Gruber, T., Muller, M. N., Strimling, P., Wrangham, R., & Zuberbuhler, K. (2009). Wild chimpanzees rely on cultural knowledge to solve an experimental honey acquisition task. *Current Biology, 19*, 1806–1810.

Gruber, T., Reynolds, V., & Zuberbuhler, K. (2010). The knowns and unknowns of chimpanzee culture. *Communicative and Integrative Biology, 3*, 221–223.

Guerin, B. (2003). Language use as social strategy: A review and an analytic framework for the social sciences. *Review of General Psychology, 7*, 251–298.

Guerra, L. G., & Silva, M. T. (2010). Learning processes and the neural analysis of conditioning. *Psychology and Neuroscience, 3*, 195–208.

Guess, D. (1969). A functional analysis of receptive language and productive speech: Acquisition of the plural morpheme. *Journal of Applied Behavior Analysis, 2*, 55–64.

Guggisberg, A. G., Mathis, J., Schnider, A., & Hess, C. W. (2010). Why do we yawn? *Neuroscience and Biobehavioral Reviews, 34*, 1267–1276.

Guillette, L. M., Farrell, T. M., Hoeschele, M., Nickerson, C. M., Dawson, M. R., & Sturdy, C. B. (2010). Mechanisms of call note-type perception in black-capped chickadees (*Poecile atricapillus*): Peak shift in a note-type continuum. *Journal of Comparative Psychology, 124*, 109–115.

Guinther, P. M., & Dougher, M. J. (2015). The clinical relevance of stimulus equivalence and relational frame theory in influencing the behavior of verbally competent adults. *Current Opinion in Psychology, 2*, 21–25.

Gully, K. J., & Dengerink, H. A. (1983). The dyadic interaction of persons with violent and nonviolent histories. *Aggressive Behavior, 7*, 13–20.

Gustafson, R. (1989). Frustration and successful vs. unsuccessful aggression: A test of Berkowitz' completion hypothesis. *Aggressive Behavior, 15*, 5–12.

Guttman, A. (1977). Positive contrast, negative induction, and inhibitory stimulus control in the rat. *Journal of the Experimental Analysis of Behavior, 27*, 219–233.

Guttman, N., & Kalish, H. I. (1956). Discriminability and stimulus generalization. *Journal of Experimental Psychology, 51*, 79–88.

Hackenberg, T. D. (2009). Token reinforcement: A review and analysis. *Journal of the Experimental Analysis of Behavior, 91*, 257–286.

Hackenberg, T. D., & Hineline, P. N. (1987). Remote effects of aversive contingencies: Disruption of appetitive behavior by adjacent avoidance sessions. *Journal of the Experimental Analysis of Behavior, 48*, 161–173.

Hackenberg, T. D., & Joker, V. R. (1994). Instructional versus schedule control of humans' choices in situations of diminishing returns. *Journal of the Experimental Analysis of Behavior, 62*, 367–383.

Hadamitzky, M., Engler, H., & Schedlowski, M. (2013). Learned immunosuppression: Extinction, renewal, and the challenge of reconsolidation. *Journal of Neuroimmune Pharmacology, 8*, 180–188.

Haggbloom, S. J. (2002). The 100 most eminent psychologists of the 20th century. *Review of General Psychology, 6*, 139–152.

Hall, G. (2009). Preexposure to the unconditioned stimulus in nausea-based aversion learning. In S. Reilly & T. R. Schachtman (Eds.), *Conditioned taste aversion: Behavioral and neural processes* (pp. 58–73). New York: Oxford University Press.

Hall, G., & Sundberg, M. L. (1987). Teaching mands by manipulating conditioned establishing operations. *The Analysis of Verbal Behavior, 5*, 41–53.

Hall, R. V., Cristler, C., Cranston, S. S., & Tucker, B. (1970). Teachers and parents as researchers using multiple baseline designs. *Journal of Applied Behavior Analysis, 3*, 247–255.

Hammack, S. E., Cooper, M. A., & Lezak, K. R. (2012). Overlapping neurobiology of learned helplessness and conditioned defeat: Implications for PTSD and mood disorders. *Neuropharmacology, 62*, 565–575.

Hammer, M., & Menzel, R. (1998). Multiple sites of associative odor learning as revealed by local brain microinjections of octopamine in honeybees. *Learning and Memory, 5*, 146–156.

Hammond, J. L., Iwata, B. A., Fritz, J. N., & Dempsey, C. M. (2011). Evaluation of fixed momentary DRO schedules under signaled and unsignaled arrangements. *Journal of Applied Behavior Analysis, 44*, 69–81.

Hand, D. J., Heil, S. H., Sigmon, S. C., & Higgins, S. T. (2014). Improving Medicaid health incentives programs: Lessons from substance abuse treatment research. *Preventive Medicine, 63*, 87–89.

Hanley, G. P., Piazza, C. C., Fisher, W. W., & Maglieri, K. A. (2005). On the effectiveness of and preference for punishment and extinction components of function-based interventions. *Journal of Applied Behavior Analysis, 38*(1), 51–65.

Hanna, E., & Meltzoff, A. N. (1993). Peer imitation by toddlers in laboratory, home and day care contexts: Implications for social learning and memory. *Developmental Psychology, 29*, 701–710.

Hansen, S. D., & Lignugaris/Kraft, B. (2005). Effects of a dependent group contingency on the verbal interactions of middle school students with emotional disturbance. *Behavioral Disorders, 30*, 170–184.

Hanson, H. M. (1959). Effects of discrimination training on stimulus generalization. *Journal of Experimental Psychology, 58*, 321–334.

Hardin, G. (1968). The tragedy of the commons. *Science, 162*, 1243.

Harlow, H. F., & Zimmerman, R. R. (1959). Affectional responses in the infant monkey. *Science, 130*, 421–432.

Harper, D. N., & McLean, A. P. (1992). Resistance to change and the law of effect. *Journal of the Experimental Analysis of Behavior, 57*, 317–337.

Harris, J. L., Bargh, J. A., & Brownell, K. D. (2009). Priming effects of television food advertising on eating behavior. *Health Psychology, 28*(4), 404–413.

Harris, M. (1974). *Cows, pigs, wars, and witches*. New York: Vintage Books.

Harris, M. (1979). *Cultural materialism*. New York: Random House.

Harris, S., Sheth, S., & Cohen, M. (2007). Functional neuroimaging of belief, disbelief, and uncertainty. *Annals of Neurology, 63*, 141–147.

Hasazi, J. E., & Hasazi, S. E. (1972). Effects of teacher attention on digit-reversal behavior in an elementary school child. *Journal of Applied Behavior Analysis, 5*, 157–162.

Hastad, O., Ernstdotter, E., & Odeen, A. (2005). Ultraviolet vision and foraging in dip and plunge birds. *Biological Letters, 1*, 306–309.

Hatch, J. P. (1980). The effects of operant reinforcement schedules on the modification of human heart rate. *Psychophysiology, 17*, 559–567.

Hausmann, F., Arnold, K. E., Marshall, N. J., & Owens, I. P. F. (2003). Ultraviolet signals in birds are special. *Proceedings of the Royal Society B, 270*, 61–67.

Havermans, R. C., & Jansen, A. T. M. (2003). Increasing the efficacy of cue exposure treatment in preventing relapse of addictive behavior. *Addictive Behaviors, 28*, 989–994.

Havermans, R. C., Salvy, S., & Jansen, A. (2009). Single-trial exercise-induced taste and odor aversion learning in humans. *Appetite, 53*, 442–445.

Haw, J. (2008). Random-ratio schedules of reinforcement: The role of early wins and unreinforced trials. *Journal of Gambling Issues, 21*, 56–67.

Hayes, S. C. (1987). A contextual approach to therapeutic change. In N. Jacobson (Ed.), *Psychotherapists in clinical practice: Cognitive and behavioral perspectives* (pp. 329–383). New York: Guilford Press.

Hayes, S. C. (1989a). Nonhumans have not yet shown stimulus equivalence. *Journal of the Experimental Analysis of Behavior, 51*, 385–392.

Hayes, S. C. (1989b). *Rule-governed behavior: Cognition, contingencies, and instructional control*. New York: Plenum Press.

Hayes, S. C., & Ju, W. (1997). The applied implications of rule-governed behavior. In W. T. O'Donohue (Ed.), *Learning and behavior therapy* (pp. 374–391). Needham Heights, MA: Allyn & Bacon.

Hayes, S. C., & Wilson, K. G. (1994). Acceptance and commitment therapy: Altering the verbal support for experiential avoidance. *The Behavior Analyst, 17*(2), 289–303.

Hayes, S. C., Barnes-Holmes, D., & Roche, B. (2001). *Relational frame theory: A post-Skinnerian account of human language and cognition*. New York: Plenum.

Hayes, S. C., Brownstein, A. J., Haas, J. R., & Greenway, D. E. (1986). Instructions, multiple schedules, and extinction: Distinguishing rule-governed from schedule-controlled behavior. *Journal of the Experimental Analysis of Behavior, 46*, 137–147.

Hayes, S. C., Strosahl, K. D., & Wilson, K. G. (2012). *Acceptance and commitment therapy: The process and practice of mindful change*. New York: The Guilford Press.

Hayes, S.C. (2004). Acceptance and commitment therapy, relational frame theory, and the third wave of behavioral and cognitive therapies. *Behavior Therapy, 35*, 639–665. https://doi.org/10.1016/S0005-7894(04)80013-3.

Hayes, S.C., Strosahl, K.D., & Wilson, K. (2009). *Acceptance and commitment therapy: The process and practice of mindful change*. Guilford Press: New York.

Hearst, E. (1961). Resistance-to-extinction functions in the single organism. *Journal of the Experimental Analysis of Behavior, 4*, 133–144.

Hearst, E., & Jenkins, H. M. (1974). *Sign tracking: The stimulus—reinforcer relation and directed action*. Austin, TX: The Psychonomic Society.

Heerey, E. A. (2014). Learning from social rewards predicts individual differences in self-reported social ability. *Journal of Experimental Psychology: General, 143*, 332–339.

Heil, S. H., Johnson, M. W., Higgins, S. T., & Bickel, W. K. (2006). Delay discounting in currently using and currently abstinent cocaine-dependent outpatients and non-drug-using matched controls. *Addictive Behaviors, 31*, 1290–1294.

Heinrich, B., & Bugnyar, T. (2007). Just how smart are ravens? *Scientific American, 296*, 64–71.

Hendrickson, K. & Rasmussen, E. B. (2013). Effects of mindful eating training on delay and probability discounting for food and money in obese and healthy-weight individuals. *Behaviour Research & Therapy, 51*, 399–409.

Hendrickson, K., & Rasmussen, E.B. (2017). Mindful eating reduces impulsive food choice in adolescents and adults. *Health Psychology, 36*, 226–235.

Hendrickson, K., Rasmussen, E.B., & Lawyer, S.R. (2015). Measurement and validation of measures for impulsive food choice in obese and healthy weight humans. *Appetite, 90*, 254–263. https://doi.org/10.1016/j.appet.2015.03.015

Hendrik, G. R., Boulogne, J. J., van Tulder, M. W., van den Brink, W., De Jong, C. A. J., & Kerkhof, A. J. F. M. (2004). A systematic review of the effectiveness of the community reinforcement approach in alcohol, cocaine and opioid addiction. *Drug and Alcohol Dependence, 74*, 1–13.

Herman, R. L., & Azrin, N. H. (1964). Punishment by noise in an alternative response situation. *Journal of the Experimental Analysis of Behavior, 7*, 185–188.

Hermann, P. M., de Lange, R. P. J., Pieneman, A. W., ter Maat, A., & Jansen, R. F. (1997). Role of neuropeptides encoded on CDCH-1 gene in the organization of egg-laying behavior in the pond snail, *Lymnaea stagnalis*. *Journal of Neurophysiology, 78*, 2859–2869.

Herrnstein, R. J. (1961a). Stereotypy and intermittent reinforcement. *Science, 133*, 2067–2069.

Herrnstein, R. J. (1961b). Relative and absolute strength of responses as a function of frequency of reinforcement. *Journal of the Experimental Analysis of Behavior, 4*, 267–272.

Herrnstein, R. J. (1964a). Aperiodicity as a factor in choice. *Journal of the Experimental Analysis of Behavior, 7*, 179–182.

Herrnstein, R. J. (1964b). Secondary reinforcement and the rate of primary reinforcement. *Journal of the Experimental Analysis of Behavior, 7*, 27–36.

Herrnstein, R. J. (1970). On the law of effect. *Journal of the Experimental Analysis of Behavior, 13*, 243–266.

Herrnstein, R. J. (1974). Formal properties of the matching law. *Journal of the Experimental Analysis of Behavior, 21*, 159–164.

Herrnstein, R. J. (1979). Acquisition, generalization, and reversal of a natural concept. *Journal of Experimental Psychology: Animal Behavior Processes, 5*, 116–129.

Herrnstein, R. J. (1997). Melioration as behavioral dynamics. In H. Rachlin & D. I. Laibson (Eds.), *The matching law: Papers in psychology and economics by Richard J. Herrnstein* (pp. 74–99). Cambridge, MA: Harvard University Press.

Herrnstein, R. J., & de Villiers, P. A. (1980). Fish as a natural category for people and pigeons. In G. H. Bower (Ed.), *The psychology of learning and motivation* (pp. 60–95). New York: Academic Press.

Herrnstein, R. J., & Hineline, P. N. (1966). Negative reinforcement as shock frequency reduction. *Journal of the Experimental Analysis of Behavior, 9*, 421–430.

Herrnstein, R. J., & Loveland, D. H. (1964). Complex visual concept in the pigeon. *Science, 146*, 549–551.

Herrnstein, R. J., & Loveland, D. H. (1975). Maximizing and matching on concurrent ratio schedules. *Journal of the Experimental Analysis of Behavior, 24*, 107–116.

Herrnstein, R. J., Loveland, D. H., & Cable, C. (1976). Natural concepts in pigeons. *Journal of Experimental Psychology: Animal Behavior Processes, 2*, 285–302.

Hesse-Biber, S., Leavy, P., Quinn, C. E., & Zoino, J. (2006). The mass marketing of disordered eating and eating disorders: The social psychology of women, thinness and culture. In *Women's studies international forum*, 29(2) (pp. 208–224). Pergamon.

Heyes, C., Bird, G., Johnson, H., & Haggard, P. (2005). Experience modulates automatic imitation. *Cognitive Brain Research, 22*, 233–240.

Heyman, G. M. (2009). *Addiction: A disorder of choice*. Cambridge, MA: Harvard University Press.

Heyman, G. M. (2014). Drug addiction is a matter of difficult choices. *The New York Times*, Retrieved from www.nytimes.com/roomfordebate/2014/02/10/what-is-addiction/drug-addictionis-a-matter-of-difficult-choices.

Hickok, G. (2014). *The myth of mirror neurons: The real neuroscience of communication and cognition*. New York: Norton.

Higgins, S. T., & Katz, J. L. (1998). *Cocaine abuse: Behavior, pharmacology, and clinical applications*. San Diego, CA: Academic Press.

Higgins, S. T., & Petry, N. M. (1999). Contingency management: Incentives for sobriety. *Alcohol Research and Health, 23*, 122–127.

Higgins, S. T., Bickel, W. K., & Hughes, J. R. (1994). Influence of an alternative reinforcer on human cocaine self-administration. *Life Sciences, 55*, 179–187.

Higgins, S. T., Budney, A. J., Bickel, W. K., Hughes, J. R., Foerg, F., & Badger, G. J. (1993). Achieving cocaine abstinence with a behavioral approach. *American Journal of Psychiatry, 150*, 763–769.

Higgins, S. T., Delaney, D., Budney, A., Bickel, W., Hughes, J., & Foerg, F. (1991). A behavioral approach to achieving initial cocaine abstinence. *American Journal of Psychiatry*, *148*, 1218–1224.

Higgins, S. T., Silverman, K., & Heil, S. H. (Eds.) (2007). *Contingency management in substance abuse treatment*. Guilford Press.

Higgins, S. T., Washio, Y., Heil, S. H., Solomon, L. J., Gaalema, D. E., Higgins, T. M., et al. (2012). Financial incentives for smoking cessation among pregnant and newly postpartum women. *Preventive Medicine*, *55* (Suppl.), S33–S40.

Hinde, R. A., & Stevenson-Hinde, J. (1973). *Constraints on learning: Limitations and predispositions*. New York: Academic Press.

Hineline, P. N. (1970). Negative reinforcement without shock reduction. *Journal of the Experimental Analysis of Behavior*, *14*, 259–268.

Hineline, P. N., & Rosales-Ruiz, J. (2013). Behavior in relation to aversive events: Punishment and negative reinforcement. In G. J. Madden (Ed.), *APA handbook of behavior analysis: Vol. 1. Methods and principles* (pp. 483–512). Washington, DC: American Psychological Association.

Hinson, R. E., Poulos, C. X., & Cappell, H. (1982). Effects of pentobarbital and cocaine in rats expecting pentobarbital. *Pharmacology, Biochemistry, and Behavior*, *16*, 661–666.

Hiroto, D. S., & Seligman, M. E. P. (1975). Generality of learned helplessness in man. *Journal of Personality and Social Psychology*, *31*, 311–327.

Hirsch, J., & McCauley, L. (1977). Successful replication of, and selective breeding for, classical conditioning in the blowfly (*Phormia regina*). *Animal Behaviour*, *25*, 784–785.

Hockett, C. F. (1958). *A course in modern linguistics*. New York: Macmillan.

Hockett, C. F. (1968). *The state of the art*. The Hague: Mouton.

Hodos, W. (1961). Progressive ratio as a measure of reward strength. *Science*, *134*, 943–944.

Hofmann, W., De Houwer, J., Perugini, M., Baeyens, F., & Crombez, G. (2010). Evaluative conditioning in humans: A meta-analysis. *Psychological Bulletin*, *136*, 390–421.

Hogg, C., Neveu, M., Stokkan, K., Folkow, L., Cottrill, P., Douglas, P., et al. (2011). Arctic reindeer extend their visual range into the ultraviolet. *Journal of Experimental Biology*, *214*, 2014–2019.

Holding, E., Bray, M. A., & Kehle, T. J. (2010). Does speed matter? A comparison of the effectiveness of fluency and discrete trial training for teaching noun tables to children with autism. *Psychology in the Schools*, *48*, 166–188.

Holland, J. G. (1978). Behaviorism: Part of the problem or part of the solution? *Journal of Applied Behavior Analysis*, *11*, 163–174.

Hollis, J. H. (1973). "Superstition": The effects of independent and contingent events on free operant responses in retarded children. *American Journal of Mental Deficiency*, *77*, 585–596.

Horne, P. J., & Erjavec, M. (2007). Do infants show generalized imitation of gestures? *Journal of the Experimental Analysis of Behavior*, *87*, 63–88.

Horne, P. J., & Lowe, C. F. (1993). Determinants of human performance on concurrent schedules. *Journal of the Experimental Analysis of Behavior*, *59*, 29–60.

Horne, P. J., & Lowe, C. F. (1996). On the origins of naming and other symbolic behavior. *Journal of the Experimental Analysis of Behavior*, *65*, 185–241.

Horne, P. J., Greenhalgh, J., Erjavec, M., Lowe, C. F., Viktor, S., & Whitaker, C. J. (2011). Increasing preschool children's consumption of fruit and vegetables: A modeling and rewards intervention. *Appetite*, *56*, 375–385.

Horner, R. H., Carr, E. G., Strain, P. S., Todd, A. W., & Reed, H. K. (2002). Problem behavior interventions for young children with autism: A research synthesis. *Journal of Autism and Developmental Disorders, 32,* 423–446.

Horner, R. H., Day, H. M., & Day, J. R. (1997). Using neutralizing routines to reduce problem behavior. *Journal of Applied Behavior Analysis, 30,* 601–614.

Hothersall, D. (1990). *History of psychology.* New York: McGraw-Hill.

Houston, A. (1986). The matching law applies to wagtails' foraging in the wild. *Journal of the Experimental Analysis of Behavior, 45,* 15–18.

Howard, J. S., & Rice, D. E. (1988). Establishing a generalized autoclitic repertoire in preschool children. *The Analysis of Verbal Behavior, 6,* 45–60. https://doi.org/10.1016/S0006-3223(96)00059-5.

Huffstetter, M., King, J. R., Onwuegbuzie, A. J., Schneider, J. J., & Powell-Smith, K. A. (2010). Effects of a computer-based early reading program on the early reading and oral language skills of at-risk preschool children. *Journal of Education for Students Placed at Risk, 15,* 279–298.

Hughes, S., & Barnes-Holmes, D. (2016). Relational frame theory: The basic account. In R. D. Zettle, S. C. Hayes, D. Barnes-Holmes, & A. Biglan (Eds.), *The Wiley handbook of contextual behavioral science* (pp. 129–178). Wiley Blackwell.

Hughes, S. C., & Boakes, R. A. (2008). Flavor preferences produced by backward pairing with wheel running. *Journal of Experimental Psychology: Animal Behavior Processes, 34*(2), 283. https://doi.org/10.1037/0097-7403.34.2.283

Hull, D. L., Langman, R. E., & Glenn, S. S. (2001). A general account of selection: Biology, immunology, and behavior. *Behavioral and Brain Sciences, 24,* 511–573.

Hunt, G. M., & Azrin, N. H. (1973). A community-reinforcement approach to alcoholism. *Behavior Research and Therapy, 11,* 91–104.

Hursh, S. R. (1991). Behavioral economics of drug self-administration and drug abuse policy. *Journal of the Experimental Analysis of Behavior, 56,* 377–393.

Hursh, S. R., Galuska, C. M., Winger, G., & Woods, J. H. (2005). The economics of drug abuse: a quantitative assessment of drug demand. *Molecular interventions, 5*(1), 20. doi:10.1124/mi.5.1.6MI February 2005, *5*(1) 20–28.

Hursh, S. R., Navarick, D. J., & Fantino, E. (1974). "Automaintenance": The role of reinforcement. *Journal of the Experimental Analysis of Behavior, 21,* 112–124.

Hutchinson, R. R. (1977). By-products of aversive control. In W. K. Honig & J. E. R. Staddon (Eds.), *Handbook of operant behavior* (pp. 415–431). Englewood Cliffs, NJ: Prentice-Hall.

Hutchinson, R. R., Azrin, N. H., & Hunt, G. M. (1968). Attack produced by intermittent reinforcement of a concurrent operant response. *Journal of the Experimental Analysis of Behavior, 11,* 489–495.

Hutsell, B. A., Negus, S. S., & Banks, M. L. (2015). A generalized matching law analysis of cocaine vs. food choice in rhesus monkeys: Effects of candidate 'agonist-based' medications on sensitivity to reinforcement. *Drug and Alcohol Dependence, 146,* 52–60.

Iacoboni, M. (2009). Neurobiology of imitation. *Current Opinion in Neurobiology, 19,* 661–665.

Iacoboni, M., Woods, R. P., Brass, M., Bekkering, H., Mazziota, J. C., & Rizzolatti, G. (1999). Cortical mechanisms of human imitation. *Science, 286,* 2526–2528.

Ingvarsson, E. T., Kahng, S., & Hausman, N. L. (2008). Some effects of noncontingent positive reinforcement on multiply controlled problem behavior and compliance in a demand context. *Journal of Applied Behavior Analysis, 41*(3), 435–440. https://doi.org/10.1901/jaba.2008.41-435

Iovannone, R., Dunlap, G., Huber, H., & Kincaid, D. (2003). Effective educational practices for students with autism spectrum disorders. *Focus on Autism and Other Developmental Disabilities, 18*(3), 150–165. https://doi.org/10.1177/10883576030180030301

Iriki, A. (2006). The neural origins and implications of imitation, mirror neurons and tool use. *Current Opinion in Neurobiology, 16*, 660–667.

Isaacs, C. D. (1982). Treatment of child abuse: A review of the behavioral interventions. *Journal of Applied Behavior Analysis, 15*(2), 273–294. https://doi.org/10.1901/jaba.1982.15-273

Ishikawa, D., Matsumoto, N., Sakaguchi, T., Matsuki, N., & Ikegaya, Y. (2014). Operant conditioning of synaptic and spiking activity patterns in single hyppocampal neurons. *The Journal of Neuroscience, 34*, 5044–5053.

Israel, M. L., Blenkush, N. A., von Heyn, R. E., & Rivera, P. M. (2008). Treatment of aggression with behavioral programming that includes supplementary contingent skin-shock. *The Journal of Behavior Analysis of Offender and Victim Treatment and Prevention, 1*(4), 119.

Ivy, J. W., Meindl, J. N., Overley, E., & Robson, K. M. (2017). Token economy: A systematic review of procedural descriptions. *Behavior Modification, 41*(5), 708–737. https://doi.org/10.1177/0145445517699559

Iwata, B. A., Dorsey, M. F., Slifer, K. J., Bauman, K. E., & Richman, G. S. (1994). Toward a functional analysis of self-injury. *Journal of Applied Behavior Analysis, 27*, 197–209.

Iwata, B. A., Pace, G. M., Kalsher, M. J., Cowdery, G. E., & Cataldo, M. F. (1990). Experimental analysis and extinction of self-injurious escape behavior. *Journal of Applied Behavior Analysis, 23*, 11–27.

Jablonka, E., & Lamb, M. J. (2002). The changing concept of epigenetics. *Annals of the New York Academy of Sciences, 981*, 82–96.

Jablonka, E., & Raz, G. (2009). Transgenerational epigenetic inheritance: Prevalence, mechanisms, and implications for the study of heredity and evolution. *The Quarterly Review of Biology, 84*, 131–176.

Jackson, R. L., Alexander, J. H., & Maier, S. F. (1980). Learned helplessness, inactivity, and associative deficits: Effects of inescapable shock on response choice escape learning. *Journal of Experimental Psychology: Animal Behavior Processes, 6*, 1–20.

Jacobs, E. A., Borrero, J. C., & Vollmer, T. R. (2013). Translational applications of quantitative choice models. In G. J. Madden (Ed.), *APA handbook of behavior analysis: Vol. 2. Translating principles into practice* (pp. 165–190). Washington, DC: American Psychological Association. doi:10.1037/13938–007.

Jaffe, Y., Shapir, N., & Yinon, Y. (1981). Aggression and its escalation. *Journal of Cross-Cultural Psychology, 12*, 21–36.

James, W. (1890). *Principles of psychology*. New York: Holt, Rinehart & Winston.

Jannetto, P. J., Helander, A., Garg, U., Janis, G. C., Goldberger, B., & Ketha, H. (2019). The fentanyl epidemic and evolution of fentanyl analogs in the United States and the European Union. *Clinical chemistry, 65*(2), 242–253. https://doi.org/10.1373/clinchem.2017.281626

Janssen, M. A., Holahan, R., Lee, A., & Ostrom, E. (2010). Lab experiments for the study of social-ecological systems. *Science, 328*, 613–617.

Jarmolowicz, D. P., Reed, D. D., DiGennaro Reed, F. D., & Bickel, W. K. (2016). The behavioral and neuroeconomics of reinforcer pathologies: Implications for managerial and health decision making. *Managerial and Decision Economics, 37*(4–5), 274–293.

Jason, L. A. (1998). Tobacco, drug, and HIV preventive media interventions. *American Journal of Community Psychology, 26*, 151–187.

Jenkins, H. M., & Moore, B. R. (1973). The form of the auto-shaped with food or water reinforcers. *Journal of the Experimental Analysis of Behavior, 20*, 163–181.

Jenkins, H. M., Barrera, F. J., Ireland, C., & Woodside, B. (1978). Signal-centered action patterns of dogs in appetitive classical conditioning. *Learning and Motivation, 9*, 272–296.

Johansen, J. P., Cain, C. K., Ostroff, L. E., & LeDoux, J. E. (2011). Molecular mechanisms of fear learning and memory. *Cell, 147*, 509–524.

Johnson, P. M., Kenny, P. J. (2010) Dopamine D2 receptors in addiction-like reward dysfunction and compulsive eating in obese rats. *Nature Neuroscience*, *13*(5), 635–41. doi:10.1038/nn.2519. Epub March 28, 2010. Erratum in: *Nature Neuroscience*, (2010) *13*(8) 1033. PMID: 20348917; PMCID: PMC2947358.

Johnson, K. R., & Layng, T. V. J. (1994). The Morningside Model of generative instruction. In R. Gardner III, D. M. Sainato, J. O. Cooper, T. E. Heron, W. L. Heward, J. Eshleman et al. (Eds.), *Behavior analysis in education: Focus on measurably superior instruction* (pp. 173–197). Monterey, CA: Brooks/Cole.

Johnson, R. N., Oldroyd, B. P., Barron, A. B., & Crozier, R. H. (2002). Genetic control of honey bee (*Apis mellifera*) dance language: Segregating dance forms in a backcrossed colony. *Journal of Heredity*, *93*, 170–173.

Johnston, J. M., & Pennypacker, H. S. (1993). *Strategies and tactics of human behavioral research*. Hillsdale, NJ: Lawrence Erlbaum Associates.

Jones, K. M., & Friman, P. C. (1999). A case study of behavioral assessment and treatment of insect phobia. *Journal of Applied Behavior Analysis*, *32*, 95–98.

Jones, S. S. (2009). The development of imitation in infancy. *Philosophical Transactions of the Royal Society B: Biological Sciences*, *364*, 2325–2335.

Juujaevari, P., Kooistra, L., Kaartinen, J., & Pulkkinen, L. (2001). An aggression machine V: Determinants in reactive aggression revisited. *Aggressive Behavior*, *27*, 430–445.

Kalat, J. W. (2014). *Introduction to psychology* (10th ed.). Stamford, CT: Cengage Learning.

Kamin, L. J. (1969). Predictability, surprise, attention, and conditioning. In B. A. Campbell & R. M. Church (Eds.), *Punishment and aversive behavior* (pp. 279–296). New York: Appleton-Century-Crofts.

Kana, R. K., Wadsworth, H. M., & Travers, B. G. (2011). A systems level analysis of the mirror neuron hypothesis and imitation impairments in autism spectrum disorders. *Neuroscience and Biobehavioral Reviews*, *35*, 894–902.

Kandel, E. R. (2006). *In search of memory*. New York: W. W. Norton & Co.

Kangas, B. D. (2007). Cultural materialism and behavior analysis. *The Behavior Analyst*, *30*, 37–47.

Karsina, A., Thompson, R. H., & Rodriguez, N. M. (2011). Effects of a history of differential reinforcement on preference for choice. *Journal of the Experimental Analysis of Behavior*, *95*, 189–202.

Kastak, C. R., Schusterman, R. J., & Kastak, D. (2001). Equivalence classification by California sea lions using class specific reinforcers. *Journal of the Experimental Analysis of Behavior*, *76*, 131–158.

Kastak, D., & Schusterman, R. J. (1994). Transfer of visual identity matching to sample in two California sea lions (*Zalophus californianus*). *Animal Learning and Behavior*, *22*, 427–435.

Kaufman, A., Baron, A., & Kopp, R. E. (1966). Some effects of instructions on human operant behavior. *Psychonomic Monograph Supplements*, *11*, 243–250.

Kawa, S., & Giordano, J. (2012). A brief historicity of the Diagnostic and Statistical Manual of Mental Disorders: issues and implications for the future of psychiatric canon and practice. *Philosophy, ethics, and humanities in medicine: PEHM*, *7*, 2. https://doi.org/10.1186/1747-5341-7-2

Kawai, M. (1965). Newly acquired pre-cultural behavior of the natural troop of Japanese monkeys on Koshima Islet. *Primates*, *6*, 1–30.

Kawamura, S. (1959). The process of sub-culture propagation among Japanese macaques. *Primates*, *2*, 43–60.

Kazdin, A. E. (1977). *The token economy: A review and evaluation*. New York: Plenum Press.

Kazdin, A. E. (1983). Failure of persons to respond to the token economy. In E. B. Foa & P. M. G. Emmelkamp (Eds.), *Failures in behavior therapy* (pp. 335–354). New York: John Wiley & Sons.

Kazdin, A. E. (1989). *Behavior modification in applied settings* (4th ed.). Belmont, CA: Brooks/Cole.

Kazdin, A. E. (2013). *Behavior modification in applied settings* (7th ed.). Long Grove, IL: Waveland Press.

Kazdin, A. E., & Klock, J. (1973). The effect of nonverbal teacher approval on student attentive behavior. *Journal of Applied Behavior Analysis, 6*, 643–654.

Keehn, J. D. (1986). *Animal models for psychiatry.* London: Routledge & Kegan Paul.

Keehn, J. D., & Jozsvai, E. (1989). Induced and noninduced patterns of drinking by food-deprived rats. *Bulletin of the Psychonomic Society, 27*, 157–159.

Keith-Lucas, T., & Guttman, N. (1975). Robust single-trial delayed backward conditioning. *Journal of Comparative and Physiological Psychology, 88*, 468–476.

Kelleher, R. T. (1956). Intermittent conditioned reinforcement in chimpanzees. *Science, 124*, 679–680.

Kelleher, R. T. (1958a). Concept formation in chimpanzees. *Science, 128*, 777–778.

Kelleher, R. T. (1958b). Fixed-ratio schedules of conditioned reinforcement with chimpanzees. *Journal of the Experimental Analysis of Behavior, 1*, 281–289. doi: 10.1901/jeab.1958.1-281.

Kelleher, R. T. (1966). Conditioned reinforcement in second-order schedules. *Journal of the Experimental Analysis of Behavior, 9*, 475–485.

Kelleher, R. T., & Gollub, L. R. (1962). A review of positive conditioned reinforcement. *Journal of the Experimental Analysis of Behavior, 5*, 543–597.

Keller, F. S. (1968). "Good-bye, teacher …" *Journal of Applied Behavior Analysis, 1*, 79–89.

Kelley, H. H. (1987). Attribution in social interaction. In E. E. Jones, D. E. Kanouse, H. H. Kelley, R. E. Nisbett, S. Valins, & B. Weiner (Eds.), *Attribution: Perceiving the causes of behavior* (pp. 1–26). Hillsdale, NJ: Lawrence Erlbaum Associates.

Keysers, C. (2015). The straw man in the brain. *Science, 347*, 240.

Khatchadourian, R. (2015). We know how you feel. *The New Yorker.* Retrieved January 2019 from www.newyorker.com/magazine/2015/01/19/know-feel

Killeen, P. R. (2015). The logistics of choice. *Journal of the Experimental Analysis of Behavior, 104*, 72–92.

Killeen, P. R., & Pellón, R. (2013). Adjunctive behaviors are operants. *Learning & Behavior, 41*, 1–24.

Killeen, P. R., Posadas-Sanchez, D., Johansen, E. B., & Thrailkill, E. A. (2009). Progressive ratio schedules of reinforcement. *Journal of Experimental Psychology: Animal Behavior Processes, 35*, 35–50.

Killeen, P. R., Wald, B., & Cheney, C. (1980). Observing behavior and information. *The Psychological Record, 30*, 181–190.

Killen, M., & Rutland, A. (2011). *Children and social exclusion: Morality, prejudice and group identity.* West Sussex: UK, Wiley-Blackwell.

King, L. E., Douglas-Hamilton, I., & Vollrath, F. (2011). Beehive fences as effective deterrents for crop-raiding elephants: Field trials in northern Kenya. *African Journal of Ecology, 49*, 431–439.

Kinloch, J. M., Foster, T. M., & McEwan, J. S. (2009). Extinction-induced variability in human behavior. *The Psychological Record, 59*(3), 347–369. https://doi.org/10.1007/BF03395669

Kirby, F. D., & Shields, F. (1972). Modification of arithmetic response rate and attending behavior in a seventh-grade student. *Journal of Applied Behavior Analysis, 5*, 79–84.

Kirsch, I. (2014). Antidepressants and the Placebo Effect. *Zeitschrift fur Psychologie, 222*(3), 128–134. https://doi.org/10.1027/2151-2604/a000176

Klin, A., Chawarska, K., Rubin, E., & Volkmar, F. (2004). Clinical assessment of young children at risk for autism. In R. DelCarmen-Wiggins & A. Carter (Eds.), *Handbook of infant, toddler and preschool mental health assessment* (pp. 311–336). New York: Oxford University Press.

Knutson, J. (1970). Aggression during the fixed-ratio and extinction components of a multiple schedule of reinforcement. *Journal of the Experimental Analysis of Behavior, 13*, 221–231.

Ko, M. C., Terner, J., Hursh, S., Woods, J. H., & Winger, G. (2002). Relative reinforcing effects of three opioids with different durations of action. *Journal of Pharmacology and Experimental Therapeutics, 301*(2), 698–704. https://doi.org/10.1124/jpet.301.2.698

Kobayashi, S., Schultz, W., & Sakagami, M. (2010). Operant conditioning of primate prefrontal neurons. *Journal of Neurophysiology, 103*, 1843–1855.

Kodak, T., & Bergmann, S. (2020). Autism spectrum disorder: Characteristics, associated behaviors, and early intervention. *Pediatric Clinics, 67*(3), 525–535.

Kohler, W. (1927). *The mentality of apes* (2nd revised ed., E. Winter, trans.). London: Routledge & Kegan Paul.

Kollins, S. H., Newland, M. C., & Critchfield, T. S. (1997). Human sensitivity to reinforcement in operant choice: How much do consequences matter?. *Psychonomic Bulletin & Review, 4*(2), 208–220.

Konkel, L. (2016). Positive reinforcement helps surgeons learn. *Scientific American*. Retrieved from www.scientificamerican.com/article/positive-reinforcement-helps-surgeons-learn/?WT.mc_id=send-to-friend

Kratochwill, T. R., Hitchcock, J., Horner, R. H., Levin, J. R., Odom, S. L., Rindskopf, D. M., et al. (2010). Single-case designs technical documentation. Retrieved from *What Works Clearinghouse website*, http://ies.ed.gov/ncee/wwc/pdf/wwc_scd.pdf

Krebs, J. R., & Davies, N. B. (Eds.) (1978). *Behavioural ecology: An evolutionary approach*. Oxford: Blackwell Scientific Publications.

Kristof, N. (2015). The dangers of vaccine denial. *The New York Times*. Retrieved from http://nyti.ms/1yU8vM5.

Kruger, B., Bischoff, M., Blecker, C., Langhanns, C., Kindermann, S., Sauerbier, I., et al. (2014). Parietal and premotor cortices: Activation reflects imitation accuracy during observation, delayed imitation and concurrent imitation. *NeuroImage, 100*, 39–50.

Kubina, R. M. Jr. (2005). The relations among fluency, rate building, and practice. A response to Doughty, Chase, and O'Shields (2004). *The Behavior Analyst, 28*, 73–76.

Kuhl, P. K., & Meltzoff, A. N. (1996). Infant vocalizations in response to speech: Vocal imitation and developmental change. *Journal of the Acoustical Society of America, 100*, 2425–2438.

Kulik, C. C., Kulik, J. A., & Cohen, P. A. (1980). Instructional technology and college teaching. *Teaching of Psychology, 7*, 199–205.

Kulubekova, S. (2012). Computational model of selection by consequences: Patterns of preference change on concurrent schedules. Ph.D. dissertation of the Department of Psychology, Emory University.

Kunkel, J. H. (1997). The analysis of rule-governed behavior in social psychology. *The Psychological Record, 47*, 698–716.

Kurti, A. N., & Dallery, J. (2012). Review of Heyman's addiction: A disorder of choice. *Journal of Applied Behavior Analysis, 45*, 229–240.

Kwapis, J. L., & Wood, M. A. (2014). Epigenetic mechanisms in fear conditioning: Implications for treating post-traumatic stress disorder. *Trends in Neurosciences, 37*, 706–720.

Kysers, C., & Gazzola, V. (2014). Hebbian learning and predictive mirror neurons for actions, sensations and emotions. *Philosophical Transactions of the Royal Society B, 369*, 20130175. doi:10.1098/rstb.2013.0175.

Lamal, P. A. (Ed.) (1997). *Cultural contingencies: Behavior analytic perspectives on cultural practices*. Westport, CT: Praeger.

Lamal, P. A., & Greenspoon, J. (1992). Congressional metacontingencies. *Behavior and Social Issues, 2*, 71–81.

Lamarre, J., & Holland, J. G. (1985). The functional independence of mands and tacts. *Journal of the Experimental Analysis of Behavior, 43*, 5–19.

Langer, E., Djikic, M., Pirson, M., Madenci, A., & Donohue, R. (2010). Believing is seeing: Using mindlessness (mindfully) to improve visual acuity. *Psychological Science*, *21*, 661–666.

Langthorne, P., & McGill, P. (2009). A tutorial on the concept of the motivating operation and its importance to application. *Behavior Analysis and Practice*, *2*, 22–31.

Laraway, S., Snycerski, S., Michael, J., & Poling, A. (2003). Motivating operations and terms to describe them: Some further refinements. *Journal of Applied Behavior Analysis*, *36*, 407–414.

Lashley, R. L., & Rossellini, R. (1980). Modulation of schedule-induced polydipsia by Pavlovian conditioned states. *Physiology & Behavior*, *24*, 411–414.

Lasiter, P. S. (1979). Influence of contingent responding on schedule-induced activity in human subjects. *Physiology and Behavior*, *22*, 239–243.

Latham, G. (1994). *The power of positive parenting*. North Logan, UT: P&T.

Lattal, D. (2012). *Vigilance: Behaving safely during routine, novel and rare events*. Aubrey Daniels International, Inc. Retrieved from http://aubreydaniels.com/pmezine/vigilance-behaving-safely-during-routine-novel-and-rare-eventshttp

Lattal, K. A., & Doepke, K. J. (2001). Correspondence as conditional stimulus control: Insights from experiments with pigeons. *Journal of Applied Behavior Analysis*, *34*, 127–144.

Lattal, K. A., Reilly, M. P., & Kohn, J. P. (1998). Response persistence under ratio and interval reinforcement schedules. *Journal of the Experimental Analysis of Behavior*, *70*, 165–183.

Lattal, K. M., & Lattal, K. A. (2012). Facets of Pavlovian and operant extinction. *Behavioural Processes*, *90*, 1–8.

LaVigna, G. W., & Donnellan, A. W. (1986). *Alternatives to punishment: Solving behavior problems with non-aversive strategies*. New York: Irvington.

Layng, T. V. J., Twyman, J. S., & Stikeleather, G. (2004). Selected for success: How Headsprout Reading Basics™ teaches beginning reading. In D. J. Moran & R. W. Malott (Eds.), *Evidence-based educational methods* (pp. 171–195). San Diego, CA: Elsevier Inc.

Lazar, R. (1977). Extending sequence-class membership with matching to sample. *Journal of the Experimental Analysis of Behavior*, *27*, 381–392.

Leal, M., & Powell, B. J. (2012). Behavioural flexibility and problem-solving in a tropical lizard. *Biological Letters*, *8*, 28–30.

Lechago, S. A., Carr, J. E., Grow, L. L., Love, J. R., & Almason, S. M. (2010). Mands for information generalize across establishing operations. *Journal of Applied Behavior Analysis*, *43*, 381–395. doi:10.1901/jaba.2010.43-381.

Lee, K., Talwar, V., McCarthy, A., Ross, I., Evans, A., & Arruda, C. (2014). Can classic moral stories promote honesty in children? *Psychological Science*, *25*, 1630–1636.

Lee, R., Sturmey, P., & Fields, L. (2007). Schedule-induced and operant mechanisms that influence response variability: A review and implications for future investigations. *The Psychological Record*, *57*, 429–455.

Lee, V. L. (1981a). Prepositional phrases spoken and heard. *Journal of the Experimental Analysis of Behavior*, *35*, 227–242.

Lee, V. L. (1981b). Terminological and conceptual revision in the experimental analysis of language development: Why? *Behaviorism*, *9*, 25–53.

Lehman, P. K., & Geller, E. S. (2004). Behavior analysis and environmental protection: Accomplishments and potential for more. *Behavior and social issues*, *13*(1), 13–33. https://doi.org/10.5210/bsi.v13i1.33

Lemley, S. M., Kaplan, B. A., Reed, D. D., Darden, A. C., & Jarmolowicz, D. P. (2016). Reinforcer pathologies: Predicting alcohol related problems in college drinking men and women. *Drug and Alcohol Dependence*, *167*, 57–66.

LeMoyne, T., & Buchanan, T. (2011). Does "hovering" matter? Helicopter parenting and its effect on well-being. *Sociological Spectrum*, *31*(4), 399–418.

Leotti, L. A., & Delgado, M. R. (2011). The inherent reward of choice. *Psychological Science*, *22*, 1310–1318.

Lerman, D. C., Iwata, B. A., & Wallace, M. D. (1999). Side effects of extinction: Prevalence of bursting and aggression during the treatment of self-injurious behavior. *Journal of Applied Behavior Analysis*, *32*, 1–8.

Lerman, D. C., Iwata, B. A., Zarcone, J. R., & Ringdahl, J. (1994). Assessment of stereotypic and self-injurious behavior as adjunctive responses. *Journal of Applied Behavior Analysis*, *27*, 715–728.

Lesaint, F., Sigaud, O., & Khamassi, M. (2014). Accounting for negative automaintenance in pigeons: A dual learning systems approach and factored representations. *PloS One*, *9*, e111050. doi:10.1371/journal.pone.0111050.

Lerman, D. C., & Vorndran, C. M. (2002). On the status of knowledge for using punishment: Implications for treating behavior disorders. *Journal of Applied Behavior Analysis*, *35*(4), 431–464. https://doi.org/10.1901/jaba.2002.35-431

Leslie, J. C. (2011). Animal models of psychiatric disorders: Behavior analysis perspectives. *European Journal of Behavior Analysis*, *12*, 27–40.

Lett, B. T., & Grant, V. L. (1996). Wheel running induces conditioned taste aversion in rats trained while hungry and thirsty. *Physiology and Behavior*, *59*, 699–702.

Lett, B. T., Grant, V. L., Byrne, M. J., & Koh, M. T. (2000). Pairings of a distinctive chamber with the aftereffect of wheel running produce conditioned place preference. *Appetite*, *34*, 87–94.

Levenson, R. M., Krupinski, E. A., Navarro, V. M., & Wasserman, E. A. (2015). Pigeons (*Columba livia*) as trainable observers of pathology and radiology breast cancer images. *PLoS One*. doi:10.1371/journal.pone.0141357.

Levitin, D. (2014). *The organized mind*. New York: Penguin.

Levy, I. M., Pryor, K. W., & McKeon, T. R. (2016). Is teaching simple surgical skills using an operant learning program more effective than teaching by demonstration? Clinical *Orthopedics and Related Research*, *474*, 945–955. doi:10.1007/s11999-015-4555-8.

Lewis, M. (2010). *The big short: Inside the doomsday machine*. New York: W. W. Norton & Company.

Li, B., Piriz, J., Mirrione, M., Chung, C., Proulx, C. D., Schulz, D., et al. (2011). Synaptic potentiation onto habenula neurons in the learned helplessness model of depression. *Nature*, *470*, 535–539.

Libbrecht, R., Oxley, P. R., Keller, L., & Kronauer, D. J. C. (2015). Robust DNA methylation in the clonal raider ant brain. *Current Biology*, *26*, 391–395.

Lieberman, D. A., Cathro, J. S., Nichol, K., & Watson, E. (1997). The role of S⁻ in human observing behavior: Bad news is sometimes better than no news. *Learning and Motivation*, *28*, 20–42.

Lieberman, P. (2007). The evolution of human speech: Its anatomical and neural bases. *Current Anthropology*, *48*, 39–66.

Lieberman, P. (2012). Vocal tract anatomy and the neural bases of talking. *Journal of Phonetics*, *40*, 608–622.

Lieberman, P. (2014). Genes and the evolution of language. In P. Brambilla & A. Marini (Eds.), *Brain evolution, language and psychopathology in schizophrenia* (pp. 7–21). East Sussex, UK: Routledge.

Lieberman, P. (2015). Language did not spring forth 100,000 years ago. *PLoS Biology*. doi:10.1371/journal.pbio.1002064.

Lieberman, P., Laitman, J. T., Reidenberg, J. S., & Gannon, P. J. (1992). The anatomy, physiology, acoustic and perception of speech: Essential elements in analysis of the evolution of human speech. *Journal of Human Evolution*, *23*, 447–467.

Lima, E. L., & Abreu-Rodrigues, J. (2010). Verbal mediating responses: Effects on generalization of say-do correspondence and noncorrespondence. *Journal of Applied Behavior Analysis*, *43*, 411–424.

Lin, J. Y., Roman, C., Arthurs, J., & Reilly, S. (2012). Taste neophobia and c-Fos expression in the rat brain. *Brain Research*, *1448*, 82–88.

Lind, O., Mitkus, M., Olsson, P., & Kelber, A. (2014). Ultraviolet vision in birds: The importance of transparent eye media. *Proceedings of the Royal Society B*, *281*. Retrieved from http://dx.doi.org/10.1098/rspb.2013.2209.

Lindberg, J. S., Iwata, B. A., Kahng, S., & DeLeon, I. G. (1999). DRO contingencies: An analysis of variable-momentary schedules. *Journal of Applied Behavior Analysis*, *32*, 123–135.

Lindsley, O. R. (1972). From Skinner to precision teaching: The child knows best. In J. B. Jordan & L. S. Robbins (Eds.), *Let's try something else kind of thing: Behavioral principles of the exceptional child* (pp. 1–11). Arlington, VA: The Council for Exceptional Children.

Lindsley, O. R. (1990a). Our aims, discoveries, failures, and problem. *Journal of Precision Teaching*, *7*, 7–17.

Lindsley, O. R. (1990b). Precision teaching: By teachers for children. *Teaching Exceptional Children*, *22*, 10–15.

Lindsley, O. R. (1991). Precision teaching's unique legacy from B. F. Skinner. *Journal of Behavioral Education*, *1*, 253–266.

Linehan, C., Roche, B., & Stewart, I. (2010). A derived relations analysis of computer gaming complexity. *European Journal of Behavior Analysis*, *11*, 69–77.

Lionello-DeNolf, K. M. (2009). The search for symmetry: 25 years in review. *Learning and Behavior*, *37*, 188–203.

Lippman, L. G., & Meyer, M. E. (1967). Fixed-interval performance as related to instructions and to the subject's vocalizations of the contingency. *Psychonomic Science*, *8*, 135–136.

Lloyd, D. R., Hausknecht, K. A., & Richards, J. B. (2014). Nicotine and methamphetamine disrupt habituation of sensory reinforcer effectiveness in male rats. *Experimental and Clinical Psychopharmacology*, *22*(2), 166–175. https://doi.org/10.1037/a0034741

Lloyd, D. R., Medina, D. J., Hawk, L. W., Fosco, W. D., & Richards, J. B. (2014). Habituation of reinforcer effectiveness. *Frontiers in integrative neuroscience*, *7*, 107. https://doi.org/10.3389/fnint.2013.00107

Lloyd, K. E., & Lloyd, M. E. (1992). Behavior analysis and technology in higher education. In R. P. West & L. A. Hamerlynck (Eds.), *Designs for excellence in education: The legacy of B. F. Skinner* (pp. 147–160). Longmont, CO: Sopris West, Inc.

Locey, M. L., & Rachlin, H. (2012). Commitment and self-control in a prisoner's dilemma game. *Journal of the Experimental Analysis of Behavior*, *98*, 88–103.

Lofdahl, K. L., Holliday, M., & Hirsch, J. (1992). Selection for conditionability in *Drosophila melanogaster*. *Journal of Comparative Psychology*, *106*, 172–183.

Loftus, E. F., & Zanni, G. (1975). Eyewitness testimony: The influence of the wording of a question. *Bulletin of the Psychonomic Society*, *5*, 86–88.

Logue, A. W. (1985). Conditioned food aversion learning in humans. In N. S. Braveman & P. Bronstein (Eds.), *Experimental assessments and clinical applications of conditioned food aversions* (pp. 316–329). New York: New York Academy of Sciences.

LoLordo, V. M., & Overmier, J. B. (2011). Trauma, learned helplessness, its neuroscience and implications for posttraumatic stress disorder. In T. R. Schachtman & S. Reilly (Eds.), *Associative learning and conditioning theory: Human and non-human applications* (pp. 121–151). New York: Oxford University Press.

Lopez-Perez, R., & Spiegelman, E. (2013). Why do people tell the truth? Experimental evidence for pure lie aversion. *Experimental Economics*, *16*, 233–247.

Lorenzetti, F. D., Baxter, D. A., & Byrne, J. H. (2011). Classical conditioning analogy enhanced acetylcholine responses but reduced excitability of an identified neuron. *The Journal of Neuroscience*, *31*, 14789–14793.

Lovaas, O. I. (1961). Interaction between verbal and nonverbal behavior. *Child Development, 32*, 329–336.

Lovaas, O. I. (1966). A program for the establishment of speech in psychotic children. In J. K. Wing (Ed.), *Early childhood autism* (pp. 115–144). Elmsford, NY: Pergamon.

Lovaas, O. I. (1977). *The autistic child: Language development through behavior modification.* New York: Irvington.

Lovaas, O. I. (1987). Behavioral treatment and normal educational and intellectual functioning in young autistic children. *Journal of Consulting and Clinical Psychology, 55*, 3–9.

Lovaas, O. I. (1993). The development of a treatment-research project for developmentally disabled and autistic children. *Journal of Applied Behavior Analysis, 26*, 617–630.

Lovaas, O. I., & Simmons, J. Q. (1969). Manipulation of self-destruction in three retarded children. *Journal of Applied Behavior Analysis, 2*, 143–157.

Lovaas, O. I., Newsom, C., & Hickman, C. (1987). Self-stimulatory behavior and perceptual reinforcement. *Journal of Applied Behavior Analysis, 20*, 45–68. doi:10.1901/jaba.1987.20–45.

Lowe, C. F. (1979). Determinants of human operant behavior. In M. D. Zeiler & P. Harzem (Eds.), *Reinforcement and the organization of behaviour* (pp. 159–192). New York: John Wiley & Sons.

Lowe, C. F., Beasty, A., & Bentall, R. P. (1983). The role of verbal behavior in human learning: Infant performance on fixed-interval schedules. *Journal of the Experimental Analysis of Behavior, 39*, 157–164.

Lowenkron, B. (1998). Some logical functions of joint control. *Journal of the Experimental Analysis of Behavior, 69*, 327–354.

Lowenkron, B. (1999). Joint control of rule following: An analysis of purpose. *Annual Meeting of the Association for Behavior Analysis.* Chicago.

Lowenkron, B. (2004). Meaning: A verbal behavior account. *The Analysis of Verbal Behavior, 20*, 77–97.

Lowenkron, B. (2006). Joint control and the selection of stimuli from their description. *The Analysis of Verbal Behavior, 22*, 129–151.

Lubinski, D., & Thompson, T. (1987). An animal model of the interpersonal communication of interoceptive (private) states. *Journal of the Experimental Analysis of Behavior, 48*(1), 1–15. https://doi.org/10.1901/jeab.1987.48-1

Lubow, R. E. (1974). High-order concept formation in the pigeon. *Journal of the Experimental Analysis of Behavior, 21*, 475–483.

Lubow, R. E. (2009). Conditioned taste aversion and latent inhibition: A review. In S. Reilly & T. R. Schachtman (Eds.), *Conditioned taste aversion: Behavioral and neural processes* (pp. 37–57). New York: Oxford University Press.

Lubow, R. E., & Moore, A. U. (1959). Latent inhibition: The effect of non-reinforced preexposure to the conditioned stimulus. *Journal of Comparative and Physiological Psychology, 52*, 415–419.

Lucas, G. A., Deich, J. D., & Wasserman, E. A. (1981). Trace autoshaping: Acquisition, maintenance, and path dependence at long trace intervals. *Journal of the Experimental Analysis of Behavior, 36*, 61–74.

Luciano, M. C., Herruzo, J., & Barnes-Holmes, D. (2001). Generalization of say-do correspondence. *The Psychological Record, 51*, 111–130.

Luiselli, J. K., Ricciardi, J. N., & Gilligan, K. (2005). Liquid fading to establish milk consumption by a child with autism. *Behavioral Interventions, 20*, 155–163.

Lukas, K. E., Marr, M. J., & Maple, T. L. (1998). Teaching operant conditioning at the zoo. *Teaching of Psychology, 25*, 112–116.

Lussier, J. P., Heil, S. H., Mongeon, J. A., Badger, G. J., & Higgins, S. T. (2006). A meta-analysis of voucher-based reinforcement therapy for substance use disorders. *Addiction, 101*, 192–203.

Lutz, A. (2012). Ultra successful astrologer Susan Miller works 20-hour days and doesn't have time for haters. *Business Insider*. Retrieved from www.businessinsider.com/astrologer-susan-miller-shares-her-secrets-2012-6.

Lynn, S. K., & Barrett, L. F. (2014). "Utilizing" signal detection theory. *Psychological Science, 25*, 1663–1673.doi:10.1177/0956797614541991.

MacAleese, K. R., Ghezzi, P. M., & Rapp, J. T. (2015). Revisiting conjugate schedules. *Journal of the Experimental Analysis of Behavior, 104*, 63–73.

MacCorquodale, K. (1970). On Chomsky's review of Skinner's *Verbal Behavior. Journal of the Experimental Analysis of Behavior, 13*, 83–99.

Mace, F. C., & Critchfield, T. S. (2010). Translational research in behavior analysis: Historical traditions and imperative for the future. *Journal of the Experimental Analysis of Behavior, 93*, 293–312.

Mace, F. C., Neef, N. A., Shade, D., & Mauro, B. C. (1994). Limited matching on concurrent schedule reinforcement of academic behavior. *Journal of Applied Behavior Analysis, 27*, 585–596.

Machado, A. (1989). Operant conditioning of behavioral variability using a percentile reinforcement schedule. *Journal of the Experimental Analysis of Behavior, 52*, 155–166.

Machado, A. (1992). Behavioral variability and frequency dependent selection. *Journal of the Experimental Analysis of Behavior, 58*, 241–263.

Machado, A. (1997). Increasing the variability of response sequences in pigeons by adjusting the frequency of switching between two keys. *Journal of the Experimental Analysis of Behavior, 68*, 1–25.

MacKillop, J., Amlung, M. T., Few, L. R., Ray, L. A., Sweet, L. H., & Munafo, M. R. (2011). Delayed reward discounting and addictive behavior: A meta-analysis. *Psychopharmacology, 216*, 305–321.

MacLarnon, A. M., & Hewitt, G. P. (2004). Increased breathing control: Another factor in the evolution of human language. *Evolutionary Anthropology, 13*, 181–197.

MacPhail, E. M. (1968). Avoidance responding in pigeons. *Journal of the Experimental Analysis of Behavior, 11*, 629–632.

Madden, G. J., & Bickel, W. K. (2009). *Impulsivity: The behavioral and neurological science of discounting*. Washington, DC: American Psychological Association.

Madden, G. J., & Johnson, P. S. (2011). A delay-discounting primer. In G. J. Madden & W. K. Bickel (Eds.), *Impulsivity: The behavioral and neurological science of discounting* (pp. 11–38). Washington, DC: American Psychological Association.

Madden, G. J., Peden, B. F., & Yamaguchi, T. (2002). Human group choice: Discrete-trial and free-operant tests of the ideal free distribution. *Journal of the Experimental Analysis of Behavior, 78*, 1–15.

Madden, G. J., Petry, N. M., Badger, G. J., & Bickel, W. K. (1997). Impulsive and self-control choices in opioid-dependent patients and non-drug-using control participants: Drug and monetary rewards. *Experimental and Clinical Psychopharmacology, 5*, 256–262.

Maddox, S. A., Watts, C. S., Doyere, V., & Schafe, G. E. (2013). A naturally-occurring histone acetyltransferace inhibitor derived from Garcinia indica impairs newly acquired and reactivated fear memories. *PloS One, 8*, e54463. doi:10.1371/journal.pone.0054463.

Mahoney, A. M., Lalonde, K., Edwards, T., Cox, C., Weetjens, B., & Poling, A. (2014). Landmine-detection rats: An evaluation of reinforcement procedures under simulated operational conditions. *Journal of the Experimental Analysis of Behavior, 101*, 450–458.

Mahoney, A. M., Miguel, C. F., Ahearn, W. H., & Bell, J. (2011). The role of common motor responses in stimulus categorization by preschool children. *Journal of the Experimental Analysis of Behavior, 95*, 237–262.

Maier, S. F., & Seligman, M. E. P. (1976). Learned helplessness: Theory and evidence. *Journal of Experimental Psychology: General, 105*, 3–46.

Maier, S. F., Seligman, M. E. P., & Solomon, R. L. (1969). Pavlovian fear conditioning and learned helplessness. In B. A. Campbell & R. M. Church (Eds.), *Punishment and aversive behavior* (pp. 299–342). New York: Appleton-Century-Crofts.

Malone, J. C. (2014). Did John B. Watson really "found" behaviorism? (2014). *The Behavior Analyst, 37*, 1–12.

Malone, J. C., & Garcia-Penagos, A. (2014). When a clear strong voice was needed: A retrospective review of Watson's (1924/1930) behaviorism. *Journal of the Experimental Analysis of Behavior, 102*, 267–287. doi:10.1002/jeab.98.

Malott, M. E., & Glenn, S. S. (2006). Targets of intervention in cultural and behavioral change. *Behavior and Social Issues, 15*, 31–56.doi:10.5210/bsi.v15i1.344.

Malott, R. W. (1988). Rule-governed behavior and behavioral anthropology. *The Behavior Analysts, 11*, 181–203.

Maple, T. L., & Perdue, B. M. (2013). Chapter 7: Behavior analysis and training. *Zoo animal welfare* (pp. 119–132). Heidelberg: Springer-Verlag.

Markowitz, H. (1981). *Behavioral enrichment in the zoo*. New York: Van Nostrand Reinhold.

Markowitz, H., Schmidt, M., Nadal, L., & Squier, L. (1975). Do elephants ever forget? *Journal of Applied Behavior Analysis, 8*, 333–335.

Marr, M. J. (1977). Behavioral pharmacology: Issues of reductionism and causality. *Advances in behavioral pharmacology, 7*, 1–12.

Marr, J. M., & Zilio, D. (2013). No island entire of itself: Reductionism and behavior analysis. *European Journal of Behavior Analysis, 14*(2), 241–257. https://doi.org/10.1080/15021149.2013.11434458

Marsh, G., & Johnson, R. (1968). Discrimination reversal learning without "errors." *Psychonomic Science, 10*, 261–262.

Martens, B. K., Lochner, D. G., & Kelly, S. Q. (1992). The effects of variable-interval reinforcement on academic engagement: A demonstration of matching theory. *Journal of Applied Behavior Analysis, 25*, 143–151.

Martin, G. L., & Hrycaiko, D. (1983). *Behavior modification and coaching: Principles, procedures and research*. Springfield, IL: Charles C. Thomas.

Martin, G. L., & Pear, J. (2006). *Behavior modification: What it is and how to do it* (8th ed.). Upper Saddle River, NJ: Prentice-Hall.

Martin, G. L., & Pear, J. (2015). *Behavior modification: What it is and how to do it* (10th ed.). New York, NY: Psychology Press/Routledge.

Martin, S., & Friedman, S. G. (2011). *Blazing clickers*. Paper presented at Animal Behavior Management Alliance conference, Denver, CO. Retrieved from http://susanfriedman.net/files/journals/Blazing%20Clickers.pdf

Martinez, E. (2010). DA: Kevin and Elizabeth Schatz killed daughter with "religious whips" for mispronouncing word. *CBSNEWS Crimesider*. Retrieved from www.cbsnews.com/8301–504083_162–6009742–504083.html

Martinez-Harms, J., Marquez, N., Menzel, R., & Vorobyev, M. (2014). Visual generalization in honeybees: Evidence of peak shift in color discrimination. *Journal of Comparative Physiology A, 200*, 317–325.

Marzo, V. D., Goparaju, S. K., Wang, L., Liu, J., Batkai, S., Jara, Z., et al. (2001). Leptin-regulated endocannabinoids are involved in maintaining food intake. *Nature, 410*, 822–825.

Masaki, T., & Nakajima, S. (2008). Forward conditioning with wheel running causes place aversion in rats. *Behavioural Processes, 79*, 43–47.

Matson, J. L., & Boisjoli, J. A. (2009). The token economy for children with intellectual disability and/or autism: A review. *Research in Developmental Disabilities, 30*, 240–248.

Matson, J. L., Hattier, M. A., & Belva, B. (2012). Treating adaptive living skills of persons with autism using applied behavior analysis: A review. *Research in Autism Spectrum Disorders, 6*, 271–276.

Matsuda, K., Garcia, Y., Catagnus, R., & Brandt, J. A. (2020). Can behavior analysis help us understand and reduce racism? A review of the current literature. *Behavior Analysis in Practice, 13*(2), 336–347.

Mattaini, M. A., & McGuire, M. S. (2006). Behavioral strategies for constructing nonviolent cultures with youth: A review. *Behavior Modification, 30*, 184–224.

Matthews, B. A., Shimoff, E., & Catania, A. C. (1987). Saying and doing: A contingency-space analysis. *Journal of Applied Behavior Analysis, 20*, 69–74.

Matthews, L. R., & Temple, W. (1979). Concurrent schedule assessment of food preference in cows. *Journal of the Experimental Analysis of Behavior, 32*, 245–254. https://doi.org/10.1901/jeab.1979.32-245

Maurice, C. (1993). *Let me hear your voice*. New York: Knopf.

Maurice, C., Green, G., & Luce, S. C. (1996). *Behavioral intervention for young children with autism—A manual for parents and professionals*. Sarasota, FL: Pro-Ed.

May, J. G., & Dorr, D. (1968). Imitative pecking in chicks as a function of early social experience. *Psychonomic Science, 11*, 109–129.

Mazur, J. E. (1983). Steady-state performance on fixed-, mixed-, and random-ratio schedules. *Journal of the Experimental Analysis of Behavior, 39*, 293–307.

Mazur, J. E. (1987). An adjusting procedure for studying delayed reinforcement. In M. L. Commons, J. E. Mazur, J. A. Nevin, & H. Rachlin (Eds.), *Quantitative analyses of behavior. Vol. 5. The effect of delay and of intervening events on reinforcement value* (pp. 55–73). Hillsdale, NJ: Lawrence Erlbaum Associates.

Mazur, J. E., & Fantino, E. (2014). Choice. In F. K. McSweeney & E. S. Murphy (Eds.), *The Wiley Blackwell handbook of operant and classical conditioning* (pp. 195–220). West Sussex, UK: Wiley/Blackwell.

McAndrew, F. T. (2009). The interacting roles of testosterone and challenges to status in human male aggression. *Aggression and Violent Behavior, 14*, 330–335.

McCarty, K. F. (2012). Twin Oaks: A case study of an intentional egalitarian community. *Capstone Collection*. Paper 2494. Retrieved from http://digitalcollections.sit.edu/capstones/2494

McClung, C. A., & Nestler, E. J. (2008). Neuroplasticity mediated by altered gene expression. *Neuropsychopharmacology Reviews, 33*, 3–17.

McDevitt, M. A., & Williams, B. A. (2010). Dual effects on choice of conditioned reinforcement frequency and conditioned reinforcement value. *Journal of the Experimental Analysis of Behavior, 93*, 147–155.

McDonald, J. S. (1988). Concurrent variable-ratio schedules: Implications for the generalized matching law. *Journal of the Experimental Analysis of Behavior, 50*, 55–64.

McDougall, W. (1908). *An introduction to social psychology*. London: Methuen.

McDowell, J. J. (1981). On the validity and utility of Herrnstein's hyperbola in applied behavior analysis. In C. M. Bradshaw, E. Szabadi, & C. F. Lowe (Eds.), *Quantification of steady-state operant behaviour* (pp. 311–324). Amsterdam: Elsevier/North-Holland.

McDowell, J. J. (1982). The importance of Herrnstein's mathematical statement of the law of effect for behavior therapy. *American Psychologist, 37*, 771–779.

McDowell, J. J. (1988). Matching theory in natural human environments. *The Behavior Analyst, 11*, 95–109.

McDowell, J. J. (2004). A computational model of selection by consequences. *Journal of the Experimental Analysis of Behavior, 81*, 297–317.

McDowell, J. J. (2010). Behavioral and neural Darwinism: Selectionist function and mechanism in adaptive behavior dynamics. *Behavioural Processes, 84*, 358–365.

McDowell, J. J. (2013). A quantitative evolutionary theory of adaptive behavior dynamics. *Psychological Review, 120*, 731–750.

McDowell, J. J., & Ansari, Z. (2005). The quantitative law of effect is a robust emergent property of an evolutionary algorithm for reinforcement learning. *Advances in Artificial Life: Lecture Notes in Computer Science, 3630*, 413–422.

McDowell, J. J., & Caron, M. L. (2010). Bias and undermatching in delinquent boys' verbal behavior as a function of their level of deviance. *Journal of the Experimental Analysis of Behavior, 93*, 471–483.

McDowell, J. J., & Popa, A. (2010). Toward a mechanics of adaptive behavior: Evolutionary dynamics and matching theory statics. *Journal of the Experimental Analysis of Behavior, 94*, 242–260.

McEachin, J. J., Smith, T., & Lovaas, I. O. (1993). Long-term outcome for children with autism who received early intensive behavioral treatment. *American Journal on Mental Retardation, 97*, 359–372.

McGaha-Mays, N., & Heflin, L. J. (2011). Increasing independence in self-care tasks for children with autism using self-operated auditory prompts. *Research in Autism Spectrum Disorders, 5*, 1351–1357.

McGill, P., & Langthorne, P. (2011). Gene-environment interactions and the functional analysis of challenging behavior in children with intellectual and developmental disabilities. *Behavioral Development Bulletin, 11*, 20–25.

McIntire, K. D., Cleary, J., & Thompson, T. (1987). Conditional relations by monkeys: Reflexivity, symmetry, and transitivity. *Journal of the Experimental Analysis of Behavior, 47*, 279–285.

McSweeney, F. K., & Murphy, E. S. (2009). Sensitization and habituation regulate reinforcer effectiveness. *Neurobiology of Learning and Memory, 92*, 189–198.

McSweeney, F. K., & Murphy, E. S. (2014). Characteristics, theories, and implications of dynamic changes in reinforcer effectiveness. In F. K. McSweeney & E. S. Murphy (Eds.), *The Wiley Blackwell handbook of operant and classical conditioning* (pp. 339–368). West Sussex, UK: John Wiley & Sons.

McSweeney, F. K., & Weatherly, J. N. (1998). Habituation to the reinforcer may contribute to multiple-schedule behavioral contrast. *Journal of the Experimental Analysis of Behavior, 69*, 199–221.

McSweeney, F. K., Ettinger, R. A., & Norman, W. D. (1981). Three versions of the additive theories of behavioral contrast. *Journal of the Experimental Analysis of Behavior, 36*, 285–297.

McSweeney, F. K., Melville, C. L., & Higa, J. (1988). Positive behavioral contrast across food and alcohol reinforcers. *Journal of the Experimental Analysis of Behavior, 50*, 469–481.

Mechner, F. (2010). Chess as a behavioral model for cognitive skill research: Review of *Blindfold Chess* by Eliot Hearst and John Knott. *Journal of the Experimental Analysis of Behavior, 94*, 373–386.

Mehus, C. J. & Patrick, M. E. (2021). Prevalence of spanking in US national samples of 35-year-old parents from 1993 to 2017. *JAMA Pediatrics, 175*(1):92–93. doi:10.1001/jamapediatrics.2020.2197.

Meltzoff, A. N. (1988a). Imitation of televised models by infants. *Child Development, 59*, 1221–1229.

Meltzoff, A. N. (1988b). Infant imitation after a 1-week delay: Long-term memory for novel acts and multiple stimuli. *Developmental Psychology, 24*, 470–476.

Meltzoff, A. N. (1988c). Infant imitation and memory: Nine-month-olds in immediate and deferred tests. *Child Development, 59*, 217–225.

Meltzoff, A. N. (1999). Born to learn: What infants learn from watching us. In N. Fox & J. G. Worhol (Eds.), *The role of early experience in infant development* (pp. 1–10). Skillman, NJ: Pediatric Institute Publications.

Meltzoff, A. N., & Moore, M. K. (1977). Imitation of facial and manual gestures by human neonates. *Science, 198*, 75–78.

Meltzoff, A. N., & Moore, M. K. (1983). Newborn infants imitate adult facial gestures. *Child Development, 54*, 702–709.

Meltzoff, A. N., & Moore, M. K. (1999). Resolving the debate about early imitation. In A. Slater & D. Muir (Eds.), *Reader in developmental psychology* (pp. 151–155). Oxford: Blackwell Science.

Mendelson, J., & Chillag, D. (1970). Schedule-induced air licking in rats. *Physiology and Behavior, 5,* 535–537.

Mendoza, E., Colomb, J., Ryabak, J., Pfluger, H. J., Zars, C. S., & Brembs, B. (2014). Drosophila FoxP mutants are deficient in operant self-learning. *PLoS ONE, 9.* doi:10.1371/journal.pone.0100648.

Mendres, A. E., & Borrero, J. C. (2010). Development and modification of a response class via positive and negative reinforcement: A translational approach. *Journal of Applied Behavior Analysis, 43,* 653–672.

Merola, I., Prato-Previde, E., & Marshall-Pescini, S. (2012). Social referencing in dog-owner dyads. *Animal Cognition, 15,* 175–185.

Mery, F., & Kawecki, T. J. (2002). Experimental evolution of learning ability in fruit flies. *Proceedings of the National Academy of Sciences, 22,* 14274–14279.

Mesoudi, A. (2016). Cultural evolution: Integrating psychology, evolution and culture. *Current Opinion in Psychology, 7,* 17–22.

Meyer, D. R., Cho, C., & Wesemann, A. F. (1960). On problems of conditioning discriminated lever-press avoidance responses. *Psychological Review, 67,* 224–228.

Meyer, J. (2007). Acoustic strategy, phonetic comparison and perceptive cues of whistled languages. *HAL Archives, version 1.* Retrieved from http://halshs.archives-ouvertes.fr/halshs-00133192.

Meyer, J. (2015). *Whistled languages: A worldwide inquiry on human whistled speech.* Heidelberg: Springer-Verlag Berlin.

Meyer, L. H., & Evans, I. M. (1989). *Non-aversive intervention for behavior problems: A manual for home and community.* Baltimore, MD: Paul H. Brookes.

Michael, J. (1983). Evocative and repertoire-altering effects of an environmental event. *The Analysis of Verbal Behavior, 2,* 19–21.

Michael, J. L. (1982a). Distinguishing between discriminative and motivational functions of stimuli. *Journal of the Experimental Analysis of Behavior, 37,* 149–155.

Michael, J. L. (1982b). Skinner's elementary verbal relations: Some new categories. *The Analysis of Verbal Behavior, 1,* 1–3.

Michael, J. L. (1988). Establishing operations and the mand. *The Analysis of Verbal Behavior, 6,* 3–9.

Michael, J. L. (1993). Establishing operations. *The Behavior Analyst, 16,* 191–206.

Michael, J. L. (2000). Implications and refinements of the establishing operation concept. *Journal of Applied Behavior Analysis, 33,* 401–410.

Michael, J. L., Whitley, P., & Hesse, B. E. (1983). The pigeon parlance project. *The Analysis of Verbal Behavior, 1,* 6–9.

Miguel, C. F., Frampton, S. E., Lantaya, C. A., LaFrance, D. L., Quah, K., Meyer, C. S., et al. (2015). Effects of tact training on the development of analogical reasoning. *Journal of the Experimental Analysis of Behavior, 104,* 96–118.

Milgram, S. (1974). *Obedience to authority.* New York: Harper & Row.

Millenson, J. R., & Hendry, D. P. (1967). Quantification of response suppression in conditioned anxiety training. *Canadian Journal of Psychology/Revue canadienne de psychologie, 21*(3), 242–252. https://doi.org/10.1037/h0082981

Miller Jr, H. L. (1976). Matching-based hedonic scaling in the pigeon. *Journal of the Experimental Analysis of Behavior, 26*(3), 335–347.

Miller, J. R., Lerman, D. C., & Fritz, J. N. (2010). An experimental analysis of negative reinforcement contingencies for adult-delivered reprimands. *Journal of Applied Behavior Analysis, 43,* 769–773.

Miller, K. B., Lund, E., & Weatherly, J. (2012). Applying operant learning to the stay-leave decision in domestic violence. *Behavior and Social Issues, 21,* 135–151.

Miller, N. E., & Banuazizi, A. (1968). Instrumental learning by curarized rats of a specific visceral response, intestinal or cardiac. *Journal of Comparative and Physiological Psychology, 65*, 1–7.

Miller, N. E., & Carmona, A. (1967). Modification of a visceral response, salivation in thirsty dogs, by instrumental training with water reward. *Journal of Comparative and Physiological Psychology, 63*, 1–6.

Miller, N. E., & DiCara, L. (1967). Instrumental learning of heart rate changes in curarized rats: Shaping and specificity to discriminative stimulus. *Journal of Comparative and Physiological Psychology, 63*, 12–19.

Miller, N. E., & Dollard, J. (1941). *Social learning and imitation.* New Haven, CT: Yale University Press.

Miller, N. E., & Dworkin, B. R. (1974). Visceral learning: Recent difficulties with curarized rats and significant problems for human research. In P. A. Obrist, A. H. Black, J. Brener, & L. V. DiCara (Eds.), *Cardiovascular psychophysiology: Current issues in response mechanisms, biofeedback and methodology* (pp. 295–331). Chicago, IL: Aldine.

Miller, S. A., Rodriguez, N. M., & Rourke, A. J. (2015). Do mirrors facilitate acquisition of motor imitation in children diagnosed with autism? *Journal of Applied Behavior Analysis, 48*, 194–198.

Miltenberger, R. G., & Fuqua, R. W. (1981). Overcorrection: A review and critical analysis. *The Behavior Analyst, 4*(2), 123–141.

Miltenberger, R. G., & Gross, A. C. (2011). Teaching safety skills to children. In W. W. Fisher, C. C. Piazza, & H. S. Roane (Eds.), *Handbook of applied behavior analysis* (pp. 417–432). New York: The Guilford Press.

Miltenberger, R. G., Gatheridge, B. J., Satterlund, M., Egemo-Helm, K. R., Johnson, B. M., Jostad, C., et al. (2005). Teaching safety skills to children to prevent gun play: An evaluation of *in situ* training. *Journal of Applied Behavior Analysis, 38*, 395–398.

Minshawl, N. F. (2008). Behavioral assessment and treatment of self-injurious behavior in autism. *Child and Adolescent Psychiatric Clinics of North America, 17*, 875–886.

Minster, S. T., Elliffe, D., & Muthukumaraswamy, S. D. (2011). Emergent stimulus relations depend on stimulus correlation and not on reinforcement contingencies. *Journal of the Experimental Analysis of Behavior, 95*, 327–342.

Mirrione, M. M., Schulz, D., Lapidus, K. A. B., Zhang, S., Goodman, W., & Henn, F. A. (2014). Increased metabolic activity in the septum and habenula during stress is linked to subsequent expression of learned helplessness behavior. *Frontiers of Human Neuroscience, 8*. doi:10.3389/fnhum.2014.00029.

Mitchell, D., Kirschbaum, E. H., & Perry, R. L. (1975). Effects of neophobia and habituation on the poison-induced avoidance of exteroceptive stimuli in the rat. *Journal of Experimental Psychology: Animal Behavior Processes, 104*, 47–55.

Modaresi, H. A. (1990). The avoidance barpress problem: Effects of enhanced reinforcement and an SSDR-congruent lever. *Learning and Motivation, 21*, 199–220.

Moerk, E. L. (1990). Three-term contingency patterns in mother-child verbal interactions during first-language acquisition. *Journal of the Experimental Analysis of Behavior, 54*, 293–305.

Moerk, E. L. (2000). *The guided acquisition of first language skills.* Stamford, CT: Ablex Publishing.

Molm, L. D., Takahashi, N., & Peterson, G. (2000). Risk and trust in social exchange: An experimental test of a classical proposition. *American Journal of Sociology, 105*, 1396–1427.

Moore, J. (2003). Some further thoughts on the pragmatic and behavioral conception of private events. *Behavior and Philosophy, 31*, 151–157.

Moran, D. J., & Malott, R. W. (2004). *Evidence-based educational methods.* San Diego, CA: Elsevier Academic Press.

Morgan, C. L. (1894). *An introduction to comparative psychology.* London: W. Scott.

Morgan, D., Carter, C., DuPree, J. P., Yezierski, R. P., & Vierck, C. J., Jr. (2008). Evaluation of prescription opioids using operant-based pain measures in rats. *Experimental and Clinical Psychopharmacology, 16*, 367–375.

Morgan, L., & Neuringer, A. (1990). Behavioral variability as a function of response topography and reinforcement contingency. *Animal Learning and Behavior, 18*, 257–263.

Morris, E. K. (1988). Contextualism: The world view of behavior analysis. *Journal of Experimental Child Psychology, 46*, 289–323.

Morse, W. H. (1966). Intermittent reinforcement. *Operant behavior: Areas of research and application.* New York: Appleton-Century-Crofts, pp. 52–108.

Moseley, J. B., O'Malley, K., Petersen, N. J., Menke, T. J., Brody, B. A., Kuykendall, D. H., et al. (2002). A controlled trial of arthroscopic surgery for osteoarthritis of the knee. *New England Journal of Medicine, 347*, 81–88.

Moser, E., & McCulloch, M. (2010). Canine scent detection of human cancers: A review of methods and accuracy. *Journal of Veterinary Behavior: Clinical Applications and Research, 5*, 145–152.

Moxley, J. H., Ericsson, K. A., Charness, N., & Krampe, R. T. (2012). The role of intuition and deliberative thinking in experts' superior tactical decision-making. *Cognition, 124*, 72–78.

Mueller, M. M., Palkovic, C. M., & Maynard, C. S. (2007). Errorless learning: Review and practical application for teaching children with pervasive developmental disorders. *Psychology in the Schools, 44*, 691–700.

Mukherjee, S. (2016). *The gene: An intimate history.* New York: Scribner.

Muller, P. G., Crow, R. E., & Cheney, C. D. (1979). Schedule-induced locomotor activity in humans. *Journal of the Experimental Analysis of Behavior, 31*, 83–90.

Murphy, M. S., & Cook, R. G. (2008). Absolute and relational control of a sequential auditory discrimination by pigeons (*Columba livia*). *Behavioral Processes, 77*, 210–222.

Myerson, J., & Hale, S. (1984). Practical implications of the matching law. *Journal of Applied Behavior Analysis, 17*, 367–380.

Nader, M. A., & Woolverton, W. L. (1992). Effects of increasing response requirement on choice between cocaine and food in rhesus monkeys. *Psychopharmacology, 108*, 295–300.

Nagy, E., Compagne, H., Orvos, H., Pal, A., Molnar, P., Janszky, I., et al. (2005). Index finger movement imitation by human neonates: Motivation, learning and left-hand preference. *Pediatric Research, 58*, 749–753.

Nakajima, S., & Katayama, T. (2014). Running-based pica in rats: Evidence for the gastrointestinal discomfort hypothesis of running-based taste aversion. *Appetite, 83*, 178–184.

Naqvi, N. H., Gaznick, N., Tranel, D., & Bechara, A. (2014). The insula: A critical neural substrate for craving and drug seeking under conflict and risk. *Annals of the New York Academy of Sciences, 1316*, 53–70.

Naqvi, N. H., Rudrauf, D., Damasio, H., & Bechara, A. (2007). Damage to the insula disrupts addiction to cigarette smoking. *Science, 315*, 531–534.

National Institute on Drug Abuse (2021). Overdose death rate. www.drugabuse.gov/drug-topics/trends-statistics/overdose-death-rates

Neal, D. T., Wood, W., & Quinn, J. M. (2006). Habits—A repeat performance. *Current Directions in Psychological Science, 15*, 198–202.

Neel, J. V. (1962). Diabetes mellitus: A "thrifty" genotype rendered detrimental by "progress"? *American Journal of Human Genetics, 14*, 353–362.

Neely, L. C., Ganz, J. B., Davis, J. L., Boles, M. B., Hong, E. R., Ninci, J., & Gilliland, W. D. (2016). Generalization and maintenance of functional living skills for individuals with autism spectrum disorder: A review and meta-analysis. *Review Journal of Autism and Developmental Disorders, 3*(1), 37–47. https://doi.org/10.1007/s40489-015-0064-7

Nergardh, R., Ammar, A., Brodin, U., Bergstrom, J., Scheurink, A., & Sodersten, P. (2007). Neuropeptide Y facilitates activity-based anorexia. *Psychoneuroendocrinology, 32*, 493–502.

Neuringer, A. (2004). Reinforced variability in animals and people. *American Psychologist, 59*, 891–906.

Neuringer, A. (2009). Operant variability and the power of reinforcement. *The Behavior Analyst Today, 10*, 319–343.

Neuringer, A. J. (1986). Can people behave "randomly?" The role of feedback. *Journal of Experimental Psychology: General, 115*, 62–75.

Neuringer, A. J. (2002). Operant variability: Evidence, function, and theory. *Psychonomic Bulletin and Review, 9*, 672–705.

Neuringer, A. J., & Jensen, G. (2013). Operant variability. In G. J. Madden (Ed.), *APA handbook of behavior analysis: Vol. 1. Methods and principles* (pp. 513–546). Washington, DC: American Psychological Association.

Nevin, J. A. (1969). Signal detection theory and operant behavior: A review of David M. Green and John A. Swets' Signal Detection Theory and Psychophysics. *Journal of the Experimental Analysis of Behavior, 12*, 475–448.

Nevin, J. A. (1974). Response strength in multiple schedules. *Journal of the Experimental Analysis of Behavior, 21*, 389–408.

Nevin, J. A. (1988). Behavioral momentum and the partial reinforcement effect. *Psychological Bulletin, 103*, 44–56.

Nevin, J. A. (1992). An integrative model for the study of behavioral momentum. *Journal of the Experimental Analysis of Behavior, 57*, 301–316.

Nevin, J. A. (2012). Resistance to extinction and behavioral momentum. *Behavioural Processes, 90*, 89–97.

Nevin, J. A., & Grace, R. C. (2000). Behavior momentum and the law of effect. *Behavioral and Brain Sciences, 23*, 73–130.

Nevin, J. A., Grace, R. C., Holland, S., & McLean, A. P. (2001). Variable-ratio versus variable-interval schedules: Response rate, resistance to change, and preference. *Journal of the Experimental Analysis of Behavior, 76*, 43–74.

Nieuwenhuis, S., de Geus, E. J., & Aston-Jones, G. (2011). The anatomical and functional relationship between the P3 and autonomic components of the orienting response. *Psychophysiology, 48*, 162–175.

Norman, W. D., & McSweeney, F. K. (1978). Matching, contrast, and equalizing in the concurrent leverpress responding of rats. *Journal of the Experimental Analysis of Behavior, 29*, 453–462.

Normand, M. P., Machado, M. A., Hustyi, K. M., & Morely, A. J. (2011). Infant sign training and functional analysis. *Journal of Applied Behavior Analysis, 44*, 305–314.

Norton, W. (1997). Human geography and behavior analysis: An application of behavior analysis to the explanation of the evolution of human landscapes. *The Psychological Record, 47*, 439–460.

Notterman, J. M. (1959). Force emission during bar pressing. *Journal of Experimental Psychology, 58*, 341–347.

Nuzzolo-Gomez, R., & Greer, R. D. (2004). Emergence of untaught mands and tacts of novel adjective-object pairs as a function of instructional history. *The Analysis of Verbal Behavior, 20*, 63–67.

Nyhan, B., Reifler, J., Richey, S., & Freed, G. L. (2014). Effective messages in vaccine promotion: A randomized trial. *Pediatrics, 133*, e835–e842.

O'Brien, R. M., & Simek, T. C. (1983). A comparison of behavioral and traditional methods for teaching golf. In G. L. Martin & D. Harycaiko (Eds.), *Behavior modification and coaching: Principles, procedures, and research* (pp. 175–183). Springfield, IL: Charles C. Thomas.

O'Heare, J. (2009). On the permanence of punishment. Retrieved from www.jamesoheare.com/weblog/permanenceofpunish.html.

O'Hora, D., Barnes-Holmes, D., & Stewart, I. (2014). Antecedent and consequential control of derived instruction-following. *Journal of the Experimental Analysis of Behavior, 102*, 66–85.

O'Kelly, L. E., & Steckle, L. C. (1939). A note on long enduring emotional responses in rats. *Journal of Psychology, 8*, 125–131.

O'Leary, M. R., & Dengerink, H. A. (1973). Aggression as a function of the intensity and pattern of attack. *Journal of Experimental Research in Personality, 7*, 61–70.

O'Regan, L. M., Farina, F. R., Hussey, I., & Roche, R. A. P. (2015). Event-related brain potentials reveal correlates of the transformation of stimulus functions through derived relations in healthy humans. *Brain Research, 1599*, 168–177.

O'Rourke, T. J. (1978). *A basic vocabulary: American Sign Language for parents and children*. Silver Spring, MD: T. J. Publishers.

Oah, S., & Dickinson, A. M. (1989). A review of empirical studies of verbal behavior. *The Analysis of Verbal Behavior, 7*, 53–68.

Oberman, L. M., & Ramachandran, V. S. (2007). The stimulating social mind: The role of the mirror neuron system and simulation in the social and communicative deficits of autism spectrum disorders. *Psychological Bulletin, 133*, 310–327.

Oberman, L. M., Hubbard, E. M., McCleery, J. P., Altschuler, E. L., Ramachandran, V. S., & Pineda, J. A. (2005). EEG evidence for mirror neuron dysfunction in autism spectrum disorders. *Cognitive Brain Research, 24*, 190–198.

Odum, A. L. (2011a). Delay discounting: Trait variable? *Behavioural Processing, 83*, 1–9.

Odum, A. L. (2011b). Delay discounting: I'm a k, you're a k. *Journal of the Experimental Analysis of Behavior, 96*, 427–439.

Odum, A. L., Becker, R. J., Haynes, J. M., Galizio, A., Frye, C. C., Downey, H., ... & Perez, D. M. (2020). Delay discounting of different outcomes: Review and theory. *Journal of the Experimental Analysis of Behavior, 113*(3), 657–679.

Odum, A. L., Ward, R. D., Barnes, C. A., & Burke, K. A. (2006). The effects of delayed reinforcement on variability and repetition of response sequences. *Journal of the Experimental Analysis of Behavior, 86*, 159–179.

Okouchi, H., & Songmi, K. (2004). Differential reinforcement of human self-reports about schedule performance. *The Psychological Record, 54*, 461–478.

Oostenbroek, J., Suddendorf, T., Nielson, M., Kennedy-Costantini, S., Davis, J., et al. (2016). Comprehensive longitudinal study challenges the existence of neonatal imitation in humans. *Current Biology, 26*, 1334–1338. Retrieved from http://dx.doi.org/10.1016/j.cub.2016.03.047.

Orne, M. T., & Evans, F. J. (1965). Social control in the psychology experiment: Antisocial behavior and hypnosis. *Journal of Personality and Social Psychology, 1*, 189–200.

Ortu, D., Becker, A. M., Woelz, T. A. R., & Glenn, S. S. (2012). An iterated four-player prisoner's dilemma game with an external selecting agent: A metacontingency experiment. *Revista Latinoamericana de Psicología, 44*, 111–120.

Osgood, C. E. (1953). *Method and theory in experimental psychology*. New York: Oxford University Press.

Overmier, J. B., & Seligman, M. E. P. (1967). Effects of inescapable shock upon subsequent escape and avoidance responding. *Journal of Comparative and Physiological Psychology, 63*, 28–33.

Overskeid, G. (1992). Is any human behavior schedule-induced? *The Psychological Record, 42*, 323–340.

Page, S., & Neuringer, A. J. (1985). Variability is an operant. *Journal of Experimental Psychology: Animal Behavior Processes, 11*, 429–452.

Palya, W. L., & Zacny, J. P. (1980). Stereotyped adjunctive pecking by caged pigeons. *Animal Learning & Behavior, 8*(2), 293–303. https://doi.org/10.3758/BF03199609

Paniagua, F. A. (1989). Lying by children: Why children say one thing, do another? *Psychological Reports, 64*, 971–984.

Paniagua, F. A., & Baer, D. M. (1982). The analysis of correspondence as a chain reinforceable at any point. *Child Development, 53*, 786–798.

Papachristos, E. B., & Gallistel, C. R. (2006). Autoshaped head poking in the mouse: A quantitative analysis of the learning curve. *Journal of the Experimental Analysis of Behavior, 85*, 293–308.

Papini, M. R., & Bitterman, M. E. (1990). The role of contingency in classical conditioning. *Psychological Review, 97*, 396–403.

Parece, T. E., Grossman, L., & Geller, E. S. (2013). Reducing carbon footprint of water consumption: a case study of water conservation at a university campus. In *Climate change and water resources* (pp. 199–218). Berlin, Heidelberg: Springer. doi:10.1007/698_2013_227.

Park, R. D. (2002). Punishment revisited—science, values, and the right question: Comment on Gershoff (2002). *Psychological Bulletin, 128*, 596–601.

Parkinson, J. A., Crofts, H. S., McGuigan, M., Tomic, D. L., Everitt, B. J., & Roberts, A. C. (2001). The role of the primate amygdala in conditioned reinforcement. *Journal of Neuroscience, 21*, 7770–7780.

Partington, J. W., & Bailey, J. S. (1993). Teaching intraverbal behavior to preschool children. *Analysis of Verbal Behavior, 11*, 9–18.

Partington, J. W., Sundberg, M. L., Newhouse, L., & Spengler, S. M. (1994). Overcoming an autistic child's failure to acquire a tact repertoire. *Journal of Applied Behavior Analysis, 27*, 733–734.

Patterson, A. E., & Boakes, R. A. (2012). Interval, blocking and marking effects during the development of schedule-induced drinking in rats. *Journal of Experimental Psychology: Animal Behavior Processes, 38*, 303–314.

Patterson, G. R. (1976). The aggressive child: Victim and architect of a coercive system. In E. J. Mash, L. A. Hamerlynck, & L. H. Hendy (Eds.), *Behavior modification and families* (pp. 269–316). New York: Brunner/Mazel.

Patterson, G. R. (1982). *Coercive family processes*. Eugene, OR: Castalia.

Patterson, G. R. (2002). Etiology and treatment of child and adolescent antisocial behavior. *The Behavior Analysts Today, 3*, 133–144.

Paul, G. L. (2006). Myth and reality in Wakefield's assertions regarding Paul and Lentz (1977). *Behavior and Social Issues, 15*, 244–252.

Pauley, P. J. (1987). *Controlling life: Jacques Loeb and the engineering ideal in biology*. New York: Oxford University Press.

Pavlov, I. P. (1960). *Conditioned reflexes: An investigation of the physiological activity of the cerebral cortex* (G. V. Anrep, trans.). New York: Dover (original work published in 1927).

Pear, J. J., Schnerch, G. J., Silva, K. M., Svenningsen, L., & Lambert, J. (2011). Web-based computer-aided personalized system of instruction. *New Directions for Teaching and Learning, 128*, 85–94.

Pedersen, W. C., Gonzales, C., & Miller, N. (2000). The moderating effect of trivial triggering provocation on displaced aggression. *Journal of Personality and Social Psychology, 78*, 913–927.

Pelaez, M. (2013). Dimensions of rules and their correspondence to rule-governed behavior. *European Journal of Behavior Analysis, 14*, 259–270.

Pelaez, M., Virues-Ortega, J., & Gewirtz, J. L. (2012). Acquisition of social referencing via discrimination training in infants. *Journal of Applied Behavior Analysis, 45*, 23–36. https://doi.org/10.1901/jaba.2012.45-23

Pellón, R., & Pérez-Padilla, A. (2013). Response-food delay gradients for lever-pressing and schedule-induced licking in rats. *Learning & Behavior, 41*, 218–227.

Pepperberg, I. M. (1981). Functional vocalizations by an African gray parrot (*Psittacus erithacus*). *Zeitschrift fur Tierpsychologie, 58,* 193–198.

Pepperberg, I. M. (2000). *The Alex studies: Cognitive and communicative abilities of grey parrots.* Cambridge, MA: Harvard University Press.

Perdue, B. M., Evans, T. A., Washburn, D. A., Rumbaugh, D. M., & Beran, M. J. (2014). Do monkeys choose to choose? *Learning & Behavior, 42,* 164–175. doi:10.3758/s13420-014-0135-0.

Pereira, S., & van der Kooy, D. (2013). Entwined engrams: The evolution of associative and non-associative learning. *Worm, 2,* e22725, 1–5.

Perone, M., & Galizio, M. (1987). Variable interval schedules of timeout from avoidance. *Journal of the Experimental Analysis of Behavior, 47,* 97–113.

Perone, M., & Hursh, D. E. (2013). Single-case experimental designs. In G. J. Madden (Ed.), *APA handbook of behavior analysis: Vol. 1. Methods and principles* (pp. 107–126). Washington, DC: American Psychological Association.

Perry, G. (2012). *Behind the shock machine: The untold story of the notorious Milgram psychology experiments.* New York: The New Press.

Peterson, C., & Seligman, M. E. P. (1984). Causal explanations as a risk factor for depression: Theory and evidence. *Psychological Review, 91,* 347–374.

Peterson, G. B. (2004). A day of great illumination: B. F. Skinner's discovery of shaping. *Journal of the Experimental Analysis of Behavior, 82,* 317.

Peterson, N. (1978). *An introduction to verbal behavior.* Grand Rapids, MI: Behavior Associates Inc.

Peters-Scheffer, N., Didden, R., Korzilius, H., & Sturmey, P. (2011). A meta-analytic study on the effectiveness of comprehensive ABA-based intervention programs for children with autism spectrum disorders. *Research in Autism Spectrum Disorders, 5,* 60–69.

Petry, N. M., & Madden, G. J. (2011). Discounting and pathological gambling. In G. J. Madden & W. K. Bickel (Eds.), *Impulsivity: The behavioral and neurological science of discounting* (pp. 273–294). Washington, DC: American Psychological Association.

Petry, N. M., Martin, B., Cooney, J. L., & Kranzler, H. R. (2000). Give them prizes, and they will come: Contingency management for treatment of alcohol dependence. *Journal of Consulting and Clinical Psychology, 68,* 250–257.

Petscher, E. S., Rey, C., & Bailey, J. S. (2009). A review of empirical support for differential reinforcement of alternative behavior. *Research in Developmental Disabilities, 30,* 409–425.

Pettit, B., & Western, B. (2004). Mass imprisonment and the life course: Race and class inequality in US incarceration. *American Sociological Review, 69*(2), 151–169.

Petursdottir, A. I., Carr, J. E., & Michael, J. (2005). Emergence of mands and tacts of novel objects among preschool children. *The Analysis of Verbal Behavior, 21,* 59–74.

Phelps, B. J. (2015). Behavioral perspectives on personality and self. *The Psychological Record, 65,* 557–565.

Pierce, W. D. (1991). Culture and society: The role of behavioral analysis. In P. A. Lamal (Ed.), *Behavioral analysis of societies and cultural practices* (pp. 13–37). New York: Hemisphere.

Pierce, W. D. (2001). Activity anorexia: Biological, behavioral, and neural levels of selection. *Behavioral and Brain Sciences, 24,* 551–552.

Pierce, W. D., & Epling, W. F. (1983). Choice, matching, and human behavior: A review of the literature. *The Behavior Analyst, 6,* 57–76.

Pierce, W. D., & Epling, W. F. (1988). Biobehaviorism: Genes, learning and behavior. *Working Paper No. 88–5,* Center for Systems Research, University of Alberta, Edmonton, Alberta.

Pierce, W. D., & Epling, W. F. (1997). Activity anorexia: The interplay of culture, behavior, and biology. In P. A. Lamal (Ed.), *Cultural contingencies: Behavior analytic perspectives on cultural practices* (pp. 53–85). Westport CT: Praeger.

Pierce, W. D., Diane, A., Heth, C. D., Russell, J. C., & Proctor, S. D. (2010). Evolution and obesity: Resistance of obese-prone rats to a challenge of food restriction and wheel running. *International Journal of Obesity, 34*, 589–592.

Pierce, W. D., Epling, W. F., & Boer, D. P. (1986). Deprivation and satiation: The interrelations between food and wheel running. *Journal of the Experimental Analysis of Behavior, 46*, 199–210.

Pierce, W. D., Heth, C. D., Owczarczyk, J., Russell, J. C., & Proctor, S. D. (2007). Overeating by young obese-prone and lean rats caused by tastes associated with low energy foods. *Obesity, 15*, 1069–1079.

Pierrel, R., Sherman, G. J., Blue, S., & Hegge, F. W. (1970). Auditory discrimination: A three-variable analysis of intensity effects. *Journal of the Experimental Analysis of Behavior, 13*, 17–35.

Pietras, C. J., & Hackenberg, T. D. (2005). Response-cost punishment via token loss with pigeons. *Behavioural Processes, 69*, 343–356.

Pjetri, E., Adan, R. A., Herzog, H., de Hass, R., Oppelaar, H., et al. (2012). NPY receptor subtype specification for behavioral adaptive strategies during limited food access. *Genes, Brain, and Behavior, 11*, 105–112.

Platt, J. (1973). Social traps. *American Psychologist, 28*, 641–651.

Plaud, J. J., & Newberry, D. E. (1996). Rule-governed behavior and pedophilia. *Sexual Abuse: Journal of Research and Treatment, 8*, 143–159.

Ploog, B. O., & Zeigler, H. P. (1996). Effects of food-pellet size on rate, latency, and topography of autoshaped keypecks and gapes in pigeons. *Journal of the Experimental Analysis of Behavior, 65*, 21–35. https://doi.org/10.1901/jeab.1996.65-21 https://doi.org/10.1901/jeab.1973.20-163.

Pocock, T. L., Foster, M., & McEwan, J. S. (2010). Precision teaching and fluency: The effects of charting and goal-setting on skaters' performance. *Journal of Behavioral Health and Medicine, 1*, 93–118.

Pokorny, J., Hatt, N. V., Colombi, C., Vivanti, G., Rogers, S. J., & Rivera, S. M. (2015). The action observation system when observing hand actions in autism and typical development. *Autism Research, 8*, 284–296.

Poling, A. (1978). Performance of rats under concurrent variable-interval schedules of negative reinforcement. *Journal of the Experimental Analysis of Behavior, 30*, 31–36.

Poling, A., Edwards, T. L., Weeden, M., & Foster, T. M. (2011). The matching law. *The Psychological Record, 61*, 313–322.

Poling, A., Nickel, M., & Alling, K. (1990). Free birds aren't fat: Weight gain in captured wild pigeons maintained under laboratory conditions. *Journal of the Experimental Analysis of Behavior, 53*, 423–424.

Poling, A., Weetjens, B., Cox, C., Beyene, N. W., Bach, H., & Sully, A. (2011). Using trained pouched rats to detect land mines: Another victory for operant conditioning. *Journal of Applied Behavior Analysis, 44*, 351–355.

Pollard, K., & Blumstein, D. T. (2012). Evolving communicative complexity: Insights from rodents and beyond. *Philosophical Transactions of the Royal Society B, 367*, 1869–1878.

Pomerleau, O. F. (1979). Behavioral medicine: The contribution of the experimental analysis of behavior to medical care. *American Psychologist, 34*, 654–663.

Porter, J. H., Brown, R. T., & Goldsmith, P. A. (1982). Adjunctive behavior in children on fixed interval food reinforcement schedules. *Physiology and Behavior, 28*, 609–612.

Porter, J. H., Young, R., & Moeschl, T. (1978). Effects of water and saline preloads on schedule-induced polydipsia in the rat. *Physiology & Behavior, 21*, 333–338.

Potter, J. N., Hanley, G. P., Augustine, M., Clay, C. J., & Phelps, M. C. (2013). Treating stereotypy in adolescents diagnosed with autism by refining the tactic of "using stereotypy as reinforcement". *Journal of Applied Behavior Analysis, 46*(2), 407–423.

Poulos, C. X., Wilkinson, D. A., & Cappell, H. (1981). Homeostatic regulation and Pavlovian conditioning intolerance to amphetamine-induced anorexia. *Journal of Comparative and Physiological Psychology, 95*, 735–746.

Poulson, C. L., Kymmissis, E., Reeve, K. F., Andreatos, M., & Reeve, L. (1991). Generalized vocal imitation in infants. *Journal of Experimental Child Psychology, 51*, 267–279.

Poulson, C. L., Kyparissos, N., Andreatos, M., Kymmissis, E., & Parnes, M. (2002). Generalized imitation within three response classes in typically developing infants. *Journal of Experimental Child Psychology, 81*, 341–357.

Powell, R. A., Digdon, N., Harris, B., & Smithson, C. (2014). Correcting the record on Watson, Rayner and Little Albert: Albert Barger as "psychology's lost boy". *American Psychologist, 69*, 600–611.

Powell, R. W. (1968). The effect of small sequential changes in fixed-ratio size upon the post-reinforcement pause. *Journal of the Experimental Analysis of Behavior, 11*, 589–593.

Powers, R., Cheney, C. D., & Agostino, N. R. (1970). Errorless training of a visual discrimination in pre-school children. *The Psychological Record, 20*, 45–50.

Prather, J. F., Peters, S., Nowicki, S., & Mooney, R. (2008). Precise auditory-vocal mirroring in neurons for learned vocal communication. *Nature, 451*, 305–310. doi:10.1038/nature06492.

Premack, D. (1959). Toward empirical behavioral laws: 1. Positive reinforcement. *Psychological Review, 66*, 219–233.

Premack, D. (1962). Reversability of the reinforcement relation. *Science, 136*, 235–237.

Prendergast, M., Podus, D., Finney, J., Greenwell, L., & Roll, J. (2006). Contingency management for treatment of substance use disorders: A meta-analysis. *Addiction, 101*, 1546–1560.

Provenza, F. D., Gregorini, P., & Carvalho, P. C. F. (2015). Synthesis: Foraging decisions link plants, herbivores and human beings. *Animal Production Science, 55*, 411–425.

Provenza, F. D., Lynch, J. J., & Nolan, J. V. (1994). Food aversion conditioned in anesthetized sheep. *Physiology and Behavior, 55*, 429–432.

Provenza, F. D., Villalba, J. J., Cheney, C. D., & Werner, S. J. (1998). Self-organization of foraging behavior: From simplicity to complexity without goals. *Nutrition Research Review, 11*, 199–222.

Provine, R. R. (2005). The yawn is primal, unstoppable and contagious, revealing the devolutionary and neural basis of empathy and unconscious behavior. *American Scientist, 93*, 532–539.

Provine, R. R., & Hamernik, H. B. (1986). Yawning: Effects of stimulus interest. *Bulletin of the Psychonomic Society, 24*, 437–438.

Pryor, K. W. (1999). *Don't shoot the dog.* New York: Bantam.

Pryor, K. W., Haag, R., & O'Reilly, J. (1969). The creative porpoise: Training for novel behavior. *Journal of the Experimental Analysis of Behavior, 12*, 653–651.

Puig, M. V., Rose, J., Schmidt, R., & Freund, N. (2014). Dopamine modulation of learning and memory in the prefrontal cortex: Insights from studies in primates, rodents and birds. *Frontiers in Neural Circuits, 8*. doi:10.3389/fncir.2014.00093.

Rachlin, H. (1969). Autoshaping of key pecking in pigeons with negative reinforcement. *Journal of the Experimental Analysis of Behavior, 12*, 521–531.

Rachlin, H. (1970). *Introduction to modern behaviorism.* San Francisco, CA: W. H. Freeman.

Rachlin, H. (1974). Self-control. *Behaviorism, 2*, 94–107.

Rachlin, H. (2000). *The science of self-control.* Cambridge, MA: Harvard University Press.

Rachlin, H., & Green, L. (1972). Commitment, choice and self-control. *Journal of the Experimental Analysis of Behavior, 17*, 15–22.

Rachlin, H., Arfer, K. B., Safin, V., & Yen, M. (2015). The amount effect and marginal value. *Journal of the Experimental Analysis of Behavior, 104*, 1–6.

Rachlin, H., Green, L., Kagel, J. H., & Battalio, R. C. (1976). Economic demand theory and psychological studies of choice. In G. H. Bower (Ed.), *The psychology of learning and motivation* (Vol. 10, pp. 129–154). New York: Academic Press.

Rachlin, H., & Laibson, D. I. (1997). *The matching law: Papers in psychology and economics.* New York: Harvard University Press.

Rachlin, H., Raineri, A., & Cross, D. (1991). Subjective probability and delay. *Journal of the Experimental Analysis of Behavior, 55*, 233–244.

Rankin, C. H., Abrams, T., Barry, R. J., Bhatnagar, S., Clayton, D. F., Colombo, J., et al. (2009). Habituation revisited: An update and revised description of the behavioral characteristics of habituation. *Neurobiology of Learning and Memory, 92*, 135–138.

Rapp, J. T. (2008). Conjugate reinforcement: A brief review and suggestions for applications to the assessment of automatically reinforced behavior. *Behavioral Interventions, 23*, 113–136. doi:10.1002/bin.259.

Rasmussen, E. B., & Newland, M. C. (2008). Asymmetry of reinforcement and punishment in human choice. *Journal of the Experimental Analysis of Behavior, 89*(2), 157–167.

Rasmussen, E. B., & Newland, M. C. (2009). Quantification of ethanol's antipunishment effect in humans using the generalized matching equation. *Journal of the Experimental Analysis of Behavior, 92*(2), 161–180. https://doi.org/10.1901/jeab.2009.92-161

Rasmussen, E. B., & Hillman, C. (2011). Naloxone and rimonabant reduce the reinforcing properties of exercise in rats. *Experimental and Clinical Psychopharmacology, 19*(6), 389. https://doi.org/10.1037/a0024142

Rasmussen, E. B., & Huskinson, S. L. (2008). Effects of rimonabant on behavior maintained by progressive ratio schedules of sucrose reinforcement in obese Zucker (fa/fa) rats. *Behavioural Pharmacology, 19*, 735–742.

Rasmussen, E. B., Lawyer, S. R., & Reilly, W. (2010). Percent body fat is related to delay and probability discounting for food in humans. *Behavioral Processes, 83*, 23–30.

Ray, E., & Heyes, C. (2011). Imitation in infancy: The wealth of the stimulus. *Developmental Science, 14*, 92–105.

Raybuck, J. D., & Lattal, K. M. (2014). Bridging the interval: Theory and neurobiology of trace conditioning. *Behavioural Processes, 101*, 103–111.

Reed, D. D., Kaplan, B. A., & Brewer, A. T. (2012). Discounting the freedom to choose: Implications for the paradox of choice. *Behavioural Processes, 90*, 424–427.

Reed, K., Duncan, J. M., Lucier-Greer, M., Fixelle, C., & Ferraro, A. J. (2016). Helicopter parenting and emerging adult self-efficacy: Implications for mental and physical health. *Journal of Child and Family Studies, 25*(10), 3136–3149.

Rehfeldt, R. A. (2011). Toward a technology of derived stimulus relations: An analysis of articles published in the Journal of Applied Behavior Analysis, 1992–2009. *Journal of Applied Behavior Analysis, 44*, 109–119.

Reicher, S. D., Haslam, S. A., & Smith, J. R. (2012). Working toward the experimenter: Reconceptualizing obedience within the Milgram paradigm as identification-based followership. *Perspectives on Psychological Science, 7*, 315–324.

Reilly, S., & Schachtman, T. R. (2009). *Conditioned taste aversion.* New York: Oxford University Press.

Rendall, D., Vokey, J. R., & Nemeth, C. (2007). Lifting the curtain on the *Wizard of Oz*: Biased voice-based impressions of speaker size. *Journal of Experimental Psychology and Human Perception Performance*, *33*, 1208–1219.

Rendell, L., Boyd, R., Enquist, M., Feldman, M. W., Fogarty, L., & Laland, K. N. (2011). How copying affects the amount, evenness and persistence of cultural knowledge: Insights from the social learning strategies tournament. *Philosophical Transactions of the Royal Society B*, *366*, 1118–1128.

Repacholi, B. M., & Meltzoff, A. N. (2007). Emotional eavesdropping: Infants selectively respond to indirect emotional signals. *Child Development*, *78*, 503–521.

Rescorla, R. A. (1966). Predictability and number of pairings in Pavlovian fear conditioning. *Psychonomic Science*, *4*, 383–384.

Rescorla, R. A. (1988). Pavlovian conditioning: It's not what you think it is. *American Psychologist*, *43*, 151–160.

Rescorla, R. A., & Wagner, A. R. (1972). A theory of Pavlovian conditioning: Variations in the effectiveness of reinforcement and nonreinforcement. In A. H. Black & W. F. Prokasy (Eds.), *Classical conditioning II: Current research and theory* (pp. 64–69). New York: Appleton-Century-Crofts.

Revillo, D. A., Arias, C., & Spear, N. E. (2013). The unconditioned stimulus pre-exposure effect in preweanling rats in taste aversion learning: Role of the training context and injection cues. *Developmental Psychobiology*, *55*, 193–204.

Revusky, S. H., & Bedarf, E. W. (1967). Association of illness with prior ingestion of novel foods. *Science*, *155*, 219–220.

Revusky, S. H., & Garcia, J. (1970). Learned associations over long delays. In G. H. Bower (Ed.), *The psychology of learning and motivation: Advances in research and theory* (Vol. 4, pp. 1–84). New York: Academic Press.

Reynolds, G. S. (1961). An analysis of interactions in a multiple schedule. *Journal of the Experimental Analysis of Behavior*, *4*, 107–117.

Rhodes, G. (2006). The evolutionary psychology of facial beauty. *Annual Review of Psychology*, *57*, 199–226.

Ribes, E. M., & Martinez, C. (1998). Second-order discrimination in humans: The roles of explicit instructions and constructed verbal responding. *Behavioural Processes*, *42*, 1–18.

Ribes, E. M., & Rodriguez, M. E. (2001). Correspondence between instructions, performance, and self-descriptions in a conditional discrimination task: The effects of feedback and type of matching response. *Psychological Record*, *51*, 309–333.

Richards, R. W. (1988). The question of bidirectional associations in pigeons' learning of conditional discrimination tasks. *Bulletin of the Psychonomic Society*, *26*, 577–579.

Richardson, J. V., & Baron, A. (2008). Avoidance of timeout from response-independent food: Effects of delivery rate and quality. *Journal of the Experimental Analysis of Behavior*, *89*, 169–181.

Richardson, R., Riccio, D. C., Jamis, M., Cabosky, J., & Skoczen, T. (1982). Modification of reactivated memory through "counterconditioning." *American Journal of Psychology*, *95*, 67–84.

Risley, T. R., & Hart, B. (1968). Developing correspondence between the nonverbal and verbal behavior of preschool children. *Journal of Applied Behavior Analysis*, *1*, 267–281.

Risley, T. R., & Hart, B. (2006). Promoting early language development. In N. F. Watt, C. Ayoub, R. H. Bradley, J. E. Puma, & W. A. LeBoeuf (Eds.), *The crisis in youth mental health: Critical issues and effective programs, Volume 4, Early intervention programs and policies* (pp. 83–88). Westport, CT: Praeger.

Rispoli, M. J., O'Reilly, M. F., Sigafoos, J., Lang, R., Kang, S., Lancioni, G., & Parker, R. (2011). Effects of presession satiation on challenging behavior and academic engagement for children with autism during classroom instruction. *Education and Training in Autism and Developmental Disabilities*, 607–618. www.jstor.org/stable/24232370

Ritz, T., Rosenfield, D., Steele, A., Millard, M., & Meuret, A. (2014). Controlling asthma by training capnometry-assisted hypoventilation (CATCH) vs. slow breathing: A randomized controlled trial. *Chest, 146*, 1237–1247.

Rizley, R. C., & Rescorla, R. A. (1972). Associations in second-order conditioning and sensory preconditioning. *Journal of Comparative and Physiological Psychology, 81*, 1–11.

Rizzolatti, G. (2014). Imitation: Mechanisms and importance for human culture. *Rendiconti Lincei, 25*, 285–289.

Rizzolatti, G., & Craighero, L. (2004). The mirror-neuron system. *Annual Review of Neuroscience, 27*, 169–192.

Roane, H. S. (2008). On the applied use of progressive-ratio schedules of reinforcement. *Journal of Applied Behavior Analysis, 41*, 155–161.

Roberts, A. C., Reekie, Y., & Braesicke, K. (2007). Synergistic and regulatory effects of orbitofrontal cortex on amygdala-dependent appetitive behavior. *Annals of the New York Academy of Sciences, 1121*, 297–319.

Robertson, S. H., & Rasmussen, E. B. (2017). Effects of a cafeteria diet on delay discounting in adolescent and adult rats: Alterations on dopaminergic sensitivity. *Journal of Psychopharmacology, 31*(11), 1419–1429. https://doi.org/10.1177/0269881117735750

Robles, D. S. (2009). Thinness and beauty: when food becomes the enemy. *International Journal of Research & Review, 2*.

Rodriguez, L. R., Rasmussen, E. B., Kyne-Rucker, D., Wong, M., & Martin, K. S. (2021). Delay discounting and obesity in food insecure and food secure women. *Health Psychology, 40*(4), 242.

Rodzon, K., Berry, M. S., & Odum, A. L. (2011). Within-subject comparison of degree of delay discounting using titrating and fixed sequence procedures. *Behavioural Processes, 86*, 164–167.

Rogers, S. J., & Vismara, L. A. (2008). Evidence-based comprehensive treatments for early autism. *Journal of Clinical Child & Adolescent Psychology, 37*(1), 8–38. https://doi.org/10.1080/15374410701817808

Rohsenow, D. J., Tidey, J. W., Martin, R. A., Colby, S. M., Sirota, A. D., Swift, R. M., et al. (2015). Contingent vouchers and motivational interviewing for cigarette smokers in residential substance abuse treatment. *Journal of Substance Abuse Treatment, 55*, 29–38.

Roll, J. M., & Higgins, S. T. (2000). A within-participant comparison of three different schedules of reinforcement of drug abstinence using cigarette smoking as an exemplar. *Drug and Alcohol Dependence, 58*, 103–109.

Roll, J. M., Higgins, S. T., & Badger, G. J. (1996). An experimental comparison of three different schedules of reinforcement of drug abstinence using cigarette smoking as an exemplar. *Journal of Applied Behavior Analysis, 29*, 495–505.

Roll, J. M., Madden, G. J., Rawson, R., & Petry, N. M. (2009). Facilitating the adoption of contingency management for the treatment of substance use disorders. *Behavior Analysis in Practice, 2*, 4–13.

Roper, T. J., & Posadas-Andrews, A. (1981). Are schedule-induced drinking and displacement activities causally related? *The Quarterly Journal of Experimental Psychology Section B, 33*, 181–193.

Rosenberg, E. L., & Ekman, P. (1995). *Motivation and Emotion, 19*, 111–138.

Rost, K. A., Hemmes, N. S., & Alvero, A. M. (2014). Effects of the relative values of alternatives on preference for free-choice in humans. *Journal of the Experimental Analysis of Behavior, 102*, 241–251.

Roth, T. L. (2012). Epigenetics of neurobiology and behavior during development and adulthood. *Developmental Psychobiology, 54*, 590–597.

Roth, T. L., & Sweatt, J. D. (2011). Annual research review: Epigenetic mechanisms and environmental shaping of the brain during sensitive periods of development. *The Journal of Child Psychology and Psychiatry, 52*, 398–408.

Roth, W. J. (2002). Teaching dolphins to select pictures in response to recorded dolphin whistles with few errors. *Dissertation Abstracts International: Section B: The Sciences & Engineering, 62*(10-B), 95008.

Routtenberg, A., & Kuznesof, A. W. (1967). Self-starvation of rats living in activity wheels on a restricted feeding schedule. *Journal of Comparative and Physiological Psychology, 64*, 414–421.

Rubene, D., Hastad, O., Tauson, R., Wall, H., & Odeen, A. (2010). The presence of UV wavelengths improves the temporal resolution of the avian visual system. *Journal of Experimental Biology, 213*, 3357–3363.

Ruggles, T. R., & LeBlanc, J. M. (1982). Behavior analysis procedures in classroom teaching. In A. S. Bellack, M. Hersen, & A. E. Kazdin (Eds.), *International handbook of behavior modification and therapy* (pp. 959–996). New York: Plenum Press.

Rung, J. M., & Madden, G. J. (2018). Experimental reductions of delay discounting and impulsive choice: A systematic review and meta-analysis. *Journal of Experimental Psychology: General, 147*(9), 1349.

Russell, C. L., Bard, K. A., & Adamson, L. B. (1997). Social referencing by young chimpanzees (Pan troglodytes). *Journal of Comparative Psychology, 111*, 185–193.

Rutherford, A. (2009). *Beyond the box: B. F. Skinner's technology of behavior from laboratory to life, 1950s-1970s*. Toronto: University of Toronto Press.

Saeed, S., Quintin, J., Kerstens, H. H. D., Rao, N. A., Aghajanirefah, A., Matarese, F., et al. (2014). Epigenetic programming of monocyte-to-macrophage differentiation and trained innate immunity. *Science, 345*. doi:10.1126/science.1251086.

Salamone, J. D., Correa, M., Mingote, C. S., & Weber, S. M. (2003). Nucleus accumbens dopamine and the regulation of effort in food-seeking behavior: Implications for studies of natural motivation, psychiatry, and drug abuse. *Journal of Pharmacology and Experimental Therapeutics, 305*, 1–8.

Salvy, S. J., Mulick, J. A., Butter, E., Bartlett, R. K., & Linscheid, T. R. (2004). Contingent electric shock (SIBIS) and a conditioned punisher eliminate severe head banging in a preschool child. *Behavioral Interventions, 19*, 59–72.

Salvy, S. J., Pierce, W. D., Heth, D. C., & Russell, J. C. (2004). Taste avoidance induced by wheel running: Effects of backward pairings and robustness of conditioned taste aversion. *Physiology and Behavior, 82*, 303–308.

Sanders, G. A. (1974). Introduction. In D. Cohen (Ed.), *Explaining linguistic phenomena* (pp. 1–20). Washington, DC: Hemisphere.

Sanders, S. J., Murtha, M. T., Gupta, J. D., Murdoch, J. D., Raubeson, M. J., Willsey, A. J., et al. (2012). De novo mutations revealed by whole-exome sequencing are strongly associated with autism. *Nature, 485*, 237–241.

Santana, L. H. (2015). Two challenges of a selectionist analogy to the theory of selection by consequences. *Revista Perspectivas, 6*, 40–47.

Sargisson, R. J., & White, K. G. (2001). Generalization of delayed matching-to-sample performance following training at different delays. *Journal of the Experimental Analysis of Behavior, 75*, 1–14.

Saunders, K. J. (1989). Naming in conditional discrimination and stimulus equivalence. *Journal of the Experimental Analysis of Behavior, 51*, 379–384.

Saunders, K. J., & Spradlin, J. E. (1989). Conditional discrimination in mentally retarded adults: The effect of training the component simple discriminations. *Journal of the Experimental Analysis of Behavior, 52*, 1–12.

Sautter, R. A., & LeBlanc, L. A. (2006). Empirical applications of Skinner's analysis of verbal behavior with humans. *Analysis of Verbal Behavior, 22*, 35–48.

Savage-Rumbaugh, S. E. (1984). Verbal behavior at a procedural level in the chimpanzee. *Journal of the Experimental Analysis of Behavior, 41*, 223–250.

Savage-Rumbaugh, S. E. (1986). *Ape language: From conditioned response to symbol*. New York: Columbia University Press.

Savage-Rumbaugh, S. E., Murphy, J., Sevcik, R. A., Brakke, K. E., Williams, S. L., & Rumbaugh, D. M. (1993). Language comprehension in ape and child. *Monographs of the Society for Research in Child Development*, *58* (233).

Savage-Rumbaugh, S. E., Shanker, S. G., & Taylor, T. J. (1998). *Apes, language, and the human mind*. New York: Oxford University Press.

Sayette, M. A., Creswell, K. G., Dimoff, J. D., Fairbairn, C. E., Cohn, J. F., Heckman, B. W., et al. (2012). Alcohol and group formation: A multimodal investigation of the effects of alcohol on emotion and social bonding. *Psychological Science*, *23*, 869–878.

Schaal, D. W. (2013). Behavioral neuroscience. In G. J. Madden (Ed.), *APA handbook of behavior analysis: Vol. 1. methods and principles* (pp. 339–359). Washington, DC: American Psychological Association.

Schaefer, H. H., & Martin, P. L. (1966). Behavior therapy for "apathy" of hospitalized patients. *Psychological Reports*, *19*, 1147–1158.

Schafer, R. J., & Moore, T. (2011). Selective attention from voluntary control of neurons in the prefrontal cortex. *Science*, *332*, 1568–1571.

Scheller, R. H., & Axel, R. (1984). How genes control innate behavior. *Scientific American*, *250*, 54–62.

Schiffrin, H. H., Liss, M., Miles-McLean, H., Geary, K. A., Erchull, M. J., & Tashner, T. (2014). Helping or hovering? The effects of helicopter parenting on college students' well-being. *Journal of child and family studies*, *23*(3), 548–557.

Schlichting, C. D., & Wund, M. (2014). Phenotypic plasticity and epigenetic marking: An assessment of evidence for genetic accommodation. *Evolution*, *68*, 656–672.

Schlinger, H. D., & Blakely, E. (1987). Function-altering effects of contingency-specifying stimuli. *The Behavior Analyst*, *10*, 41–45.

Schlinger, H. D., Derenne, A., & Baron, A. (2008). What 50 years of research tell us about pausing under ratio schedules of reinforcement. *The Behavior Analyst*, *31*, 39–60.

Schlinger, H. D., Jr. (2011). Skinner as missionary and prophet: A review of *Burrhus F. Skinner: Shaper of Behaviour*. *Journal of Applied Behavior Analysis*, *44*, 217–225.

Schlund, M. W., Cataldo, M. F., & Hoehn-Saric, R. (2008). Neural correlates of derived relational responding on tests of stimulus equivalence. *Behavioral and Brain Functions*. doi:10.1186/1744-9081-4-6.

Schlund, M. W., Hoehn-Saric, R., & Cataldo, M. F. (2007). New knowledge derived from learned knowledge: Functional-anatomic correlates of stimulus equivalence. *Journal of the Experimental Analysis of Behavior*, *87*, 287–307.

Schmidt, A. C., Hanley, G. P., & Layer, S. A. (2009). A further analysis of the value of choice: Controlling for the illusory discriminative stimuli and evaluating the effects of less preferred items. *Journal of Applied Behavior Analysis*, *42*, 711–716.

Schmitt, D. R. (2001). Delayed rule following. *The Behavior Analyst*, *24*, 181–189.

Schoenfeld, W. N., Antonitis, J. J., & Bersh, P. J. (1950). A preliminary study of training conditions necessary for conditioned reinforcement. *Journal of Experimental Psychology*, *40*, 40–45.

Schopler, E., & Mesibov, G. B. (1994). *Behavioral issues in autism*. New York: Plenum.

Schrier, A. M., & Brady, P. M. (1987). Categorization of natural stimuli by monkeys (*Macaca mulatta*): Effects of stimulus set size and modification of exemplars. *Journal of Experimental Psychology: Animal Behavior Processes*, *13*, 136–143.

Schusterman, R. J., & Kastak, D. (1993). A California sea lion (*Zalophus californianus*) is capable of forming equivalence relations. *The Psychological Record*, *43*, 823–839.

Schwartz, B. (1980). Development of complex stereotyped behavior in pigeons. *Journal of the Experimental Analysis of Behavior, 33*, 153–166.

Schwartz, B. (1982a). Failure to produce response variability with reinforcement. *Journal of the Experimental Analysis of Behavior, 37*, 171–181.

Schwartz, B. (1982b). Reinforcement-induced stereotypy: How not to teach people to discover rules. *Journal of Experimental Psychology: General, 111*, 23–59.

Schwartz, B. (2004). *The paradox of choice: Why more is less.* New York: HarperCollins.

Schwartz, B., & Williams, D. R. (1972a). The role of response reinforcer contingency in negative automaintenance. *Journal of the Experimental Analysis of Behavior, 18*, 351–357.

Schwartz, B., & Williams, D. R. (1972b). Two different kinds of key peck in the pigeon: Some properties of responses maintained by negative and positive response-reinforcer contingencies. *Journal of the Experimental Analysis of Behavior, 18*, 201–216.

Segal, E. F. (1962). Effects of *dl*-amphetamine under concurrent VI DRL reinforcement. *Journal of the Experimental Analysis of Behavior, 5*, 105–112.

Seligman, M. E. P. (1970). On the generality of the laws of learning. *Psychological Review, 77*, 406–418.

Seligman, M. E. P. (1975). *Helplessness: On depression, development, and death.* San Francisco, CA: Freeman.

Seligman, M. E. P. (1991). *Learned optimism.* New York: Knopf.

Seligman, M. E. P., & Maier, F. F. (1967). Failure to escape traumatic shock. *Journal of Experimental Psychology, 74*, 1–9.

Shabani, D. B., & Fisher, W. W. (2006). Stimulus fading and differential reinforcement for the treatment of needle phobia in a youth with autism. *Journal of Applied Behavior Analysis, 39*, 449–452.

Shabani, D. B., Carr, J. E., & Petursdottir, A. I. (2009). A laboratory model for studying response-class hierarchies. *Journal of Applied Behavior Analysis, 42*, 105–121.

Shahan, T. A., & Cunningham, P. (2015). Conditioned reinforcement and information theory reconsidered. *Journal of the Experimental Analysis of Behavior, 103*, 405–418,

Shahan, T. A., & Sweeney, M. M. (2013). A model of resurgence based on behavioral momentum theory. *Journal of the Experimental Analysis of Behavior, 95*, 91–108.

Shapiro, M. M. (1960). Respondent salivary conditioning during operant lever pressing in dogs. *Science, 132*, 619–620.

Shaver, K. G. (1985). *The attribution of blame.* New York: Springer-Verlag.

Shaw, J., Seldomridge, N., Dunkle, D., Nugent, P., Spangler, L., Bromenshenk, J., et al. (2005). Polarization lidar measurements of honey bees in flight for locating land mines. *Optics Express, 13*, 5853–5863.

Shearer, J., Tie, H., & Byford, S. (2015). Economic evaluations of contingency management in illicit drug misuse programmes: A systematic review. *Drug and Alcohol Review, 34*, 289–298.

Shearn, D. W. (1962). Operant conditioning of heart rate. *Science, 137*, 530–531.

Sherman, J. A. (1965). Use of reinforcement and imitation to reinstate verbal behavior in mute psychotics. *Journal of Abnormal Psychology, 70*, 155–164.

Sherman, J. G., Ruskin, G., & Semb, G. B. (Eds.) (1982). *The personalized system of instruction: 48 seminal papers.* Lawrence, KS: TRI Publications.

Sherrington, C. (1906). *The integrative action of the nervous system* (2nd ed., 1947). New Haven, CT: Yale University Press.

Shizgal, P., & Arvanitogiannis, A. (2003). Gambling on dopamine. *Science, 299*, 1856–1858.

Shull, R. L. (1979). The postreinforcement pause: Some implications for the correlational law of effect. In M. D. Zeiler & P Harzem (Eds.), *Reinforcement and the organization of behaviour* (pp. 193–221). New York: John Wiley & Sons.

Shull, R. L., & Pliskoff, S. S. (1967). Changeover delay and concurrent schedules: Some effects on relative performance measures. *Journal of the Experimental Analysis of Behavior, 10,* 517–527.

Shull, R. L., Gaynor, S. T., & Grimer, J. A. (2002). Response rate measured as engagement bouts: Resistance to extinction. *Journal of the Experimental Analysis of Behavior, 77,* 211–231.

Sidman, M. (1953). Avoidance conditioning with brief shock and no exteroceptive warning signal. *Science, 118*(3058), 157–158. doi:10.1126/science.118.3058.157.

Sidman, M. (1960). *Tactics of scientific research.* New York: Basic Books.

Sidman, M. (1962). Reduction of shock frequency as reinforcement for avoidance behavior. *Journal of the Experimental Analysis of Behavior, 5,* 247–257.

Sidman, M. (1993). Reflections on behavior analysis and coercion. *Behavior and Social Issues, 3,* 75–85.

Sidman, M. (1994). *Equivalence relations and behavior: A research story.* Boston, MA: Authors Cooperative, Inc.

Sidman, M. (2000). Equivalence relations and the reinforcement contingency. *Journal of the Experimental Analysis of Behavior, 74,* 127–146.

Sidman, M. (2001). *Coercion and its fallout.* Boston, MA: Authors Cooperative, Inc.

Sidman, M. (2006). The distinction between positive and negative reinforcement: Some additional considerations. *The Behavior Analyst, 29,* 135–139. doi:10.1007/BF03392126.

Sidman, M., & Cresson, O., Jr. (1973). Reading and crossmodal transfer of stimulus equivalences in severe retardation. *American Journal of Mental Deficiency, 77,* 515–523.

Sidman, M., & Tailby, W. (1982). Conditional discrimination vs. matching-to-sample: An expansion of the testing paradigm. *Journal of the Experimental Analysis of Behavior, 37,* 5–22.

Sidman, M., Cresson, O., Jr., & Wilson-Morris, M. (1974). Acquisition of matching to sample via mediated transfer. *Journal of the Experimental Analysis of Behavior, 22,* 261–273.

Sidman, M., Rauzin, R., Lazar, R., Cunningham, S., Tailby, W., & Carrigan, P. (1982). A search for symmetry in the conditional discriminations of rhesus monkeys, baboons, and children. *Journal of the Experimental Analysis of Behavior, 37,* 23–44.

Siegel, S. (1975). Conditioning insulin effects. *Journal of Comparative and Physiological Psychology, 89,* 189–199.

Siegel, S. (2005). Drug tolerance, drug addiction, and drug anticipation. *Current Directions in Psychological Science, 14,* 296–300.

Siegel, S., Hinson, R. E., Krank, M. D., & McCully, J. (1982). Heroin "overdose" death: The contribution of drug-associated environmental cues. *Science, 216,* 436–437.

Sigafoos, J., Doss, S., & Reichle, J. (1989). Developing mand and tact repertoires in persons with severe developmental disabilities using graphic symbols. *Research in Developmental Disabilities, 10,* 183–200.

Sigafoos, J., Reichle, J., Doss, S., Hall, K., & Pettitt, L. (1990). Spontaneous transfer of stimulus control from tact to mand contingencies. *Research in Developmental Disabilities, 11,* 165–176.

Silbert, L. J., Honey, C. J., Simony, E., Poeppei, D., & Hasson, U. (2014). Coupled neural systems underlie the production and comprehension of naturalistic narrative speech. *Proceedings of the National Academy of Sciences,* e4687–e4696. doi:10.1073/pnas.1323812111.

Silva, S. P., & Lattal, K. A. (2010). Why pigeons say what they do: Reinforcer magnitude and response requirement effects on say responding in say-do correspondence. *Journal of the Experimental Analysis of Behavior, 93,* 395–413.

Silverman, K., Roll, J. M., & Higgins, S. T. (2008). Introduction to the special issue on the behavior analysis and treatment of drug addiction. *Journal of Applied Behavior Analysis, 41,* 471–480.

Simic, J., & Bucher, B. (1980). Development of spontaneous manding in nonverbal children. *Journal of Applied Behavior Analysis, 13*, 523–528.

Simola, D. F., Graham, R. J., Brady, C. M., Enzmann, B. L., Desplan, C., Anandasankar, R., et al., (2015). Epigenetic (re)programming of caste-specific behavior in the ant Camponotus floridanus. *Science, 351*, 42–49.

Simon, S. J., Ayllon, T., & Milan, M. A. (1982). Behavioral compensation: Contrastlike effects in the classroom. *Behavior Modification, 6*, 407–420.

Sitharthan, G., Hough, M. J., Sitharthan, T., & Kavanagh, D. J. (2001). The alcohol helplessness scale and its prediction of depression among problem drinkers. *Journal of Clinical Psychology, 57*, 1445–1457.

Skinner, B. F. (1935). Two types of conditioned reflex and a pseudo type. *Journal of General Psychology, 12*, 66–77.

Skinner, B. F. (1938). *The behavior of organisms*. New York: Appleton-Century-Crofts.

Skinner, B. F. (1945). Baby in a box. *Ladies Home Journal*. Retrieved from www.uni.edu/~maclino/cl/skinner_baby_in_a_box.pdf.

Skinner, B. F. (1948a). *Walden two*. New York: Macmillan.

Skinner, B. F. (1948b). "Superstition" in the pigeon. *Journal of Experimental Psychology, 38*, 168–172.

Skinner, B. F. (1950). Are theories of learning necessary? *Psychological Review, 57*, 193–216.

Skinner, B. F. (1953). *Science and human behavior*. New York: Free Press.

Skinner, B. F. (1957). *Verbal behavior*. New York: Appleton-Century-Crofts.

Skinner, B. F. (1960). Pigeons in a pelican. *American Psychologist, 15*, 28–37.

Skinner, B. F. (1968). *The technology of teaching*. New York: Appleton-Century-Crofts.

Skinner, B. F. (1969). *Contingencies of reinforcement: A theoretical analysis*. New York: Appleton-Century-Crofts.

Skinner, B. F. (1971). *Beyond freedom and dignity*. New York: Alfred A. Knopf.

Skinner, B. F. (1976). *Particulars of my life*. New York: McGraw-Hill.

Skinner, B. F. (1979). *The shaping of a behaviorist*. New York: McGraw-Hill.

Skinner, B. F. (1981). Selection by consequences. *Science, 213*, 501–504.

Skinner, B. F. (1983). *A matter of consequences*. New York: Alfred A. Knopf.

Skinner, B. F. (1984). The evolution of behavior. *Journal of the Experimental Analysis of Behavior, 41*, 217–222.

Skinner, B. F. (1986). The evolution of verbal behavior. *Journal of the Experimental Analysis of Behavior, 45*(1), 115–122. doi:10.1901/jeab.1986.45-115.

Skinner, B. F., & Vaughan, M. E. (1983). *Enjoy old age: A program of self-management*. New York: W. W. Norton.

Smeets, P. M., Barnes-Holmes, D., & Nagle, M. (2000). Transfer and stimulus equivalence classes derived from simultaneously presented S+ and S− stimuli. *European Journal of Behavior Analysis, 1*, 33–49.

Smirnova, A., Zorina, Z., Obozova, T., & Wasserman, E. (2015). Crows spontaneously exhibit analogical reasoning. *Current Biology, 25*, 256–260.

Smith, S. L., & Rasmussen, E. B. (2010). Effects of 2-AG on the reinforcing properties of wheel activity in obese and lean Zucker rats. *Behavioural Pharmacology, 21*, 292–300.

Smith, T. (2001). Discrete trial training in the treatment of autism. *Focus on Autism and Other Developmental Disabilities, 16*(2), 86–92.

Smith, T. R., & Jacobs, E. A. (2015). Concurrent token production in rats. *Psychological Record, 65*, 101–113.

Smith, T., & Iadarola, S. (2015). Evidence base update for autism spectrum disorder. *Journal of Clinical Child & Adolescent Psychology, 44*(6), 897 922.

Sobell, L. C., Ellingstad, T. P., & Sobell, M. B. (2000). Natural recovery from alcohol and drug problems: Methodological review of the research with suggestions for future directions. *Addiction, 95*(5), 749–764. https://doi.org/10.1046/j.1360-0443.2000.95574911.x

Sobsey, D. (1990). Modifying the behavior of behavior modifiers: Arguments for countercontrol against aversive procedures. In A. C. Repp & N. N. Singh (Eds.), *Perspectives on the use of nonaversive and aversive interventions for persons with developmental disabilities* (pp. 421–433). Sycamore, IL: Sycamore Publishing Co.

Solomon, P. R., Blanchard, S., Levine, E., Velazquez, E., & Groccia-Ellison, M. (1991). Attenuation of age-related deficits in humans by extension of the interstimulus interval. *Psychology of Aging, 6*, 36–42.

Solomon, R. L., & Brush, E. S. (1956). Experimentally derived conceptions of anxiety and aversion. In M. R. Jones (Ed.), *Nebraska symposium on motivation* (pp. 212–305). Lincoln, NE: University of Nebraska Press.

Solomon, R. L., & Wynne, L. C. (1953). Traumatic avoidance learning: Acquisition in normal dogs. *Psychological Monographs: General and Applied, 67*(4), 1–19. https://doi.org/10.1037/h0093649

Sossin, W. S., Kirk, M. D., & Scheller, R. H. (1987). Peptidergic modulation of neuronal circuitry controlling feeding in *Aplysia. Journal of Neuroscience, 7*, 671–681.

Specter, D. (2014). Drool: Ivan Pavlov's real quest. *The New Yorker*, 123–126.

Spetch, M. L., & Friedman, A. (2006). Pigeons see correspondence between objects and their pictures. *Psychological Science, 17*, 966–972.

Spetch, M. L., Cheng, K., & Clifford, C. W. G. (2004). Peak shift but not range effects in recognition of faces. *Learning and Motivation, 35*, 221–241.

Spetch, M. L., Wilkie, D. M., & Pinel, J. P. (1981). Backward conditioning: A reevaluation of the empirical evidence. *Psychological Bulletin, 89*, 163–175.

Spiegelman, B. (2008). The skinny fat. *The Scientist, 22*, 28. Retrieved from http://classic.the-scientist.com/article/display/54033/

Springer, C. R., & Pear, J. J. (2008). Performance measures in courses using computer-aided personalized system of instruction. *Computers & Education, 51*, 829–835.

Squires, N., & Fantino, E. (1971). A model for choice in simple concurrent and concurrent-chains schedules. *Journal of the Experimental Analysis of Behavior, 15*, 27–38.

Staddon, J. E. R. (1977). Schedule-induced behavior. In *Handbook of operant behavior* (pp. 125–152). Abingdon, Oxford: Routledge.

Staddon, J. E. R., & Simmelhag, V. L. (1971). The "superstition" experiment: A re-examination of its implications for the principles of adaptive behavior. *Psychological Review, 78*, 3–43.

Stafford, D., & Branch, M. (1998). Effects of step size and break-point criterion on progressive-ratio performance. *Journal of the Experimental Analysis of Behavior, 70*, 123–138.

Stein, L., & Belluzzi, J. D. (1988). Operant conditioning of individual neurons. In M. L. Commons, R. M. Church, J. R. Stellar, & A. R. Wagner (Eds.), *Quantitative analyses of behavior. Vol. 7. Biological determinants of reinforcement and memory* (pp. 249–264). Hillsdale, NJ: Lawrence Erlbaum Associates.

Stein, L., & Belluzzi, J. D. (2014). Operant conditioning of individual neurons. In Michael L. Commons, Russell M. Church, James R. Stellar, & Allan R. Wagner (Eds.), *Quantitative analysis of behavior: Biological determinants of reinforcement* (Vol. VII, pp. 249–264). New York: Psychology Press.

Stein, L., Xue, B. G., & Belluzzi, J. D. (1994). *In vitro* reinforcement of hippocampal bursting: A search for Skinner's atoms of behavior. *Journal of the Experimental Analysis of Behavior, 61*, 155–168.

Stenseng, F., Belsky, J., Skalicka, V., & Wichstrom, L. (2014). Preschool social exclusion, aggression, and cooperation: A longitudinal evaluation of the need-to-belong and the social-reconnection hypotheses.

Personality and Social Psychology Bulletin, 40, 1637–1647. Retrieved from http://psp.sagepub.com/content/early/2014/10/10/0146167214554591.

Stevens, J. R., & Stephens, D. W. (2011). The adaptive nature of impulsivity. In G. J. Madden & W. K. Bickel (Eds.), *Impulsivity: The behavioral and neurological science of discounting* (pp. 361–388). Washington, DC: American Psychological Association.

Stewart, I., McElwee, J., & Ming, S. (2013). Language generativity, response generalization, and derived relational responding. *Journal of Applied Behavior Analysis, 29*, 137–155.

Stiers, M., & Silberberg, A. (1974). Lever-contact responses in rats: Automaintenance with and without a negative response-reinforcer dependency. *Journal of the Experimental Analysis of Behavior, 22*, 497–506.

Stokes, P. D. (2001). Variability, constraints, and creativity: Shedding light on Claude Monet. *American Psychologist, 36*, 355–359.

Stokes, P. D., Mechner, F., & Balsam, P. D. (1999). Effects of different acquisition procedures on response variability. *Animal Learning and Behavior, 27*, 28–41.

Stokes, T. F., & Baer, D. M. (1977). An implicit technology of generalization. *Journal of Applied Behavior Analysis, 10*, 349–367.

Stokes, T. F., Fowler, S. A., & Baer, D. M. (1978). Training preschool children to recruit natural communities of reinforcement. *Journal of Applied Behavior Analysis, 11*, 285–303.

Stoops, W. W. (2008). Reinforcing effects of stimulants in humans: Sensitivity of progressive-ratio schedules. *Experimental Clinical Psychopharmacology, 16*, 503–512.

Stoops, W. W., Glaser, P. E. A., Fillmore, M. T., & Rush, C. R. (2004). Reinforcing, subject-rated, performance and physiological effects of methylphenidate and d-amphetamine in stimulant-abusing humans. *Journal of Psychopharmacology, 18*, 534–543.

Storms, L. H., Boroczi, G., & Broen, W. E., Jr. (1962). Punishment inhibits an instrumental response in hooded rats. *Science, 135*, 1133–1134.

Straus, M. A. (2001). *Beating the devil out of them: Corporal punishment in American families and its effects on children* (2nd ed.). New Brunswick, NJ: Transaction Publishers.

Strickland, J. C., & Johnson, M. W. (2021). Rejecting impulsivity as a psychological construct: A theoretical, empirical, and sociocultural argument. *Psychological Review, 128*(2), 336.

Stroop, J. R. (1935). Studies of interference in serial verbal reactions. *Journal of Experimental Psychology, 18*, 643–662.

Strosahl, K. D., & Robinson, P. J. (2016). Acceptance and commitment therapy: Applications to the treatment of clinical depression. In A. Wells & P. Fisher (Eds.), *Treating depression: MCT, CBT and third wave therapies* (pp. 319–343). West Sussex, UK: John Wiley & Sons, Ltd.

Stuttgen, M. C., Yildiz, A., & Gunturkun, O. (2011). Adaptive criterion setting in perceptual decision making. *Journal of the Experimental Analysis of Behavior, 96*, 155–176.

Substance Abuse and Mental Health Services Administration (2014). National Survey on Drug Use and Health: Summary of Methodological Studies, 1971–2014 [Internet]. Rockville, MD: Substance Abuse and Mental Health Services Administration (US). Available from www.ncbi.nlm.nih.gov/books/NBK519735/

Sullivan, R. M., & Hall, W. G. (1988). Reinforcers in infancy: Classical conditioning using stroking or intraoral infusions of milk as UCS. *Developmental Psychobiology, 21*, 215–223.

Sulzer-Azaroff, B. (1986). Behavior analysis and education: Crowning achievements and crying needs. *Division 25 Recorder, 21*, 55–65.

Sun, B., & Tsai, P. S. (2011). A gonadotropin-releasing hormone-like molecule modulates the activity of diverse central neurons in a gastropod mollusk, *Aplysia californica. Frontiers in Endocrinology, 2*, 1–8.

Sundberg, M. L. (1985). Teaching verbal behavior to pigeons. *The Analysis of Verbal Behavior, 3*, 11–17.

Sundberg, M. L. (1996). Toward granting linguistic competence to apes: A review of Savage-Rumbaugh et al.'s *Language Comprehension in Ape and Child*. *Journal of the Experimental Analysis of Behavior*, *65*, 477–492.

Sundberg, M. L., & Michael, J. (2001). The benefits of Skinner's analysis of verbal behavior for children with autism. *Behavior Modification*, *25*, 698–724.

Sundberg, M. L., & Sundberg, C. A. (2011). Intraverbal behavior and verbal conditional discriminations in typically developing children and children with autism. *The Analysis of Verbal Behavior*, *27*, 23–43.

Sundberg, M. L., Endicott, K., & Eigenheer, P. (2000). Using intraverbal prompts to establish tacts for children with autism. *The Analysis of Verbal Behavior*, *17*, 89–104.

Svartdal, F. (1992). Operant modulation of low-level attributes of rule-governed behavior by nonverbal contingencies. *Learning and Motivation*, *22*, 406–420.

Sweatt, D. (2009). Experience-dependent epigenetic modifications in the central nervous system. *Biological Psychiatry*, *65*, 191–197.

Swedo, E., Idaikkadar, N., Leemis, R., et al. (2020). Trends in U.S. Emergency Department Visits Related to Suspected or Confirmed Child Abuse and Neglect Among Children and Adolescents Aged <18 Years Before and During the COVID-19 Pandemic — United States, January 2019–September 2020. *Morbidity and Mortality Weekly Report*, *69*, 1841–1847. doi: http://dx.doi.org/10.15585/mmwr.mm6949a1externalicon.

Swisher, M., & Urcuioli, P. J. (2015). Symmetry in the pigeon with sample and comparison stimuli in different locations, II. *Journal of the Experimental Analysis of Behavior*, *104*, 119–132.

Swithers, S. E. (2015). Not so sweet revenge: Unintended consequences of high-intensity sweeteners. *The Behavior Analyst*, *38*, 1–17.

Swithers, S. E., & Davidson, T. L. (2008). A role for sweet taste: Calorie predictive relations in energy regulation by rats. *Behavioral Neuroscience*, *122*, 161–173.

Swithers, S. E., Doerflinger, A., & Davidson, T. L. (2006). Consistent relationships between sensory properties of savory snack foods and calories influence food intake in rats. *International Journal of Obesity*, *30*, 1685–1692.

Taghert, P. H., & Nitabach, M. N. (2012). Peptide neuromodulation in invertebrate model systems. *Neuron*, *76*, 82–97.

Takamori, K., Yoshida, S., & Okuyama, S. (2001). Repeated treatment with imipramine, fluvoxamine, and tranylcypromine decreases the number of escape failures by activating dopaminergic systems in a rat learned helplessness test. *Life Sciences*, *69*, 1919–1926.

Takemoto, H. (2008). Morphological analyses and 3D modeling of the tongue musculature of the chimpanzee (*Pan troglodytes*). *American Journal of Primatology*, *70*, 966–975.

Talwar, V., Arruda, S., & Yachison, S. (2015). The effect of punishment and appeals for honesty on children's truth-telling behavior. *Journal of Experimental Child Psychology*, *130*, 209–217.

Tammen, S. A., Friso, S., & Choi, S. W. (2013). Epigenetics: The link between nature and nurture. *Molecular Aspects of Medicine*, *34*, 753–764.

Taylor, I., & O'Reilly, M. F. (1997). Toward a functional analysis of private verbal self-regulation. *Journal of Applied Behavior Analysis*, *30*, 43–58.

Taylor, S. P., & Pisano, R. (1971). Physical aggression as a function of frustration and physical attack. *Journal of Social Psychology*, *84*, 261–267.

Terrace, H. S. (1963). Discrimination learning with and without "errors." *Journal of the Experimental Analysis of Behavior*, *6*, 1–27.

Terrace, H. S. (1966). Stimulus control. *Operant behavior: Areas of research and application*. New York: Appleton-Century-Crofts, pp. 271–344.

Thompson, R. F., & Spencer, W. A. (1966). Habituation: A model phenomenon for the study of neuronal substrates of behavior. *Psychological Review, 73*, 16–43.

Thompson, R. H., Bruzek, J. L., & Cotnoir-Bichelman, N. M. (2011). The role of negative reinforcement in infant caregiving: An experimental simulation. *Journal of Applied Behavior Analysis, 44*, 295–304.

Thompson, T. (2007). Relations among functional systems in behavior analysis. *Journal of the Experimental Analysis of Behavior, 87*, 423–440.

Thorndike, E. L. (1898). Animal intelligence. *Psychological Review Monograph Supplements (Serial No. 8)*.

Thorpe, W. H. (1963). *Learning and instinct in animals*. Cambridge, MA: Harvard University Press.

Tiger, J. H., Hanley, G. P., & Hernandez, E. (2006). An evaluation of the value of choice with preschool children. *Journal of Applied Behavior Analysis, 39*, 1–16.

Tighe, T. J., & Leaton, R. N. (1976). *Habituation*. Hillsdale, NJ: Lawrence Erlbaum Associates.

Timberlake, W. (1983). Rats' responses to a moving object related to food or water: A behavior-systems analysis. *Animal Learning and Behavior, 11*, 309–320.

Timberlake, W., & Grant, D. L. (1975). Auto-shaping in rats to the presentation of another rat predicting food. *Science, 190*, 690–692.

Tinbergen, N. (1951). *The study of instinct*. Oxford: Oxford University Press.

Tinbergen, N., & Kuenen, D. J. (1957). Feeding behavior in young thrushes. In C. H. Schiller (Ed.), *Instinctive behavior: Development of a modern concept* (pp. 209–236). London: Methuen.

Tobler, P. N., Fiorillo, C. D., & Schultz, W. (2005). Adaptive coding of reward value by dopamine neurons. *Science, 307*, 1642–1645.

Todd, J. T. (2015). Old horses in new stables: Rapid prompting, facilitated communication, science, ethics, and the history of magic. In R. M. Foxx & J. A. Mulick (Eds.), *Controversial therapies for developmental disabilities: Fad fashion, and science in professional practice* (2nd ed.) (pp. 372–409). Mahwah, NJ: Routledge.

Todd, T. P. (2013). Mechanisms of renewal after extinction of instrumental behavior. *Journal of Experimental Psychology: Animal Behavior Processes, 39*, 193–207.

Tomasello, M. (2009). *The question of chimpanzee culture, plus postscript (Chimpanzee culture, 2009)*. In K. N. Laland & B. G. Galef (Eds.), *The question of animal culture* (pp. 198–221). Cambridge, MA: Harvard University Press.

Tourinho, E. Z. (2006). Private stimuli, covert responses and private events: Conceptual remarks. *The Behavior Analyst, 29*, 13–31.

Toussaint, K. A. (2011). Teaching tactual discrimination of Braille characters to beginning Braille readers. *Dissertation submitted to the Department of Psychology*, Louisiana State University, Agricultural and Mechanical College.

Towe, A. L. (1954). A study of figural equivalence in the pigeon. *Journal of Comparative and Physiological Psychology, 47*, 283–287.

Travis, C. (2013). Book review: 'Behind the shock machine' by Gina Perry. *The Wall Street Journal*, Retrieved from http://on.wsj.com/1Jf7baNcom/1Jf7baN.

Tuckman, G. (2011). CNN Video: Spare the rod, spoil the child. Retrieved from http://ac360.blogs.cnn.com/2011/08/15/video-spare-the-rod-spoil-the-child/

Tulving, E. (1983). *Elements of episodic memory*. New York: Oxford University Press.

Twenge, J. M., Baumeister, R. F., Tice, D. M., & Stucke, T. S. (2001). If you can't join them, beat them: Effects of social exclusion on aggressive behavior. *Journal of Personality and Social Psychology, 81*, 1058–1069.

Twyman, J. S. (1996). The functional independence of impure mands and tacts of abstract stimulus properties. *The Analysis of Verbal Behavior, 13*, 1–19.

Twyman, J. S. (2014). Behavior analysis in education. In F. K. McSweeney & E. S. Murphy (Eds.), *The Wiley Blackwell handbook of operant and classical conditioning* (pp. 533–558). West Sussex, UK: Wiley.

Ulrich, R. E., & Azrin, N. H. (1962). Reflexive fighting in response to aversive stimulation. *Journal of the Experimental Analysis of Behavior, 5*, 511–520.

Ulrich, R. E., Hutchinson, R. R., & Azrin, N. H. (1965). Pain-elicited aggression. *The Psychological Record, 15*, 111–126.

Ulrich, R. E., Wolff, P. C., & Azrin, N. H. (1964). Shock as an elicitor of intra- and inter-species fighting behavior. *Animal Behaviour, 12*, 14–15.

Van Hest, A., van Haaren, F., & van de Poll, N. E. (1989). Operant conditioning of response variability in male and female Wistar rats. *Physiology and Behavior, 45*, 551–555.

Van Houten, R., Axelrod, S., Bailey, J. S., Favell, J. E., Foxx, R. M., Iwata, B. A., et al. (1988). The right to effective behavioral treatment. *Journal of Applied Behavior Analysis, 21*, 381–384.

Van Houten, R., Malenfant, J. E. L., Reagan, I., Sifrit, K., Compton, R., & Tenenbaum, J. (2010). Increasing seat belt use in service vehicle drivers with a gearshift delay. *Journal of Applied Behavior Analysis, 43*, 369–380.

Van Houten, R., Reagan, I. J., & Hilton, B. W. (2014). Increasing seat belt use: Two field experiments to test engineering-based behavioral interventions. *Transportation Research Part F: Traffic Psychology and Behavior, 23*, 133–146.

Vargas, E. A. (1998). Verbal behavior: Implications of its mediational and relational characteristics. *Analysis of Verbal Behavior, 15*, 149–151.

Vargas, J. S. (1990). B. F. Skinner: Fact and fiction. *The International Behaviorology Association Newsletter, 2*, 8–11.

Vasquez, E. A., Denson, T. F., Pedersen, W. C., Stenstrom, D. M., & Miller, N. (2005). The moderating effect of trigger intensity on triggered displaced aggression. *Journal of Experimental Social Psychology, 41*, 61–67.

Vaughan, M. E., & Michael, J. L. (1982). Automatic reinforcement: An important but ignored concept. *Behaviorism, 10*, 217–227.

Vaughn, W., Jr. (1988). Formation of equivalence sets in pigeons. *Journal of Experimental Psychology: Animal Behavior Processes, 14*, 36–42.

Ventura, R., Morrone, C., & Puglisi-Allegra, S. (2007). Prefrontal/accumbal catecholamine system determines motivational salience attribution to both reward- and aversion-related stimuli. *Proceedings of the National Academy of Sciences, 104*, 5181–5186.

Vergoz, V., Roussel, E., Sandoz, J. C., & Giurfa, M. (2007). Aversive learning in honeybees revealed by the olfactory conditioning of the stinger extension reflex. *PLoS One, 2*, e288. Retrieved from http://dx.doi.org/10.1371/journal.pone.0000288.

Verhave, T. (1966). The pigeon as a quality control inspector. *American Psychologist, 21*, 109–115.

Vichi, C., Andery, M. A., & Glenn, S. S. (2009). A metacontingency experiment: The effects of contingent consequences on patterns of interlocking contingencies of reinforcement. *Behavior and Social Issues, 18*, 41–57.

Villareal, J. (1967). Schedule-induced pica. *Physiology and Behavior, 6*, 577–588.

Vits, S., & Schedlowski, M. (2014). Learned placebo effects in the immune system. *Zeitschrift fur Psychologie, 222*, 148–153.

Vits, S., Cesko, E., Enck, P., Hillen, U., Schadendorf, D., & Schedlowski, M. (2011). Behavioural conditioning as the mediator of placebo responses in the immune system. *Philosophical Transactions of the Royal Society B, 366*, 1799–1807. doi:10.1098/rstb.2010.0392.

Vivanti, G., & Rogers, S. (2014). Autism and the mirror neuron system: Insights from learning and teaching. *Philosophical Transactions of the Royal Society B, 369*. Retrieved from http://dx.doi.org/10.1098/rstb.2013.0184.

Vogelstein, J. T., Park, Y., Ohyama, T., Kerr, R., Truman, J. W., Priebe, C. E., et al. (2014). Discovery of brainwide neural-behavioral maps via multiscale unsupervised structure learning. *Sciencexpress*. doi:10.1126:science.1250298.

Voineagu, I., Wang, X., Johnston, P., Lowe, J. K., Tian, Y., Horvath, S., et al. (2011). Transcriptomic analysis of autistic brain reveals convergent molecular pathology. *Nature, 474*, 380–384.

Volkmar, F., Carter, A., Grossman, J., & Klin, A. (1997). Social development in autism. In D. J. Cohen & F. R. Volkmar (Eds.), *Handbook of autism and pervasive developmental disorders* (2nd ed.) (pp. 171–194). New York: John Wiley & Sons.

Volkow, N. D., Fowler, J. S., & Wang, G.-J. (2003). The addicted human brain: Insights from imaging studies. *Journal of Clinical Investigation, 111*, 1444–1451.

Vollmer, T. R., & Hackenberg, T. D. (2001). Reinforcement contingencies and social reinforcement: Some reciprocal relations between basic and applied research. *Journal of Applied Behavior Analysis, 34*, 241–253.

Voltaire, M., Gewirtz, J. L., & Pelaez, M. (2005). Infant responding under conjugate- and continuous-reinforcement schedules. *Behavioral Development Bulletin, 1*, 71–79.

Vuchinich, R. E., & Simpson, C. A. (1999). Delayed reward discounting in alcohol abuse. In F. J. Chaloupka, M. Grossman, W. K. Bickel, & H. Saffer (Eds.), *The economic analysis of substance use and abuse* (pp. 103–122). Chicago: University of Chicago Press.

Wacker, D. P., Harding, J. W., Berg, W. K., Lee, J. F., Schieltz, K. M., Padilla, Y. C., et al. (2011). An evaluation of persistence of treatment effects during long-term treatment of destructive behavior. *Journal of the Experimental Analysis of Behavior, 96*, 261–282.

Waddell, J., Anderson, M. L., & Shors, T. J. (2011). Changing rate and hyppocampal dependence of trace eyeblink conditioning: Slow learning enhances survival of new neurons. *Neurobiology of Learning and Memory, 95*, 159–165.

Wakefield, J. C. (2006). Is behaviorism becoming a pseudo-science? Power versus scientific rationality in the eclipse of token economies by biological psychiatry in the treatment of schizophrenia. *Behavior and Social Issues, 15*, 202–221.

Waller, M. B. (1961). Effects of chronically administered chlorpromazine on multiple-schedule performance. *Journal of the Experimental Analysis of Behavior, 4*, 351–359.

Walsh, J. J., Friedman, A. K., Sun, H., Heller, E. A., Ku, S. M., Juarez, B., et al. (2014). Stress and CRF gate neural activation of BDNF in the mesolimbic reward pathway. *Nature Neuroscience, 17*, 27–29.

Wanchisen, B. A., Tatham, T. A., & Mooney, S. E. (1989). Variable-ratio conditioning history produces high- and low-rate fixed-interval performance in rats. *Journal of the Experimental Analysis of Behavior, 52*, 167–179.

Wansink, B., & Sobal, J. (2007). Mindless eating: The 200 daily food decisions we overlook. *Environment and Behavior, 39*(1):106–123. doi:10.1177/0013916506295573.

Ward, T. A., Eastman, R. L., & Ninness, C. (2009). An experimental analysis of cultural materialism: The effects of various modes of production on resource sharing. *Behavior and Social Issues, 18*, 58–80.

Warmington, M., & Hitch, G. J. (2014). Enhancing the learning of new words using an errorless learning procedure: Evidence from typical adults. *Memory, 22*, 582–594.

Warren, Z., McPheeters, M. L., Sathe, N., Foss-Feig, J. H., Glasser, A., & Veenstra-VanderWeele, J. (2011). A systematic review of early intensive intervention for autism spectrum disorders. *Pediatrics, 127*, e1303–e1311. doi:10.1542/peds.2011-0427.

Wasserman, E. A. (1973). Pavlovian conditioning with heat reinforcement produces stimulus-directed pecking in chicks. *Science, 181*, 875–877.

Wasserman, E. A., & Young, M. E. (2010). Same–different discrimination: The keel and backbone of thought and reasoning. *Journal of Experimental Psychology: Animal Behavior Processes, 36*, 3–22.

Watanabe, S., Sakamoto, J., & Wakita, M. (1995). Pigeons' discrimination of paintings by Monet and Picasso. *Journal of the Experimental Analysis of Behavior, 63*(2), 165–174.

Watson, J. B. (1913). Psychology as the behaviorist views it. *Psychological Review, 20*, 158–177.

Watson, J. B., & Rayner, R. (1920). Conditioned emotional reactions. *Journal of Experimental Child Psychology, 3*, 1–14.

Wawrzyncyck, S. (1937). Badania and parecia *Spirostomum ambiguum major*. *Acta Biologica Experimentalis (Warsaw), 11*, 57–77.

Weiner, H. (1969). Controlling human fixed-interval performance. *Journal of the Experimental Analysis of Behavior, 12*, 349–373.

Weisberg, P., & Rovee-Collier, C. (1998). Behavioral processes of young infants and young children. In K. A. Lattal & M. Perone (Eds.), *Handbook of research methods in human operant behavior* (pp. 325–370). New York: Plenum Press.

Weiss, B., & Gott, C. T. (1972). A microanalysis of drug effects on fixed-ratio performance in pigeons. *Journal of Pharmacology and Experimental Therapeutics, 180*, 189–202.

West, M. J., & King, A. P. (1988). Female visual displays affect the development of male song in the cowbird. *Nature, 334*, 244–246.

West, R. P., & Young, K. R. (1992). Precision teaching. In R. P. West & L. A. Hamerlynck (Eds.), *Designs for excellence in education: The legacy of B. F. Skinner* (pp. 113–146). Longmont, CO: Sopris West, Inc.

West, R. P., Young, R., & Spooner, F. (1990). Precision teaching: An introduction. *Teaching Exceptional Children, 22*, 4–9.

White, A. J., & Davison, M. C. (1973). Performance in concurrent fixed-interval schedules. *Journal of the Experimental Analysis of Behavior, 19*, 147–153.

White, C. T., & Schlosberg, H. (1952). Degree of conditioning of the GSR as a function of the period of delay. *Journal of Experimental Psychology, 43*, 357–362.

White, K. G. (2002). Psychophysics of remembering: The discrimination hypothesis. *Current Directions in Psychological Science, 11*, 141–145.

White, K. G. (2013). Remembering and forgetting. In G. J. Madden (Ed.), *APA handbook of behavior analysis: Vol. 1. methods and principles* (pp. 411–437). Washington DC: American Psychological Association.

White, K. G., & Brown, G. S. (2011). Reversing the course of forgetting. *Journal of the Experimental Analysis of Behavior, 96*, 177–189.

White, K. G., & Sargisson, R. J. (2011). Maintained generalization of delay-specific remembering. *Behavioural Processes, 87*, 310–313.

White, O. R. (1986). Precision teaching—Precision learning. *Exceptional Children, 52*, 522–534.

Whiten, A., & Boesch, C. (2001). The cultures of chimpanzees. *Scientific American, 284*, 60–68.

Whiten, A., Hinde, R. A., Laland, K. N., & Stringer, C. B. (2011). Culture evolves. *Philosophical Transactions of the Royal Society B, 366*, 938–948.

Whiten, A., McGuigan, N., Marshall-Pescini, S., & Hopper, L. M. (2009). Emulation, imitation, over-imitation and the scope of culture for child and chimpanzee. *Philosophical Transactions of the Royal Society B, 364*, 2417–2428.

Wiesler, N. A., Hanson, R. H., Chamberlain, T. P., & Thompson, T. (1988). Stereotypic behavior of mentally retarded adults adjunctive to a positive reinforcement schedule. *Research in Developmental Disabilities, 9*, 393–403.

Wilkes, G. (1994). *A behavior sampler*. North Bend, WA: Sunshine Books.

Williams, A. R. (1997). Under the volcano: Montserrat. *National Geographic, 192*, 59–75.

Williams, C. D. (1959). The elimination of tantrum behavior by extinction procedures. *Journal of Abnormal and Social Psychology, 59*, 269.

Williams, D. R., & Williams, H. (1969). Automaintenance in the pigeon: Sustained pecking despite contingent non-reinforcement. *Journal of the Experimental Analysis of Behavior, 12*, 511–520.

Williams, G., Carnerero, J. J., & Perez-Gonzalez, L. A. (2006). Generalization of tacting actions in children with autism. *Journal of Applied Behavior Analysis, 39*, 233–237. doi:10.1901/jaba.2006.175-04.

Williams, J. L., & Lierle, D. M. (1986). Effects of stress controllability, immunization, and therapy on the subsequent defeat of colony intruders. *Animal Learning and Behavior, 14*, 305–314.

Williams, K. D., & Nida, S. A. (2011). Ostracism: Consequences and coping. *Current Directions in Psychological Science, 20*, 71–75.

Wilson, D. S., Hayes, S. C., Biglan, A., & Embry, D. D. (2014). Evolving the future: Toward a science of intentional change. *Behavioral and Brain Sciences, 37*, 395–416. doi:10.1017/S0140525X13001593.

Wilson, L., & Rogers, R. W. (1975). The fire this time: Effects of race of target, insult, and potential retaliation on black aggression. *Journal of Personality and Social Psychology, 32*, 857–864.

Wing, V. C., & Shoaib, M. (2010). A second-order schedule of food reinforcement in rats to examine the role of the CB1 receptors in the reinforcement-enhancing effects of nicotine. *Addiction Biology, 15*, 380–392.

Winger, G., & Woods, J. H. (2013). Behavioral pharmacology. In G. J. Madden (Ed.), *APA handbook of behavior analysis: Vol. 1. methods and principles* (pp. 547–567). Washington, DC: American Psychological Association.

Winstanley, C. A. (2011). The neural and neurochemical basis of delay discounting. In G. J. Madden & W. K. Bickel (Eds.), *Impulsivity: The behavioral and neurological science of discounting* (pp. 95–122). Washington, DC: American Psychological Association.

Wisniewski, M. G., Church, B. A., & Mercado, E., III. (2009). Learning-related shifts in generalization gradients for complex sounds. *Learning and Behavior, 37*, 325–335.

Witnauer, J. E., & Miller, R. R. (2011). Some determinants of second-order conditioning. *Learning & Behavior, 39*, 12–26.

Witte, K., & Allen, M. (2000). A meta-analysis of fear appeals: Implications for effective public health campaigns. *Health Education & Behavior, 27*, 591–615.

Wolfe, B. M., & Baron, R. A. (1971). Laboratory aggression related to aggression in naturalistic social situations: Effects of an aggressive model on the behavior of college students and prisoner observers. *Psychonomic Science, 24*, 193–194.

Wolfe, J. B. (1936). Effectiveness of token rewards for chimpanzees. *Comparative Psychology Monographs, 12*, 1–72.

Wolpe, J. (1961). The systematic desensitization treatment of neuroses. *Journal of Nervous and Mental disease*.

World Health Organization (2021). Corporal punishment and health. Retrieved May 22, 2022, from www.who.int/news-room/fact-sheets/detail/corporal-punishment-and-health

World Health Organization (2022). WHO Coronavirus (COVID-19) Dashboard.

Wyckoff, L. B., Jr. (1952). The role of observing responses in discrimination learning. Part 1. *Psychological Review, 59*, 431–442.

Wyckoff, L. B., Jr. (1969). The role of observing responses in discrimination learning. In D. P. Hendry (Ed.), *Conditioned reinforcement* (pp. 237–260). Homewood, IL: Dorsey Press.

Xue, B. G., Belluzzi, J. D., & Stein, L. (1993). *In vitro* reinforcement of hippocampal bursting activity by the cannabinoid receptor agonist (-)- CP-55,940. *Brain Research, 626*, 272–277.

Yamamoto, T. (2007). Brain regions responsible for the expression of conditioned taste aversion in rats. *Chemical Senses*, *32*, 105–109.

Yang, Z., Bertolucci, F., Wolf, R., & Heisenberg, M. (2013). Flies cope with uncontrollable stress by learned helplessness. *Current Biology*, *23*, 799–803.

Yi, J. I., Christian, L., Vittimberga, G., & Lowenkron, B. (2006). Generalized negatively reinforced manding in children with autism. *Analysis of Verbal Behavior*, *22*, 21–33.

Yoon, S., & Bennett, G. M. (2000). Effects of a stimulus-stimulus pairing procedure on conditioning vocal sounds as reinforcers. *Analysis of Verbal Behavior*, *17*, 75–88.

Zarcone, J. R., Mullane, M. P., Langdon, P. E. & Brown, I. (2020), Contingent Electric Shock as a Treatment for Challenging Behavior for People with Intellectual and Developmental Disabilities: Support for the IASSIDD Policy Statement Opposing Its Use. *Journal of Policy and Practice in Intellectual Disabilities*, *17*, 291–296. https://doi.org/10.1111/jppi.12342

Zelinski, E. L., Hong, N. S., Tyndall, A. V., Halsall, B., & McDonald, R. J. (2010). Prefrontal cortical contributions during discriminative fear conditioning, extinction, and spontaneous recovery in rats. *Experimental Brain Research*, *203*, 285–297.

Zentall, T. R. (2006). Imitation: Definitions, evidence and mechanisms. *Animal Cognition*, *9*, 335–353.

Zentall, T. R. (2011). Social learning mechanisms: Implications for a cognitive theory of imitation. *Interaction Studies*, *12*, 233–261.

Zentall, T. R., Wasserman, E. A., & Urcuioli, P. J. (2014). Associative concept learning in animals. *Journal of the Experimental Analysis of Behavior*, *101*, 130–151.

Zettle, R. D., & Hayes, S. C. (1982). Rule-governed behavior: A potential theoretical framework for cognitive-behavior therapy. In P. C. Kendall (Ed.), *Advances in cognitive behavioral research and therapy*, 1 (pp. 73–118). New York: Academic Press.

Zielinska, E. (2012). Treating fat with fat. *The Scientist*, *26*, 65.

Zimmerman, J., & Ferster, C. B. (1963). Intermittent punishment of S^Δ responding in matching-to-sample. *Journal of the Experimental Analysis of Behavior*, *6*, 349–356.

AUTHOR INDEX

Abbott, B. 371
Abramson, C. I. 292
Abramson, L. Y. 218
Abreu-Rodrigues, J. 394
Ackert, A. M. 263
Adamson, L. B. 278
Ader, R. 96
Afifi, T. O. 203
Agostino, N. R. 292
Ahearn, W. H. 450
Ainslie, G. W. 342, 344
Albanos, K. 252
Alberto, P. A. 469, 477
Alcock, J. 516
Aleksejev, R. M. 221
Alexander, J. H. 216
Alexander, J. L. 500
Alferink, L. A. 321, 366
Allen, C. 537
Allen, G. L. 385
Allen, M. 212
Allen, R. 486
Alling, K. 125
Almason, S. M. 438
Alsiö, J. 334
Altemus, M. 264
Alvarez, L. W. 511
Alvero, A. M. 331
Alvord, J. R. 385
Amat, J. 221
Amlung, M. 339
Amtzen, E. 456
Anderson, C. A. 225
Anderson, C. D. 284
Anderson, M. L. 91
Anderson, N. D. 292
Andery, M. A. 533
Andre, J. 252
Andreatos, M. 409
Andrew, S. C. 287
Andrews, D. 477
Anger, D. 180
Anokhin, A. P. 341
Anrep, G. V. 82, 84, 85
Ansari, Z. 325
Antonitis, J. J. 137, 147, 157, 158, 368
Aoyama, K. 126–127

Appel, J. B. 196, 197
Arantes, J. 292
Arcediano, F. 92
Arfer, K. B. 342
Arias, C. 89
Armitage, K. B. 529
Arnold, K. E. 278
Arnold, M. L. 469
Aronsen, G. 223
Arruda, S. 395
Arthurs, J. 252
Arvanitogiannis, A. 7
Asaro, F. 511
Aston-Jones, G. 240, 334
Atalayer, D. 335
Austin, J. 14, 31, 469
Autor, S. M. 364
Avargues-Weber, A. 86
Axel, R. 514, 515
Ayllon, T. 284, 385, 470
Ayres, K. M. 500
Azar, B. 305
Azrin, N. H. 138, 193–194, 196, 197–198, 221,
    222, 223, 227–228, 232, 255, 385, 483
Azrin, R. D. 193–194

Badger, G. J. 176, 177, 339, 485
Badia, P. 371
Baer, D. M. 41, 43, 136, 393, 394, 407, 408, 409,
    410–412, 427, 470, 474, 475, 495
Baeyens, F. 5, 93
Bailey, J. S. 14, 444, 472
Baker, A. G. 142
Baker, T. B. 94
Baldwin, J. M. 396
Balsam, P. D. 136
Bandura, A. 396, 407, 412–416, 419–420, 495
Banko, K. M. 116, 118
Banks, M. L. 353
Banna, K. M. 321
Banuazizi, A. 247
Barbera, M. L. 498
Bard, K. A. 278, 402
Bargh, J. A. 45
Barnes, C. A. 136
Barnes-Holmes, D. 393, 421, 441, 450, 462, 474
Barnes-Holmes, D. 462

Barnes-Holmes, Y. 441, 450, 474
Baron, A. 161, 166, 182, 204, 210, 419
Baron, R. A. 224
Barot, S. K. 252
Barrera, F. J. 238
Barreto, F. 488
Barrett, L. F. 65
Barron, A. B. 515
Bartlett, S. M. 195
Battalio, R. C. 332
Baum, M. 208
Baum, W. M. 180, 182, 212, 319, 320, 321, 326, 328, 345, 346, 347, 352, 546
Bauman, K. E. 47
Baumeister, R. F. 227
Baxter, D. A. 234, 524
Baxter, M. G. 368
Beasty, A. 167
Beavers, G. A. 41, 48, 49
Bechara, A. 252
Beck, H. P. 21–22
Becker, A. M. 534
Bedarf, E. W. 251
Beeby, E. 344
Belke, T. W. 262, 263, 284, 335, 384
Bell, J. 450
Bellack, A. S. 500
Belles, D. 480, 493, 508
Belluzzi, J. D. 122, 529
Belsky, J. 227
Bem, D. J. 394, 436
Benbassat, D. 292
Benitti, V. 488
Bennett, A. T. D. 278
Bennett, G. M. 446
Benthall, R. P. 167
Bentzley, B. S. 334
Beran, M. J. 330
Bereznak, S. 500
Berg, M. L. 278
Bergmann, S. 469
Berkowitz, L. 224
Bernard, C. 49–52, 57
Bernstein, I. L. 252
Berry, M. S. 338
Bersh, P. J. 368
Bertaina-Anglade, V. 219
Bertolucci, F. 216
Berton, O. 220
Besalel-Azrin, V. A. 193–194
Betts, K. R. 227
Bewley-Taylor, D. 486
Bezzina, C. W. 338

Bhatnagar, S. 78
Bickel, W. K. 25, 176, 338, 339, 340, 341, 485
Bierley, C. 81
Biesmeijer, J. C. 515
Biglan, A. 14, 509
Bijou, S. 43
Binder, C. 490, 492
Bird, G. 404
Bitterman, M. E. 86, 90
Bjork, D. W. 24
Blackman, D. 457
Blakely, E. 422
Blanchard, S. 89
Blass, E. M. 81
Bloom, P. 394
Bloom, S. E. 138
Blough, D. S. 136, 278, 294, 297
Blue, S. 280
Blumstein, D. T. 529
Boakes, R. A. 253, 259, 263
Boer, D. P. 261, 262
Boesch, C. 538
Boisjoli, J. A. 386
Boisseau, R. P. 130
Bolin, B. L. 172
Bolívar, H. A. 483
Bolles, R. C. 205
Bonasio, R. 97, 522
Bondy, A. 497
Bonem, E. J. 155, 160, 284
Bonem, M. K. 155, 160
Boomhower, S. 339
Borba, A. 535
Borden, R. J. 225
Boren, J. J. 181
Boren, M. C. P. 159, 175, 304
Boroczi, G. 196
Borrero, J. C. 41, 321, 322
Bostow, D. E. 14
Bouton, M. E. 85, 142, 571
Boutot, E. A. 469
Bowdle, B. F. 226
Bowen, R. 225
Bower, G. H. 85
Boyce, T. E. 488
Boyle, J. 284
Bradlyn, A. S. 480, 493, 508
Bradshaw, C. A. 167
Bradshaw, C. M. 321
Brady, P. M. 302
Braesicke, K. 368
Brainard, M. S. 129, 130
Branch, M. 175

Brandt, J. A. 14
Bratcher, M. S. 9, 353
Bray, M. A. 492
Bray, S. 82
Brechner, K. C. 535
Breland, K. 235–237
Breland, M. 235–237
Brembs, B. 113, 524
Brettel, L. 86
Brewer, A. T. 332
Breyer, N. L. 385
Brody, H. 96
Broen, W. E., Jr. 196
Brooks, D. C. 85
Brooks, K. R. 288
Brown, D. L. 288
Brown, G. S. 299, 300, 301
Brown, J. L. 294
Brown, M. P. 371
Brown, P. L. 239–240, 242, 264, 265, 268
Brown, R. T. 257, 461, 529
Browne, C. 305
Brownell, K. D. 45
Brownstein, A. J. 319, 421
Bruce, S. 39
Brunner, E. 292
Brush, E. S. 208
Bruzek, J. L. 206, 207
Buccino, G. 405
Buchanan, T. 219
Bucher, B. 437
Buckley, J. L. 321, 322, 345, 354
Buckley, K. B. 23
Buckley, K. E. 225
Budney, A. J. 483
Bugnyar, T. 303
Bullock, C. E. 266, 382
Burch, A. E. 483, 485
Burger, J. M. 419
Burgess, R. L. 49, 50, 51
Burke, K. A. 136
Burkholder, E. O. 479
Bushell, D., Jr. 49, 50, 51
Buske-Kirschbaum, A. 96
Buskist, W. F. 421
Butler, B. K. 278
Byford, S. 486
Byrne, J. H. 234, 524
Byrne, M. J. 253, 254

Cable, C. 302
Cabosky, J. 200
Cain, C. K. 198

Call, J. 538
Cameron, J. 116–117, 118, 149, 229
Can, D. D. 461
Cancel-Tassin, G. 306
Cappell, H. 93
Carmona, A. 245–246, 269
Carnagey, N. L. 225
Carnerero, J. J. 440
Carnett, A. 386
Carnine, D. 488, 492
Caron, M. L. 321
Caroni, P. 123
Carr, D. 457
Carr, E. G. 201, 326
Carr, J. E. 41, 436, 438, 439, 440, 442, 444, 479, 494, 495, 497
Carrigan, P. F., Jr. 457
Carroll, M. E. 333
Carroll, R. J. 437, 439, 440
Carter, A. 495
Carter, C. 56
Carton, J. S. 385
Cartwright, W. S. 486
Carvalho, L. S. 278
Carvalho, P. C. F. 470
Case, D. A. 373, 374
Castilla, J. L. 259
Catagnus, R. 14
Cataldo, M. F. 456, 477
Catania, A. C. 16, 192, 312, 316, 324, 329, 393, 417, 452, 454, 455
Cate, C. T. 287
Cathro, J. S. 374
Catmur, C. 404
Cautela, J. R. 194, 471
Cerutti, D. 424
Chamberlain, T. P. 258
Chance, P. 24
Chandler, R. K. 486
Chang, S. 242
Charlop-Christy, M. H. 495
Charness, N. 31, 32
Chase, P. N. 460, 492
Chawarska, K. 495
Chen, G. 152
Chen, X. 402
Cheney, C. D. 155, 165, 212, 250, 259–260, 265, 284, 292, 293, 366, 371, 385, 470
Cheng, J. 369
Cheng, K. 287
Cheng, S. C. 97
Cheng, T. D. 91
Cherek, D. R. 224, 257

Chillag, D. 255, 256
Chittka, A. 11
Chittka, L. 11, 515
Cho, C. 208
Choi, S. W. 11
Chomsky, N. 427
Christ, T. J. 477
Christian, L. 438
Christophersen, E. R. 501
Church, B. A. 287
Ciano, P. D. 369
Classe, A. 531
Cleary, J. 457
Clifford, C. W. G. 287
Cockburn, J. 330
Cohen, D. 226
Cohen, M. 166
Cohen, N. 96
Cohen, P. A. 488
Cohen, P. S. 196
Cohen, S. L. 170, 171
Cohn, S. L. 169
Coker, C. C. 371
Collinger, J. L. 245
Collins, A. G. E. 330
Colquitt, G. 488
Conger, R. 321
Conyers, C. 195
Cook, D. 488
Cook, R. G. 289
Cooney, J. L. 484
Cooper, J. O. 194
Cooper, M. A. 221
Corballis, M. C. 529
Coren, S. 404
Cornu, J. N. 306
Correa, M. 369
Cotnoir-Bichelman, N. M. 206, 207
Courage, M. L. 402
Courtney, K. 214
Cowdery, G. E. 477
Cowles, J. T. 381
Craighero, L. 403
Craik, F. I. 292
Cranston, S. S. 477, 478, 479, 480
Cresson, O., Jr. 460
Cristler, C. 477, 478, 479, 480
Critchfield, T. S. 395, 470, 471
Crombez, G. 93
Cross, D. 338
Crossman, E. K. 160, 366
Crow, R. E. 165, 256
Crozier, R. H. 515

Culbertson, S. 159, 175, 304
Cullinan, V. 441
Cumming, W. W. 305
Cummings, A. R. 497
Cunningham, P. 389
Cunningham, S. 457
Cussenot, O. 306
Cuzzocrea, F. 488

D'Amato, M. R. 457
Dale, R. H. I. 144
Dallery, J. 485
Damasio, H. 252
Dapcich-Miura, E. 385
Dardig, J. 473, 474
Darley, J. M. 63
Dasiewicz, P. 203
Davidson, T. L. 505
Davies, N. B. 327, 517
Davis, D. R. 176
Davis, J. L. 498
Davison, M. C. 321, 345, 346, 351, 352, 365
Dawkins, R. 512
Day, H. M. 46
Day, J. J. 12
Day, J. R. 46
de Boer, B. 430
De Brugada, I. 89
de Geus, E. J. 240
De Houwer, J. 5, 93
de Lange, R. P. J. 514
de Villiers, P. A. 283, 302, 315, 316, 320, 325
Deacon, J. R. 394
deCharms, R. C. 252
Deci, E. L. 116
DeFulio, A. 210
Deguchi, H. 410
DeHart, W. B. 25, 340
DeHouwer, J. 474
Deich, J. D. 91
Deisseroth, K. 242
Delaney, P. F. 31
DeLeon, I. G. 471
Delgado, M. R. 330
Dempsey, C. M. 471
Demuru, E. 74
Dengerink, H. A. 224, 225
Denson, T. F. 224
Derenne, A. 161, 182, 287
Desmond, J. E. 91
Dessalles, J. L. 428
Detrich, R. 394
DeWall, C. N. 227

Dewey, R. A. 237
DeWulf, M. J. 284
Diane, A. 264, 501, 503
Dias, B. G. 12
DiCara, L. 246
Dickerson, F. B. 385, 386
Dickins, D. W. 456, 528
Dickins, T. E. 528
Dickinson, A. M. 434
Didden, R. 495
Digdon, N. 22
DiMarzo, V. 174
Dinsmoor, J. A. 212, 280, 371
Disterhoft, J. F. 91
Dixon, D. R. 47
Dixon, P. D. 263
Djikic, M. 65
Dobek, C. 253, 254
Doepke, K. J. 393
Doerflinger, A. 505
Dollard, J. 407, 469
Domire, S. C. 500
Domjan, M. 40
Donahoe, J. W. 526
Donato, F. 123
Donnellan, A. W. 202
Donnerstein, E. 224
Donohue, B. C. 469
Donohue, R. 65
Donovan, W. I. 206
Donovan, W. J. 278
Dornhaus, A. 515
Dorr, D. 399
Dorrance, B. R. 400
Dorsey, M. F. 47
Dos Santos, C. V. 219
Doss, S. 440
Dotto-Fojut, K. M. 294
Dougan, J. D. 9, 353
Dougher, M. J. 456, 461
Doughty, A. H. 136
Doughty, S. S. 492
Douglas-Hamilton, I. 188
Doupe, A. J. 129, 130
Dove, L. D. 256
Doyere, V. 199
Doyle, T. A. 257
Draganski, B. 8
Du, L. 410, 412
Dube, W. V. 170
Duhigg, C. 112
Dukas, R. 519
Dulany, D. E. 420

Duncan, I. D. 262, 263, 335
Dunlap, A. S. 250
Dunlap, G. 497
DuPree, J. P. 56
Durand, V. M. 475
Dussutour, A. 130
Dutra, L. 485
Dworkin, B. R. 247
Dwyer, D. M. 263
Dymond, D. 471

Eastman, R. L. 538
Eckel, L. A. 263
Eckerman, D. A. 158
Edelman, G. M. 523
Edling, T. 470
Edwards, T. L. 344
Egan, L. C. 394
Egger, M. D. 369, 370
Eibl-Eibesfeldt, I. 72
Eigenheer, P. 444
Eisenberger, R. 116, 118, 136
Ekman, P. 431
Elder, S. T. 170
Ellenwood, D. W. 460
Elliffe, D. 460
Ellis, D. A. 91
Elsmore, T. F. 320
Embry, D. D. 14, 509
Endicott, K. 444
Engelmann, S. 292, 488
Engler, H. 96
Enns, J. T. 404
Epling, W. F. 252, 259–260, 261, 262, 264, 289, 321, 322, 334, 471, 492, 500, 501, 509, 527, 530
Epstein, L. H. 127
Epstein, R. 159, 398–400, 426
Ericsson, K. A. 31, 32
Erjavec, M. 409, 410
Erlich, P. R. 428
Ernstdotter, E. 278
Estes, W. K. 100
Etkowitz, H. 526
Ettinger, R. A. 283
Ettinger, R. H. 284
Evans, F. J. 418
Evans, I. M. 202
Evans, R. I. 415
Evans, T. A. 330

Fadiga, L. 403
Fagot, J. 296

Falck-Ytter, T. 403
Falk, J. L. 255, 256, 257, 258, 269, 270n2
Fantino, E. 267, 352, 365, 368, 373, 374, 375–376, 379, 387–389, 553
Farina, F. R. 456
Farmer-Dougan, V. 9, 353
Fawcett, T. W. 338
Febbo, S. 170
Feenstra, M. G. P. 369
Fehr, E. 535
Feldman, M. A. 201
Fender, K. M. 334
Ferguson, A. 321
Ferland, R. J. 284
Fernandez-Hermida, J. 484
Ferrari, P. F. 405
Ferster, C. B. 131, 149, 152, 159, 160, 163, 165, 168, 175, 197, 304, 312, 362
Festinger, L. 394
Field, D. P. 522
Fields, L. 135, 423, 456
Fields, R. D. 8
Fietz, A. 86
Figlewicz, D. P. 125
Filby, Y. 197
Fillmore, M. T. 172, 173
Finney, J. W. 486, 501
Fiorillo, C. D. 9
Fischer, J. L. 492
Fischer, M. 226
Fisher, E. B. 343
Fisher, W. W. 327, 469
Fishman, S. 469
Fixsen, D. L. 385
Fletcher, B. W. 486
Floresco, S. B. 369
Flyn, L. 488
Fogassi, L. 403, 405
Follman, M. 223
Fordham, R. 305
Foreman, A. M. 214
Fosco, W. D. 127
Foster, M. 488
Foster, T. A. 383
Foster, T. M. 137, 322, 344, 470
Fowler, J. S. 86
Fowler, S. A. 475
Foxx, R. M. 193, 494, 495
Francis, G. 12
Frank, A. J. 458
Frank, M. J. 330
Frederiksen, L. W. 223
Freed, D. E. 335

Freed, G. L. 212
Freedman, D. H. 504
Freeman, J. H. 89
Freund, N. 9
Fridlund, A. J. 21
Friedman, A. 302
Friedman, S. G. 305, 359–360, 470
Friman, P. C. 469, 501
Frisch, K. von 515
Friso, S. 11
Fritz, J. N. 138, 204, 471
Frost, L. 497
Fryer, R. G., Jr. 117
Fuhrman, D. 397
Fuqua, R. W. 194

Galef, B. G., Jr. 537
Galizio, M. 166, 204, 214, 396, 420–421
Gallese, V. 403
Gallistel, C. R. 267
Gallup, A. C. 74
Galuska, C. M. 334
Gamba, J. 436, 441, 443
Gamble, E. H. 170
Ganchrow, J. R. 81
Gannon, P. J. 429
Ganz, J. B. 498
Gapp, K. 12
Garcia, J. 249–50, 251
Garcia, Y. 14
Garcia-Fernandez, G. 483, 484
Garcia-Penagos, A. 20
Garcia-Rodriguez, O. 484
Gardner, E. T. 212
Gardner, R. M. 288
Garris, P. A. 9, 353
Gast, D. L. 493
Gaston, S. 487
Gaynor, S. T. 169
Gaznick, N. 252
Gazzola, V. 404
Gear, A. 118
Geen, R. G. 226
Gehm, T. 219
Geiger, B. 226
Geller, E. S. 14, 469
Geller, I. 56
Gellman, M. 501
Gendall, K. A. 264
Gerard, H. B. 394
Gershoff, E. T. 189
Gewirtz, J. L. 156, 275, 276, 277, 409
Ghazanfar, A. A. 429

Ghezzi, P. M. 156, 494, 495
Gillespie-Lynch, K. 529
Gilligan, G. 293
Ginsburg, S. 519, 520
Gintis, H. 535
Giordano, J. 26
Giorno, K. G. 136
Girardet, C. 306
Gitter, S. A. 227
Giurfa, M. 86
Glaser, P. E. A. 172, 173
Glasscock, S. T. 501
Glenn, S. S. 14, 509, 522, 524, 532, 533, 534, 535
Glenwick, D. 476
Glucksberg, S. 63
Godfrey, A. 306
Goetz, E. M. 136
Goldiamond, I. 65
Goldie, W. D. 21
Goldin-Meadow, S. 529
Goldsmith, P. A. 257
Goldsmith, T. H. 278
Goldstein, M. H. 530
Golinkoff, R. M. 461
Gollub, L. R. 152, 361, 365, 366, 368
Gonzales, C. 226
Goodwyn, F. D. 498
Gott, C. T. 180
Gould, S. J. 512
Goyos, C. 436, 441, 443
Grace, R. C. 169, 498
Grant, D. L. 241, 244
Grant, D. S. 297
Grant, J. D. 341
Grant, V. L. 253, 254, 269
Gravina, N. 469
Gredeback, G. 403
Green, D. M. 65
Green, E. J. 301
Green, G. 498
Green, L. 332, 335, 343, 344
Green-Paden, L. D. 385
Greene, W. A. 247
Greenfield, P. M. 529
Greenspoon, J. 534
Greenway, D. E. 421
Greenwell, L. 486
Greer, R. D. 410, 412, 443, 449
Gregorini, P. 470
Gregory, M. K. 41
Griffiths, R. R. 161
Griggio, M. 278
Grimer, J. A. 169

Grindle, A. C. 14
Grissom, N. 78
Groccia-Ellison, M. 89
Groopman, J. 52
Gross, A. C. 469
Gross, L. 403
Grossman, J. 495
Grossman, L. 14
Grow, L. L. 438
Gruber, T. 538–539
Guerin, B. 433, 434
Guerra, L. G. 526
Guess, D. 427, 439, 440
Guggisberg, A. G. 74
Guillette, L. M. 287
Guinther, P. M. 456
Gully, K. J. 224
Gunturkun, O. 67
Gustafson, R. 224
Guttman, A. 283
Guttman, N. 91, 285, 286, 287, 289

Haag, R. 136
Haas, J. R. 421
Hackenberg, T. D. 195, 210, 212–213, 379, 382–383, 419, 580
Hackett, S. 469
Hadamitzky, M. 96
Haddox, P. 292
Haggard, P. 404
Haggbloom, S. J. 17
Hake, D. F. 138, 197–198, 221
Hale, S. 322, 323
Hall, G. 89, 437
Hall, K. 440
Hall, R. V. 477, 478, 479, 480
Hall, W. G. 81
Hallam, C. 486
Halsall, B. 85
Hamernik, H. B. 73
Hammack, S. E. 221
Hammer, M. 86
Hammond, J. L. 138, 471
Hand, D. J. 469
Hanley, G. P. 201, 330
Hanna, E. 402
Hansen, S. D. 475
Hanson, H. M. 286, 287, 289
Hanson, R. H. 258
Hardin, G. 535, 536
Harlow, H. F. 527
Harnard, S. 16
Harper, D. N. 156

Harris, B. 22
Harris, J. A. 259
Harris, J. L. 45
Harris, M. 532, 537, 538
Harris, S. 166
Harsh, J. 371
Hart, B. 393, 530
Harte, C. 462
Hasazi, J. E. 53, 54, 57
Hasazi, S. E. 53, 54, 57
Haslam, S. A. 419
Hasson, U. 432
Hastad, O. 278
Hatch, J. P. 170
Hatfield, D. B. 14
Hausknecht, K. A. 127
Hausman, N. L. 472
Hausmann, F. 278
Havermans, R. C. 143, 253
Haw, J. 163
Hawk, L. W. 127
Hayes, S. C. 14, 25, 26, 215, 296, 417, 421, 432, 441, 450, 457, 474, 509
Healy, O. 488
Hearst, E. 157, 243
Heath, A. C. 341
Heerey, E. A. 379
Heflin, L. J. 469, 477
Hegge, F. W. 280
Heidenreich, B. A. 9, 353
Heil, S. H. 339, 469, 470, 485
Heinrich, B. 303
Heisenberg, M. 216
Hellhammer, D. 96
Hemmes, N. S. 331
Hendrickson, K. 339, 341
Hendrik, G. R. 483, 485
Herman, R. L. 198
Hermann, P. M. 514
Hernandez, E. 330
Heron, T. E. 194
Herrnstein, R. J. 157, 212, 301, 302, 314, 316, 317, 318, 319, 321, 323, 324, 325, 326, 328, 356, 365, 562
Herruzo, J. 393
Hersen, M. 500
Hess, C. W. 74
Hesse, B. E. 437, 439, 440
Hesse-Biber, S. 289
Heth, C. D. 253, 254, 501, 503, 505
Heth, D. C. 253
Heward, W. L. 194, 473, 474
Hewitt, G. P. 430

Heyes, C. 404, 405, 406
Heyman, G. M. 312, 334, 340, 485
Hickman, C. 156
Hickok, G. 404, 405, 496
Higa, J. 284
Higgins, S. T. 176, 177, 339, 469, 483, 484, 485
Hilgard, E. R. 85
Hillman, C. 9
Hilton, B. W. 469
Hinde, R. A. 235, 527, 538
Hindz, V. B. 227
Hineline, P. N. 192, 212–213, 488
Hinson, R. E. 93, 94, 95
Hiroto, D. S. 216
Hirsch, J. 518, 519
Hirsh-Pasek, K. 461
Hitch, G. J. 292
Hockett, C. F. 427
Hodos, W. 171
Hoehn-Saric, R. 456
Hofmann, W. 93
Hogg, C. 278
Hoi, H. 278
Holahan, R. 535
Holding, E. 492
Holland, J. G. 439, 440, 441, 442, 443, 476, 481
Holland, P. C. 242
Holland, S. 169, 498
Holliday, M. 518
Hollis, J. H. 258
Holz, W. C. 197–198, 222, 223, 227–228
Honey, C. J. 432
Hong, N. S. 85
Hopper, L. M. 538
Horne, P. J. 409, 410, 421, 441, 448, 504, 563
Horner, R. H. 46, 201
Hothersall, D. 76
Hough, M. J. 218
Houston, A. 321, 338
Hovell, M. F. 385
Howard, J. S. 446, 447, 448, 492
Howe, M. L. 402
Hrycaiko, D. 488
Huber, H. 497
Huffstetter, M. 489
Hughes, J. R. 176
Hughes, S. C. 253, 462, 474
Hull, D. L. 509
Hume, K. 469
Hunt, D. M. 278
Hunt, G. M. 255, 483
Hunziker, M. H. L. 219
Hursh, D. E. 55

Hursh, S. R. 160, 267, 334
Huskinson, S. 173
Hussey, I. 456
Hustyi, K. M. 443
Hutchinson, N. R. 222
Hutchinson, R. R. 138, 221, 222, 255
Hutsell, B. A. 353

Iacoboni, M. 403, 404
Iadarola, S. 497
Ikegaya, Y. 123
Ingvarsson, E. T. 472
Iovannone, R. 497
Ireland, C. 238
Iriki, A. 404
Irons, G. 21
Isaacs, C. D. 14
Ishikawa, D. 123
Israel, M. L. 201
Ivy, J. W. 498
Iwata, B. A. 41, 47, 48, 49, 138, 256, 471, 477

Jabaij, L. 96
Jablonka, E. 12, 513, 519, 520
Jackson, R. L. 216
Jacobs, E. A. 321, 382
Jaffe, Y. 225
James, W. 396
Jamis, M. 200
Jannetto, P. J. 160
Jansen, A. T. M. 143, 253
Jansen, R. F. 514
Janssen, M. A. 535
Jarmolowicz, D. P. 340
Jason, L. A. 476
Jenkins, H. M. 238, 239–240, 242, 243, 264, 265, 266, 268
Jensen, G. 136
Johansen, E. B. 171
Johansen, J. P. 198, 199
Johnson, H. 404
Johnson, K. R. 492
Johnson, M. W. 339, 342
Johnson, P. M. 25
Johnson, P. S. 338
Johnson, R. 292
Johnson, R. N. 515
Johnston, J. M. 49, 481
Joker, V. R. 419
Jones, J. H. 225
Jones, K. M. 469
Jones, S. S. 402
Jozsvai, E. 256

Ju, W. 421
Juujaevari, P. 226

Kaartinen, J. 226
Kagel, J. H. 332
Kahng, S. 471, 472
Kalat, J. W. 21
Kalish, H. I. 285, 286, 287, 289
Kalsher, M. J. 477
Kamin, L. J. 99–100
Kana, R. K. 495, 496
Kandel, E. R. 8
Kangas, B. D. 538
Kaplan, B. A. 332
Karmely, J. 469
Karsina, A. 330, 331
Kastak, C. R. 458
Kastak, D. 295, 458
Katayama, T. 253, 255
Katz, J. L. 483
Kaufman, A. 419
Kavanagh, D. J. 218
Kawa, S. 26
Kawai, M. 397
Kawamura, S. 538
Kawecki, T. J. 9, 518
Kaye, W. H. 264
Kazdin, A. E. 379, 385, 470, 481, 500
Keehn, J. D. 256, 500
Kehle, T. J. 492
Keith-Lucas, T. 91
Kelber, A. 278
Kelleher, R. T. 302, 366–367, 382, 383
Keller, F. S. 488, 489–490, 508, 566
Keller, L. 521
Kelley, H. H. 433
Kelly, S. Q. 327
Kelso, S. E. 495
Kendig, M. D. 259
Kenny, P. J. 25
Keysers, C. 405, 496
Khamassi, M. 266, 267
Khatchadourian, R. 154
Killeen, P. R. 171, 172, 259, 321, 345, 346, 371
Killen, M. 227
Kincaid, D. 497
Kinchla, R. A. 63
King, A. P. 530
King, J. R. 489
King, L. E. 188
Kinloch, J. M. 137
Kirby, F. D. 379
Kirk, M. D. 514

Kirsch, I. 96
Kirschbaum, C. 96
Kirschbaum, E. H. 251
Klebez, J. 170
Klin, A. 495
Klock, J. 379
Knott, B. 278
Knutson, J. 290
Ko, M. C. 160
Kobayashi, S. 122, 123
Kodak, T. 469
Koelling, R. A. 249–250
Koestner, R. 116
Koffamus, M. N. 341
Koh, M. T. 253, 254
Kohler, W. 282
Kohn, J. P. 169
Kollins, S. H. 321, 345
Konarski, E. A., Jr. 394
Konkel, L. 360
Kooistra, L. 226
Kopp, R. E. 419
Korzilius, H. 495
Krampe, R. T. 31, 32
Krank, M. D. 94, 95
Krantz, P. J. 294
Kranzler, H. R. 484
Kratochwill, T. R. 60
Krebs, J. R. 327, 517
Kristof, N. 210
Kronauer, D. J. C. 521
Kruger, B. 405
Krupinski, E. A. 306
Kubina, R. M., Jr. 492
Kuenen, D. J. 71
Kuhl, P. K. 402
Kulak, J. T., Jr. 56
Kulik, C. C. 488
Kulik, J. A. 488
Kulubekova, S. 523
Kunkel, J. H. 396
Kurti, A. N. 485
Kuznesof, A. W. 259, 334
Kwapis, J. L. 199, 200
Kymmissis, E. 409
Kyparissos, N. 409
Kysers, C. 404

La Rochelle, C. D. 219
La Via, M. C. 264
Lac, S. T. 333
Laibson, D. I. 344
Laitman, J. T. 429

Laland, K. N. 538
Lamal, P. A. 13, 171, 476, 486, 533, 534
Lamarre, J. 439, 440, 441, 442, 443
Lamb, M. J. 513
Lambert, J. 469, 488
Langer, E. 65
Langman, R. E. 509
Langone, J. 500
Langthorne, P. 47, 496
Lanson, R. N. 158
Laraway, S. 45
Lashley, R. L. 259
Lasiter, P. S. 257
Latham, G. 476
Lattal, D. 66
Lattal, K. A. 139, 169, 393
Lattal, K. M. 91, 139
Lau, C. E. 255
LaVigna, G. W. 202
Lawrence, C. E. 371
Layer, S. A. 330
Layng, T. V. J. 489, 492
Lazar, R. 457, 460
Leal, M. 517
Leaton, R. N. 78
Lebbon, A. 469
LeBlanc, J. M. 379
LeBlanc, L. A. 434, 444, 497
Lechago, S. A. 438
LeDoux, J. E. 198
Lee, A. 535
Lee, K. 395
Lee, R. 135
Lee, V. L. 427, 439, 440
Lehman, P. K. 14
Lemley, S. M. 340
LeMoyne, T. 219
Leotti, L. A. 330
Lerman, D. C. 48, 49, 138, 202, 204, 256
Lesaint, F. 266, 267
Leslie, J. C. 56
Lett, B. T. 253, 254, 269
Levenson, R. M. 306
Levine, E. 89
Levinson, S. 21
Levitin, D. 68
Levy, I. M. 360
Lewis, M. 344
Lewis, P. 212
Lezak, K. R. 221
Li, B. 220
Libbrecht, R. 521
Lieberman, D. A. 374, 375, 376

Lieberman, P. 427, 429, 430
Lierle, D. M. 219
Lignugaris/Kraft, B. 475
Lima, E. L. 394
Lin, J. Y. 252
Lind, O. 278
Lindberg, J. S. 471
Lindsley, O. R. 488, 490, 492
Linehan, C. 450
Lionello-DeNolf, K. M. 458
Lippman, L. G. 419
Lloyd, D. R. 127
Lloyd, K. E. 490
Lloyd, M. E. 490
Locey, M. L. 343
Lochner, D. G. 327
Locke, E. 416
Lofdahl, K. L. 518
Loftus, E. F. 203
Logan, C. A. 376
Logue, A. W. 251
LoLordo, V. M. 219
Lopez-Perez, R. 395
Lorenzetti, F. D. 234, 524
Loukas, E. 457
Lovaas, O. I. 156, 201, 393, 471, 488, 494, 495, 497, 498, 553
Love, J. R. 438
Loveland, D. H. 301, 302, 314, 323
Lovett, V. E. 409
Lowe, C. F. 165–166, 167, 421, 441, 448, 504, 564
Lowenkron, B. 423, 438, 453, 456
Lubinski, D. 394
Lubow, R. E. 88, 89, 302
Lucas, G. A. 91
Luce, S. C. 498
Luciano, M. C. 393
Luiselli, J. K. 293
Lukas, K. E. 359
Lund, E. M. 190, 498
Lussier, J. P. 485
Lutz, A. 2
Lyn, H. 529
Lynch, J. J. 250
Lynn, S. I. 65

MacAleese, K. R. 156
MacCorquodale, K. 434
Mace, F. C. 321, 470
Machado, A. 136, 292
Machado, M. A. 443
MacKillop, J. 339
MacLarnon, A. M. 430

MacMillan, H. L. 203
Macphail, E. M. 208
Madden, G. J. 321, 338, 339, 341, 460, 485, 486
Maddox, S. A. 199
Madenci, A. 65
Mahoney, A. M. 305, 450
Maier, F. F. 500
Maier, S. F. 216, 219, 221
Malone, J. C. 20
Malott, M. E. 535
Malott, R. W. 422, 469
Maple, T. L. 359, 470
Markowitz, H. 144, 470
Marquez, N. 287
Marr, M. J. 359
Marsh, G. 292
Marshall, N. J. 278
Marshall-Pescini, S. 278, 397, 538
Martens, B. K. 327
Martin, B. 484
Martin, G. L. 93, 470, 474, 488
Martin, P. L. 385
Martin, S. 86, 305, 359–360
Martinez, C. 421
Martinez, E. 189
Martinez-Harms, J. 287
Marzo, V. D. 354
Masaki, T. 255
Mathis, J. 74
Matson, J. L. 386, 495
Matsuda, K. 14
Matsuki, N. 123
Matsumoto, N. 123
Mattaini, M. A. 476
Matthews, B. A. 393, 417
Matthews, L. R. 321, 322
Maugard, A. 296
Maurice, C. 488, 495, 498
Mauro, B. C. 321
May, J. G. 399
Maynard, C. S. 292
Mazur, J. E. 162, 327, 336, 337, 338, 352, 558
McAndrew, F. T. 226
McBride, S. A. 320
McCarthy, D. 345
McCarty, K. F. 469
McCauley, L. 518, 519
McClannahan, L. C. 294
McClung, C. A. 8
McCollum, S. 488
McConaha, C. W. 264
McCulloch, M. 306
McCully, J. 94, 95

McDevitt, M. A. 389
McDonald, J. S. 314
McDonald, R. J. 85
McDougall, W. 396
McDowell, J. J. 317, 320, 321, 322, 323, 324, 325, 326, 327, 345, 509, 523–524
McEachin, J. J. 494
McElwee, J. 460
McEwan, J. S. 488
McGaha-Mays, N. 469
McGill, P. 47, 496
McGuigan, N. 538
McGuire, M. S. 476
McIlvane, W. J. 170, 457
McIntire, K. D. 457
McKeon, T. R. 360
McKewan, J. S. 137
McLean, A. P. 156, 169
McNamara, J. M. 338
McSweeney, F. K. 81, 126–127, 283, 284, 321
Mechling, L. C. 500
Mechner, F. 31, 33, 136
Medina, D. J. 127
Mehus, C. J. 190
Meltzoff, A. N. 400–402, 404, 405, 406, 412, 426
Melville, C. L. 284
Mendelson, J. 255, 256
Mendoza, E. 114, 524
Mendres, A. E. 41
Menzel, R. 86, 287
Mercado, E., III 287
Merola, I. 278
Mery, F. 9, 518
Mesibov, G. B. 495
Mesoudi, A. 433
Meuret, A. 245
Meyer, D. R. 208
Meyer, J. 531, 532
Meyer, L. H. 202
Meyer, M. E. 419
Michael, J. L. 44, 45, 284, 365, 422, 434, 437, 439, 440, 442, 444, 448, 462, 470
Michel, H. V. 511
Miguel, C. F. 436, 439, 440, 444, 450, 458
Milan, M. A. 284
Milgram, S. 418–419
Militello, J. 74
Millard, M. 245
Millenson, J. R. 84
Miller, H. L., Jr. 136, 322, 421
Miller, J. R. 204, 215
Miller, K. B. 190, 191, 192

Miller, N. E. 224, 226, 245–246, 247, 269, 369, 370, 407, 469
Miller, R. R. 92, 93
Miller, S. A. 2, 410
Miltenberger, R. G. 194, 469
Ming, S. 460
Mingote, C. S. 369
Minshawl, N. F. 201
Minster, S. T. 460
Mirrione, M. M. 219
Mitchell, D. 251
Mitkus, M. 278
Modaresi, H. A. 208
Moerk, E. L. 461
Moeschl, T. 256
Molm, L. D. 537
Mond, J. M. 288
Mongeon, J. A. 485
Moody, L. 341
Mooney, R. 129
Mooney, S. E. 167
Moore, A. G. 492
Moore, A. U. 88
Moore, B. R. 266
Moore, J. 30
Moore, M. K. 400–402, 404, 406, 412, 426
Moore, T. 123
Moran, D. J. 469
Morasco, B. J. 483
Morely, A. J. 443
Morford, M. 529
Morgan, C. L. 396
Morgan, D. 56
Morgan, L. 136
Morris, E. K. 10
Morrone, C. 369
Morse, W. H. 179, 180
Moseley, J. B. 96
Moser, E. 306
Mota, N. P. 203
Moxley, J. H. 31, 32
Mueller, M. M. 292
Muhammad, Z. 39
Mukherjee, S. 10, 513
Muller, D. 123
Muller, M. N. 539
Muller, P. G. 165, 256
Mulligan, R. C. 341
Murdaca, A. M. 488
Murphy, E. S. 126, 283
Murphy, M. S. 289
Murray, E. A. 368
Muthukumaraswamy, S. D. 460

Myers, T. W. 266
Myerson, J. 322, 323

Nadal, L. 144
Nader, M. A. 333
Nagy, E. 402
Nakajima, S. 253, 255
Naqvi, N. H. 252
Nartey, R. K. 456
Navarick, D. J. 267
Navarro, V. M. 306
Neal, D. T. 153
Neef, N. A. 321
Neel, J. V. 501
Neely, L. C. 498
Negus, S. S. 353
Nergardh, R. 264
Nestler, E. J. 8
Neuringer, A. J. 129, 134, 135, 136, 159
Nevin, J. A. 65, 139, 140, 169, 498
Newberry, D. E. 421
Newhouse, L. 440
Newland, M. C. 45, 313, 321
Newsom, C. 156
Nichol, K. 374
Nickel, M. 125
Nida, S. A. 1
Nieuwenhuis, S. 240
Ninness, C. 538
Nisbett, R. E. 226
Nitabach, M. N. 514
Nolan, J. V. 250
Norman, W. D. 283, 321
Normand, M. P. 443
Norton, W. 533
Notterman, J. M. 137
Nowicki, S. 129
Nuzzolo-Gomez, R. 443
Nygaard, S. L. 333
Nyhan, B. 212

O'Brien, R. M. 363–364
O'Doherty, J. 82
O'Heare, J. 202
O'Hora, D. 421
O'Kelly, L. E. 221
O'Leary, M. R. 225
O'Regan, L. M. 456
O'Reilly, J. 136
O'Reilly, M. F. 417
O'Rourke, T. J. 431
O'Shields, E. 492
Oah, S. 434

Oberman, L. M. 404, 496
Obozova, T. 296, 458
Odeen, A. 278
Odum, A. L. 136, 338, 339, 341
Okouchi, H. 394
Okuyama, S. 219
Oldroyd, B. P. 515
Oliva, P. 488
Olsson, P. 278
Ondet, V. 306
Oostenbroek, J. 406
Orne, M. T. 418
Ortu, D. 534
Osgood, C. E. 530
Ostroff, L. E. 198
Ostrom, E. 535
Overmier, J. B. 216, 219
Overskeid, G. 258
Owczarczyk, J. 505
Owens, I. P. F. 278
Oxley, P. R. 521

Pace, G. M. 477
Page, S. 134
Palagi, E. 74
Palkovic, C. M. 292
Pan, D. 223
Paniagua, F. A. 393
Papachristos, E. B. 267
Papini, M. R. 90
Parece, T. E. 14
Park, R. D. 189
Parkinson, J. A. 368
Parnes, M. 409
Partington, J. W. 440, 444
Patrick, M. E. 190
Patterson, A. E. 259
Patterson, G. R. 225, 380
Paul, E. 221
Paul, G. L. 386
Pauley, P. J. 20
Pavlov, I. P. 17–19, 79–80, 83, 84, 85, 88, 575
Pear, J. J. 93, 469, 470, 474, 488, 490
Peden, B. F. 321
Pedersen, W. C. 224, 226
Pelaez, M. 156, 275–276, 277, 416, 421
Péllon, R. 259
Pennypacker, H. S. 49, 481
Pepperberg, I. M. 302, 446
Perdue, B. M. 330, 470
Pereira, S. 79
Perez-Gonzalez, L. A. 440
Pérez-Padilla, A. 259

Perlow, S. 343
Perone, M. 55, 214, 395
Perry, G. 419
Perry, R. L. 251
Perugini, M. 93
Peters, L. C. 206
Peters, S. 129
Peters-Scheffer, N. 495
Peterson, C. 218
Peterson, G. 537
Peterson, G. B. 359
Peterson, G. L. 223
Peterson, N. J. 196, 447
Peterson, R. F. 407, 410, 495
Petker, T. 339
Petry, N. M. 176, 339, 483, 484, 486
Petscher, E. S. 472
Pettit, B. 487
Pettitt, L. 440
Petursdottir, A. I. 41, 436, 441, 442, 443
Phelps, B. J. 154
Phillips, E. A. 385
Phillips, E. L. 385
Pieneman, A. W. 514
Pierce, W. D. 9, 116, 118, 229, 252, 253, 254, 260, 261, 262, 263, 264, 284, 289, 321, 322, 334, 335, 384, 471, 472, 500, 501, 503, 505, 509, 527, 530, 537
Pierrel, R. 280
Pietras, C. J. 195
Pilastro, A. 278
Pinel, J. P. 91
Pirson, M. 65
Pisano, R. 225
Pistoljevic, N. 449
Pjetri, E. 264
Platt, J. 535
Plaud, J. J. 421
Pliskoff, S. S. 316, 319
Ploog, B. O. 266
Pocock, T. L. 488
Podus, D. 486
Poehlmann, R. J. 250
Poeppei, D. 432
Pokorny, J. 495
Poling, A. 45, 125, 305, 321, 344, 470
Pollard, K. 529
Pomerleau, O. F. 501
Popa, A. 523
Porter, J. H. 256, 257
Posadas-Andrews, A. 256
Posadas-Sanchez, D. 171
Potter, J. N. 497

Poulos, C. X. 93
Poulson, C. L. 294, 409, 445
Powell, B. J. 517
Powell, R. A. 22, 384
Powell, R. W. 181
Power, J. M. 91
Powers, R. 292
Prather, J. F. 129
Prato-Previde, E. 278
Premack, D. 118–119, 573
Prendergast, M. 486
Prior, K. W. 360
Pritchard, T. 488
Proctor, S. D. 501, 503, 505
Progar, P. R. 294
Provenza, F. D. 250, 470
Provine, R. R. 73, 74
Pryor, K. W. 136, 305, 359
Puglisi-Allegra, S. 369
Puig, M. V. 9
Pulkkinen, L. 226

Quinn, J. M. 153

Rachlin, H. C. 241, 319, 332, 338, 342, 343, 344
Raineri, A. 338
Rakoczy, H. 402
Ramachandran, V. S. 496
Rankin, C. H. 78
Rapp, J. T. 156
Rasmussen, E. B. 9, 25, 45, 124, 173, 313, 321, 322, 339, 341, 345, 354
Rauzin, R. 457
Ravignani, A. 397
Rawson, R. 486
Ray, E. 404, 405, 406
Raybuck, J. D. 91
Rayner, R. 21, 22
Raz, G. 12
Reagan, I. J. 469
Reed, D. D. 332
Reed, H. K. 201
Reed, K. 219
Reed, P. 167
Reekie, Y. 368
Reeve, K. F. 294
Rehfeldt, R. A. 458
Reicher, S. D. 419
Reichle, J. 440
Reidenberg, J. S. 429
Reifler, J. 212
Reilly, M. P. 169
Reilly, S. 251, 252

Reinberg, D. 97
Rendall, D. 429
Rendell, L. 540
Repacholi, B. M. 405
Rescorla, R. A. 81, 90, 93, 102
Ressler, K. 12
Revillo, D. A. 89
Revusky, S. H. 251
Rey, C. 472
Reyes, F. D. 524
Reynolds, A. R. 172
Reynolds, G. S. A. 283, 324
Reynolds, V. 539
Rhodes, G. 82
Ribes, E. M. 421
Ricciardi, J. N. 293
Riccio, D. C. 200
Rice, E. E. 446, 447, 448
Richards, J. B. 127
Richards, R. W. 457
Richardson, D. R. 224
Richardson, J. V. 170, 210
Richardson, R. 200
Richey, S. 212
Richman, G. S. 47
Ringdahl, J. 256
Rispoli, M. J. 46
Ritz, T. 245
Rizley, R. C. 93
Rizzolatti, G. 403, 405
Roane, H. S. 172
Roberts, A. C. 368, 369
Robertson, S. H. 25
Robinson, P. J. 220, 221
Robles, D. S. 289
Roche, B. 441, 450, 462
Roche, R. A. P. 456
Rodriguez, L. R. 338, 341
Rodriguez, M. E. 421
Rodriguez, N. M. 330, 331, 410
Rodzon, K. 338
Rogers, R. W. 226
Rogers, S. J. 495, 497
Rohsenow, D. J. 485
Roll, J. M. 176–177, 483, 486
Roman, C. 252
Romanowich, P. 389
Roper, T. J. 256
Rosales-Ruiz, J. 192, 212
Rose, J. 9
Rosenberg, E. L. 431
Rosenfield, D. 245

Ross, D. 412, 413
Ross, S. A. 412, 413
Rossellini, R. 259
Rost, K. A. 331
Roth, T. L. 11, 12
Roth, W. J. 292
Rourke, A. J. 410
Roussel, E. 86
Routtenberg, A. 259, 334
Rovee-Collier, C. 156
Rowe, C. 287
Rowland, N. E. 335
Rozzi, S. 405
Rubene, D. 278
Rubin, E. 495
Ruddle, H. V. 321
Rudrauf, D. 252
Ruggles, T. R. 379
Rumbaugh, D. M. 330
Rung, J. M. 341
Rush, C. R. 172, 173
Ruskin, G. 489
Russell, C. L. 278
Russell, J. C. 253, 501, 503, 505
Rutherford, A. 16, 25
Rutland, A. 227
Ryan, R. M. 116

Sacket, S. 74
Saeed, S. 97
Safin, V. 342
Sagvolden, T. 329
Sakagami, M. 122
Sakaguchi, T. 123
Salamone, J. D. 369
Sallery, R. D. 222
Salmon, D. P. 457
Salvy, S. J. 202, 253
Samson, H. H. 257
Sanchez-Hervas, E. 484
Sanders, G. A. 428
Sanders, S. J. 495
Sandoz, J. C. 86
Santana, L. H. 509
Santos, L. R. 394
Sareen, J. 203
Sargisson, R. J. 298, 299, 301
Saunders, K. J. 457
Sautter, R. A. 434, 444
Savage-Rumbaugh, S. E. 437, 438, 439, 529
Sayette, M. A. 46
Schaal, D. W. 8
Schachtman, T. R. 251

Schaefer, H. H. 385
Schafe, G. E. 199
Schafer, R. J. 123
Schafer, S. 86
Schedlowski, M. 96
Scheller, D. K. 219
Scheller, R. H. 514, 515
Schiffrin, H. H. 219
Schlichting, C. D. 519
Schlinger, H. D. 161, 162, 422
Schlosberg, H. 90
Schlund, M. W. 456, 457
Schmidt, A. C. 330
Schmidt, M. 144
Schmidt, R. 9
Schmitt, D. R. 422
Schnerch, G. J. 469, 488
Schnider, A. 74
Schoenfeld, W. N. 368
Schopler, E. 495
Schrier, A. M. 302
Schultz, W. 9, 122
Schusterman, R. J. 295, 458
Schwartz, B. 133–134, 265–266, 332
Schwarz, N. 226
Schweitzer, J. B. 385
Secades-Villa, R. 484
Seeley, T. D. 515
Segal, E. F. 169
Seifter, J. 56
Seligman, M. E. P. 216, 218, 219, 250, 500
Semb, G. B. 489
Shabani, D. B. 41, 469
Shade, D. 321
Shahan, T. A. 159, 389
Shanker, S. G. 437, 439
Shanock, L. 118, 136
Shapir, N. 225
Shapiro, M. M. 234
Shaver, K. G. 282
Shaw, J. 86
Shearer, J. 486
Shearn, D. W. 246
Sherman, G. J. 280
Sherman, J. A. 293, 407, 408, 409, 410, 495
Sherman, J. G. 489
Sherman, L. 343
Sherrington, C. 75, 76
Sheth, S. 166
Shields, F. 379
Shimoff, E. H. 393, 417
Shizgal, P. 7
Shoaib, M. 366

Shors, T. J. 91
Shull, R. L. 169, 182, 316
Sidman, M. 17, 49, 57, 152, 169, 175, 189, 209–210, 218, 228, 229, 232, 451, 457, 460, 547
Siegel, S. 93, 94, 95
Sigafoos, J. 440
Sigaud, O. 266, 267
Sigmon, S. C. 469
Silberberg, A. 241, 375–376
Silbert, L. J. 432
Silva, K. M. 469, 488
Silva, M. T. 526
Silva, S. P. 393
Silverman, K. 470, 483
Simek, T. C. 363–364
Simic, J. 437
Simmelhag, V. L. 182, 255
Simmons, J. Q. 201, 471
Simola, D. F. 521, 522
Simon, S. J. 284
Simony, E. 432
Simpson, C. A. 339
Simpson, R. L. 498
Sipols, A. 125
Sitharthan, G. 218
Sitharthan, T. 218
Skalicka, V. 227
Skinner, B. F. 7, 13, 14–17, 24, 64–65, 100, 111, 120–121, 123, 131, 136, 141, 143, 149, 152, 159, 160–161, 163, 165, 168, 176, 178, 196, 214–215, 226, 259, 282, 305, 312, 359, 362, 379, 380, 391, 395, 406, 415, 416, 417, 419, 428, 430, 432, 433, 434, 436, 440, 443, 446, 448, 461, 462, 463, 464, 469, 488, 492, 509, 511, 517, 518, 526, 528, 530, 532, 544
Skloot, R. 225
Skoczen, T. 200
Slaughter, V. 406
Slifer, K. J. 47
Smartwood, L. 74
Smirnova, A. 296, 458
Smith, T. H. 494, 497, 498
Smith, J. R. 419
Smith, S. L. 9
Smith, T. R. 382
Smithson, C. 22
Snycerski, S. 45
Sobell, L. C. 340
Sobsey, D. 201
Soderstrom, M. 461
Solomon, P. R. 89
Solomon, R. L. 208, 209, 210, 212, 216
Songmi, K. 394

Author Index

Sossin, W. S. 514
Sparkman, C. R. 492
Spear, J. 423
Spear, N. E. 89
Specter, D. 19
Spencer, W. A. 78
Spengler, S. M. 440
Spetch, M. L. 91, 287, 302
Spiegelman, B. 504
Spiegelman, E. 395
Spooner, F. 492
Spradlin, J. E. 457
Springer, C. R. 488
Squier, L. 144
Squires, N. 376
Staddon, J. E. R. 182, 255, 256, 259, 270N2
Stafford, D. 175
Stafford, K. 305
Stechle, L. C. 221
Steele, A. 245
Stefan, L. 260, 334, 501
Stein, L. 122, 526
Steiner, J. E. 81
Steinmetz, A. B. 89
Steinmetz, J. E. 152
Steinwald, H. 142
Stenseng, F. 227
Stenstrom, D. M. 224
Stephen, I. D. 288
Stephens, D. W. 250, 341
Stevens, J. R. 341
Stevenson, R. J. 288
Stevenson-Hinde, J. 235, 527
Stewart, I. 421, 450, 460
Stierle, H. 96
Stiers, M. 241
Stikeleather, G. 489
Stokes, P. D. 136
Stokes, T. F. 474, 475
Stolfi, L. 449
Stoops, W. W. 172, 173
Storms, L. H. 196
Strada, M. J. 469
Strain, P. S. 201
Straus, M. A. 189
Striano, T. 402
Strickland, J. C. 342
Strimling, P. 539
Stringer, C. B. 538
Stroop, J. R. 62
Strosahl, K. D. 220, 221, 474
Stucke, T. S. 227
Sturmey, P. 135, 495

Stuttgen, M. C. 67
Sullivan, R. M. 81
Sulzer-Azaroff, B. 488
Sun, B. 514
Sundberg, C. A. 444
Sundberg, M. L. 437, 439, 440, 444, 448, 498
Sutor, L. T. 247
Svartdal, F. 421
Svenningsen, L. 469, 488
Sweatt, D. 97
Sweatt, J. D. 11, 12
Swedo, E. 189
Sweeney, M. M. 159
Swets, J. A. 65
Swisher, M. 458
Swithers, S. E. 505
Symonds, M. 89
Szabadi, E. 321

Taghert, P. H. 514
Tailby, W. 451, 457
Takahashi, N. 537
Takamori, K. 219
Takemoto, H. 430
Talwar, V. 395
Tam, V. 293
Tammen, S. A. 11, 12
Tarbox, J. 47
Tatham, T. A. 167
Tauson, R. 278
Taylor, I. 417
Taylor, S. P. 225
Taylor, T. J. 437, 439
Teasdale, J. D. 218
Temple, W. 321, 322, 470
Tenhula, W. N. 385
Tennie, C. 538
ter Maat, A. 514
Terrace, H. S. 286, 291, 293
Thomas, S. 5
Thompson, R. F. 78
Thompson, R. H. 206, 207, 330, 331
Thompson, T. 161, 258, 394, 457, 496
Thorndike, E. L. 23–24, 120, 121, 398, 426, 581
Thorpe, W. H. 396
Thrailkill, E. A. 171
Tice, D. M. 227
Tie, H. 486
Tierney, J. 487
Tiffany, S. T. 94
Tiger, J. H. 330
Tighe, T. J. 78
Timberlake, W. 241, 244

Tinbergen, N. 71, 72, 513
Tobler, P. N. 9
Todd, A. W. 201
Todd, J. T. 495
Todd, T. P. 141
Tomasello, M. 538
Tomie, A. 457
Tourinho, E. Z. 30, 535
Toussaint, K. A. 292
Towe, A. L. 302
Townsend, D. B. 294
Tranel, D. 252
Trapp, N. L. 160
Travers, B. G. 495
Travis, C. 419
Troutman, A. C. 469
Tsai, P. S. 514
Tu, S. 97
Tucker, B. 477, 478, 479, 480
Tucker, D. 170
Tuckman, G. 189
Tulving, E. 297
Turner, J. R. 501
Twenge, J. M. 227
Twyman, J. S. 442, 488, 489, 492
Tyndall, A. V. 85

Ulrich, R. E. 221, 222, 232
Urcuioli, P. J. 458

Vaidya, M. 383
van de Poll, N. E. 136
van der Kooy, D. 79
van der Wall, S. B. 250
van Haaren, F. 136
van Hest, A. 136
Van Houten, R. 201, 469
Vannieuwkerk, R. 81
Vargas, E. A. 430, 432
Vargas, J. S. 16
Vasquez, E. A. 224
Vaughan, M. E. 17, 176, 284
Vaughn, W., Jr. 457
Velazquez, E. 89
Ventura, R. 369
Vergoz, V. 86
Verhave, T. 304, 305
Vichi, C. 533
Vierck, C. J., Jr. 56
Villalba, J. J. 470
Villareal, J. 255
Virues-Ortega, J. 275, 276, 277
Vismara, L. A. 497

Vits, S. 96, 97
Vittimberga, G. 438
Vivanti, G. 495
Vogel, T. 47
Vogelstein, J. T. 8, 13
Voineagu, I. 495
Volkmar, F. 495
Volkow, N. D. 86, 486
Vollmer, T. R. 321, 322, 379
Vollrath, F. 182
Voltaire, M. 156
von Hofsten, C. 403
Vorndran, C. M. 202
Vorobyev, M. 287
Vuchinich, R. E. 339

Wacker, D. P. 170
Waddell, J. 91
Wadsworth, H. M. 495
Wagner, A. R. 81, 102
Wagner, S. M. 529
Wakefield, J. C. 386
Wald, B. 371
Wall, H. 278
Wallace, M. D. 138
Waller, M. B. 169
Walsh, J. J. 221
Walsh, V. 404
Wanchisen, B. A. 167
Wang, G.-J. 86
Wansink, B. 311
Ward, L. M. 404
Ward, R. D. 136
Ward, T. A. 538
Warmington, M. 292
Warren, Z. 495
Washburn, D. A. 330
Wasserman, E. A. 91, 241, 244, 290, 296, 306, 458
Watanabe, S. 302
Watkins, C. L. 490, 492
Watkins, L. R. 221
Watson, E. 374
Watson, J. B. 12–13, 19–23
Watts, C. S. 199
Wawrzyncyck, S. 76
Weatherly, J. N. 190, 283
Weber, S. M. 369
Weeden, M. 344
Weigel, J. W. 501
Weiner, H. 166
Weisberg, P. 156
Weiss, B. 180

Werner, S. J. 470
Wesemann, A. F. 208
West, M. J. 530
West, R. P. 488, 490, 491, 492
West, S. A. 517
Western, B. 487
Wheeler, D. S. 242
White, A. J. 351
White, C. T. 90
White, K. G. 297–298, 299, 300, 301, 344, 571
White, O. R. 492
Whiten, A. 397, 538
Whitley, P. 439
Wichstrom, L. 227
Wiesler, N. A. 258
Wilkes, G. 359
Wilkie, D. M. 91
Wilkins, E. E. 252
Wilkinson, D. A. 93
Wilkinson, K. M. 457
Williams, A. R. 77
Williams, B. A. 389
Williams, C. D. 144
Williams, D. R. 265–266
Williams, G. 440
Williams, H. 265, 266
Williams, J. L. 219
Williams, K. D. 1
Williams, M. D. 284
Williams, W. L. 494, 495
Wilson, A. G. 341
Wilson, D. S. 14, 509
Wilson, K. G. 215, 474
Wilson, L. 226
Wilson-Morris, M. 460
Wing, V. C. 366
Winger, G. 55, 334
Winstanley, C. A. 341
Wisniewski, M. G. 287
Witnauer, J. E. 93
Witte, K. 212
Wixted, J. 374
Woelz, T. A. R. 534
Wolf, M. M. 385, 470

Wolf, R. 216
Wolfe, B. M. 224
Wolfe, J. B. 381
Wolfe, P. 500
Wolff, P. C. 221
Wolpe, J. 215
Wood, M. A. 199, 200
Wood, W. 153
Woods, J. H. 55, 334
Woodside, B. 238
Woolverton, W. L. 333
Wrangham, R. 539
Wund, M. 519
Wyckoff, L. B., Jr. 370–371
Wynne, L. C. 209, 210, 212

Xue, B. G. 122

Yachison, S. 395
Yamaguchi, T. 321
Yamamoto, T. 252
Yang, Z. 216
Yen, M. 342
Yezierski, R. P. 56
Yi, J. I. 438
Yildiz, A. 67
Yinon, Y. 225
Yoon, S. 446
Yoshida, S. 219
Young, K. R. 488, 490, 491
Young, M. E. 290
Young, R. 256, 492

Zanni, G. 203
Zarcone, J. R. 201, 202, 256
Zelinski, E. L. 85
Zentall, T. R. 400, 402, 458, 460
Zettle, R. D. 474
Ziegler, H. P. 266
Zielinska, E. 504
Zimmerman, J. 197
Zimmerman, R. R. 527
Zorina, Z. 296, 458
Zuberbuhler, K. 539

SUBJECT INDEX

Note: Page numbers in *italics* indicate figures; page numbers in **bold** indicate glossary entries.

5-hydroxytryptamine (5-HT) 219

A-B-A-B design 52–55, *54*, 476–477, **543**
abolishing operation (AO) 45, 46, 47, **543**
absolute rate of response 323
absolute stimulus control 289, **543**
abstinence, reinforcement of 483–485, *484*
abuse, domestic violence 190–192
Acceptance and Commitment Therapy (ACT) 25–26, 215, 474, **543**
acquisition 112; Rescorla–Wagner equation and 102–103, *102*; respondent conditioning 82–83, *83*, 84, 87, 88–89; *see also* respondent conditioning
active intermodal mapping 402–405, *404*
activity anorexia: defined **543**; experimental analysis of 259–264, *261*, *262*; substitutability of food and physical activity and 334–335
ad libitum weight 131, **543**
adaptive hypothesis of obesity 502
addiction, contingency management of 143, 176, 483–488, **549**–**550**; *see also* drug use and abuse
adjunctive behavior: defined 270n2, **543**; experimental analysis of 255–259
adjusting-amount procedure 338
adjusting-delay procedure 337
adventitious reinforcement 282
aggression: Bobo doll experiment 412–414, *413*; cycle of 224–226; displaced 221–223, *222*, **553**; functional analysis of 48–49, *49*; operant **565**; as response to social exclusion 227; schedule-induced **574**; submissive behavior and 380
Ainslie-Rachlin principle 344
air crib 16
alternative behavior: differential reinforcement of (DRA) 138, 193–194, 471–472, **552**; matching 319–320
American Psychiatric Association 26
American Psychological Association 26
American Sign Language (ASL) 431, *431*
animals: clicker training for 359–360; puzzle box for 120–121, *121*; signal detection by 304–306, *304*, *306*; *see also* birds

anorexia nervosa, peak shift in 288–289, *288*
antecedent stimulus (S) **544**; types of 310n1
ants, epigenetic reprogramming of 521–522, *521*
applied behavior analysis (ABA): for autism 494–500; challenging behavior reduction 472; characteristics of 470–476; defined 4, 470, **544**; in education 488–492; neurodiversity movement 496–497; for obesity 500–505; research strategies in 476–482; for self-control 492–493, *493*; for substance use disorders 483–488
Association for Behavior Analysis International (ABAI) 24, 26
associative learning, evolution 519–520
associative strength 101, **544**
assumption of generality 165–167, **544**
attention deficit hyperactivity disorder (ADHD) 170, 172, 257
autism/children with autism: applied behavior analysis for 494–500; defined **544**; mirror-neuron system and 404
autoclitic relation (verbal behavior) 446–448, **544**
automaintenance, negative 264–265, **564**
autoshaping: defined 239–242, *240*, **544**; as operant-respondent interrelationships 264–267, *265*
aversive control of behavior 188–192
aversive stimuli: aggression as side effect of 221–227; behavioral persistence as side effect of 215–216; conditioned 187, **548**; crying babies 205–206, *207*; defined 187, **544**; frequency, and avoidance behavior 212; learned helplessness as side effect of 216–220; social disruption as side effect of 227–228; *see also* punishment
avoidance 203–204, 212–214; discriminated 208–209, **552**; and escape, difference between 206, *208*; experiential 215; as impending doom 212–213; nondiscriminated 209–212, **565**; *see also* negative reinforcement
avoidance behavior: aversive stimulus frequency and 212; extinction of 215–216
avoidance learning 206–209
avoidance paradox 209–210

backward chaining 363–364, *364*, **544**
backward conditioning *90*, 91–92, *92*, 253, **544**
bad news, effect of 40, 370–376
Baer, Donald 407, 410–412
Bandura, Albert 412–416, *413*
Barger, Albert (Little Albert) 21–22, *22*
baseline: defined 52, **544**; operant, for behavioral neuroscience 55–56; *see also* multiple baseline designs
baseline level 84
baseline sensitivity 56, **545**
Baum, William 346, *347*
behavior: adjunctive 255–259, 270n2, **543**; avoidance 212–214; context of 10, 44; contingency-shaped 417–422, **550**; defined 1, **545**; displacement 257, **553**; elicited 39, 75, 109, **553**; emitted 40, 109, **553**; functional analysis of 37–41; genetic regulation of 513–520; impulsive 338, 341–344; interim 255, 270n2; intraverbal *441*, 443–444, **560**; mediating 165; perceiving as 62–65; phylogenetic 72–78, 399; precurrent 417, **567–568**; private 30–31, **568**; recording 151–152, *151–152*, 481–482, *482*; reflexive 75, 205, 245–246; respondent 39–41; schedule-induced 182, 255; selection by consequences and 509, *510*, 523–527; sequences of 72–75; social, epigenetic and reprogramming of 521–522, *521*; stereotyped 258; submissive 380; superstitious 282, **578–579**; terminal 255, **579**; textual 446, **579–580**; theories of 2; thinking as 29–31, 33; *see also* challenging behavior; ontogenetic behavior; operant behavior; rule-governed behavior; self-injurious behavior; verbal behavior
behavior analysis: culture and 13–14; defined 3–4, **546**; neuroscience and 8–9; Pavlov and 17–19; as progressive science 150–151; Skinner and 24–27; Thorndike and 23–24; Watson and 19–23; *see also* applied behavior analysis; experimental analysis of behavior
Behavior Analysis Certification Board (BACB) 499–500
behavior analysts 24, **545**
behavior chains and social referencing 274–278, *275*, *277*
behavior-consequence learning (BCL) 112–114, *113*
behavior-feedback stream 235
behavior maintenance 475, **546**
Behavior of Organisms, The (Skinner) 13, 359
behavior sequences, stimulus control of 274
behavior system 244, **546**
behavior therapy 25
behavior trapping 475, 476, **546**
behavioral assessment 473

behavioral contract 473–474, *474*, **545**
behavioral contrast 283–284, **545**
behavioral economics: choice, addiction, and 332–344; choice, foraging, and 327–332; defined 332, **545**
behavioral engineering *see* applied behavior analysis
behavioral flexibility 517–520, *519*, *520*, **545**
behavioral medicine 500–501, **545**
behavioral momentum 169–170, 498
behavioral neuroscience: activity anorexia and neuropeptide Y 263–264; conditioned reinforcement 368–369; defined 9, **545**; derived stimulus relations and 456–457; helplessness 219–220; honeybees 86; matching, sensitivity, and 353–354; operant baselines for 55–56; operant conditioning of neuron 121–123; PR schedules 172–174; of social defeat 220–221; stimulus control 85, 278; taste aversion 252; *see also* neuroscience
behavioral research *see* research
behavioral rigidity 513–514, 515
behavioral variability 129, **545**
behaviorism 20, **546**
bias, matching law 345, 346–350, **546**
biofeedback 170–171, 244–245
biological context: of behavior 10; of conditioning 248–255; defined **546**
biology, selection by consequences and 509–513, *510*
birds: begging reflex 75; concept formation in 301–303; delayed imitation in 398–400; discrimination by 208, 278, 290–291; imprinting by 10; quality control by 304–305, *304*; shaping of song of 129–130; vision of 278
bivalent effect (wheel running) 253–256, *254*
Black Applied Behavior Analyst, Inc. 26
"blaming the victim" 282
blocked-response CEO 437
blocking 99–100, **546**
Bobo doll experiment 412–414, *413*
brain-derived neurotrophic factor (BDNF) 220–221
break-and-run pattern 153, 160, *160*, 161, *165*, 165, **546**
breakpoint 171, **546**
Breland demonstration 235–237, *236*, *237*

carpenter ants 521–522, *521*
cats, puzzle box for 120–121, *121*
celeration 490–492, *491*, **546**
chain schedule of reinforcement 360–363, *361*, *362*; concurrent 376–379, *377*
chains: backward 363–364, *364*; reaction 74–75, *74*

challenging behavior: applied behavior analysis 472–474; direct treatment of 472–474, *473*; as focus of applied behavior analysis 471; functional analysis of 47–49, *49*; see also self-injurious behavior (SIB)
change in associative strength 101, **546**
change in level (baseline to treatment) 60, *61*, **546**
changeover delay (COD) 316, **546–547**
changeover response 315–316, **547**
changing criterion design 480, 493, *493*, **547**
chess moves, thinking aloud about 31–33
child rearing: helicopter parenting 219; infant caregiving and escape learning 205–206, *207*; use of punishment in 189–190, *190*, 203
children: dropping out 228–229, *229*; education 228–229, *229*, 290, 292, 492; generalized social reinforcement 379–380; imitation 396, 400–402, *400*, *401*, 404–405, 407, 408, 409–410; infant caregiving 205–206, *207*; naming relation 448–450; obesity in 503–504; object permanence 37–38; physical punishment 203; verbal behavior 441–443, 444, 445–446, 447–448; see also autism/children with autism
choice: defined 311–312, **547**; experimental analysis of 311–316; foraging, behavioral economics, and 327–332; quantification of 344–353
"chunking," in formation of habits 112
clarity of definition 481
classical conditioning 21, 79–80, 90, 239, 365, 366; see also respondent conditioning
clicker training 305, 306, 359–360
coercion 228–229, **547**
Coercion and Its Fallout (Sidman) 209–210, 228–229
cognitive psychology 28
commitment response 343–344, **547**
common resource problem 535
communication: effects of, and costly use of punishment 535–537; matching and 321–322, *322*; see also verbal behavior
community-health programs and avoidance behavior 210–212
community reinforcement approach (CRA) 483–485, *484*
completeness of definition 481
complex conditioning 98–100
compound stimuli 98–100, **547**
concept formation in pigeons 301–303
conceptualization 33
concurrent-chain schedule 376–379, *377*, **547**
concurrent-interval schedule 314–315
concurrent-ratio schedule 313–314, *314*

concurrent schedules of reinforcement 312–313, *313*, **547**
concurrent superstition 316
concurrent variable-interval schedule 315
conditional discrimination 303, **547**
conditioned aversive stimulus 187, **547**
conditioned compensatory response 96
conditioned establishing operation (CEO) 437, **548**
conditioned immunosuppression 96–97
conditioned overeating 505, **548**
conditioned place aversion (CPA) 255
conditioned place preference (CPP) 81, 254–255, **548**
conditioned reflex 18–19; see also conditioned response; conditioned stimulus
conditioned reinforcement: chain schedules and 360–363, *361*, *362*; clicker training 359–360; defined 358–359, **548**; delay reduction and 376–379, 386–389, *389*; effectiveness of 364–365; experimental analysis of 366–368; generalized 379–384; information and 369–376; neuroscience and 368–369
conditioned reinforcer 128, 358, **548**
conditioned response (CR): defined 5, 80, **548**; seeing as 64
conditioned stimulus (CS) 80, **548**
conditioned-stimulus function 5, 42, **548**
conditioned suppression 100, 212–213, **548**
conditioned taste aversion (CTA) 80–81, 88–89, **548–549**
conditioned taste preference (CTP) 253, 254
conditioned withdrawal 95, **549**
conditioning 4; in applied behavior analysis 471–472; backward *90*, 91–92, *92*, 253, **544**; biological context of 248–255; complex 98–100; delayed 89, *90*, **551**; first-order 92, **555**; forward 253; Rescorla–Wagner model 100–104; respondent 4–5, 79–80, *79*, 93–94, 233–235, **572**; second-order 92–97, **575**; simultaneous *90*, 90, **576**; temporal relations 89–92, *90*; trace *90*, 90–91, **580**; see also fear conditioning; operant conditioning; respondent conditioning
confirmation bias 2, **549**
confounding variable 52
conjugate reinforcement 156
constant k assumption, of matching 326
construction of S^Ds 417, **549**
context for conditioning 248–255, **549**
context of behavior 10, 44, **549**
contextual stimuli 85, **549**
contiguity, role in conditioning 5, 81–82, 90, 247
contingencies of reinforcement *510*

contingencies of survival 399, *510*, 511–512, 518, 526–527, **549**
contingency (respondent) 8, 81–82, **549**
contingency management: defined **549**; of substance use disorders 176, 483–488
contingency of reinforcement 45, *46*; contingencies of survival compared to 526–527; defined 110–112, *111*, **549–550**; instructions and 419–422; interlocking 464–465, 532–53, 534, **559**; response patterns and 152; in single-subject research 57; types of 114–116, *115*; *see also* extinction
contingency-shaped behavior 417–422, **550**
contingency-specifying stimuli 416, **550**
contingent electric skin stimulation (CESS) 201–202
contingent response **550**
continuous reinforcement (CRF) *132*, 132, *133*, 155–159, *156*, *158*, **550**
control 3
conversion therapy 26
correlation, among events and behavior 4, **550**
correspondence relations: defined 393, **550**; human behavior and 393–396; *see also* imitation; rule-governed behavior
countercontrol 223–224, **550**
covariation, to establish causation 52
COVID-19 pandemic: child abuse 189; rule following 421–422, *422*; vaccination programs 211
CRA (community reinforcement approach) 483–485, *484*
craving, drugs 95, 96, 241, 252
cross-commodity discounting 341
crying of infants and caregiving behavior 205–206, *207*
CS-pre-exposure effect 88, **550**
CS–US disparity 101
cultural evolution 14, 538–540, *539*, **550**
cultural practice 532–534, **550**
culture: behavior analysis and 13–14; defined 14, **551**; origin, transmission, and evolution of 537–540; selection by consequences and 509, *510*, 532–534
cumulative effect, of macrocontingency for group 534
cumulative record 151–152, *152*, **550–551**
cumulative recorder 151, *151*, **551**

Darwin, Charles 510–511, *511*
delay discounting 335–336, **551**; drug abuse, gambling, and obesity 338–340; trait vs. state 340–341

delay reduction and conditioned reinforcement 376–379, 386–389, *389*
delay-reduction hypothesis 376, **551**
delayed conditioning 89, *90*, **551**
delayed imitation: behavior analysis of 405–407; defined 397, **551**; by infants 402; studies with pigeons 398–400
delayed matching to sample (DMTS) 297–299, *298*, **551**
delayed punishment 197
demand curve 332–333, *332*, **551**
dependent variable 51–52, *51*, **551**
depression and learned helplessness 218–220
deprivation operation 125, **551**
derived stimulus relations 456–457
differential imitation 415
differential reinforcement 4, 128; of alternative behavior (DRA) 471–472, **551**; defined 110, **551–552**; extinction and 137; of other behavior (DRO) 282, 471, **552**; response chains and 274; social referencing and 274–278, *275*, *277*; as three-term contingency 272–274; use of 53–54
differential response 43, 272, *273*, **552**
diffusion of responsibility 225
direct replication 57, **552**
discounted value 336
discounting curve 336, *337*
discounting rate 337
discriminated avoidance 208–209, **552**
discriminated extinction 138–139, **552**
discrimination 4; conditional 303, **547–548**; defined **552**; differential reinforcement and 272–278; errorless 291–292, *291*, **554**; in pigeons 302–303; respondent 88, **572**; simultaneous 290, **576–577**; successive 290, **578**
discrimination index (ID) 280–281, *280*, **552**
discriminative function 43, **552–553**
discriminative stimulus (S^D): conditioned reinforcement 362; defined 43, 109–110, 271, **552**; extinction and 138, 140, *141*; rules as 415, 418, 432
discriminative-stimulus account of conditioned reinforcement 368, **553**
dishabituation 126–127
displaced aggression 221–223, *222*, **553**
displacement behavior 257, **553**
diversity 26
domestic violence and aversive control 190–192
doom, impending, avoidance as 212–213
dopamine, role in neural reward system 56, 82, 86
dropping out 228–229, *229*
Drosophila, operant learning in 112–114, *113*

drug use and abuse: adjunctive behavior and 25, 257; applied behavior analysis for 483–488; behavioral economics, choice, and 333–334, 338–340; contingency management for abstinence 143, 176–178, *177*, 483–487, *484*; cravings, neural activity, and taste aversion 252; extinction and 142–143; fixed-ratio schedules and 160; heroin overdose 94–95, *95*; PR schedules 172–173; respondent conditioning and 93–94
duration recording 482, **553**

early intensive behavioral intervention (EIBI) 494, 495, 497, 498, **553**
eating, motivational interrelations between physical activity and 252–255
echoic relation *445*, 445–446, **553**
ecological costs of resource depletion 534–537
education, applied behavior analysis in 488–492
elasticity of commodities 332–333
elephants, aversive buzz of bees 188–189
elicited (behavior) 39, 75, 109, **553**
embedded operant contingencies 244–248
embedded respondent contingencies 235–244
emitted (behavior) 40, 109, **553**
emotional response 138, **553**
Enjoy Old Age (Skinner & Vaughan) 17
environment: defined 41, **553**; functional analysis of 41–49; interaction between phenotype and, in obesity 501–504, *502*; ontogenetic behavior 78–79; phenotype, genotype, and 512, *512*; social focus on 476
epigenetic marking, DNA methylation and histone acetylation 10–11
epigenetic "memory" and trained immunity 97–98
epigenetics: in reprogramming of social behavior 521–522, *521*; retention of early learning and 10–12, *11*; in retention of fear conditioning 198–200
Epstein, Robert 398–400
equivalence relations *see* stimulus equivalence
errorless discrimination 291–292, *291*, **553–554**
escape 203; and avoidance, difference between 206, *208*; *see also* negative reinforcement
escape learning 204–206, *205*
established-response method 366, **554**
establishing operation (EO) 44–45, 46, 47, **554**
event recording 481
evolution: behavioral dynamics and 523–524; behavioral flexibility and 517–520, *519*, *520*; cultural 14, 538–540, *539*, **550**; defined **554**; of learning 9–10; natural selection and 510–513; reinforcement, verbal behavior, and 527–532; for vocal behavior 428–430, *429*, 529–530

exclusive preference, with concurrent FR schedules 314
experiential avoidance 215
experimental analysis of behavior 2–3, 37, **554**; *see also* behavior analysis
exposure therapy 215
external validity 224, **554**
extinction: of avoidance behavior 215–216; behavioral side effects 136–139; defined 136, **554**; discriminated 138–139, **552**; forgetting and 143–144, *143*; negative punishment compared to 194; operant selection and 526; Rescorla–Wagner equation and 103–104, *104*; respondent 84, **572–573**; of temper tantrums 144–145, *144*; use of 53–54; *see also* resistance to extinction
extinction burst 137, **555**
extinction-induced variability 137
extinction stimulus 272, **554**
extraneous sources of reinforcement 324, **555**

fading 293–294, **555**
falsifiability **555**
Fantino, Edmund 387–388, *388*
fear conditioning 21–22, 198–200
feelings: real, but not causes 28–29; reports of 29
finger spelling 431, *431*
first-order conditioning 92, **555**
fixed-action pattern (FAP): defined 72, **555**; genetic control of 514–517, *515*, *516*, *517*
fixed-interval schedule: adjunctive behavior and 257–258; defined *159*, 164–165, *165*, **555**; instructions, effect on 165–167, 419; postreinforcement pause on 181–183, *182*; scallop pattern on 361–362, *362*; token production and 382
fixed-ratio schedule: defined 159–162, *159*, **555**; postreinforcement pause on 181–183, *182*; token production and 195, 382–383, *383*
fixed-time schedule 164
fluency 490, **555**
Food Dudes Healthy Eating Program 504
force of response 137–138, **556**
forgetting: extinction and 143–144, *143*; reversal of 299–301, *300*
formal similarity 445, **555**
forward conditioning 253
free-operant method 123–124, **556**
function, as characteristic effect of behavior 37
function-altering event 422–423, **556**
functional analysis 8; of behavior 37–41; of challenging behavior 47–49, *49*; defined 37, **556**; of environment 41–49
functional independence: of basic verbal classes 440–443; defined 436–437, **556**

functional response classes *40*
functional stimulus classes *42*

Galizio, Mark *420*, 420–421
gambling and delay discounting 339–340
gender identity 26
generality: in applied behavior analysis 474–475; assumption 165–167, **544**; defined 57–58, **556**; of matching 320–321
generalization: aspects of 285–287, 289–290; defined 86–88, *87*, 284–285, **556**
generalization gradient 87, *87*, 285–286, *286*, **556**
generalized conditioned reinforcer 379, 384, **556**
generalized imitation: defined 407–409, *408*, **557**; by infants 409–410; naming relation and 448–450; training 410–412
generalized matching equation 320, 328, 346, 353–354
generalized matching relation 317, 320–322, 344–353, **557**
generalized matching-to-sample 452
generalized response chain 277
generalized social reinforcement 379–381, **557**
generalized trained immunity 97
genetic variation: in natural selection 129; sources of 512–513
genotype 512, *512*, 515, 523, **557**
Glenn, Sigrid 533, *533*
good news, effect of 370–376

habituation 75, 76–78, **557**; reinforcer 126–127, *127*
Headsprout™ 488–489
HealthyWage.com 487–488
hedonic value 368
helicopter parenting 219
helplessness, learned 216–220, **561**
heroin overdose 94–95, *95*
heterogeneous chain schedule 362–363, **557**
higher-order operant class 448
history of reinforcement 38–39, **557–558**
homeostasis 95–96, **558**
homogenous chain schedule 362–363, **558**
honeybees 86, 515–517, *516*
hyperbolic curve 324–326, *325*
hyperbolic discounting equation 336–338, *337*, **558**
hypothetical construct 62, **558**

identity matching 294–296, *295*, *296*, 451–452, **558**
imitation: action understanding, mirror neurons, and 403–405, *404*; behavior analysis of 405–407; complex observational learning and 412–415; defined 396, **558**; delayed 402; differential 415; in laboratory 398–400, *398*; operant and generalized 407–412, *408*; spontaneous 396–397, 400–402, *400*, *401*; see also generalized imitation
immediacy of change (baseline to treatment) 60, *61*, **558–559**
immediacy of punishment 196–197
immediate causation 7, **558–559**
immunity, trained 97–98
immunosuppression, conditioned 96–97
impulsive behavior: defined 341–344; delay discounting and 338
in-vitro reinforcement (IVR) 122, **560**
incentive salience 241–242, **559**
inclusion 26
independent variable 51–52, *51*, **559**
indifference point 336
infant caregiving and escape learning 205–206, *207*
information account of conditioned reinforcement 374, **559**
information and conditioned reinforcement 369–376
initial-link schedules 378, 386, 388
innate imitation 397; see also spontaneous imitation
instinctive drift 236, *237*, **559**
Institutional Review Boards (IRBs) 225
instructions: contingencies of reinforcement and 419–422; naming relation and 449–450
instrumental response **559**
intensity of punishment 196
intensive behavioral intervention 498–499, **559**
interim behavior 255, 270n2; see also adjunctive behavior
interlocking contingencies 462, 464–465, 532–533, 534, **559**
intermittent reinforcement effect **559**
intermittent schedule of reinforcement 139–140, **559**
intermodal mapping 402–403
internal validity 52, 476, **559**
interreinforcement interval (IRI) 175, 181, *182*, **559–560**
interresponse time (IRT) 178, **560**
interval recording *482*, 482, **560**
interval schedule: defined 159, **560**; fixed 164–165, *165*; variable 167–170, *168*
intraverbal behavior *441*, 443–444, **560**
intrinsic motivation 116–118, *117*
Introduction to the Study of Experimental Medicine, An (Bernard) 50

joint control 423–424, *423*, 456, **560**
Journal of Applied Behavior Analysis 470
Journal of the Experimental Analysis of Behavior 150

Keller Plan, personalized system of instruction (PSI) 489–490
Khan Academy 488

language 427–428; *see also* verbal behavior
latency 120–121, **560**; law of 76, 84, **561**
latent inhibition 88–89, **560**
law of effect 24, 120, **560**
law of intensity–magnitude 76, **560**
law of latency 76, 84, **561**
law of the threshold 76, **560**
learned helplessness 216–220, **561**
learning: associative 519–520; avoidance 206–209; behavior-consequence 112–114, *113*; defined 1, 78, **561**; in *Drosophila* 112–114, *113*; early, retention of 10–12; escape 204–206, *205*; evolution of 9–10; in honeybees 86; stimulus-relation 112–114, *113*; taste aversion 248–252, *249–250*, *254*, **579**; theories of 2; trial-and-error 23, 24, **581**; *see also* observational learning
leptin, hormone in obesity 125, 173, *174*, 264, 354, 501
limited hold 168, **561**
listening and speaking 432–433
lithium in taste aversion learning 80–81, 89, 249, 252
Little Albert study 21–22, *22*
log-linear matching equation 352, **561**
Los Horcones 469
Lovaas, Ivar 494, 497

macrocontingency 534–537, **561**
magazine training 128, **561**
main effect of schedule 152, 153
manding 435–436, *436*, 437–438, 440–443, 462–464, *463*, **561–562**
matching (relation) 316–317, 322–323; behavioral neuroscience and 353–354; communication and 321–322, *322*; defined 319, **561**; generality of 320–321; on more than two alternatives 320; proportional 317–319, *319*, 388; relational 295–296; on single-operant schedules 323–326, *325*; symbolic 452, **579**; time on alternative 319–320; *see also* generalized matching relation
matching to sample 294–296, *295*, *296*, **562**
maximization 328, **562**
maximum association strength 101, **562**
McDowell, Jack 326
McSweeney, Frances 126, *126*
measurement in applied behavior analysis 480–481
mediating behavior 165
melioration 328, **562**

memory: rehearsal 297, 424; traces 297
metacontingency *510*, 533–534, **562**
microaggression 226
Milgram, Stanley 418–419, *418*
Miller experiments 246–248, *246*
mirror neurons 402–405, *404*, 495–496
mixed schedule of reinforcement 370, **562**
mobile technology in treatment of autism 500
modal action pattern (MAP) 72–73, **562**
molar account of schedule performance: defined **563**; rate differences and 180–181; rate of response and 178; shock frequency avoidance behavior, and 212
molecular account of schedule performance: defined **563**; rate of response and 178–180, *179*; shock frequency avoidance behavior, and 212
money, as economic reinforcement 381–382, 384
motivational operation (MO) 45–47, *46*, **563**
multiple baseline across behaviors 478–480, *480*, **563**
multiple baseline across participants 477–478, *479*, **563**
multiple baseline across settings 477, *478*, **563**
multiple baseline designs 477–480, **563**
multiple-exemplar instructions (MEIs) 449
multiple schedules 279–284, *279*, **563–564**
mutation 513, **563–564**

naming relation 448–450, **564**
natural contingency and schedule of reinforcement 154–155
natural selection: defined **564**; evolution and 510–513; for vocal behavior 428–430
negative automaintenance 264–265, **564**
negative contrast 283; *see also* behavioral contrast
negative punishment *115*, 116, *193*, 194–195, **564**
negative reinforcement: contingencies of *193*, 203–214; defined 114–115, *115*, **564**; in schools 228–229
negative reinforcer 44, 204, **564**
neural activity, taste aversion, and drug cravings 252
neural basis of reward 9
Neuringer, Allen 134, *135*
neurodiversity 26
neurodiversity movement 496–497
neuron: mirror 129, 495–496; operant conditioning of 121–123
neuropeptide Y (NPY) and activity anorexia 263–264
neuroplasticity 1, 123, 404, **564**

neuroscience: autoshaping, sign tracking, and 241–242; behavioral analysis and 8–9; conditioned reinforcement and 368–369; depression, learned helplessness, and 219–220; learning in honeybees and 86; progressive-ratio schedule and 172–174; shaping of birdsong and 129–130; stimulus control and 278; *see also* behavioral neuroscience
neutral stimulus (NS) 79
new-response method for conditioned reinforcement 359, 366, **564**
non-contingent reinforcement (NCR) 472
nondiscriminated avoidance 209–212, **565**
nonspecific reinforcement 438
nonverbal discriminative stimulus 438, 439
novel behavior 68, 397, 526, 538

obesity 25; applied behavior analysis for 500–505; delay discounting and 339–340
object permanence 37–38
objectivity 481
observational learning: abstract rules and 415; behavioral interpretation of 414–415; cognitive theory of 414; complex 412–415; defined 396, **564–565**
observing behavior and conditioned reinforcement 372–376
observing response 370, **565**
octopamine 86
odor detection by dogs 305–306
omission procedure *265*, 265; *see also* negative automaintenance
online help resources and contingency management 487–488
ontogenetic **565**
ontogenetic behavior: conditioned and unconditioned responses 83–84; contiguity and contingency of stimuli 81–82; discrimination 88; generalization 86–88, *87*; overview 78–79; pre-exposure to stimuli 88–89; relative nature of stimuli 80–81; respondent acquisition 82–83, *83*; respondent conditioning 79–80; respondent extinction 84; spontaneous recovery 84–85, *85*
ontogenetic selection **565**
opaque imitation 402
operant 5, 109, **565**
operant aggression **565**
operant baselines for behavioral neuroscience 55–56
operant behavior: contingencies of reinforcement 110–112, *111*, 114–116, *115*; defined 39–41; discriminative stimuli 109–110; identification of reinforcing stimuli 118–120; overview 109; recording 151–152, *151–152*; selection of 12–13
operant chamber 124–125, *124*, **566**
operant class 128; defined 109, **565–566**
operant conditioning: defined 5–7, *6*, 108, **566**; model experiment in 130–132, *131*, *132*, *133*; of neuron 121–123; overview 120–121; procedures in 123–129; respondent conditioning and 233–235
operant imitation 407, **566**
operant level **566**; extinction and 131–132, 140, 157
operant rate 121, 123; *see also* rate of response
operant–respondent interrelationships: activity anorexia 259–264, *261*, *262*; adjunctive behavior 255–259; analysis of contingencies 234–248; autoshaping 264–267, *265*; biological context of conditioning 248–255; overview 233–234, *234*
operant selection 524–526
operant variability 137, **566**
optimal foraging 327–328
overcorrection 193–194, **566**
overeating, conditioned, and childhood obesity 505
overmatching in generalized matching equation 345, **566**
overshadowing 98–99, **566**

paradoxical effects of punishment **566**
partial reinforcement effect (PRE) 139–140, **566**
pause-and-run pattern *see* break-and-run pattern
Pavlov, Ivan P. 5, 17–19, *18*
payoff matrix 65–68, *66*
peak shift 286–289, *287*, *288*, **566**
Pelaez, Martha 275–277, *275*
perceiving, as behavior 62–65
perception, signal detection, and payoff matrix 65–68, *66*
permanence of punishment 202, **566**
personalized system of instruction (PSI) 489–490, **566–567**
Perspectives on Behavior Science 24
phenotype 512, *512*, **566–567**
phonation 429
phylogenetic **567**
phylogenetic behavior 72–78, 399
phylogeny 510, **567**
physical activity: activity anorexia, and substitutability of food and 334–335; motivational interrelations between eating and 259–264, *261*, *262*; taste conditioning induced by 252–255
physical restraint procedures 201

Picture Exchange Communication System (PECS) 497–498, *499*
placebo effect 96, **567**
plasticity, neural 404; *see also* neuroplasticity
point-to-point correspondence 443–444
polydipsia **567**; induced by interval schedule of reinforcement 255, 256–257, *256*, 258–259
positive contrast 283; *see also* behavioral contrast
positive counterconditioning 200
positive practice 193–194
positive punishment: defined *115*, 115–116, 192–194, *193*, **567**; effectiveness of 197–198
positive reinforcement: defined 114, *115*, **567**; schedules of 155–159, *156*, *158*
positive reinforcer 44, 110, 114, 118, **567**
postreinforcement pause (PRP): defined **567**; fixed-interval schedules and 164; fixed-ratio schedules and 160–161; on fixed schedules 181–183, *182*; variable-ratio schedules and 162–163, *162*
power law for matching 345; *see also* generalized matching law
precision teaching 490–492, *491*, **567**
precurrent behavior 417, **567**
preference 311, **568**
preference for choice 329–332, *331*, **568**
preference reversal 336, 341–344, *342*, *343*, **568**
Premack principle 118–120, *119*, **568**
preparedness 248–251, *249–250*, **568**
preratio pause 161, **568**
primary aversive stimulus 187, **568**
primary laws of the reflex 76, 84, **568**
private behavior 30–31, **568**
private event 27–28
probability of response 123, **568**
probe trial 285
problem behavior *see* challenging behavior
problem solving and reinforcement 133–136, *134*
programming for generality 474–475
progressive-delay procedure 337–338
progressive-ratio (PR) schedule: defined 171–172, **568**; neuroscience and 172–174
proportional matching 317–319, *319*, 388
proportional rate of response 317–318, 319
provocation, aggression as response to 224–226
psychiatric disorders and punishment 203
punisher 192, **568**
punishment: contingencies of 192–198, *192*; to control human behavior 189–190, *190*; debate over use of 201–202, **581**; defined 192, **568–569**; effectiveness of 195–198; immediacy of 196–197; intensity of 196; macrocontingency, and costly use of 534–537; negative *115*, 116, *193*, 194–195, **564**; paradoxical effects of **566**; permanence of 202, **566**; psychiatric disorders and 203; relativity of **571**; schedule of 197; use in treatment 200–202; *see also* aversive stimuli; positive punishment
puzzle box for cats in trial and error learning 120–121, *121*

quality control and signal detection 304–306, *304*, *306*
quantitative law of effect 324, **568–569**

random-ratio (RR) schedule 163, *331*, 331
range of variability (in assessment) 60, *61*, **569**
rate of response: absolute 323; defined 121, **569**; extinction and 136, 139, 140, *141*; matching relation and 317, 320, 328; proportional 317–318, 319; punishment and 192, 196, 197; relative 317, 318, **570**; on schedules 178–181, *179*
ratio schedule: concurrent 313–314, *314*; defined 159, **569**; fixed 159–162; variable 162–164, *162*
ratio strain 161, 175, **569**
rats, as subjects of experiments 270n1
reaction chain 74–75, *74*, **569–570**
recombinant DNA technology 514
recording behavior 151–152, *151–152*, 481–482, *482*
reflex: conditioned 18–19; defined 4, 39, 71, 75, **570**; primary laws of the 76, 84, **568**
reflexive behavior: negative reinforcement and 205; overview 75; reinforcement of 245–246
reflexivity 450–451, 452–453, *452*, **569–570**
reinforcement 3; adventitious 282; behavioral selection by 522–527; biofeedback, robotic limbs, and 244–245; chain schedule of **546**; concurrent schedules of **547**; conjugate 156; contingency of 57; continuous *132*, 132, *133*, 155–159, *156*, *158*, **550**; defined **570**; dynamics 44–47; extraneous sources of 324, **555**; generalized social 379–381, **557**; history of 38–39, **558**; *in-vitro* 122, **560**; non-contingent (NCR) 472; nonspecific 438; in operant conditioning 6; perception and 64–65; problem solving and 133–136, *134*; of reflexive behavior 245–246; response stereotypy and 133–134, *134*; response variability and 134–135; susceptibility to 526–527; unconditioned 364–365, 376; *see also* conditioned reinforcement; contingency of reinforcement; differential reinforcement; negative reinforcement; positive reinforcement; postreinforcement pause; schedule of reinforcement

reinforcement efficacy 172, **570**
reinforcement function 43, **570**
reinforcer 109
reinforcer habituation 126–127, *127*
reinforcer pathologies 25, 340, **570**
reinforcing stimulus, identification of 118–120
reinstatement (of response) 142–143, **570**
relational frame 450
relational frame theory (RFT) 26, 474, **570–571**
relational frames 461–462
relational matching 295–296
relative rate of reinforcement: behavioral contrast and 283–284; defined **570**; delay reduction and 389; proportional matching and 318
relative rate of response 317, 318, **570**
relative stimulus control 289–290, **570**
releasing stimuli 72
reliability of observation 482, **571**
remembering 297–299, **571**
remote causation 7–8, **571**
renewal (of responding) **571**
repertoire (of behavior) 129, **571**
replication (of results) 57–58, **571**
Rescorla–Wagner equation 102–104
Rescorla–Wagner model 100–104, **571**
research: A-B-A-B design 52–55, *54*, 476–477, **543**; in applied behavior analysis 470–471; changing criterion design 480, 493, *493*, **547**; measurement issues in 480–481; multiple baseline designs 477–480, **563**; recording behavior for 481–482, *482*; reliability of observations in 482; tactics of 49–55; *see also* single-subject research
resistance to extinction: continuous reinforcement and 156–157; defined 139, **572**; discriminative stimuli and 140, *141*; partial reinforcement effect 139–140; reinstatement of responding and 142–143; spontaneous recovery and 140–142; type of conditioned response and 84
resource depletion, ecological costs of 534–537
respondent 4, 40, 80, **572**
respondent acquisition 82–83, *83*, **572**
respondent behavior 39–41
respondent conditioning: defined 4–5, *6*, 79–80, *79*, **572**; drug use and abuse and 93–94; operant conditioning and 233–235
respondent discrimination 88, **572**
respondent extinction 84, **572–573**
respondent generalization 86–88, *87*, **573**
response: assessment of experimental control of 59–62; changeover 315–316, **547**; commitment 343–344, **547**; conditioned 5; contingent **550**; defined 39; definition and measurement of 58–59; differential 43, 272, *273*, **552**; echoic *445*, 445–446, **553**; emotional 138, **553**; force of 137–138, **555**; instrumental **559**; observing 370, **565**; probability of 123, **568**; reinstatement of 142–143, **570**; run of responses 160, 182, **574**; schedules of reinforcement and patterns of 152–154, *153*; unconditioned 4–5, 75, **581**; *see also* conditioned response; rate of response
response alternatives 198
response bias 345, 350
response chain 274, **572-3**
response class 41, **573**
response cost 195, **573**
response deprivation 120, **573**
response deprivation hypothesis **573**
response differentiation 137, **573**
response functions 39–41
response generalization 475, **573**
response hierarchy 41, 120, **573**
response-independent schedule 164
response–shock interval (RSI) 210, **573**
response stereotypy 133–134, *134*, 157–159, *158*
response variability 59, 134–135
responsibility, diffusion of 225
restitution 193
restraint procedures 201
resurgence 159, **573**
retention: of early learning 10–12; of fear conditioning 198–200
retention interval 297–298, *298*, *299*, **574**
retroactive interference 299–301
reversal design (A-B-A-B) 52–55, *54*, 476–477, **543**
reversal test 453–454, *454*, **573–574**
reverse forgetting 299–301, *300*
reward and intrinsic motivation 116–118, *117*
reward contingency 117–118
robotic limbs 244–245
rule-governed behavior: contingency-shaped behavior and 417–419; defined 416–417, **574**; function-altering events and 422–423; instructions and 419–422; joint control and 423–424, *423*; listening and 432
run of responses 160, 182, **574**

S-delta (S^Δ) as extinction stimulus 110, 272, **575**
S–S account of conditioned reinforcement 366, **577**
salience 102, **574**
satiation **574**
Savage-Rumbaugh, Sue 438, *439*
saying and doing, consistency between 393–396

scalloping 164, *165*, 165, **574**
Schatz, Lydia, murder of 189
schedule-induced behavior 182, 255; *see also* adjunctive behavior
schedule of punishment 197
schedule of reinforcement: assumption of generality and 165–167; biofeedback and 170–171; cigarettes and 176–178, *177*; concurrent 312–313, *313*; concurrent-chain 376–379, *377*, **547**; defined 149, **575**; importance of 149–150; intermittent 139–140, **559**; interval 164–165, *165*, 167–170, *168*; matching, modification, and 323; mixed 370, **562**; motivation and 153; natural contingencies and 154–155; patterns of response and 152–154, *153*; positive reinforcement 155–159, *156*, *158*; postreinforcement pause on 181–183, *182*; progressive-ratio schedules 171–174; rate of response on 178–181, *179*; ratio and interval schedules 159–165, *159*; second-order 366–368, *367*, **575**; side effects of **576**; token 382–384, *383*, **580**
schedule performance: postreinforcement pause on fixed schedules 181–183, *182*; rate of response on schedules 178–181, *179*; in transition 174–176
Science and Human Behavior (Skinner) 16
science of behavior 3; *see also* behavior analysis
second-order conditioning 92–97, **575**
second-order schedule of reinforcement 366–368, *367*, **575**
selection: as causal process 7–8; by consequences 7, 235, 509, *510*, **575–576**; cultural 7, 509; ontogenetic **565**; operant 524–526; of operant behavior 12–13; for operant processes 524–526; sexual 75, **576**; *see also* natural selection
self-administration (of drug) 142
self-control: defined 341–344, **576**; training 492–493, *493*
self-efficacy, statements of 416
self-injurious behavior (SIB): aversive therapy for 200–202; extinction of 138; functional analysis of 47–49, *49*; single-operant equation and 326–327, *327*
sensitivity parameter in matching equation 345–350
sensory preconditioning **576**
sequences of behavior 72–75
sexual orientation 26
sexual selection 75, **576**
shaping 128–130, *128*, **576**
shock–shock interval (SSI) 210, **576**

side-effect schedule pattern 152–153
side effects of schedules of reinforcement **576**
Sidman avoidance *see* nondiscriminated avoidance
sign tracking 237–239, *238*, 241–242, **576**
signal detection 65–68, *66*, 304–306, *304*, *306*
simultaneous conditioning *90*, 90, **576**
simultaneous discrimination 290, **576–577**
single exemplar instructions (SEIs) 449–450
single-operant rate equation 326–327, *327*
single-operant schedule, matching on 323–326, *325*
single-subject research: assessment of control and behavior change 58–62; defined 57, **577**; generality and 57–58; reversal design in 52–55
Skinner, B. F.: *The Behavior of Organisms* 13, 359; biography and career of 14–17; *Enjoy Old Age* 17; photograph *15*; punishment and 196, 214–215; rise of behavior analysis and 24–27; *Science and Human Behavior* 16; *The Technology of Teaching* 16–17; *Verbal Behavior* 434, 462; *Walden Two* 469
smoking: during pregnancy 485; schedule of reinforcement and 176–178, *177*
social behavior, epigenetic reprogramming of 521–522, *521*
social defeat and behavioral neuroscience 220–221
social disruption 227–228, **577**
social environment, focus on 476
social episode 462, **576–577**
social exclusion, aggression as response to 227
social referencing and behavior chains 274–278, *275*, *277*
social reinforcement, generalized 379–381, **557**
social signals 527–529
social use of words 433–434
speaking: ecology, contingencies, and 531–532, *531*; listening and 432–433; *see also* verbal behavior
species-specific behavior and contingencies 242–244, *243*
spontaneous imitation: behavior analysis of 405–407; defined 396–397, **577**; by infants 396, 400–402, *400*, *401*; studies with pigeons 398–400, *398*
spontaneous recovery: defined 84–85, *85*, **577**; extinction and 140–142; habituation and 78
steady-state performance 55–56, 152–153, 174–175, **577**
stereotyped behavior 258
stereotypy, functional analysis of 48–49, *49*
Stickk.com 487

stimuli 4; compound 98–100, **547**; conditioned 80, **548**; contextual 85, **549**; contiguity and contingency of 81–82; contingency-specifying 416, **550**; extinction 272, **555**; information value of 369–372; multiple functions of 464, **563**; neutral 79; pre-exposure to 88–89; primary aversive **568**; reinforcing 118–120; relative nature of 80–81; releasing 72; unconditioned 4–5, 75, **581**; *see also* aversive stimuli; discriminative stimulus; stimulus control
stimulus class 43–44, **577**
stimulus control: absolute and relative 289–290; of behavior sequences 274; complex 294–296; defined 273, **577**; fading of 293–294; multiple schedules and 279–281, *279*; neuroscience and 278; three-term contingency 272–274; *see also* generalization
stimulus enhancement 396, 400
stimulus equivalence: application of 458–460; defined 450, **577–578**; derived stimulus relations 456–457; experimental analysis of 451–452, *451*, *452*; reflexivity 452–453, *452*; relational frames 461–462; research on 457–458, *259*; symmetry 453–454, *454*; transitivity 454–456, *455*; types of 450–451
stimulus function 42–43, 273, **577**
stimulus generalization 285, **578**; *see also* generalization
stimulus-relation learning (SRL) 112–114, *113*
stimulus–response theory of behavior 12–13
stimulus substitution 243–244, **578**
stimulus-to-stimulus (S-S) conditioning 39, 81
Stockholm syndrome 380
Stroop effect 62–64, *63*
structural approach 37–38, **578**
submissive behavior 380
substance use and abuse *see* drug use and abuse
substitutability 333–334, **579**
successive approximation 128–130, *128*, **578**
successive discrimination 290, **578**
superstitious behavior 282, **578–579**
survival contingencies *see* contingencies of survival
susceptibility to reinforcement 526–527
symbolic behavior and stimulus equivalence 450–460
symbolic matching 452, **579**
symmetry 450–451, 453–454, *454*, **578–579**
systematic replication 57, **579**

taboo, cultural 537–538
tacting 435–436, 438–443, *439*, *464*, 464–465, **579**
tandem schedule 361–362, *362*, **579**
taste aversion induced by physical activity 252–255

taste aversion learning 248–252, *249–250*, *254*, **579**
teaching: clicker training and 360; coercion in 228–229; discrimination and 4, 282; errorless procedures for 292; precision 490–492, *491*, **567**; Skinner and 16–17; sports 363–364, *364*; token economies for 384–386; *see also* education, applied behavior analysis in; learning
Technology of Teaching, The (Skinner) 16–17
temper tantrums 144–145, *144*
temporal pairing 5, **579**
temporal relations and conditioning 89–92, *90*
terminal behavior **579**
Terzich, Brenda, ABC School of Autism 499
textual behavior 446, **579–580**
thinking, as behavior 29–31, 33
Thorndike, Edward L. 23–24, *23*
three-term contingency 45, 111–112, *111*, 180, 238, 272–274
threshold, law of 76, **561**
thrifty gene theory 501
time sampling 482, **579–580**
timeout: from avoidance 213–214, **580**; from positive reinforcement 194–195, 210, **580**; from reinforcement 134
token economy 384–386, **580**
token reinforcement 381–384, *381*, *383*
token schedule of reinforcement 382–384, *383*, **580**
tolerance (to drug) 93–95, 96, **580**
tongue, human vocalization 430
topography of response 8, 38, 109, 157, 239, 362, 424, 435, **580**
trace conditioning *90*, 90–91, **580**
tragedy of commons 535
trained immunity 97–98
training: clicker training 305, 306, 359–360; generalized imitation 410–412; magazine training 128, **561**; self-control 492–493, *493*; verbal operants 436–440
trans-disease process 339–340
transition state 175, **580–581**
transitivity 451, 454–456, *455*, **580**
translational research 470
trend (in baseline) 60, *61*, 62, **581**
trial-and-error learning 23, 24, **581**
Twin Oaks 469
two-key concurrent-chains procedure 376–378, *377*
two-key procedure 312–316, *313*, *314*, **581**

unconditioned reflex 5
unconditioned reinforcement 364–365, 376
unconditioned reinforcer 358, **581**
unconditioned response (UR) 4–5, 75, **581**
unconditioned stimulus (US) 4–5, 75, **581**

undermatching 345, 350, 353, **581**
unfalsifiable predictions 2
use of punishment debate 201–202, **581**
US-pre-exposure effect 89, **581**
UV vision of birds 278

vaccination programs and avoidance behavior 210–212
variability: operant 135–136, 137, 157–158; range of 60, *61*; response 134–135, 157–158
variable-interval (VI) schedule *159*, 167–170, *168*, **581**; choice and 315, 324, *325*; instructions and 419; proportional 317
variable-ratio (VR) schedule *159*, 162–164, *162*, **582**
verbal behavior: autoclitic relations 446–448; defined 395, 428, **581–582**; echoic relations *445*, 445–446; effects of 430–431; evolution of vocal apparatus 428–430; evolution, reinforcement, and 527–532; functional independence of basic verbal classes 440–443; intraverbal relations *441*, 443–444; language and 427–428; mand relation, analysis of 462–464, *463*; naming relations 448–450; operant functions of 430, 434–436; range of 431–432; social use of words 433–434; speaking, listening, and verbal community 432–433; symbolic behavior and 450–460; tact relation, analysis of *464*, 464–465; textual relations 446; three-term contingencies and natural speech 461; training of verbal operants 436–440
Verbal Behavior (Skinner) 434, 462
verbal community 431–434, **582**
video modeling 500
vision of birds 278
vocal tract: evolution of 430; in speech production 429–430, *429*
vocalization and speech sounds 529–530
volcanic eruption, habituation of 77, 77–78
voucher-based contingency management (CM) 483–487, *484*

Walden Two (Skinner) 469
Watson, John B. 19–23, *20*
whistling, speaking by 531–532, *531*
words, social use of 433–434
writing, as verbal behavior 430

yawning, as modal action pattern (MAP) *73*, 73–74

9781032065144